Genetics of Subpolar Fish and Invertebrates

Developments in environmental biology of fishes 23

Series Editor
DAVID L.G. NOAKES

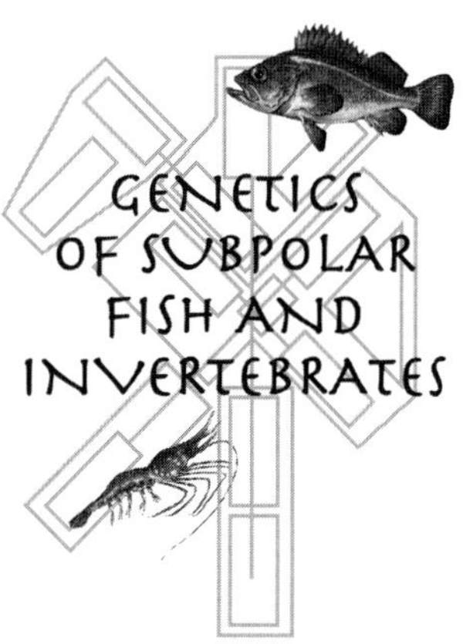

Genetics of Subpolar Fish and Invertebrates

Guest Editors:

Anthony J. Gharrett, Richard G. Gustafson, Jennifer L. Nielsen, James E. Seeb, Lisa W. Seeb, William W. Smoker, Gary H. Thorgaard and Richard L. Wilmot

Reprinted from *Environmental Biology of Fishes*, Volume 69 (1–4), 2004

KLUWER ACADEMIC PUBLISHERS

DORDRECHT / BOSTON / LONDON

A C.I.P. Catalogue record for this book is available from the library of Congress

ISBN 1-4020-2033-3

Published by Kluwer Academic Publishers,
P.O. Box 17, 3300 AA Dordrecht, The Netherlands

Sold and distributed in North, Central and South America
by Kluwer Academic Publishers,
101 Philip Drive, Norwell, MA 02061, U.S.A.

In all other countries, sold and distributed
by Kluwer Academic Publishers,
P.O. Box 322, 3300 AH Dordrecht, The Netherlands

Printed on acid-free paper

CONTENTS

Genetics of populations: methods and applications

Genetic variability: fitness and adaptation

Genetic variability: phenotype and maintenance of variation

Aquaculture genetics: interactions of cultured and wild fish

Environmental Biology of Fishes **69:** ix, 2004.

Acknowledgements

Genetics of Subpolar Fish and Invertebrates, the 20th Lowell Wakefield Fisheries Symposium, was organized and coordinated by Brenda Baxter and Sherri Pristash, University of Alaska Fairbanks, Alaska Sea Grant College Program; A.J. Gharrett (Chair), University of Alaska Fairbanks, Fisheries Division; Rick Gustafson, National Marine Fisheries Service, Northwest Fisheries Science Center; Jennifer Nielsen, U.S. Geological Survey, Biological Resources Division; Jim Seeb and Lisa Seeb, Alaska Department of Fish and Game, Gene Conservation Laboratory; W.W. Smoker, University of Alaska Fairbanks, Fisheries Division; Gary Thorgaard, Washington State University, School of Biological Sciences; and Richard Wilmot, National Marine Fisheries Service, Alaska Fisheries Science Center.

Sponsors of the meeting were Alaska Sea Grant College Program, University of Alaska Fairbanks; National Marine Fisheries Service; North Pacific Fishery Management Council; and the Wakefield Endowment, University of Alaska Foundation.

Many thanks to the following people who reviewed one or more paper for the proceedings book: Syuiti Abe, Bill Ardren, Meredith L. Bartron, Terry D. Beacham, P. Bentzen, Louis Bernatchez, Craig Busack, Mike Canino, Dmitri Churikov, Joe Cloud, Penny Crane, R.A. Curry, Roy Danzmann, Bobette R. Dickerson, John Emlen, John Epifanio, Dylan J. Fraser, Tony Gharrett, Stewart Grant, Andy Gray, Bruno Guinand, Christopher Habicht, Eric M. Hallerman, Michael M. Hansen, Jeffrey J. Hard, Lorenz Hauser, Daniel Heath, Dennis Hedgecock, Ron Heintz, Anne Henderson-Arzapalo, Pamela C. Jensen, Knut Jørstad, Ji-Eun Kim, Timothy King, Kathy L. Knudsen, Christine Kondzela, Khai D. Le, Erica Leder, Anne R. Marshall, Jeff Marliave, Christopher Martyniuk, Makoto P. Matsuoka, Niall McKeown, Kristi M. Miller, Tor Naesje, James J. Nagler, Kerry Naish, Krista Nichols, Marc A. Noakes, Jeffrey B. Olsen, Patrick T. O'Reilly, Ken Overturf, Jerry Pella, Guy Perry, Ruth Phillips, Paulo A. Prodöhl, Maureen Purcell, Tom Quinn, Axayacatl Rocha-Olivares, D. Rogers, Sean Rogers, Daniel E. Ruzzante, Marjatta Saisa, Kim T. Scribner, James B. Shaklee, Christian T. Smith, Adrian Spidle, Paul Spruell, William D. Templin, Fred Utter, Robin Waples, John K. Wenburg, Paul Wheeler, Peter Wimberger, Gary Winans, Ruth Withler, Chris Wood, and Sewall Young. Thanks to Sue Keller, Alaska Sea Grant, who coordinated the review process.

This book is Alaska Sea Grant College Program publication number AK-SG-03-02.

Environmental Biology of Fishes **69**: 1–5, 2004.

Introduction to genetics of subpolar fish and invertebrates

A.J. Gharrett[a], S. Keller[b], R.G. Gustafson[c], P. Johnson[d], J.L. Nielsen[e], J.E. Seeb[f] , L.W. Seeb[f],
W.W. Smoker[a], G.H. Thorgaard[g] & R.L. Wilmot[d]
[a]*Fisheries Division, School of Fisheries and Ocean Sciences, University of Alaska Fairbanks,
11120 Glacier Highway, Juneau, AK 99801, U.S.A. (e-mail: ffajg@uaf.edu)*
[b]*Alaska Sea Grant College Program, School of Fisheries and Ocean Sciences, University of
Alaska Fairbanks, Fairbanks, AK 99775-5040, U.S.A.*
[c]*National Marine Fisheries Service, Northwest Fisheries Science Center, Conservation Biology Division,
2725 Montlake Blvd. E, Seattle, WA 98112, U.S.A.*
[d]*National Marine Fisheries Service, Auke Bay Laboratory, Alaska Fisheries Science Center,
11305 Glacier Highway, Juneau, AK 99801, U.S.A.*
[e]*U.S. Geological Survey, Alaska Science Center, 1011 E. Tudor Road, Anchorage, AK 99503, U.S.A.*
[f]*Alaska Department of Fish and Game, Commercial Fisheries Division, 333 Raspberry Road,
Anchorage, AK 99518, U.S.A.*
[g]*School of Biological Sciences and Center for Reproductive Biology, Washington State University,
Pullman, WA 99164-4236, U.S.A.*

Key words: quantitative genetics, population genetics, molecular genetics, aquaculture, systematics, wild–hatchery
interactions

This 20th Wakefield Symposium, 'The genetics of subpolar fish and invertebrates', is the successor of the 11th Wakefield Symposium, 'Genetics of subarctic fish and shellfish', which was held in Juneau, Alaska in 1993. In the introduction to that symposium (Gharrett & Smoker 1994), it was noted that: 'beginning in the 1960s, modern tools of genetic analysis began to be broadly applied in fisheries science', and that 'within the past decade (referring to the 1980s), fisheries genetics had entered the mainstream of fisheries resource utilization'. That observation may be an understatement in today's world of fisheries science. Once-vigorous fisheries in many parts of the world have failed, growing demand for fisheries products has led to full utilization of many remaining capture resources and is driving an increase in aquaculture productivity, and the role of aquaculture has increased dramatically (FAO 2002). Looming over concerns of lost stocks and persistent erosion of genetic variability are predictions of global warming, which may further tax genetic resources. One of the consequences of these developments is an increased interest in and reliance on genetic applications to many aspects of fisheries management, aquaculture, and conservation.

In addressing those concerns, fisheries scientists have increased their attention to the genetics of fish and fish populations; the number of fish genetics citations has increased fourfold in the last decade (Figure 1). In addition to the increased attention, the application of sophisticated genetic analysis tools, such as studies of mitochondrial DNA (mtDNA) and microsatellites, have nearly caught up with the more traditional allozyme studies in annual citations (Figure 1).

One of the predominant themes of both symposia was the descriptive study of population structure. Descriptive studies are an important step in developing management or conservation plans because they can provide markers for use in stock identification programs and because they can identify distinct productivity units (the geographical scale of such units) in species that do not have convenient or obvious geographical boundaries, such as many marine species. Analyses of allozyme variation, which dominated early fisheries genetics research, still provide valuable

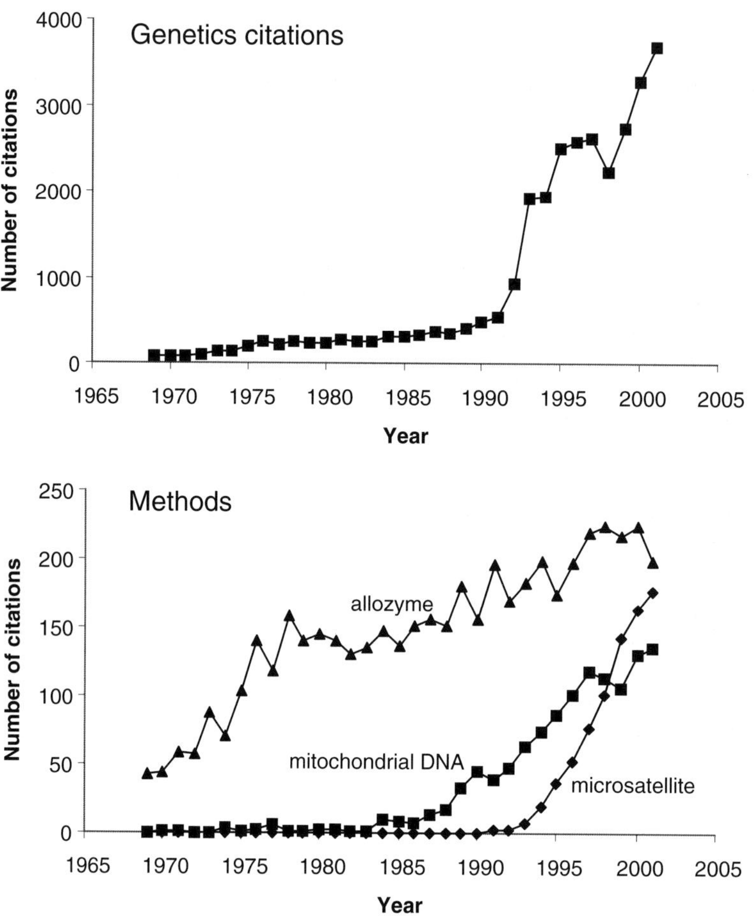

Figure 1. The number of genetics citations from a search of BIOSIS® using combinations of the words and terms (pisces, osteicthyes, or fish) and (genetics or cytogenetics or mtDNA or allozyme or electrophoresis or microsatellite) for papers published each year from 1969 to 2001 (upper). The number of citations for specific molecular methods from a search of BIOSIS Previews (BIOSIS 2001) using combinations of the words and terms (pisces, osteicthyes, or fish) with (1) mtDNA or mitochondrial DNA for mtDNA analyses, (2) allozyme or electrophoresis for allozyme analyses, or (3) microsatelite for microsatellite analyses in papers published each year from 1969 to 2001 (lower). BIOSIS® Previews, BIOSIS, Philadelphia, PA. 20 May 2001. http://www.biosis.org.

information and are represented here by studies on Pacific salmon (*Oncorhynchus* spp.; Seeb et al. 2004; Guthrie & Wilmot 2004), herring (*Clupea herengus* and *Clupea pallasi*; Jorstad 2004) and Pacific hake (*Merluccius productus*; Iwamoto et al. 2004). However, descriptive studies increasingly use analyses of DNA variation (Figure 1), particularly mtDNA polymorphisms (Brykov et al. 2004; Sato et al. 2004) and microsatellite variation (Beacham et al. 2004; Curry et al. 2004; Habicht et al. 2004; Matala et al. 2004; Shubina et al. 2004; Wennevik et al. 2004; Young et al. 2004a,b). Some studies combine analyses of allozyme, mtDNA, and microsatellite variation (Jorstad et al.

2004; Thrower et al. 2004). In the last decade, there has also been increased attention to marine species (Brykov et al. 2004; Jorstad 2004; Jorstad et al. 2004; Iwamoto et al. 2004; Matala et al. 2004; Shubina et al. 2004) and focus on finer levels of genetic structure (Curry et al. 2004; Habicht et al. 2004; Thrower et al. 2004; Young et al. 2004a,b).

As geneticists address more complicated – and more interesting – questions, descriptive studies are becoming increasingly sophisticated. A.R. Templeton's keynote address presented an approach to descriptive studies that uses the temporal information inherent in gene and haplotype trees in combination with the geographical distribution of the genes to make rigorous inferences about historical demographic events that shaped the current distribution. The marriage of phylogenetic information to population genetics promises to provide important insights from studies of fish and invertebrate resources.

One of the genetic applications to fisheries management is the resolution of stock mixtures. These methods have proved useful in a variety of applications including forensics (Withler et al. 2004a). Preliminary studies of high seas distributions of chum salmon (*Oncorhynchus keta*) is extending our knowledge of their biology (Seeb et al. 2004). The analytical basis of this process continues to be developed and refined (Guinand et al. 2004; Olsen et al. 2004; Reynolds & Templin 2004), and data from new descriptive studies must be examined for their suitability for application to specific management issues (Beacham et al. 2004).

The genetic composition of a population is molded from the evolutionary forces and stochastic events that have acted on it. Population differences can emerge from subtle local differences and from random drift, and isolated populations may accrue substantially different genetic architectures, which can be studied using genetic analysis. Consequently, genetics can be used to address a variety of biological questions ranging from elucidating complex mating systems (Seamons et al. 2004; Withler et al. 2004b) to resolving the phenotypic differences among coho populations that involve embryo development and timing (Granath et al. 2004). The differences in genetic architecture among populations take on additional implications in the context of stock transfers that can accompany hatchery, stock enhancement, and aquaculture programs. In addition, artificial culture practices can reduce variability in cultured stocks (Kim et al. 2004). One of the consequences of introgression of transplanted fish is outbreeding depression, which was documented in pink salmon (*Oncorhynchus gorbuscha*) (Gilk et al. 2004). It is likely that steelhead (*Oncorhynchus mykiss*) repeatedly stocked into the Great Lakes contribute their genes to naturally spawning populations (Bartron & Scribner 2004); but, in contrast, the reproductive success of hatchery steelhead in Washington appears to be lower than that of naturally spawning steelhead (McLean et al. 2004). Various measures, including supplementation programs, have been initiated to address the drastic declines in chinook salmon returns that the Columbia River system has experienced in the last century. Detecting stressors while they are occurring,

rather than after the fact, would reduce the time needed by managers to detect and possibly circumvent negative effects. One approach may be detection of decreased embryological developmental stability, which might be indexed by increases in asymmetry of paired structures (like left and right pectoral fin rays) in fish (Johnson et al. 2004). One result of the stressors may be sex conversion, as suggested by the large number of Columbia River system chinook salmon females that carry what were previously presumed to be male-specific markers (Chowen & Nagler 2004).

Several reports involve diverse aquacultural applications. Among the challenges to aquacultural enterprises are disease outbreaks and a need for nutritious, inexpensive diets. In some instances, domesticated stocks can be selected for fish that possess characteristics that are desirable in intensive culture. Resistance to infectious hematopoietic necrosus virus was examined in genes of the major histocompatibility complex of Atlantic salmon (*Salmo salar*) (Miller et al. 2004); and the potential of using variation in gene expression (detected by real-time PCR) to direct selection for growth was also explored (Overturf et al. 2004). In addition, success in increasing the sperm motility of sex-reversed male rainbow trout (*O. mykiss*) will improve the culture of all-female lines (Kobayashi et al. 2004). In what undoubtedly reflects an important future direction for many aspects of fisheries biology and genetics, quantitative trait loci for several meristic traits were detected and examined in rainbow trout (Nichols et al. 2004).

Genetic differences within and between species provide phylogenetic characters that can be used to learn about the demographic history of a species (Templeton 2004), resolve subspecies or species differences (Frolov & Frolova 2004; Oleinik et al. 2004; Templeton 2004), or develop phylogenies for taxa that include numerous species (Phillips et al. 2004). Several different genetic methods were used in phylogenetic applications, including karyotypes (Frolov & Frolova 2004) and mtDNA (Oleinik et al. 2004) in chars (*Salvelinus* spp.), DNA sequences of growth hormone introns in Salmoninae (Phillips et al. 2004), transferrin sequences in brown trout (*Salmo trutta*) species complex (Templeton 2004), and microsatellite flanking sequences in rockfish (*Sebastes* spp.) (Asahida et al. 2004).

In aggregate, these papers reflect questions that are important to modern fisheries science and genetics and, in comparison to the science presented at the 11th Wakefield Symposium, illustrate the evolution of the field over the past decade. The improved technology will continue to provide tools to address increasingly complicated problems not only in traditional applications but also in ecological and behavioral studies. The union between molecular and quantitative genetics, where many of the major questions about population structure and evolution remain unanswered, will also benefit from the new technologies.

References

Asahida, T.A., K. Gray & A.J. Gharrett. 2004. Use of microsatellite locus flanking regions for phylogenetic analysis? A preliminary study of *Sebastes* subgenera. Environ. Biol. Fish. 69: 461–470.

Bartron, M.L. & K.T. Scribner. 2004. Temporal comparisons of genetic diversity in Lake Michigan steelhead, *Oncorhynchus mykiss*, populations: effects of hatchery supplementation. Environ. Biol. Fish. 69: 395–407.

Beacham, T.D., K.D. Le & J.R. Candy. 2004. Population structure and stock identification of steelhead trout, *Oncorhynchus mykiss*, in British Columbia and the Columbia River based on microsatellite variation. Environ. Biol. Fish. 69: 95–109.

Brykov, V.A., N.E. Polyakova, T.F. Priima & O.N. Katugin. 2004. Mitochondrial DNA variation in northwestern Bering Sea walleye pollock, *Theragra chalcogramma* (Pallas). Environ. Biol. Fish. 69: 167–175.

Chowen, T.R. & J.J. Nagler. 2004. Temporal and spatial occurrence of female chinook salmon carrying a male-specific genetic marker in the Columbia River watershed. Environ. Biol. Fish. 69: 427–432.

Curry, R.A., S.L. Currie, L. Bernatchez & R. Saint-Laurent. 2004. The rainbow smelt, *Osmerus mordax*, complex of Lake Utopia: threatened or misunderstood? Environ. Biol. Fish. 69: 153–166.

FAO (Fisheries Department, Food and Agriculture Organization of the United Nations). 2002. The state of world fisheries and aquaculture 2002. FAO, Rome. 150 pp.

Frolov, S.V. & V.N. Frolova. 2004. Karyological differentiation of northern Dolly Varden and sympatric chars of the genus *Salvelinus* in northeastern Russia. Environ. Biol. Fish. 69: 441–447.

Gharrett, A.J. & W.W. Smoker. 1994. Introduction to genetic and subarctic fish and shellfish. Can. J. Fish. Aquat. Sci. 94(Suppl. 1): 1–3.

Gilk, S.E., I.A. Wang, C.L. Hoover, W.W. Smoker, S.G. Taylor, A.K. Gray & A.J. Gharrett. 2004. Outbreeding depression in hybrids between spatially separated pink salmon (*Oncorhynchus gorbuscha*) populations: marine survival, homing ability, and variability in family size. Environ. Biol. Fish. 69: 287–297.

Granath, K.L., W.W. Smoker, A.J. Gharrett & J.J. Hard. 2004. Effects on embryo development time and survival of intercrossing three geographically separate populations of Southeast Alaska coho salmon, *Oncorhynchus kisutch*. Environ. Biol. Fish. 69: 299–306.

Guinand, B., K.T. Scribner, A. Topchy, K.S. Page, W. Punch & M.K. Burnham-Curtis. 2004. Sampling issues affecting accuracy of likelihood-based classification using genetical data. Environ. Biol. Fish. 69: 245–259.

Guthrie III, C.M. & R.L. Wilmot. 2004. Genetic structure of wild chinook salmon populations of Southeast Alaska and northern British Columbia. Environ. Biol. Fish. 69: 81–93.

Habicht, C., J.B. Olsen, L. Fair & J.E. Seeb. 2004. Smaller effective population sizes evidenced by loss of microsatellite alleles in tributary-spawning populations of sockeye salmon from the Kvichak River, Alaska drainage. Environ. Biol. Fish. 69: 51–62.

Iwamoto, E., M.J. Ford & R.G. Gustafson. 2004. Genetic population structure of Pacific hake, *Merluccius productus,* in the Pacific Northwest. Environ. Biol. Fish. 69: 187–199.

Johnson, O., K. Neely & R. Waples. 2004. Lopsided fish in the Snake River Basin—fluctuating asymmetry as a way of assessing impact of hatchery supplementation in chinook salmon, *Oncorhynchus tshawytscha*. Environ. Biol. Fish. 69: 379–393.

Jørstad, K.E. 2004. Evidence for two highly differentiated herring groups at Goose Bank in the Barents Sea and the genetic relationship to Pacific herring, *Clupea pallasi*. Environ. Biol. Fish. 69: 211–221.

Jørstad, K.E., P.A. Prodöhl, A.-L. Agnalt, M. Hughes, A.P. Apostolidis, A. Triantafyllidis, E. Farestveit, T.S. Kristiansen, J. Mercer & T. Svåsand. 2004. Sub-arctic populations of European lobster, *Homarus gammarus*, in northern Norway. Environ. Biol. Fish. 69: 223–231.

Kim, J.E., R.E. Withler, C. Ritland & K.M. Cheng. 2004. Genetic variation within and between domesticated chinook salmon, *Oncorhynchus tshawytscha*, strains and their progenitor populations. Environ. Biol. Fish. 69: 371–378.

Kobayashi, T., S. Fushiki & K. Ueno. 2004. Improvement of sperm motility of sex-reversed male rainbow trout, *Oncorhynchus mykiss*, by incubation in high-pH artificial seminal plasma. Environ. Biol. Fish. 69: 419–425.

Matala, A.P., A.K. Gray, J. Heifetz & A.J. Gharrett. 2004. Population structure of Alaskan shortraker rockfish, *Sebastes borealis*, inferred from microsatellite variation. Environ. Biol. Fish. 69: 201–210.

McLean, J.E., P. Bentzen & T.P. Quinn. 2004. Differential reproductive success of sympatric, naturally spawning hatchery and wild steelhead, *Oncorhynchus mykiss*. Environ. Biol. Fish. 69: 359–369.

Miller, K.M., J.R. Winton, A.D. Schulze, M.K. Purcell & T.J. Ming. 2004. Major histocompatibility complex loci are associated with susceptibility of Atlantic salmon to infectious hematopoietic necrosis virus. Environ. Biol. Fish. 69: 307–316.

Nichols, K.M., P.A. Wheeler & G.H. Thorgaard. 2004. Quantitative trait loci analyses for meristic traits in *Oncorhynchus mykiss*. Environ. Biol. Fish. 69: 317–331.

Oleinik, A.G., L.A. Skurikhina, S.V. Frolov, V.A. Brykov & I.A. Chereshnev. 2004. Differences between two subspecies of Dolly Varden, *Salvelinus malma*, revealed by RFLP–PCR analysis of mitochondrial DNA. Environ. Biol. Fish. 69: 449–459.

Olsen, J.B., C. Habicht, J. Reynolds & J.E. Seeb. 2004. Moderately and highly polymorphic microsatellites provide discordant estimates of population divergence in sockeye salmon, *Oncorhynchus nerka*. Environ. Biol. Fish. 69: 261–273.

Overturf, K., D. Bullock, S. LaPatra & R. Hardy. 2004. Genetic selection and molecular analysis of domesticated rainbow trout for enhanced growth on alternative diet sources. Environ. Biol. Fish. 69: 409–418.

Phillips, R.B., M.P. Matsuoka, N.R. Konkol & S. McKay. 2004. Molecular systematics and evolution of the growth hormone introns in the Salmoninae. Environ. Biol. Fish. 69: 433–440.

Reynolds, J.H. & W.D. Templin. 2004. Detecting specific populations in mixtures. Environ. Biol. Fish. 69: 233–243.

Sato, S., H. Kojima, J. Ando, H. Ando, R.L. Wilmot, L.W. Seeb, V. Efremov, L. LeClair, W. Buchholz, D.-H. Jin, S. Urawa, M. Kaeriyama, A. Urano & S. Abe. 2004. Genetic population structure of chum salmon in the Pacific Rim inferred from mitochondrial DNA sequence variation. Environ. Biol. Fish. 69: 37–50.

Seamons, T.R., P. Bentzen & T.P. Quinn. 2004. The mating system of steelhead, *Oncorhynchus mykiss*, inferred by molecular analysis of parents and progeny. Environ. Biol. Fish. 69: 333–344.

Seeb, L.W., P.A. Crane, C.M. Kondzela, R.L. Wilmot, S. Urawa, N.V. Varnavskaya & J.E. Seeb. 2004. Migration of Pacific Rim chum salmon on the high seas: insights from genetic data. Environ. Biol. Fish. 69: 21–36.

Shubina, E.A., M.N. Mel'nikova, A.I. Glubokov & B.M. Mednikov. 2004. Analysis of the genetic structure of northwestern Bering Sea walleye pollock *Theragra chalcogramma*. Environ. Biol. Fish. 69: 177–185.

Templeton, A.R. 2004. Using haplotype trees for phylogeographic and species inference in fish populations. Environ. Biol. Fish. 69: 7–20.

Thrower, F., C. Guthrie III, J. Nielsen & J. Joyce. 2004. A comparison of genetic variation between an anadromous steelhead, *Oncorhynchus mykiss*, population and seven derived populations sequestered in freshwater for 70 years. Environ. Biol. Fish. 69: 111–125.

Wennevik, V., Ø. Skaala, S.F. Titov, I. Studyonov & G. Nævdal. 2004. Microsatellite variation in populations of Atlantic salmon from North Europe. Environ. Biol. Fish. 69: 143–152.

Withler, R.E., J.R. Candy, T.D. Beacham & K.M. Miller. 2004a. Forensic DNA analysis of Pacific salmonid samples for species and stock identification. Environ. Biol. Fish. 69: 275–285.

Withler, R.E., J.R. King, J.B. Marliave, B. Beaith, S. Li, K.J. Supernault & K.M. Miller. 2004b. Polygamous mating and high levels of genetic variation in lingcod, *Ophiodon elongatus*, of the Strait of Georgia, British Columbia. Environ. Biol. Fish. 69: 345–357.

Young, S.F., M.R. Downen & J.B. Shaklee. 2004a. Microsatellite DNA data indicate distinct native populations of kokanee, *Oncorhynchus nerka*, persist in the Lake Sammamish basin, Washington. Environ. Biol. Fish. 69: 63–79.

Young, S.F., J.G. McLellan & J.B. Shaklee. 2004b. Genetic integrity and microgeographic population structure of westslope cutthroat trout, *Oncorhynchus clarki lewisi*, in the Pend Oreille Basin in Washington. Environ. Biol. Fish. 69: 127–142.

Environmental Biology of Fishes **69**: 7–20, 2004.
© 2004 *Kluwer Academic Publishers. Printed in the Netherlands.*

Using haplotype trees for phylogeographic and species inference in fish populations

Alan R. Templeton
Department of Biology, Washington University, St. Louis, MO 63130-4899, U.S.A.
(e-mail: temple_a@biology.wustl.edu)

Received 17 April 2003   Accepted 19 April 2003

Key words: hybridization, history, diversity

Synopsis

Genetic variation is now routinely screened at the DNA sequence level in many studies. If the DNA region being screened has not experienced excessive amounts of recombination, it is often possible to reconstruct the evolutionary history of the genetic variation in the form of a haplotype tree. This tree estimates the evolutionary pathway that interconnects all the different haplotypes (sequence variants) observed in the sample. This haplotype tree can be used to define a series of nested branches (clades) that reflects the relative temporal history of the haplotypes and groups of haplotypes. Geographical information can then be overlaid upon this temporal series to test for significant associations between geography and temporal position in the haplotype tree. This allows a reconstruction of how the genetic variation arose and spread in both space and time. Such reconstructions can yield many insights into the joint roles of recurrent events such as gene flow and of historical events such as fragmentation or range expansion. These points are illustrated with studies on the chub, *Leuciscus cephalus*. There is also a need to extend such nested phylogeographic analyses to a phylo/reticulate geographic analysis that incorporates both assortment and recombination between and within DNA regions. A preliminary phylo/reticulate geographic analysis is presented of the transferrin locus in the brown trout, *Salmo trutta*, species complex that reveals the importance of hybridization in the recent evolutionary history of this group. This example shows the inadequacy of a strictly phylogenetic approach and illustrates the need to incorporate reticulate evolution. The results of nested clade phylogeographic analysis and the new phylo/reticulate geographic analysis are then used for inferring species status of the marbled trout. The results indicate that an old hybridization event may have played a role in the origin of the marbled trout. Currently the marbled trout is primarily endangered by hybridization with introduced brown trout. These results show both the positive and negative impacts of hybridization upon biodiversity. Such phylo/reticulate geographic studies will challenge both our concepts of species and our conservation management strategies.

Introduction

Many management decisions for fish populations ideally should be based upon a knowledge of the recent evolutionary history of the populations as well as upon an accurate and reliable assessment of their taxonomic status. One genetic tool for investigating both of these issues is a haplotype tree. When a region of homologous DNA is sequenced in many individuals sampled from one or more populations, many polymorphic nucleotide sites and small insertions and

deletions are typically encountered. A haplotype is a distinct genetic state of a single DNA molecule as simultaneously defined for all polymorphic sites within the sequenced region. Because all the DNA molecules being surveyed today are homologous, they must all be derived from a single common ancestral DNA molecule that existed in the past. The currently observed array of site variation arose as mutations occurred during some of the DNA replication events that interconnect the ancestral molecule to the current molecules. If the sequenced DNA region experiences

8

no recombination, the accumulation of these mutations through time completely determines the array of
current haplotypic variation. A haplotype tree is the
evolutionary sequence of mutations that defines the
current haplotypes through the DNA lineages that interconnect the current DNA molecules to the common
ancestral DNA molecule. A variety of algorithms can
be used to estimate this evolutionary history, including statistical parsimony (Templeton et al. 1992) and
maximum parsimony (Swofford 2002).

Some DNA regions are subject to recombination
and/or gene conversion, and these genetic processes
can generate further haplotypic diversity by creating
new combinations of the polymorphic sites. If recombination and gene conversion are common and uniformly
distributed throughout the sequenced region, most of
the observed haplotypic variation would have arisen
from recombination or gene conversion, making the
concept of a haplotype tree biologically meaningless.
However, when recombination is rare or when it is
concentrated at a hotspot, the haplotype variation can
be described by a set of haplotype trees that represent
the phylogenetic accumulation of mutational variation
that has either not been affected by recombination
or has arisen after recombination plus a finite number of estimated recombination/gene conversion events
(Templeton et al. 2000a,c). This paper will consider
only those cases in which recombination is absent or in
which it is sufficiently rare or concentrated as to allow
a partitioning of haplotypic variation into phylogenetic
and recombinational components.

The purpose of this paper is to show how haplotype
trees, with and without recombination, can be used as
a tool to address the recent evolutionary history of fish
populations and to make inference about species status.
A major theme of this paper is that by using multiple DNA regions or regions subject to recombination,
both phylogenetic history and historical reticulation
(hybridization) can be incorporated into an integrated
analysis which offers insights that go beyond a strict
phylogeographic analysis.

*The nested clade analysis for phylogeographic
inference*

The first use of haplotype trees to be discussed is the
inference of the recent evolutionary history of populations when recombination can be ignored. Such trees
are an important tool in phylogeography; that is, how
the populations spread through space and time to create

the observed geographical distributions of the current
populations. One method for using haplotype trees for
phylogeographic inference is the nested clade analysis
(Templeton et al. 1995). With this approach, the haplotype tree is used to define a series of hierarchically
nested clades (branches within branches). Such nested
hierarchies are commonly used in comparative evolutionary analyses of species or higher taxa, but can also
be applied to the haplotype variation found within a
species if that variation can be placed into a haplotype
tree (Templeton et al. 1987).

A set of rules is used to produce a nested series
of haplotypes and clades (Templeton et al. 1987,
Templeton & Sing 1993). To achieve the first level of
nesting, one starts at the tips of the haplotype tree. A tip
simply refers to a haplotype that is connected to the tree
by only one branch. In contrast, haplotypes with more
than one connecting branch in the tree represent interior
nodes of the tree. These haplotypes are therefore called
interior haplotypes. To create the first level of nested
haplotypes, move one mutational step into the interior
from the tips, and place all haplotypes that are intercon-
necting by this procedure into a single clade to produce
'1-step clades'. There may be many interior haplotypes
that are more than one mutational step from any tip
haplotype and were therefore not placed into a 1-step
clade by this procedure. In those cases, the initial set
of 1-step clades are pruned off the haplotype tree, and
the same nesting procedure is then applied to the more
interior portions of the pruned tree. Additional rounds
of pruning and nesting are repeated as needed until all
haplotypes have been placed into 1-step clades. Additional nesting rules are needed in case some haplotypes
are still left un-nested and to deal with ambiguities in
the topology of the haplotype tree (Templeton & Sing
1993). The second level of nesting uses the same rules,
but the rules are now applied to 1-step clades rather than
haplotypes and result in '2-step clades'. The nesting
procedure is repeated using 2-step clades as its units,
and so on until a nesting level is reached such that the
next higher nesting level would result in only a single
clade spanning the entire original haplotype network.

Nested haplotype trees contain much temporal information. When rooted, we know which clade is the
oldest one in any given nested category and which
clades are the younger mutational derivatives. Even if
the tree were unrooted, coalescent theory predicts that
tips are highly likely to be younger than the interiors
to which they are connected (Castelloe & Templeton
1994), so both rooted and unrooted trees contain temporal information in their nested clade hierarchies. This

relative temporal information extends to the higher nesting levels. In this manner, turning a haplotype tree into a series of nested clades captures much information about relative temporal orderings, although some aspects of time are left undefined. This partial information about temporal ordering can be used to analyze the spread of haplotypes and clades through space and time in a manner that does not depend upon a molecular clock or dating.

A nested clade analysis also requires a quantification of the spatial distribution of haplotypes and clades of haplotypes. The geographical data are quantified in two main fashions (Templeton et al. 1995). The first is the clade distance, D_c, which measures how widespread the clade is spatially. The clade distance is determined by calculating the average latitude and longitude for all observations of the clade in the sample, weighted by the local frequencies of the clade at each location. This estimates the geographical center for the clade. Next, the great circle distance from a location containing one or more members of the clade to the geographical center is calculated, and these distances are averaged over all locations containing the clade of interest, once again weighted by the frequency of the clade in the local sample. Sometimes geographical distance is not the most appropriate measure of space. For example, suppose a sample is taken of a riparian fish species. Because rivers do not flow in straight lines and because the fish are confined in their movements to the river, the geographical distance between two sample sites in the river is not relevant to the fish; rather, the important distance in this case is the distance between the two points going only along the river. In cases such as these, the investigator should define the distances between any two sample points in the most biologically relevant fashion, and the clade distance is now calculated as the average pairwise distance between all observations of the clade, once again weighted by local frequencies.

The second measure of geographical distribution of a haplotype or clade is the nested clade distance, D_n. The nested clade distance quantifies how far away a haplotype or clade is located from those haplotypes or clades to which it is most closely related evolutionarily; that is, the clades with which it is nested into a higher level clade. For geographical distance, the first step in calculating the nested clade distance is to find the geographical center for all individuals bearing members not only of the clade of interest, but also bearing any other clades that are nested with the clade of interest at the next higher level of nesting. This is the geographical center of the nesting clade. The nested clade distance is then calculated as the average distance that an individual bearing a haplotype from the clade of interest lies from the geographical center of the nesting clade. Once again, all averages are weighted by local frequencies. When the investigator defines the distances between sample locations, the nested clade distance is the average pairwise distance between an individual bearing a haplotype from the clade of interest to individuals bearing any haplotype from the nesting clade that contains the clade of interest.

Because of sampling artifacts, it is dangerous to make biological inferences from a visual overlay of geography upon a haplotype tree or from just the observed values of quantitative distance measurements. To adjust for sampling, the nested clade analysis first quantifies the degree of confidence in the quantitative distance measures by testing the null hypothesis that the haplotypes or clades nested within a high-level nesting clade show no geographical associations given their overall sample numbers. This null hypothesis is tested by randomly permuting the observations within a nesting clade across geographical locations in a manner that preserves the overall clade frequencies and sample sizes per locality (Templeton et al. 1995). After each random permutation, the clade and nested clade distances can be recalculated. By doing this a thousand or more times, the distribution of these distances under the null hypothesis of no geographical associations for a fixed frequency can be simulated. The observed clade and nested clade distances can then be contrasted to this null distribution, and we can infer which distances are statistically significant.

Because our biological interest in haplotype trees centers around how space and time are associated, some statistical power can be enhanced within a nesting clade by taking the average of the clade and nested clade distances for all the tips pooled together and subtracting the tip average from the corresponding average for the older interiors. The average interior-tip difference still captures the temporal contrast of old *versus* young within a nesting clade, but often has greater power to reject the null hypothesis of no geographical association.

Statistical significance is not the same as biological significance. Statistical significance tells us that the measures we are calculating are based upon a sufficient number of observations such that we can be confident that geographical associations exist with the haplotype tree. However, statistical significance alone does not tell us how to interpret those geographical associations. To arrive at biological significance, we must examine

10

how various types of recurrent gene flow or historical events can create specific patterns of geographical association.

For example, restricted gene flow creates associations between genetic variation and geography. Because restricted gene flow implies only limited movement by individuals during any given generation, it takes time for a newly arisen haplotype to spread geographically. Obviously, when a mutation first occurs, the resulting new haplotype is found only in its area of origin. With each passing generation, a haplotype lineage that persists has a greater and greater chance of spreading to additional locations via restricted gene flow. Hence, the clade distances should increase with time under a model of restricted gene flow. One of the more common types of restricted gene flow is isolation-by-distance (Wright 1943). Under recurrent gene flow restricted through isolation by distance, the spread of a haplotype through space occurs via small geographical movements in any given generation, resulting in a strong correlation between how widespread a haplotype (or clade) is (as measured by D_c) and its temporal position in the haplotype tree. The older the haplotype, the more widespread it is expected to be. Moreover, newer haplotypes are found within the geographical range of the haplotype from which they were derived (taking into account sampling error), and since geographical centers move slowly under this model, the clade and nested distances should yield similar patterns of statistical significance. The expectations under isolation by distance are illustrated by some of the results from a nested clade analysis of mtDNA from the chub, *Leuciscus cephalus* (Durand et al. 1999). Figure 1 shows the geographical ranges of the interior and tip clades found in two different nested clades. In both cases, the geographical ranges of the clades increase as we go from younger to older clades, and these changes were associated with statistically significant distance measures.

Historical events can also create strong associations between haplotypes and geography. One such event is past fragmentation followed by complete or nearly complete genetic isolation. Because of genetic isolation, the haplotypes or clades that arose after fragmentation but in the same population will show concordant restricted spatial distributions that correspond to the geographical area occupied by the isolates in which they arose. If the fragmentation event lasts longer than the typical time to coalescence to a common ancestral molecule, the isolate will be marked as a monophyletic clade in the haplotype tree. If the fragmentation event

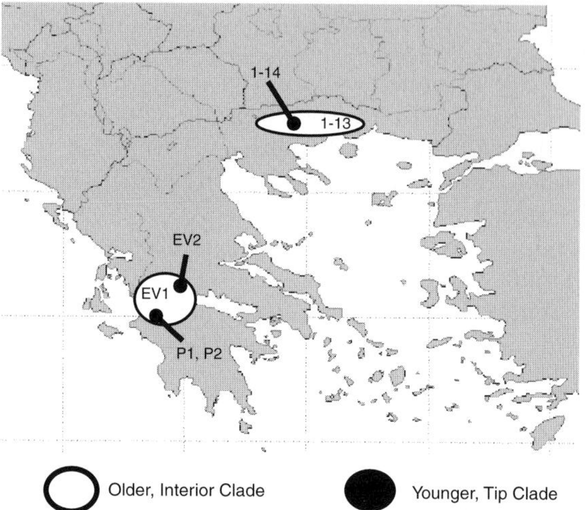

Figure 1. The expectations under isolation by distance as illustrated by some of the results from a nested clade analysis of mtDNA from the chub, *L. cephalus* (Durand et al. 1999). The map shows the areas containing the sampled rivers that drain into the Adriatic, Ionian, Aegean and Black Seas. Ovals show the approximate geographic distributions of haplotypes or clades. The geographic distributions found in two different nesting clades are shown that yield a statistically significant inference of gene flow restricted by isolation by distance. The first case contains the older interior haplotype EV1 and three younger tip haplotypes derived from it, EV2, P1 and P2. The second case contains the older interior clades 1-13 and the younger tip clades derived from it, 1-14.

is much older than the coalescent time, many mutations should accumulate, resulting in the clades that mark the different isolates being interconnected with branch lengths that are much longer than the average branch length in the tree. This pattern is illustrated in Figure 2, which also comes from the nested clade analysis of mtDNA from the chub, *L. cephalus* (Durand et al. 1999). However, not all cases of fragmentation and isolation are marked by strict monophyly of haplotype tree clades, and strict monophyly can also be destroyed by subsequent admixture events (Templeton 2001). Therefore, a strict monophyletic correspondence of clades with geography is a strong but not necessary indicator of fragmentation. Regardless of whether there is monophyly or not, haplotypes or clades that arose after the fragmentation event cannot spread beyond the confines of the isolate in which they arose. This means that the clade distance cannot increase beyond the geographical ranges of the fragmented isolates. Even if this clade had been introduced to another isolate by some rare admixture or dispersal event, the frequency of the clade in the other isolate will generally be rare

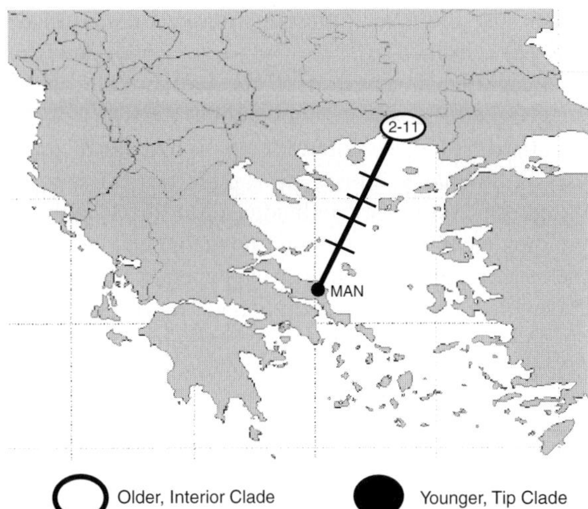

Older, Interior Clade   Younger, Tip Clade

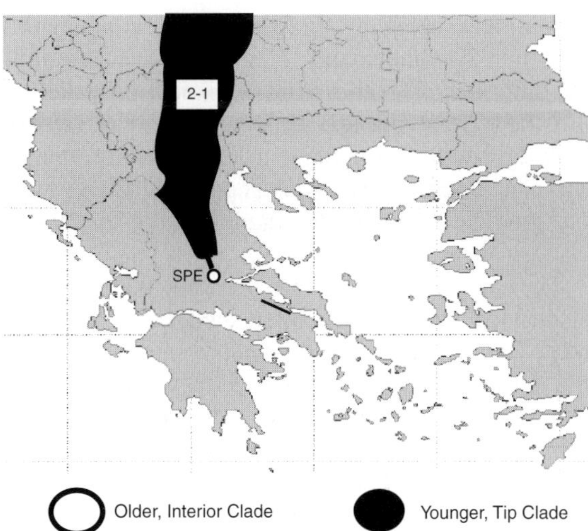

Older, Interior Clade   Younger, Tip Clade

Figure 2. The expectations under population fragmentation as illustrated by some of the results from a nested clade analysis of mtDNA from the chub, *L. cephalus* (Durand et al. 1999). The map shows the areas containing the sampled rivers that drain into the Adriatic, Ionian, Aegean and Black Seas. Ovals show the approximate geographic distributions of two clades from a nesting clade that yields a statistically significant inference of fragmentation. The older interior clades 2–11 and the younger tip haplotype dereived from it is MAN. Tic marks on the line connecting these two clades indicate that four mutational changes have accumulated between them.

Figure 3. The expectations under population range expansion as illustrated by some of the results from a nested clade analysis of mtDNA from the chub, *L. cephalus* (Durand et al. 1999). The map shows the areas containing the sampled rivers that drain into the Adriatic, Ionian, Aegean and Black Seas. Ovals show the approximate geographic distributions of two clades from a nesting clade that yields a statistically significant inference of range expansion. The older interior clade is SPE and the younger tip clade derived from it is 2-1.

as long as isolation is the norm. Such rarely occurring admixture events therefore have little impact on clade distance. Although the magnitude of the clade distance is severely restricted by fragmentation, the same is not true for the nested clade distances. The nested clade distances can suddenly become much larger than the clade distances when the nesting clade contains haplotypes or clades found in other isolates. This is also shown in Figure 2, where the limited geographical ranges of clades 2–11 and MAN indicate that they have small clade distances, but the nested clade distances are large because the isolates are located in distant geographical regions within the area sampled.

Another type of historical event that can create strong geographical associations is range expansion. When range expansion occurs, those haplotypes found in the ancestral population/s that were the source of the range expansion can become widespread geographically (large clade distances), including relatively young haplotypes or clades that are globally rare but that were present in the expanding population. Young, rare haplotypes in the ancestral source population that are carried along with the population range expansion will have clade distances that are large for their frequency and temporal position. Another pattern of range expansion is created by those haplotypes or clades that arise by mutation in the newly colonized areas. Such new haplotypes or clades tend to be tips, may have small clade distances, but will often be located far from the geographical center of their ancestral range, resulting in large nested clade distances. Figure 3 shows the approximate geographical distributions of two clades that contributed to an inference of range expansion in the chub (Durand et al. 1999). As can be seen from that figure, the younger clade has a much more widespread geographical distribution than its immediate ancestral clade – one of the patterns associated with range expansion and exactly the opposite of that associated with isolation by distance (Figure 1).

No single test statistic discriminates between recurrent gene flow, past fragmentation, and past range expansion in the nested clade analysis. Rather, it is a pattern formed from several statistics that allows discrimination. Also, as indicated in the above discussion, many different patterns can sometimes lead to the same biological conclusion. Finally, sometimes the pattern associated with significant clade and nested clade distances is an artifact of inadequate geographical sampling. In light of these complexities (which reflect the reality of evolutionary possibilities and sampling constraints), an inference key is provided as an appendix to Templeton et al. (1995), with the latest version being available at http://bioag.byu.edu/zoology/crandall_lab/geodis.htm along with the program GEODIS for implementing the nested clade analysis.

The use of this inference key not only protects against making biological inference affected by inadequate geographical sampling, but is essential in searching out multiple, overlaying patterns within the same data set. As Figures 1–3 show, the chub populations were influenced by a mixture of isolation by distance, fragmentation, and range expansion. There is nothing about the evolutionary factors of restricted gene flow, fragmentation events, or range expansion events that make them mutually exclusive alternatives. One of the great strengths of the nested clade inference procedure is that it explicitly searches for the combination of factors that best explains the current distribution of genetic variation and does not make *a priori* assumptions that certain factors should be excluded or ignored. Moreover, by using the temporal polarity inherent in a nested design (or by outgroups when available), the various factors influencing current distributions of genetic variation are reconstructed as a dynamic process through time. Hence, nested clade analysis does not merely identify and geographically localize the various factors influencing the spatial distribution of genetic variation, rather it brings out the dynamical structure and temporal juxtaposition of these evolutionary factors.

Limitations of nested clade analysis for phylogeographic inference

Although the nested clade approach to phylogeographic inference has many strengths, it does have limitations. In particular, inference is limited by (1) sample size and sample sites, (2) insufficient genetic resolution to detect an event or process that actually occurred, (3) false inferences arising from the evolutionary stochasticity of the coalescent process itself or by the haplotype tree being skewed or otherwise altered by natural selection, (4) the failure of the original inference key and statistical structure to incorporate important evolutionary processes, such as secondary contact and hybridization and (5) the dependence of the nested clade analysis upon a strictly phylogenetic structure for the analyzed haplotypic variation.

Because biological interpretation in a nested clade analysis is limited to those distance statistics that result in a significant rejection of the null hypothesis of no geographical associations within a nesting clade, the ability to make inference in nested clade analysis is obviously limited by sample size. A sample based upon only a few individuals has little chance of yielding meaningful inference, no matter how dramatic the resulting genetic patterns may appear. In addition, even when significant geographical associations are detected, the inference key may lead to the conclusion that there has been an inadequate geographical sampling for unambiguous biological interpretation. When these sampling limitations are encountered, an investigator can only circumvent them by additional sampling, either additional individuals per site and/or additional sites. Such sampling limitations can be regarded as a strength of the nested clade approach rather than a weakness. The sampling considerations ensure that when inference is made, it is based upon adequate sampling of individuals and sites. Moreover, when the inference key results in the conclusion of inadequate sampling, the regions that need to be sampled are identified. Hence, the nested clade analysis provides specific guidance for future sampling efforts.

The next two limitations are the failure to detect events or processes because of insufficient genetic resolution (as opposed to inadequate sampling) and the danger of false inferences. Both of these are real possibilities. For example, Templeton (1998a) validated the original inference criteria for range expansion by examining 12 actual biological examples for which strong prior evidence existed that range expansion had indeed occurred, all with good sampling, and four examples for which there was no prior evidence for range expansion. The nested clade analysis correctly identified 11 of the 12 known range expansions, but one range expansion yielded no significant statistics and was undetected. The failure to detect this known range expansion was shown to be not due to inadequate sampling, but rather due to the fact that an appropriate mutation had not occurred in the right place and time to mark the expansion event. This shows that no one locus or DNA region can capture the totality of a species' population structure and recent evolutionary history.

Of the four cases with no prior evidence for range expansion, one yielded a significant inference of range expansion (Templeton 1998a). This inference was not necessarily wrong, just not known from prior evidence. However, this inference could have been a false positive. The processes of mutation and genetic drift, which shape the haplotype tree upon which the nested analysis is based, are both random processes, so sometimes the expected pattern will not arise just by chance alone. Moreover, natural selection can skew both the shape of the haplotype tree and the geographical distribution of certain haplotypes, thereby creating patterns that do not necessarily reflect phylogeographic processes. This can yield false inferences. Regarding the one inferred range expansion without prior evidence as a false inference, the resulting two-by-two contingency table of prior evidence for range expansion and no prior evidence *versus* inferred range expansion and no inferred range expansion is significant at the 0.03 level with a Fisher's exact test (Templeton 1998a). This result shows that the inference criteria do well most of the time, but that failure to detect known events and false inferences are both possible.

Both of these limitations can be circumvented by performing nested clade analyses on many loci or gene regions. By studying multiple DNA regions, an investigator can obtain a more complete evolutionary history, thereby reducing the danger of missing an event or process due to the lack of an appropriately placed mutation in time and space in any one DNA region. The chances of making a false inference can be reduced with multiple loci by cross validating inferences across DNA regions. This multi-locus approach with cross validation was used in a nested clade analysis of recent human evolutionary history using ten different DNA regions; the human mitochondrial genome and nine nuclear genome regions, including Y-, X-linked and autosomal regions (Templeton 2002). What was most remarkable about the cross validated inferences in this case was the high degree of incompleteness found in the analysis of any one DNA region. This illustrates that failure to detect events or processes is a common phenomenon, so any analysis based upon a single DNA region should be regarded as incomplete. Interestingly, most inferences were cross validated by two or more DNA regions, thereby indicating that the problem of false inferences may not be so common an occurrence as the failure to make an inference.

The fourth limitation, the failure to incorporate into the inference key such important evolutionary events and processes such as secondary contact and hybridization, is addressed in Templeton (2001). New statistics were presented in that paper that allow the detection of secondary contact, although not necessarily hybridization. The multi-locus approaches used in the study of human evolution also provide a powerful tool for examining the roles of secondary contact and hybridization.

Hybridization makes possible assortment or recombination between DNA regions and recombination within nuclear DNA regions.

The possibility of recombination relates to the final limitation of the nested clade analysis; namely that it is a phylogeographic analysis that is limited to that portion of the haplotypic variation that arose through phylogenetic processes. If recombination is rare, the recombinant haplotypes can be excluded from the analysis with little erosion of power, as was done in the analysis of the nuclear DNA regions of humans (Templeton 2002). However, sometimes recombination is sufficiently common that it represents a major contributor to current haplotype variation, yet it is still sufficiently rare that much phylogenetic structure remains (Templeton et al. 2000c). To incorporate both phylogeny and recombination requires a new type of analysis that goes beyond phylogeography; a phylo/reticulate geographic analysis. The next section presents an example of an attempt at such a phylo/reticulate analysis in the species complex of brown trout, *Salmo trutta*.

The phylo/reticulate geography of brown trout

The brown trout, *S. trutta* L., is the most widely distributed freshwater fish native to the Palearctic region, being found from Norway to North Africa and from Iceland to Afganistan (Bernatchez 2001). The brown trout is a polytypic species or a species complex with much phenotypic diversity and life history variation, including anadromous, fluviatile and lacustrine modes of life (Behnke 1972). Over the last two decades, geographical patterns of extreme genetic differentiation were observed with allozymes (e.g. Bouza et al. 1999) and mtDNA (e.g. Bernatchez 2001). Bernatchez (2001) performed a nested clade analysis of mtDNA haplotype variation in 1 794 trout from 174 populations. This nested clade analysis indicated the existence of five major evolutionary lineages that evolved in geographic isolation during the Pleistocene and that have remained largely allopatric since then, as well as finer phylogeographic structuring within some of these major lineages. The five major evolutionary lineages are the Danubian (DA), the Adriatic (AD), the marbled trout (MA), the Mediterranean (ME) and the Atlantic (AT) (Bernatchez 2001), with the Atlantic lineage being the first to be fragmented from the others. The distribution of allozyme variants is partially congruent with these lineages (García-Marín et al. 1996).

14

One of the allozyme loci frequently used to discriminate salmonid populations is the Transferrin (*TF*) locus that codes for an iron binding protein found in vertebrate blood serum and interstitial spaces. The brown trout shows considerable population genetic differentiation for the electromorphs at this locus (Antunes et al. 2002). The Atlantic populations of the brown trout are often fixed for the electromorph *TF*100*, although southwestern Atlantic populations have moderate to high frequencies of the *TF*95* electromorph. The Mediterranean brown trout is often fixed for the *TF*102* electromorph, but some populations also have the *TF*80* allele. The marble trout, *S. trutta marmoratus*, is often fixed for the *TF*75* allele (except in cases of hybridization with introduced stock), and the Italian carpione, *S. trutta carpio*, for the *TF*78* allele.

Antunes et al. (2002) sequenced 3 696 bp from 62 copies of the *TF* gene in individuals representing the above electromorph categories, in addition to four copies from the Atlantic salmon (*S. salar*) as an outgroup. Two major complications exist when attempting to perform a phylogeographic analysis with nuclear DNA such as the *TF* locus. The first complication is haplotype determination. Because *TF* is an autosomal locus, individuals bear two copies of the gene. When an individual is heterozygous for two or more polymorphic sites, the phase of these sites is ambiguous. For *TF*, haplotypes were determined by the haplotype-substraction method (Clark 1990) coupled with some molecular cloning. However, molecular cloning and other molecular methods for phasing are currently expensive and time consuming, so it was not feasible to resolve all ambiguities. As a result, the resulting haplotype tree has loops that represent ambiguity in determining the haplotypes.

The second complication when studying nuclear DNA is recombination. Antunes et al. (2002) used the method of Crandall & Templeton (1999) as modified by Templeton et al. (2000a) to detect several statistically significant recombination and/or gene conversion events that have helped shaped the haplotypic variation at this locus. Antunes et al. (2002) then used the procedures outlined in Templeton et al. (2000c) to partition the evolutionary history of the *TF* haplotypic variation into a portion that reflects only the phylogenetic accumulation of mutations uninfluenced by recombination or gene conversion events (Figure 4) and a portion that was created by recombination and/or gene conversion events followed by post-recombinant phylogenetic accumulation of mutations (Figure 5).

A nested clade analysis cannot be applied to the totality of the haplotypic variation shown in Figures 4 and 5. For example, the nesting rules do not apply to recombinant clades because there is no nested phylogenetic hierarchy in this case. Excluding the haplotypic variation that has been influenced by recombination (Figure 5) would in this case exclude a substantial portion of the data and throw away much potential information. One of the great needs in working with nuclear DNA subject to recombination is the development of a formal statistical framework analogous to nested clade analysis that explicitly incorporates both phylogeny and reticulation through recombination and/or gene conversion. Such a framework does not yet exist, but the potential of an integrated phylo/reticulate geographic analysis can be illustrated with the *TF* data.

Although nested hierarchies cannot be defined for the total dataset, the relative temporal relationship of recombinant to non-recombinant clades of haplotypes can still be determined. A recombinant clade must be younger than the youngest parental type involved in its origin. Hence, in lieu of a formal statistical analysis, it is possible to go through a temporal hierarchy that integrates Figures 4 and 5 with geographical distributions. Figure 6 shows the oldest portion of the *TF* evolutionary history, a portion of Figure 4 near the outgroup root that has never experienced detectable recombination. Figure 6 indicates that the brown trout originated in the Black, Caspian and Aral Sea drainages (as marked by the TF-BCA clade and largely corresponding to the DA mtDNA lineage). Next was an expansion into the Adriatic and Mediterranean, as marked by clades such as TF-102 that are derived from the node labeled 'T-1' in Figure 4.

Westward expansion and subsequent phylogenetic evolution continued with the spread of the TF-95 clade from the Adriatic and Mediterranean into the Atlantic (Figure 7). Figure 7 also shows a gene conversion event that occurred between parental types in the TF-BCA clade and the Adriatic/Mediterranean clade. Hence, a hybridization event must have occurred between fish originally separated by the major expansion illustrated in Figure 6. This recombinant lineage is an old one, as indicated by the large number of mutations that have accumulated since it originally occurred (Figure 5). This recombinant lineage is the one found in present-day populations of marbled trout (the TF-75 clade).

Figure 8 shows the most recent portions of the *TF* phylogeny and recombination/gene conversion events.

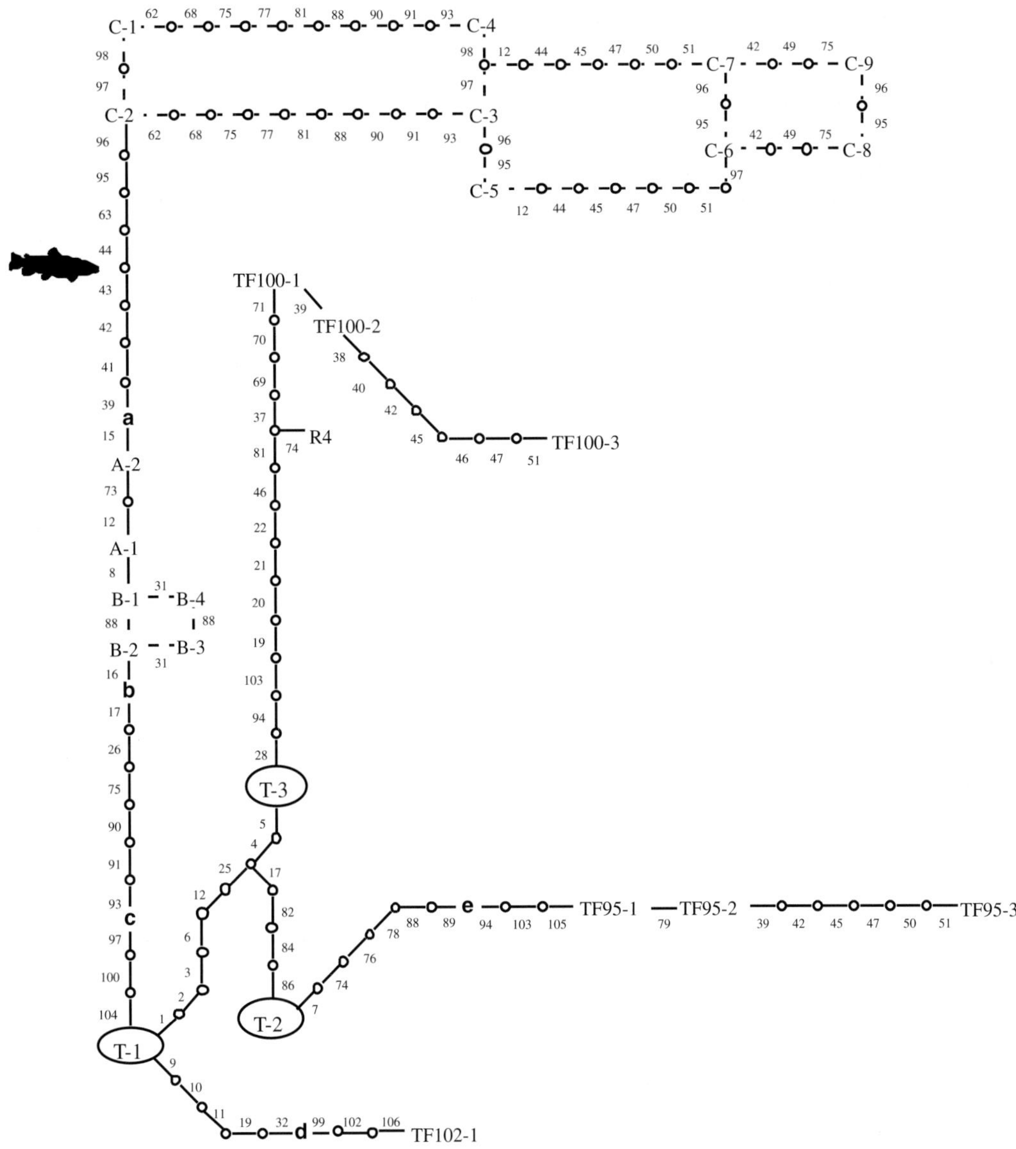

Figure 4. The estimated haplotype tree for the Transferrin haplotypes unaffected by recombination or gene conversion. Small circles indicate nodes in the tree that represent intermediate haplotype states not found in the sample. Each line (solid or dashed) represents a single mutational event. The site involved in the mutation is indicated near the line by a small boldface number, with the numbers corresponding to the variable site numbers given in Antunes et al. (2002). Dashed lines indicate where loops or alternative haplotype phasing create ambiguity in the topology of the tree. Nodes that define three major clades are indicated by an oval containing T-i, where i can be 1, 2 or 3. Other nodes involved in recombination events are indicated by the boldface lowercase letters a–e. The black salmonid fish indicates the connection with the outgroup (*S. salar*) and hence indicates the rooting of the tree.

Figure 8 reveals a range expansion of the Atlantic populations of the brown trout (the TF-100 clade) and another important recombination event involving parents within the Adriatic/Mediterranean populations that produced through subsequent phylogenetic evolution the TF-80 clade found in Mediterranean populations and the TF-78 clade found in the Italian carpione. These findings of repeated hybridization and recombination giving rise to major *TF* clades and population lineages show the inadequacy of a

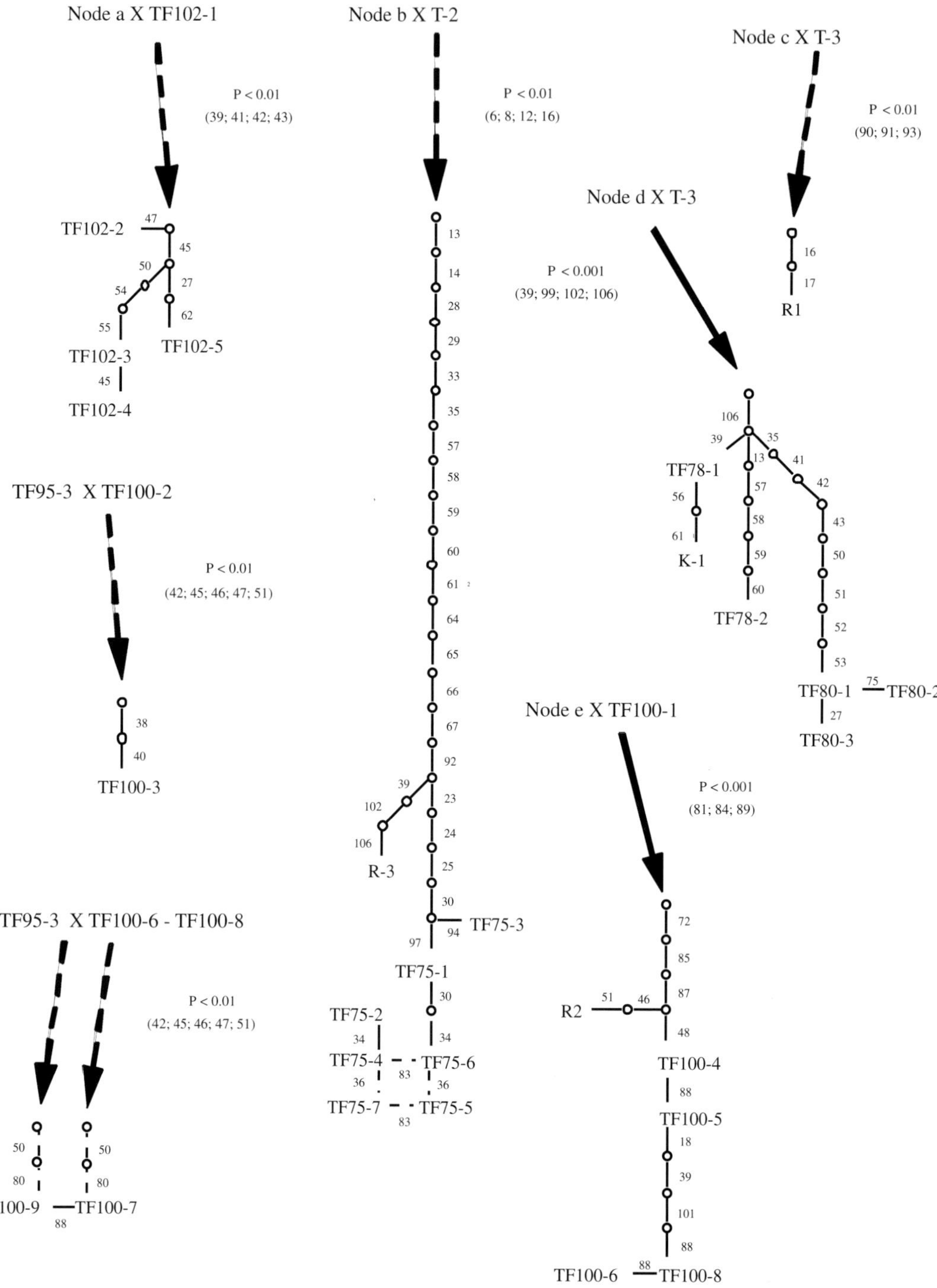

Figure 5. The estimated recombinant clades for Transferrin. Solid and dashed arrows represent recombination and gene conversion events, respectively. The p-value near the arrow corresponds to the tail probability from the test given in Templeton et al. (2000a) of the null hypothesis of no recombination or gene conversion, and the spanned variable sites involved in the crossover event are indicated in parentheses. The post-recombinational phylogenetic evolution is estimated through statistical parsimony, with the layout being the same as that given in Figure 4.

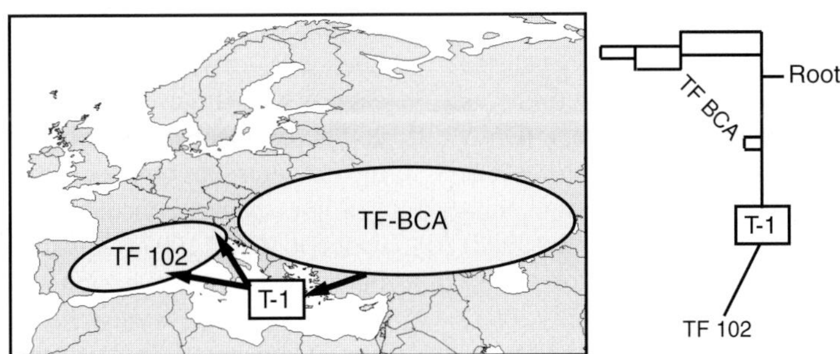

Figure 6. The early phylogeography of the brown trout species complex as inferred from the *TF* haplotype tree. Ovals show approximate geographic distributions of clades, and solid arrows show phylogenetic transitions between clades. The basic topology of the haplotype phylogeny is shown to the right, as shown in more detail in Figure 4.

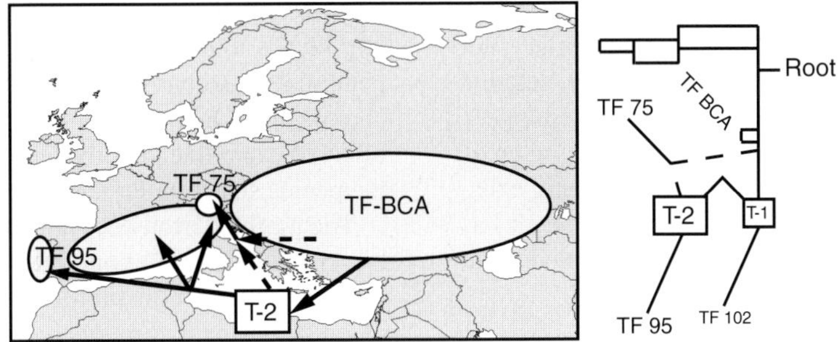

Figure 7. The middle phylo/reticulate geography of the brown trout species complex as inferred from the TF haplotypes. Ovals show approximate geographic distributions of clades, solid arrows show phylogenetic transitions between clades, dashed arrows show recombination/gene conversion events. The basic topology of the haplotype phylogeny is shown to the right, as shown in more detail in Figure 4, and the dashed lines indicate recombination and/or gene conversion events as shown in Figure 5.

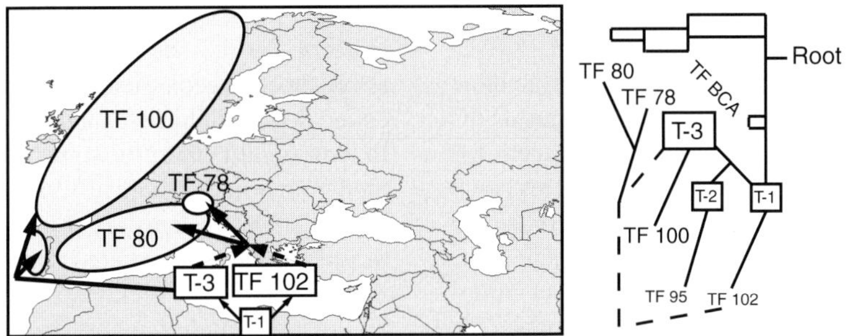

Figure 8. The more recent phylo/reticulate geography of the brown trout species complex as inferred from the TF haplotypes. Ovals show approximate geographic distributions of clades, solid arrows show phylogenetic transitions between clades, dashed arrows show recombination/gene conversion events. The basic topology of the haplotype phylogeny is shown to the right, as shown in more detail in Figure 4, and the dashed lines indicate recombination and/or gene conversion events as shown in Figure 5.

strictly phylogenetic approach and illustrate the need to incorporate reticulate evolution.

The evolutionary history of the brown trout as inferred from Figures 6 to 8 may at first seem somewhat incompatible with the results of the nested clade analysis of mtDNA (Bernatchez 2001). For example, Figures 6–8 imply that the brown trout complex originated in the Black, Caspian and Aral Sea drainages, followed by westward expansion with the Atlantic populations being the most recently established. In contrast, Bernatchez (2001) inferred that the Atlantic populations represent the oldest fragmented lineage.

18

However, there is no incompatibility here. Basic coalescent theory predicts that mtDNA should coalesce to a common ancestral from roughly four times as rapidly as autosomal DNA. As a consequence, mtDNA in general is sampling a different and more recent time period of evolutionary history than autosomal DNA (Templeton 2002). Indeed, Antunes et al. (2002) date the origin and westward expansion of brown trout to pre-Pleistocene times, whereas Bernatchez (2001) dates the fragmentation of brown trout populations to the Pleistocene. There is no incompatibility of a pre-Pleistocene origin of the brown trout in western Asia followed by a pre-Pleistocene westward expansion to the Atlantic, which in turn is followed by Pleistocene fragmentation events going from west to east. The mtDNA and *TF* data are therefore complementary, not contradictory, and together provide a more detailed history of the brown trout over a broader time range.

The mtDNA and *TF* data also result in many cross validated results, such as the marbled trout being a distinct evolutionary lineage. This concordant result has implications for the species status of the marbled trout, as will now be discussed.

Species inference

All species inference depends upon the species concept being used. Hybridization and introgression among brown trout lineages has obviously occurred in the past and continues today (Bernatchez et al. 1992, Povz 1995, Antunes et al. 2002). Accordingly, the biological species concept that defines species in terms of reproductive isolating mechanisms does not seem to be applicable to this group. An alternative species concept is the cohesion species concept (Templeton 1989). A cohesion species is an evolutionary lineage that maintains its cohesiveness as a lineage over time because it is a reproductive community capable of exchanging gametes and/or an ecological community sharing a derived adaptation or adaptations needed for successful reproduction. One advantage of this concept is that it can be implemented as a set of testable null hypotheses (Templeton 1998b, 1999, 2001, Templeton et al. 2000b).

The cohesion species is first and foremost an evolutionary lineage. Therefore, the first null hypothesis to be tested is that the organisms sampled are derived from a single evolutionary lineage. This null hypothesis was formally rejected for the brown trout by the nested clade analysis of mtDNA by Bernatchez (2001). The mtDNA inference of the marbled trout as a distinct evolutionary lineage is cross validated by its being a distinct *TF* lineage as well (Figure 7). Hence, the first null hypothesis is clearly rejected, one of the several lineages within the brown trout complex is the marbled trout.

Some species concepts (e.g. the diagnostic species concept, Hull 1997) equate species to evolutionary lineages, but the cohesion species concept does not. Cohesion species must be at least a single evolutionary lineage, but they can also contain more than one evolutionary lineage if those lineages have not significantly diverged with respect to the attributes that define cohesion mechanisms (Templeton 1989). Hence, when the first null hypothesis is rejected, the inference chain must proceed to testing a second null hypothesis that the previously identified lineages are genetically exchangeable and/or ecologically interchangeable. Only when this second null hypothesis is rejected can a lineage or group of lineages be elevated to species status under the cohesion concept.

The marbled trout does not appear to be well defined by reproductive isolation; indeed, this failure of reproductive isolation constitutes its greatest current source of endangerment through hybridization with continuously and massively introduced brown trout from other lineages. Therefore, this second null hypothesis cannot be rejected on the basis of a lack of genetic exchangeability. Reproductive isolation is not necessary (although it is sufficient given an evolutionary lineage) for species status under the cohesion species concept, so the second null hypothesis can also be tested through ecological interchangeability (originally called demographic exchangeability, Templeton 1989). In this regard, there is some evidence that the marbled trout is not ecologically interchangeable with other brown trout lineages. First, the marbled trout lives in upland streams, whereas other brown trout live in middle and lower sections (Povz 1995). Second, not all upland streams support marbled trout, which seem to be sensitive to the geological composition of the ground as well (Sommani 1960). These results suggest that the second null hypothesis can be rejected on the basis of a lack of ecological interchangeability of marble and brown trout. If ecological interchangeability is indeed rejected, then the distinct evolutionary lineage of marble trout should be elevated to species status under the cohesion species concept. However, the evidence is only suggestive at present. Detailed quantitative or qualitative ecological data on the marbled trout and other brown trout lineages needs to be gathered in order to execute a formal statistical test of the null hypothesis of ecological interchangeability before the species status of the marbled trout can be more definitively inferred.

Phylo/reticulate geography and the role of hybridization

Implementing the cohesion concept through testable null hypotheses requires much data, but the data rich inference procedure of the cohesion concept automatically insures much insight into likely contributors to the speciation process. This is evident for the marbled trout example. The nested clade analysis of Bernatchez (2001) and the phylo/reticulate geographic analysis of Antunes et al. (2002) on *TF* jointly indicate that the marbled trout lineage arose after a hybridization event between the DA lineage and the Mediterranean/Adriatic lineage followed by allopatric fragmentation. Neither analysis definitely indicates whether hybridization or fragmentation or both is the major contributor to this likely speciation event, but these analyses do indicate that hybridization and fragmentation were potentially involved in the process of speciation. Note that the nested clade analysis of mtDNA alone did not – indeed, cannot – identify ancient hybridization as a factor in the evolutionary history of marbled trout. Only by going to nuclear genes subject to recombination or to multi-locus systems (this would include using morphology as a proxy for a multi-locus system, as in inferring hybridization through discrepancies between mtDNA haplotypes and morphology) can phylogeography be extended to phylo/reticulate geography. This extension is a critical one for assessing the role of hybridization in speciation and microevolution.

Botanists have long recognized the importance of hybridization in plant evolution (Rieseberg 1997), but hybridization has received less attention by zoologists. Part of the reason for this difference has been the dominance of the biological species concept among zoologists. The biological species concept downplays the role of hybridization, and indeed treats it at best as an inconvenience and at worst as something negative and destructive to genotypic integrity (O'Brien & Mayr 1991). There is no doubt that hybridization can sometimes play such a destructive role. Indeed, the primary danger for the marbled trout at present is hybridization with continuously and massively introduced populations of brown trout (Povz 1995). However, the phylo/reticulate geographic

study of Antunes et al. (2002) shows that an ancient hybridization event was present at the origin of the lineage that is now called the marbled trout. Thus, this taxon owes its origin in part or in whole to hybridization. Hybridization therefore plays a dual role in both evolution and conservation; it can trigger speciation and thereby enhance biodiversity, or it can diminish biodiversity. Both roles need to be appreciated.

The ability to perform phylo/reticulate geographic analyses provides a tool for studying the creative role of hybridization in animal speciation and evolution. As shown by the example of the marbled trout, such phylo/reticulate geographic studies will challenge both our concepts of species and our conservation management strategies.

Acknowledgements

I thank A.J. Gharrett and the University of Alaska Sea Grant program for the opportunity to present this material at the 20th Lowell Wakefield Fisheries Symposium. This work also greatly benefitted from comments and input from Agostinho Antunes, Louis Bernatchez and an anonymous reviewer. This work supported in part by NIH grant R01 GM60730.

References

Antunes, A., A.R. Templeton, R. Guyomard & P. Alexandrino. 2002. The role of nuclear genes in intraspecific evolutionary inference: Genealogy of the transferrin gene in the brown trout. Mol. Biol. Evol. 19: 1272–1287.

Behnke, R.J. 1972. The systematics of salmonid fishes of recently glaciated lakes. J. Fish. Res. Board Can. 29: 639–671.

Bernatchez, L. 2001. The evolutionary history of brown trout (*Salmo trutta* L.) inferred from phylogeographic, nested clade, and mismatch analyses of mitochondrial DNA variation. Evolution 55: 351–379.

Bernatchez, L., R. Guyomard & R. Bonhomme. 1992. DNA sequence variation of the mitochondrial control region among geographically and morphologically remote European brown trout *Salmo trutta* populations. Mol. Ecol. 1: 161–173.

Bouza, C., J. Arias, J. Castro, L. Sánchez & P. Martínez. 1999. Genetic structure of brown trout, *Salmo trutta* L., at the southern limit of the distribution range of the anadromous form. Mol. Ecol. 8: 1991–2002.

Castelloe, J. & A.R. Templeton. 1994. Root probabilities for intraspecific gene trees under neutral coalescent theory. Mol. Phylogen. Evol. 3: 102–113.

Clark, A.G. 1990. Inference of haplotypes from PCR-amplified samples of diploid populations. Mol. Biol. Evol. 7: 111–122.

Crandall, K.A. & A.R. Templeton. 1999. Statistical approaches to detecting recombination. pp. 153–176. *In*: K.A. Crandall (ed.)

The Evolution of HIV, The Johns Hopkins University Press, Baltimore.

Durand, J.D., A.R. Templeton, B. Guinand, A. Imsiridou & Y. Bouvet. 1999. Nested clade and phylogeographic analyses of the chub, *Leuciscus cephalus* (Teleostei, Cyprinidae) in Greece: Implications for Balkan Peninsula biogeography. Mol. Phylogen. Evol. 13: 566–580.

García-Marín, J.L., F.M. Utter & C. Pla. 1996. Origins and relationships of native populations of *Salmo trutta* (brown trout) in Spain. Heredity 77: 313–323.

Hull, D.L. 1997. The ideal species concept – and why we can't get it. pp. 357–380. *In*: M.F. Claridge, H.A. Dawah & M.R. Wilson (ed.) Species: The Units of Biodiversity, Chapman & Hall, London.

O'Brien, S.J. & E. Mayr. 1991. Bureaucratic mischief: Recognizing endangered species and subspecies. Science 251: 1187–1188.

Povz, M. 1995. Status of freshwater fishes in the Adriatic catchment of Slovenia. Biol. Conserv. 72: 171–177.

Rieseberg, L.H. 1997. Hybrid origins of plant species. Ann. Rev. Ecol. Syst. 28: 359–389.

Sommani, E. 1960. II *Salmo marmoratus* Cuv. sua origine e distribuzione nell'Italia settentrionale. Bolletino di Pesca, Piscicoltura e Idrobiologia 15: 40–47.

Swofford, D. 2002. PAUP*: Phylogenetic Analysis Using Parsimony (and other methods). Version 4. Sinauer, Sunderland, MA.

Templeton, A.R. 1989. The meaning of species and speciation: A genetic perspective. pp. 3–27. *In*: D. Otte & J.A. Endler (ed.) Speciation and its Consequences, Sinauer, Sunderland, Massachusetts.

Templeton, A.R. 1998a. Nested clade analyses of phylogeographic data: Testing hypotheses about gene flow and population history. Mol. Ecol. 7: 381–397.

Templeton, A.R. 1998b. Species and speciation: Geography, population structure, ecology, and gene trees. pp. 32–43. *In*: D.J. Howard & S.H. Berlocher (ed.) Endless Forms: Species and Speciation, Oxford University Press, Oxford.

Templeton, A.R. 1999. Using gene trees to infer species from testable null hypothesis: Cohesion species in the *Spalax ehrenbergi* complex. pp. 171–192. *In*: S.P. Wasser (ed.) Evolutionary Theory and Processes: Modern Perspectives, Papers in Honour of Eviatar Nevo, Kluwer Academic Publishers, Dordrecht.

Templeton, A.R. 2001. Using phylogeographic analyses of gene trees to test species status and processes. Mol. Ecol. 10: 779–791.

Templeton, A.R. 2002. Out of Africa again and again. Nature 416: 45–51.

Templeton, A.R. & C.F. Sing. 1993. A cladistic analysis of phenotypic associations with haplotypes inferred from restriction endonuclease mapping. IV. Nested analyses with cladogram uncertainty and recombination. Genetics 134: 659–669.

Templeton, A.R., E. Boerwinkle & C.F. Sing. 1987. A cladistic analysis of phenotypic associations with haplotypes inferred from restriction endonuclease mapping. I. Basic theory and an analysis of Alcohol Dehydrogenase activity in *Drosophila*. Genetics 117: 343–351.

Templeton, A.R., K.A. Crandall & C.F. Sing. 1992. A cladistic analysis of phenotypic associations with haplotypes inferred from restriction endonuclease mapping & DNA sequence data. III. Cladogram estimation. Genetics 132: 619–633.

Templeton, A.R., E. Routman & C. Phillips. 1995. Separating population structure from population history: A cladistic analysis of the geographical distribution of mitochondrial DNA haplotypes in the Tiger Salamander, *Ambystoma tigrinum*. Genetics 140: 767–782.

Templeton, A.R., A.G. Clark, K.M. Weiss, D.A. Nickerson, J. Stengård, E. Boerwinkle & C.F. Sing. 2000a. Recombinational and mutational hotspots within the human *Lipoprotein Lipase* gene. Amer. J. Hum. Gen. 66: 69–83.

Templeton, A.R., S.D. Maskas & M.B. Cruzan. 2000b. Gene trees: A powerful tool for exploring the evolutionary biology of species and speciation. Plant Species Biol. 15: 211–222.

Templeton, A.R., K.M. Weiss, D.A. Nickerson, E. Boerwinkle & C.F. Sing. 2000c. Cladistic structure within the human Lipoprotein lipase gene and its implications for phenotypic association studies. Genetics 156: 1259–1275.

Wright, S. 1943. Isolation by distance. Genetics 28: 114–138.

Environmental Biology of Fishes **69**: 21–36, 2004.
© 2004 *Kluwer Academic Publishers. Printed in the Netherlands.*

Migration of Pacific Rim chum salmon on the high seas: insights from genetic data

Lisa W. Seeb[a], Penelope A. Crane[a,e], Christine M. Kondzela[b], Richard L. Wilmot[b], Shigehiko Urawa[c], Natalya V. Varnavskaya[d] & James E. Seeb[a]
[a]*Gene Conservation Laboratory, Alaska Department of Fish and Game, 333 Raspberry Road, Anchorage, AK 99518, U.S.A. (e-mail: lisa_seeb@fishgame.state.ak.us)*
[b]*National Marine Fisheries Service, Auke Bay Laboratory, 11305 Glacier Highway, Juneau, AK 99801, U.S.A.*
[c]*National Salmon Resources Center, Fisheries Agency of Japan, 2-2 Nakanoshima, Toyohira-ku, Sapporo 062-0922, Japan*
[d]*Kamchatka Research Institute of Fisheries and Oceanography, 683600, Petropavlovski-Kamchatsky, Naberejnaya 18, Russia*
[e]*Current address: Conservation Genetics Laboratory, U.S. Fish & Wildlife Service, 1011 E. Tudor Road, Anchorage, AK 99503, U.S.A.*

Received 7 April 2003   Accepted 27 April 2003

Key words: Oncorhynchus keta, mixed stock analyses, allozyme electrophoresis

Synopsis

Wild stocks of chum salmon, *Oncorhynchus keta*, have experienced recent declines in some areas of their range. Also, the release of hatchery chum salmon has escalated to nearly three billion fish annually. The decline of wild stocks and the unknown effects of hatchery fish combined with the uncertainty of future production caused by global climate change have renewed interest in the migratory patterns of chum salmon on the high seas. We studied the composition of high-seas mixtures of maturing and immature individuals using baseline data for 20 allozyme loci from 356 populations from throughout the Pacific Rim. Composition estimates were made from three time series. Two of these time series were from important coastal migratory corridors: the Shumagin Islands south of the Alaska Peninsula and the east coast of the Kamchatka Peninsula. The third was from chum salmon captured incidentally in the Bering Sea trawl fishery for walleye pollock. We also analyzed geographically dispersed collections of chum salmon captured in the month of July. The time series show dynamic changes in stock composition. The Shumagin Island corridor was used primarily by Northwest Alaskan and Asian populations in June; by the end of July stocks from the Alaska Peninsula and southern North America dominated the composition. The composition along the Kamchatka coast changed dramatically from primarily Russian stocks in May to primarily Japanese stocks in August; the previously undocumented presence of stocks from the Alaska Peninsula and Gulf of Alaska was also demonstrated. Immature chum salmon from throughout the Pacific Rim, including large proportions of southern North American stocks, contributed to the Bering Sea bycatch during the months of September and October. The migration routes of North American stocks is far more widespread than previously observed, and the Bering Sea is an important rearing area for maturing and immature chum salmon from throughout the species' range.

Introduction

Recent fluctuations in abundance of major stock assemblages have prompted interest in the migratory patterns of chum salmon, *Oncorhynchus keta*, in the North Pacific Ocean and Bering Sea. In North America, chum salmon from Norton Sound, the Yukon River, and the Kuskokwim River have recently declined

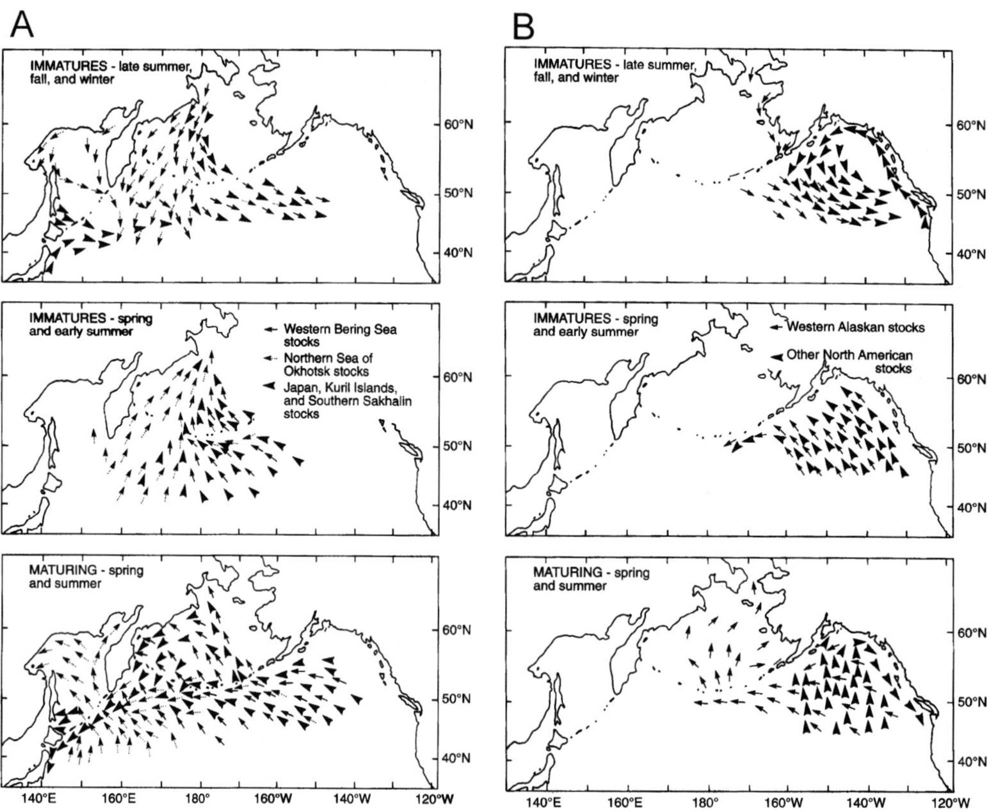

Figure 1. Migration routes of chum salmon in the North Pacific based on studies conducted by NMFS.[2] (A) Asian chum salmon, (B) North American chum salmon.

to historical lows, prompting the State of Alaska to restrict commercial and subsistence fisheries and declare northwestern Alaska a disaster area (Figure 1). Chum salmon catches in East Kamchatka are now consistently one third of the historical average (Radchenko 1998). Coincident with these declines of wild stocks, hatcheries in Asia and North America have increased the release of chum salmon into the North Pacific Ocean to nearly three billion fry annually (Mahnken et al. 1998); many of these hatchery fish migrate into and graze in the Gulf of Alaska and Bering Sea. The level of interactions and effects of these hatchery releases on wild stocks are largely unknown, although Kaeriyama (1998) concluded that the decreasing body size and increasing age at maturity of hatchery chum salmon from Hokkaido was due to increased intraspecific competition on the high seas. Further, changing ocean conditions since the 1970s may have contributed to fluctuations in abundance (Beamish & Bouillon 1993).

Chum salmon are anadromous, spawning in freshwater and growing and maturing during long oceanic migrations lasting 2–5 years. Much is known of the freshwater life history and ecology of chum salmon; much less is known of the oceanic migration patterns and relative marine survival of specific stocks. While some stocks in eastern Bering Sea drainages are disastrously low, others prosper and provide robust commercial and subsistence fisheries (e.g., stocks inhabiting Kotzebue Sound and Bristol Bay). Largely untested hypotheses argue that spatial and temporal migratory patterns of salmonids are under strict genetic control (Brannon 1984), suggesting that migration along stock-specific migration corridors could lead to differing marine survival and varying rates of return among stocks in a fluctuating marine environment.

Chum salmon are distributed in Asia from Korea to the Arctic coast of Russia and in North America from central Oregon to the Arctic coast east of the Mackenzie River on the Beaufort Sea (modified after Salo 1991). The abundance of Asian stocks is greater than that of North American stocks. The early high-seas tagging studies of the International

Pacific Salmon Commission[1] and the National Marine Fisheries Service (NMFS) (Figure 1)[2] serve as the basis of our understanding of migration patterns. Those studies showed that the known ocean ranges of Asian and North American chum salmon are broadly overlapping though Asian chum salmon are more widely distributed. Asian chum salmon extend eastward to at least 140°W, a more distant migration than North American chum salmon that extend westward to 175°E (Salo 1991). Historical data also suggest that North American stocks originating south of the Alaska Peninsula do not enter the Bering Sea[1]. However, more recent high-seas tagging studies[3] and studies using thermal marks (Farley & Munk 1997) and scale pattern analysis (Patton et al. 1998) have refined our understanding of chum salmon migration and document that chum salmon from Southeast Alaska, British Columbia, and Washington do move into the Bering Sea.

Simultaneously with projects utilizing tagging, thermal, or scale marks, researchers developed a Pacific Rim-wide database of gene markers to use in high-seas studies. Research began in the mid-1980s to identify the origin of chum salmon harvested illegally in the high seas and incidentally caught in the high-seas drift-net fisheries for flying squid, *Ommastrephes bartramii* (Smouse et al. 1990). In the 1990s, efforts focused on collecting a common and standardized set of allozyme loci and alleles, and large data sets of allele frequencies from Washington, British Columbia, Alaska, Yukon Territory, Russia, and Japan were completed (Kondzela et al. 1994, Phelps et al. 1994, Wilmot et al. 1994, Winans et al. 1994, Seeb & Crane 1999a).

Using 164 population collections from this database, Seeb & Crane (1999b) developed a model to estimate the origin of chum salmon harvested in fisheries occurring along the south side of the Alaska Peninsula using data from 20 allozyme loci. Identifiable stock groups were selected from heterogeneity and multidimensional scaling (MDS) analyses and refined using simulation studies of artificial mixtures. Proportionate contributions to south Alaska Peninsula fisheries were estimated for chum salmon of eight stock groups: Japan, Russia, Northwest Alaska summer, Yukon River fall run, Alaska Peninsula-Kodiak Island, Southeast Alaska, British Columbia, and Washington (Seeb & Crane 1999b).

As new population data became available, researchers independently updated the original baseline and expanded applications. The database was used to estimate the origin of chum salmon sampled from the Bering Sea trawl fishery for walleye pollock, *Theragra chalcogramma* (Wilmot et al. 1998), high-seas test fisheries (Winans et al. 1998, Urawa et al. 2000), and cargoes seized from illegal harvests.[4,5] Later work refined the estimates into 10 stock groups, splitting the large Russian group and realigning the Gulf of Alaska and southern populations resulting in the following groups: Japan, China/Southern Russia, Northern Russia, Northwest Alaska summer, Yukon River fall run, Alaska Peninsula/Kodiak Island, Susitna River, Prince William Sound, Southeast Alaska/Northern British Columbia, and Southern British Columbia/Washington.[6] The allozyme baseline was recently significantly enlarged and currently includes 356 individual populations ranging from the Columbia River to Kotzebue Sound in North America and from Honshu to the Anadyr River in Asia (Table 1).[7]

---

[1] Neave, F., T. Yonemori & R.G. Bakkala. 1976. Distribution and origin of chum salmon in offshore waters of the North Pacific Ocean. Bulletin Number 35, International North Pacific Fisheries Commission, Vancouver, BC.

[2] Fredin, R.A., R.L. Major, R.G. Bakkala & G. Tanonaka. 1977. Pacific salmon and the high seas salmon fisheries of Japan (Processed report). Northwest and Alaska Fisheries Center, National Marine Fisheries Service, Seattle.

[3] Myers, K.W., K.Y. Aydin, R.V. Walker, S. Fowler & M.L. Dahlberg. 1996. Known ocean ranges of stocks of Pacific salmon and steelhead as shown by tagging experiments, 1956–1995. (North Pacific Anadromous Fish Commision Doc. 192). FRI-UW-9614. University of Washington, Fisheries Research Institute, Box 357980, Seattle, WA 98195-7980.

[4] Wilmot, R.L., C.M. Kondzela, C.M. Guthrie III, A. Moles, E. Martinson & J.H. Helle. 1999. Origins of sockeye and chum salmon seized from the Chinese vessel *Ying Fa*. (North Pacific Anadromous Fish Commission NPAFC Doc. 410). Auke Bay Fisheries Laboratory, Alaska Fisheries Science Center, NMFS, NOAA, 11305 Glacier Highway, Juneau, AK 99801-8626

[5] Wilmot, R.L., C.M. Kondzela, C.M. Guthrie III, A. Moles, J.J. Pella & M. Masuda. 2000. Origins of salmon seized from the F/V *Arctic Wind*. (North Pacific Anadromous Fish Commission NPAFC Doc. 471). Auke Bay Fisheries Laboratory, Alaska Fisheries Science Center, NMFS, NOAA, 11305 Glacier Highway, Juneau, AK 99801-8626.

[6] Crane, P.A. & L.W. Seeb. 2000. Genetic analysis of chum salmon harvested in the South Peninsula, post June fishery, 1996–1997. Regional Information Report No. 5J00-05, Alaska Department of Fish and Game, Anchorage.

[7] Kondzela, C.M., P.A. Crane, S. Urawa, N.V. Varnavskaya, V. Efremov, X. Luan, W.B. Templin, K. Hayashizaki, R.L. Wilmot & L.W. Seeb. 2002. Development of a comprehensive allozyme baseline for Pacific Rim chum salmon. (North Pacific Anadromous Fish Commission, NPAFC Doc 629). Alaska Department of Fish and Game, 333 Raspberry Road, Anchorage, AK 99518, U.S.A.

Table 1. Geographic origins of Pacific Rim chum salmon populations included in this study. Major reporting groups and regions within those reporting groups are given. Abbreviations for major reporting groups are given in upper case. Regional numbers are listed in parentheses following the regional name.

Major reporting group	Region	Geographic area
1. NW AK (Northwest Alaska)	Northwest Alaska Summer (1)	Kotzebue Sound Norton Sound Yukon River Summer Kuskokwim Bay and Lower Kuskokwim River Bristol Bay
	Yukon River Fall (2)	
	Kuskokwim River-Upper (3)	
2. AK PEN/GOA (Alaska Peninsula, Gulf of Alaska)	Alaska Peninsula and Kodiak Island (4)	North Alaska Peninsula South Alaska Peninsula Kodiak Island
	Susitna River (5)	
	Prince William Sound (6)	
3. SE AK/N BC (Southeast Alaska, Northern British Columbia)	Southeast Alaska, Northern British Columbia (7)	Northern Southeast Alaska Mainland Southeast Alaska Northern British Columbia Georgia Strait
	Prince of Wales Island (8)	
	Queen Charlotte Island (9)	
4. S BC/PNW (Southern British Columbia, Pacific Northwest)	Puget Sound, Southern British Columbia (10)	Fraser River West Coast Vancouver Island Strait of Juan de Fuca Northern Puget Sound Southern Puget Sound Hood Canal
	Hood Canal Summer (11)	
	Coastal Washington, Columbia River (12)	
5. JAPAN (Japan)	Honshu (13)	Pacific Coast of Honshu Japan Sea Coast of Honshu
	Hokkaido (14)	Pacific Coast of Hokkaido Nemuro Coast Okhotsk Coast of Hokkaido Japan Sea Coast Hokkaido
6. RUSSIA/CHINA (Russia, China)	Amur River China (15)	
	Kuril Islands (16)	
	Premorye/Suifen (17)	
	Sakhalin Island (18)	
	Northern Russia (19)	Anadyr River Eastern Kamchatka Peninsula Western Kamchatka Peninsula Magadan

In this study, we describe the enlarged database and use it to estimate stock contributions to temporally and spatially structured collections from mixed aggregations of immature and maturing chum salmon. Our objectives were to use the contribution estimates to infer stock-specific migration patterns and to compare these patterns to historic data from tagging studies and recent information from thermal marks. We evaluated mixtures sampled through time from the Shumagin Islands off the South Alaska Peninsula, the east coast of the Kamchatka Peninsula, and the southeastern Bering Sea. The Shumagin Islands and the east coast of the

Kamchatka Peninsula are important migration corridors while the southeastern portion of the Bering Sea is an important high-seas rearing area for chum salmon. We also analyzed a set of geographically dispersed collections taken during the month of July across the North Pacific Ocean. The data indicate that the migration of North American stocks is far more widespread than previously observed and that the Bering Sea is an important rearing area for immature and maturing chum salmon from throughout the species' range.

Materials and methods

Baseline construction

A total of 356 populations representing over 42 000 individuals was included. Populations were scored for a common set of 20 allozyme loci: *ALAT**; *mAAT-1**; *sAAT-1,2**; *mAH-3**; *ESTD**; *G3PDH-2**; *GPI-A**; *GPI-B1,2**; *mIDHP-1**; *sIDHP-2**; *LDH-A1**; *LDH-B2**; *sMDH-A1**; *sMDH-B1,2**; *mMEP-2**; *sMEP-1**; *MPI**; *PEPA**; *PEPB-1**; and *PGDH*.[7] Alleles were standardized and pooled as necessary to insure consistency. Samples collected in more than a single year at a location were pooled if no significant differences were found (p < 0.001) between years. Unlike previous versions of the chum salmon baseline, no large regional pooling was conducted, and individual populations rather than representative populations for Southeast Alaska, British Columbia, and Washington were included.

Population structure

Genetic distances between all pairwise combinations of populations were estimated using Cavalli-Sforza & Edwards' (1967) chord distances. Genetic distances were used in a MDS analysis that clusters populations so that expected distances between populations closely match the observed interpopulational distances in multidimensional space. A gene diversity analysis (Nei 1973) was performed to partition variation among and within major groups identified in the MDS analysis and simulation analyses (see below). G_{st}, the proportion of gene diversity due to an among population component, was calculated for each major group and over all populations. We tested for allele frequency heterogeneity to interpret G_{st}; heterogeneity tests can be considered an indirect test of the null hypothesis ($G_{st} = 0$; Chakraborty and Leimar 1987).

Mixture collections

Samples from mixed aggregations of immature and maturing chum salmon were collected from processing plants, onboard high-seas research cruises, and by NMFS observers onboard walleye pollock fishery vessels (Table 2). Temporal samples from the Shumagin Islands were collected in 1996 from test and commercial fisheries.[6] Temporal collections from the east coast of the Kamchatka Peninsula were collected within the 200-mile Russian Economic Zone in the Bering Sea and northwestern Pacific using drift nets during research cruises in 1998. Bycatch samples were collected by NMFS observers during the pollock fishery in the southeastern Bering Sea (cf., Wilmot et al. 1998). Geographically dispersed samples originated from Japanese high-seas research cruises conducted in 1998 (Urawa et al. 2000).

Maximum likelihood model

Stock assemblages to be estimated were determined though MDS analyses and simulation studies. Simulations were performed using the Statistics Program for Analyzing Mixtures (SPAM ver. 3.5, Debevec et al. 2000). In each simulation, baseline and mixture genotypes were randomly generated from the baseline allele frequencies using Hardy–Weinberg expectations. Each simulated mixture (N = 400) was composed of 100% of the region under study, with each population in the region contributing equally to the mixture. Average estimates of mixture proportions were derived from 1 000 simulations. Individual population estimates were first calculated and then summed into regions (allocate-sum procedure; Wood et al. 1987). Based on geography and the MDS analyses, 40 regions were initially evaluated. Regions were enlarged until ~90% of the mixture on an average was allocated to the correct group. Ninety percent confidence intervals for each regional estimate were computed using 1 000 bootstrap resamples and the symmetric percentile method. For ease of discussion, individual regions were summed into larger regional aggregations using the allocate-sum procedures. To avoid confusion, larger regional aggregations will be termed 'reporting groups', and discussions including individual regions will include a regional numeric designation in parentheses.

Table 2. Collections of chum salmon from the North Pacific Ocean and Bering Sea for mixture analyses and temporal and geographic comparisons. Collection dates, sample size (N), maturity level (I = immature, M = maturing), vessel or location, and source are given.

Sample	Dates	N	Maturity	Vessel or location	Source
Temporal comparisons					
Shumagin Islands					
June Test Fishery Period 1	6–10 June 1996	395	M	Sand Point, AK	ADF& G[13]
June Test Fishery Period 2	11–16 June 1996	400	M	Sand Point, AK	ADF& G[13]
June Fishery Period 1	18–20 June 1996	398	M	Sand Point, AK	ADF& G[14]
June Fishery Period 2	21–25 June 1996	398	M	Sand Point, AK	ADF& G[14]
June Fishery Period 3	26–30 June 1996	399	M	Sand Point, AK	ADF& G[14]
Post June Test Fishery	7–18 July 1996	507	M	Sand Point, AK	ADF& G[13]
Post June Fishery Period 1	20–26 July 1996	394	M	Sand Point, AK	ADF& G[13]
Post June Fishery Period 2	27 July–2 August 1996	399	M	Sand Point, AK	ADF& G[13]
Post June Fishery Period 3	3–10 August 1996	400	M	Sand Point, AK	ADF& G[13]
Kamchatka Peninsula					
Kamchatka Period 1	24 May 1998	98	M	*Altair*	ADF& G[15]
Kamchatka Period 2	9 June 1998	98	M	*Altair*	ADF& G[15]
Kamchatka Period 3	10 July 1998	100	M	*Altair*	ADF& G[15]
Kamchatka Period 4	1 August 1998	100	M	*Altair*	ADF& G[15]
Bering Sea Bycatch					
Bering Sea Bycatch Period 1	1–7 September 1996	487	I	NMFS observers	NMFS[16]
Bering Sea Bycatch Period 2	8–14 September 1996	384	I	NMFS observers	NMFS[16]
Bering Sea Bycatch Period 3	15–21 September 1996	365	I	NMFS observers	NMFS[16]
Bering Sea Bycatch Period 4	22–28 September 1996	684	I	NMFS observers	NMFS[16]
Bering Sea Bycatch Period 5	29 September–5 October 1996	519	I	NMFS observers	NMFS[16]
Bering Sea Bycatch Period 6	6–17 October 1996	378	I	NMFS observers	NMFS[16]
Bering Sea Bycatch all periods	1 September–12 October 1996	95	M	NMFS observers	NMFS[16]
Geographic comparisons					
Central Gulf of Alaska (49.00°–56.00°N, 145.00°W)	4–11 July 1998	336	I/M	*T/V Oshoro-maru*	Urawa et al. (2000)
Bering Sea (52.30°–57.30°N, 177.30°E–177.30°W)	3–17 July 1998	615	I/M	*T/V Oshoro-maru*	FAJ[17]
Western North Pacific (50–51°N, 165°E)	3–6 July 1998	180	I/M	*T/V Hokko-maru*	FAJ[17]
Central North Pacific (43.00°–51.30°N, 179.30°W–180°)	23 June–2 July 1998	161	I/M	*I/V Wakatake-maru*	Urawa et al. (2000)

[13]Crane, P.A. & L.W. Seeb. 2000. Genetic analysis of chum salmon harvested in the South Peninsula, post June fishery, 1996–1997. Regional Information Report No. 5J00-05, Alaska Department of Fish and Game, Anchorage, Alaska.

[14]Seeb, L.W., P.A. Crane & E.M. Debevec. 1997. Genetic analysis of chum salmon harvested in the South Unimak and Shumagin Islands June Fisheries, 1993–1996. Regional Information Report No. 5J97-17, Alaska Department of Fish and Game, Anchorage, AK.

[15]Alaska Department of Fish and Game, unpublished.

[16]NMFS, Auke Bay Laboratory, Juneau, Alaska, USA, unpublished.

[17]National Salmon Resources Center, Fisheries Agency of Japan, Sapporo, Japan, unpublished.

Conditional maximum likelihood estimates for the mixture collections were generated in a similar manner using SPAM ver. 3.5 and the allocate-sum procedure. Symmetric 90% confidence intervals for each regional contribution estimate were computed as above using 1 000 bootstrap resamples of the baseline and mixture.

Results

Baseline evaluation

Regional groups for the mixture analyses were identified through gene diversity analyses, MDS, and simulation studies (Figures 2 and 3; Table 3). Forty regions

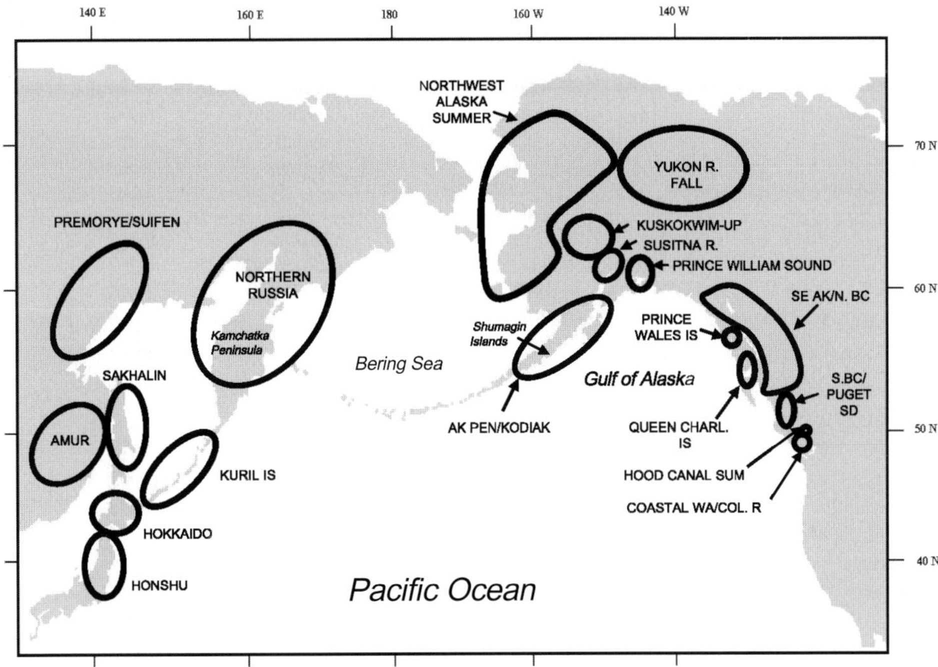

Figure 2. Map of North Pacific Ocean, Gulf of Alaska, and Bering Sea showing 19 regional groups of populations of chum salmon used in mixture analyses.

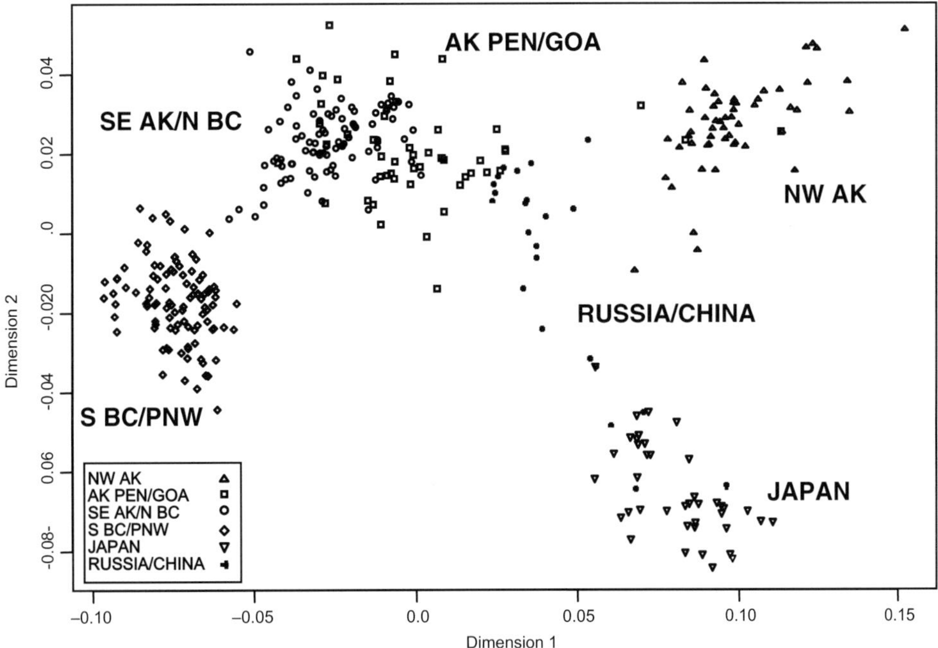

Figure 3. Multidimensional scaling of Pacific Rim chum salmon.

Table 3. Gene diversity analysis (Nei 1973). Log-likelihood tests were used as an indirect test of the null hypothesis ($G_{st} = 0$).

(A) Hierarchical analysis for regions and reporting groups

Locus	H_T	H_S	Within populations	Among populations within regions	Among regions within reporting group	Among reporting groups
sAAT-1,2*	0.148	0.144	0.978	0.011	0.002	0.010
mAAT-1*	0.339	0.291	0.860	0.018	0.012	0.111
mAH-3*	0.496	0.462	0.931	0.021	0.016	0.032
ALAT*	0.271	0.263	0.968	0.018	0.007	0.006
ESTD*	0.273	0.191	0.700	0.035	0.019	0.245
G3PDH-2*	0.217	0.207	0.953	0.016	0.010	0.022
GPI-B1,2*	0.002	0.002	0.950	0.000	0.000	0.000
GPI-A*	0.004	0.004	1.000	0.000	0.000	0.000
mIDHP-1*	0.175	0.162	0.927	0.040	0.010	0.023
sIDHP-2*	0.627	0.600	0.957	0.019	0.005	0.019
LDH-A1*	0.150	0.134	0.892	0.019	0.009	0.081
LDH-B2*	0.008	0.008	0.964	0.012	0.012	0.012
sMDH-A1*	0.108	0.103	0.956	0.017	0.010	0.016
sMDH-B1,2*	0.030	0.030	0.980	0.013	0.003	0.003
mMEP-2*	0.265	0.250	0.945	0.018	0.012	0.025
sMEP-1*	0.014	0.013	0.948	0.030	0.007	0.015
MPI*	0.215	0.205	0.955	0.019	0.013	0.014
PEPA*	0.002	0.002	1.000	0.000	0.000	0.000
PEPB1*	0.381	0.355	0.932	0.020	0.011	0.037
PGDH*	0.037	0.036	0.970	0.016	0.008	0.005
Mean	0.188	0.173	0.938	0.017**	0.008**	0.034**

(B) G_{st} for major reporting groups

Reporting group	Group abbreviation	G_{st}
Northwest Alaska	NW AK	0.016**
Alaska Peninsula and Gulf of Alaska	AK PEN/GOA	0.036**
Southeast Alaska and N. British Columbia	SE AK/N BC	0.020**
S British Columbia and Pacific Northwest	S BC/PNW	0.022**
Japan	JAPAN	0.022**
Russia and China	RUSSIA/CHINA	0.055**

**p < 0.001.

were initially evaluated; 19 composite regions had an acceptable level of accuracy and precision in the simulation studies (Table 1, Figure 4A,B).[8] The 19 regions were summed into six major reporting groups. A gene diversity analysis indicated that 94% of the total genetic diversity is due to within population variation, 2% to variation among populations within regions, 1% to variation among regions within reporting groups, and 3% to variation among reporting groups (Table 3(A)).

Three of the major reporting groups, Southern British Columbia and the Pacific Northwest (S BC/PNW), Northwest Alaska (NW AK), and Japan (JAPAN) each formed a distinct cluster within the MDS analysis (Figure 3). G_{st} values for these major reporting groups ranged from 0.016 to 0.022 (Table 3(B)). Results from 100% simulations exceeded 0.90 for S BC/PNW, NW AK, and JAPAN, and simulations for regions within these reporting groups were also near or above 0.90 with the exception of Hokkaido (14) in JAPAN (Figure 4A,B). The two regions within JAPAN, Honshu (13) and Hokkaido (14), allocate to each other in the 100% simulations, indicative of close similarity and possible gene flow between these island groups.

---

[8] Detailed results of the simulation studies and the mixture analyses for the 19 regions are available from the Alaska Department of Fish and Game website at: http://www.genetics.cf.adfg.state.ak.us

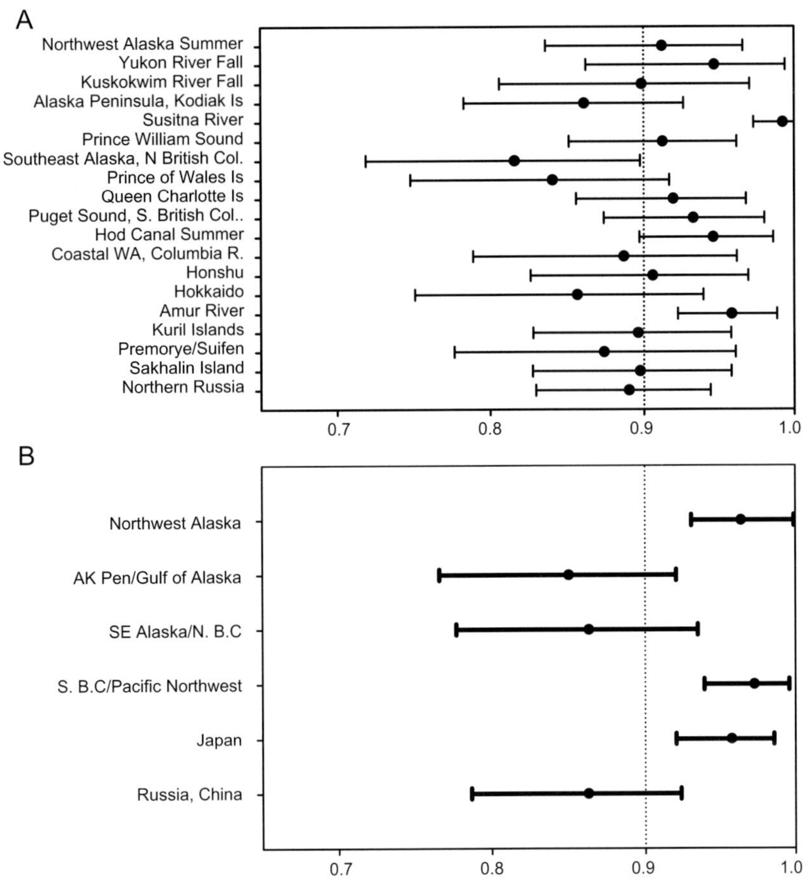

Figure 4. Simulation results for chum salmon. Each simulated mixture (N = 400) was composed of 100% of the region under study. Mean estimates and 90% bootstrap confidence intervals are given. Regions and reporting groups are from Table 1. (A) 19 regions, (B) 6 major reporting groups.

Gene flow may be a result of natural straying and historical colonization patterns as well as recent hatchery practices and stock transfers within Japan.

The other three major reporting groups are also apparent in the MDS analysis (Figure 3), Alaska Peninsula and Gulf of Alaska (AK PEN/GOA), Russia and China (RUSSIA/CHINA), and Southeast Alaska and Northern British Columbia (SE AK/N BC). These reporting groups are more heterogeneous than those described above with generally larger G_{st} values: 0.020 for SE AK/N BC, 0.036 for the AK PEN/GOA, and 0.055 for RUSSIA/CHINA. RUSSIA/CHINA is the most heterogeneous with the largest G_{st} value. It encompasses the largest geographic area being composed of five regions: Amur River in China and Russia (15), Kuril Islands (16), Premorye/Suifen (17), Sakhalin Island (18), and Northern Russia (19). Kuril

Island populations show affinities to some Hokkaido (14) populations while Northern Russia (19) population are closely aligned with some Alaska Peninsula and Kodiak Island populations (4) in Figure 3.

Temporal distribution

Shumagin Islands. A dynamic shift in stock composition through time was apparent in the test and commercial fisheries in the Shumagin Islands (Figure 5A). Chum salmon from NW AK were the largest contributors in the early periods with estimates as high as 0.69 in the early June test fisheries. The NW AK contribution declined through June, and a final sharp decrease was observed between the end of June and mid-July, decreasing from 0.32 to 0.05 The contribution from the

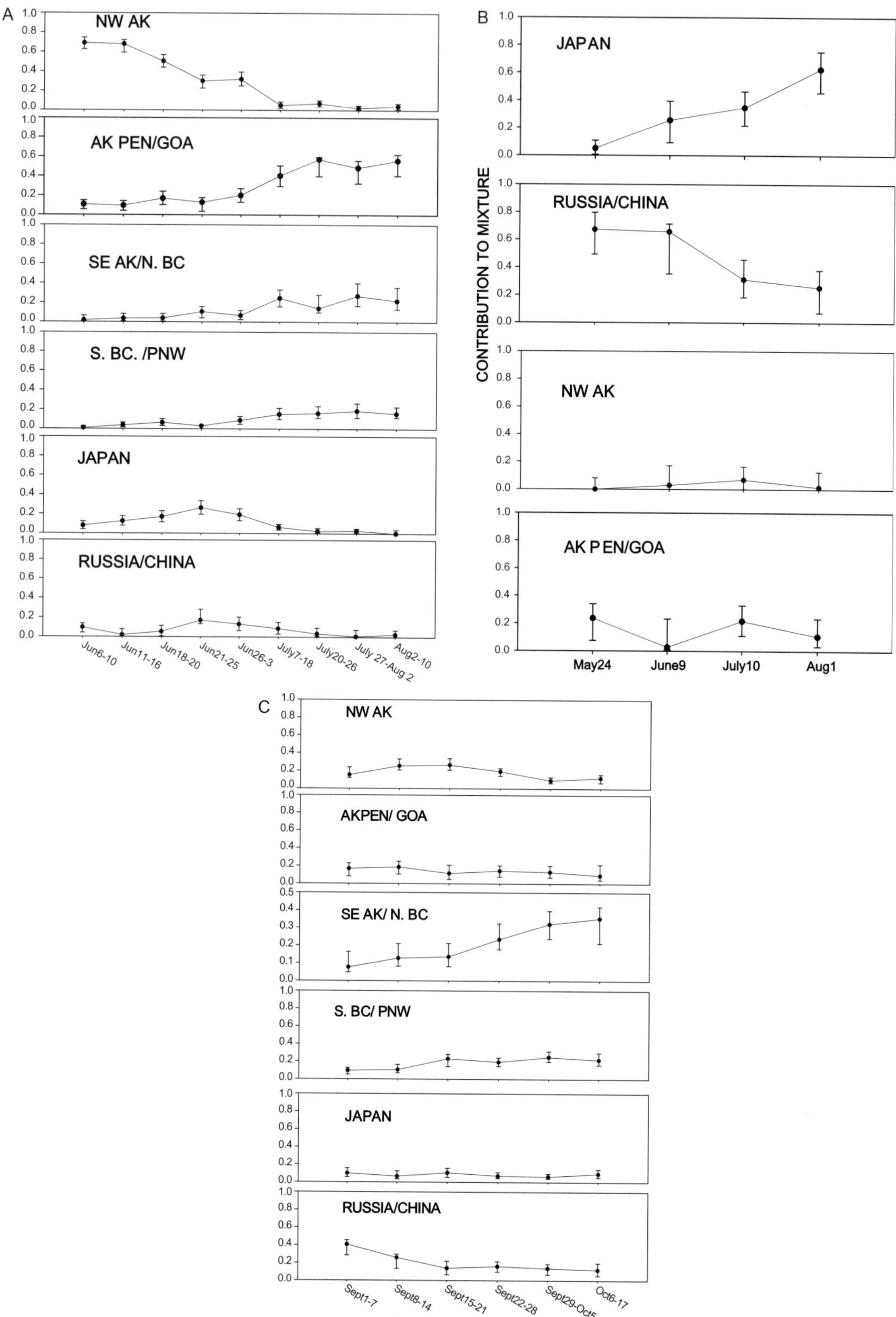

Figure 5. Temporal mixture estimates for Pacific Rim chum salmon. (A) Shumagin Islands from the south Alaska Peninsula. (B) East coast of the Kamchatka Peninsula. (C) Bycatch from the walleye pollock trawl fishery from the southeastern Bering Sea.

Asian continent (JAPAN and RUSSIA/CHINA) was low in early June totaling less than 0.20, increased to a high of 0.43 for the two groups combined during 21–25 June, and declined between the end of June and mid-July.

Stocks from the AK PEN/GOA increased from June to early August. Contributions ranged from 0.10 to 0.20 in June to ~0.56 in late July and August (Figure 5A). Chum salmon from more southerly areas in North America (SE AK/N BC and S BC/PNW) were low in June, but increased through July and August. The combined estimates for the two reporting groups ranged from 0.30 to 0.46 for the three periods in the Post June fishery.

East coast of Kamchatka. A shift in stock composition through time was also seen in the migration corridor along the east coast of the Kamchatka Peninsula from the end of May through the beginning of August. In May, mixtures were composed predominantly of RUSSIA/CHINA and nearly exclusively of populations originating in the Northern Russian (19) group (Figure 5B). August samples were dominated by JAPAN while the RUSSIA/CHINA contribution dropped markedly. Contributions from JAPAN increased through time from an estimated 0.05 in the May sample to 0.62 in the August sample. Japanese populations originated primarily from the Hokkaido (14) region.

North American populations from Alaska also use this corridor. Estimates of chum salmon from AK PEN/GOA ranged from 0.03 to 0.24 and were highest in the May 24 and July 10 samples (Figure 5B). NW AK populations were small contributors, but were likely present as 0.000 is not included in the confidence interval. The largest estimate for NW AK was 0.07 in the July 10 sample. The contributions of the other reporting groups from North America were low or absent.

Bering Sea trawl bycatch. The Bering Sea trawl bycatch samples were collected from immature chum salmon over six sampling periods from September through mid-October. Unlike mixtures of maturing chum salmon from the Shumagin Islands and the east coast of Kamchatka, all reporting groups contributed to every period (Figure 5C), and no confidence interval for any group in any period included 0.000. Similar to other temporal samples, there was a shift in composition during the sampling period. RUSSIA/CHINA declined from a high of 0.41 in the first period to 0.12 in the

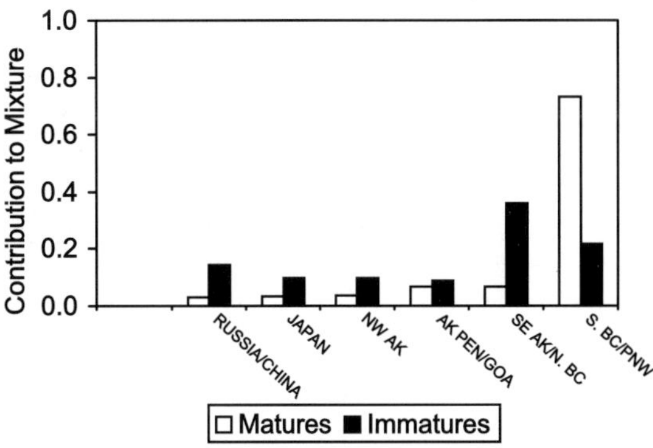

Figure 6. Estimated composition of immature and mature chum salmon caught in the walleye pollock fishery in the southeastern Bering Sea, September to October, 1998.

final period. A simultaneous increase was seen in populations from the SE AK/N BC and S BC/PNW with contribution of these two southern groups exceeding 0.50 in the final period.

A small sample of maturing fish (N = 95) was also collected over the time series and compared to a pooled sample of immature fish over the period (N = 2 897). Whereas the pooled sample of immature fish was composed of all groups, the mature sample was heavily weighted towards S BC/PNW with a contribution estimated at 0.73 (Figure 6).

Spatial distribution

Four additional mixtures were analyzed from high-seas collections taken in the Bering Sea and North Pacific in July (Table 2). These mixtures were compared to two mixtures from July periods described above, the Post June Fishery Period 1 from the Shumagin Islands and Kamchatka mixture Period 3, to provide a 'snapshot' of chum salmon distributions during the month of July (Figure 7).

The most northern collection, Bering Sea, was taken at 180° longitude. The sample was composed primarily of chum salmon from the Asian continent. NW AK populations were likely farther east in coastal waters or had already entered freshwater at the time of collection. Only a small proportion (<0.05) of the chum salmon originated from the two southern North American groups, SE AK/N BC and S BC/PNW.

Two mixtures were composed primarily of Asian chum salmon. In the western North Pacific sample (165°E), the contribution of RUSSIA/CHINA was

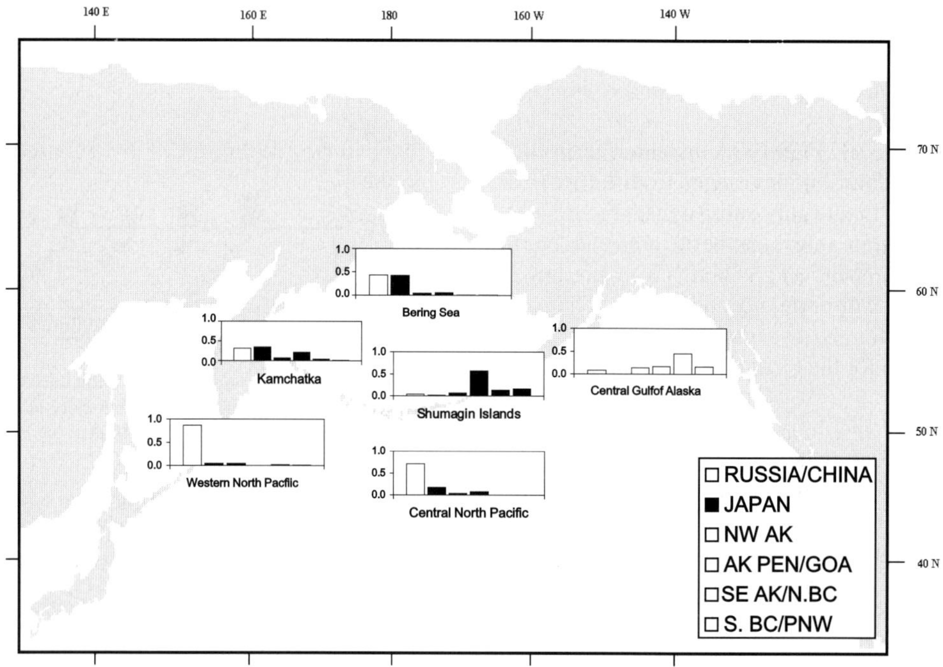

Figure 7. Estimated composition of spatially dispersed mixtures of chum salmon.

estimated at 0.87 with JAPAN likely contributing the small remainder. RUSSIA/CHINA populations were also present in high proportions (0.71) in the Central North Pacific sample taken at ~180° longitude. The remainder of the central North Pacific sample was composed of JAPAN (0.17) with lesser contributions of AK PEN/GOA (0.08) and NW AK (0.04).

The most easterly sample was taken from the central Gulf of Alaska (145°W) and was composed primarily of North American chum salmon. The SE AK/N BC group was the largest contributor with at 0.45 with contributions of ~0.15 from each of the other three North American reporting groups. The contribution of RUSSIA/CHINA was estimated at 0.082.

Discussion

Temporal distribution

Shumagin Islands. Migrating chum salmon are harvested incidentally in fisheries for sockeye salmon on major capes of the Shumagin Islands. The fishery has been prosecuted since the early 1900s with record harvests of chum salmon reaching 1.1 million fish,[9] although since 1996 the harvests have been limited by a conservation ceiling. Genetic studies were initiated in the 1990s as a result of declines in northwestern Alaska chum salmon (Seeb & Crane 1999b). The Shumagin Islands time series indicates a dramatic change in stock composition for chum salmon through time. In June, chum salmon migrating through the Shumagin Islands were mainly of NW AK origin, followed by chum salmon from JAPAN and RUSSIA/CHINA. A fairly abrupt change occurred in July, when most chum salmon originated from the AK PEN/GOA or the southern North American groups.

These results were concordant with earlier physical tagging studies conducted off the South Alaska Peninsula. Recoveries of salmon tagged in June were made in western Alaska and Asia, while chum salmon

---

[9] Eggers, D.M., K. Rowell & B. Barrett. 1991. Stock composition of sockeye and chum salmon catches in southern Alaska Peninsula fisheries in June. Fishery Research Bulletin No. 91-01, Alaska Department of Fish and Game, Division of Commercial Fisheries, Juneau, AK, 49 pp.

tagged in July were more often recovered in local areas (Gilbert & Rich 1926). [9,10]

The Shumagin Islands data were also similar to the model of chum salmon migration derived from high-seas tagging studies for Asian and western Alaskan stocks. Historic tagging information demonstrated that chum salmon from western Alaska move rapidly from the Gulf of Alaska to the Bering Sea in June.[1] Asian stocks move from the northern Pacific Ocean and Gulf of Alaska during this time as well and are also common in the eastern Aleutian Islands in summer.

These early studies reported that maturing chum salmon bound for central Alaska, Southeast Alaska, British Columbia, and Washington tend to move to the northern Gulf of Alaska after the majority of western Alaska chum salmon have migrated through.[1,2] Data indicated that these North American populations were not commonly found west of 155°W. However, the Shumagin Islands are farther west at 160°W, and our estimates indicate that chum salmon from these areas were the primary contributors to the July fisheries in the Shumagin Islands.

East coast of Kamchatka. Research along the east coast of the Kamchatka Peninsula was initiated to determine the composition through time of fisheries prosecuted within the 200-mile Russian Economic Zone. Similar to the Shumagin Islands, this area is an important migration corridor for chum salmon. Stocks from RUSSIA/CHINA were the major contributors in May and June, but a dramatic shift in composition occurred in June and July when chum salmon from JAPAN predominated in the fishery. The high proportion of North American chum salmon, principally from AK PEN/GOA, estimated along this coast is of considerable interest. Historical tagging studies provide little information for comparison. A single chum salmon, tagged along the Russian coast north of the Kamchatka Peninsula, returned to general area encompassing the Arctic coast of Alaska, Yukon, and Kuskokwim rivers.[3] Additional research into the stock composition of salmon using this important corridor is definitely warranted.

Bering Sea bycatch. Historic tagging data suggest that chum salmon from areas in Southeast Alaska and further south spend most of their ocean life history in the Gulf of Alaska and are seldom found west of 180°W.[1] This again contrasts with results of this study and other recent studies. Wilmot et al. (1998) used an allozyme baseline to analyze samples from the 1994 and 1995 Bering Sea fishery for walleye pollock from late August to early October. They also detected a broad representation of stocks of immature chum salmon in the Bering Sea and large proportions of maturing chum salmon originating from the British Columbia and the Pacific Northwest. The results for maturing chum salmon were concordant with results from scale pattern analysis (Patton et al. 1998). Thermally marked chum salmon from Southeast Alaska and British Columbia have also been recovered in the Bering Sea.[11] The differences between these studies and the older tagging studies for chum salmon from southern North American populations may be due either to decadal changes in the marine environment or changes in abundance.

Spatial distribution

The geographically disbursed mixtures provide an overview of chum salmon distributions in the month of July. Combined with the temporal series, several important trends are evident. North American chum salmon populations originating from Northwest Alaska and the Gulf of Alaska eastward to Prince William Sound are likely present along the east coast of Kamchatka. Populations from the Gulf of Alaska and south to the Pacific Northwest migrate along the South Alaska Peninsula in July along with local stocks from the Alaska Peninsula. Fish from all regions, including southern North American stocks, are present as immatures in the Bering Sea.

Results from this study can also be compared to Urawa et al. (2000) who used allozymes to analyze samples of chum salmon from the central Gulf of Alaska and central North Pacific. Their findings were generally consistent with those derived in this

---

[10] Thorsteinson, F.V. & T.R.J. Merrell. 1964. Salmon tagging experiments along the south shore of Unimak Island and the southwestern shore of the Alaska Peninsula. Special Scientific Report 486, U.S. Fish and Wildlife Service, Washington, DC.

[11] Ignell, S.E., C.M. Guthrie, J.H. Helle & K. Munk. 1997. Incidence of thermally marked chum salmon in the 1994–1996 Bering Sea pollock B-season trawl fishery. (North Pacific Anadromous Fish Commission, NPAFC Doc. 246) Auke Bay Laboratory, Alaska Fisheries Science Center, NMFS, NOAA, 11305 Glacier Highway, Juneau, AK 99801-8626.

34

study; North American chum salmon predominated in the central Gulf of Alaska (145°W) while Asian chum predominated along the central North Pacific (165°W). In addition, they presented thermal mark information showing that the contribution of Prince William Sound and Southeast Alaska hatchery stocks was high among immature chum salmon in the central Gulf of Alaska.

Lineage-specific migration routes

The data may also provide insights into whether lineages of close genetic similarity follow similar migration paths or, alternatively, if migration paths of geographically proximal lineages, regardless of their genetic similarities, follow similar routes. Chum salmon provide an interesting case study because some rather divergent genetic lineages are geographically proximate, allowing a test of the hypothesis that genetically similar lineages follow similar migration routes. For example, a major contact zone between the Northwest Alaska and the Alaska Peninsula/Gulf of Alaska groups exists between Meshik River and Lawrence Valley Creek on the northern Alaska Peninsula (Seeb & Crane 1999a). These populations are separated by ~150 km and yet show significant frequency differences at half the loci sampled. A second contact zone exists in Cook Inlet of the Gulf of Alaska. Populations of the Susitna River, which drains into Cook Inlet, are genetically more similar to Northwest Alaska populations than to other Cook Inlet and Gulf of Alaska populations. It is possible that post-glacial colonization of these populations occurred through inland dispersal via a connection of northern tributaries of the Yukon River (Seeb & Crane 1999a).

Appropriate data to evaluate lineage-specific routes are limited in this data set, and any comparisons need to be conducted using the regional groups rather than the larger reporting groups. The first and third periods from the Kamchatka coast do provide some insights into the occurrence of the Susitna River (5) region relative to the Northwest Alaska summer (1) and Alaska Peninsula, Kodiak Island (4) regions. In the first sample, the Northwest Alaska summer (1) region was absent; the Alaska Peninsula, Kodiak Island (4) and Susitna River (5) regions were estimated at 0.040 and 0.061, respectively. In the third period, the Northwest Alaska summer (1) region was estimated at 0.037; the other two regions were 0.134 and 0.083, respectively. These data sets, though clearly limited, support the hypothesis that the geographically, but not

genetically, close populations of the Alaska Peninsula, Kodiak Island (4) and Susitna River (5) regions follow similar migration routes.

Importance of high-seas research

There is renewed interest on the migratory patterns of salmonids in part because of declines of major stock assemblages and the uncertainty of future production caused by global climate change. Climate in the Gulf of Alaska and Bering Sea play an important role in salmon production (Beamish & Bouillon 1993). Decadal changes have introduced unanticipated impacts on ecology, ocean productivity, and fishing economies of the Pacific Rim.

The Bering Sea ecosystem in particular is experiencing extraordinary change (see reviews in Loughlin & Ohtani (1999)). Major coccolithophore blooms (*Emilania huxley*), once rare, have occurred annually since 1997. Of special interest to some ecologists are dramatic declines of charismatic species including both marine mammals and seabirds (NRC 1996). The red king crab (*Paralithodes camtschatica*) population in the eastern Bering Sea increased an order of magnitude during the 1970s, only to crash to a historically low level where it has remained (Wooster 1992). In contrast, walleye pollock numbers steadily increased since the 1960s to become the dominant marine species.

These recent dynamics coupled with the temporal and spatial patchiness of nutrient and plankton resources (Sapozhnikov 1999, Sukhanova et al. 1999) have lead to international concern over the future of salmon resources of the Bering Sea. In the summer of 2002, member nations of the North Pacific Anadromous Fish Commission (NPAFC: Canada, Japan, Russia, and United States) initiated an aggressive program for long-term coordinated study of the ecology and migration of chum, sockeye, and Chinook salmon in the Bering Sea.

This Bering-Aleutian Salmon International Survey (BASIS) program[12] calls for four one-month surveys (spring, summer, winter, and fall) at 105 sampling stations across the Bering Sea each year for five years. Detailed biological and stock identification analyses will attempt to determine migration and growth

---

[12] North Pacfic Anadromous Fish Commission (NPAFC). 2001. Draft plan for NPAFC Bering-Aleutian Salmon International Survey (BASIS). (NPAFC Doc. 579) 27 pp.

of regional stock aggregations. Genetic data from mixtures of fish collected at the sampling stations will be linked to oceanographic and other biological data to improve our understanding of the productivity of salmon populations and to test hypotheses of stock-specific migration routes.

The current allozyme database provides exhaustive coverage of chum salmon spawning populations throughout the Pacific Rim. A variety of applications are now possible throughout the species' range including analysis of near-shore and high-seas mixtures of chum salmon in all life history stages. The results presented here suggest a high degree of overlap in the areas of the Pacific Ocean and Bering Sea utilized by Asian and North American stocks of chum salmon. Stock composition along important migration corridors changes dramatically over relatively short periods of time. Further, the Bering Sea is an important rearing area for stocks from throughout the range including those as far south in North America as the Pacific Northwest.

Acknowledgements

This study is a collaboration of the Working Committee on Stock Identification of the North Pacific Anadromous Fish Commission. We thank Joel Reynolds and Bill Templin from the Alaska Department of Fish and Game for their help with the construction of the baseline and with the simulation studies. We thank Doug Eggers, Alaska Department of Fish and Game, for his support of high-seas studies and for arranging funding from the State of Alaska for this project. Conversations with Kate Myers, Trey Walker, and Nancy Davis of the High-Seas Research Program, University of Washington, and Jack Helle and Loh-Lee Low, National Marine Fisheries Service, were instrumental in developing ideas presented in this study. The manuscript benefited from reviews by Sue Merkouris, Alaska Department of Fish and Game, and two anonymous reviewers. This manuscript is dedicated to the memory of Steve Phelps who devoted 22 years of his life to teasing apart the genetic relationships of chum salmon populations. This study would not have been possible without Steve's extensive data from Washington and Southern British Columbia populations. Contribution PP-230 of the Alaska Department of Fish and Game, Commercial Fisheries Division.

References

Beamish, R.J. & D.R. Bouillon. 1993. Pacific salmon production trends in relation to climate. Can. J. Fish. Aquat. Sci. 50: 1002–1016.

Brannon, E.L. 1984. Influence of stock origin of salmon migratory behavior. pp. 103–111. *In*: J.D. McCleave (ed.) Mechanisms of Migration in Fishes. NATO Conference Series IV, Marine Sciences Vol. 14, Plenum Press, New York.

Cavalli-Sforza, L.L. & A.W.F. Edwards. 1967. Phylogenetic analysis: Models and estimation procedures. Evolution 21: 550–570.

Chakraborty, R. & O. Leimar. 1987. Genetic variation within a subdivided population. pp. 89–120. *In*: N. Ryman & F.M. Utter (ed.) Population Genetics and Fishery Management, University of Washington Press, Seattle.

Debevec, E.M., R.B. Gates, M. Masuda, J. Pella, J. Reynolds & L.W. Seeb. 2000. SPAM (Version 3.2): Statistics program for analyzing mixtures. J. Hered. 91: 509–511.

Farley, E.V. Jr. & K. Munk. 1997. Incidence of thermally marked pink and chum salmon in the coastal water of the Gulf of Alaska. Alaska Fish. Res. Bull. 4: 181–187.

Gilbert, C.H. & W.H. Rich. 1926. Second experiment in tagging salmon in the Alaska Peninsula fisheries reservation, summer of 1923. Bull. U.S. Bureau Fish. 42: 27–75.

Kaeriyama, M. 1998. Dynamics of chum salmon, *Oncorhynchus keta*, populations released from Hokkaido, Japan. N. Pac. Anadr. Fish Comm. Bull. No. 1: 90–102.

Kondzela, C.M., C.M. Guthrie, S.L. Hawkins, C.D. Russell, J.H. Helle & A.J. Gharrett. 1994. Genetic relationships among chum salmon populations in southeast Alaska and northern British Columbia. Can. J. Fish. Aquat. Sci. 51: 50–64.

Loughlin, T.R. & K. Ohtani. 1996. Dynamics of the Bering Sea. University of Alaska Sea Grant, AK-SG-99-03. Fairbanks. 825 pp.

Mahnken, C., G. Ruggerone, W. Waknize & T. Flagg. 1998. A historical perspective on salmonid production from Pacific Rim hatcheries. N. Pac. Anadr. Fish Comm. Bull. 1: 38–53.

National Research Council (NRC). 1996. The Bering Sea Ecosystem, National Academy Press, Washington, DC.

Nei, M. 1973. Analysis of gene diversity in subdivided populations. Proc. Natl. Acad. Sci. U.S.A. 70: 3321–3323.

Patton, W.S., K.W. Myers & R.V. Walker. 1998. Origins of chum salmon caught incidentally in the eastern Bering Sea walleye pollock trawl fishery as estimated from scale pattern analysis. N. Am. J. Fish. Management 18: 704–712.

Phelps, S.R., L.L. Leclair, S. Young & H.L. Blankenship. 1994. Genetic diversity patterns of chum salmon in the Pacific Northwest. Can. J. Fish. Aquat. Sci. 51: 65–83.

Radchenko, V.I. 1998. Historical trends of fisheries and stock condition of Pacific salmon in Russia. N. Pac. Anadr. Fish. Comm. Bull. No. 1: 28–37.

Salo, E.O. 1991. Life history of chum salmon (*Oncorhynchus keta*). pp. 231–309. *In*: C. Groot & L. Margolis (ed.) Pacific Salmon Life Histories, UBC Press, University of British Columbia, Vancouver.

Sapozhnikov, V.V. 1999. Mesoscale anticyclonic eddies at the shelf break and their impact on the formation of hydrochemical structures of the Bering Sea. pp. 251–259. *In*: T.R. Loughlin & K. Ohtani (ed.) Dynamics of the Bering Sea, University of Alaska Sea Grant, AK-SG-03, Fairbanks.

Seeb, L.W. & P.A. Crane. 1999a. High genetic heterogeneity in chum salmon in Western Alaska, the contact zone between northern and southern lineages. Trans. Am. Fish. Soc. 128: 58–87.

Seeb, L.W. & P.A. Crane. 1999b. Allozymes and mitochondrial DNA discriminate Asian and North American populations of chum salmon in mixed-stock fisheries along the south coast of the Alaska Peninsula. Trans. Am. Fish. Soc. 128: 88–103.

Smouse, P.E., R.S. Waples & J.A. Tworek. 1990. A genetic mixture analysis for use with incomplete source population data. Can. J. Fish. Aquat. Sci. 47: 620–634.

Sukhanova, I.R., H.J. Semina & M.V. Venttsel. 1999. Spatial distribution and temporal variability of phytoplankton in the Bering Sea. pp. 453–483. *In*: T.R. Loughlin & K. Ohtani (ed.) Dynamics of the Bering Sea, University of Alaska Sea Grant, AK-SG-99-03, Fairbanks.

Urawa, S., M. Kawana, G. Anma, Y. Kamei, T. Shoji, M. Fukuwaka, K.M. Munk, K.W. Myers & E.V. Farley Jr. 2000. Geographic origin of high-seas chum salmon determined by genetic and thermal otolith markers. N. Pac. Anadr. Fish Comm. Bull. 2: 283–290.

Wilmot, R.L., R.J. Everett, W.J. Spearman, R. Baccus, N.V. Varnavskaya & S.V. Putivkin. 1994. Genetic stock structure of Western Alaska chum salmon and a comparison with Russian Far East stocks. Can. J. Fish. Aquat. Sci. 51: 84–94.

Wilmot, R.L., C.M. Kondzela, C.M. Guthrie & M. Masuda. 1998. Genetic stock identification of chum salmon harvested incidentally in the 1994 and 1995 Bering Sea trawl fishery. N. Pac. Anadr. Fish Comm. Bull. No. 1: 285–299.

Winans, G.A., P.B. Aebersold, Y. Ishida & S. Urawa. 1998. Genetic stock identification of chum salmon in highseas test fisheries in the Western North Pacific Ocean and Bering Sea. N. Pac. Anadr. Fish Comm. Bull. No. 1: 220–226.

Winans, G.A., P.B. Aebersold, S. Urawa & N.V. Varnavskaya. 1994. Determining continent of origin of chum salmon (*Oncorhynchus keta*) using genetic stock identification techniques: Status of allozyme baseline in Asia. Can. J. Fish. Aquat. Sci. 51: 95–113.

Wood, C.C., S. McKinnell, T.J. Mulligan & D.A. Fournier. 1987. Stock identification with the maximum-likelihood mixture model: Sensitivity analysis and application to complex problems. Can. J. Fish. Aquat. Sci. 44: 866–881.

Wooster, W. 1992. King crab dethroned. pp. 15–30. *In*: M.H. Glanz (ed.) Climate Variability, Climate Change, and Fisheries, Cambridge University Press, Cambridge.

Environmental Biology of Fishes **69**: 37–50, 2004.
© 2004 *Kluwer Academic Publishers. Printed in the Netherlands.*

Genetic population structure of chum salmon in the Pacific Rim inferred from mitochondrial DNA sequence variation

Shunpei Sato[a], Hiroyuki Kojima[b], Junko Ando[a], Hironori Ando[a], Richard L. Wilmot[c], Lisa W. Seeb[d], Vladimir Efremov[e], Larry LeClair[f], Wally Buchholz[g], Deuk-Hee Jin[h], Shigehiko Urawa[i], Masahide Kaeriyama[b], Akihisa Urano[a,j] & Syuiti Abe[k,l]

[a]*Division of Biological Science, Graduate School of Science, Hokkaido University, Sapporo 060-0810, Japan*
[b]*Graduate School of Science and Engineering, Hokkaido Tokai University, Sapporo 005-8601, Japan*
[c]*Auke Bay Laboratory, Alaska Fisheries Science Center, NOAA, Juneau, U.S.A.*
[d]*Alaska Department of Fish and Game, Anchorage, U.S.A.*
[e]*Russian Academy of Science, Vladivostok, Russia*
[f]*Washington Department of Fish and Wildlife, Olympia, Washington, U.S.A.*
[g]*U.S. Fish and Wildlife Service, Anchorage, AK, U.S.A.*
[h]*Kangnung National University, Kangnung, Korea*
[i]*Salmon Resources Center, Sapporo 062-0922, Japan*
[j]*Field Science Center, Hokkaido University, Sapporo 060-0811, Japan*
[k]*Laboratory of Animal Cytogenetics, Center for Advanced Science and Technology, Hokkaido University, Sapporo 060-0810, Japan (e-mail: sabe@ees.hokudai.ac.jp)*
[l]*Laboratory of Breeding Science, Graduate School of Fisheries Sciences, Hokkaido University, Hakodate 041-8611, Japan*

Received 17 April 2003   Accepted 27 April 2003

Key words: mtDNA control region, Pacific salmon, haplotype genealogy, genetic divergence

Synopsis

We examined the genetic population structure of chum salmon, *Oncorhynchus keta*, in the Pacific Rim using mitochondrial (mt) DNA analysis. Nucleotide sequence analysis of about 500 bp in the variable portion of the 5′ end of the mtDNA control region revealed 20 variable nucleotide sites, which defined 30 haplotypes of three genealogical clades (A, B, and C), in more than 2,100 individuals of 48 populations from Japan (16), Korea (1), Russia (10), and North America (21 from Alaska, British Columbia, and Washington). The observed haplotypes were mostly associated with geographic regions, in that clade A and C haplotypes characterized Asian populations and clade B haplotypes distinguished North American populations. The haplotype diversity was highest in the Japanese populations, suggesting a greater genetic variation in the populations of Japan than those of Russia and North America. The analysis of molecular variance and contingency χ^2 tests demonstrated strong structuring among the three geographic groups of populations and weak to moderate structuring within Japanese and North American populations. These results suggest that the observed geographic pattern might be influenced primarily by historic expansions or colonizations and secondarily by low or restricted gene flow between local groups within regions. In addition to the analysis of population structure, mtDNA data may be useful for constructing a baseline for stock identification of mixed populations of high seas chum salmon.

Introduction

Chum salmon, *Oncorhynchus keta*, has the widest natural geographical distribution among all Pacific salmon species in the North Pacific Rim, ranging from Korea and Japan northward to the Arctic coasts of Russia and North America and then southward to Oregon (Salo 1991). Spawning adults, like other Pacific salmons, are anadromous to the natal river. Such restricted homing behavior will lead geographically distinct populations to partial genetic isolation. Estimation of genetic variation among and within the Pacific Rim populations of chum salmon is therefore important for addressing the population history, the patterns of ocean migration, and the stock composition in high seas aggregations and coastal commercial fisheries.

The genetic variation of chum salmon has been examined by allozyme analysis (Kondzela et al. 1994, Phelps et al. 1994, Wilmot et al. 1994, Winans et al. 1994), which has been used for stock identification of population mixtures in the ocean (Wilmot et al. 1998, Seeb & Crane 1999b). However, analysis of allozymes requires careful collection and handling of tissues (Brown et al. 1979), and resolution of allozymes remains mostly at the regional- to continental-levels (Brown et al. 1979, Wilmot et al. 1998). Moreover, allozyme data are often inadequate for discriminating causal factors of population divergence (Zhivotovsky et al. 1994).

Recently developed molecular techniques are likely to provide a powerful means to observe genetic variation in salmon populations with increased accuracy and resolution (Ferguson et al. 1995). Maternally inherited mitochondrial (mt) DNA has higher sequence variability than most single copy nuclear genes (Brown et al. 1979). The non-coding control region has been recomended for assessing intraspecific genetic variation in the species of interest, because it often has higher variability than the coding regions (Moritz et al. 1987, Meyer 1993), although this situation is not always the case for some whale (Hoelzel et al. 1991) and fish species (Bernatchez et al. 1992, Pigeon et al. 1998) including salmon (Churikov et al. 2001). Thus, analysis of mtDNA has become a method of choice in many phylogenetic, population genetic, and evolutionary studies (Moritz et al. 1987, Meyer 1993).

Previous studies on restricted fragment length polymorphisms (RFLPs) showed low levels of sequence variation in limited mtDNA segments of salmonids including chum salmon (Wilson et al. 1987, Cronin et al. 1993, Park et al. 1993, Bickham et al. 1995). Although a genome-wide RFLP study using a number of restriction enzymes detected a few hypervariable coding regions of mtDNA in chum and other Pacific salmon species (Churikov et al. 2001), such an extensive analysis is not easy to apply for stock identification using a large number of fish. Moreover, allozyme and mtDNA RFLP analyses provided similar estimates in stock identification of mixed fisheries of chum salmon

(Seeb & Crane 1999b), although that study did not survey the entire mtDNA genome.

Recently, Sato et al. (2001) detected greater amount of variation in the mtDNA control region by nucleotide sequence analysis than the variation observed by a previous RFLP analyses (Park et al. 1993). This finding indicates an increased potential of mtDNA sequence analysis to estimate the genetic variation of chum salmon populations. In this collaborative study, nucleotide sequencing of the mtDNA control region was conducted to examine its potential use in analyzing the genetic variation and population structure of chum salmon, using more than 2,100 individuals from 48 populations in Japan, Korea, Russia, and North America.

Materials and methods

Samples

Liver, blood, or muscle samples of chum salmon were collected from 1,617 individuals of 36 populations from Japan (four populations), Korea (one population), Russia (10 populations), and North America including Alaska (13 populations), British Columbia (three populations), and Washington (five populations) from 1988 to 2000 (Table 1 and Figure 1). All samples,

Table 1. Sampling locations, date of collection, and the numbers of chum salmon samples (N) used for mtDNA analysis.

Sampling location	Date of collection	N
Japan		
Hokkaido Island		
1 Chitose River*	14 Oct. 1996	51
2 Tokushibetsu River*	23 Sep. 1997	51
3 Tokoro River (late-run)*	20 Nov. 1998	44
4 Tokoro River (early-run)*	13 Oct. 1999	49
5 Nishibetsu River*	25 Sep. 1997	41
6 Kushiro River	22 Oct. 1998	49
7 Tokachi River*	17 Oct. 1996	46
8 Yurappu River*	17 Nov. 1998	40
Honshu Island		
9 Tsugaruishi River (late-run), Iwate Pref.*	10 Dec. 1997	44
10 Tsugaruishi River (early-run), Iwate Pref.	Oct. 1999	47
11 Otsuchi River, Iwate Pref.*	8 Apr. 1999	49
12 Koizumi River, Miyagi Pref.*	21 Nov. 1996	47
13 Kawabukuro River, Akita Pref.*	18 Nov. 1997	30
14 Gakko River, Yamagata Pref.*	10 Dec. 1996	45
15 Uono River, Niigata Pref.	23–24 Oct. 1996	49
16 Jintsu River, Toyama Pref.	7 Nov. 1995	49

Table 1. (*Continued*)

Sampling location	Date of collection	N
Korea		
17 Namadae River	13 Nov. 2000	46
Russia		
Anadyr		
18 Anadyr River	1990	43
Kamchatka Peninsula		
19 Hairsova River	1993	41
20 Kamchatka River	1991	46
21 Vorovskaya River	1990	32
22 Kol River	1991	44
Sakhalin island		
23 Kalininka River	1994	42
Magadan		
24 Ola River	1990	33
25 Magadan River	1991	37
Nikolaevsk-na-Amure		
26 Amur River	9 Sep. 2000	50
Primorye		
27 Avakumovka River	1994	30
North America		
Northwest Alaska		
28 Salmon River	1991	45
29 Sheenjek River (fall-run)	1992	45
30 Andreafsky River (summer-run)	1993	48
31 Togiak River	1993	49
Alaska Peninsula		
32 Belkofski River	1992	44
Southcentral Alaska		
33 Kizhuyak River	1992	46
34 Olsen Creek	1992	45
Southeast Alaska		
35 Sawmill Creek, Berner's Bay	28 July. 1993	50
36 Long Bay, Chichigof Island	25–26 Aug. 1991	49
37 Whale Bay, Baranof Island	12 Aug. 1993	48
38 Port Beauclerc, Kuiu Island	20 Aug. 1995	45
39 Fish Creek, Portland Canal	25 Sep. 1988	49
40 Disappearance Creek, POW Island	25 Sep. 1998	50
British Columbia		
41 Ecstall River, Skeena River area	12 Sep. 1988	45
42 Bag Harbor, QCI	Mid-Oct. 1989	50
43 Nekite Channel	15 Sep. 1989	33
Washington		
44 Nooksack River	1998	47
45 Quilcene Bay	1998	49
46 Blackjack Creek	1998	50
47 Satsop River	1998	49
48 Hamilton Creek	1998	43

*Cited from Sato et al. (2001).

except for those from three populations (Tsugaruishi River early-run in Japan, Namadae River in Korea, and Amur River in Russia), were from archives. The fish used in this study were captured when they returned to their natal rivers. Archived samples had been previously used for allozyme and/or mtDNA RFLP analyses (Kondzela et al. 1994, Wilmot et al. 1994, Winans et al. 1994, Seeb & Crane 1999b). Several populations were sampled in more than one year and we assumed that the genetic composition was temporally stable, as has been suggested by previous allozyme studies (Wilmot et al. 1994, Seeb & Crane 1999a,b). A possible genetic difference, between summer- and fall-run fish was examined for fish collected in the Yukon River. Early- and late-run populations from two Japanese rivers also were included in the analysis (Tokoro and Tsugaruishi Rivers), as in the previous study (Sato et al. 2001). Liver and blood samples were stored at $-80°C$ and muscle samples were stored in 100% ethanol at room temperature until DNA extraction.

DNA extraction

DNA was isolated from the stored specimens following the phenol–chloroform method (Sambrook et al. 1989). Prior to extraction of DNA, the muscle samples were washed twice in 500 µl sodium–Tris–EDTA buffer (STE; 0.1 M NaCl, 10 mM Tris–HCl, and 1 mM EDTA, pH 8.0). The frozen liver samples were immediately homogenized in the same solution. About 50 µl of whole blood and homogenates of liver or muscle were added to 500 µl STE buffer containing 500 µg/ml proteinase K and 0.5% SDS, and incubated at 37°C overnight. DNA was extracted three times with a mixture of phenol (250 µl) and 24 : 1 chloroform : isoamylalcohol (250 µl), and then twice with 500 µl of the chloroform–isoamylalcohol alone. DNA in aqueous phase was recovered by ethanol precipitation, dried in air, and dissolved in Tris–EDTA buffer (TE; 10 mM Tris–HCl, 1 mM EDTA, pH 7.5).

PCR amplification and nucleotide sequence analysis

The control region of mtDNA was amplified by PCR as in the previous study (Sato et al. 2001). The PCR products were purified by the QIAquick PCR Purification Kit (QIAGEN, Hilden, Germany) after confirmation of their sizes by gel-electrophoresis. Approximately 500 bp in the variable position of the 5′ end of the

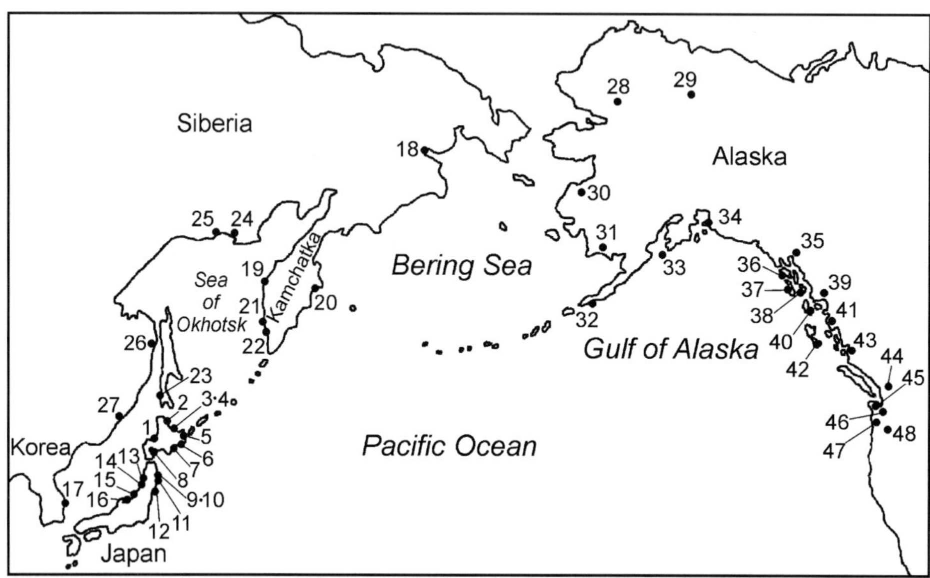

Figure 1. Geographical locations of sampling sites (see Table 1 for the site names).

mtDNA control region was sequenced with a Hitachi SQ-5500L DNA Sequencer (Hitachi, Tokyo) (Sato et al. 2001).

Nucleotide sequence data analysis

For data analysis, the nucleotide sequences of 537 previously examined individuals from 12 Japanese collections (11 populations) were also included. Thus, a total of 2,154 individuals from 48 Pacific Rim populations was analyzed in this study (Table 1 and Figure 1). The sequence data of the 5′ end of the mtDNA control region were aligned by GENETIX-WIN version 4.0.6 (Software Development Co. Ltd, Tokyo) to identify nucleotide variations, from which the haplotypes were defined. Haplotypes were connected into the parsimony network using two different programs: the Network version 3.111 program (Bandelt et al. 1999, available at the web site http://www.fluxus-engineering.com) and the TCS version 1.13 software (Posada & Crandall 2001, available at the web site http://zoology.byu.edu/crandall_lab/programs.htm).

Population genetic data analysis

Haplotype and nucleotide diversity within populations, and nucleotide divergence between populations were estimated according to Nei (1973) and Nei & Tajima (1981) using the DA program in REAP (McElroy et al. 1993). The heterogeneity of the haplotype frequencies within and between geographic regions was evaluated using the contingency χ^2 test (Roff & Bentzen 1989), with 10,000 Monte Carlo simulations by CHIRXC program (Zaykin & Pudovkin 1993). Populations were grouped by the neighbor-joining method (Saitou & Nei 1987) using pairwise nucleotide divergences. The topology obtained was tested for its stability by a consensus analysis using 1,000 replicates of the original population divergence matrix obtained by bootstrap resampling of individuals from each population. A neighbor-joining tree was constructed for each replicate with NEIGHBOR and the consensus tree was generated using CONSENSUS in PHYLIP version 3.5c software[1]. In order to assess the extent of genetic divergence at different levels of geographic hierarchy, the overall molecular variance was partitioned into components corresponding to the population divergence within and among regions by the analysis of molecular variance (AMOVA) model (Excoffier et al. 1992) using the Arequin version 2.000 program package[2].

[1] Felsenstein, J. 1993. PHYLIP (Phylogeny inference package), Version 3.5c. Department of Genetics, University of Washington, Seattle: Available at the web site http://evolution.genetics.washington.edu/phylip.html.
[2] Schneider, S., D. Roessli & L. Excoffier. 2000. Arlequin, Version 2.000 University of Geneva, Geneva: Available at the web site http://lgb.unige.ch/arlequin/.

Results

mtDNA control region haplotypes in chum salmon

Haplotype sequence variation. Sequence analysis of the 481 bp 5′ variable position of the mtDNA control region disclosed 20 variable sites in a total of 2,154 individuals from 48 populations, which defined a total of 30 haplotypes, which we designated as A-1 to C-5 (Table 2). The nucleotide sequence variations observed included one nucleotide insertion, one nucleotide deletion, and 18 nucleotide substitutions including 11 transitions and seven transversions (see Table 2). Designation of the 12 haplotypes reported in the previous study (Sato et al. 2001) were changed as follows: OKDL-1 to A-1, OKDL-2 to A-5, OKDL-3 to A-6, OKDL-4 to A-7, OKDL-5 to A-8, OKDL-6 to B-1, OKDL-7 to B-3, OKDL-8 to B-4, OKDL-9 to C-1, OKDL-10 to C-2, OKDL-11 to C-4, and OKDL-12 to C-5 (GenBank accession numbers AB039890–AB039901). The sequences of 18 newly identified mtDNA control region haplotypes were registered in the DDBJ/EMBL/GenBank with accession numbers AB091514–AB091531.

Haplotype genealogy. The observed haplotypes of chum salmon were grouped into three clades based on the nucleotide variation shown in Table 2, i.e., A-1 to A-8 in clade A, B-1 to B-17 in clade B, and C-1 to C-5 in clade C. The T to C transition at nucleotide 30 separated clade C from clade A, and a deletion at nucleotide 386 and C to A transversion at nucleotide 395 discriminated clade B from clades A and C, respectively, as shown in Table 2. Two different algorithms following Templeton et al. (1992) and Bandelt et al. (1999) created essentially the same parsimony network connecting the 30 control region haplotypes (data not shown). A single minimum spanning tree (Figure 2) was then produced

Table 2. Variable nucleotide positions in the 5′ half of mtDNA control region of chum salmon.

Haplotype	10	30	42	57	70	78	96	108	154	194	231	242	250	260	339	340	386	395	401	471
A-1	T	T	A	A	T	T	-	A	C	A	T	C	T	A	T	C	G	C	T	A
A-2	C	.	.	.	.	.	.	.	.	.	.	.	.	.	.	.	.	.	.	.
A-3	.	.	G	.	.	.	.	.	.	.	.	.	.	.	.	.	.	.	.	.
A-4	.	.	.	.	.	.	.	C	.	.	.	.	.	.	.	.	.	.	.	.
A-5	.	.	.	.	.	.	.	.	.	T	.	.	.	.	.	.	.	.	.	.
A-6	.	.	.	.	.	.	.	.	.	.	C	.	.	.	.	.	.	.	.	.
A-7	.	.	.	.	.	.	.	.	.	.	.	.	.	.	.	.	.	.	.	C
A-8	.	.	.	.	.	.	A	.	.	.	.	.	.	.	.	.	.	.	.	.
B-1	.	.	.	.	.	.	.	.	.	.	.	.	.	.	.	.	-	A	.	.
B-2	.	C	.	.	.	.	.	.	.	.	.	.	.	.	.	.	-	A	.	.
B-3	.	.	.	.	.	.	.	G	.	.	.	.	.	.	.	.	-	A	.	.
B-4	.	.	.	.	.	.	.	.	.	.	C	.	.	.	.	.	-	A	.	.
B-5	C	.	.	.	.	.	.	G	.	.	.	.	.	.	.	.	-	A	.	.
B-6	.	.	.	.	C	.	.	G	.	.	.	.	.	.	.	.	-	A	.	.
B-7	.	.	.	.	.	C	.	G	.	.	.	.	.	.	.	.	-	A	.	.
B-8	.	.	.	.	.	.	C	G	.	.	.	.	.	.	.	.	-	A	.	.
B-9	.	.	.	.	.	.	.	G	.	.	C	.	.	.	.	.	-	A	.	.
B-10	.	.	.	.	.	.	.	G	.	.	.	T	.	.	.	.	-	A	.	.
B-11	.	.	.	.	.	.	.	G	.	.	.	.	C	.	.	.	-	A	.	.
B-12	.	.	.	.	.	.	.	G	.	.	.	.	.	G	.	.	-	A	.	.
B-13	.	.	.	.	.	.	.	G	.	.	.	.	.	.	A	.	-	A	.	.
B-14	.	.	.	.	.	.	.	G	.	.	.	.	.	.	.	.	-	A	C	.
B-15	.	.	.	.	.	.	.	G	.	.	.	.	.	.	.	.	-	A	.	C
B-16	.	.	.	.	.	.	.	G	.	.	.	.	.	.	A	T	-	A	.	.
B-17	.	.	.	.	.	.	.	G	.	.	.	.	.	.	A	.	-	A	C	.
C-1	.	C	.	.	.	.	.	.	.	.	.	.	.	.	.	.	.	.	.	.
C-2	.	C	.	T	.	.	.	.	.	.	.	.	.	.	.	.	.	.	.	.
C-3	.	C	.	.	C	.	.	.	.	.	.	.	.	.	.	.	.	.	.	.
C-4	.	C	.	.	.	.	T	.	.	.	.	.	.	.	.	.	.	.	.	.
C-5	.	C	.	.	.	.	.	.	.	.	C	.	.	.	.	.	.	.	.	.

The nucleotide at each position is given for A-1. The hyphen represents the deletion and dot represents the same nucleotide at the same position as in the A-1.

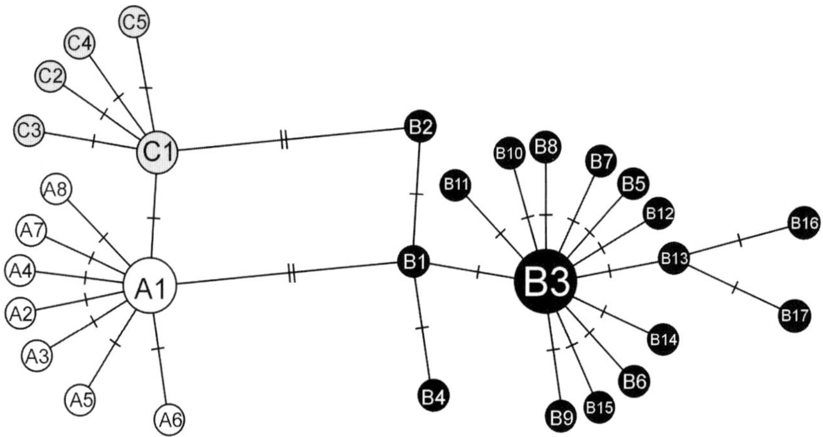

Figure 2. A single minimum spanning tree for the 30 mtDNA control region haplotypes (481 bp sequences) of chum salmon presented in Table 2. Circle sizes reflect haplotype abundances.

by assuming that transitions were more probable than transversions, that ambiguously positioned haplotypes were more likely to be descended from other haplotypes in their geographic neighborhood and from more abundant and interior haplotypes (Castelloe & Templeton 1994). A-1, B-3, and C-1 were focal haplotypes in their respective clades, from which rarer haplotypes were radiated, thus showing three star-like genealogies. However, the connection between these focal haplotypes remained ambiguous based on these data (Figure 2).

Haplotype distribution in the Pacific Rim populations

The distribution of 30 haplotypes among 48 populations of chum salmon is presented in Table 3. The frequency of haplotypes across the Pacific Rim was nonrandom, and mostly associated with regional structure.

Asian populations. The Japanese populations had the most haplotypes, 16 of 30, in the regions examined. All three haplotype clades were represented in every population except for the Tsugaruishi late-run (9) and the Otsuchi (11) populations both of which lacked clade B haplotypes (see Table 3 and Figure 1). The A-1 and C-1 haplotypes were common in all the populations of Japan; the B-3 was common in most Japanese populations. The frequency and composition of the haplotypes differed between early (10) and late (9) runs from the Tsugaruishi River on the Pacific coast of Honshu,

whereas the Tokoro River early (4) and late (3) runs in Hokkaido showed similar haplotype distributions (Sato et al. 2001).

Thirteen haplotypes occurred in 10 Russian populations and four haplotypes occurred in a single Korean population (Table 3). All three haplotype clades were observed in the Namadae River (17) in Korea and the Avakumovka River (27) in Russia, both of which are on the Sea of Japan coast (see Table 3 and Figure 1). Nine other Russian populations contained clade B and C haplotypes, although the clade C haplotypes were less abundant than clade B haplotypes (Table 3). Haplotype B-3 was predominant in most of the Russian populations (Table 3).

North American populations. The number of haplotypes in North American populations was typically less than that of populations in Japan and Russia. The North American populations exhibited no clade A, 10 clade B, and one clade C haplotypes (Table 3). Clade C haplotypes were rare (<8%) and occurred in only two populations, one from the Alaska Peninsula (Belkofski River, 32) and the other from Kodiak Island (Kizhuyak River, 33), both areas of Southcentral Alaska (see Table 3 and Figure 1). Among the observed clade B haplotypes, B-3 and B-13 were common in most of the North American populations, although the latter was less frequent than the former. Four populations in western Alaska, Salmon River (28), fall-run (Sheenjak River, 29) and summer-run (Andreafsky River, 30) of the Yukon River, and Togiak River (31), lacked the B-13 haplotype and were fixed or nearly fixed for the

Table 3. Distribution of mtDNA control region haplotypes among 48 populations of chum salmon in the Pacific Rim.

Population no.	A1	A2	A3	A4	A5	A6	A7	A8	B1	B2	B3	B4	B5	B6	B7	B8	B9	B10	B11	B12	B13	B14	B15	B16	B17	C1	C2	C3	C4	C5
Japan																														
Hokkaido Island																														
1	22	0	0	0	1	0	0	0	0	0	14	2	0	0	0	0	0	0	0	0	0	0	0	0	0	10	0	0	0	2
2	30	0	0	0	0	0	0	0	0	0	13	0	0	0	0	0	0	0	0	0	0	0	0	0	0	8	0	0	0	0
3	26	0	0	0	0	0	1	0	0	0	8	0	0	0	0	0	0	0	0	0	0	0	0	0	0	8	0	0	0	1
4	21	0	0	0	0	1	0	0	1	0	16	0	0	0	0	0	0	0	0	0	0	0	0	0	0	9	0	0	0	1
5	12	0	0	0	0	0	0	0	0	0	18	0	0	0	0	0	0	0	0	0	0	0	0	0	0	11	0	0	0	0
6	23	1	4	0	0	0	0	0	1	0	8	0	0	0	0	0	0	0	0	0	0	0	0	0	0	10	0	1	0	1
7	18	0	0	0	2	0	4	0	0	0	12	0	0	0	0	0	0	0	0	0	0	0	0	0	0	8	1	0	1	0
8	24	0	0	0	0	0	0	0	0	0	6	0	0	0	0	0	0	0	0	0	0	0	0	0	0	10	0	0	0	0
Honshu Island																														
9	25	0	0	0	0	0	0	0	0	0	0	0	0	0	0	0	0	0	0	0	0	0	0	0	0	19	0	0	0	0
10	20	0	0	0	0	4	0	1	0	0	7	0	0	0	0	0	0	0	0	0	0	0	0	0	0	15	0	0	0	0
11	26	0	0	0	0	2	0	0	0	0	0	0	0	0	0	0	0	0	0	0	0	0	0	0	0	21	0	0	0	0
12	24	0	0	0	0	1	0	0	0	0	1	0	0	0	0	0	0	0	0	0	0	0	0	0	0	21	0	0	0	0
13	19	0	0	0	0	0	0	0	0	0	5	0	0	0	0	0	0	0	0	0	0	0	0	0	0	5	0	0	0	1
14	26	0	0	0	0	0	0	0	0	0	4	1	0	0	0	0	0	0	0	0	0	0	0	0	0	14	0	0	0	0
15	29	0	0	2	0	0	0	0	0	0	8	0	0	0	0	0	0	0	0	0	0	0	0	0	0	9	0	0	0	1
16	37	0	0	0	0	0	0	0	0	0	2	0	0	0	0	0	0	0	0	0	0	0	0	0	0	10	0	0	0	0
Korea																														
17	36	0	0	0	0	0	0	0	0	0	6	0	0	0	0	0	0	0	0	0	0	0	0	0	0	3	0	0	0	1
Russia																														
Anadyr																														
18	0	0	0	0	0	0	0	0	0	0	33	0	1	0	0	0	0	0	0	0	0	0	0	0	0	9	0	0	0	0
Kamchatka Peninsula																														
19	0	0	0	0	0	0	0	0	0	0	32	0	0	0	0	0	1	0	0	0	0	0	0	0	0	8	0	0	0	0
20	0	0	0	0	0	0	0	0	0	0	31	0	0	0	1	0	0	0	0	0	0	0	0	0	0	0	0	0	0	0
21	0	0	0	0	0	0	0	0	0	0	38	0	0	0	0	0	1	0	0	1	0	0	0	0	0	6	0	0	0	0
22	0	0	0	0	0	0	0	0	0	0	38	0	0	0	0	0	2	0	0	1	0	0	0	0	0	3	0	0	0	0
Sakhalin Island																														
23	0	0	0	0	0	0	0	0	0	0	20	0	0	0	0	6	16	0	0	0	0	0	0	0	0	7	0	0	0	1
Magadan																														
24	0	0	0	0	0	0	0	0	1	0	20	0	0	1	0	0	3	0	0	0	0	0	0	0	0	7	0	0	0	1
25	0	0	0	0	0	0	0	0	0	4	31	0	0	0	0	0	1	0	0	0	0	0	0	0	0	0	1	0	0	0

Table 3. (*Continued*)

Population no.	A1	A2	A3	A4	A5	A6	A7	A8	B1	B2	B3	B4	B5	B6	B7	B8	B9	B10	B11	B12	B13	B14	B15	B16	B17	C1	C2	C3	C4	C5
Nikolaevsk-na-Amure																														
26	0	0	0	0	0	0	0	0	2	0	45	0	0	0	0	0	2	0	0	0	0	0	0	0	0	1	0	0	0	0
Primorye																														
27	7	0	0	0	0	0	0	0	0	0	9	0	0	0	0	1	0	0	0	0	0	0	0	0	0	6	0	0	0	7
North America																														
Northwest Alaska																														
28	0	0	0	0	0	0	0	0	0	0	48	0	0	0	0	0	0	0	0	0	0	0	0	0	0	0	0	0	0	0
29	0	0	0	0	0	0	0	0	0	0	45	0	0	0	0	0	0	0	0	0	0	0	0	0	0	0	0	0	0	0
30	0	0	0	0	0	0	0	0	0	0	45	0	0	0	0	0	0	0	0	0	0	0	0	0	0	0	0	0	0	0
31	0	0	0	0	0	0	0	0	0	0	48	0	0	0	0	0	0	0	0	0	0	0	1	0	0	0	0	0	0	0
Alaska Peninsula																														
32	0	0	0	0	0	0	0	0	0	0	37	0	0	0	0	0	0	0	0	0	5	0	0	0	0	4	0	0	0	0
Southcentral Alaska																														
33	0	0	0	0	0	0	0	0	0	0	36	0	0	0	0	0	1	0	0	0	6	0	0	0	0	1	0	0	0	0
34	0	0	0	0	0	0	0	0	0	0	35	0	0	0	0	0	0	0	2	0	6	0	0	0	2	0	0	0	0	0
Southeast Alaska																														
35	0	0	0	0	0	0	0	0	0	0	39	0	0	0	0	0	1	0	5	0	5	0	0	0	0	0	0	0	0	0
36	0	0	0	0	0	0	0	0	0	0	40	0	0	0	0	0	1	0	1	0	7	0	0	0	0	0	0	0	0	0
37	0	0	0	0	0	0	0	0	0	0	33	0	0	0	0	0	2	0	0	0	13	0	0	0	0	0	0	0	0	0
38	0	0	0	0	0	0	0	0	0	0	40	0	0	0	0	0	4	0	0	0	1	0	0	0	0	0	0	0	0	0
39	0	0	0	0	0	0	0	0	0	0	45	0	0	0	0	0	0	0	0	0	3	0	0	0	1	0	0	0	0	0
40	0	0	0	0	0	0	0	0	0	0	33	0	0	0	0	0	5	0	0	0	12	0	0	0	0	0	0	0	0	0
British Columbia																														
41	0	0	0	0	0	0	0	0	0	0	29	0	0	0	0	0	1	0	0	0	15	0	0	0	0	0	0	0	0	0
42	0	0	0	0	0	0	0	0	0	0	32	0	0	0	0	0	0	0	1	0	17	0	0	0	0	0	0	0	0	0
43	0	0	0	0	0	0	0	0	0	0	25	0	0	0	0	0	0	0	0	0	8	0	0	0	0	0	0	0	0	0
Washington																														
44	0	0	0	0	0	0	0	0	0	0	39	0	0	0	0	0	0	0	0	0	8	0	0	0	0	0	0	0	0	0
45	0	0	0	0	0	0	0	0	0	0	41	0	0	0	0	0	0	3	0	0	5	0	0	0	0	0	0	0	0	0
46	0	0	0	0	0	0	0	0	0	0	45	0	0	0	0	0	0	0	0	0	3	0	0	2	0	0	0	0	0	0
47	0	0	0	0	0	0	0	0	0	0	23	0	0	0	0	0	1	0	0	0	17	8	0	0	0	0	0	0	0	0
48	0	0	0	0	0	0	0	0	0	0	23	0	6	0	0	0	0	0	0	0	12	2	0	0	0	0	0	0	0	0
Total	425	1	4	2	1	10	1	5	5	4	1206	3	7	1	1	7	42	3	9	2	143	10	1	2	3	236	2	1	1	16

B-3 haplotype (Table 3 and Figure 1). Other North American populations usually included one or two clade B haplotypes in addition to the B-3 and B-13 haplotypes.

Inter-regional haplotype distribution. Among the observed haplotypes, 22 were region-specific, including seven clade A (A-2 to A-8), one clade B (B-4) and two clade C (C-3 and C-4) of 16 haplotypes in Japan; seven clade B (B-10, B-11, B-13 to B-17) of 11 haplotypes in North America; and five clade B (B-2, B-6 to B-8, and B-12) of 13 haplotypes in Russia (Table 3). The B-3 and C-1 haplotypes occurred in all three geographic regions, although the haplotype C-1 was rare and its occurrence was limited in the North American populations. The A-1, B-1, C-2, and C-5 haplotypes occurred in both the Japanese and Russian populations, and the B-5 and B-9 were shared in the Russian and North American populations. Therefore, it is reasonable to conclude that clade A and C haplotypes characterize the Asian populations, and clade B haplotypes predominate in the North American populations.

Population genetic analysis

Haplotype diversity was highest in the populations of Japan (0.63 ± 0.01), followed by those of Russia (0.43 ± 0.03) and North America (0.34 ± 0.02), whereas nucleotide diversity was similar in the Japanese (0.0028) and Russian populations (0.0025), but lower in the North American populations (Table 4). These findings suggest greater genetic variation in the populations of Japan than those of Russia and North America.

Population clustering. The populations examined were clustered using the neighbor-joining method (Figure 3). The population consensus tree clearly separated Japan/Korea from the other geographic groups with high bootstrap support (99%), although one Russian population (Avakumovka) on the Sea of Japan coast was included in the Japan/Korea cluster. Interestingly, western Alaskan populations were separated from the other North American groups with more than 50% of the bootstrap replicates, but included two Russian populations (the Kamchatka River in eastern Kamchatka Peninsula and the Kalininka River in Sakhalin Island) in the same cluster. Other Russian populations formed a separate cluster with more than

95% bootstrap support. Thus, four major population clusters of Japan/Korea, Russia, Northwest Alaska, and the rest of the North American groups were apparent on the consensus tree.

Heterogeneity in the haplotype distribution. The contingency χ^2 test showed highly significant heterogeneity ($p < 0.001$) in the haplotype frequencies for the entire set of populations (Table 5). Significant regional heterogeneity ($p < 0.001$) was also observed between each set of populations from Japan and Russia, Japan and North America, and Russia and North America, respectively. Furthermore, significant heterogeneity was observed among North American populations ($p < 0.005$), which suggested five sub-regional North American groups: western Alaska, the Alaska Peninsula/Southcentral Alaska, Southeast Alaska, British Columbia, and Washington (Table 5). A similar level of heterogeneity ($p < 0.005$) was also apparent among Hokkaido, the Pacific coast of Honshu, and the Sea of Japan coast in the Japanese populations (Table 5, Sato et al. 2001). No significant heterogeneity was observed among the populations in the Yukon summer- and fall-runs in western Alaska or the late- and early-runs in the Tokoro River in Hokkaido (Sato et al. 2001), whereas significant heterogeneity was observed between the Tsugaruishi late- and early-runs on the Pacific coast of Honshu ($p < 0.01$).

Geographic hierarchy in the Pacific Rim populations. Using AMOVA to partition of the molecular variance AMOVAs (Table 6) revealed the following population structure in chum salmon: (i) very strong geographic structuring among Japan, Russia, and North America (56.2% of the total variance, Analysis I), as compared with the average extent of structuring among populations within each geographic group (4.3% of the total variance); (ii) weak to moderate structuring among western Alaska, Alaska Peninsula/Southcentral Alaska, Southeast Alaska, British Columbia, and Washington (4.9% of the variance, Analysis IV); (iii) similar level of intra-regional structuring among Hokkaido, the Pacific coast of Honshu and the Sea of Japan coast of Honshu in Japan (7.3% of the variance, Analysis II) as described previously (Sato et al. 2001); and (iv) unclear geographic structuring among six regional groups (Tables 1 and 3) (18.2% of the variance, $p = 0.075$, Analysis III) and very weak structuring among local populations within groups in Russia (3.0% of the variance). The latter is likely associated

46

Table 4. Haplotype diversity (h, ±SD) and nucleotide diversity (π, in parentheses) within 48 populations calculated from mtDNA haplotype frequencies.

Population	h (π)	Population	h (π)	Population	h (π)
Japan	0.63 ± 0.01 (0.0028)	*Russia*	0.43 ± 0.03 (0.0025)	Sawmill Cr.	0.38 ± 0.08 (0.00085)
Chitose	0.71 ± 0.04 (0.0038)	Anadyr	0.38 ± 0.08 (0.0029)	Long Bay	0.32 ± 0.08 (0.00069)
Tokushibetsu	0.58 ± 0.05 (0.0030)	Hairusova	0.36 ± 0.08 (0.0028)	Whale Bay	0.46 ± 0.06 (0.0010)
Tokoro	0.60 ± 0.07 (0.0028)	Kamchatka	0.06 ± 0.06 (0.00013)	Port Beauclere	0.21 ± 0.08 (0.00044)
Tokoro-E	0.69 ± 0.04 (0.0037)	Vorovskaya	0.31 ± 0.08 (0.0021)	Fish Cr.	0.16 ± 0.07 (0.00040)
Nishibetsu	0.67 ± 0.03 (0.0040)	Kol	0.25 ± 0.08 (0.0014)	Disappearance Cr.	0.51 ± 0.06 (0.0012)
Kushiro	0.72 ± 0.05 (0.0032)	Kalininka	0.63 ± 0.04 (0.0016)	Ecstall	0.48 ± 0.05 (0.0010)
Tokachi	0.75 ± 0.04 (0.0039)	Ola	0.59 ± 0.08 (0.0038)	Bag Harbor	0.48 ± 0.05 (0.0010)
Yurappu	0.57 ± 0.06 (0.0024)	Magadan	0.33 ± 0.09 (0.0019)	Nekite Channel	0.38 ± 0.08 (0.00079)
Tsugaruishi	0.50 ± 0.03 (0.0010)	Amur	0.19 ± 0.07 (0.00065)	Nooksack	0.29 ± 0.07 (0.00060)
Tsugaruishi-E	0.70 ± 0.04 (0.0030)	Avakumovka	0.79 ± 0.03 (0.0048)	Quilcene Bay	0.29 ± 0.08 (0.00063)
Otsuchi	0.54 ± 0.03 (0.0012)			Blackjack Cr.	0.19 ± 0.07 (0.00055)
Koizumi	0.55 ± 0.03 (0.0014)	*N. America*	0.34 ± 0.02 (0.00083)	Satsop	0.65 ± 0.04 (0.0016)
Kawabukuro	0.56 ± 0.09 (0.0026)	Salmon	0.00 ± 0.00 (0.0000)	Hamilton Cr.	0.63 ± 0.05 (0.0016)
Gakko	0.57 ± 0.05 (0.0022)	Sheenjek	0.00 ± 0.00 (0.0000)		
Uono	0.60 ± 0.06 (0.0027)	Andreafsky	0.00 ± 0.00 (0.0000)	Total	0.63 ± 0.01 (0.0037)
Jintsu	0.39 ± 0.07 (0.0012)	Togiak	0.04 ± 0.04 (0.000085)		
		Belkofski	0.34 ± 0.08 (0.0018)		
Korea		Kizhuyak	0.32 ± 0.08 (0.00097)		
Namadae	0.37 ± 0.08 (0.0019)	Olsen Cr.	0.38 ± 0.08 (0.00099)		

with an insufficient number of populations in the study, i.e., a single population in four of the locales.

Discussion

This mtDNA analysis resolved 20 variable positions in the approximately 500 bp nucleotide sequences at the 5′ end of the control region (Table 2), which define a total of 30 haplotypes among more than 2,100 individuals from 48 populations of chum salmon in the Pacific Rim (Table 3). Analysis of the variation among populations demonstrated; (i) substantial genetic divergence among three geographic groups of chum salmon, i.e., Japan, Russia, and North America; (ii) greatest genetic variation between Japanese and North American populations; and (iii) weak to moderate genetic isolation within Japanese and North American populations.

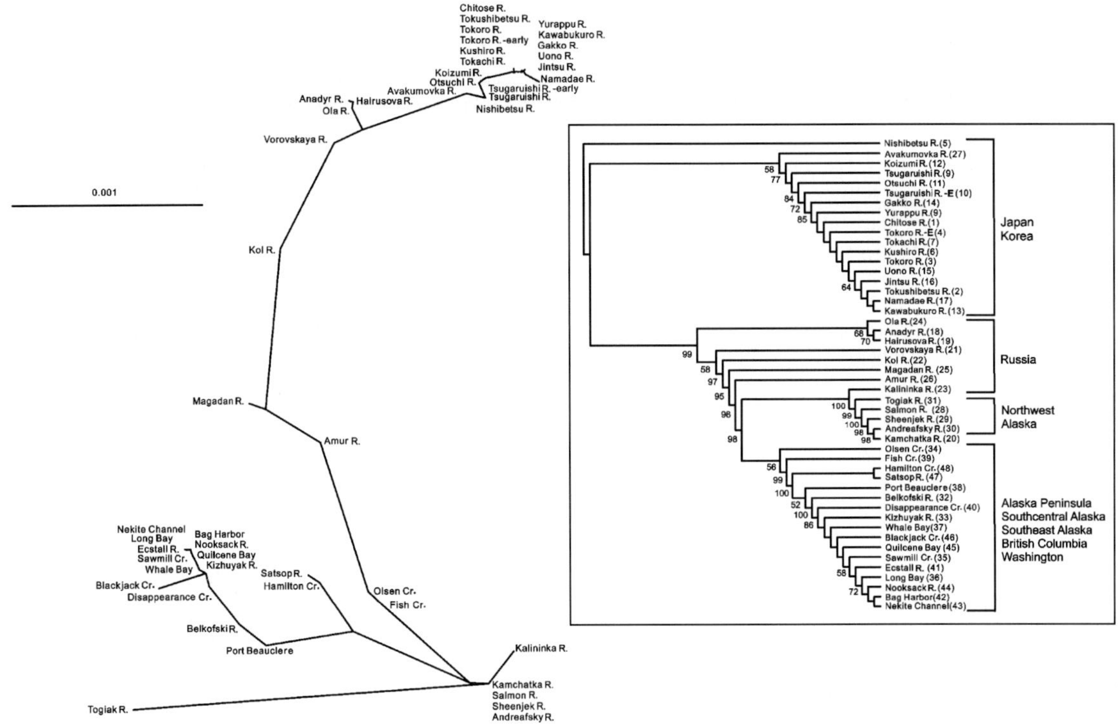

Figure 3. Unrooted neighbor-joining tree showing net nucleotide divergence among 48 chum salmon populations. The scale corresponds to 0.1% of nucleotide divergence. In the inset, the topology of the consensus tree (not scaled) is shown with nodal values for bootstrap support over 50% of the 1,000 replicated trees. The number in parenthesis indicates the geographical location of each population shown in Figure 1.

Table 5. Results of homogeneity test for pairwise geographic regions with indicated number of chum salmon populations in parenthesis. The probability of homogeneity for pairwise geographic regions was given below diagonal, which was calculated using contingency χ^2 test with 10,000 Monte Carlo simulations (Roff & Bentzen 1989). The probability of homogeneity within regions was given on diagonal.

Region	Japan		Russia	North America				
	HOK	HON	RUS	NWA	AP/SCLA	SEA	BCL	WSG
Japan								
Hokkaido (8)	<0.001							
Honshu (8)	<0.001	<0.001						
Russia								
Russia (10)	<0.001	<0.001	<0.001					
North America								
Northwest Alaska (4)	<0.001	<0.001	<0.001	1				
Alaska Peninsula/ Southcentral Alaska (3)	<0.001	<0.001	<0.001	<0.001	0.069			
Southeast Alaska (6)	<0.001	<0.001	<0.001	<0.001	<0.001	<0.001		
British Columbia (3)	<0.001	<0.001	<0.001	<0.001	<0.005	<0.001	0.75	
Washington (5)	<0.001	<0.001	<0.001	<0.001	<0.001	<0.001	<0.001	<0.001

Our nucleotide sequence analysis detected similar or higher levels of variation than that found at the 3′ end of the control region in other *Oncorhynchus* species, including *O. mykiss*, *O. kisutch*, and *O. tshawytscha* (Nielsen et al. 1994). Such inter-specific comparisons must be interpreted with care, because the extent of sequence variation in the control region is site-specific and species-specific in fish (Meyer 1993). However, the

Table 6. Results of the hierarchical analyses of molecular variance for chum salmon.

Variance component	%	P	Φ
Analysis I			
Among three regional groups	56.2	<0.001	0.56
(Japan, Russia, and North America)			
Among populations within groups	4.3	<0.001	0.098
Within populations	39.5	<0.001	0.60
Analysis II			
Among three regional groups in Japan	7.3	<0.001	0.073
Among populations within groups	1.5	<0.05	0.017
Within populations	91.2	<0.001	0.088
Analysis III			
Among six regional groups in Russia	18.2	0.075	0.18
Among populations within groups	3.0	<0.001	0.03
Within populations	78.8	<0.001	0.21
Analysis IV			
Among five regional groups	4.9	<0.005	0.049
in North America			
Among populations within groups	3.7	<0.001	0.038
Within populations	91.4	<0.001	0.085

The percentage of variance (%), probability estimated from permutation (P), and the F-statistics (Φ) are given at hierarchical level (Exoffier et al. 1992).

variation found herein was higher than that detected in previous RFLP analyses of chum salmon mtDNA in terms of the number of the observed haplotypes (Cronin et al. 1993, Park et al. 1993, Seeb & Crane 1999b, Churikov et al. 2001). These observations indicate the potential of mtDNA sequence analysis to estimate the genetic variation and to examine the population structure of chum salmon.

The thirty haplotypes observed were genealogically connected in the three clades, although their relationships are ambiguous (Figure 2). The A-1, B-3, and C-1 haplotypes are presumably ancestral to other haplotypes within clades A, B, and C, respectively, given their abundance and centrality in each genealogy. The three star-like genealogies suggest that most of the rarer haplotypes radiating from the central ones may have evolved after colonization of chum salmon in each of the three geographic regions. This is supported by the observation that the radiated haplotypes include most of the region-specific haplotypes. The Japanese populations have some features not found in other geographic groups: the largest number of haplotypes including the region-specific ones, the greatest haplotype diversity, and the presence of all three haplotype clades. These results suggest that the Japanese populations

have longer histories than Russian and North American populations.

The neighbor-joining tree, contingency χ^2 test, and AMOVA revealed clear geographic structuring in the Pacific Rim chum salmon populations, with distinct genetic divergence among Japan, Russia, and North America. Genetic divergence among the three regional groups of chum salmon was also observed in previous studies using variation of allozyme loci (Okazaki 1983, Wilmot et al. 1998, Seeb & Crane 1999b), mtDNA RFLPs (Seeb & Crane 1999a), and minisatellite DNA (Taylor et al. 1994). Furthermore, several studies have also inferred geographic structuring within regions (Kondzela et al. 1994, Phelps et al. 1994, Wilmot et al. 1994, 1998, Seeb & Crane 1999b). In this study, subregional structure was observed between populations within Japan and North America, although the extent of structure within regions was weak to moderate as compared with inter-regional structure (Table 6). In Japan, low or restricted gene flow between Hokkaido and Honshu was implied by higher F_{ST} estimates between Hokkaido and Honshu populations than within Hokkaido or Honshu populations (Sato et al. 2001). Low gene flow between the two regions in Japan may result from differences in the route of spawning migration, run timing, and distance between populations, in addition to possible geological or historical factors (Sato et al. 2001). Over the past few decades, however, contemporary forces such as extensive human-mediated transplantation of stocks may have eroded an even stronger previously existing structure. The Hokkaido and Honshu populations have undergone extensive hatchery operations and transplantation of eggs and fry from one river population to another in the history of commercial salmon production (Kijima & Fujio 1982). In this study, we attempted to minimize such a contemporary influence by sampling late-run fish, particularly in Honshu, because introduced Hokkaido populations migrate earlier than native Honshu populations (Salo 1991).

The population structure within North America, i.e., western Alaska, Alaska Peninsula/Southcentral Alaska, Southeast Alaska, British Columbia, and Washington, is also intriguing. Although no study exists that encompasses all of North America, some of the regional structure was observed in previous allozyme studies (Kondzela et al. 1994, Wilmot et al. 1998, Seeb & Crane 1999b). Contingency χ^2 tests of our mtDNA data indicate low gene flow among these five regions, particularly between western Alaska and other North American locales (Table 5). Although all the factors involved in the structure observed within North American salmon have not been determined, the glacial history of this region has likely had an influence (Utter et al. 1980, Gharrett et al. 1987, Varnavskaya & Beacham 1992, Cronin et al. 1993, Varnavskaya et al. 1994, Bickham et al. 1995, Seeb & Crane 1999a). Two refugia, i.e., Beringia and Cascadia, have been suggested as a Pleistocene cradle for North American salmon species including chum salmon; whereas salmonid populations of western Alaska and Russia were probably recolonized from Beringia, the populations of the Gulf of Alaska, British Columbia, and Washington were recolonized from Cascadia (Seeb & Crane 1999a). The geographic grouping of North American chum salmon populations in this mtDNA analysis may reflect the glacial history, although this inference needs confirmation by further studies including additional collections, particularly from Russia, and analytical methods that estimate population history, such as nested cladistic analysis (Templeton et al. 1992).

The distinct genetic population structure that we observed in Pacific Rim chum salmon revealed herein also suggests that mtDNA sequence analysis may provide good regional estimates of stocks in mixed-stock ocean aggregates. Additional populations from Russia and North America are needed to establish the mtDNA baseline for accurate stock identification. As mentioned above, our mtDNA sequence data may be useful for analyzing the evolutionary mechanisms that shaped the current geographic distribution of chum salmon in the Pacific Rim. Such studies are now ongoing in our laboratories.

Acknowledgements

We would like to thank Dmitri Churikov and Anthony Gharrett, Juneau Center, School of Fisheries and Ocean Sciences, University of Alaska, Fairbanks, for valuable advice on statistical analysis. We also would like to thank the Tsugaruishi fisherman's cooperative association for their help in sampling. This study was supported in part by Grants-in-Aid from the Fisheries Agency, Northern Advancement Center for Science and Technology, and the Ministry of Education, Culture, Sports, Science and Technology, Japan.

References

Bandelt, H.J., P. Forster & A. Rohl. 1999. Median-joining networks for inferring intraspecific phylogenies. Mol. Biol. Evol. 16: 37–48.

Bernatchez, L., R. Guyomard & F. Bonhomme. 1992. DNA sequence variation of the mitochondrial control region among geographically and morphologically remote European brown trout *Salmo trutta* populations. Mol. Ecol. 1: 161–173.

Bickham, J.W., C.C. Wood & J.C. Patton. 1995. Biogeographic implications of cytochrome *b* sequences and allozymes in sockeye (*Oncorhynchus nerka*). J. Hered. 86: 140–144.

Brown, W.M., M. George, Jr. & A.C. Wilson. 1979. Rapid evolution of animal mitochondrial DNA. Proc. Natl Acad. Sci. U.S.A. 76: 1967–1971.

Castelloe, J., & A.R. Templeton. 1994. Root probabilities for intraspecific gene trees under neutral coalescent theory. Mol. Phylogen. Evol. 3: 102–113.

Churikov, D., M. Matsuoka, X. Luan, A.K. Gray, V.A. Brykov & A.J. Gharrett. 2001. Assessment of concordance among genealogical reconstructions from various mtDNA segments in three species of Pacific salmon (genus *Oncorhynchus*). Mol. Ecol. 10: 2329–2339.

Cronin, M.A., W.J. Spearman, R.L. Wilmot, J.C. Patton & J.W. Bickham. 1993. Mitochondrial DNA variation in chinook (*Oncorhynchus tshawytscha*) and chum salmon (*O. keta*) detected by restriction enzyme analysis of polymerase chain reaction (PCR) products. Can. J. Fish. Aquat. Sci. 50: 708–715.

Excoffier, L., P.E. Smouse & J.M. Quattro. 1992. Analysis of molecular variance inferred from metric distances among DNA haplotypes: Application to human mitochondrial DNA restriction data. Genetics 131: 479–491.

Ferguson, A., J.B. Taggart, P.A. Prodohl, O. Mcmeel, C. Thompson, C. Stone, P. Mcginnity & R.A. Hynes. 1995. The application of molecular markers to the study and conservation of fish populations, with special reference to *Salmo*. J. Fish. Biol. 47(Suppl. A): 103–126.

Gharrett, A.J., S.M. Shirley & G.R. Tromble. 1987. Genetic relationships among populations of Alaskan chinook salmon (*Oncorhynchus tshawyscha*). Can. J. Fish. Aquat. Sci. 44: 765–774.

Hoelzel, A.R., J.M. Hancock & G.A. Dover. 1991. Evolution of the cetacean mitochondrial D-loop region. Mol. Biol. Evol. 8: 475–493.

Kijima, A. & Y. Fujio. 1982. Correlation between geographic distance and genetic distance in populations of chum salmon *Oncorhynchus keta*. Bull. Jpn. Soc. Sci. Fish. 48: 1703–1709

Kondzela, C.M., C.M. Guthrie, S.L. Hawkins, C.D. Russell, J.H. Helle & A.J. Gharrett. 1994. Genetic relationship among chum salmon populations in southeast Alaska and northern British Columbia. Can. J. Fish. Aquat. Sci. 51: 50–64.

McElroy, D., P. Moran, E. Bermingham & I. Kornfield. 1993. REAP: An integrated environment for the manipulation and phylogenetic analysis of restriction data. J. Hered. 83: 157–158.

Meyer, A. 1993. Evolution of mitochondrial DNA in fish. pp. 1–38. *In*: P.W. Hochachka & T.P. Mommsen (ed.) Biochemistry and Molecular Biology of Fishes, Vol. 2, Elsevier, Amsterdam.

Moritz, C., T.E. Dowling & W.M. Brown. 1987. Evolution of animal mitochondrial DNA: Relevance for population biology and systematics. Ann. Rec. Ecol. Syst. 18: 269–292.

Nei, M. 1973. Analysis of gene diversity in subdivided populations. Proc. Natl Acad. Sci. U.S.A. 70: 3321–3323.

Nei, M. & F. Tajima. 1981. DNA polymorphism detectable by restriction endonucleases. Genetics 97: 145–163.

Nielsen, J.L., C. Gan & W.K. Thomas. 1994. Differences in genetic diversity for mitochondrial DNA between hatchery and wild population of *Oncorhynchus*. Can. J. Fish. Aquat. Sci. 51: 290–297.

Okazaki, T. 1983. Genetic structure of chum salmon *Oncorhynchus keta* river populations. Bull. Jpn. Soc. Sci. Fish. 49: 189–196.

Park, L.K., M.A. Brainard, D.A. Dightman & G.A. Winans. 1993. Low levels of intraspecific variation in the mitochondrial DNA of chum salmon (*Oncorhynchus keta*). Mol. Mar. Biol. Biotech. 2: 362–370.

Phelps, S.R., L.L. Leclair, S. Young & H.L. Iankenship. 1994. Genetic diversity patterns of chum salmon in the Pacific northwest. Can. J. Fish. Aquat. Sci. 51(Suppl. 1): 65–83.

Pigeon, D., J.J. Dodson & L. Bernatchez. 1998. A mtDNA analysis of spatiotemporal distribution of two sympatric larval populations of rainbow smelt (*Osmerus mordax*) in the St. Lawrence River estuary, Quebec, Canada. Can. J. Fish. Aquat. Sci. 55: 1739–1747.

Posada, D. & K.A. Crandall. 2001. Intraspecific gene genealogies: Trees grafting into networks. Trends. Ecol. Evol. 16: 37–45.

Roff, D.A. & P. Bentzen. 1989. The statistical analysis of mitochondrial DNA polymorphisms: Chi 2 and the problem of small samples. Mol. Biol. Evol. 6: 539–545.

Saitou, N. & M. Nei. 1987. The neighbor-joining method: A new method for reconstructing phylogenetic trees. Mol. Biol. Evol. 4: 406–425.

Salo, E.O. 1991. Life history of chum salmon (*Oncorhynchus keta*). pp. 231–309. *In*: C. Groot & L. Margolis (ed.) Pacific Salmon Life Histories. University of British Columbia Press, Vancouver.

Sambrook, J., E. F. Fritsch & T. Maniatis. 1989. Molecular Cloning: A Laboratory Manual, 2nd edition, Cold Spring Harbor Laboratory Press, New York. 9.16–9.23.

Sato, S., J. Ando, H. Ando, S. Urawa, A. Urano & S. Abe. 2001. Genetic variation among Japanese populations of chum salmon inferred from the nucleotide sequences of the mitochondrial DNA control region. Zool. Sci. 18: 99–106.

Seeb, L.W. & P.A. Crane. 1999a. High genetic heterogeneity in chum salmon in western Alaska, the contact zone between northern and southern lineages. Trans. Am. Fish. Soc. 128: 58–87.

Seeb, L.W. & P.A. Crane. 1999b. Allozymes and mitochondrial DNA discriminate Asian and north American populations of chum salmon in mixed-stock fisheries along the south coast of the Alaska Peninsula. Trans. Am. Fish. Soc. 128: 88–103.

Taylor, E.B., T.D. Beacham & M. Kaeriyama. 1994. Population structure and identification of north Pacific Ocean chum salmon (*Oncorhynchus keta*) revealed by an analysis of minisatellite DNA variation. Can. J. Fish. Aquat. Sci. 51: 1430–1442.

Templeton, A.R., K.A. Crandall & C.F. Sing. 1992. A cladistic analysis of phenotypic associations with haplotypes inferred from restriction endonuclease mapping and DNA sequence data. III. Cladogram estimation. Genetics 132: 619–633.

Utter, F.M., Campton, D., Grant, S., Milner, G., Seeb, J. & L. Wishard. 1980. Population structure of indigenous salmonid species of the Pacific northwest. pp. 285–304. *In*: W.J. McNeil & D.C. Himsworth (ed.) Salmonid ecosystems of the North Pacific. Oregon State University Press, Corvallis.

Varnavskaya, N.V. & T.D. Beacham. 1992. Biochemical genetic variation in odd-year pink salmon (*Oncorhynchus gorbuscha*) from Kamchatka. Can. J. Zool. 70: 2115–2120.

Varnavskaya, N.V., C.C. Wood, R.J. Everett, R.L.Wilmot, V.S.Varnavsky, V.V. Midanaya & T.P. Quinn. 1994. Genetic differentiation of subpopulations of sockeye salmon (*Oncorhynchus nerka*) within lakes of Alaska, British Columbia, and Kamchatka, Russia. Can. J. Fish. Aquat. Sci. 51(Suppl. 1): 147–157.

Wilmot, R.L., R.J. Everett, W.J. Spearman, R. Baccus, N.V. Varnavskaya & S.V. Putivkin. 1994. Genetic stock structure of western Alaska chum salmon and a comparison with Russia far east stocks. Can. J. Fish. Aquat. Sci. 51(Suppl. 1): 84–94.

Wilmot, R.L., C.M. Kondzela, C.M. Guthrie & M.M. Masuda. 1998. Genetic stock identification of chum salmon harvested incidentally in the 1994 and 1995 Bering Sea trawl fishery. N. Pac. Anadr. Fish. Comm. Bull. 1: 285–299.

Wilson, G.M., W.K. Thomas & A.T. Beckenbach. 1987. Mitochondrial DNA analysis of Pacific northwest populations of *Oncorhynchus tshawytscha*. Can. J. Fish. Aquat. Sci. 44: 1301–1305.

Winans, G.A., P.B. Aebersold, S. Urawa & N.V. Varnavskaya. 1994. Determining continent of origin of chum salmon (*Oncorhynchus keta*) using genetic stock identification techniques: Status of allozyme baseline in Asia. Can. J. Fish. Aquat. Sci. 51: 95–113.

Zaykin, D.V. & A.I. Pudovkin. 1993. Two programs to estimate significance of χ^2 values using pseudo-probability tests. J. Hered. 84: 152.

Zhivotovsky, L.A., A.J. Gharrett, A.J. MacGregor, M.K. Glubokovsky & M.W. Feldman. 1994. Gene differentiation in Pacific salmon (*Oncorhynchus* sp.): Facts and models with reference to pink salmon (*O. gorbuscha*). Can. J. Fish. Aquat. Sci. 51: 223–232.

Environmental Biology of Fishes **69**: 51–62, 2004.
© 2004 *Kluwer Academic Publishers. Printed in the Netherlands.*

Smaller effective population sizes evidenced by loss of microsatellite alleles in tributary-spawning populations of sockeye salmon from the Kvichak River, Alaska drainage

Christopher Habicht[a], Jeffrey B. Olsen[a,b], Lowell Fair[a] & James E. Seeb[a]
[a]*Alaska Department of Fish and Game, Commercial Fisheries Division, 333 Raspberry Road, Anchorage, AK 99518, U.S.A. (e-mail: chris_habicht@fishgame.state.ak.us)*
[b]*Current address: U.S. Fish and Wildlife Service, 1011 E. Tudor Road, Anchorage, AK 99503, U.S.A.*

Received 17 April 2003   Accepted 19 April 2003

Key words: bottleneck, Naknek River, Bristol Bay, Alaska, *Oncorhynchus nerka*, commercial fishery, ecotype, M, heterozygosity excess, gametic disequilibrium, straying, migration obstacles, beach spawners

Synopsis

We tested signals of historical reductions in effective population size within populations of sockeye salmon *Oncorhynchus nerka* returning to Bristol Bay, Alaska, to examine the roles that ecotype, migration obstacles, and drainage might play in the highly variable production of the Kvichak River drainage. We collected data for eight microsatellite loci from ~100 fish at each of 16 locations within the Kvichak River drainage and five locations within the more productively stable Naknek River drainage. Pair-wise exact tests were used to group similar collections within ecotype, within drainage, and above and below migration obstacles. After grouping, collections represented independent populations for further analyses. We examined the number of alleles per locus, mean ratio of the number of alleles to the range in allele size, heterozygosity excess, and gametic disequilibrium as measures of reduction-in-population-size events. Number of alleles per locus revealed the largest number of significant differences. Tributary populations showed a stronger signal consistent with reduced effective population size than did beach populations within the Kvichak River drainage. Kvichak River drainage populations showed a stronger signal consistent with reduced effective population size than did the Naknek River drainage populations. Populations above migration obstacles showed signals consistent with reduction in historical population sizes in multiple measures indicating some of these reductions may be severe enough to qualify as demographic bottlenecks.

Introduction

Sockeye salmon *Oncorhynchus nerka* returning to Bristol Bay, Alaska, support the most valuable commercial salmon fishery in North America. Over the past 20 years, fish from the Kvichak River drainage (Figure 1) have supported a quarter of this fishery. However, unlike other drainages in Bristol Bay, the number of sockeye salmon returning to the Kvichak River drainage varied greatly, ranging from 400 000 to 48 million fish per year. The average deviations of the long-term mean of adult production of Kvichak River drainage is about twice that of the neighboring Naknek River drainage (Figure 2).

Sockeye salmon that spawn in the Kvichak River drainage may be classified into two ecotypes: river spawners and lake (beach) spawners. The annual census of the two ecotypes is similar in magnitude. River spawners are larger at maturation, females are more fecund with smaller eggs, and males are shallower-bodied with relatively shorter lower jaws than beach spawners (Blair et al. 1993). Some of the variables that affect survival during the freshwater stage differ systematically between the two ecotypes. For example, embryos from beach-spawning populations may be susceptible to low wind conditions, which reduces intragravel water flow (Leonetti 1997), and to low lake water levels and ice scouring during winter

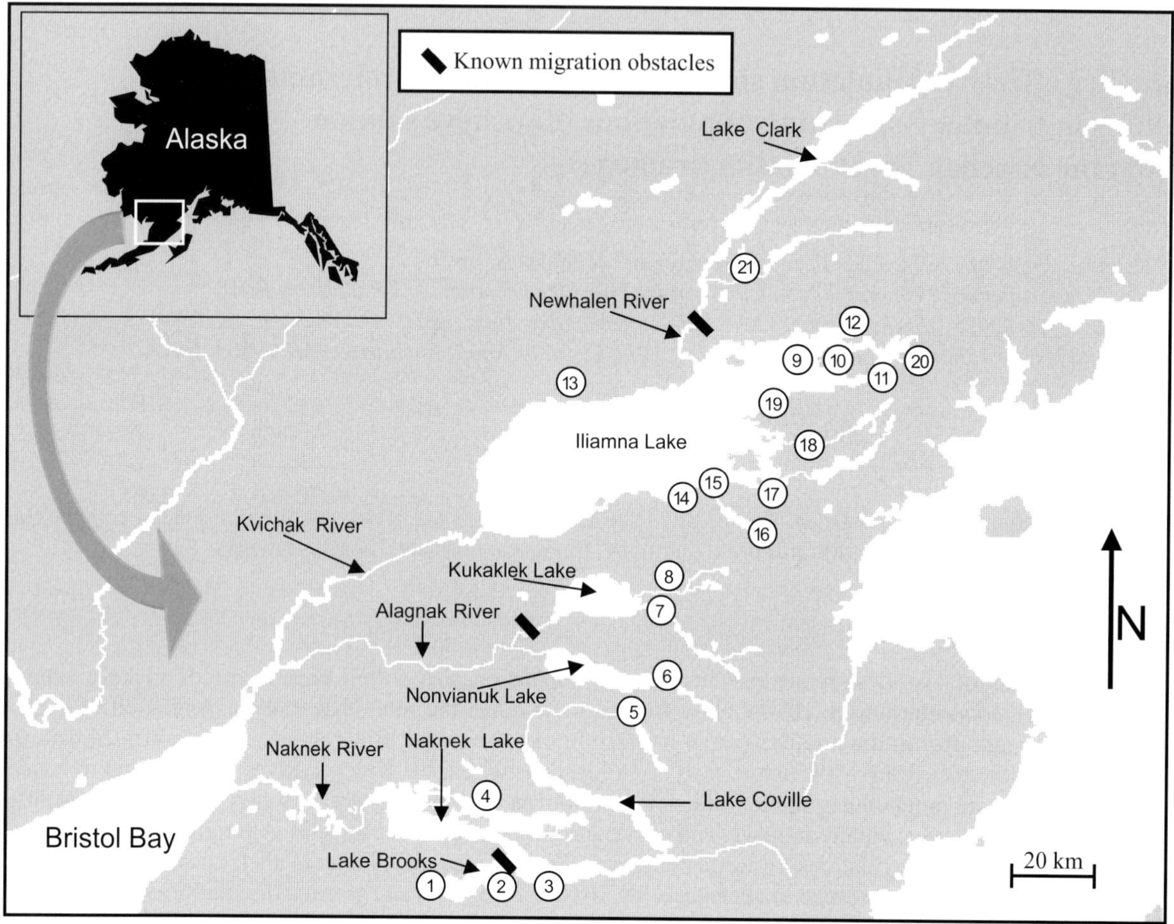

Figure 1. Numbers denote sampling locations for sockeye salmon (see Table 2) and known migration obstacles are identified within the Kvichak and Naknek River drainages.

(Blair et al. 1993). Embryos from river spawners are more susceptible to severe flow changes caused by drought or floods (Forester 1968, Selifonov 1987) and adults more susceptible to bear predation (Quinn & Kinnison 1999). Another confounding variable that may also have a systematic effect on survival within sockeye salmon populations is the presence of migration obstacles such as falls or rapids (see data in Allendorf & Seeb 2000). Environmental effects of the highly variable subarctic climate of Bristol Bay could drastically reduce the effective population sizes (N_e) in different ecotypes or exacerbate the effects of migration obstacles in different years.

Genetic bottlenecks arise from drastic reductions in effective population sizes (i.e., $N_e < 100$, see Cornuet & Luikart 1996) and produce genetic signals that can be detected in extant populations. Less drastic reduction in effective population sizes, such as persistently low population sizes, produce a different set of signals (Table 1). Identifying patterns of these signals between ecotypes may provide insight into the mechanisms responsible for the large fluctuations in productivity in the Kvichak River drainage. Very little is known about intragravel survival of embryos of the two ecotypes within the Kvichak River drainage. No previous examinations of genetic demographic signals associated with reductions in effective populations sizes have been performed in these systems.

The large number of potentially discrete spawning aggregates and the enormous annual variation in numbers of fish returning to spawn in the Kvichak River system make it possible for local reductions in effective population sizes to occur despite the large numbers of fish entering the system to spawn. There

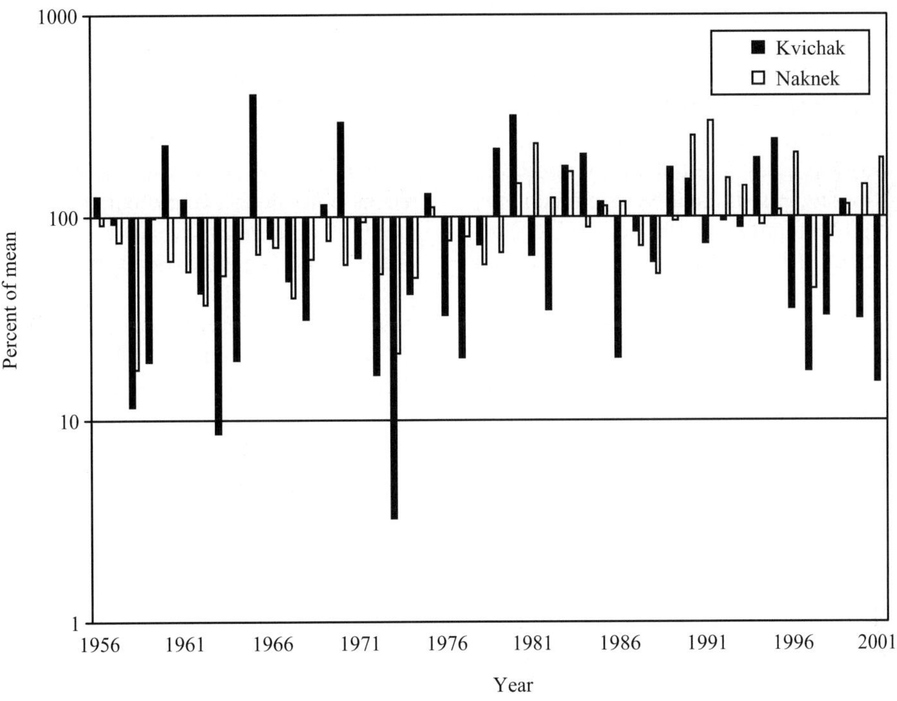

Figure 2. Percent of mean number of sockeye salmon returning to the Kvichak and Naknek River drainages from 1956 to 2001.

Table 1. Characteristics for genetic bottleneck/low-N_e tests used to examine populations of sockeye salmon from the Kvichak and Naknek River drainages.

Test	Characteristics	Citations
Number of alleles per locus	Suitable for markers with large numbers of alleles Good at detecting reductions in N_e that may not be as drastic as genetic bottlenecks	Nei et al. (1975), Allendorf (1986), Spencer et al. (2000), Garza & Williamson (2001), Spong & Hellborg (2002)
M	M = number of alleles/range of alleles Powerful for detecting very recent, single bottleneck events	Garza & Williamson 2001
Heterozygosity excess	Suitable for loci with high heterozygosity Can detect recent, low N_e bottlenecks	Cornuet & Luikart (1996), Spencer et al. (2000), Spong & Hellborg (2002), Luikart et al. (1998)
Gametic disequilibrium	Suitable for very recent, low N_e bottlenecks	Waples & Smouse 1990

are 119 creeks and rivers and 38 lakes and ponds that support spawning by sockeye salmon within the Kvichak drainage.[1] Iliamna Lake and Lake Clark support spawning at multiple beach locations. This

geographic complexity and the observation that the fish home to their natal spawning area precisely (Hendry et al. 1995) suggests that there may be a large number of genetically discrete populations (e.g. Seeb et al. 2000). In addition to the geographic population structure there also may be temporal genetic structure, with sockeye salmon spawning early and late in the season. Taken all together, there are possibly 200–300 discrete spawning aggregates within the system.

---

[1] ADF&G. 1998. Catalog of waters important for the spawning, rearing or migration of anadromous fishes. Alaska Department of Fish and Game, Habitat and Restoration Division, 333 Raspberry Road, Anchorage, AK. http://gis. habitat.adfg.state.ak.us/AWC_IMS/viewer.htm.

In addition, cryptic reductions in N_e, situations where estimated population size (N) do not represent N_e (see citations in Luikart et al. 1998), likely occur within these spawning aggregates. Among the factors that may produce cryptic reductions in N_e for salmonids are polygamy and polyandry (Bentzen et al. 2001), variations in redd quality within ecotype (e.g. Leonetti 1997), family-specific survival differences (Geiger et al. 1997, Pimm et al. 1989) and fluctuations in population sizes (Frankham 1995). For example, Heath et al. (2002) found large differences between N and estimated N_e for steelhead trout (N_e/N ranged from 0.06 to 0.29).

In other wildlife species, fluctuation in population size is the most important variable producing cryptic reductions in N_e (Frankham 1995). The reason for this large effect is because N_e is the harmonic mean of the number of spawners per year and therefore years with low numbers of returns have a disproportionately large impact on N_e (Wright 1978). Aerial surveys of the number of fish observed on spawning grounds commonly vary by three orders of magnitude from year-to-year within the same locations in the Kvichak drainage.[2] Some of this variability in counts may be due to inter-annual variation in run timing, weather conditions, and water conditions (e.g. turbidity, tannin load, flood events) or variation in observer bias, but high levels of variation have been observed on the ground (T. Quinn, University of Washington, Seattle, WA, personal communication). These fluctuations are not always consistent among spawning locations within years indicating that survival may vary among incubation locations.

Different measures using molecular genetic data have been developed to test for bottlenecks and for less drastic reductions in effective populations size (Table 1, also see Cornuet & Luikart 1996). Because these measures are affected differently by the intensity and duration of bottlenecks, using a suite of tests provides insight into the character of reduction in N_e experienced by the population. Number of alleles per locus is only meaningful as a comparative measure testing between groups of populations because the number of alleles observed varies by locus. This measure has been observed to decrease even in cases where reductions in effective population sizes that were not as drastic as genetic bottlenecks (Garza & Williamson 2001,

Spong & Hellborg 2002). M-value can also be used as a comparative measure, as it is in this study, but generally average values for multiple loci below 0.68 have been observed in populations known to have undergone reduced effective population sizes (Garza & Williamson 2002). M-value is powerful at detecting recent bottleneck events, but loses power following multiple bottleneck events because the number of alleles remaining is drastically reduced. Both heterozygosity excess and gametic disequilibrium are measures testable within populations and are well suited for detecting recent bottleneck events. Heterozygosity excess is most powerful when using allelic-rich loci such as microsatellites while gametic disequilibrium is most powerful when using loci with high heterozygosity (Waples & Smouse 1990, Cornuet & Luikart 1996). Hetrozygosity excess is not the same as excess heterozygosity relative to Hardy–Weinberg equilibrium. Instead, heterozygosity excess is the excess in heterozygosity relative to the expected heterozygosity given a set number of alleles in a population at equilibrium between gains of alleles due to mutation and loss of alleles due to drift. Under this equilibrium, a large proportion of the alleles are expected to have low frequencies which produces low heterozygosity given the number of alleles present. After a bottleneck and before the population reaches equilibrium, fewer but higher frequency alleles remain (lower frequency alleles are more likely to be eliminated during a bottleneck) which produces higher heterozygosity (heterozygosity excess).

In this paper, we test for evidence that the signals of a reduction in N_e vary with ecotypes within the Kvichak River drainage. We also compare the signals of a reduction in N_e between collections from the Kvichak River drainage and the more productively stable Naknek River drainage (Figure 1) to test whether reductions in N_e occurred in the spawning aggregates from Kvichak River drainage after accounting for ecotype effects. If reductions in N_e occurred more frequently within one of the ecotypes, then we would expect to see a consistent signal within one of the ecotypes. If both ecotypes are affected by reductions in N_e associated with production cycles in the Kvichak River drainage, then we would expect stronger signals within collections of the Kvichak River drainage than within collections from the Naknek River drainage. No signals of N_e reductions would indicate that either no N_e reduction occurred or the N_e reduction was too minor or short-lasting to be detectable with these methods.

---

[2] Regnart, J.R. 1996. Kvichak River sockeye salmon spawning ground surveys, 1955–1996. Alaska Department of Fish and Game, 333 Raspberry Road, Anchorage AK, RIR 2A96-42, 25 pp.

Table 2. Sampling locations, year, and map numbers in Figure 1 for collections from sockeye salmon spawning aggregates in the Naknek and Kvichak River drainages used in genetic analyses. Ecotypes include beach (B) and tributary (T) spawners. 'Lake' denotes the lake that the tributary drains into or where the beach spawning occurs. 'Obstacle' indicates known types of migration obstacles along migratory route. With the exception of Finger Beach (N = 85), all collections consisted of 100 fish.

Location name	Year	Map no.	Ecotype	Lake	Obstacle
Naknek River					
Above obstacles					
Headwaters Creek	2001	1	T	Brooks	Falls
Up-a-tree Creek	2000	2	T	Brooks	Falls
Below obstacles					
Margot Creek	2000	3	T	Naknek	—
Idavain Creek	2000	4	T	Naknek	—
American Creek	2000	5	T	Coville	—
Kvichak River					
Above obstacles					
Battle River	2001	7	T	Kukaklek	Rapids
Moraine Creek	2001	8	T	Kukaklek	Rapids
Tazimina River	2001	21	T	Six Mile	Rapids
Below obstacles					
Flat Island Beach	2000	9	B	Iliamna	—
Triangle Island Beach	2000	10	B	Iliamna	—
Finger Beach	2000	11	B	Iliamna	—
Knutson Bay Beach	2000	12	B	Iliamna	—
Kulik River	2001	6	T	Nonvianuk	—
Lower Talarik Creek	2000	13	T	Iliamna	—
Dennis Creek	2000	14	T	Iliamna	—
Gibraltar River	2000	15	T	Iliamna	—
Southeast Creek	2000	16	T	Iliamna	—
Nick N. Creek	2000	17	T	Iliamna	—
Copper River	2000	18	T	Iliamna	—
Tommy Creek	2000	19	T	Iliamna	—
Chinkelyes Creek	2000	20	T	Iliamna	—

We include collections from spawning aggregates both above and below potential migration obstacles (Table 2, Figure 1). We first test for relative presence of reduction-in-N_e signals for populations above and below potential migration obstacles. If reduction-in-N_e signals are associated with position of the collections relative to migration obstacles, then to avoid confounded tests, collections above migration obstacles were excluded when testing for reduction-in-N_e signals between ecotypes and between drainages.

Methods

Sockeye salmon were sampled on spawning grounds at 21 locations in the Kvichak and Naknek River drainages in 2000 and 2001 (Table 2; Figure 1). Heart tissue was collected on wet ice and frozen within 12 h. Samples were stored at $-80°$C prior to analysis.

Five of these collections were made above three potential migration obstacles. Two were made above Brooks Falls, a 2 m waterfall between Lake Brooks and Naknek Lake, which flows at about 11 m^3 s^{-1} during migration season. A fish ladder was installed in 1949 to help sockeye salmon ascend the falls during adverse water conditions.[3] The fish pass has been intermittently used by migrating salmon and has not been systematically maintained. Two collections were made above Alagnak River rapids, a 50 m, Class 4 rapids which flows at 74 m^3 s^{-1} during migration season and is located upstream of the confluence with the Nonvianuk River. Little is known about its potential to restrict salmon migration. One collection was made above the

[3] Wilmot, R., C. Burger & P. Steinbach. 1986. Genetics of sockeye salmon (*Oncorhynchus nerka*) in Katmai National Park and Preserve, Alaska. Report to U.S. Department of Interior, National Park Service. Interagency agreement no. IA-9700-4-8010.

Newhalen River which contains a series of three rapids over 15 km, starting about 3 km from the river mouth, including Petrof Falls, a 200 m, Class 5 rapid that flows at about 300 m^3 s^{-1} during migration season.[4] This section has been reported to become a velocity barrier to sockeye salmon migration in years with high summer temperatures that produce rapid glacial melt rates.[5]

Eight microsatellite loci were screened for variation: *Omy77**, *One102**, *One108**, *One109**, *One111**, *Ots107**, *Ots3**, and *uSat60**. DNA was amplified using polymerase chain reaction according to Olsen et al. (2000). Microsatellites were size fractionated on acrylamide gel using an Applied Biosystems Prism 377 DNA sequencer, alleles for each locus were scored, and data were tabulated for importing into statistical software according to the methods of Olsen et al. (2004). Loci were tested for deviations from Hardy–Weinberg expectations (H–W) using GENEPOP (version 3.3, updated version of Raymond & Rousset 1995, Rousset & Raymond 1995). Within each of the five spawning categories (defined by the drainage, ecotype, and position relative to migration obstacles) we performed pair-wise exact tests for genic differentiation (Goudet et al. 1996) calculated in GENEPOP with Markov chain parameters: 5000 as the dememorization number; 1000 batches; and 1000 iterations per batch. If the exact tests indicated homogeneity between a pair of collections, we investigated the geographic and life history relationships among these collections. If the homogeneity appeared to be clustered within life history or geographic characteristics, then we grouped collections and recalculated pair-wise exact tests among remaining collections. These tests of allele frequency homogeneity can be considered as a test of the null hypothesis $F_{st} = 0$ between collections (Chakraoborty & Leimar 1987), therefore the grouped and ungrouped collections represented independent populations.

We used four genetic bottleneck measures: (1) number of alleles per locus; (2) M (mean ratio of the number of alleles to the range in allele size; Garza & Williamson 2001); (3) heterozygosity excess; and (4) gametic

disequilibrium (Table 1). Because both the number of alleles per locus and M are influenced by collection size (Garza & Williamson 2001, Leberg 2002), we calculated these measures for each collection and averaged them in cases where collections were grouped. Each collection contained ~100 fish.

We used t-tests to detect differences in the average number of alleles per locus and average M among collections as follows: first we tested between collections from tributaries above and below migration obstacles within the Kvichak and the Naknek Rivers separately. If we found evidence of historic reductions in N_e within the collections above migration obstacles, these collections were eliminated from further analyses. Then we tested between collections from beach and tributary spawners within the Kvichak River drainage. If we found differences in reduction-in-N_e signals between beach and tributary spawners, we eliminated beach spawners from further analyses. Finally we tested between remaining collections from the Kvichak and Naknek River drainages.

Heterozygosity excess and deficiency was tested within the original collections in BOTTLENECK (version 1.1.03, Cornuet & Luikart 1996) using 1000 iterations assuming both the stepwise-mutation model (SMM) and the infinite-allele model (IAM) and using the Wilcoxon sign-rank test to determine significance. Gametic disequilibrium was tested within collections using an omnibus permutation test of correlation across locus pairs (Smouse & Neel 1977, Waples & Smouse 1990; locally written function in S-Plus 2000, Insightful Corporation). Original, ungrouped collections were used in these tests to avoid erroneous conclusions due to potential type II errors in the exact tests. Significance levels were adjusted for multiple tests using a sequential Bonferroni adjustment (Rice 1989).

Results

We analyzed a total of 2085 fish, ~100 from each of 21 locations (Table 2). The 16 locations from the Kvichak River drainage included 12 tributaries, three of which were above migration obstacles, and four beaches. The five locations from the Naknek River drainage included only tributaries, two of which were above migration obstacles. All collections were in H–W equilibrium for all loci.

Of the 49 exact tests made, 41 were significant, indicating that most collections within each spawning

---

[4] Demory, R.L., R.F. Orrell & D.R. Heinle. 1964. Spawning ground catalogue of the Kvichak River system, Bristol Bay, Alaska. Special Scientific Report – Fisheries No. 488, U.S. Fish and Wildlife Service, Washington, DC.

[5] Poe, P.H. & O.A. Mathisen. 1980. The 1979 and 1980 enumeration of sockeye salmon runs to the Newhalen River–Lake Clark system. Preliminary Report. Fisheries Research Institute, College of Fisheries, University of Washington, Seattle, WA 98195, 27 pp.

category differed genetically from one another, and six of the loci contributed most to these differences: *One109**, *Omy77**, *One108**, *One102**, *One111** and *uSat60**. The eight non-significant tests revealed two cases where homogeneity among collections could not be excluded. First, by combining #11-Finger Beach and #12-Knutson Bay Beach (the only two mainland beaches) into one collection, all pair-wise exact tests were significant within this category. Second, by combining #15-Dennis Creek, #16-Gibraltar River and #17-Southeast Creeks (adjacent drainages into Iliamna Lake), all pair-wise exact tests were significant within this category. These groupings allow each resulting collection to represent independent populations and decrease the potential for pseudoreplication.

The number of alleles per locus ranged from 10 to 39, and the lowest frequencies of the common allele within collections ranged from 8% to 70% (raw data available from primary author). *One111** had the largest number of alleles per population with a range of 22–30; followed by *One102**, *One108** and *One109** with a range of 10–19 alleles per locus; and *Omy77**, *Ots107**, *Ots3** and *uSat60** with a range of 3–10 alleles per locus (Table 3). M-values varied greatly both within loci across collections and within collections across loci (Table 4). Collections above the migration obstacles in both drainages showed a significantly reduced numbers of alleles per locus but not reduced M-values (Tables 5 and 6). Within the Kvichak drainage, tributary spawners showed a lower number of alleles per locus compared to the beach spawners, but again, no difference was observed for the M-values (Table 6). Within tributary spawners below migration obstacles, the Kvichak River drainage spawners showed lower numbers of alleles per locus relative to the Naknek River drainage spawners (Table 6). All loci appeared to contribute to these observed reductions in mean numbers of alleles

Table 3. Number of alleles for each collection of sockeye salmon from the Naknek and Kvichak Rivers at the eight microsatellite loci surveyed.

Collection	Locus							
	*Omy77**	*One102**	*One108**	*One109**	*One111**	*Ots107**	*Ots3**	*uSat60**
Naknek River								
Above obstacles								
Headwaters Creek	5.0	11.0	14.0	12.0	24.0	4.0	4.0	5.0
Up-a-tree Creek	5.0	13.0	15.0	13.0	25.0	5.0	6.0	5.0
Below obstacles								
Margot Creek	7.0	15.0	17.0	12.0	27.0	5.0	7.0	8.0
Idavain Creek	7.0	16.0	19.0	13.0	28.0	6.0	7.0	9.0
American Creek	6.0	19.0	17.0	12.0	26.0	6.0	5.0	6.0
Kvichak River								
Above obstacles								
Moraine Creek	5.0	12.0	10.0	11.0	22.0	3.0	5.0	4.0
Battle River	5.0	12.0	10.0	11.0	23.0	3.0	7.0	4.0
Tazimina River	7.0	15.0	15.0	13.0	27.0	5.0	4.0	10.0
Below obstacles								
Flat Island Beach	6.0	19.0	17.0	13.0	30.0	4.0	6.0	9.0
Triangle Island Beach	5.0	17.0	15.0	12.0	29.0	6.0	6.0	8.0
Finger/Knutson Beaches	6.5	16.0	16.0	13.0	29.0	4.0	5.5	7.0
Finger Beach	7.0	16.0	17.0	13.0	28.0	4.0	6.0	7.0
Knutson Bay Beach	6.0	16.0	15.0	13.0	30.0	4.0	5.0	7.0
Kulik River	6.0	16.0	16.0	13.0	28.0	6.0	4.0	3.0
Lower Talarik Creek	6.0	14.0	14.0	12.0	26.0	5.0	6.0	6.0
Dennis/Gibraltar/SE	6.3	16.3	14.3	14.7	27.3	5.7	6.0	6.3
Dennis Creek	7.0	17.0	13.0	15.0	24.0	5.0	5.0	5.0
Gibraltar River	7.0	15.0	15.0	14.0	30.0	6.0	7.0	5.0
Southeast Creek	5.0	15.0	15.0	14.0	28.0	5.0	6.0	9.0
Nick N. Creek	8.0	15.0	14.0	14.0	26.0	5.0	5.0	7.0
Copper River	6.0	14.0	15.0	13.0	24.0	7.0	6.0	9.0
Tommy Creek	8.0	17.0	15.0	11.0	28.0	4.0	6.0	10.0
Chinkelyes Creek	6.0	14.0	14.0	13.0	26.0	5.0	6.0	8.0

Table 4. M (Garza & Williamson 2001) for each collection of sockeye salmon from the Naknek and Kvichak Rivers at the eight microsatellite loci surveyed.

Collection	Locus							
	*Omy77**	*One102**	*One108**	*One109**	*One111**	*Ots107**	*Ots3**	*uSat60**
Naknek River								
Above obstacles								
Headwaters Creek	0.83	0.79	0.82	0.75	0.96	0.36	0.80	0.45
Up-a-tree Creek	1.00	0.87	0.94	1.00	0.89	0.45	0.46	0.63
Below obstacles								
Margot Creek	1.00	0.68	0.94	0.92	0.82	0.83	0.54	0.47
Idavain Creek	0.70	0.94	0.73	0.93	0.90	0.86	0.54	0.56
American Creek	0.86	0.86	0.94	0.92	0.81	0.50	0.71	0.67
Kvichak River								
Above obstacles								
Battle River	1.00	0.92	1.00	0.85	0.68	1.00	0.78	0.44
Moraine Creek	0.83	0.86	1.00	0.92	0.67	1.00	1.00	0.44
Tazimina River	0.70	0.83	0.94	0.93	0.93	0.50	0.57	0.83
Below obstacles								
Flat Island Beach	0.86	1.00	0.81	0.72	0.94	1.00	0.43	0.82
Triangle Island Beach	0.83	0.89	0.88	0.80	0.88	0.55	0.40	0.73
Finger/Knutson Beaches	0.94	0.83	0.91	0.87	0.74	0.90	0.46	0.64
Finger Beach	0.88	0.89	0.81	0.87	0.70	1.00	0.50	0.58
Knutson Bay Beach	1.00	0.76	1.00	0.87	0.77	0.80	0.42	0.70
Kulik River	0.86	0.94	1.00	1.00	0.80	0.50	1.00	0.33
Lower Talarik Creek	1.00	0.93	0.93	0.92	0.93	1.00	0.46	0.60
Dennis/Gibraltar/SE	0.90	0.94	0.76	0.98	0.77	0.52	0.55	0.53
Dennis Creek	1.00	1.00	0.62	1.00	0.75	0.50	0.56	0.56
Gibraltar River	1.00	0.94	0.71	0.93	0.73	0.55	0.58	0.50
Southeast Creek	0.71	0.88	0.94	1.00	0.82	0.50	0.50	0.53
Nick N. Creek	0.80	0.83	0.67	0.93	0.84	1.00	0.83	0.44
Copper River	0.60	0.78	0.68	1.00	0.80	1.00	0.50	0.75
Tommy Creek	0.73	0.77	1.00	0.92	0.80	1.00	0.50	0.59
Chinkelyes Creek	0.86	0.88	0.67	0.87	0.79	1.00	0.50	0.67

(Table 3). Both collections above migration obstacles in the Naknek drainage showed heterozygosity excess when assuming the IAM (Table 5). No other significant values, after adjusting for multiple tests, were detected within collections for either heterozygosity excess (using the IAM) or gametic disequilibrium (Table 5). Assuming the SMM, all collections yielded significant heterozygosity deficiency (data not shown).

Discussion

Our results were consistent with the hypothesis that tributary-spawning populations have had lower effective population sizes than beach-spawning populations. However, these lower effective population sizes either did not fall into the category of demographic bottlenecks, in the strict sense, or they occurred a long time ago because no other measures of bottlenecks were significant. Our results were consistent with the hypothesis that at least two of the three migration obstacles produced reductions in N_e, and in one case, this reduction may have qualified as a demographic bottleneck because of the concordance of the heterozygosity excess signals within the collections above migration obstacles in the Naknek River drainage. Finally, our results were also consistent with the hypothesis that the populations from the Kvichak River drainage have undergone more drastic or more recent reductions in N_e than have the populations from the Naknek River drainage.

These findings are notable because straying among populations and the overlapping generation structure of sockeye salmon may mute the signals of a reduction in N_e. Although sockeye salmon home precisely, low

Table 5. Results from tests for a reduction in N_e within collections of sockeye salmon from the Kvichak and Naknek River drainages (see Table 1 for test citations). Bolded p-values under heterozygosity excess using the IAM were significant after adjusting for multiple tests within strata.

Collection	No. alleles (mean)	M (mean)	Heterozygosity excess (p-value)	Gametic disequilibrium (p-value)
Naknek River				
Above obstacles				
Headwaters Creek	9.88	0.72	**0.006**	0.063
Up-a-tree Creek	10.88	0.78	**0.037**	0.543
Below obstacles				
Margot Creek	12.25	0.78	0.422	0.161
Idavain Creek	13.13	0.77	0.578	0.874
American Creek	12.13	0.79	0.422	0.244
Kvichak River				
Above obstacles				
Battle River	9.00	0.83	0.230	0.087
Moraine Creek	9.38	0.84	0.156	0.783
Tazimina River	12.00	0.78	0.371	0.323
Below obstacles				
Flat Island Beach	13.00	0.82	0.527	0.234
Triangle Island Beach	12.25	0.75	0.527	0.090
Finger/Knutson Beaches	12.13	0.79	N/A	N/A
Finger Beach	12.25	0.78	0.422	0.368
Knutson Bay Beach	12.00	0.79	0.371	0.648
Kulik River	11.50	0.80	0.371	0.156
Lower Talarik Creek	11.13	0.85	0.320	0.580
Dennis/Gibraltar/Southeast	12.00	0.74	N/A	N/A
Dennis Creek	11.38	0.75	0.371	0.818
Gibraltar River	12.38	0.74	0.473	0.487
Southeast Creek	12.25	0.74	0.578	0.311
Nick N. Creek	11.88	0.79	0.628	0.388
Copper River	11.75	0.76	0.578	0.528
Tommy Creek	12.38	0.79	0.578	0.191
Chinkelyes Creek	11.50	0.78	0.578	0.780

Table 6. Student's t-test results comparing the average number of alleles per locus and M for microsatellite data collected from populations of sockeye salmon in the Kvichak and Naknek River drainages (sample sizes were similar among collections). Mean values (m_1 and m_2) for the first and second category identified in the 'Test' column are given. Probability (p) is significant for number of alleles per locus in all comparisons (in bold).

Test	Number alleles per locus			M		
	m_1	m_2	p	m_1	m_2	p
Obstacles vs. no obstacles						
Kvichak tributaries	10.13	11.73	**0.015**	0.82	0.79	0.121
Naknek tributaries	10.38	12.50	**0.015**	0.75	0.78	0.146
Tributary vs. beach						
Kvichak, below barriers	11.73	12.46	**0.038**	0.79	0.79	0.985
Kvichak vs. Naknek						
Tributaries, below barriers	11.73	12.50	**0.018**	0.79	0.78	0.371

levels of straying among populations have been documented and are thought to be evolutionarily adaptive (Wood 1995). In addition, fish return to spawn in Bristol Bay at ages 4 (23%), 5 (63%) and 6 (14%). This overlapping generation structure could obscure bottleneck signals due to drastic single-year reductions in population sizes (Waples 1990). Over the last 50 years, escapement data indicate that low numbers within the drainage as a whole seldom occur in consecutive years (Figure 2). The propensity of these life history characteristics to mute the signals of a reduction in N_e indicates that these signals are conservative measures of past bottleneck events (Withler et al. 2000) and that only the most sensitive statistics can detect genetic patterns consistent with reductions in N_e that are not as severe as demographic bottlenecks.

Most genetic measures produce a significant signal only after a drastic reduction to a very small effective population size to (i.e. $N_e < 100$ individuals; see Table 1 and references therein). In our study, the only statistic that consistently indicated reductions in N_e was the number of alleles per locus. This finding is consistent with previous studies on other species when the reduction in N_e has not been to the level of a bottleneck (Spong & Helborg 2002) or in cases where low N_e has been present for an extended period of time (Waldick et al. 2002). Our findings of significant reduction in the number of alleles per locus, but not in M, was consistent with an earlier reduction in N_e such as a founder effect or with repeated periods of low N_e (Garza & Williamson 2001). In addition, some of the tests for reduction in N_e are sensitive to the distribution of allele frequencies and in particular gaps in this distribution. The significant heterozygosity deficiency tests yielded by the SMM analysis were likely an artifact of the gaps in distribution of allelic lengths (Figure 3) as described in Cornuet & Luikart (1996) because no heterozygosity deficiency tests were significant when using the IAM. These gaps in distribution of allelic lengths also produced highly variable M-values as epitomized by *Ots107** (Table 4, Figure 3).

The most pronounced difference in the number of alleles per locus that we detected was between collections above and below migration obstacles (Table 3). The four collections with the lowest number of alleles per locus were all above migration obstacles (Table 5). In addition, the populations above migration obstacles in the Naknek River were the only populations to show heterozygosity excess using the IAM (Table 5). Annual

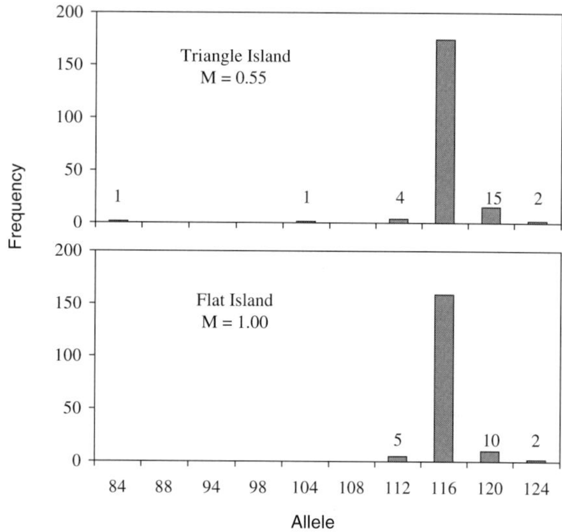

Figure 3. Allele frequencies for *Ots107** from beach spawners collected from Flat Island and Triangle Island (allele counts provided for alleles with counts lower than 20). Note the gaps in allele categories which can produce highly variable M-values and type I error in tests for heterozygosity deficiency when assuming the SMM (Cournet & Luikart 1996). Within the observed range of alleles for all collections, 12 allele categories were never seen.

migrations of sockeye salmon above known migration obstacles do not account for a high enough proportion of the escapement into the Kvichak drainage to explain the large variation in production, but they do provide insight into the biogeography and genetic relationships among populations within each drainage. The evidence of a reduction in N_e within the populations above migration obstacles compared with populations below the obstacles indicated that straying between these population groups is low. Low straying rates between these population groups may have been maintained as a consequence of adaptations by the upstream population groups either to bypass the migration obstacles or to utilize habitats above the migration obstacles. These adaptations would reduce the chances of recolonization by neighboring population groups following an extirpation event.

In addition to reduced numbers of alleles per locus, significant heterozygosity excess were detected within both populations sampled from Lake Brooks indicating that, of all the reductions in N_e identified, this may have been the most recent or most drastic and may indicate that a true demographic bottleneck occurred.

Wilmot et al.[6] document a large genetic divergence between the Lake Brooks tributary populations and the Naknek Lake tributary populations which is also consistent with a bottleneck event affecting the Lake Brooks tributary populations. A fish pass was installed in 1949 based on speculation that the falls was a barrier to migration. There is no empirical documentation over the last 100 years that the falls created such a barrier.[7] However, Wilmot et al.[6] document similar allele frequencies between collections from the Brooks River above and below the falls. This similarity was attributed to possible recent increases in straying as a result of the fish pass between lower and upper river fish. Our findings of a bottleneck signal in Lake Brooks tributary spawners, in combination with the genetic dissimilarity between the Lake Brooks and Naknek Lake tributary spawners, indicate that the fish pass has not increased straying, to the level required for homogenization, between tributary spawners from the two lakes.

Finally, although the Tazimina River collection was grouped together with the other collections above migration obstacles in the Kvichak River drainage for statistical analyses, it did not demonstrate a reduction in allele numbers (Tables 3 and 5). Removing it from the analysis provides strong evidence of a reduction in allele numbers in collections above the Alagnak River rapids. The Tazimina River population demonstrates that not all suspected migration obstacles produce significant reductions in N_e.

In conclusion, the observed patterns of reduced-N_e signals between sockeye salmon populations spawning in the Kvichak River drainage and the Naknek River drainage and between those spawning above and below migration obstacles are concordant with expectations based on the life history of these populations. Escapements to the Kvichak River drainage are highly variable and N_e is driven by years of low escapement. Obstacles to migration are also thought to increase variation in escapement because annual fluctuations in environmental variables may block or severely restrict passage. Our finding of a stronger reduction in N_e in tributary spawners relative to beach spawners within the Kvichak River drainage may provide insight into the mechanisms responsible for the large fluctuations in productivity.

Acknowledgements

We thank S. Wilson, N. DeCovich and Z. Grauvogel for collecting the data in the laboratory. A. Gharrett and three anonymous reviewers greatly improved this manuscript and S. Merkouris provided editorial guidance. T. Hamon provided insights into migration obstacles in the Naknek and Kvichak River drainages. Sample site selection and collection could not have been done without the support of S. Morstad and other staff from ADF&G King Salmon office, and T. Quinn and other faculty and students from the University of Washington Fisheries Research Institute. The Tazimina River collection was shared by C. Woody and K. Ramstad. This project was funded by grant NA96FW0196 from the National Marine Fisheries Service to the Alaska Department of Fish and Game.

References

Allendorf, F.W. 1986. Genetic drift and the loss of alleles *versus* heterozygosity. Zool. Biol. 5: 181–190.

Allendorf, F.W. & L.W. Seeb. 2000. Concordance of genetic divergence among sockeye salmon populations at allozyme, nuclear DNA, and mitochondrial DNA markers. Evolution 54: 640–651.

Bentzen, P., J.B. Olsen, J.E. McLean, T.R. Seamons & T.P. Quinn. 2001. Kinship analysis of Pacific salmon: Insights into mating, homing, and timing of reproduction. J. Hered. 92: 127–136.

Blair, G.R., D.E. Rogers & T.P. Quinn. 1993. Variation in life history characteristics and morphology of sockeye salmon in the Kvichak River system, Bristol Bay, Alaska. Trans. Am. Fish. Soc. 122: 550–559.

Chakraoborty, R. & O. Leimar. 1987. Genetic variation within a subdivided population. pp. 89–120. *In*: N. Ryman & R. Utter (ed.) Population Genetics and Fishery Management, University of Washington Press, Seattle.

Cornuet, J.M. & G. Luikart. 1996. Description and power analysis of two tests for detecting recent population bottlenecks from allele frequency data. Genetics 144: 2001–2014.

Forester, R.E. 1968. The sockeye salmon, *Oncorhynchus nerka*. Fish. Res. Board Can. Bull. 162: 422 pp.

Frankham, R. 1995. Effective population size/adult population size ratios in wildlife: A review. Genet. Res. 66: 95–107.

Garza, J.C. & E.G. Williamson. 2001. Detection of reduction in population size using data from microsatellite loci. Mol. Ecol. 10: 305–318.

---

[6] Wilmot, R., C. Burger & P. Steinbach. 1986. Genetics of sockeye salmon (*Oncorhynchus nerka*) in Katmai National Park and Preserve, Alaska. Report to U.S. Department of Interior, National Park Service. Interagency agreement no. IA-9700-4-8010.

[7] Hamon, T., Katmai National Park, 2002, personal communication.

Geiger, H.J., W.W. Smoker, L.A. Zhivotovsky & A.J. Gharrett. 1997. Variability of family size and marine survival in pink salmon (*Oncorhynchus gorbuscha*) has implications for conservation biology and human use. Can. J. Fish. Aquat. Sci. 54: 2684–2690.

Goudet, J., M. Raymond, T. de Meeus & F. Rousset. 1996. Testing differentiation in diploid populations. Genetics 144: 1933–1940.

Heath, D.D., C. Busch, J. Kelly & D.Y. Atagi. 2002. Temporal change in genetic structure and effective population size in steelhead trout (*Oncorhynchus mykiss*). Mol. Ecol. 11: 197–214.

Hendry, A.P., F.E. Leonetti & T.P. Quinn. 1995. Spatial and temporal isolating mechanisms – the formation of discrete breeding aggregations of sockeye salmon (*Oncorhynchus nerka*). Can. J. Zool. 73: 339–352.

Leberg, P.L. 2002. Estimating allelic richness: Effects of sample size and bottlenecks. Mol. Ecol. 11: 2445–2449.

Leonetti, F.E. 1997. Estimation of surface and intragravel water flow at sockeye salmon spawning beaches in Iliamna Lake, Alaska. N. Am. J. Fish. Manag. 17: 194–201.

Luikart, G., F.W. Allendorf, J.-M. Cornuet & W.B. Sherwin. 1998. Distortion of allele frequency distributions provides a test for recent population bottlenecks. J. Hered. 89: 238–247.

Nei, M., T. Maruyama & R. Chakraborty. 1975. The bottleneck effect and genetic variability in populations. Evolution 29: 1–10.

Olsen, J.B., S.L. Wilson, E.J. Kretschmer, K.C. Jones & J.E. Seeb. 2000. Characterization of 14 tetranucleotide microsatellite loci derived from sockeye salmon. Mol. Ecol. 9: 2185–2287.

Olsen, J.B., C. Habicht & J.E. Seeb. 2004. Moderately and highly polymorphic microsatellites provide discordant estimates of population divergence in sockeye salmo, *Oncorhynchus nerka*. Environ. Biol. Fish. 69: 261–273.

Pimm, S.L., J.L. Gittleman & G.F. McCracken. 1989. Plausible alternatives to bottlenecks to explain reduced genetic diversity. Trends Ecol. Evol. 4: 176–178.

Quinn, T.P. & M.T. Kinnison. 1999. Size-selective and sex-selective predation by brown bears on sockeye salmon. Oecologia 121: 273–282.

Raymond, M. & F. Rousset. 1995. GENEPOP (version 1.2): Population genetics software for exact tests and ecumenicism. J. Hered. 86: 248–249.

Rice, W.R. 1989. Analyzing tables of statistical tests. Evolution 4: 223–225.

Rousset, F. & M. Raymond. 1995. Testing heterozygote excess and deficiency. Genetics 140: 1413–1419.

Seeb, L.W., C. Habicht, W.D. Templin, K.E. Tarbox, R.Z. Davis, L.K. Brannian & J.E. Seeb. 2000. Genetic diversity of sockeye salmon of Cook Inlet, Alaska, and its application to management of populations affected by the *Exxon Valdez* oil spill. Trans. Am. Fish. Soc. 129: 1223–1249.

Selifonov, M.M. 1987. Influence of environment on the abundance of sockeye salmon (*Oncorhynchus nerka*) from the Ozernaya and Kamchatka Rivers. pp. 125–128. *In*: H.D. Smith, L. Margolis & C.C. Wood (ed.) Sockeye Salmon (*Oncorhynchus nerka*) Population Biology and Future Management, Can. Spec. Publ. Fish. Aquat. Sci. 96.

Smouse, P.E. & J.V. Neel. 1977. Multivariate analysis of gametic disequilibrium in the Yanomama. Genetics 85: 733–752.

Spencer, C.C., J.E. Neigel & P.L. Leberg. 2000. Experimental evaluation of the usefulness of microsatellite DNA for detecting demographic bottlenecks. Mol. Ecol. 9: 1517–1528.

Spong, G. & L. Hellborg. 2002. A near-extinction in lynx: Do microsatellite data tell the tale? Conservation Ecology 6: 15 [online] URL: http://www.consecol.org/vol6/iss1/art15.

Waldick, R.C., S. Kraus, M. Brown & B.N. White. 2002. Evaluating the effects of historic bottleneck events: An assessment of microsatellite variability in the endangered, North Atlantic right whale. Mol. Ecol. 11: 2241–2249.

Waples, R.S. 1990. Conservation genetics of Pacific salmon. II. Effective population size and the rate of loss of genetic variability. J. Hered. 81: 267–276.

Waples, R.S. & P.E. Smouse. 1990. Gametic disequilibrium analysis as a means of identifying mixtures of salmon populations. Am. Fish. Soc. Symp. 7: 439–458.

Withler, R.E., K.D. Le, R.J. Nelson, K.M. Miller & T.D. Beacham. 2000. Intact genetic structure and high levels of genetic diversity in bottlenecked sockeye salmon (*Oncorhynchus nerka*) populations of the Fraser River, British Columbia, Canada. Can. J. Fish. Aquat. Sci. 57: 1985–1998.

Wood, C.C. 1995. Life history variation and population structure in sockeye salmon. Am. Fish. Soc. Symp. 17: 195–216.

Wright, S. 1978. Evolution and the Genetics of Populations. Volume 4: Variability Within and Among Natural Populations, The University of Chicago Press, Ltd, London.

Environmental Biology of Fishes **69**: 63–79, 2004.
© 2004 *Kluwer Academic Publishers. Printed in the Netherlands.*

Microsatellite DNA data indicate distinct native populations of kokanee, *Oncorhynchus nerka*, persist in the Lake Sammamish Basin, Washington

Sewall F. Young[a], Mark R. Downen[b] & James B. Shaklee[a]
[a]*Washington Department of Fish and Wildlife, 600 Capitol Way N, Olympia, WA 98501, U.S.A.*
(e-mail: youngsfy@dfw.wa.gov)
[b]*Washington Department of Fish and Wildlife, P.O. Box 100, La Conner, WA 98257, U.S.A.*

Received 17 April 2003   Accepted 19 April 2003

Key words: genetics, hatchery introductions, divergence

Synopsis

Large-scale introductions of resident and anadromous salmonids from exogenous sources and urbanization have led to major changes in, and concern for the fate of, indigenous fish populations of the Lake Sammamish/Lake Washington Basin. Specifically, introductions of kokanee (the resident form of *Oncorhynchus nerka*) from the Lake Whatcom Hatchery and sockeye (the anadromous form of *O. nerka*) from Baker Lake have caused uncertainty about the ancestry of the kokanee that currently spawn in the basin. We used nine microsatellite loci to investigate the inter-relationships of kokanee populations that spawn in streams in the Sammamish sub-basin, sockeye salmon populations that share spawning areas with the kokanee, Lake Whatcom Hatchery kokanee and Baker Lake sockeye, and an out-group, Meadow Creek kokanee, from Lake Kootenay which drains into the upper Columbia River. We observed high levels of genetic variation (5–49 alleles per locus). Explicit tests of population sub-division revealed that collections from most spawning aggregations differed from each other. Observed allele frequency distributions strongly suggest that natural spawning kokanee in the basin are not descended from recent Lake Whatcom stock introductions. We found no compelling evidence to suggest that the kokanee sampled from spawning areas within the Lake Sammamish sub-basin have resulted from, or been altered substantially by, past introductions of non-native kokanee or sockeye.

Introduction

The Lake Washington/Lake Sammamish Basin (Figure 1), which includes the Seattle metropolitan area, is one of five watersheds in Washington that historically supported native kokanee, *Oncorhynchus nerka*, populations[1] (Hendry et al. 1996). It also is one of seven watersheds in the state that supports lake rearing populations of sockeye salmon, the anadromous form of, *O. nerka*.[2] Kokanee typically have an adfluvial

life history, migrating from natal streams to rear and mature in a lake, avoiding the ocean migration that characterizes sockeye salmon. Despite their recognized conspecific status, sympatric populations of sockeye and kokanee are biologically and genetically distinct (Foote et al. 1989).

The Lake Washington/Lake Sammamish watershed supported only small populations of sockeye but large populations of kokanee in the period between 1890 and 1920.[3] Historically, the Bear Creek system

---

[1] Pfeifer, B. 1995. Decision document for the management and restoration of indigenous kokanee of the Lake Sammamish/ Sammamish River Basins with special emphasis on the Issaquah Creek stock. Draft Report, Washington Department of Fish and Wildlife, Mill Creek, WA.
[2] Shaklee, J.B., J. Ames & L. Lavoy. 1996. Genetic diversity units and major ancestral lineages for sockeye salmon in Washington. Chapter E *In*: C. Busack & J.B. Shaklee (ed.).

Genetic Diversity Units and Major Ancestral Lineages of Salmonid Fishes in Washington. Tech. Rept. RAD 95-02/96, Washington Department of Fish and Wildlife, Olympia, Washington.
[3] Kerwin, J. 2001. Salmon and steelhead habitat limiting factors report for Cedar–Sammamish Basin (Water Resource Inventory Area 8). Washington Conservation Commission, Olympia, WA.

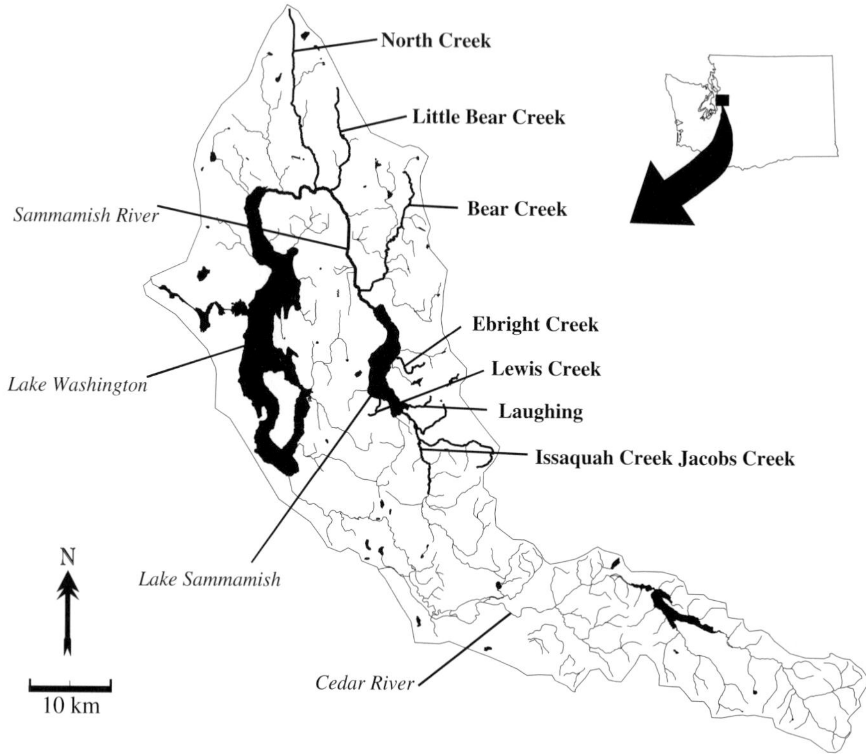

Figure 1. The study area is in the Puget Sound Basin in Washington State. Lake Sammamish lies ~16 km east of the City of Seattle.

supported September–October spawning kokanee populations estimated to number in the tens of thousands in the 1930s, based on annual egg take records of the Washington Department of Game.[3] Prior to 1982, the August–September spawning kokanee in Issaquah Creek numbered from 400 to over 1000 fish annually.[1] Other tributaries to Lake Sammamish have supported November–December runs of kokanee at least since the 1930s, however, little is known about these fish and they commonly have been assumed to have descended from substantial fry plants of Lake Whatcom Hatchery kokanee prior to 1978[1] (Hendry 1995).

Few records exist regarding the historical status of anadromous sockeye in Bear Creek or Issaquah Creek. Large numbers of Baker River (a tributary of the Skagit River in northwestern Washington) and Cultus River (a tributary of the lower Fraser River in British Columbia) stocks were introduced in the first half of the 20th century suggesting native sockeye abundance was relatively low (Hendry et al. 1996). However, several investigations of genetic relationships among Lake Washington/Sammamish sockeye have revealed the uniqueness of Bear Creek sockeye with respect to

Baker River and Cultus stocks and pointed to a strong native component in these fish[4] (Hendry et al. 1996, Gustafson et al. 1997).

In recent years, kokanee spawning stocks in Issaquah Creek,[1] in tributaries to Lake Washington and the Sammamish River,[5,6] and November–December runs into creeks that flow into Lake Sammamish have experienced declining or relatively low abundance. The abundance of Lake Washington beach spawning

[4] Seeb, J. & L. Wishard. 1977. Genetic marking and mixed fishery analysis: The use of biochemical genetics in the management of Pacific salmon stocks. Unpublished Final Report submitted to Washington Department of Fisheries for Service Contract 792. 65 pp.

[5] Ostergaard, E. 1996. 1995 status report: Abundance of spawning kokanee in the Sammamish River Basin. Addendum to the 1994 status report. King County Surface Water Management Division, Seattle, WA.

[6] Ostergaard, E. 1998. Salmon spawning surveys in the Lake Washington watershed: Results from the 1996 salmon spawning survey program and kokanee spawner survey program. King County Department of Natural Resources, Seattle, WA.

sockeye also has declined recently.[7] While the causes for these declines have not been rigorously established, they likely include myriad effects due to extensive urbanization in the region, loss and degradation of habitat, fishery harvests, introduction and proliferation of non-native fish species, and former hatchery management practices aimed at eliminating kokanee as a means of controlling infectious hematopoetic necrosis (IHN) disease.[8]

Particular concern has been focused on the early-run kokanee population in Issaquah Creek for three reasons. First, the abundance of this population has declined to extremely low numbers in recent years. Second, this population's very early spawn timing with peak activity in mid-August differs from that of other kokanee and sockeye populations in the basin, which spawn from late September through early January.[1,4,9] Third, the genetic distinctiveness of the early-run Issaquah Creek kokanee population has been recognized previously based on both allozyme electrophoretic data[4] (Hendry 1995, Hendry et al. 1996) and, more recently, microsatellite DNA data.[9] In contrast, little attention has been paid to the late-run kokanee of the Sammamish Basin, primarily because managers assumed that they originated from Lake Whatcom Hatchery strain kokanee planted in the basin from 1940 to 1978. Prior to this study no genetic evidence existed to support or refute this assumption.

We undertook the present study to elucidate genetic characteristics of kokanee and sockeye populations in the Lake Washington/Lake Sammamish Basin, to assess their inter-relationships, and to ascertain the likely origins of the extant populations with particular attention to the assertion that some or all are descended from plants of hatchery fish from other areas. We used microsatellite DNA markers (Wright & Bentzen 1994) because earlier work done on sockeye and kokanee using allozyme electrophoresis (e.g. Wood et al. 1994, Winans et al. 1996) revealed relatively low levels of detectable genetic variation with relatively few variable

loci, and we believed that more variable genetic markers would provide greater statistical power to this study. A number of recent investigations have demonstrated the power of microsatellite DNA markers to elucidate population structure (Small et al. 1998, Beacham et al. 1999, Beacham & Wood 1999, Banks et al. 1999, 2000, Olsen et al. 2000b, Young et al. 2004).

Materials and methods

Spawner surveys and sample collection

Washington Department of Fish and Wildlife (WDFW) and King County Department of Natural Resources (KCDNR) personnel conducted weekly spawner escapement surveys on index reaches of Sammamish Basin streams from mid-August 2000 through early January 2001. Index reaches were designated to contain known spawning habitat and aggregations of kokanee and sockeye. Conservative escapement and temporal abundance estimates (Figure 2) were based on live fish counts from the weekly surveys assuming an average stream life of 12 days. During the surveys, and on other occasions, fin tissue samples for subsequent DNA analysis were collected from both carcasses and live fish, generally in proportion to the estimated numbers of fish in the stream (Table 1).

DNA samples and extractions

We analyzed a total of 874 *O. nerka* from 12 collections in this study (Table 1). Nine of the 12 collections were obtained from natural spawning areas in the basin during fall 2000. Paul Bentzen and Ingrid Spies, University of Washington School of Fisheries provided 13 Issaquah Creek early kokanee that were collected by Andrew Hendry, McGill University Department of Biology. In addition to the samples representing local spawners, we included four collections from outside the basin. Two collections of Lake Whatcom Hatchery strain kokanee, one from the Lake Whatcom Fish Hatchery in Whatcom County, Washington and one from the Spokane Fish Hatchery in Stevens County, Washington represent a source of kokanee introductions to the basin estimated at >40 million fish between 1940 and 1978.[9] A collection of Baker River sockeye represents the predominant source of introduced sockeye during the early 1900s. The fourth exogenous collection is of Meadow Creek kokanee young from the Spokane Hatchery. Those young were produced

---

[7] Washington Department of Fish and Wildlife and Western Washington Treaty Tribes. 1994. 1992 Washington State Salmon and Steelhead Stock Inventory. Appendix 1 Puget Sound Stocks– South Puget Sound Volume. WDFW, Olympia, WA. 371 pp.

[8] Hopley, C.W. Washington Department of Fish and Wildlife, pers. comm.

[9] Bentzen, P. & I. Spies. 2000. Investigation of genetic variability within and between Lake Washington sockeye salmon populations using microsatellite markers. Unpublished Report submitted to City of Seattle. 20 pp.

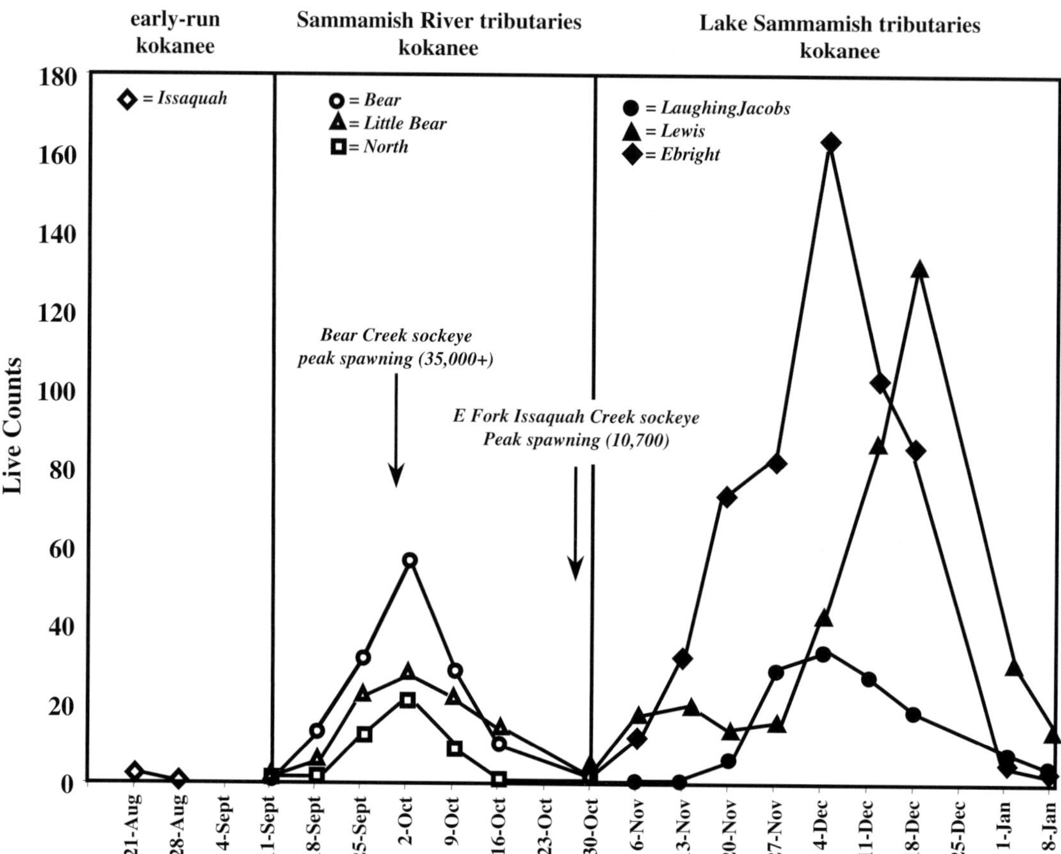

Figure 2. Kokanee spawning ground surveys during 2000–2001 revealed a tri-modal temporal distribution of spawning activity. The peak of presumed kokanee spawner abundance in the tributaries to the Sammamish River during September and October coincided with the peak of sockeye spawner abundance.

Table 1. Sample collection details.

Collection name	Type	N	Spawn timing	Collection site
Issaquah Creek[a]	Kokanee	13	Aug.	rkm 0.0–8.0
Issaquah Creek	Sockeye	61	Oct.	rkm 0.0–8.0
Bear Creek[b]	Kokanee	48	Sept.–Oct.	rkm 0.3–8.8
Bear Creek	Sockeye	52	Sept.–Oct.	rkm 0.3–8.8
Little Bear Creek[b]	Kokanee	25	Sept.–Oct.	rkm 3.2–5.6
Lewis Creek	Kokanee	100	Nov.–Dec.	rkm 0.0–0.8
Laughing Jacobs Creek	Kokanee	55	Nov.–Dec.	rkm 0.0–1.6
Ebright Creek	Kokanee	100	Nov.–Dec.	rkm 0.0–0.3
Baker Lake	Sockeye	100	Oct.–Dec.	Baker Spawning channel
Whatcom at Whatcom Hatchery	Kokanee	100		Lake Whatcom Fish Hatchery
Whatcom at Spokane Hatchery	Kokanee	100		Spokane Tribal Fish Hatchery
Meadow Creek	Kokanee	100		Kootenay Lake tributary (BC)

[a]Included in analyses despite small sample size due to suspected uniqueness.
[b]Judged to be kokanee based on size at maturity. Collected during peak of sockeye spawning.

from adults taken from Meadow Creek, a tributary to the North Arm of Lake Kootenay, British Columbia in the upper Columbia River Basin. The Meadow Creek kokanee population was included as an outgroup to provide perspective to the relationships of the other collections. All collections except the Whatcom Hatchery strain at the Spokane Hatchery and the Meadow Creek stock were of adult fish. Samplers removed ∼0.25 cm^2 of fin tissue or operculum from adults or collected whole juveniles. Tissue samples were preserved in 100% ethanol until DNA was extracted.

DNA extractions from three sets of samples (Ebright Creek, Lewis Creek, and Meadow Creek kokanee) were done using an ammonium acetate precipitation purification following proteinase K digestion of the tissue samples (procedure modified from Gentra Systems–Puregene). DNA extractions for all other samples were done using commercially available, 96-well silica membrane based kits (Macherey–Nagel Nucleospin multi-96 tissue kits or Qiagen DNeasy 96 tissue kits).

DNA amplification

We amplified nine microsatellite DNA loci (Table 2) via the polymerase chain reaction (PCR) using fluorescent-labeled primers. Eight of the loci were isolated from sockeye salmon (Olsen et al. 2000a) and one (*Ots-103*) was isolated from chinook salmon, *Oncorhynchus tshawytscha* (Small et al. 1998). We combined the loci into three multiplexes of three loci each, to allow coamplification of the constituent loci (Olsen et al. 1996) and provide efficiency in PCR and allele detection. The specific PCR amplification protocols we used were:

PCR multiplex OneA: *One-108* at 0.06 μM; *One-110* at 0.1 μM; and *One-100* at 0.4 μM

PCR multiplex OneB: *One-102* at 0.075 μM; *One-114* at 0.1 μM; and *One-115* at 0.06 μM

PCR multiplex OneC: *One-105* at 0.04 μM; *Ots-103* at 0.2 μM; and *One-101* at 0.06 μM

The three multiplex PCRs were optimized with a standard thermal profile: an initial 3 min denaturation at 92°C, followed by 38 cycles of 15 s denaturation at 92°C, 30 s annealing at 50°C, and 60 s extension at 72°C; and then a final 30 min extension at 72°C to encourage uniform adenylation.

Microsatellite DNA data collection

Microsatellite DNA analysis followed procedures established in our laboratory. We collected microsatellite data using a 96-lane, ABI-377 automated DNA sequencer utilizing in-lane size standards (GeneScan-500 rox; Applied Biosystems). Raw data from the DNA sequencer was processed using GeneScan (v. 3.0) and Genotyper (v. 2.5) software (Applied Biosystems). Two biologists scored the microsatellite DNA mobility patterns of all samples independently and all scoring discrepancies were reviewed and resolved. Allele length estimates from Genotyper were imported into Microsoft Excel, where we defined allele size bins based on modes and gaps in the distribution of allele length estimates. Allele bins were centered on median length estimates within the clusters between gaps, with inter-bin spaces equal to one-half the interval between cluster medians. This insured that all alleles within a bin were more similar to each other than to alleles in neighboring bins. Genotypes with at least one raw size estimate that fell between bins were considered ambiguous and zeroed at that locus. Those ambiguous alleles comprised 3.6% (567) of our 15,768 allele observations.

Table 2. Locus summary (N = number of fish successfully scored; H$_e$ = expected heterozygosity; H$_o$ = observed heterozygosity).

Locus	N	# Alleles	H$_e$	H$_o$	Repeat class (bp)		Allelic range (bp)	
					Published as	Scored as	Smallest	Largest
One-100	592	46	0.950	0.905	4	Tetranucleotide	244	484
One-108	675	25	0.921	0.859	4	Dinucleotide	179	261
One-110	682	33	0.934	0.894	4	Dinucleotide	211	299
One-102	594	19	0.897	0.823	4	Tetranucleotide	198	274
One-114	744	28	0.928	0.898	4	Tetranucleotide	204	316
One-115	743	19	0.909	0.871	4	Tetranucleotide	176	256
One-101	739	47	0.958	0.936	4	Tetranucleotide	173	357
Ots-103	768	23	0.922	0.882	4	Tetranucleotide	134	234
One-105	727	6	0.735	0.664	4	Tetranucleotide	130	150

Analyses of DNA data

We used the software FSTAT (Goudet 1995) v. 2.9.3.2[10] to construct allele frequency tables (Appendix A) and to identify population structure among the collections with explicit genotypic tests of population differentiation. We used genotypic permutation tests because they do not rely on Hardy–Weinberg equilibrium conditions (Goudet et al. 1996).

We tested our data for non-random associations of alleles within populations by conducting tests of Hardy–Weinberg equilibrium and linkage disequilibrium using the software, GDA v. 1.1.[11] This implementation is a permutation approach that emulates Fisher's exact test. Evidence of non-random associations, manifested in significant test results, can suggest that loci are inappropriate for use in subsequent analyses or that sample collections were drawn from heterogeneous groups.

We compared pairwise estimates of coancestry distance, $d = -\ln(1 - \theta)$ (Weir 1996), to investigate the inter-relationships between collections. We used the software, Microsat,[12] to bootstrap alleles at each locus and estimate pairwise coancestry distances and their standard errors. We then used Neighbor and Consense, components of the phylogenetic analysis software package Phylip,[13] to construct neighbor-joining trees of the coancestry distances and to find a majority rule consensus tree. We used the results of the tests of population differentiation and the analysis of coancestry distances to combine individual collections into groups.

We constructed scatter plots of all genotypes at each locus to visualize allelic distribution patterns among collection groups. Although the distribution patterns can be deduced from allele frequency tables, and can be quantified only from those tables, it is often difficult to detect patterns in large tables of allele frequencies.

---

[10] Goudet, J. 2001. FSTAT, a program to estimate and test gene diversities and fixation indices (version 2.9.3). Available from http://www.unil.ch/izea/softwares/fstat.html. Updated from Goudet (1995).

[11] Lewis, P.O. & D. Zaykin. 2001. Genetic Data Analysis: Computer program for the analysis of allelic data. Version 1.0 (d16c). Free program distributed by the authors over the internet from http://lewis.eeb.uconn.edu/lewishome/software.html.

[12] Available from the Human Population Genetics Laboratory at Stanford University (http://hpgl.stanford.edu/projects/microsat).

[13] Felsenstein, J. 1993. PHYLIP (Phylogeny Inference Package) version 3.5c. Distributed by the author. Department of Genetics, University of Washington, Seattle.

Graphical presentation is a much more efficient way to reveal patterns in the data than tabular presentation.

Results

Distribution and abundance of spawners

Weekly spawner escapement surveys conducted between August 2000 and January 2001 suggested three major groups of kokanee (early-run, north lake tributaries, and late-run) and two major groups of sockeye (Bear Creek and East Fork Issaquah Creek) spawning in the Lake Sammamish Basin based on distribution and timing (Figure 2).

Surveys for early-run kokanee were conducted in Issaquah Creek and many of its tributaries. Two presumed early-run kokanee were observed in the Issaquah Creek drainage during the third and fourth weeks of August, one of which was responsible for the construction of a redd immediately downstream of the Issaquah Creek Hatchery. Surveys of other Lake Sammamish tributaries and of several Sammamish River tributaries also conducted between 25 July and 31 August yielded only one additional kokanee observation; in Little Bear Creek on 31 August.

Spawning surveyors observed a second group of presumed kokanee in the north tributaries of the Sammamish River from early September through late October. These fish were dull olive to brown in color and ranged from 250 to 360 mm in fork length (FL). Of the estimated escapement of 170 kokanee in these creeks, most were males. These observations in Bear, Little Bear, and North creeks occurred while a run of over 35,000 sockeye spawned in the same reaches of these three tributaries (Figure 2).

A third group of kokanee entered east and south Lake Sammamish tributaries from October through early January (Figure 2). These fish were morphologically distinct from the kokanee mentioned above, with heavy spotting along their entire dorsal surface and both lobes of their caudal fins and with varying degrees of red coloration laterally. They ranged from 340 to 520 mm (FL) with an estimated total escapement of 620 fish. Only three adult sockeye were observed in the tributaries used by these kokanee. Kokanee spawning in these Lake Sammamish tributaries was later than that of the second major group of sockeye in the basin, the East Fork Issaquah sockeye. That sockeye run peaked 26th October with a total estimated escapement of 10,700 fish. Kokanee were not observed in East Fork Issaquah Creek during this time.

Population genetics

Field sampling efforts yielded nine kokanee and 11 sockeye from North Creek. We excluded those collections from our analyses because the sample sizes were very small. However, we included 13 Issaquah Creek early-run kokanee that were collected in 1993 in our analyses because we believed that they are the only available remnants of this distinctive native population (Table 1). Considering this small sample, inferences regarding the Issaquah early kokanee should be treated with caution. Also, we combined the data from the kokanee collections from Bear Creek and Little Bear Creek for our analyses based on temporal proximity of the two and because genotypic tests of differentiation (see below) failed to reveal significant divergence between them.

*One-100** and *One-102** had substantially more missing scores than the other loci, however, analyses run without those loci produced results that were similar to those obtained with all nine loci. The results we report here are based on all nine loci to increase statistical power.

Tests of Hardy–Weinberg equilibrium at each locus in each of the 11 collections included in the analysis revealed significant departures in three cases (at a nominal p-value of 0.05 after correction for multiple tests): *One-100** in the Bear Creek sockeye collection and both *One-101** and *Ots-103** in the Lewis Creek kokanee collection. We saw no evidence from these tests that any of the nine loci are inappropriate for further analyses; however, significant tests at two loci in the Lewis Creek collection could indicate a departure from random mating in that population. Tests for linkage disequilibrium produced eight significant results (p<0.05 after correction for multiple simultaneous tests) among 396 pairwise tests. No locus pair showed evidence of linkage in more than one population so we see no evidence that any of these loci are closely linked. The Laughing Jacobs Creek and the Meadow Creek kokanee collections each had one significant test (*One-102*/Ots-103** and *One-100*/One-102**, respectively) whereas the Lewis Creek sample had six significant tests (*One-100*/One-114**, *One-110*/One-101**, *One-114*/One-101**, *One-114*/Ots-103**, *One-115*/One-101**, and *One-101*/Ots-103**), four of which involved *One-101** and *Ots-103**,

loci that deviated from Hardy–Weinberg expectations. When we consider jointly the six tests that suggested non-random associations between locus pairs in the Lewis Creek collection and the departures from Hardy–Weinberg expected genotypic proportions noted above, we cannot dismiss the possibility that the Lewis Creek sample was drawn from an admixture of fish from two or more distinct populations or perhaps was biased in some other way.

Genotypic tests of differentiation provided evidence for significant genetic divergence among all study populations except for four pairs of collections: (1) Lake Whatcom Fish Hatchery kokanee vs. Lake Whatcom kokanee at Spokane Fish Hatchery; (2) Bear Creek kokanee vs. Little Bear Creek kokanee; (3) Bear Creek sockeye vs. Bear Creek kokanee; and (4) Bear Creek sockeye vs. Little Bear Creek kokanee.

The consensus neighbor-joining tree of pairwise coancestry distances (Figure 3) illustrates the distinctiveness of the Lake Sammamish tributaries kokanee populations (Lewis Creek, Laughing Jacobs Creek, and Ebright Creek) with 98% bootstrap support and of the Lake Whatcom kokanee with 100% bootstrap support. No other population groups were supported with at least 80% bootstrap support so no other relationships were convincingly resolved by this analysis.

Pairwise coancestry distances (Table 3) revealed non-zero distances between all pairs of collections, except the two collections of Whatcom kokanee from successive brood years. Distances of ∼0.020 among the Lewis Creek, Laughing Jacobs Creek, and Ebright Creek kokanee populations contrast with distances of ∼0.070 between each of those populations and the two Whatcom kokanee collections. The pairwise distances among the Sammamish River tributary kokanee populations (Bear Creek and Little Bear Creek), the Bear Creek sockeye, and the Issaquah Creek sockeye populations are ∼0.025 while the pairwise distances between the Sammamish River kokanee population and the Lake Sammamish tributaries kokanee populations are ∼0.060.

Based on the tests of population differentiation and the clustering by genetic distances, we pooled the genetic data for Lewis, Ebright, and Laughing Jacobs creeks into a Lake Sammamish tributaries group and we grouped the data for the two Lake Whatcom Hatchery strain kokanee collections (from Lake Whatcom Hatchery and from Spokane Hatchery) into a Whatcom group to examine several scenarios that might explain their divergence.

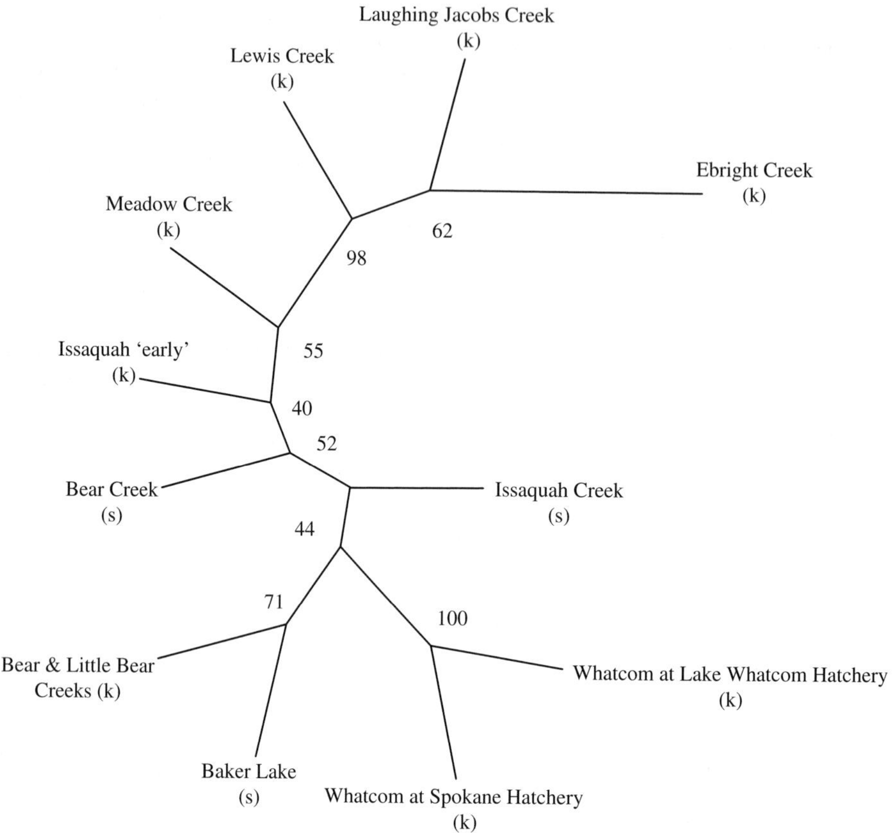

Figure 3. Consensus neighbor-joining tree of coancestry distances. Numbers at the nodes indicate the number of trees out of 100 replicates that supported the more distal grouping. Internal branch lengths are proportional to bootstrap values.

Table 3. Pairwise coancestry distances. All values except that for the Lake Whatcom vs. Whatcom at Spokane comparison are significantly greater than zero based on bootstrap analysis. k = kokanee, s = sockeye.

System	Issaquah (k)	Lewis (k)	Laughing Jacobs (k)	Ebright (k)	Whatcom at Spokane (k)	Lake Whatcom (k)	Issaquah Cr (s)	Sammamish R tribs (k*)	Bear (s)	Baker Lake (s)
Lewis Cr (k)	0.083									
Laughing Jacobs Cr (k)	0.113	0.018								
Ebright Cr (k)	0.106	0.015	0.019							
Whatcom–Spokane (k)	0.082	0.067	0.072	0.064						
Whatcom–Whatcom (k)	0.085	0.069	0.078	0.067	0.004					
Issaquah Cr (s)	0.071	0.047	0.065	0.050	0.026	0.027				
Sammamish R tribs (k*)	0.016	0.075	0.056	0.056	0.069	0.025	0.030			
Bear Cr (s)	0.018	0.068	0.049	0.057	0.069	0.015	0.030	0.033		
Baker Lake (s)	0.083	0.041	0.122	0.050	0.088	0.101	0.033	0.050	0.051	
Meadow Cr (k)	0.066	0.059	0.036	0.036	0.086	0.051	0.058	0.035	0.033	0.040

*Although this collection was originally thought to be kokanee, we now consider it to be residualized sockeye (see text).

We observed large differences in allele frequency distributions between the Lake Sammamish tributaries kokanee and the Whatcom kokanee. Fifty-one percent (141) of the 278 alleles observed at *One-100** in the pooled Whatcom kokanee collections were shorter than 340 bp whereas 97% (335) of the 346 *One-100** alleles observed in the pooled Lake Sammamish tributaries collections were 340 bp or longer (Appendix A). Similarly, at *One-108**, 77% (251) of the 324 alleles observed in the Whatcom kokanee were 219 bp or shorter while 72% (297) of 416 alleles from the Lake Sammamish tributaries were longer than 219 bp.

We observed a strong non-random pattern of allele presence/absence at *One-110** among collections (Figure 4). In all collections except the two Whatcom kokanee collections, we observed predominantly 4 bp intervals between allele classes. One thousand one hundred and fifty of 1340 alleles (86%) observed at *One-110** in the whole data set occurred in allele classes that were consistent with a tetranucleotide repeat model for the locus whereas 170 of the 190 alleles (89%) that were inconsistent with a simple tetranucleotide repeat model were observed in the two Whatcom kokanee collections. Within the Lake Whatcom kokanee collections we observed a strong 2 bp interval between allele classes: 50% of the alleles observed (170 of 342) at *One-110** were offset 2 bp from the tetranucleotide repeat pattern that predominated in the entire data set. We observed alleles *237, *249, *253, *257, *265, *269, *273, *277, *281, *285, and *289 in both Whatcom kokanee collections but not in the Lewis Creek, Laughing Jacobs Creek, or Ebright Creek collections. We observed three *261 alleles in the Lewis Creek sample but none in the Laughing Jacobs Creek or Ebright Creek samples. The 2 bp pattern is not unique to the Whatcom kokanee strain in our data set – we observed seven of the allele classes that are out-of-phase with the predominant 4 bp interval in the Meadow Creek kokanee collection and four in the *O. nerka* from the Sammamish River tributaries, but those allele classes accounted for more than 4% of the observations in only the Lake Whatcom and Meadow Creek samples.

Discussion

The data reveal significant population structure among the naturally spawning resident *O. nerka* in the Lake

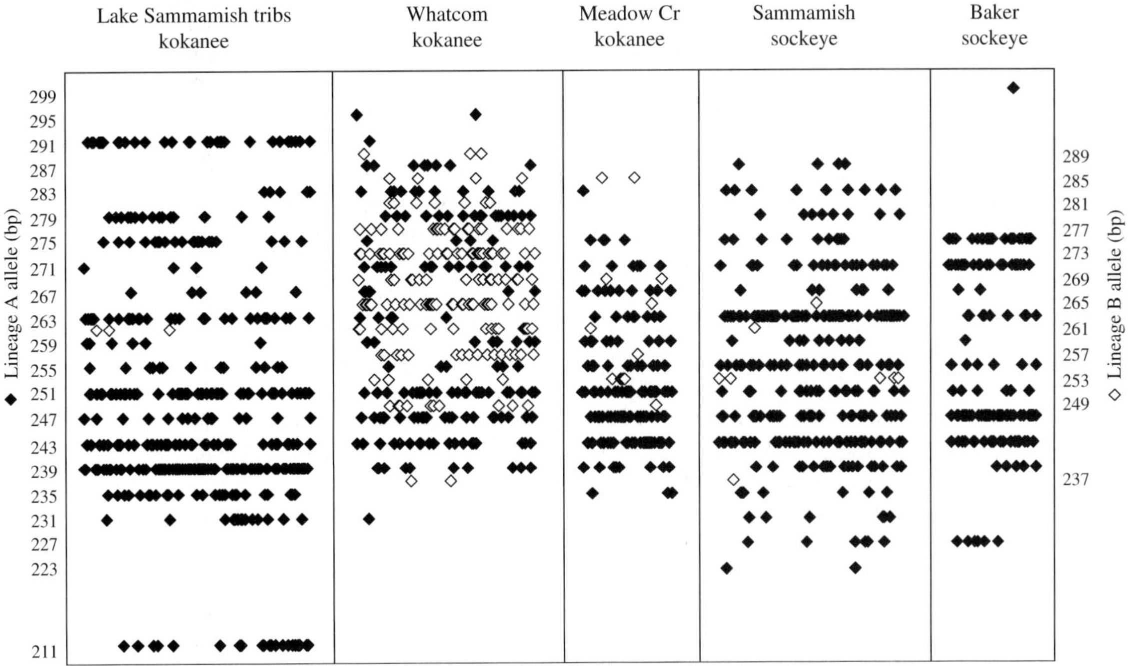

Figure 4. Scatter plot of the allelic distribution at locus *One-110**. This plot displays the genotypes of 682 kokanee and sockeye that we scored at this locus for this study and reveals two allelic lineages.

Sammamish system. Samples collected from spawning, presumed non-anadromous *O. nerka* in Bear Creek and Little Bear Creek were genetically undifferentiated from each other and from anadromous adults that spawned contemporaneously in Bear Creek, but all three groups were significantly differentiated from the non-anadromous adults that spawned approximately a month later in three Lake Sammamish tributaries – Lewis Creek, Laughing Jacobs Creek, and Ebright Creek. The lack of significant differentiation among the non-anadromous *O. nerka* and the more abundant sockeye that were collected during spawning in tributaries to the Sammamish River leads us to conclude that the non-anadromous spawners are sockeye that failed to migrate to saltwater. Elemental analyses of Sr:Ca ratios in otolith primordia that suggested the non-anadromous spawners had anadromous mothers[14] support this conclusion. The differentiation we observed between the sockeye that spawn in tributaries to the Sammamish River and the Lake Sammamish tributaries kokanee populations is qualitatively similar to differentiation between sympatric sockeye and kokanee populations reported elsewhere based on minisatellites and mitochondrial DNA restriction sites (Taylor et al. 1996), and allozymes (Foote et al. 1989, Wood & Foote 1996). However, direct, quantitative comparisons between the results of the present study and those of these earlier investigations were not attempted because of the different marker classes used in the various studies.

Although tests of population differentiation show that the Lewis Creek, Laughing Jacobs Creek, and Ebright Creek collections are significantly differentiated, there is strong bootstrap support for grouping these three populations into a Lake Sammamish tributaries aggregate population with respect to all other collections in this study. In fact, the consensus neighbor-joining tree of coancestry distances supports only this group and the Whatcom kokanee aggregate. All other relationships are unresolved by that analysis (i.e., bootstrap values were less than 75%).

The divergence of the Lake Sammamish tibutaries kokanee populations and of the Whatcom kokanee does not by itself clearly resolve the ancestry of the Lake Sammamish tributaries kokanee. We imagine three possible scenarios that would lead to the observed divergence: (1) the Lake Sammamish tributaries kokanee might be indigenous to the basin and show little or no signs of introgression of Lake Whatcom alleles; (2) the Lake Sammamish tributaries kokanee might have descended from some of the more than 45 million Lake Whatcom kokanee fry that were planted into the basin during the mid-1900s but in the intervening years either the Lake Sammamish line, the Lake Whatcom Hatchery strain, or both have gone through an effective size bottleneck, with associated divergence; or (3) the Lake Sammamish tributaries kokanee are an introgressed line with both indigenous and Lake Whatcom ancestry.

The allele distributions that we observed suggest that the divergence between the Lake Sammamish tributaries kokanee and the Lake Whatcom kokanee is not the result of a bottleneck in either population aggregate. Divergence of the two groups from a common gene pool over fewer than 10 generations since Whatcom strain releases into the basin stopped would have required either a simultaneous decrease in frequency of the eight *One-100** alleles shorter than 340 bp in the Lake Sammamish populations or a simultaneous increase in frequency of those alleles in the Whatcom strain. Under random genetic drift, neighboring allele classes at highly variable loci would be expected to increase or decrease more or less independently, so the concerted frequency divergence in the under 340 bp and over 340 bp allele classes at *One-100**, is unexpected. The same holds for the allelic distributions we observed at *One-108** in the two groups. Similar logic argues against the observed allelic distributions having arisen from a recent founder effect due to introductions of Whatcom kokanee into the basin.

The allele distributions at *One-110** in the two groups also provide compelling evidence that the population divergence we observed is not due to random genetic drift following a recent bottleneck or founding event. Half of the alleles that we observed in the Whatcom kokanee were in allele classes that were 2 bp out-of-phase with the 4 bp repeat pattern we observed in the Lake Sammamish tributaries aggregate. We attribute that pattern to the presence of two allele lineages (lineages A and B in Figure 4) at *One-110**. Lineage A is predominant in all populations we studied except the Whatcom Hatchery strain kokanee; lineage B is virtually absent from the Lake Sammamish tributaries kokanee. Both A and B lineages occur at frequencies of ~0.5 in the Whatcom kokanee and the B lineage accounts for 11 of 33 *One-110** allele

---

[14] Eric Volk, Washington Department of Fish and Wildlife, pers. comm.

classes observed in that population. Karhu et al. (2000) described the divergence of microsatellite allele lineages (within several pine species) by expansion of different repeat motifs.

Derivation of the Lake Sammamish tributaries kokanee from the Whatcom strain would have required either the loss in the Lake Sammamish tributaries kokanee of 10 lineage B alleles or the rapid evolution of the B lineage in the Whatcom kokanee in the 25 years since Whatcom kokanee introductions to the Sammamish Basin ceased. Random genetic drift is an extremely unlikely explanation for the absence of allele lineage B in the Lake Sammamish tributaries kokanee and we doubt that in the last 25 years the B lineage evolved 10 new alleles that account for 50% of the alleles in the current Whatcom Hatchery kokanee strain. Instead, we interpret these differences as evidence that these populations represent evolutionarily distinct lineages.

The allele distributions at *One-110** also suggest that the Lake Sammamish tributaries kokanee are not substantially introgressed with Whatcom strain alleles. Only three of 346 *One-110** alleles sampled from the Lake Sammamish tributaries kokanee aggregate were in allele lineage B.

The high frequency of a distinctive allele lineage in the Lake Whatcom kokanee that is nearly absent in all other populations examined in this study highlights the value of examining microsatellite size variation for evidence of anomolous allele distribution patterns in studies of the inter-relationships of populations. Although tests of population differentiation and genetic distances demonstrated that the Whatcom kokanee differed substantially from other *O. nerka* populations in this study, they did not reveal the separate ancestries of the Whatcom group and the Lake Sammamish tributaries kokanee. The presence/absence of distinctive allele lineages is an important feature of a data set and it can provide important insights into ancestral relationships among populations.

Considering the magnitude of Lake Whatcom kokanee fry plants into the Lake Washington/Lake Sammamish Basin during the 1900s, the observed dissimilarity between the extant populations and the putative donor stock is surprising. We note that the Lake Washington/Lake Sammamish Basin is an IHN positive environment and that Lake Whatcom is IHN free.[15]

We speculate that IHN vulnerability might explain the apparent lack of success of the Lake Whatcom kokanee introductions, however, confirmation or refutation would require further study.

The persistence of native kokanee populations in the Lake Sammamish Basin following massive infusions of Lake Whatcom Hatchery kokanee fry over many years demonstrates that large, historical plants of exogenous stocks do not necessarily result in the loss of native gene pools. The implications are two-fold: (1) even in areas of intensive historical supplementation, considerable, indigeneous genetic diversity might persist; and (2) where the goal of such hatchery plants is supplementation of a naturally spawning population, genetic monitoring can be an informative evaluation tool.

Acknowledgements

Cherril Bowman and Alice Pichahchy provided technical assistance in the laboratory. Ken Warheit provided helpful advice on both technical and analytical matters. Various agency staff (KCDNR, Northwest Indian Fisheries Commission, and WDFW) and volunteers collected samples. Paul Bentzen and Ingrid Spies (University of Washington) shared purified DNA samples from the 13 Issaquah Creek early-run kokanee samples with us. Steve Foley (WDFW) provided the sockeye count, escapement, and timing data depicted in Figure 2. Jim Matilla (KCDNR), whose life-long experience in the creeks in the region was instrumental in the success of the field portions of this project and numerous dedicated volunteers who initially discovered a number of kokanee populations in the Sammamish Basin and who continue to monitor these fish today. We thank Richard LeCaire, the Colville Confederated Tribes, and BPA for allowing us to use the data for Meadow Creek kokanee. We also thank Jed Varney and Larry Sissan (WDFW) for providing the Baker Lake sockeye and the Lake Whatcom Hatchery kokanee samples, respectively. Eric Volk (WDFW) kindly shared the results of elemental analyses of otolith primordia. The data collection and analyses described herein were funded by the King Conservation District and by State of Washington General Funds for wild salmonid genetics work. The editor and two anonymous reviewers provided many suggestions that greatly improved this manuscript.

---

[15] Unpublished data; WDFW Fish Health Unit, Olympia, WA.

References

Banks, M.A., M.S. Blouin, B.A. Baldwin, V.K. Rashbrook, H.A. Fitzgerald, S.M. Blankenship & D. Hedgecock. 1999. Isolation and inheritance of novel microsatellites in chinook salmon (*Oncorhynchus tshawytscha*). J. Hered. 90: 281–288.

Banks, M.A., V.K. Rashbrook, M.J. Calavetta, C.A. Dean & D. Hedgecock. 2000. Analysis of microsatellite DNA resolves genetic structure and diversity of chinook salmon (*Oncorhynchus tshawytscha*) in California's central valley. Can. J. Fish. Aquat. Sci. 57: 915–927.

Beacham, T.D., S. Pollard & K.D. Lee. 1999. Population structure and stock identification of steelhead in southern British Columbia, Washington, and the Columbia River based on microsatellite DNA variation. Trans. Amer. Fish. Soc. 128: 1068–1084.

Beacham, T.D. & C.C. Wood. 1999. Application of microsatellite DNA variation to estimation of stock composition and escapement of Nass River sockeye salmon (*Oncorhynchus nerka*). Can. J. Fish. Aquat. Sci. 56: 297–310.

Foote, C.J., C.C. Wood & R.E. Withler. 1989. Biochemical comparison of sockeye salmon and kokanee, the anadromous and nonanadromous forms of *Oncorhynchus nerka*. Can. J. Fish. Aquat. Sci. 46: 149–158.

Goudet, J. 1995. FSTAT (ver. 1.2): A computer program to calculate F-statistics. J. Hered. 86: 485–486.

Goudet, J., M. Raymond, T. de Meeüs & F. Rousset. 1996. Testing differentiation in diploid populations. Genetics 144: 1933–1940.

Gustafson, R.G., T.C. Wainwright, G.A. Winans, F.W. Waknitz, L.T. Parker & R.S. Waples. 1997. Status review of sockeye salmon from Washington and Oregon. NOAA technical memorandum NMFS-NWDSC-33. 282 pp.

Hendry, A.P. 1995. Sockeye salmon (*Oncorhynchus nerka*) in Lake Washington: An investigation of ancestral origins, population differentiation and local adaptation. MS thesis, University of Washington, Seattle, WA. 159 pp.

Hendry, A.P., T.P. Quinn & F.M. Utter. 1996. Genetic evidence for the persistence and divergence of native and introduced sockeye salmon (*Oncorhynchus nerka*) within Lake Washington, Washington. Can. J. Fish. Aquat. Sci. 53: 823–832.

Karhu, A., J.-H. Dieterich & O. Savolainen. 2000. Rapid expansion of microsatellite sequences in pines. Mol. Biol. Evol. 17: 259–265.

Olsen, J.B., J.K. Wenburg & P. Bentzen. 1996. Semiautomated multilocus genotyping of Pacific salmon (*Oncorhynchus* spp.) using microsatellites. Mol. Mar. Biol. Technol. 5: 259–272.

Olsen, J.B., S.L. Wilson, E.J. Kretschmer, K.C. Jones & J.E. Seeb. 2000a. Characterization of 14 tetranucleotide microsatellite loci derived from sockeye salmon. Mol. Ecol. 9: 2185–2187.

Olsen, J.B., P. Bentzen, M.A. Banks, J.B. Shaklee & S. Young. 2000b. Microsatellites reveal population identity of individual pink salmon to allow supportive breeding of a population at risk of extinction. Trans. Amer. Fish. Soc. 129: 232–242.

Small, M.P., T.D. Beacham, R.E. Withler & R.J. Nelson. 1998. Discriminating coho salmon (*Oncorhynchus kisutch*) populations within the Fraser River, British Columbia. Mol. Ecol. 7: 141–155.

Taylor, E.B., C.J. Foote & C.C. Wood. 1996. Molecular genetic evidence for parallel life-history evolution within a Pacific salmon (sockeye salmon and kokanee, *Oncorhynchus nerka*). Evolution 50: 401–416.

Weir, B.S. 1996. Genetic Data Analysis II. Sinauer Associates, Inc. Sunderland, MA. 445 pp.

Winans, G.A., P.B. Aebersold & R.S. Waples. 1996. Allozyme variability of *Oncorhynchus nerka* in the Pacific Northwest, with special consideration to populations of Redfish Lake, Idaho. Trans. Amer. Fish. Soc. 125: 645–663.

Wood, C.C. & C.J. Foote. 1996. Evidence for sympatric genetic divergence of anadromous and nonanadromous morphs of sockeye salmon (*Oncorhynchus nerka*). Evolution 50: 1265–1279.

Wood, C.C., B.E. Riddell, D.T. Rutherford & R.W. Withler. 1994. Biochemical genetic survey of sockeye salmon (*Oncorhynchus nerka*) in Canada. Can. J. Fish. Aquat. Sci. 51(Suppl 1): 114–131.

Wright, J.M. & P. Bentzen. 1994. Microsatellites: Genetic markers for the future. Rev. Fish Biol. Fish. 4: 384–388.

Young, S.F., J.G. McLellan & J.B. Shaklee. 2004. Genetic integrity and microgeographic population structure of westslope cutthroat trout, *Oncorhynchus clarki lewisi*, in the Pend Oreille Basin in Washington. Environ. Biol. Fish. 69: 127–142.

Appendix A. Allele frequencies ('k' = kokanee; 's' = sockeye; N = number of fish successfully scored; '—' indicates allele was not observed in that collection).

Locus alleles	Early Issaquah Cr (k)	Lewis Cr (k)	Laughing Jacobs Cr (k)	Ebright Cr (k)	Whatcom at Whatcom (k)	Whatcom at Spokane (k)	Meadow Cr (k)	Bear & L Bear Cr (k)	Issaquah Cr (s)	Bear Cr (s)	Baker Lake (s)
*One-100**											
N	10	61	39	73	68	71	78	45	40	38	77
244	—	—	—	—	—	—	0.006	—	—	—	—
252	—	—	—	—	—	—	0.019	—	—	—	—
256	—	—	—	—	—	—	0.038	—	—	—	—
260	—	—	—	—	—	—	0.019	—	—	—	—
264	—	—	—	—	—	—	0.013	—	—	—	—

Appendix A. Continued

Locus alleles	Early Issaquah Cr (k)	Lewis Cr (k)	Laughing Jacobs Cr (k)	Ebright Cr (k)	Whatcom at Whatcom (k)	Whatcom at Spokane (k)	Meadow Cr (k)	Bear & L Bear Cr (k)	Issaquah Cr (s)	Bear Cr (s)	Baker Lake (s)
268	—	—	—	—	—	—	0.006	—	—	—	—
292	—	—	—	—	—	—	—	—	0.025	—	—
296	—	—	0.013	—	—	—	—	—	0.038	—	—
300	—	—	0.013	—	0.007	0.007	—	0.011	0.038	—	—
304	—	—	—	—	0.007	0.007	0.013	—	0.025	0.013	0.013
308	0.050	—	—	—	0.029	0.007	0.006	—	—	—	0.078
312	—	0.025	—	0.007	0.044	0.007	0.013	0.011	0.013	0.079	0.013
316	—	—	0.013	0.014	0.059	0.063	0.051	0.011	0.013	—	—
320	—	—	—	—	—	—	—	—	0.013	—	—
324	0.100	—	—	—	0.088	0.085	0.026	0.011	0.138	0.026	0.013
328	—	—	—	—	—	—	—	—	—	0.013	—
332	0.100	—	—	—	0.162	0.183	0.019	0.067	0.025	0.026	0.097
336	0.150	—	0.013	0.007	0.140	0.120	0.045	0.089	0.063	0.105	0.104
340	0.200	0.090	0.038	0.034	0.096	0.113	0.038	0.078	0.013	0.184	0.006
344	0.050	0.066	0.115	0.089	0.066	0.099	0.064	0.145	0.050	0.118	0.117
348	0.250	0.016	0.026	—	0.022	0.056	0.032	0.211	0.013	0.066	0.097
352	0.100	0.131	0.103	0.062	0.022	0.042	0.038	0.089	0.063	0.053	0.052
356	—	0.049	0.090	0.027	0.037	0.042	0.019	0.044	0.088	0.066	0.032
360	—	0.156	0.179	0.205	0.066	0.014	0.045	0.022	0.038	0.013	0.006
364	—	0.172	0.154	0.110	0.074	0.042	0.032	0.011	0.013	0.013	—
368	—	0.016	0.051	0.034	0.037	0.035	0.083	0.033	0.050	0.026	—
372	—	0.082	0.077	0.130	0.037	0.035	0.071	0.011	0.025	0.026	0.013
376	—	0.008	—	0.137	—	0.021	0.051	0.022	0.013	0.026	—
380	—	0.025	0.064	0.014	—	—	0.026	—	0.038	—	—
384	—	0.016	0.013	0.027	—	0.007	0.032	0.033	0.050	0.039	—
388	—	—	—	0.014	—	0.007	0.038	—	0.025	—	—
392	—	0.057	0.013	0.068	0.007	0.007	0.026	0.011	0.013	0.013	—
396	—	—	—	—	—	—	0.013	—	—	—	—
400	—	0.016	—	—	—	—	0.019	—	—	—	—
404	—	0.074	0.026	0.021	—	—	0.006	—	—	—	—
408	—	—	—	—	—	—	0.006	—	—	—	—
412	—	—	—	—	—	—	0.006	—	—	—	—
416	—	—	—	—	—	—	0.019	—	—	—	—
428	—	—	—	—	—	—	—	0.033	0.025	0.026	0.071
436	—	—	—	—	—	—	0.019	0.033	0.025	0.039	0.123
440	—	—	—	—	—	—	0.019	—	—	—	0.013
444	—	—	—	—	—	—	0.006	0.022	0.025	—	0.097
448	—	—	—	—	—	—	—	—	0.038	—	0.045
452	—	—	—	—	—	—	—	—	0.013	0.026	0.006
472	—	—	—	—	—	—	0.006	—	—	—	—
484	—	—	—	—	—	—	0.006	—	—	—	—

*One-108**

	Early Issaquah Cr (k)	Lewis Cr (k)	Laughing Jacobs Cr (k)	Ebright Cr (k)	Whatcom at Whatcom (k)	Whatcom at Spokane (k)	Meadow Cr (k)	Bear & L Bear Cr (k)	Issaquah Cr (s)	Bear Cr (s)	Baker Lake (s)
N	7	74	53	81	79	83	82	47	51	39	79
179	—	—	—	—	0.006	—	—	—	—	—	—
183	0.071	—	—	—	0.013	0.024	—	0.042	—	0.064	—
187	—	—	—	—	0.076	0.060	—	0.021	0.010	0.013	—
191	—	—	—	—	0.089	0.090	0.091	0.011	0.059	0.026	0.044
195	—	—	0.009	0.012	0.101	0.096	0.061	0.075	0.020	0.064	0.304
197	—	—	—	—	—	—	—	—	0.010	—	—
199	0.143	—	—	—	0.025	0.024	0.091	0.096	0.069	—	0.234
203	—	0.054	—	0.031	0.051	0.072	0.110	0.064	0.049	0.064	0.070
205	—	—	—	—	—	—	—	—	0.010	—	—

Appendix A. Continued

Locus Alleles	Early Issaquah Cr (k)	Lewis Cr (k)	Laughing Jacobs Cr (k)	Ebright Cr (k)	Whatcom at Whatcom (k)	Whatcom at Spokane (k)	Meadow Cr (k)	Bear & L Bear Cr (k)	Issaquah Cr (s)	Bear Cr (s)	Baker Lake (s)
207	0.143	0.014	0.028	0.006	0.095	0.120	0.128	0.064	0.059	0.090	0.082
209	—	—	—	—	—	—	—	0.032	—	—	—
211	—	0.128	0.170	0.074	0.127	0.108	0.116	0.170	0.078	0.167	0.025
215	—	0.047	0.066	0.074	0.070	0.114	0.110	0.117	0.186	0.179	0.019
217	—	—	0.075	0.037	—	—	—	—	—	—	—
219	—	—	0.038	0.025	0.082	0.108	0.085	0.106	0.167	0.051	0.082
223	—	0.128	0.123	0.173	0.025	0.024	0.067	0.043	0.059	0.115	—
227	0.286	0.061	0.028	0.049	0.038	0.042	0.085	0.021	0.049	0.026	—
231	0.286	0.385	0.255	0.346	0.051	0.030	0.024	0.096	0.137	0.051	0.133
237	—	0.007	0.028	0.031	0.019	—	—	0.011	—	0.026	—
241	0.071	0.047	0.057	0.074	0.051	0.048	0.006	0.021	0.010	0.038	0.006
245	—	—	—	0.006	0.025	0.012	0.018	—	0.020	0.026	—
249	—	—	—	—	0.032	0.006	0.006	—	0.010	—	—
253	—	—	—	—	0.006	—	—	0.011	—	—	—
257	—	—	—	—	0.019	0.018	—	—	—	—	—
261	—	0.128	0.123	0.062	—	—	—	—	—	—	—

*One-110**

Locus Alleles	Early Issaquah Cr (k)	Lewis Cr (k)	Laughing Jacobs Cr (k)	Ebright Cr (k)	Whatcom at Whatcom (k)	Whatcom at Spokane (k)	Meadow Cr (k)	Bear & L Bear Cr (k)	Issaquah Cr (s)	Bear Cr (s)	Baker Lake (s)
N	12	77	53	88	84	87	69	49	52	41	70
211	—	0.026	0.019	0.102	—	—	—	—	—	—	—
223	—	—	—	—	—	—	—	0.011	—	0.012	—
227	—	—	—	—	—	—	—	—	0.010	0.049	0.043
231	—	0.013	—	0.091	0.006	—	—	0.011	0.010	0.037	—
235	0.208	0.065	0.047	0.063	—	—	0.014	0.041	0.010	0.024	—
237	—	—	—	—	0.006	0.006	—	0.011	—	—	—
239	0.250	0.214	0.274	0.273	0.030	0.029	0.065	0.031	0.096	0.073	0.043
243	0.042	0.123	0.085	0.068	0.071	0.052	0.203	0.214	0.125	0.122	0.150
247	—	0.039	0.066	0.023	0.083	0.069	0.167	0.113	0.067	0.122	0.257
249	—	—	—	—	0.048	0.017	0.007	—	—	—	—
251	0.083	0.117	0.208	0.131	0.065	0.080	0.159	0.041	0.096	0.061	0.043
253	0.083	—	—	—	0.018	0.011	0.022	0.020	—	0.024	—
255	0.125	0.039	0.019	0.034	0.024	0.017	0.051	0.112	0.087	0.073	0.043
257	—	—	—	—	0.030	0.063	0.007	—	—	—	—
259	—	0.032	—	0.006	0.012	0.034	0.065	0.011	0.048	0.024	0.007
261	—	0.019	—	—	0.036	0.075	0.007	0.011	—	—	—
263	—	0.091	0.038	0.063	0.030	—	0.058	0.245	0.183	0.171	0.064
265	—	—	—	—	0.107	0.086	0.007	—	0.010	—	—
267	0.208	0.006	0.019	0.017	0.024	0.011	0.065	0.011	0.029	0.049	0.014
269	—	—	—	—	0.042	0.063	0.014	—	—	—	—
271	—	0.006	0.019	0.006	0.054	0.052	0.043	0.041	0.067	0.110	0.171
273	—	—	—	—	0.089	0.098	—	—	—	—	—
275	—	0.052	0.113	0.017	0.006	0.017	0.029	0.031	0.058	—	0.157
277	—	—	—	—	0.054	0.063	—	—	—	—	—
279	—	0.071	0.019	0.011	0.030	0.075	—	0.011	0.048	0.024	—
281	—	—	—	—	0.024	0.017	—	—	—	—	—
283	—	—	—	0.028	0.048	0.023	0.007	0.031	0.038	0.024	—
285	—	—	—	—	0.012	0.011	0.007	—	—	—	—
287	—	—	—	—	0.036	0.011	—	0.011	0.019	—	—
289	—	—	—	—	0.006	0.011	—	—	—	—	—
291	—	0.084	0.075	0.068	0.006	—	—	—	—	—	—
295	—	—	—	—	0.006	0.006	—	—	—	—	—
299	—	—	—	—	—	—	—	—	—	—	0.007

Appendix A. Continued

Locus / Alleles	Early Issaquah Cr (k)	Lewis Cr (k)	Laughing Jacobs Cr (k)	Ebright Cr (k)	Whatcom at Whatcom (k)	Whatcom at Spokane (k)	Meadow Cr (k)	Bear & L Bear Cr (k)	Issaquah Cr (s)	Bear Cr (s)	Baker Lake (s)
*One-102**											
N	**9**	**46**	**52**	**39**	**93**	**91**	**84**	**45**	**52**	**37**	**81**
198	—	—	—	—	—	—	0.030	—	—	—	—
202	—	—	—	—	—	—	0.006	0.033	0.038	—	0.105
206	—	—	—	—	—	—	0.155	—	—	—	—
210	—	—	—	—	0.075	0.088	0.012	0.067	—	0.014	—
214	—	—	—	—	0.043	0.066	—	0.011	—	0.014	—
218	—	0.022	—	0.013	0.027	0.033	0.012	—	—	—	—
222	—	0.087	0.038	0.077	0.032	0.022	0.060	0.034	0.019	—	0.006
226	0.056	0.011	0.058	0.038	0.140	0.159	0.149	0.122	0.010	0.068	—
230	—	0.196	0.433	0.256	0.065	0.071	0.113	0.033	0.029	0.108	—
234	0.389	0.087	0.058	0.090	0.102	0.110	0.077	0.123	0.106	0.162	—
238	0.444	0.120	0.087	0.026	0.134	0.154	0.089	0.200	0.260	0.216	0.093
240	—	—	—	0.038	—	—	0.006	—	—	—	—
242	—	0.304	0.260	0.385	0.113	0.082	0.065	0.122	0.240	0.176	0.259
244	—	0.011	—	—	—	—	—	—	—	—	—
246	—	0.120	0.048	0.077	0.161	0.126	0.071	0.089	0.135	0.162	0.272
248	—	0.011	—	—	—	—	—	—	—	—	—
250	—	0.022	0.019	—	0.048	0.055	0.065	0.100	0.087	0.027	0.136
254	0.111	0.011	—	—	0.038	0.027	0.030	0.033	0.019	0.027	—
258	—	—	—	—	0.022	0.005	0.030	0.023	0.058	0.014	0.130
262	—	—	—	—	—	—	0.024	0.011	—	0.014	—
274	—	—	—	—	—	—	0.006	—	—	—	—
*One-114**											
N	**10**	**73**	**54**	**75**	**97**	**93**	**95**	**58**	**48**	**45**	**96**
204	—	—	—	—	0.005	—	—	—	—	—	—
208	—	0.021	0.046	0.053	0.005	—	—	0.009	0.010	—	—
212	—	—	—	—	—	—	—	—	0.010	—	—
216	—	0.021	0.019	—	0.010	0.011	—	0.034	—	0.011	—
220	0.300	0.055	0.065	0.027	0.026	0.016	0.005	0.017	0.042	0.044	0.005
224	0.050	0.123	0.130	0.153	0.005	0.022	0.011	0.035	0.021	0.011	—
228	0.050	—	—	—	—	0.005	0.037	0.061	—	0.022	0.026
232	—	0.055	0.028	0.033	0.021	0.022	0.005	0.017	0.042	0.078	0.010
234	—	0.007	—	—	—	—	—	—	—	—	—
236	—	—	—	0.007	0.041	0.011	0.047	0.043	0.042	0.044	0.005
240	—	0.075	0.083	0.060	0.072	0.070	0.047	0.009	0.021	0.011	0.063
244	0.150	0.027	0.019	0.020	0.088	0.124	0.047	0.086	0.073	0.078	0.234
248	0.100	0.116	0.028	0.107	0.093	0.070	0.142	0.103	0.104	0.111	0.083
252	0.100	0.048	0.009	0.033	0.139	0.086	0.116	0.043	0.083	0.122	0.141
256	0.050	0.240	0.157	0.213	0.057	0.070	0.132	0.069	0.042	0.011	0.146
258	—	0.007	—	—	—	—	—	—	—	—	—
260	0.100	0.027	0.111	0.087	0.093	0.065	0.079	0.052	0.042	0.022	0.068
264	0.050	0.164	0.259	0.160	0.072	0.140	0.079	0.086	0.104	0.078	0.042
268	0.050	—	0.009	0.027	0.103	0.124	0.058	0.129	0.198	0.133	0.057
272	—	0.007	0.019	0.007	0.052	0.070	0.021	0.060	0.073	0.067	—
276	—	—	0.009	—	0.015	0.038	0.026	0.086	0.021	0.033	0.016
280	—	—	—	0.007	0.062	0.027	0.047	0.017	0.031	0.033	0.073
284	—	0.007	0.009	0.007	0.026	0.016	0.011	0.018	—	0.067	0.026
288	—	—	—	—	0.010	0.011	0.042	0.009	—	0.011	0.005
292	—	—	—	—	0.005	0.005	0.026	—	—	—	—
296	—	—	—	—	—	—	0.011	0.018	—	—	—
300	—	—	—	—	—	—	0.005	—	0.042	0.011	—
316	—	—	—	—	—	—	0.005	—	—	—	—

Appendix A. Continued

Locus Alleles	Early Issaquah Cr (k)	Lewis Cr (k)	Laughing Jacobs Cr (k)	Ebright Cr (k)	Whatcom at Whatcom (k)	Whatcom at Spokane (k)	Meadow Cr (k)	Bear & L Bear Cr (k)	Issaquah Cr (s)	Bear Cr (s)	Baker Lake (s)
*One-115**											
N	10	65	53	69	97	92	96	59	57	48	97
176	—	—	—	—	0.010	0.016	0.016	0.008	—	0.021	—
180	—	0.008	—	—	0.036	0.038	0.031	0.025	0.061	0.021	—
184	—	0.015	0.019	0.036	0.036	0.022	0.052	0.008	0.079	—	0.026
188	—	0.015	0.028	0.022	0.046	0.043	0.078	0.051	0.026	0.125	—
192	—	0.038	0.009	—	0.057	0.087	0.104	0.042	0.044	0.115	0.005
196	0.150	0.146	0.132	0.036	0.062	0.027	0.167	0.127	0.132	0.125	0.119
200	0.050	0.115	0.019	0.203	0.124	0.141	0.167	0.195	0.158	0.156	0.155
204	0.150	0.015	0.028	0.058	0.206	0.245	0.115	0.043	0.088	0.042	0.098
208	—	0.085	0.075	0.058	0.222	0.179	0.115	0.094	0.070	0.104	0.124
212	0.350	—	0.019	0.022	0.072	0.060	0.073	0.068	0.096	0.063	0.082
216	—	0.100	0.047	0.065	0.021	0.038	0.042	0.144	0.132	0.125	0.253
220	0.150	0.054	0.075	0.014	0.031	0.038	0.010	0.068	0.053	0.021	0.134
224	0.050	0.292	0.434	0.304	0.041	0.033	0.005	0.043	0.018	0.021	—
226	—	0.008	—	—	—	—	—	—	—	—	—
228	0.100	0.085	0.085	0.116	0.031	0.022	0.016	0.059	0.035	0.031	0.005
230	—	0.008	—	—	—	—	—	—	—	—	—
232	—	0.015	0.028	0.065	0.005	0.011	—	0.008	—	0.010	—
236	—	—	—	—	—	—	0.005	0.017	0.009	0.021	—
256	—	—	—	—	—	—	0.005	—	—	—	—
*One-101**											
N	12	77	50	80	95	97	95	56	61	37	79
173	0.042	—	—	—	—	—	0.005	—	—	—	—
177	—	—	—	—	0.005	0.005	—	—	—	—	—
181	—	—	0.010	—	0.021	0.005	—	—	—	—	—
185	—	—	—	—	—	0.015	0.011	—	—	—	0.013
189	—	—	—	0.019	0.021	0.010	0.026	0.044	0.016	—	—
193	—	—	—	—	0.037	0.015	—	0.036	—	0.014	0.013
197	0.167	—	—	—	0.063	0.041	—	—	—	—	—
201	—	—	0.010	—	0.021	0.077	0.026	0.009	0.008	—	—
205	—	—	0.030	0.013	0.047	0.052	0.032	0.018	0.016	0.027	—
209	0.042	0.019	0.020	0.006	0.058	0.046	0.005	0.036	0.008	—	—
213	0.125	0.097	0.090	0.063	0.021	0.036	—	—	0.025	0.041	—
217	—	0.039	0.080	0.063	0.037	0.026	0.005	0.036	0.008	0.027	0.038
221	—	—	—	—	0.047	0.067	0.005	0.009	0.016	—	—
225	—	—	—	—	0.037	0.015	—	0.009	0.008	—	0.013
229	—	—	—	0.006	0.053	0.046	0.005	0.009	0.016	0.014	—
233	—	—	—	—	0.021	0.010	—	0.027	—	—	0.006
237	—	—	—	—	0.026	0.021	0.021	0.045	—	—	—
241	—	—	—	—	0.011	0.031	—	—	—	—	—
245	—	—	—	—	0.005	0.021	0.011	—	—	—	—
249	—	—	—	—	0.011	0.010	—	0.018	—	—	0.006
253	—	—	—	—	0.005	0.010	0.011	0.009	—	0.014	—
257	—	—	—	—	0.005	0.010	0.005	0.009	0.008	—	—
261	—	—	—	—	—	—	0.005	—	—	—	—
265	—	—	—	—	0.005	—	0.011	—	—	—	—
269	—	—	—	—	0.005	0.010	0.032	—	—	0.027	—
273	—	0.013	0.020	—	0.011	0.005	0.032	0.009	—	0.014	—
277	—	—	—	—	0.032	0.010	0.026	—	—	—	—
281	—	0.013	—	0.006	—	0.010	0.068	—	0.033	0.027	—
285	0.083	0.006	0.040	0.050	0.021	0.010	0.042	0.009	0.082	0.014	—

Appendix A. Continued

Locus Alleles	Early Issaquah Cr (k)	Lewis Cr (k)	Laughing Jacobs Cr (k)	Ebright Cr (k)	Whatcom at Whatcom (k)	Whatcom at Spokane (k)	Meadow Cr (k)	Bear & L Bear Cr (k)	Issaquah Cr (s)	Bear Cr (s)	Baker Lake (s)
289	—	0.097	—	0.013	0.016	0.021	0.063	0.018	—	0.054	0.057
293	—	0.019	—	0.019	0.026	0.026	0.037	0.009	0.033	0.014	0.006
297	—	0.006	0.020	—	0.047	0.031	0.042	0.054	0.098	—	0.019
301	—	—	—	0.013	0.016	0.031	0.053	0.098	0.057	0.068	0.108
305	—	0.175	0.250	0.188	0.047	0.036	0.074	0.107	0.148	0.041	0.133
309	0.083	0.156	0.040	0.044	0.053	0.036	0.079	0.054	0.098	0.108	0.063
313	0.042	0.104	0.070	0.094	0.047	0.021	0.037	0.045	0.098	0.041	0.057
317	—	—	—	—	0.011	0.052	0.026	0.098	0.049	0.135	0.108
321	0.208	0.045	0.040	0.075	0.011	0.010	0.042	0.009	0.016	0.014	—
325	—	0.019	0.020	0.056	0.037	0.026	0.032	0.045	0.066	0.041	0.190
329	0.083	—	0.040	0.006	0.032	0.031	0.032	0.054	—	0.176	0.139
333	—	0.052	0.120	0.156	0.021	0.031	0.042	—	0.025	0.014	0.019
337	—	0.039	0.040	0.038	0.005	0.031	0.005	0.027	0.016	0.068	—
341	—	0.058	0.030	0.019	—	—	0.021	0.026	0.049	—	—
345	—	0.019	0.010	0.006	—	—	0.005	0.026	—	0.014	—
349	—	0.013	0.020	—	—	—	0.005	—	—	—	—
353	0.042	0.006	—	—	0.005	—	0.016	—	—	—	0.013
357	0.083	—	—	0.050	—	—	0.005	—	—	—	—

*One-105**

Locus Alleles	Early Issaquah Cr (k)	Lewis Cr (k)	Laughing Jacobs Cr (k)	Ebright Cr (k)	Whatcom at Whatcom (k)	Whatcom at Spokane (k)	Meadow Cr (k)	Bear & L Bear Cr (k)	Issaquah Cr (s)	Bear Cr (s)	Baker Lake (s)
N	11	76	53	76	96	97	95	55	60	30	78
130	—	0.007	—	0.007	—	—	—	0.036	0.008	0.050	0.013
134	0.636	0.493	0.349	0.362	0.120	0.149	0.258	0.273	0.342	0.450	0.199
138	0.182	0.342	0.443	0.296	0.177	0.268	0.595	0.191	0.150	0.150	0.167
142	0.136	0.125	0.179	0.316	0.344	0.366	0.089	0.445	0.342	0.283	0.500
146	0.045	0.033	0.028	0.020	0.354	0.206	0.042	0.055	0.158	0.067	0.122
150	—	—	—	—	0.005	0.010	0.016	—	—	—	—

*Ots-103**

Locus Alleles	Early Issaquah Cr (k)	Lewis Cr (k)	Laughing Jacobs Cr (k)	Ebright Cr (k)	Whatcom at Whatcom (k)	Whatcom at Spokane (k)	Meadow Cr (k)	Bear & L Bear Cr (k)	Issaquah Cr (s)	Bear Cr (s)	Baker Lake (s)
N	10	87	53	94	91	86	87	64	55	42	99
134	—	—	—	—	—	0.023	0.006	—	—	—	—
142	—	—	—	—	—	—	0.006	—	—	—	—
150	—	—	—	—	0.016	—	0.034	—	—	—	—
154	—	0.006	—	—	0.011	0.012	0.017	—	—	—	0.005
158	—	—	—	—	0.049	0.047	0.017	—	0.009	—	—
162	—	0.017	0.019	0.027	0.093	0.081	0.057	0.070	0.018	0.036	0.005
166	—	0.092	0.057	0.112	0.099	0.128	0.069	0.055	0.118	0.083	0.015
170	—	0.040	0.142	0.032	0.099	0.140	0.034	0.047	0.127	0.095	0.076
174	0.050	0.086	0.009	0.069	0.132	0.081	0.063	0.015	0.136	0.024	0.010
178	0.050	0.029	0.066	0.085	0.044	0.041	0.075	0.117	0.145	0.095	0.303
182	—	0.190	0.292	0.271	0.033	0.052	0.075	0.094	0.027	0.036	0.217
186	0.300	0.098	0.028	0.048	0.066	0.070	0.098	0.141	0.100	0.048	0.045
190	0.350	0.149	0.132	0.059	0.110	0.116	0.040	0.071	0.064	0.071	0.030
194	0.100	0.092	0.057	0.112	0.082	0.087	0.052	0.094	0.045	0.083	0.192
198	0.100	—	0.019	—	0.038	0.041	0.121	0.086	0.073	0.119	0.086
202	—	0.011	0.019	0.011	0.082	0.064	0.086	0.094	0.036	0.131	—
206	0.050	0.046	0.094	0.027	0.038	0.006	0.029	0.023	0.009	0.071	0.010
210	—	0.109	0.066	0.128	—	—	0.029	0.071	0.009	0.060	0.005
214	—	0.034	—	0.021	—	—	0.052	0.015	0.073	0.036	—
218	—	—	—	—	0.005	0.006	0.017	0.008	0.009	0.012	—
222	—	—	—	—	—	0.006	0.006	—	—	—	—
226	—	—	—	—	—	—	0.011	—	—	—	—
234	—	—	—	—	—	—	0.006	—	—	—	—

Environmental Biology of Fishes **69**: 81–93, 2004.
© 2004 *Kluwer Academic Publishers. Printed in the Netherlands.*

Genetic structure of wild chinook salmon populations of Southeast Alaska and northern British Columbia

Charles M. Guthrie III & Richard L. Wilmot
NMFS Auke Bay Laboratory, 11305 Glacier Highway, Juneau, AK 99801, U.S.A.
(e-mail: chuck.guthrie@noaa.gov)

Received 17 April 2003   Accepted 19 April 2003

Key words: allozyme, postglacial colonization, stock identification

Synopsis

Allozyme variation was used to examine population genetic structure of adult chinook salmon, *Oncorhynchus tshawytscha*, collected between 1988 and 1993 from 22 spawning locations in Southeast Alaska and northern British Columbia. Thirty-five loci and two pairs of isoloci were variable, and of these, 25 loci and one pair of isoloci expressed the most abundant allele with a frequency of less than or equal to 0.95 in at least one collection. A neighbor-joining (NJ) tree of genetic distances defined five regional groups: (1) King Salmon River (the only island collection), which has large allelic frequency differences from other populations in this study; (2) heterogeneous coastal populations from southern southeast Alaska; (3) transmountain collections from the Taku and Stikine Rivers on the eastern side of the coastal mountain range; (4) Chilkat River in northern Southeast Alaska; and (5) northern coastal Southeast Alaska, which consists of the Situk River and the Klukshu River, a tributary of the Alsek River. A second NJ tree that included collections from the Yukon River and British Columbia did not reveal any strong genetic similarity between Southeast Alaska and the Yukon River. The data suggest that Southeast Alaska may have been colonized from both northern and southern refugia following the last glaciation – a period of sufficient time to allow for isolation by distance to occur.

Introduction

Chinook salmon, *Oncorhynchus tshawytscha*, is a valuable sport and commercial species in Southeast Alaska and northern British Columbia. It has also been the focus of Pacific Salmon Treaty negotiations because intercept fisheries on both sides of the border may harvest stocks from both countries. Recovery of coded-wire tagged chinook salmon caught as by-catch in domestic and foreign trawl fleets shows that fish from Southeast Alaska and northern British Columbia migrate well out into the North Pacific Ocean and the Bering Sea where they are vulnerable to both legal and illegal fishing (Healy 1991). Efforts are underway in both the U.S. and Canada to develop a genetic baseline for spawning populations of chinook salmon for use in determining the origins of fish harvested in interception fisheries near the U.S./Canada border (transboundary rivers, and between Southeast Alaska and northern British Columbia), and as by-catch in the North Pacific and Bering Sea trawling fleets (Teel et al.,[1] Wilmot et al.[2]).

Of the approximately 2 000 watersheds in Southeast Alaska, fewer than 40 support chinook salmon, and most of the major spawning areas are widely separated (Heard et al. 1995). Genetic similarities among populations of a species can be used to detect contemporary and historic relationships among them.

---

[1] Teel, D.J., P.A. Crane, C.M. Guthrie III, A.R. Marshall, D.M. Van Doornik, W.D. Templin, N.V. Varnavskaya & L.W. Seeb. 1999. Comprehensive allozyme database discriminates chinook salmon around the Pacific Rim. NPAFC document 440. 25 pp.

[2] Wilmot, R.L., C.M. Kondzela, C.M. Guthrie III, A. Moles, Jerome J. Pella & Michele Masuda. 2000. Origins of salmon seized from the F/V *Arctic Wind*. (NPAFC Doc. 471) Auke Bay Fisheries Laboratory, Alaska Fisheries Science Center, NMFS, NOAA, 11305 Glacier Highway, Juneau, AK 99801-8626. 18 pp.

The rivers of Southeast Alaska and the transboundary rivers that flow from northern British Columbia through Southeast Alaska support stocks of 'stream-type' chinook salmon that spend 1 year in freshwater before migrating to sea. A single 'ocean-type' population, which migrates to sea in its first year, occurs in the Situk River (Johnson et al. 1992). Teel et al. (2000) examined genetic structure of chinook salmon collections from British Columbia, and observed differences between inland stream-type and coastal ocean-type collections. Gharrett et al. (1987) found that chinook salmon populations from Southeast Alaska were genetically more heterogeneous than populations from elsewhere in Alaska, which presumably reflects the low gene flow among, or multiple founding sources for the Southeast Alaska populations. They also speculated that chinook salmon from Southeast Alaska may have originated from the Yukon River drainage through headwater capture of the transmountain rivers flowing out of northern British Columbia. There is geological evidence that the Taku River was blocked periodically and backed up into the Yukon River drainage (Kerr 1948). The Alsek River was also similarly connected to the Yukon River drainage in the past via Kluane Lake and the White River (Lindsey et al. 1981).

The objectives of this study were to use allozyme variation to describe the genetic structure of chinook salmon populations in Southeast Alaska and the transboundary rivers flowing out of northern British Columbia, and to investigate possible historic relationships among some Alaskan populations and chinook salmon from other areas.

Methods

Between 1988 and 1993, 37 collections of tissue samples were taken from adult chinook salmon at 22 spawning locations in Southeast Alaska and northern British Columbia (Table 1, Figure 1). Between late July and early September, fish were sampled at weirs, carcass weirs, stream spawning areas, and during egg takes for hatchery propagation. Ten locations were sampled in multiple years (Table 1). Whole eyes and samples of liver, cheek muscle, and heart were collected from each adult fish, cooled with frozen gel–ice, subsequently frozen at $-20°C$, and shipped to the Auke Bay Laboratory where they were stored at $-80°C$ until analyzed. Liver was not collected from the Farragut River samples.

Table 1. Group designation (letters correspond to spawning areas listed in Figures 1 and 2 & Appendices 1 and 2 available at ftp.afsc.noaa.gov/sida/chinook/) location, date of collection, and sizes of chinook salmon collections (N), used for electrophoretic analysis. Districts are Alaska Department of Fish and Game Statistical Areas.

Group	Location	Date	N
District 101–Boca De Quadra			
A.	Keta R.	Sep. 5–6, 89	15
West Behm canal			
B.	Chickamin R. (pooled)		151
	Chickamin South Fork	Aug. 24, 89	66
	Leduc R.	Aug. 24, 89	30
	Humpy C.	Aug. 24, 89	4
	Chickamin South Fork	Aug. 21, 90	15
	Leduc R.	Aug. 21, 90	25
	Chickamin R.	Aug. 21, 90	11
Unuk R. drainage			
C.	Clear C.	Aug. 30, 89	33
D.	Cripple C.	Aug. 16, 88	121
E.	Gene's Lake C.	Aug. 30, 89	67
District 107–Bradfield canal			
F.	Harding R.	Aug. 16,22,26, 89	45
District 108–Stikine R. drainage			
G./H.	Lower Stikine (pooled)		75
G.	North Arm C.	Aug. 17, 89	18
H.	Andrews C.	Aug. 17, Sep. 9, 89	57
I.	Shakes C.	1993	29
J.	Little Tahltan R. (pooled)		228
	Little Tahltan R.	Aug. 89	101
	Little Tahltan R.	Aug. 7–8, 90	50
	Little Tahltan R.	Jul., Aug. 91	77
District 110–Farragut Bay			
K.	Farragut R.	Aug. 22, 89	8
District 111–Northwest Admiralty Island			
L.	King Salmon R. (pooled)		86
	King Salmon R.	Jul. 25,30, 88	37
	King Salmon R.	Jul. 24,31, 89	31
	King Salmon R.	Jul. 24,31, 90	18
Taku R. drainage			
M.	Nakina R. (pooled)		198
	Nakina R.	Aug. 14–24, 89	104
	Nakina R.	Aug., 90	94
N.	Kowatua C.(pooled)		190
	Kowatua C.	Aug. 25,29, Sep. 1, 89	95
	Kowatua C.	Aug. 29, Sep. 6, 90	95
O.	Tatsatua C. (pooled)		228
	Tatsatua C.	Aug. 30, 89	112
	Tatsatua C.	Aug. 24, Sep. 6, 90	116
P.	Dudidontu R.	Aug. 13, 90	28
Q.	Tseta R.	Aug. 6,10, 89	81
R.	Nahlin R. (pooled)		129
	Upper Nahlin R.	Aug. 1–5, 89	81
	Upper Nahlin R.	Jul. 31, Aug. 5, 90	48

Table 1. (Continued)

Group	Location	Date	N
District 115–Chilkat R. drainage			
S.	Big Boulder C.	Aug. 5–15, 91	27
T.	Tahini R. (pooled)		83
	Tahini R.	Jul. 31, 89	26
	Tahini R.	Aug. 90	48
	Tahini R.	Aug. 2–5, 91	19
District 181–Alsek R. drainage			
U.	Klukshu R. (pooled)		250
	Klukshu R.	Aug. 89	105
	Klukshu R.	Aug. 8, 90	100
	Klukshu R.	Sep. 91	45
Situk R. drainage			
V.	Situk R. (pooled)		174
	Situk R.	Jul. 6, Sep. 90	31
	Situk R.	Jul. 12, Sep. 10, 91	65
	Situk R.	Aug. 29, 92	78

Protein electrophoresis was conducted as described by Aebersold et al. (1987). Specific enzyme activities were stained according to Harris & Hopkinson (1976) or Aebersold et al. (1987). Loci for which data were taken, the tissues in which they were expressed, and the buffer systems with which they were resolved are listed in Table 2.

Conformance to Hardy–Weinberg equilibrium expectations (e.g., Hardy 1908) was tested with Chi-square goodness-of-fit tests. For less variable loci, probabilities of the statistic were estimated using a Monte Carlo simulation [a FORTRAN program written by A.J. Gharrett (unpublished) analogous to Roff & Bentzen (1989)] using 2 000 iterations. Hardy–Weinberg tests were not conducted on co-migrating, duplicated loci (isoloci) or for three loci with dominant allele expression, $GPI\text{-}B2^*60$ allele, $GPIr^*$ and $sMEP\text{-}2^*$, which were scored as dominant–recessive phenotypes. Variation at isoloci was treated as if all the variability was at one locus and the other was monomorphic. This is a conservative treatment for isoloci having relatively low allelic variability (Gharrett & Thomason 1987). We used log-likelihood ratio analysis (Sokal & Rohlf 1981) to test homogeneity of allele frequencies at four levels of hierarchy. For tests that included low-frequency alleles, probability levels were estimated by generating distributions of the G-statistic, a Monte Carlo method analogous to that described for the Hardy–Weinberg goodness-of-fit test (FORTRAN program from A.J. Gharrett, unpubl.). The relative amounts of variation at different levels of hierarchy were test using an approximate F-statistic:

$$F_{df\,among,\,df\,within} = \frac{G_{among}/df_{among}}{G_{within}/df_{within}}$$

Relationships among collections were examined descriptively with neighbor-joining (NJ) trees (Saitou & Nei 1987) constructed from chord distances (Cavalli-Sforza & Edwards 1967) between pairs of populations. Allelic frequencies were estimated for $sMEP\text{-}2^*$ and $GPI\text{-}B2^*$ from the alternate homozygotes ($GPIr^*$ is monomorphic) assuming Hardy–Weinberg equilibrium. Significance of nodes in the NJ tree were inferred from test of heterogeneity among branches joined at the nodes (Chakraborty & Leimar 1987). Monte-Carlo simulations were conducted to assess the influence of sampling error of allele frequency estimates on the topology of trees (Hawkins et al. 2002). The entire data set was iteratively (20 000 trials) regenerated by resampling with replacement alleles from each population, and NJ trees were constructed for each iteration. A consensus tree was determined from these 20 000 trees using CONSENSE from PHYLIP (Felsenstein[3]). The stability of the topology is shown in the proportion of trees that share specific nodes. We used Arlequin (Schneider et al.[4]) to describe hierarchical genetic structure within (F_{SC}) and among (F_{CT}) regional groupings described by the NJ tree. The significance of the F-statistics was estimated from distributions of the statistics generated by 17 000 permutations. Three loci were dropped from this analysis: $GPI\text{-}2^*$, $GPIr^*$ and $sMEP\text{-}2^*$ since heterozygotes could not be detected for these loci.

A second NJ tree, using the 32 loci and three isolocus pairs available for all three data sets (Table 2; $GPI\text{-}B2^*60$ allele was pooled with the common allele) and incorporating 13 additional populations from British Columbia (Teel et al. 2000) and 12 from the upper Yukon River (Wilmot et al.,[5] Wilmot unpubl. data), was constructed to examine the genetic

[3] Felsenstein, J. 1993. PHYLIP (*Phylogeny Inference Package*) Version 3.60a, Seattle WA, Department of Genetics, University of Washington.

[4] Schneider, S., J.-M. Kueffer, D. Roseli & Laurant Excoffier. 2000. Arlequin version 2.0: A Software for Population Genetic Data Analysis, Genetics and Biometry Laboratory, University of Geneva, Switzerland.

[5] Wilmot, R.L., R.J. Everett, W.J. Spearman & R. Baccus. 1992. Genetic stock structure of Yukon River chum and chinook salmon 1987 to 1990. Progress Report, Alaska Fish and Wildlife Research Center, U.S. Fish and Wildlife Service, U.S. Department of Interior, 1011 East Tudor Road, Anchorage AK, 99503. 132 pp.

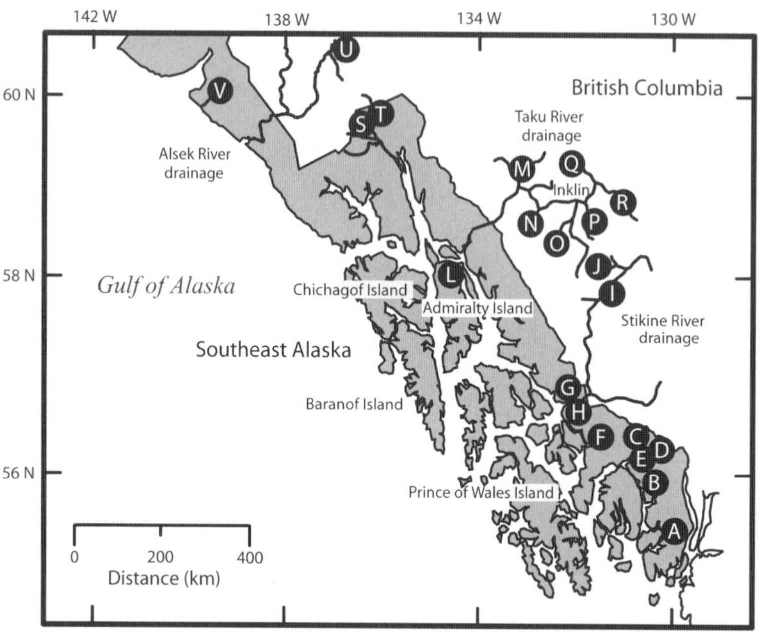

Figure 1. Sampling sites for chinook salmon in Southeast Alaska and northern British Columbia. Letters correspond to collection sites listed in Table 1.

Table 2. Protein coding loci (Shaklee et al. 1990) for enzymes resolved in this study, Enzyme Commission (EC) numbers (IUBMBNC 1992), and the tissues and buffer in which they were resolved. Peptidase loci are designated according to their substrate specificity. L = liver, H = heart, M = muscle, E = eye.

Enzyme	EC number	Locus	Tissue	Buffer[a]	Variability level[b]
Aconitate hydratase	4.2.1.3	*sAH-1**	L	B	4, 5
Adenosine deaminase	3.5.4.4	*ADA-1**	M,H	G,F	4, 5
		*ADA-2**	M,H	G,F	3, 5
Alanine aminotransferase	2.6.1.2	*ALAT**	M,H	G,E	4
Aspartate aminotransferase	2.6.1.1	*mAAT-1**	M,H	B,C,F	4, 5
		*sAAT-1,2**	M,H	B,C	4, 5
		*sAAT-3**	E	A	4
Creatine kinase	2.7.3.2	*CK-C1**	E	G,E	1
		*CK-C2**	E	G,E	4
		*CK-B**	E	G,E	1
Formaldehyde dehydrogenase (glutathione)	1.2.1.1	*FDHG**	H,E,L,M	G	4, 5
Fructose-biphosphate aldolase	4.1.1.13	*FBALD-3**	E	A	1
Glucose-6-phosphate isomerase	5.3.1.9	*GPI-B1**	H	D,G	1, 5
		*GPI-B2**	H	D,G	4, 5
		*GPI-A**	H	D,G	4, 5
		*GPI-r**	H	D,G	1
Glyceraldehyde-3-phosphate dehydrogenase	1.2.1.12	*GAPDH-2**	H	C	3
		*GAPDH-3**	H	C	3

Table 2. (Continued)

Enzyme	EC number	Locus	Tissue	Buffer[a]	Variability level[b]
Glycerol-3-phosphate	1.1.1.8	*G3PDH-1**	M,H	B,C	1
dehydrogenase		*G3PDH-2**	M,H,	B,C	1
		*G3PDH-3**	M,H,	B,C	3
		*G3PDH-4**	M,H	B,C	1
Glutathione reductase	1.6.4.2	*GR**	M,H	F	4, 5
β-N-Acetylhexosaminidase	3.2.1.30	*β-HEX**	L	F	4, 5
L-Iditol dehydrogenase	1.1.1.14	*IDDH-1**	L,H	D	4
		*IDDH-2**	L,H	D	1
Isocitrate dehydrogenase	1.1.1.42	*mIDHP-1**	M,H	B,F	2, 5
		*mIDHP-2**	M,H	B,F	3, 5
		*sIDHP-1**	M,H,L,E	A,B,F	4, 5
		*sIDHP-2**	H,L,E	A,B,F	4, 5
Lactate dehydrogenase	1.1.1.27	*LDH-B1**	M,H,E	G,D	3, 5
		*LDH-B2**	M,H,L,E	G,D	1, 5
		*LDH-C**	E	G,D	1, 5
Malate dehydrogenase	1.1.1.37	*sMDH-A1,2**	M,H	B,C	1, 5
		*sMDH-B1,2**	H,M,L	B,C	4, 5
		*mMDH-1**	H,M	B,C	1
		*mMDH-2**	H,M	B,C	4, 5
		*mMDH-3**	H,M	B,C	1
Malic enzyme	1.1.1.40	*sMEP-1**	M,H	F	4, 5
		*sMEP-2**	M,H	F	4
Mannose-6-phosphate isomerase	5.3.1.8	*MPI**	M,H,E	G,F	4, 5
Peptidases	3.4.*.*				
Dipeptidase	3.4.13.18	*PEPA**	M,H,E	D,G,C	4, 5
Glycl-leucine activity		*PEPC**	E	E	1
Tripeptide aminopeptidase	3.4.11.4	*PEPB-1**	M,H,E,L	F,G	4, 5
Leucyl-glycl-glycine activity		*PEPB-2**	M,H	F,G	3
Leucyl-tyrosine activity	3.4.11.*	*PEP-LT**	M,H	D,F	4, 5
X-Pro aminopeptidase	3.4.13.9	*PEPD-1**	M,H,E	C,D,F	1, 5
Phenylalanyl-proline activity		*PEPD-2**	M,H,E	C,D,F	1
Phosphoglucomutase	5.4.2.2	*PGM-1**	M,H	G	3, 5
		*PGM-2**	M,H	G	1, 5
6-Phosphogluconate dehydrogenase	1.1.1.44	*PGDH**	M,H,E,L	A,B	1, 5
Phosphoglycerate kinase	2.7.2.3	*PGK-1**	E,M,H	A,B	1, 5
		*PGK-2**	E,M,H	A,B	4, 5
Superoxide dismutase	1.15.1.1	*mSOD-1**	M,H	D,F,G	4, 5
		*sSOD-1**	M,H,E	D,F,G	4, 5
Triose-phosphate isomerase	5.3.1.1	*TPI-1**	M,E,H	G	1
		*TPI-2**	M,H	G	1
		*TPI-3**	M,E,H	G	4, 5
		*TPI-4**	M,E,H	G	4, 5

[a] A = CA 7.0 (Clayton and Tretiak 1972), B and C = CAME 7.0 and 7.0-NAD (Aebersold et al. 1987), D = R(Ridgeway et al. 1970), E = MF (Markert & Faulhaber 1965), F = TC-4 (buffer 'a' of Schaal and Anderson 1974), G = TG (Holmes and Masters 1970).

[b] 1 = monomorphic; 2 = rarely variable; most abundant allele ≥ 0.99; 3 = variable; most abundant allele < 0.99 but > 0.95; 4 = highly variable; most abundant allele ≤ 0.95; 5 = used in broad region analyses.

relationship with populations from Southeast Alaska. To test for isolation by distance and to assess the effects of gene flow and drift, the F_{ST} between all pairs of 10 coastal populations was calculated by the method of Weir & Cockerham (1984). The effective number of migrants ($N_e m$) per generation, was calculated between all pairs of populations from Wright's (1931) equation $F_{ST} = 1/(4N_e m + 1)$. The $\log_{10}(N_e m)$ was then regressed against the $\log_{10}$(distance) measured in kilometers between each pair of populations (Slatkin 1993). Distances were measured by water between the mouths of each drainage. For populations with g subdivisions the equation:

$$F_{ST} = \frac{1}{4N_e m \, (g/(g-1))^2} - 1$$

was used to assess divergence among subdivisions.

Results

We obtained data from 62 protein coding loci in the adult chinook salmon samples (Table 2). Variation was observed at 35 loci and two isolocus pairs, with the frequency of the most abundant allele was less than or equal to 0.95 in at least one collection at 25 loci and one isolocus pair. The frequency of the most abundant allele at four loci and one isolocus pair was between 0.95 and 0.99 in at least one collection, and was less than 0.99 in at least one collection at six loci. Twenty-one loci and one isolocus pair were monomorphic. Out of 703 tests possible for these collections, 22 did not conform to Hardy–Weinberg expectations, three of which remained significant when a sequential Bonferroni adjustment was applied (Rice 1989). These were: *LDHB1** in [Q] Tseta Creek 1989; *PEPLT** in [J] Little Tahltan River 1989; and *mAAT-1* in [D] Cripple Creek 1988. The departures from Hardy–Weinberg were due to a paucity of heterozygotes. Allele frequency appendices for polymorphic and monomorphic loci are available for download at ftp.afsc.noaa.gov/Sida/chinook/.

Multiple collections from different years were taken from 10 locations. *G*-tests indicated that seven of these locations were temporally heterogeneous (Table 3). We followed Waples (1990) recommendation for pooling temporal samples from the same location, when the differences were not too large to be attributed to sampling error or genetic drift. Temporal collections from the same location were pooled for analysis (Table 1), and we also pooled multiple, homogeneous collections within the Chickamin River drainage, as well as Andrews and North Arm Creeks, which are on opposite banks of the lower Stikine River (Tables 1 and 3).

Combined heterogeneity for among river drainages exceeded heterogeneity within drainages at 26 loci and two isolocus pairs (Table 3), indicating that divergence between rivers within drainages was less than divergence among drainages ($F_{333, 962} = 10.66$; p $= 2.11 \times 10^{-182}$). Significant heterogeneity (p < 0.05) was observed within every river drainage except the Chickamin River. The Taku, Stikine, and Chilkat rivers exhibited heterogeneity at three, two, and one additional levels of hierarchy, respectively. A complete appendix of the hierarchal log-likelihood analysis is available for download at ftp.afsc.noaa.gov/Sida/chinook/.

An unrooted tree (Figure 2) constructed from chord distances by NJ identified five regional groupings of chinook salmon collections (letters from Table 1 and Figure 1): (1) southern coastal Southeast Alaska ([A] Keta River, [B] Chickamin River, [C] Clear Creek, [D] Cripple Creek, [E] Gene's Lake, [F] Harding River, and [G] Lower Stikine River); (2) [L] King Salmon River; (3) transmountain ([I] Shakes Creek, [J] Little Tahltan River, [M] Nakina River, [N] Kowatua Creek, [O] Tatsatua Creek, [P] Dudidontu River, [Q] Tseta Creek, and [R] Upper Nahlin River; (4) Chilkat River ([S] Big Boulder Creek, and [T] Tahini River); and (5) northern coastal Southeast Alaska ([U] Klukshu River and [V] Situk River). The same five regional groupings of chinook salmon were identified in a principal component analysis (not shown). The consensus tree shared the same topology with the point estimate of the NJ tree with one exception; Chickamin River and Clear Creek populations positions were reversed on the consensus tree. The consensus tree represents the stability of the position of nodes on the trees, but lacks the genetic distances of the point NJ tree. Because the point estimate and consensus trees were nearly coincidental we were able to label the nodes with the bootstrap values. The majority of the nodes had bootstrap values >50%, which indicate stability in the tree structure. A gene diversity analysis, based on five geographic areas inferred from the NJ tree, estimated that 29.5% ($F_{SC} = 0.01692$; p $\ll 1 \times 10^{-4}$) of the genetic diversity was within regions, while 70.5% ($F_{CT} = 0.04051$; p $\ll 1 \times 10^{-4}$) was among regions, which suggests greater gene flow within than among regions.

A second NJ tree (Figure 3) was used to examine the relationships among Southeast Alaska/northern

Table 3. Log-likelihood ratio analysis of regional genetic structure of chinook salmon from Southeast Alaska and northern British Columbia for a subset of loci – the full analysis is available at ftp://ftp.afsc.noaa.gov/Sida/chinook/. Probability levels (*p < 0.05, **p < 0.01, ***p < 0.005) were based on the distribution of G-statistics for each test based on Monte-Carlo resampling. G is the log-likelihood statistic and d.f. the degrees of freedom. Values of d.f. in parentheses are in the totals for all alleles in all populations and the d.f. shown are from the alleles observed in each test.

Source	Locus															
	*ADA-1**		*GPIB-2**		*IDDH-1**		*sIDHP-1**		*sMEP-1**		*sMEP-2**		Total		Grand total	
	d.f.	G	d.f.	G	d.f.	G	d.f.	G	d.f.	G	d.f.	G	d.f.	G	d.f.	G
Within Situk River	2	5.330	2		2	2.639	2	4.370	2	15.469***	2	2.867	12	30.675***	(74) 42	64.241*
Within Klukshu River	2	6.190*	2		2	0.240	2	0.223	2	21.615***	2	5.450	12	33.718***	(74) 38	64.918***
Within Chilkat River	3		3		3	15.461**	3	1.704	3	3.293	3	2.876	18	23.334	(111) 57	92.78***
Within Tahini River					2	12.941**	2	0.828	2	1.737	2	2.687	8	18.193*	28	38.346
Between Big Boulder and Tahini					1	2.520	1	0.876	1	1.556	1	0.189	4	5.141	19	54.434***
Within Taku River	9	26.471**	9	4.531	9	34.822***	9	14.001	9	13.732	9	19.188*	54	112.745***	(333) 306	486.658***
Between Nakina and Inklin	1	0.401	1	2.976	1	0.022	1	1.171	1	1.414	1	0.016	6	6.000	34	65.774***
Within Nakina River	1	7.696**	1	1.555	1	2.547	1	0.736	1	1.059	1	3.459	6	17.052**	30	42.704
Within Inklin River	7	18.374*	7		7	32.253***	7	12.094	7	11.259	7	15.714*	42	89.694***	238	378.183***
Among Kowatua, Tatsatua, Nahlin	2	5.192	2		2	18.344**	2	0.948	2	5.256	2	3.914	12	33.654***	72	165.706***
Within Kowatua	1	1.389	1		1	1.378	1	0.002	1	0.163	1	5.312*	6	8.244	32	49.56*
Within Tatsatua River	1	0.378	1		1	5.682	1	0.635	1	3.297	1	2.614	6	12.606*	33	38.799
Within Nahlin River	3	11.415**	3		3	6.850	3	10.508*	3	2.542	3	3.874	18	35.189**	90	124.119**

Table 3. (Continued)

Source	Locus																
	$ADA\text{-}1^*$		$GPIB\text{-}2^*$		$IDDH\text{-}1^*$		$sIDHP\text{-}1^*$		$sMEP\text{-}1^*$		$sMEP\text{-}2^*$		Total		Grand total	
	d.f.	G	d.f.	G	d.f.	G	d.f.	G	d.f.	G	d.f.	G	d.f.	G	d.f.	G
Among Tseta, U. Nahlin, Dudidontu	2	10.509**	2		2	6.339*	2	10.322**	2	2.541	2	3.836	12	33.547***	60	91.522*
Within upper Nahlin River	1	0.906	1		1	0.511	1	0.186	1	0.001	1	0.039	6	1.643	29	32.597
Within King Salmon River	2	6.945*	2	3.671	2	9.609**	2	1.627	2	3.431	2		12	25.283*	(74) 26	46.801**
Within Stikine River	5	9.351	5		5	28.603***	5	21.723***	5	18.231**	5	2.027	30	79.935***	(185) 160	247.382***
Between upper and lower Stikine	1	5.335	1		1	22.533***	1	8.623	1	15.595***	1	0.030	6	52.116***	32	125.83***
Within lower Stikine	1	0.278	1		1	0.820	1	0.490	1	0.147	1	0.003	6	1.738	28	24.536
Within upper Stikine	3	3.738	3		3	5.251	3	12.610**	3	2.489	3	1.994	18	26.082	93	97.013
Between Shakes and Tahltan	1	1.209	1		1	0.531	1	2.185	1	0.038	1	1.012	6	4.975	30	32.898
Within Tahltan River	2	2.529	2		2	4.720	2	10.425**	2	2.451	2	0.983	12	21.108*	60	64.118
Within Unuk River	2	4.191	2		2	2.308	2	1.653	2	5.331	2	0.402	12	13.885	64	105.108***
Within Chickamin River	1	0.418	1		1	0.017	1	0.445	1	1.839	1	0.828	6	3.547	(37) 32	32.296
Within rivers	26	58.895***	26	8.201	26	93.699***	26	45.746**	26	82.941***	26	33.638	156	323.120***	962	1140.184***
Among rivers	9	849.843***	9	192.383***	9	168.845***	9	41.136***	9	245.704***	9	48.867***	54	1546.778***	333	4204.841***
Total (within + among)	35	908.738***	35	200.584***	35	262.544***	35	86.882***	35	328.645***	35	82.505***	210	1869.898***	1295	5345.024***

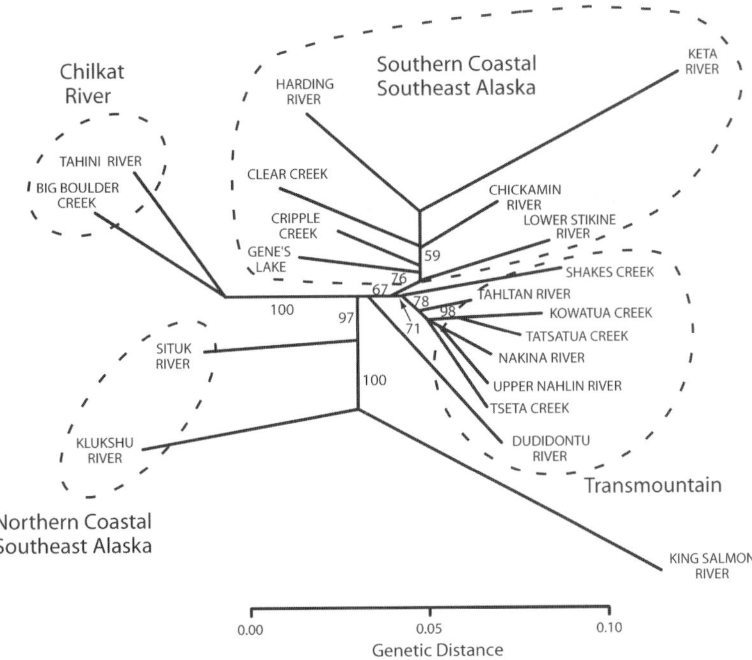

Figure 2. Neighbor-joining tree (Saitou & Nei 1987) using genetic distances (Cavalli-Sforza & Edwards 1967) estimated from allelic frequencies at 62 loci. Numbers on branches are percent consensus tree values from a bootstrap of alleles within populations.

British Columbia populations and those from the Yukon River (Wilmot et al.;[5] Wilmot unpubl. data) and other northern British Columbia populations (Teel et al. 2000). The tree illustrated eight regional groupings: (1) a coastal group which includes central coastal British Columbia (Kitimat, Atnarko, and Wannock) and southern coastal Southeast Alaska (identified from the previous tree); (2) Skeena River (Kitsumkalum, Cedar, Kitiwanga, Bulkey, Morice, Kispiox, Babine, and Bear; (3) Nass River (Damdochax and Cranberry); (4) Yukon River (North Klondike 89 & 90, McQuesten, Blind, Ross, Tatchun, Big Salmon, Little Salmon, Takhini, Stoney, Nitsutlin, and Nulato; and the four of the five previously identified regional groupings: (5) King Salmon River; (6) transmountain; (7) Chilkat River; and (8) northern coastal Southeast Alaska. Two collections are on branches outside their geographical groupings; the Lower Stikine arises on a branch with the Nass and Skeena Rivers, and the Bulkey, part of the Skeena River, is on a separate branch. Fifty percent of all nodes shared among bootstrap topologies had bootstrap values >50%. The consensus tree shared the same topology with the point estimate NJ tree on nodes with >51% bootstrap estimates.

A plot of all pairwise $N_e m$ values for central coastal British Columbia and southern coastal Southeast Alaska against geographic distance shows there is an apparent pattern of isolation by distance (Figure 4). The slope of the regression of $\log 10(N_e m)$ on $\log_{10}$(distance) is −0.299 and the regression explains 39% ($r^2 = 0.385$, p < 0.001) of the variation in the data. Estimation of the average numbers of migrants between rivers within drainages was about 14.5 fish per generation, whereas the number of migrants between drainages was about 3.8. Both estimates assume that the population systems are near a migration-drift equilibrium (Zhivotovsky et al. 1994).

Discussion

Descriptive analyses resolved five regional groups of chinook salmon in Southeast Alaska and northern British Columbia in this study. Four of the five groups remained intact when data from somewhat different sets of alleles were compared to data from other adjacent regions to the north and south. The clusters generally reflect geographic proximity, with several exceptions. The two northern coastal Southeast Alaska

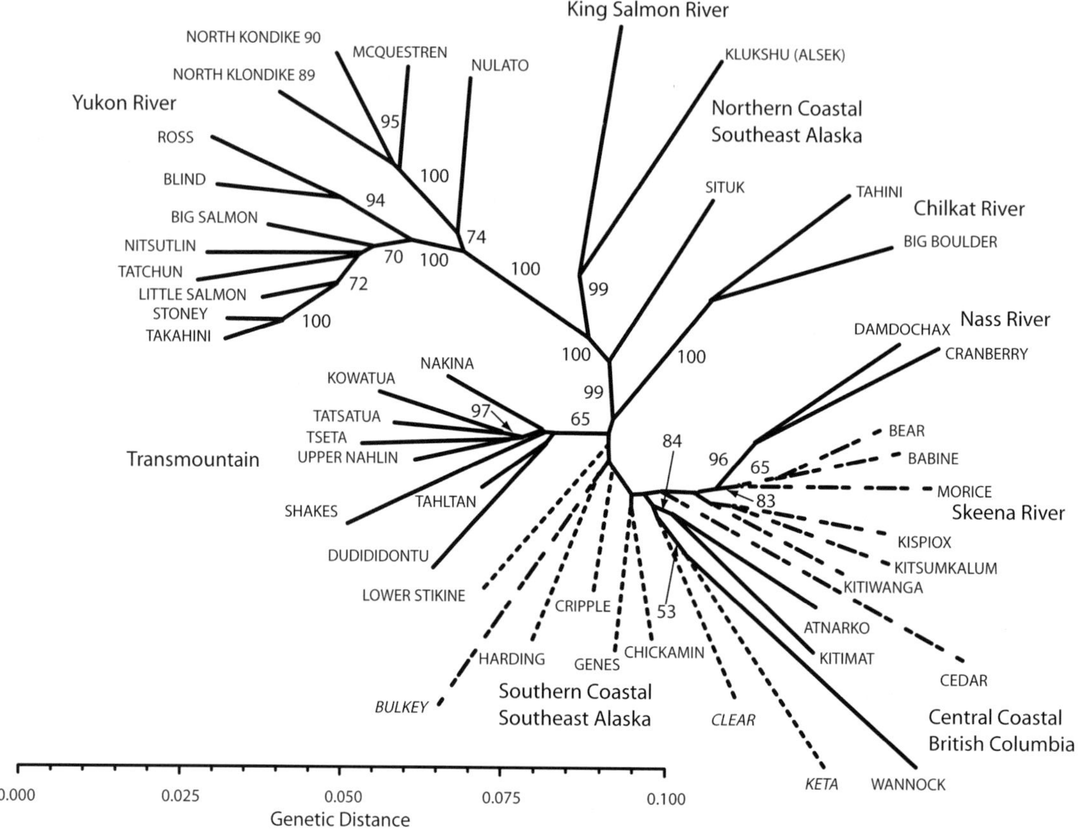

Figure 3. Neighbor-joining tree (Saitou & Nei 1987) using genetic distances (Cavalli-Sforza & Edwards 1967) estimated from allelic frequencies at 34 loci for 45 collections. Numbers on branches are percent consensus tree values from a bootstrap of alleles within populations. Collections which are italicized arise from branches outside their regional grouping.

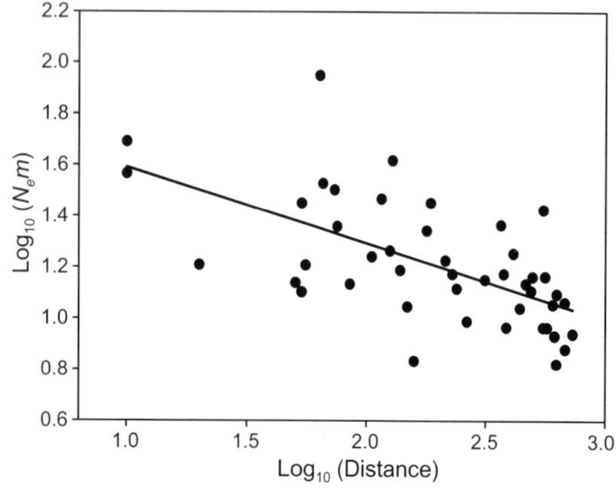

Figure 4. $\log_{10}(N_e m)$ plotted against $\log_{10}$(distance) measured in km for 10 coastal populations of chinook salmon from Southeast Alaska and central British Columbia. The linear regression equation is $\log_{10}(N_e m) = 1.892 - 0.299 * \log_{10}$(distance) with an adjusted $r^2 = 0.385$, $p < 0.001$.

collections do not form a strong cluster, which may be the result of the Situk chinook salmon having an ocean-type life-history (Johnson et al. 1992) with a relatively short distance for juvenile outmigration, whereas the chinook salmon in the Klukshu, a tributary far up the Alsek on the eastern side of the coastal mountain range, are stream-type. The King Salmon River clusters near the Klukshu River on both NJ trees (Figures 2 and 3), however they are separated by a large genetic distance. The King Salmon River which is the only significant (>100 spawners) wild island population in Southeast Alaska (Pahlke[6]), is unique in that its frequency of *GPIB-1*60 allele (58%) far exceeds the frequency in all other systems (the next highest

---

[6] Pahlke, K.A. 2000. Escapements of Chinook Salmon in Southeast Alaska and Transboundary Rivers in 1999, Alaska Department of Fish and Game, Division of Sport Fish, Fishery Data Series No. 00-34. 51 pp.

is 7.2% for Nakina River). The large genetic distance between the King Salmon River collection and the other collections may be a result of early run timing, unique habitat, small population size, and relative isolation from other chinook spawning areas. Wheeler and Greens creeks, drain western Admiralty Island, and may support small chinook salmon runs (Halupka et al.[7]). Although those in Wheeler Creek may be hatchery strays, there are anecdotal reports of historical Native harvests of chinook salmon at Wheeler Creek (Heard[8]). However, the wild population was probably eliminated by indiscriminate harvesting by a cannery at its mouth and its proximity to major fishing grounds (Pahlke[6]). The King Salmon River population is geographically isolated and may have escaped a similar demise.

The Chilkat River collections (Figures 2 and 3) cluster together on the NJ trees. However the large genetic distance between these two collections from this river is not unexpected because they are widely separated geographically (Figure 1). Big Boulder Creek [S], is a clearwater stream that flows into the Klehini River (about 15 km upriver from the confluence with the Chilkat mainstem) (Mecum & Kissner[9]), and the Tahini River [T] flows directly into the Chilkat mainstem ~25 km upriver from the Klehini. The southern Southeast Alaska coastal group clusters on a single branch on the NJ tree (Figure 2), but individual collections are separated by varying genetic distances. This group includes three collections from the Unuk River (Clear Creek, Gene's Lake, and Cripple Creek), which were significantly heterogeneous (Table 3) and could not be pooled, but cluster together in the descriptive analyses. The Keta River is an outlier of this group, but that could be due to the small sample size (N = 15).

The transmountain group consists of collections from the Taku and Stikine rivers east of the coastal mountain range. The genetic similarities of the transmountain group illustrated by the NJ tree (Figure 2) may have

resulted from a headwater capture of the Taku River causing it to flow into the Stikine River. Two geological features of the Taku River support this idea: the Taku and Stikine rivers share the drainage of the Stikine Plateau, hence the upper reaches of these rivers lie close to each other (Kerr 1948). Other biological evidence for exchange between the Taku and the Stikine has been observed in sockeye salmon, which exhibit the same transmountain regional groupings (Guthrie et al.[10]). Kowatua and Tatstua both cluster strongly together, perhaps because they are both associated with glacial fed lakes, and have a later run timing than the other populations on the Inklin branch of the Taku River. Gharrett et al. (1987) suggested that a headwater capture (Kerr 1948, Lindsey et al. 1981) of some transmountain rivers (Taku, Stikine, Alsek [Klukshu], and Chilkat) might have been a mechanism for colonization from the Yukon River. The Alsek and Chilkat rivers do not cluster with the transmountain group in this study (Figures 2 and 3). The NJ tree (Figure 3) does not show evidence of gene flow between the Yukon River and the transmountain group. A study of mitochondrial DNA haplotypes may shed light on this possibility.

All the coastal groups (southern Southeast Alaska and central British Columbia) form a geographic cluster (Figure 3), but with large genetic distances between collections. Wannock Creek in central British Columbia is the most divergent, possibly because it is the most southern of all the collections and has ocean-type (Atnarko and Kitimat are intermediate) life-history (Teel et al. 2000). The Keta River is also divergent, but this may be due to the small collection size. The Nass and Skeena river collections, with the exception of Bulkey which was also a genetic outlier in Teel et al. (2000), clustered strongly together (Figure 3), and these populations would be inland groups as described by Teel et al. (2000). There appears to be a loose grouping of inland and coastal groups. Lower Stikine cluster with the Nass/Skeena group, but is a short distance from the coastal groups with weak topology.

The absence of chinook salmon on Prince of Wales Island in southern Southeast Alaska, which may have harbored some plant and animal populations during the

---

[7] Halupka, K.C., M.D. Bryant, M.F. Willson & F.H. Everest. 2000. Biological characteristics and population status of anadromous salmon in Southeast Alaska. General Technical Report PNW-GTR-468. Portland, OR, U.S. Deptartment of Agriculture, Forest Service, Pacific Northwest Research Station. 255 pp.

[8] William Heard, NMFS Auke Bay Laboratory, 11305 Glacier Highway, Juneau, AK 99801 (pers. comm.).

[9] Mecum, R.D. & P.D. Kissner, Jr. 1989. A study of chinook salmon in Southeast Alaska, Alaska Department of Fish & Game, Fishery Data Series No. 117. 76 pp.

---

[10] Guthrie III, C.M., J.H. Helle, P. Aebersold, G.A. Winans & A.J. Gharrett. 1994. Preliminary report on the genetic diversity of sockeye salmon populations from southeast Alaska and northern British Columbia. U.S. National Marine Fisheries Service, AFSC (Alaska Fisheries Science Center) Processed Report 94-03, Seattle. 109 pp.

92

last ice age (Baichtal et al. 1997), suggests that colonization occurred from regions outside of Southeast Alaska. The overall genetic structure we observed suggests that Southeast Alaska was possibly colonized by chinook salmon from refugia to the north and the south. Their strong regional structure suggests that the Situk, Klukshu, King Salmon, and Chilkat rivers may have been populated by individual colonization events, isolated for a long period, or may have experienced several bottlenecks. Specifically, the King Salmon River may have been originally colonized via Wheeler Creek or Greens Creek on western Admiralty Island, which could explain why these fish arrive in ripe spawning condition. Both creeks may have been part of the same drainage as the King Salmon River since their headwaters are in close geographic proximity and may have been separated as the land rebounded after the last glaciation. The transmountain group may have been colonized from the south from the Nass/Skeena drainages, but has been isolated from these groups for some time. The southern coastal Southeast Alaska group may have been colonized more recently from the central coastal British Columbia populations, given the proximity of these groups on the NJ tree. Although there is genetic similarity between these two regions, there are life-history differences. The central coastal British Columbia populations have intermediate and ocean-type life histories, while the southern coastal Southeast Alaska populations have a stream-type life-history, possibly an example of the life-history plasticity in salmonids. Chinook salmon transplanted to New Zealand have shown life-history plasticity (Unwin & Quinn 1993). One could speculate that cooler northern climate may have caused the life-history change. This would be a useful adaptation to temperature regime shifts. New genetic markers (based on randomly amplified polymorphic DNA) have been developed that discriminate ocean-type and stream-type chinook salmon in the Columbia River basin (Rasmussen et al. 2003). It would be interesting to examine the southern coastal Southeast Alaska populations using these markers in future studies.

The isolation by distance analysis suggests chinook salmon in coastal areas of southern Southeast Alaska and central British Columbia area are at approximate migration-drift equilibrium with relatively high level of gene flow that is somewhat geographically restricted, and that long distance dispersal is not sufficient to prevent isolation by distance.

Acknowledgements

We would like to thank the numerous individuals who collected samples for this study from the following agencies: Alaska Department of Fish & Game, Department of Fisheries and Oceans Canada, and the NMFS Auke Bay Laboratory. We thank Claire Noll formerly of ABL, who performed the initial laboratory analysis; current and former ABL employees Sharon Hawkins, Hanhvan Nguyen, Bichhang Nguyen, John Carswell, and John Pohl who assisted in the laboratory. We also thank to Dr. A.J. Gharrett of University of Alaska Fairbanks school of Fisheries and Ocean Sciences who led the genetics program during the design of this project, and for his editorial input. We thank Christine M. Kondzela, and three anonymous reviewers for their thoughtful comments on this manuscript.

References

Aebersold, P.B., G.A. Winans, D.J. Teel, G.B. Milner & F.M. Utter. 1987. Manual for Starch Gel Electrophoresis: A Method for the Detection of Genetic Variation, U.S. Department of Commerce, NOAA Technical Report NMFS 61. 19 pp.

Baichtal, J., G. Streveler & T. Fifield. 1997. The Geological, Glacial and Cultural History of Southern Southeast, Alaska Geographic, Vol. 24, No. 1, Anchorage: Alaska Geographical Society. 96 pp.

Cavalli-Sforza, L.L. & A.W.F. Edwards. 1967. Phylogenetic analysis: Models & estimation procedures. Evolution 21: 550–570.

Chakraborty, R. & O. Leimar. 1987. Variation within a subdivided population. pp. 89–120. *In*: N. Ryman & F.M. Utter (ed.) Population Genetics and Fishery Management, University of Washington Press, Seattle.

Clayton, J.W. & D.N. Tretiak. 1972. Amino citrate buffer for pH control of starch gel electrophoresis. J. Fish. Res. Board Can. 29: 1169–1172.

Gharrett, A.J., S.M. Shirley & G.R. Tromble. 1987. Genetic relationships among populations of Alaskan chinook salmon (*Oncorhynchus tshawytscha*). Can. J. Fish. Aquat. Sci. 44: 765–774.

Gharrett, A.J. & M.A. Thomason. 1987. Genetic changes in pink salmon (*Oncorhynchus gorbuscha*) following their introduction into the Great Lakes. Can. J. Fish. Aquat. Sci. 44: 787–792.

Hardy, G.H. 1908. Mendelian proportions in a mixed population. Science 28: 49–50.

Harris, H. & D.A. Hopkinson. 1976. Handbook of Enzyme Electrophoresis in Human Genetics, American Elsevier, New York. 120 pp.

Hawkins, S.L., N.V. Varnavskaya, E.A. Matzak, V.V. Efremov, C.M. Guthrie III, R.L. Wilmot, H. Mayama, F. Yamazaki &

A.J. Gharrett. 2002. Population structure of odd-broodline Asian pink salmon and its contrast to the even-broodline structure. J. Fish Biol. 60: 370–388.

Healey, M.C. 1991. Life history of chinook salmon (*Oncorhynchus tshawytscha*). pp. 311–394. *In*: C. Groot & L. Margolis (ed.) Pacific Salmon Life Histories, University of British Columbia Press, Vancouver.

Heard, W., R. Burkett, F. Thrower & S. McGee. 1995. A review of chinook salmon resources in Southeast Alaska and development of an enhancement program designed for minimal hatchery-wild stock interaction. pp. 21–37. *In*: H.L. Schramm, Jr. & R.G. Piper (ed.) 15th International Symposium and Workshop on Uses and Effects of Cultured Fishes in Aquatic Ecosystems, Albuquerque, NM (U.S.A.), 12–17 Mar 1994, American Fisheries Society, Bethesda.

Holmes, R.S. & C.J. Masters. 1970. Epigenetic interconversion of the multiple forms of mouse liver catalase. FEBS Lett. 11: 45–48.

IUBMBNC (International Union of Biochemistry and Molecular Biology, Nomenclature Committee). 1992. Enzyme Nomenclature 1992, Academic Press, Orlando, Florida. 864 pp.

Johnson, S.W., J.F. Thedinga & K.V. Koski. 1992. Life history of juvenile ocean-type chinook salmon (*Oncorhynchus tshawytscha*) in the Situk River, Alaska. Can. J. Fish. Aquat. Sci. 49: 2621–2629.

Kerr, F.A. 1948. Taku River map-area, British Columbia. Geol. Surv. Can. Mem. 248. 84 pp.

Lindsey, C.C., K. Patalas, R.A. Bodaly & C.P. Archibald. 1981. Glaciation and the physical, chemical, and biological limnology of Yukon lakes. Can. Tech. Rep. Fish. Aquat. Sci. 966: 37 pp.

Markert, C.L. & I. Faulhauber. 1965. Lactate dehydrogenase isozyme patterns of fish. J. Exp. Zool. 159: 319–332.

Rassmussen, C., C.O. Ostberg, D.R. Clifton, J.L. Holloway & R.J. Rodriguez. 2003. Identification of a genetic marker that discriminates ocean-type and stream-type chinook salmon in the Columbia River basin. Trans. Am. Fish. Soc. 132: 131–142.

Rice, W.R. 1989. Analyzing tables of statistical tests. Evolution 43: 223–225.

Ridgway, G.J. & S.W. Sherburne & R.D. Lewis. 1970. Polymorphisms in the serum esterases of Atlantic herring. Trans. Am. Fish. Soc. 99: 147–151.

Roff, D.A. & P. Bentzen. 1989. The statistical analysis of mitochondrial DNA polymorphisms: Π^2 and the problem of small samples. Mol. Biol. Evol. 6: 539–545.

Saitou, N. & M. Nei. 1987. The neighbor-joining method: A new method for constructing phylogenetic trees. Mol. Biol. Evol. 4: 406–425.

Schaal, B.A. & W.W. Anderson. 1974. An outline of techniques for starch gel electrophoresis of enzymes from the American oyster (*Crassostrea virginica*) Gmelin. Georgia Mar. Sci. Cent. Tech. Rep. 74-3.

Shaklee, J., F. Allendorf, D. Morizot & G. Whitt. 1990. Genetic nomenclature for protein-coding loci in fish. Trans. Am. Fish. Soc. 119: 2–15.

Slatkin, M. 1993. Isolation by distance in equilibrium and non-equilibrium populations. Evolution 47: 264–279.

Sokal, R.R. & F.J. Rohlf. 1981. Biometry, 2nd edition, W.H. Freeman, San Francisco. 859 pp.

Teel, D.J., G.B. Milner, G.A. Winans & W.S. Grant. 2000. Genetic population structure and origin of life-history types in chinook salmon in British Columbia, Canada. Trans. Am. Fish. Soc. 129: 194–209.

Unwin, M.J. & T.P. Quinn. 1993. Variation in life history patterns among New Zealand chinook salmon (*Oncorhynchus tshawytscha*) populations. Can. J. Fish. Aquat. Sci. 50: 1168–1175.

Waples, R.S. 1990. Temporal changes of allele frequency in Pacific salmon: Implications for mixed-stock fishery analysis. Can. J. Fish. Aquat. Sci. 47: 968–976.

Weir, B.S. & C.C. Cockerham. 1984. Estimating F-statistics for the analysis of population structure. Evolution 38: 1358–1370.

Wright, S. 1931. Evolution in Mendelian populations. Genetics 28: 97–159.

Zhivotovsky, L.A., A.J. Gharrett, A.J. MacGregor, M.K. Glubokovsky & M.W. Feldman. 1994. Gene differentiation in Pacific salmon (*Oncorhynchus* sp.): Facts and models with respect to pink salmon (*O. gorbuscha*). Can. J. Fish. Aquat. Sci. 51(Suppl. 1): 223–232.

Environmental Biology of Fishes **69**: 95–109, 2004.
© 2004 *Kluwer Academic Publishers. Printed in the Netherlands.*

Population structure and stock identification of steelhead trout (*Oncorhynchus mykiss*) in British Columbia and the Columbia River based on microsatellite variation

Terry D. Beacham, Khai D. Le & John R. Candy
*Department of Fisheries and Oceans, Pacific Biological Station, Nanaimo, BC, Canada V9T 6N7
(e-mail: beachamt@pac.dfo-mpo.gc.ca)*

Received 17 April 2003   Accepted 24 April 2003

Key words: DNA analysis, genetic variation, individual identification

Synopsis

Variation at 13 microsatellite loci was surveyed from ∼3 800 steelhead trout, *Oncorhynchus mykiss*, from 51 populations in British Columbia, Washington, and the Columbia River drainage. Mean F_{ST} over all 13 loci and 51 populations was 0.066. Regional structuring of populations was apparent, with Thompson River, upper Fraser River, and Columbia River populations forming distinct groups. In the Nass River, winter-run populations were distinct from the summer-run populations. Significant differences in allele frequencies were observed among regional stock groups at all loci. Analysis of variance components indicated that 5.7% of the total observed variation was distributed among 11 regions, and 2.3% of the variation was among populations within regions. Analysis of simulated mixed-stock samples suggested that variation at the microsatellite loci provided relatively accurate and precise estimates of stock composition for fishery management applications, and this was confirmed by application to actual fishery samples of known origin. Within the Fraser River drainage, individual steelhead trout can be identified to one of the three regions of origin with an accuracy of 94–97%. Microsatellites provided an effective way to determine population structure, and provided reliable estimates of stock composition in mixed-stock fisheries.

Introduction

Steelhead trout, *Oncorhynchus mykiss*, the anadromous form of rainbow trout, are found in all major coastal river systems in British Columbia. Steelhead trout abundance has become of increasing concern to fisheries managers, because the status of many populations in British Columbia may require increased emphasis on conservation (Slaney et al. 1996). The current management focus is to ensure the continued viability and productivity of existing populations, and after conservation requirements for specific populations have been obtained, to allow for limited exploitation of target populations. Steelhead trout are not abundant enough to support directed commercial fisheries, although some may be incidentally caught in salmon fisheries. Directed exploitation is largely restricted to recreational fisheries, although there may be some limited aboriginal harvest in some areas.

Determination of population structure of exploited species is an essential component in successful management of fisheries. Specifically, this information can be used for applications ranging from the determination of appropriate conservation or management units to estimation of stock composition in mixed-stock fisheries. In British Columbia, surveys of genetic variation have been used to describe population structure in steelhead trout (Parkinson 1984, Taylor 1995, Beacham et al. 1999, 2000a, Heath et al. 2001, Hendry et al. 2002). All of these studies, except for Parkinson (1984), have been directed at local populations or regional groups of steelhead trout. Parkinson (1984) surveyed variation at five allozyme loci in primarily juveniles and parr from 73 presumed steelhead trout and rainbow trout populations in British Columbia. Development of highly polymorphic microsatellite loci has improved the resolution of detectable genetic variation among steelhead trout populations. No province-wide survey of steelhead trout population structure utilizing microsatellite variation has yet been reported.

Microsatellites are very useful genetic markers to survey genetic variation among salmonid populations. Non-lethal sampling and the abundance of loci make this method very effective for describing population structure in steelhead trout (Nielsen et al. 1994, 1997, Wenburg et al. 1996, Ostberg & Thorgaard 1999, Hendry et al. 2002). Microsatellite loci can also be very useful in estimating stock composition in mixed-stock fisheries that intercept steelhead trout (Beacham et al. 1999, 2000). We further surveyed microsatellite variation among steelhead trout populations in British Columbia to resolve population structure and further refine associated stock identification applications.

The objectives of this study were to expand a broad-scale survey of population structure of steelhead trout in British Columbia using microsatellite variation and to evaluate potential applications for stock identification among major watersheds (Nass, Skeena, Fraser rivers) in the province. We evaluated whether the new microsatellite variation in a more complete and extensive population survey can provide accurate and precise estimates of stock composition for steelhead trout from major watersheds when they occur together as bycatch in marine mixed-stock fisheries for salmon. We also estimated stock composition of steelhead trout caught in test fisheries in the lower portions of three major watersheds in British Columbia, and the utility of microsatellites for identification of population and region of origin for individual steelhead trout in the Fraser River drainage.

Materials and methods

Collection of DNA samples and laboratory analysis

Using a chelex resin protocol, we extracted DNA from scales, previously collected frozen samples stored at $-20°C$, or a punch of operculum or fin tissue preserved in 90% ethanol. We sampled adult fish from 42 steelhead trout populations in British Columbia, juvenile steelhead trout from two populations in coastal Washington, and from juvenile steelhead trout (except the Clearwater, Lower Salmon, and Kalama populations which were sampled as adults) in seven Columbia

River basin populations as described by Beacham et al. (1999) (Figure 1). The main method of sampling adults was by angling, although enumeration fences were used at some locations in some years. Fishery samples of returning adults in the Nass, Skeena, and Fraser rivers were collected with fish wheels (Nass River) or a gillnet test fishery (Skeena River, Fraser River). General regions, populations within each region, year of sampling, and number of fish analyzed are outlined in Table 1. Further sampling details were outlined by Beacham et al. (1999, 2000a).

For the survey of baseline populations, PCR products at 13 microsatellite loci: *Ogo4* (Olsen et al. 1998), *Oke4* (Buchholz et al. 1999[1]), *Ots1*, *Ots2*, *Ots9* (Banks et al. 1999), *Ots108* (Nelson & Beacham 1999), *Oki10* (Smith et al. 1998), *One101*, *One111*, *One114* (Olsen et al. 2000), and *Omy325* (O'Connell et al. 1997), *Ssa197* (O'Reilly et al. 1996), and *Ssa408* (Cairney et al. 2000) were separated by size on denaturing polyacrylamide gels and allele sizes were determined with the ABI 377 automated DNA sequencer in conjunction with Genescan 3.1 and Genotyper 2.5 software (PE Biosystems, Foster City, CA). Allele frequencies for all samples surveyed at all locations in this study are available at http://www-sci.pac.dfo-mpo.gc.ca/aqua/pages/bgsid.htm.

Data analysis

Each population at each locus was tested for departure from Hardy–Weinberg equilibrium using GDA (Lewis & Zaykin 2001[2]). Each sampling year was analyzed separately for each population, years with fewer than 10 individuals sampled were ignored, and a total of 94 tests conducted for each locus (Table 1). Critical significance levels for simultaneous tests were evaluated using sequential Bonferroni adjustment (Rice 1989). F_{ST} estimates for each locus were calculated with FSTAT (Goudet 1995), and the standard deviation of the estimate for an individual locus was determined by jackknifing over populations and for all loci combined by bootstrapping over loci. All annual samples available from a location were combined to estimate

---

[1] Buchholz, W., S.J. Miller & W.J. Spearman. 1999. Summary of PCR primers for salmonid genetic studies. US Fish. Wild. Serv. Alaska Fish. Prog. Rep. 99–1.

[2] Lewis, P.O. & D. Zaykin. 2001. Genetic Data Analysis: Computer program for the analysis of allelic data. Version 1.0 (d16c). Free program distributed by the authors over the internet from http://lewis.eeb.uconn.edu/lewishome/software.html.

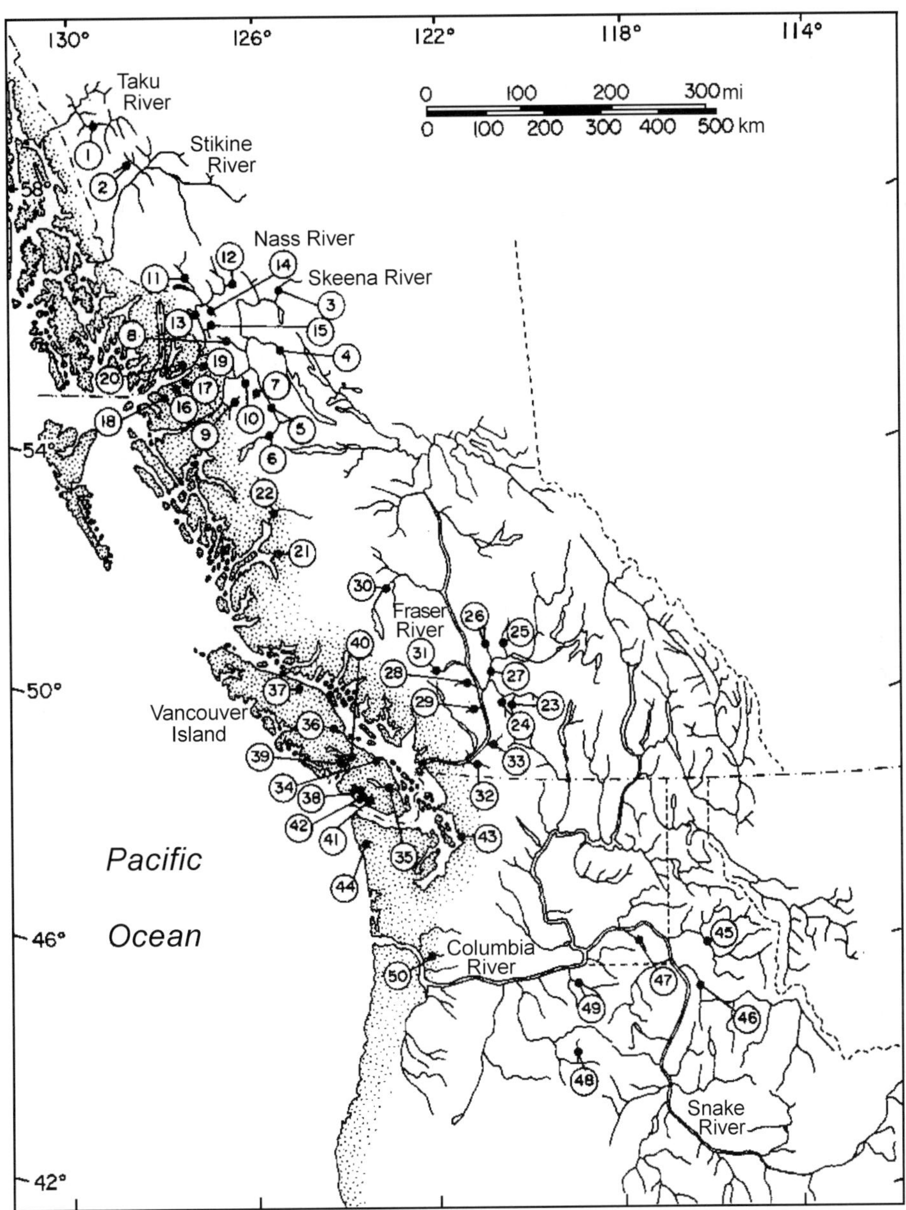

Figure 1. Locations of steelhead trout populations sampled in the survey. Numbers and locations are indicated in Table 1.

population allele frequencies, as was recommended by Waples (1989). A hierarchical gene diversity analysis using ARLEQUIN version 2.0 (Schneider et al. 2000[3]) was conducted to assess structuring of genetic variation among sampling year within a population and differences among populations (years with fewer than 10 fish sampled excluded). Tests of genetic differentiation with pairwise comparisons between all population pairs (annual samples combined) at each locus were also conducted using GENEPOP version 3.1 (Raymond & Rousset 1995). The dememorization number was set at 1 000, and 50 batches were run for each test with 1 000 iterations/batch. The Cavalli-Sforza & Edwards (1967) chord distance was used

---

[3] Schneider, S., D. Roessli & L. Excoffier. 2000. Arlequin version 2.000: A software for population genetics analysis. Genetics and Biometry Laboratory, University of Geneva, Geneva.

Table 1. Steelhead trout samples collected and analyzed from 51 populations in British Columbia, Washington, and the Columbia River. Sample sizes (N) are for years sampled.

Population	Years sampled	N	Total N
Taku River			
(1) Inklin River	1997	9	9
Stikine River			
(2) Tahltan River	1996, 1997	23, 7	30
Skeena River Summer-run			
(3) Sustut River	1994, 1996, 1997, 1998	27, 50, 84, 50	211
(4) Babine River	1991, 1992, 1995, 1996, 1997, 1998	19, 18, 22, 31, 28, 129	247
(5) Bulkley River	1995, 1996, 1997	7, 36, 20	63
(6) Morice River	1991, 1992, 1995, 1998	20, 30, 15, 46	111
(7) Toboggan Creek	1998	128	128
(8) Kispiox River	1992, 1995, 1998	20, 30, 35	85
(9) Zymoetz River	1993, 1995, 1997	16, 18, 38	72
(10) Kitseguecla River	1998	13	13
Nass River Summer-run			
(11) Bell-Irving River	1997, 1998	96, 92	188
(12) Damdochax River	1994, 1995, 1996, 1997, 1998	27, 35, 36, 20, 72	190
(13) Meziadin River	1996, 1997, 1998	9, 5, 40	55
(14) Kwinageese River	1995, 1997, 1998	11, 8, 92	111
(15) Cranberry River	1995, 1996, 1997	19, 29, 147	195
Nass River Winter-run			
(16) Chambers Creek	1997, 1998	26, 24	50
(17) Ishkheenickh River	1997, 1998	14, 19	33
(18) Kwinamass River	1997, 1998	17, 4	21
(19) Tseax River	1998	16	16
(20) Kincolith River	1996, 1997, 1998	16, 15, 10	41
Central Coast			
(21) Bella Coola	1997, 1998	29, 8	37
(22) Dean River	1994	49	49
Thompson River			
(23) Spius Creek	1986, 1987, 1988, 1989,1995, 1996,1998	17, 17, 15, 20, 50, 1,20	140
(24) Coldwater River	1995, 1996, 1997, 1998	32, 3, 31, 6	72
(25) Deadman River	1995, 1997,1998	84, 75, 3	162
(26) Bonaparte River	1994, 1995	95, 250	345
(27) Thompson River	1998	23	23
Upper Fraser River			
(28) Stein River	1976, 1978, 1980, 1997, 1998	1, 1, 2, 13, 12	29
(29) Nahatlatch River	1976, 1979, 1997, 1998	10, 3, 30, 13	56
(30) Chilko River	1995, 1996	27, 22	49
(31) Bridge River	1980, 1996, 1997, 1998	8, 9, 5, 11	33
Lower Fraser			
(32) Chilliwack River	1996, 1997	30, 6	36
(33) Coquihalla River	1994, 1995, 1996, 1997	16, 10, 18, 61	105
Vancouver Island			
(34) Nanaimo River	1997, 1998	8, 7	15
(35) Cowichan River	1997, 1998	10, 17	27
(36) Puntledge River	1996, 1997	18, 35	53
(37) Salmon River	1997, 1998	26, 13	39
(38) Caycuse River	1997	47	47
(39) Nahmint River	1997	44	44
(40) Robertson Creek	1995	50	50
(41) Harris Creek	1997	11	11
(42) Gordon River	1997	29	29
Washington Coast			
(43) Deer Creek	1993, 1995	52, 50	102
(44) Bogachiel River	1994	51	51

Table 1. (*Continued*)

Population	Years sampled	N	Total N
Upper Columbia River Basin			
(45) Clearwater River	1996	50	50
(46) Lower Salmon	1996	50	50
(47) Upper Tucannon	1991	99	99
(48) Beech River	1996	21	21
(49) Umatilla River	1994, 1996	21, 28	49
Lower Columbia River			
(50) Kalama River (winter)	1997	10	10
Kalama River (summer)	1989, 1991, 1996,	5, 5, 9	19

to estimate genetic distance among populations. An unrooted consensus neighbor-joining tree based upon 500 replicate trees was generated with the CONSENSE program from PHYLIP (Felsenstein 1993[4]). Estimates of variance components of regional differences, and population differences within regions were determined with ARLEQUIN. A hierarchical gene diversity analysis was conducted to assess geographic structuring of genetic variations among regions and populations. Eleven regions were defined for the analysis: Stikine/Taku rivers, Nass/Skeena rivers (both summer runs), Nass River winter run, Central B.C. Coast, lower Fraser River, upper Fraser River, Thompson River, Vancouver Island, Washington coast, lower Columbia River, and upper Columbia River basin (Table 1).

Estimation of stock composition

Genotypic frequencies were determined at each locus in each population and the statistical package for the analysis of mixtures software program (SPAM) (Debevec et al. 2000) was used to estimate stock composition of each simulated mixture. Each locus was assumed to be in HWE, and expected genotypic frequencies were determined from the observed allele frequencies and used as model inputs. More alleles were present at the microsatellite loci than was practical for stock identification applications. For the mixed-stock analysis only, we combined low frequency (frequency generally <0.02 in all populations) adjacent alleles to reduce the number of genotypic frequencies to be estimated with the available samples; the pooling strategy for each locus is outlined in Table 2. This was done to minimize and hopefully eliminate the occurrence of fish in the mixed sample from a specific

population having an allele not observed in the baseline populations.

Each baseline population was resampled with replacement in order to simulate random variation involved in the collection of the baseline samples before the estimation of stock composition of each test fishery or simulated mixture. Simulated mixtures composed of steelhead trout from different regions were examined in order to evaluate accuracy and precision of the estimated stock compositions. Simulated fishery samples of 150 fish were generated by randomly resampling with replacement the baseline populations in each drainage. Estimated stock composition of a simulated mixture was then determined, and the whole process was repeated 100 times to estimate the mean and standard deviation of the individual stock composition estimates. For the test fishery samples, point estimates were made of the stock composition, and standard deviations of individual stock estimates derived from 100 bootstrap resamplings of both the baseline populations and the fishery sample.

Identification of individuals to specific populations was done with the program GENECLASS 1.0 (Cornuet et al. 1999). The probabilities of individuals belonging to all populations were calculated using a Bayesian approach and each individual was assigned to the population in which it had the highest marginal probability. The individual to be classified was not included in the baseline population sample during the classification procedure.

Results

Variation within populations

As expected, high levels of variation were observed at the microsatellite loci surveyed. Mean observed

---

[4] Felsenstein, J. 1993. PHYLIP: Phylogeny Inference Package. University of Washington, Seattle.

Table 2. Method of pooling low-frequency alleles if they should occur in any population to reduce the number of genotypic frequencies to be estimated in baseline populations for mixed stock analysis. Not all allele bins considered for pooling may contain observed alleles for steelhead trout.

Pooled alleles, renumbered	Microsatellite allele numbers pooled for each locus												
	Ogo4	*Oke4*	*Ots1*	*Ots2*	*Ots9*	*Ots108*	*Oki10*	*One101*	*One111*	*One114*	*Omy325*	*Ssa197*	*Ssa408*
1	1–7	1–16	1–12	1–18	1–2	1–0	1–6	1–6	1–20	1–21	1–7	1–11	1–12
2	8	17	13–14	19–20	3–4	11–12	7	7–8	21	22	8–9	12	13–14
3	9	18	15–16	21–22	5–6	13–14	8	9–13	22	23–24	10–11	13	15
4	10–13	19	17–18	23–24	7–8	15–16	9	14–15	23–24	25–26	12–14	14–15	16
5	14–15	20–21	19–22	25–26	9–10	17–18	10–11	16–25	25	27	15–16	16–25	17
6	16–17	22–23	23	27	11	16–18	12–13	26–33	26	28	17–18	26–39	18–20
7	18–20	24–30	24	28	12	19	14–15		27	29–30	19–21		21–22
8			25	29–30	13	20	16–18		28	31–32	22–25		23–24
9			26	31–32	14–15	21–22	19–26		29–30	33–34	26–29		25–26
10			27–30	33–34	16	23	27–41		31–42	35–36	30–37		27–37
11				35–42	17–18	24–25				37–51			
12				43–47	19–20	26–27							
13					21–25	28							
14					26–31	29–31							
15						32–44							
16						45–53							

heterozygosities of the 13 loci surveyed over all populations ranged from 0.48 to 0.88, with the locus means and population ranges as follows: *Omy325* 0.82 (0.53–1.00), *Ogo4* 0.75 (0.50–0.96), *One101* 0.48 (0.12–0.80), *One111* 0.69 (0.39–0.92), *One114* 0.88 (0.72–1.00), *Ots1* 0.66 (0.35–0.96), *Ots2* 0.72 (0.50–0.94), *Ots9* 0.84 (0.70–1.00), *Ots108* 0.82 (0.67–1.00), *Ssa197* 0.54 (0.31–0.77), *Ssa408* 0.85 (0.59–1.00), *Oke4* 0.76 (0.58–1.00), *Oki10* 0.69 (0.20–1.00). Heterozygosities averaged over all loci were very similar throughout the range of populations surveyed, with mean observed heterozygosity ranging from 0.70 in the winter-run Nass River populations to 0.74 for populations in Washington and the Columbia River basin. Small populations, particularly along the east coast of Vancouver Island, were not less heterozygous than larger populations in other locations.

Genotypic frequencies observed at the 13 loci and in all 51 populations surveyed in our study were those expected for populations in Hardy–Weinberg equilibrium with one exception. Significant departures from the expected Hardy–Weinberg distribution of genotypic frequencies were observed at six loci in the upper Tucannon River sample. There was no evidence of a consistent departure of genotypic frequencies expected from Hardy–Weinberg distribution at any locus, indicating that null alleles were not abundant (Table 3).

Table 3. Number of alleles observed, expected heterozygosity (H_e), observed heterozygosity (H_o), percent significant Hardy–Weinberg equilibrium tests (HWE, N = 94 tests), and F_{ST} among 51 steelhead trout populations for 13 microsatellite loci.

Locus	Alleles	H_e	H_o	HWE	F_{ST}
Ogo4	17	0.77	0.74	3.2	0.054 (0.009)
Oke4	21	0.84	0.77	1.1	0.113 (0.019)
Ots1	24	0.78	0.66	7.4	0.083 (0.014)
Ots2	38	0.89	0.76	6.4	0.047 (0.006)
Ots9	31	0.87	0.83	2.2	0.044 (0.006)
Ots108	51	0.93	0.81	7.4	0.064 (0.013)
Oki10	35	0.75	0.60	0.0	0.110 (0.021)
One101	32	0.53	0.49	5.3	0.061 (0.015)
One111	24	0.79	0.68	5.3	0.065 (0.008)
One114	39	0.92	0.88	1.1	0.033 (0.004)
Omy325	37	0.88	0.82	3.2	0.058 (0.006)
Ssa197	33	0.56	0.51	1.1	0.080 (0.015)
Ssa408	28	0.91	0.85	1.1	0.054 (0.008)
All loci					0.066 (0.007)

Standard deviations are in parentheses.

In a hierarchical gene diversity analysis of the 51 populations surveyed, annual variation within populations accounted for 1.0% of total observed variation, whereas differences among the populations surveyed accounted for 6.0% of total observed variation. The loci with the greatest population differentiation compared with annual variation within populations were

Oke4 and *Oki10*, with population differentiation 31 times greater than annual variation at *Oke4*, and 19 times greater at *Oki10*. As differences in allele frequencies among populations were on average six times larger than annual variation within populations, all samples within populations were pooled for subsequent analysis.

Variation among populations

Large regional differences in allele frequencies were observed at some loci. For example, the frequency of *Oke4*[246] ranged from 0.18 to 0.48 in Thompson River populations, 0.13 to 0.33 in upper Fraser River populations, 0.10 to 0.21 in upper Columbia River populations, but was virtually absent in all other populations. Similarly, the frequency of *Ots108*[146] ranged from 0.28 to 0.43 in Thompson River populations, but was usually <0.10 in all other populations. Within the Nass River drainage, significant differentiation was observed between summer-run and winter-run populations (p < 0.05). For example, the frequencies of both Oke4[248] and Oke[250] were >0.25 in summer-run populations, but <0.10 in winter-run populations.

Pairwise comparisons of allele frequencies between populations were conducted to evaluate differences among all 51 populations surveyed. In the Nass River, virtually all comparisons between summer- and winter-run populations were significant (p < 0.05). Many of the comparisons between winter-run populations were not significant, likely reflecting the smaller numbers of fish sampled from these populations (Table 1) and the reduced power of the tests to detect differentiation. All comparisons between Nass River or Skeena River populations with populations in more southerly regions were significant (p < 0.01). Vancouver Island populations differed significantly from those in the Fraser River and coastal Washington. Many comparisons among Vancouver Island populations were not significant, again likely reflecting the smaller numbers of fish sampled from these populations. In the Fraser River drainage, Thompson River populations differed significantly from upper and lower Fraser River populations (p < 0.01), and differentiation at virtually all loci was observed between upper and lower Fraser River populations. Upper Columbia populations differed significantly from lower Columbia River populations, as well as from populations in British Columbia.

Population structure

We observed regional structuring among populations surveyed. In the Fraser River drainage, the five Thompson River drainage populations comprised a distinct group, clustering together 98% of the time (Figure 2). Similarly, upper Fraser River populations were distinct, and clustered together 75% of the time, as did the lower Fraser River populations. Upper Columbia River populations were also distinct, clustering together 98% of the time, and clustered separately from lower river populations. Vancouver Island populations formed a regional group, but it was not as well defined (23% bootstrap value) as regional groups in the Fraser River and Columbia River. The two British Columbia central coast populations clustered together 70% of the time. There was not a clean separation of summer-run populations in the Nass River and Skeena River drainages, but winter-run populations in the Nass River were distinct from summer-run populations (Figure 2). The single populations sampled in the Taku River and Stikine River drainages clustered together, distinct from all Nass and Skeena populations.

Within our 11 regions, hierarchical gene diversity analysis indicated that 92% of observed variation occurred within populations, with variation among populations within regions accounted for 2.3% of the observed variation, and regional differentiation accounted for 5.7% of total variation (Table 4). Thus, differences among regions were about 2.5 times greater than variation among populations within regions. The greatest differentiation among regions was observed at *Oke4* and *Oki10*, and not unexpectedly, these two loci had the largest F_{ST} values (Table 3). Little regional differentiation was observed at *One114*, as variation among populations within regions was as large as any differences in allele frequencies among regions.

The mean F_{ST} value over all 51 samples and 13 loci surveyed was 0.066 (Table 3). In pairwise comparisons of populations, Thompson River populations were the most distinct group, with many comparisons exceeding a mean F_{ST} value of 0.10 (Table 5). They were most similar to populations in the upper Fraser River drainage, their nearest neighbors, although with a mean pairwise F_{ST} of 0.066, they were still quite distinct. Similar F_{ST} values were observed between populations in the lower and upper portions of the drainage in both the Fraser River and Columbia River. In northern British Columbia, mean pairwise F_{ST} between all Nass and Skeena summer-run populations in these separate, large

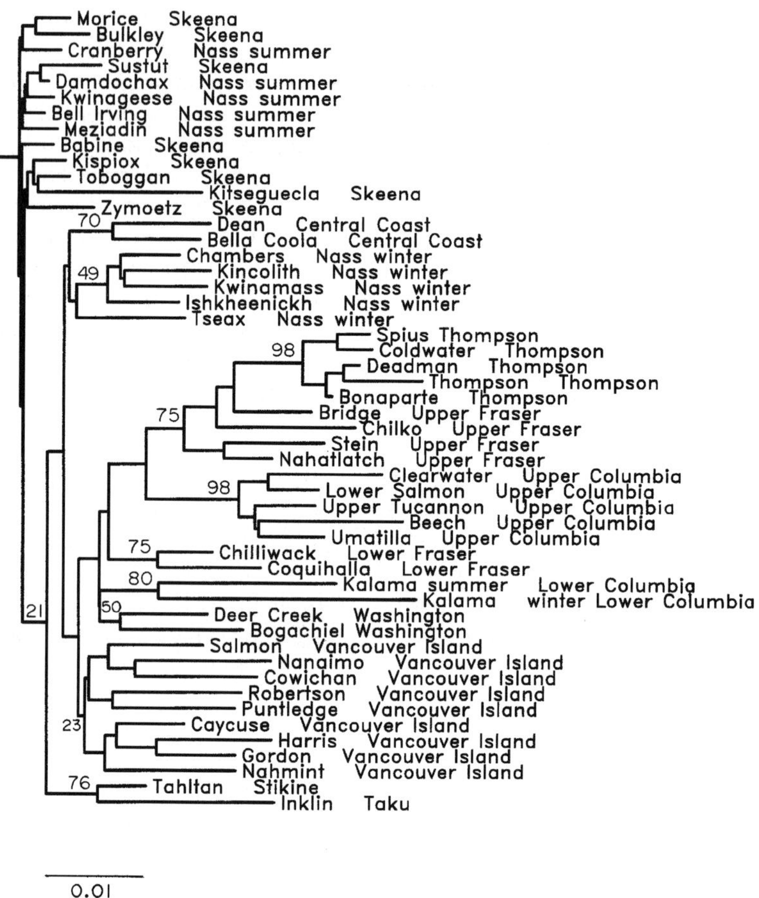

Figure 2. Unrooted neighbor-joining dendrogram based on chord distances outlining relationships of 51 steelhead trout populations. Bootstrap values at the regional nodes indicate the percentage of 500 trees where populations to the right of the node occurred together.

drainages was 0.016, indicative of weak differentiation, as was illustrated in Figure 2. The Stikine and Taku river populations were well differentiated from those in the Nass and Skeena rivers based on pairwise F_{ST}.

Comparison of individual loci

Determination of the relative power of individual loci for regional discrimination is of prime significance for practical stock identification applications. In simulations comparing the relative power of the microsatellite loci to estimate stock compositions of samples comprised of only a single regional stock, there were wide differences in the relative power of the individual loci. The mean accuracy of the estimates ranged between 37% and 79% (Table 6). Loci with fewer than 10 pooled alleles generally provided less accurate estimates of stock compositions than did loci with more alleles. Clearly, not all loci were equally effective in stock identification. The ability of a locus to resolve mixtures containing a single regional stock depended both upon the genetic distinctiveness of the regional stock and population sample sizes within the regional stock. Stocks with populations sampled for fewer than 20 individuals were estimated less accurately than stocks with larger baseline samples. The Thompson River stock was the most accurately estimated, with an average accuracy of 81% when only a single locus was used to estimate compositions of single-stock mixtures.

Estimation of stock composition

We evaluated whether the level of genetic differentiation observed among the regional stocks can be applied to practical issues of stock identification. Three simulated fishery mixtures were developed that might be representative of samples from fisheries where either northern stocks were dominant or southern stocks dominant. Populations with fewer than 20 individuals sampled were excluded from the baseline used for estimation of stock composition in the simulated mixtures, leaving 44 populations in the baseline (Table 1). Estimates of the Fraser River component were on average within 2% of the actual value for mixtures containing 20–70% Fraser River origin (Table 7). Estimated values for the Upper Columbia River and Washington coastal components were also within 2% of the actual value for mixtures containing 0–30% of these regional stocks. The Stikine River and Central Coast components were underestimated in the mixtures, likely reflecting the restricted baseline samples available for these regions, and concurrently the Skeena/Nass component was overestimated. Overall, the analysis of the simulated mixtures indicated that estimation of stock composition with these 13 loci produced results of sufficient accuracy to be of value in real fishery applications.

Accurate determination of the origins of steelhead trout in marine fisheries can be a challenging task. Accuracy of estimation can initially be evaluated by analysis of simulated fishery samples, but analysis of known-origin mixed stock samples provides a check on the reliability of the baseline data for estimation of stock composition. Known-origin mixed-stock samples can be obtained from fisheries in fresh water near the mouths of major rivers in British Columbia. Stock composition of samples collected from the Nass River and Skeena River in 1998 and the Fraser River in 1996 were estimated using the 44 populations surveyed in our study. The 96-fish Nass River sample was estimated as 98.8% (SD = 9.0%) Nass/Skeena origin, the 169-fish Skeena River sample as 99.8% (SD = 8.3%) Nass/Skeena origin, and the 168-fish Fraser River sample as 92.5% (SD = 4.7%) Fraser River origin. Accurate regional stock composition estimates

Table 4. Hierarchical gene diversity analysis incorporating populations within regions and differences among regions for 51 populations of steelhead trout in 11 geographic regions surveyed at 13 microsatellite loci.

Locus	Within populations	Among populations within regions	Among regions
Ogo4	0.9538	0.0128**	0.0334**
Oke4	0.8944	0.0279**	0.0777**
Ots1	0.9262	0.0289**	0.0415**
Ots2	0.9537	0.0178**	0.0285**
Ots9	0.9586	0.0127**	0.0287**
Ots108	0.9336	0.0182**	0.0482**
Oki10	0.9079	0.0182**	0.0739**
One101	0.9649	0.0089**	0.0262**
One111	0.9433	0.0186**	0.0381**
One114	0.9672	0.0165**	0.0163**
Omy325	0.9428	0.0149**	0.0423**
Ssa197	0.9521	0.0176**	0.0303**
Ssa408	0.9465	0.0163**	0.0372**
All loci	0.9201	0.0234**	0.0565**

**p < 0.01.

Table 5. Mean pairwise F_{ST} averaged over 13 loci for 51 populations of steelhead trout in 11 regional groups in British Columbia and the United States.

	Taku Stikine	Nass winter	Nass Skeena	Central Coast	Thompson	Upper Fraser	Lower Fraser	Van. Island	Lower Columbia	Upper Columbia	Washington
Taku, Stikine	0.026	0.065	0.055	0.065	0.136	0.106	0.079	0.064	0.081	0.100	0.066
Nass winter	0.014	0.021	0.040	0.051	0.127	0.090	0.068	0.047	0.059	0.099	0.039
Nass, Skeena	0.020	0.012	0.016	0.049	0.110	0.074	0.065	0.057	0.061	0.083	0.059
Central	0.022	0.010	0.014	0.034	0.122	0.100	0.068	0.062	0.055	0.092	0.062
Thompson	0.016	0.016	0.015	0.017	0.020	0.066	0.108	0.121	0.105	0.082	0.113
Upper Fraser	0.021	0.017	0.015	0.020	0.015	0.049	0.072	0.092	0.093	0.076	0.084
Lower Fraser	0.017	0.026	0.017	0.019	0.009	0.023	0.035	0.071	0.064	0.079	0.061
Van. Island	0.016	0.017	0.014	0.019	0.026	0.024	0.025	0.046	0.062	0.091	0.044
L. Columbia	0.015	0.011	0.015	0.008	0.014	0.024	0.021	0.022	0.021	0.072	0.050
U. Columbia	0.019	0.015	0.018	0.015	0.015	0.018	0.013	0.021	0.023	0.029	0.077
Washington coast	0.008	0.008	0.011	0.009	0.014	0.022	0.026	0.013	0.009	0.015	0.025
Within region SD	—	0.014	0.011	—	0.011	0.023	—	0.021	—	0.014	—

Comparisons were conducted between individual populations within each region and between each region, with the number of populations in each region listed in Table 1. Standard deviation is below the diagonal.

Table 6. Mean estimated stock compositions for simulated samples containing single regional stocks for each of 11 regional stocks of steelhead trout (expected answer 100%) calculated with individual microsatelite loci. Simulations were conducted using a 51 population baseline, 150 fish in the mixture sample and 100 resamplings in the mixture sample and baseline samples, with each mixture sample composed solely of steelhead trout from one region.

	Stikine	Nass winter	Skeena/ Nass	Central Coast	Thompson	Upper Fraser	Lower Fraser	Van. Island	Lower Columbia	Upper Columbia	Wash. Coast	Mean %	Number of alleles	
													Observed	Pooled
Ssa197	75	42	38	37	52	39	32	30	28	28	12	37	33	6
One101	44	31	61	15	31	36	52	50	35	42	25	39	32	6
Ogo4	12	51	69	45	84	60	71	83	30	83	27	56	17	7
Oki10	47	45	64	49	93	79	50	54	31	88	77	63	35	10
One114	50	54	82	66	88	66	72	49	75	33	46	63	39	11
Ots1	95	48	84	78	84	71	66	48	41	82	37	69	24	10
Omy325	83	41	74	61	87	70	46	75	46	92	59	69	37	10
Ots9	61	89	77	81	87	69	71	60	60	76	49	69	31	14
One111	30	63	75	69	80	86	77	58	67	85	63	69	24	11
Oke4	58	37	83	49	92	75	61	80	78	88	43	71	21	7
Ots2	66	76	74	54	87	78	81	76	48	84	71	72	38	12
Ssa408	59	70	64	62	93	72	81	71	72	86	76	74	28	10
Ots108	68	81	83	76	96	88	88	63	80	85	62	79	51	16
Mean	58	56	71	57	81	68	65	61	53	73	50			

Table 7. Estimated percentage composition of three simulated mixtures of steelhead trout using variation at 13 microsatellite loci and a 44-population baseline. Each mixture of 150 fish was generated 100 times with replacement, and stock compositions of the mixtures were estimated by resampling with replacement each baseline population.

Region	Sample 1		Sample 2		Sample 3	
	Actual	Estimated	Actual	Estimated	Actual	Estimated
Stikine	10	5.6 (2.4)		0.0 (0.2)		0.1 (0.2)
Nass winter		0.6 (0.7)		0.3 (0.5)		0.5 (0.6)
Nass/Skeena	50	60.3 (4.0)		4.8 (1.9)		5.5 (2.2)
Central Coast	20	13.7 (3.0)		0.0 (0.2)		0.1 (0.2)
Thompson River	10	11.3 (2.5)	40	42.4 (3.8)	10	11.6 (2.8)
Upper Fraser	10	7.5 (2.3)	20	16.4 (3.3)	10	7.9 (2.4)
Lower Fraser		0.1 (0.3)	10	9.5 (2.5)	10	9.8 (2.2)
Σ Fraser River drainage	20	18.9 (3.0)	70	68.8 (3.7)	30	29.3 (3.7)
Vancouver Island		0.5 (0.6)	10	6.8 (2.5)	20	17.3 (3.3)
Upper Columbia		0.3 (0.5)	10	9.5 (2.2)	30	28.8 (3.5)
Washington Coast		0.2 (0.3)	10	10.3 (2.7)	20	18.4 (3.6)

Σ is the regional sum of all populations in the region. Standard deviation is in parentheses.

should be possible from analysis of mixed-stock samples.

Identification of Individuals

Determination of the origin of an individual steelhead trout can sometimes be of interest in either enforcement or management. One such example is in the Fraser River drainage, where, in order to provide a cost effective method of studying Thompson River steelhead trout by radio telemetry, only Thompson River steelhead trout are desired for radio tagging while still in the lower Fraser River. Based on the 11 surveyed Fraser River populations, classification accuracy of individuals to specific populations in the Thompson River was 66%, but classification accuracy for the Thompson River region was 99% (Table 8). If steelhead trout migrating through the lower Fraser River are held for up to 48 h pending DNA analysis, microsatellite variation can accurately estimate an individual's region of origin and provide researchers flexibility in experimental design. Steelhead trout originating from both lower and upper Fraser River drainages can also be identified with a high level of accuracy (Table 8).

Discussion

Population analysis

In this study, we focussed on sampling adult steelhead trout when possible to obtain the DNA samples.

Table 8. Percent correct classification of individual steelhead trout to population and region of origin for 11 populations and three regions (Thompson River, Upper Fraser River, and Lower Fraser River) in the Fraser River drainage.

Population	N	Population	Region
Coldwater[1]	72	86.1	97.2
Spius[1]	140	85.1	99.3
Deadman	162	52.8	99.4
Bonaparte	345	64.1	99.4
Thompson	23	21.7	100.0
Σ Thompson	742	66.4	99.2
Chilko	49	85.7	95.9
Bridge	33	69.7	84.8
Stein	29	69.0	96.6
Nahatlatch	56	85.7	96.4
Σ Upper Fraser	167	79.6	94.0
Coquihalla	105	97.1	99.0
Chilliwack	36	75.0	91.7
Σ Lower Fraser	141	91.5	97.2

[1]Nicola River drainage incorporating both Coldwater and Spius considered as single population.

Sampling adults lessens the possibility of non-representative sampling of juveniles (Hansen et al. 1997, Wenburg et al. 1998), avoids the potential of mixing samples of juvenile rainbow trout with juvenile steelhead trout (Parkinson 1984), and avoids bias from juveniles that rear in non-natal streams. However, as steelhead trout are in generally low abundance in a river and many of the rivers are remote and difficult to access, it is possible that some of our samples were not derived from a single breeding population. Fish could have been angled from the mainstem of large rivers (such as the Bulkley and Babine rivers) prior to spawning, or sampled at enumeration fences (such as the Bonaparte or Sustut rivers) such that if multiple populations were present, they could not be identified at sampling. However, as genotypic frequencies were generally in HWE for the loci surveyed in individual samples (except for the sample of juveniles from the Tucannon River in which the excess of homozygous fish observed at six loci may have been a result of nonrandom family sampling), mixtures of discrete populations in individual samples does not appear to have been a significant problem.

Population structure of steelhead trout in British Columbia has previously been investigated by Parkinson (1984) who reported, based on the analysis of five allozyme loci, that there were three main regional groups: (1) Skeena River populations in northern British Columbia, (2) all coastal populations including those in the Fraser River (Stein and Nahatlatch rivers), and (3) an interior Fraser River group that included Thompson River and the upper Fraser River (Chilko and Bridge rivers). Our survey based upon 13 microsatellite loci indicated a finer scale structure. In northern British Columbia, steelhead trout in the Taku and Stikine rivers were distinct from those in the Nass and Skeena rivers, and steelhead trout from the central coast were distinct from those in the larger, northern rivers. In southern British Columbia, Thompson River populations were distinct from all upper and lower Fraser River populations, and Vancouver Island populations were distinct from those in the Fraser River, and from other coastal populations in either British Columbia or Washington. In contrast to the previous allozyme survey in the Fraser River drainage, the Stein River and Nahatlatch River populations were more similar to other upper Fraser River populations than they were to lower river or coastal populations. The structure observed among the southern populations is consistent with previous studies (Beacham et al. 1999).

Steelhead trout populations from the Thompson River, upper Fraser River, and upper Columbia River comprised distinct groups. Previous studies have noted that *O. mykiss* in the upper portions of both the Fraser River and Columbia River drainages were genetically distinct from the lower portions of these systems or coastal populations (Okazaki 1984, Reisenbichler et al. 1992). Substantial differentiation between upper and lower Fraser River drainage river populations has also

been observed in coho salmon, *O. kisutch*, (Small et al. 1998), sockeye salmon, *O. nerka*, (Withler et al. 2000), and chinook salmon, *O. tshawytscha*, (Beacham et al. 2003). These consistent differences within the species likely reflect the isolation of the species in two refugia during the last glaciation and subsequent colonization of the drainage by fish from these two distinct origins (McPhail & Lindsey 1986, Hewitt 1996). All Fraser River populations were also differentiated from coastal populations on Vancouver Island, a distinction not observed using allozymes (Parkinson 1984).

In our study, differentiation among populations was on average six times larger than annual variability within populations. Annual variation in allele frequencies at genetic markers in steelhead trout was observed previously with allozymes (Chilcote et al. 1980, Parkinson 1984, Reisenbichler & Phelps 1989, Reisenbichler et al. 1992) and microsatellites (Beacham et al. 1999, Heath et al. 2002). Temporal changes in microsatellite variation in samples collected 40 years apart in three Skeena River populations indicate that the variation among years within populations was about the same as the variance among those populations (Heath et al. 2002). Heath et al. (2002) interpreted these results as a contrast to other salmonid studies that have reported temporal stability in salmonid population genetic structure over time. For example, in Fraser River sockeye salmon, where samples in some populations were 15 years apart, Withler et al. (2000) reported that population differences were 7.4 times larger than annual variation within samples. As multiple independent samples are rarely collected in one year within populations, it is impossible to separate variation due to sampling from interannual variation. As the number of steelhead trout surveyed in most years by Heath et al. (2002) was small relative to the number of alleles observed at the loci, temporal or spatial sampling error may have contributed significantly to the observed annual variation.

In studies where a lack of genetic differentiation is reported between two or more groups of putative populations, it is often difficult to determine whether in fact the putative populations are genetically homogeneous, or whether the genetic markers used by investigators were appropriate for the analysis. Although differentiation at microsatellite loci was observed in other salmonids in the Nass and Skeena drainages, two sets of microsatellite loci employed for steelhead trout did not reveal the level of differentiation observed in other salmonids. The lack of differentiation would appear to be not a result of a choice of inappropriate loci, as

19 loci have been surveyed in total, but as a result of genuine genetic similarity. Genetic differentiation at neutral loci such as microsatellites is thought to reflect founder effects resulting from postglacial dispersal from different refuges (genetic differentiation among refuges assumed) and from subsequent genetic drift after colonization. Salmonids in the Nass and Skeena rivers should presumably have had a similar postglacial dispersal history, there have been no known deliberate transfers of steelhead trout between the two drainages, there is no evidence to suggest that significant straying occurs between the two watersheds, and regional genetic differentiation was observed in steelhead trout in other areas. Like Heath et al. (2001), and given the similarity of the Sustut River population in the headwaters of the Skeena drainage to the Nass River populations, the best explanation that we can offer to account for the similarity is a possible recent transfer of steelhead in the headwaters of the two drainages.

Withler (1966) noted that steelhead can enter spawning streams through the year, and categorized those as entering streams between the beginning of May and the end of September as summer-run, and those entering streams between November and April as winter-run. In the Nass River, significant differentiation was observed between summer-run and winter-run populations, but in addition to differences in run-timing, there was geographic separation of the populations, with winter-run populations spawning in lower portions of the drainage. It seems likely that though that run-timing differences contribute significantly to the variation observed in the drainage.

Mixed-stock analysis

Simulations may at times provide an optimistic view of model performance when compared with applications to actual mixed fishery samples, particularly if fish from stocks not included in the baseline occur in the mixtures. We thus wished to test model performance on actual mixed-fishery samples with a known stock composition. This was done by using mixed-fishery samples from fresh water, where presumably steelhead trout originated from the entire river drainage. These fishery samples confirmed that the microsatellite loci and populations surveyed provided reliable estimates of stock composition on a regional basis. Although an increase in baseline coverage is always desirable, the current coverage suggests that estimated stock compositions from fisheries in southern British Columbia

should be quite accurate. It would be desirable to increase baseline coverage from the Stikine River and central coast regions of British Columbia, but current representation should reliably estimate stock composition of northern fisheries.

The 13 microsatellite loci we evaluated clearly differed in their ability to provide accurate estimates of stock composition. Generally, loci with fewer bins of pooled alleles were less effective for population identification than were loci with more alleles. Whether there is a general relationship between allele number and power of the locus for stock identification needs to be evaluated for other species and applications.

Microsatellite loci provide practical markers for stock identification of steelhead trout only if there is adequate differentiation among populations in the baseline and the level of annual variation of allele frequencies within populations is minimal relative to population differentiation. Practical considerations with respect to baseline sampling and cost require that the differentiation among populations be greater than the variation within populations so that samples may be pooled over several years to obtain representative samples of populations contributing to fishery samples. In our study, population differentiation was on average about 6.0 times greater than annual variation within populations. This level of population differentiation relative to annual variation was sufficient to provide reasonable levels of accuracy and precision in stock composition estimation. However, periodic monitoring of the populations would be warranted to ensure that no substantial drift in allele frequencies has occurred.

Previous analysis of a different set of microsatellite loci did not provide enough differentiation of Nass River and Skeena River populations to provide reliable drainage-specific estimates of stock composition when applied to marine fisheries (Beacham et al. 2000a). Good drainage-specific discrimination exists among other Nass River and Skeena River salmonids that allows accurate and precise estimates of drainage specific stock composition in marine fisheries. Sockeye salmon (Beacham et al. 2000b), coho salmon (Beacham et al. 2001), and chinook salmon (Beacham et al. unpublished) populations of these two rivers possessed differences that allowed reliable estimates of stock composition. We undertook a survey of new microsatellite loci in steelhead trout on the chance that new loci would provide discrimination among Nass and Skeena steelhead trout. Unfortunately, the new loci did not provide sufficient differentiation between populations in the two drainages to enable accurate (within a few percentage) drainage-specific stock composition estimates. The loci did, however, allow for very accurate estimates of stock composition of both drainages combined, and this was confirmed by testing samples from fresh water fisheries in both drainages. Although it was not possible to provide accurate estimates of Nass River and Skeena River steelhead trout contributions in marine fisheries, microsatellite variation was effective for population-specific estimates of stock composition in freshwater fisheries in either river (Beacham et al. 2000a).

In the Fraser River drainage, the substantial differentiation observed among steelhead trout populations in the Thompson River, upper Fraser River, and lower Fraser River allowed very accurate estimates of regional stock composition for freshwater samples from the drainage, even to the extent that individual steelhead trout could be identified to region of origin with a high degree of accuracy. Identification of individuals for applications such as forensics or radio-tagging is possible even when few fish comprise a sample.

Acknowledgements

A very substantial effort was undertaken to obtain samples from adult steelhead trout sampled in this study. We would like to acknowledge staff from the provincial Ministry of Environment, Lands, and Parks (MELP) who collected adult steelhead trout samples, as well as the various agencies, organizations, and companies who collected samples in British Columbia. For the Nass River specifically, these included the Nisga'a Tribal Council, LGL Ltd., SKR Consultants, and Cascadia Natural Resource Consulting. Samples from United States populations were kindly provided by S. Young of the Washington Department of Fish and Wildlife and G. Winans of the National Marine Fisheries Service, Seattle. Two referees and A.J. Gharrett provided many constructive comments on the manuscript.

References

Banks, M.A., M.S. Blouin, B.A. Baldwin, V.K. Rashbrook, H.A. Fitzgerald, S.M. Blankenship & D. Hedgecock. 1999. Isolation and inheritance of novel microsatellites in chinook salmon (*Oncorhynchus tshawytscha*). J. Hered. 90: 281–288.

Beacham, T.D., S. Pollard, S. & K.D. Le. 1999. Population structure and stock identification of steelhead trout in southern British Columbia, Washington, and the Columbia River based on microsatellite DNA variation. Trans. Am. Fish. Soc. 128: 1068–1084.

Beacham, T.D., S. Pollard & K.D. Le. 2000a. Microsatellite DNA population structure and stock identification of steelhead trout (*Oncorhynchus mykiss*) in the Nass and Skeena Rivers in northern British Columbia. Mar. Biotech. 2: 587–600.

Beacham, T.D., C.C. Wood, R.E. Withler & K.M. Miller. 2000b. Application of microsatellite DNA variation to estimation of stock composition and escapement of Skeena River sockeye salmon (*Oncorhynchus nerka*). N. Pac. Anad. Fish. Comm. Bull. 2: 263–276.

Beacham, T.D., J.R. Candy, K.J. Supernault, T. Ming, B. Deagle, A. Schultz, D. Tuck, K. Kaukinen, J.R. Irvine, K.M. Miller & R.E. Withler. 2001. Evaluation and application of microsatellite and major histocompatibility complex variation for stock identification of coho salmon in British Columbia. Trans. Am. Fish. Soc. 130: 1116–1155.

Beacham, T.D., K.J. Supernault, M. Wetklo, B. Deagle, K. Labaree, J.R. Irvine, J.R. Candy, K.M. Miller, R.J. Nelson & R.E. Withler. 2003. The geographic basis of population structure in Fraser River chinook salmon, *Oncorhynchus tshawytscha*. Fish. Bull. 101: 229–242.

Cairney, M., J.B. Taggart & B. Hoyheim. 2000. Characterization of microsatellite and minisatellite loci in Atlantic salmon (*Salmo salar* L.) and cross-species amplification in other salmonids. Mol. Ecol. 9: 2175–2178.

Cavalli-Sforza, L.L. & A.W.F. Edwards. 1967. Phylogenetic analysis: Models and estimation procedures. Amer. J. Hum. Genet. 19: 233–257.

Chilcote, M.W., B.A. Crawford & S.A. Leider. 1980. A genetic comparison of sympatric populations of summer and winter steelhead trouts. Trans. Am. Fish. Soc. 109: 203–206.

Cornuet, J.M., S. Piry, G. Luikart, A. Estoup & M. Solignac. 1999. Comparison of methods employing multilocus genotypes to select or exclude populations as origins of individuals. Genetics 153: 1989–2000.

Debevec, E.M., R.B. Gates, M. Masuda, J. Pella, J. Reynolds & L.W. Seeb. 2000. SPAM (Version 3.2): Statistics program for analyzing mixtures. J. Hered. 91: 509–510.

Goudet, J. 1995. FSTAT A program for IBM PC compatibles to calculate Weir and Cockerham's (1984) estimators of F-statistics (version 1.2). J. Heredity 86: 485–486.

Hansen, M.M., E.E. Nielsen & K.L.D. Mensberg. 1997. The problem of sampling families rather than populations: Relatedness among individuals in samples of juvenile brown trout *Salmo trutta* L. Mol. Ecol. 6: 469–474.

Heath, D.D., S. Pollard & C. Herbinger. 2001. Genetic structure and relationships among steelhead trout (*Oncorhynchus mykiss*) populations in British Columbia. Heredity 86: 618–627.

Heath, D.D., C. Busch, J. Kelly & D.Y. Atagi. 2002. Temporal change in genetic structure and effective population size in steelhead trout (*Oncorhynchus mykiss*). Mol. Ecol. 11: 197–214.

Hendry, M.A., J.K. Wenburg, K.W. Myers & A.P. Hendry. 2002. Genetic and phenotypic variation through the season provides evidence for multiple populations of wild steelhead trout in the Dean River, British Columbia. Trans. Am. Fish. Soc. 131: 418–434.

Hewitt, G.M. 1996. Some genetic consequences of ice ages, and their role in divergence and speciation. Biol. J. Linn. Soc. 58: 247–276.

McPhail, J.D. & C.C. Lindsey. 1986. Zoogeography of the freshwater fishes of Cascadia (the Columbia system and rivers north to the Stikine). pp: 615–637. *In*: C.H. Hocutt & E.O. Wiley (ed.) Zoogeography of North American freshwater fishes. Wiley, New York.

Nelson, R.J. & T.D. Beacham. 1999. Isolation and cross species amplification of microsatellite loci useful for study of Pacific salmon. Anim. Genet. 30: 228–229.

Nielsen, J.L., C.A. Gan, J.M. Wright, D.B. Morris & W.K. Thomas. 1994. Biogeographic distributions of mitochondrial and nuclear markers for southern steelhead trout. Mol. Mar. Biol. Biotech. 3: 281–293.

Nielsen, J.L., C. Carpanzano, M.C. Fountain & C.A. Gan. 1997. Mitochondrial DNA and nuclear microsatellite diversity in hatchery and wild *Oncorhynchus mykiss* from freshwater habitats in southern California. Trans. Amer. Fish. Soc. 126: 397–417.

Olsen, J.B., P. Bentzen & J.E. Seeb. 1998. Characterization of seven microsatellite loci derived from pink salmon. Mol. Ecol. 7: 1083–1090.

Olsen, J.B., S.L. Wilson, E.J. Kretschmer, K.C. Jones & J.E. Seeb. 2000. Characterization of 14 tetranucleotide microsatellite loci derived from sockeye salmon. Mol. Ecol. 9: 2185–2187.

O'Connell, M., R.G. Danzmann, J.M. Cornuet, J.M. Wright & M.M. Ferguson. 1997. Differentiation of rainbow trout populations in Lake Ontario and the evaluation of the stepwise mutation and infinite allele mutation models using microsatellite variability. Can. J. Fish. Aquat. Sci. 54: 1391–1399.

Okazaki, T. 1984. Genetic divergence and its zoogeographic implications in closely related species *Salmo gairdneri* and *Salmo mykiss*. Jpn. J. Ichthyol. 31:297–311.

O'Reilly, P.T., L.C. Hamilton, S.K. McConnell & J.M. Wright. 1996. Rapid analysis of genetic variation in Atlantic salmon (*Salmo salar*) by PCR multiplexing of dinucleotide and tetranucleotide microsatellites. Can. J. Fish. Aquat. Sci. 53: 2292–2298.

Ostberg, C.O. & G.H. Thorgaard. 1999. Geographic distribution of chromosome and microsatellite DNA polymorphisms in *Oncorhynchus mykiss* native to western Washington. Copeia 1999: 287–298.

Parkinson, E.A. 1984. Genetic variation in populations of steelhead trout trout (*Salmo gairdneri*) in British Columbia. Can. J. Fish. Aquat. Sci. 41: 1412–1420.

Raymond, M. & F. Rousset. 1995. GENEPOP (Version 1.2): Population genetics software for exact tests and ecumenism. Heredity 86: 248–249.

Reisenbichler, R.R., J.D. McIntyre, M.F. Solazzi & S.W. Landing. 1992. Genetic variation in steelhead trout of Oregon and northern California. Trans. Am. Fish. Soc. 121: 158–169.

Reisenbichler, R.R. & S.R. Phelps 1989. Genetic variation in steelhead trout (*Salmo gairdneri*) from the north coast of Washington. *Can. J. Fish. Aquat. Sci.* 46: 66–73.

Rice, W.R. 1989. Analyzing tables of statistical tests. Evolution 43: 223–225.

Slaney, T.L., K.D. Hyatt, T.G. Northcote & R.J. Fielden. 1996. Status of anadromous salmon and trout in British Columbia and Yukon. Fisheries 21: 20–35.

Small, M.P., T.D. Beacham, R.E. Withler & R.J. Nelson. 1998. Discriminating coho salmon (*Oncorhynchus kisutch*) populations within the Fraser River, British Columbia using microsatellite DNA markers. Mol. Ecol. 7: 141–155.

Smith, C.T., B.F. Koop & R.J. Nelson. 1998. Isolation and characterization of coho salmon (*Oncorhynchus kisutch*) microsatellites and their use in other salmonids. Mol. Ecol. 7: 1614–1616.

Taylor, E.B. 1995. Genetic variation at minisatellite DNA loci among North Pacific populations of steelhead trout and rainbow trout (*Oncorhynchus mykiss*). J. Hered. 86: 354–363.

Waples, R.S. 1989. Temporal variation in allele frequencies: Testing the right hypothesis. Evolution 43: 1236–1251.

Wenburg, J.K., J.B. Olsen & P. Bentzen. 1996. Multiplexed systems of microsatellites for genetic analysis in coastal cutthroat trout (*Oncorhynchus clarki clarki*) and steelhead trout (*Oncorhynchus mykiss*). Mol. Mar. Biol. Biotech. 5: 273–283.

Wenburg, J.K., P. Bentzen & C.J. Foote. 1998. Microsatellite analysis of genetic population structure in an endangered salmonid: The coastal cutthroat trout (*Oncorhynchus clarki clarki*). Mol. Ecol. 7: 733–749.

Withler, I.R. 1966. Variability in life history characteristics of steelhead trout (*Salmo gairdneri*) along the Pacific coast of North America. J. Fish. Res. Board. Can. 23: 365–393.

Withler, R.E, K.D. Le, R.J. Nelson, K.M. Miller & T.D. Beacham. 2000. Intact genetic structure and high levels of genetic diversity in bottlenecked sockeye salmon, *Oncorhynchus nerka*, populations of the Fraser River, British Columbia, Canada. Can. J. Fish. Aquat. Sci. 57: 1985–1998.

Environmental Biology of Fishes **69**: 111–125, 2004.
© 2004 *Kluwer Academic Publishers. Printed in the Netherlands.*

A comparison of genetic variation between an anadromous steelhead, *Oncorhynchus mykiss*, population and seven derived populations sequestered in freshwater for 70 years

Frank Thrower[a], Charles Guthrie III[a], Jennifer Nielsen[b] & John Joyce[a]
[a]*National Marine Fisheries Service, Auke Bay Laboratory, 11305 Glacier Hwy, Juneau, AK 99801, U.S.A. (e-mail: frank.thrower@noaa.gov)*
[b]*USGS, Alaska Science Center, Office of Biological Science, Anchorage, AK, U.S.A.*

Received 17 April 2003   Accepted 16 June 2003

Key words: rainbow trout, DNA, microsatellites, allozymes

Synopsis

In 1926 cannery workers from the Wakefield Fisheries Plant at Little Port Walter in Southeast Alaska captured small trout, *Oncorhynchus mykiss*, from a portion of Sashin Creek populated with a wild steelhead (anadromous *O. mykiss*) run. They planted them into Sashin Lake which had been fishless to that time and separated from the lower stream by two large waterfalls that prevented upstream migration of any fish. In 1996 we sampled adult steelhead from the lower creek and juvenile *O. mykiss* from an intermediate portion of the creek, Sashin Lake, and five lakes that had been stocked with fish from Sashin Lake in 1938. Tissue samples from these eight populations were compared for variation in: microsatellite DNA at 10 loci; D-loop sequences in mitochondrial DNA; and allozymes at 73 loci known to be variable in steelhead. Genetic variability was consistently less in the Sashin Lake population and all derived populations than in the source anadromous population. The cause of this reduction is unknown but it is likely that very few fish survived to reproduce from the initial transplant in 1926. Stockings of 50–85 fish into five other fishless lakes in 1938 from Sashin Lake did not result in a similar dramatic reduction in variability. We discuss potential explanations for the observed patterns of genetic diversity in relation to the maintenance of endangered anadromous *O. mykiss* populations in freshwater refugia.

Introduction

In recent years many stocks of steelhead, *Oncorhynchus mykiss*, in the western United States have been listed as threatened or endangered under the Endangered Species Act by the National Marine Fisheries Service (Busby et al.[1]). In many cases, freshwater habitat destruction has been cited as a principal factor of population decline and, without substantial habitat restoration, these declines will probably continue. Restoration of freshwater habitats can frequently take years or decades and, in some cases, the continued risk to the remaining population requires some more drastic form of intervention to prevent extinction. In some cases (Flagg et al. 1995, Baugh & Deacon 1988) portions of the wild populations are brought into captivity while habitat restoration efforts are underway. However, the maintenance of wild populations in captivity is fraught with genetic pitfalls. The effective breeding size of these populations is frequently constrained by economics since maintaining captive populations is expensive, and this expense is directly related to the numbers maintained. However, small populations are more subject to genetic change through genetic drift (Falconer 1981), inbreeding depression (Kincaid 1983), domestication selection (Reisenbichler & Brown 1995), and founder effects (Luczynski et al. 1996).

---

[1] Busby, P.J., T.C. Wainwright, G.J. Bryant, L.J. Lierheimer, R.S. Waples, F.W. Waknitz & I.V. Lagomarsino. 1996. Status review of west coast steelhead from Washington, Idaho, Oregon and California. NOAA Tech. Memo. NMFS-NWFSC-27. 261 pp.

An alternative is to maintain endangered populations in different natural environments that allow for large breeding populations and natural reproduction (Baugh & Decon 1988). This is rarely possible especially for larger animals. For an anadromous species such as *O. mykiss*, normal mortality rates in the marine phase routinely exceed 90%. This high mortality can exceed the reproductive potential of an already endangered stock. To reduce this mortality, pumped seawater systems and marine net-pens are currently used to maintain captive populations (Shaklee et al. 1995), however, this usually involves artificial feeding and captive breeding and, concurrently, the associated genetic risks. If the life cycle of a normally anadromous fish can be completed without the marine phase – and the ability to adapt to seawater is not lost after decades of freshwater sequestration – then large, naturally reproducing populations of endangered, normally anadromous fish might be maintained in protected freshwater habitats until their native habitats are restored. This could reduce some of the genetic concerns (e.g. domestication selection, inbreeding depression) for captive populations.

Long-term genetic change within specific populations has not been studied extensively on a biochemical level because many of the tools we use today (starch gel electrophoresis and DNA sequencing) have only been developed and used extensively in the last two or three decades. Thus, while many populations of animals have been maintained in a captive state for many decades, no genetic record exists of the populations originally brought into captivity. Since most of these captive populations contain relatively small numbers of individuals, gene frequencies would most likely have changed due to founder effects, genetic drift and domestication selection over the decades. If the population has been maintained as a large naturally breeding population in a natural (although perhaps, not native) habitat that has not seen substantial disruption (either anthropogenic or natural), then it is more likely that gene frequencies of 'neutral' alleles might not have changed substantially due to genetic drift and the loss of rare alleles would be minimal. Selection would presumably alter frequencies of alleles with high selection coefficients that were favored in the new environment. While any substantial change in gene frequencies could be seen as undesirable, for some critically endangered populations, the only alternative may be extinction.

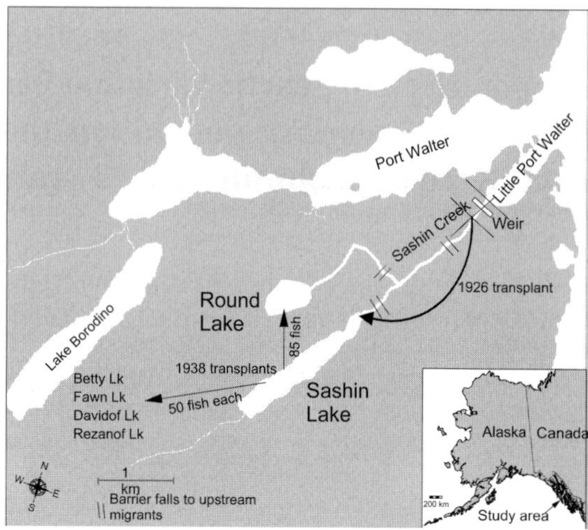

Figure 1. Map of Port Walter showing Sashin Creek study area and indicating the initial transplant (1926) from the anadromous portion of the creek to Sashin Lake and the secondary transplants (1938) to five other barren lakes.

The purpose of this study was to determine if long-term sequestration in fresh water of a normally anadromous stock of fish would result in significant changes in genetic variation that could preclude it as a useful methodology in the preservation of endangered steelhead populations. We compared genetic variation within a wild, anadromous steelhead population (Sashin Creek) in Southeast Alaska with genetic variation in a rainbow trout population from a semi-isolated lake (Sashin Lake) in the same drainage that had been established with a single transplant from the anadromous portion of Sashin Creek in 1926 (70 years earlier) (Anonymous 1939) (Figure 1). We also extended this comparison to include five other lake populations that had been stocked with fish from Sashin Lake in 1938 (approximately 60 years earlier), and a stream population in the intermediate section of Sashin Creek that is separated by barrier falls from the anadromous portion of the creek and Sashin Lake. All of the study lakes and the intermediate stream section were barren of any species of fish at the time of stocking, are above barrier falls that prevent entry of any fish from below, and have no records of subsequent transplants. Fish from all eight populations (hereafter referred to as the 'study' populations) were examined for variation at allozyme and microsatellite loci and mitochondrial DNA haplotypes.

The number of fish originally transplanted to Sashin Lake is unknown. A survey conducted in 1934 indicated the *O. mykiss* population in Sashin Lake to be large (thousands) so we assumed the initial stocking size was large or survival was quite high in the first generation. Stocking records report numbers stocked for each of the five secondary transplants (Chipperfield[2]) and indicate the maximum breeding size of the secondary transplants was small (50–85 fish). Given that the fish were stocked in July, and thus subject to natural mortality for 10 months prior to first spawning and that the sex ratios at stocking were probably unequal, we hypothesized that founder effects could have substantially altered gene frequencies through loss of rare alleles and increased genetic homozygosity. All of the watersheds in the study area remain pristine and currently support population sizes of at least several hundred to several thousand fish (Thrower, pers. observ.).

Materials and methods

A weir on Sashin Creek was used to capture all adult anadromous steelhead in 1996 and 1997. Hoop nets, minnow traps and sport fishing gear were used to capture resident fish in the study lakes and the intermediate section of Sashin Creek. Tissue samples for DNA extraction consisted of ventral fin clips of live fish. Samples from Sashin Creek and Sashin Lake in 1996 were collected from adult fish and stored in 100% ethanol, whereas those from other populations consisting of mixtures of adults and juveniles were preserved by air drying. Tissue samples for allozyme analysis were collected from a portion of the adult steelhead return in both 1996 and 1997, and from resident fish in 1997 with a separate collection from Sashin Lake made in 1996. The samples from anadromous fish were placed in $-20°C$ freezers for 2 months and transferred to $-70°C$ freezers until processed, whereas resident fish were kept alive during transit to Little Port Walter where tissues were removed and placed on ice for up to 2 h, transferred into liquid nitrogen for up to 3 months, and moved to $-70°C$ freezers until processed.

A seventh lake (Deer Lake), also initially stocked with fish from Sashin Lake, was included for contrast because it is known to have had multiple introductions of fish from outside the study area. At Deer Lake, allozyme samples in the spring of 1997, and DNA samples in the spring of 1998, were collected from fish migrating out of the lake (mostly smolts).

---

[2] Chipperfield, W.A. 1938. Memo for files, District Ranger U.S. Forest Service, July 30, Juneau, Alaska.

Laboratory analysis

Mitochondrial DNA
A total of 256 fish were examined for mtDNA haplotype variability. DNA was extracted from a small portion of dried fin tissue using Chelex 100 resin (BioRad) following methods given in Nielsen et al. (1994a). We used conserved primers (S-phe and P2) to amplify a highly variable segment of trout mtDNA, including 188 base pairs (bp) of the control region and 5 bp of the adjacent phenylalanine tRNA gene. Double- and single-stranded amplifications were performed using polymerase chain reaction (PCR). PCR products were sequenced directly and the DNA visualized on X-ray film. DNA protocols, sequence for specific primers, and the complete control region segment amplified in *O. mykiss* are given in Nielsen et al. (1994b). Sequences were aligned using MacDNASIS (Hatachi Software Engineering Company, Ltd.).

Microsatellites
Ten nuclear microsatellite loci developed in other laboratories were chosen for this study based on their high level of polymorphism in previous studies of rainbow trout and steelhead in our laboratory. The Omy-series of microsatellite loci were developed specifically for *O. mykiss*; the Oneμ-series was developed for sockeye salmon, *Oncorhynchus nerka*; Ots-series for chinook salmon, *Oncorhynchus tshawytascha*; Sfo-series for brook trout *Salvelinus fontinalis*; and the Ssa-series was developed for Atlantic salmon, *Salmo salar*. For each locus, primer B was labeled according to protocols given in Nielsen et al. (1994b). Amplification of microsatellites followed the methods given in Nielsen et al. (1997) using three fluorescent dyes and running all microsatellite gels on an ABI 373 (Applied Biosystems) adapted for microsatellite analysis. All microsatellite gels were read using ABI Prism Genotyper Software (Applied Biosystems). All loci were initially run individually as separate PCR reactions to determine allelic size distributions in the Alaska rainbow trout. PCR products were then multiplexed on the gels according to the protocol given in Table 1. The size reported here for each microsatellite allele was equal to the size of the total product amplified (including amplified primer sequence). Allelic size was determined by two methods: (1) reference to the ABI Genescan-500 size marker ladder and (2) known *O. mykiss* DNA samples that were rerun on each gel.

Table 1. Multiplex conditions used for amplifications of 10 microsatellite loci in southwest Alaska rainbow trout and steelhead.

	Anneal (°C)	Locus (primer conc.)		
		6Fam-blue	Tet-green	Hex-yellow
Mykiss A	56	One14 (0.14)	Ots1 (0.17)	One11 (0.06)
			Ssa85 (0.06)	Sfo8 (0.10)
Mykiss B	52	Omy77 (0.30)	Ssa4 (0.55)	Omy325 (0.11)
		One2 (0.055)		One8 (0.13)

Primer concentrations are given in parentheses.

Binning of alleles was performed after an analysis of variance for size distributions of each allele at each locus identified by Genotyper. To ensure consistency in both PCR reactions and scoring of microsatellites, 7.8% of all samples were run again on different gels and scored independently. Repeated runs were not included in the analysis of variance performed to establish allelic binning protocols. Alleles found in <5% of the total study population (all samples combined) were considered rare.

Allozymes
Seventy-three allozyme loci known to be variable in *O. mykiss* were screened in 612 fish (Appendix 2). Protein electrophoresis was conducted as described by Aebersold et al.[3] Specific enzyme activities were stained according to Harris & Hopkinson (1976), or Aebersold et al.[3] We followed Reisenbichler & Phelps (1989) and B. Baker (Washington Department of Fish and Wildlife, pers. commun.) for presumed loci for which data were obtained, the tissues in which they were expressed, and the buffer systems with which they were resolved.

Data analysis

To test for a recent genetic bottleneck, an analysis of allozyme and microsatellite data based on Cornuet & Luikart (1996) which examines differences between the observed heterozygosity and expected heterozygosity based on the observed number of alleles using both an infinite alleles model and a stepwise mutation model was conducted on all study populations using BOTTLENECK (version 1.2.02 (16.II.99) Piry et al.[4]).

Results

Analysis of scale samples of anadromous steelhead indicates that smolting takes place at age three or four in Sashin Creek steelhead. The smolts spend 2–3 years at sea and repeat spawners comprise 10–30% of the anadromous adults. Age validation of scale reading on resident fish in Sashin Lake by marking or tagging has not been accomplished, and reliable aging of older fish is difficult; however, resident males mature as early as age two and commonly at age three and females can mature at age three and age four. Maximum age is thought to be at least 8 and possibly substantially older (F. Thrower, unpubl. data).

Mitochondrial DNA

Sashin Lake and all lake populations derived solely from Sashin Lake, and the fish from the intermediate section of Sashin Creek (Sashin Creek residents) were monomorphic for haplotype MYS1. Only the anadromous population collected from Sashin Creek and the Deer Lake population (that had multiple transplants of different origins) showed any variation in the region of the d-loop examined (Table 2). Anadromous steelhead from Sashin Creek had four additional haplotypes and resident fish from Deer Lake had two additional haplotypes. One of the Deer Lake haplotypes (MYS10) was not found in the other study sites.

[3] Aebersold, P.B., G.A. Winans, D.J. Teel, G.B. Milner & F.M. Utter. 1987. Manual for starch gel electrophoresis: A method for the detection of genetic variation. U.S Dept. Commerce NOAA Tech. Rept. NMFS 61. 19 pp.

[4] Piry, S., G. Luikart & J.M. Cornuet. BOTTLENECK: A program for detecting recent effective population size reductions from allele data frequencies. Version 1.2.02 (16.II.1999). Available online. URL: http://www.ensam.inra.fr/URLB/bottlenect/bottleneck.html.

Table 2. Distribution of mtDNA haplotypes in Sashin Creek anadromous steelhead and seven derived landlocked populations.

Population	mtDNA haplotype						Total
	MYS1	MYS3	MYS10	MYS12	MYS21	CLA1	
Sashin Cr. Anadromous	33	3	0	14	5	1	56
Sashin Lake	26	0	0	0	0	0	26
Sashin Cr. Residents	20	0	0	0	0	0	20
Round Lake	20	0	0	0	0	0	20
Betty Lake	20	0	0	0	0	0	20
Davidof Lake	19	0	0	0	0	0	19
Fawn Lake	20	0	0	0	0	0	20
Rezanof Lake	49	0	0	0	0	0	49
Deer Lake	22	3	1	0	0	0	26
Total	229	6	1	14	5	1	256

One anadromous steelhead (designated SCS57) carried a mtDNA sequence highly divergent from the other *O. mykiss* haplotypes found in this study (Table 3). Alignment of this haplotype with other *Oncorhynchus* sequences for the same segment of the mtDNA d-loop, showed close identity between this fish and a coastal cutthroat trout, *O. clarki clarki*, from British Columbia (J. Nielsen, unpubl. data). Only a single variable site differed between the sequence derived from SCB57 and our coastal cutthroat trout. Five additional sites differentiated both the British Columbia coastal cutthroat and SCS57 from a sequence derived from an interior cutthroat (*O. clarki henshawi*) from Nevada.

Microsatellite DNA

All 10 microsatellite loci were variable in at least one of the study populations. Allelic variants ranged from a low of three per locus (One8) to a high of 15 (Ssa85). A total of 111 allelic variants for the 10 loci were detected in the eight study populations (Appendix 1). Of these, 84 were unique (present in only one population) or rare alleles (whose frequencies were less than or equal to 5% of all samples combined). The Deer Lake samples, which were used for contrast, had 17 additional unique alleles. The anadromous Sashin Creek samples contained 24 rare and 35 unique alleles, whereas the Sashin Lake resident population samples contained only 15 rare alleles and one unique allele (Figure 2). When the unique and rare alleles are pooled and adjusted for sample size, the anadromous fish had on average one unique or rare allele per fish, whereas the resident fish had only one unique or rare allele per four fish.

Differences between the Sashin Lake residents and the secondary transplant populations were far less dramatic. The Sashin Creek residents in the intermediate portion of the creek and the Round Lake population had a similar ratio of unique and rare alleles per fish as the Sashin Lake population. The four other lake populations had ratios varying from eight fish per unique or rare allele in the case of Betty Lake to five fish per allele in Rezanof and Davidof lakes to about four fish per allele in the Fawn Lake population.

Distribution of the 27 common alleles was more uniform and did not show a reduction as a result of the initial transplant to Sashin Lake. They ranged from a low of 23 in Betty Lake to highs of 27 in Sashin Lake, Sashin Creek residents and Round Lake, compared to 26 in the anadromous Sashin Creek fish. All the secondary transplant lake populations initiated with 50 fish had lost common alleles (from 1 to 4 per population) compared to the secondary source population (Sashin Lake).

Allozymes

Seventy-three loci were examined of which 18 were found to be variable in at least one of the study populations and the remainder, 52, were invariant (Appendix 3). Within these 18 loci, 40 allelic variants were found in the eight study populations. Only six of these 18 loci were polymorphic (all study populations combined). Fourteen common alleles were detected among this range of loci and populations. There were 13 unique and two rare alleles among the 18 variable loci. The anadromous steelhead sample had eight unique and two rare alleles which, when combined and adjusted for sample size, implies one

116

Table 3. Mitochondrial DNA sequence (185 base pairs) for MYS1, MYS3, MYS12, MYS21(Alaska), SCS57(Sashin Creek steelhead) and coastal (CLA1) and inland (CLA2) cutthroat trout.

Base pair

Haplotype	1									10										20									
MYS1	T	A	T	A	C	A	T	T	A	A	T	A	A	A	C	T	T	T	A	—	T	G	C	A	C	T	T	T	A
MYS3	T	A	T	A	C	A	T	T	A	A	T	A	A	A	C	T	T	T	A	—	T	G	C	A	C	T	T	T	A
MYS12	T	A	T	A	C	A	T	T	A	A	T	A	A	A	C	T	T	T	A	—	T	G	C	A	C	T	T	T	A
MYS21	T	A	T	A	C	A	T	T	A	A	T	A	A	A	C	T	T	T	A	—	T	G	C	A	C	T	T	T	A
SCS57	T	A	T	A	C	A	T	T	A	A	T	A	A	A	C	T	T	T	A	G	T	G	C	A	C	T	T	T	A
CLA1	T	A	T	A	C	A	T	T	A	A	T	A	A	A	C	T	T	T	A	G	T	G	C	A	C	T	T	T	A
CLA2	T	A	T	A	C	A	T	T	A	A	T	A	A	A	C	T	T	T	A	G	T	G	C	A	C	T	T	T	A

Haplotype	30									40										50										
MYS1	—	—	G	C	A	T	T	T	G	G	C	A	C	C	G	A	C	A	G	C	G	C	T	G	T	A	A	T	G	C
MYS3	—	—	G	C	A	T	T	T	G	G	C	A	C	C	G	A	C	A	G	C	G	C	T	G	T	A	A	T	G	C
MYS12	—	—	G	C	A	T	T	T	G	G	C	A	C	C	G	A	C	A	G	C	G	C	T	G	T	A	A	T	G	C
MYS21	—	—	G	C	A	T	T	T	G	G	C	A	C	C	G	A	C	A	G	C	G	C	T	G	T	A	A	T	G	C
SCS57	T	A	G	C	A	T	T	T	G	G	C	A	C	C	G	A	C	A	A	C	G	C	T	G	T	G	G	T	G	C
CLA1	T	A	G	C	A	T	T	T	G	G	C	A	C	C	G	A	C	A	A	C	G	C	T	G	T	G	G	T	G	C
CLA2	T	A	G	C	A	T	T	T	G	G	C	A	C	C	G	A	C	A	A	C	G	C	T	G	T	G	G	T	A	C

Haplotype	60									70										80										
MYS1	G	T	A	C	A	C	T	T	T	C	A	T	A	A	A	T	A	A	A	G	T	A	T	A	C	A	T	T	A	A
MYS3	G	T	A	C	A	C	T	T	T	C	A	T	A	A	A	T	A	A	A	G	T	A	T	A	C	A	T	T	A	A
MYS12	G	T	A	C	A	C	T	T	T	C	A	T	A	A	A	T	A	A	A	G	T	A	T	A	C	A	T	T	A	A
MYS21	G	T	A	C	A	C	T	T	T	C	A	T	A	A	A	T	A	A	A	A	T	A	T	A	C	A	T	T	A	A
SCS57	G	T	A	C	A	C	T	T	T	C	A	T	A	A	A	T	A	A	A	G	T	A	T	A	C	A	T	T	A	A
CLA1	G	T	A	C	A	C	T	T	T	C	A	T	A	A	A	T	A	A	A	G	T	A	T	A	C	A	T	T	A	A
CLA2	G	T	A	C	A	C	T	T	T	C	A	T	A	A	A	T	A	A	A	G	T	A	T	A	C	A	T	T	A	A

	90										100										110									
MYS1	T	A	A	A	C	T	T	T	C	G	A	T	C	C	A	C	T	T	T	G	T	A	G	C	A	C	C	T	A	G
MYS3	T	A	A	A	C	T	T	T	C	G	A	T	C	C	A	C	T	T	T	G	T	A	G	C	A	C	C	T	A	G
MYS12	T	A	A	A	C	T	T	T	C	G	A	C	C	C	A	C	T	T	T	G	T	A	G	C	A	C	C	T	A	G
MYS21	T	A	A	A	C	T	T	T	C	G	A	T	C	C	A	C	T	T	T	G	T	A	G	C	A	C	C	T	A	G
SCS57	T	A	A	A	C	T	T	T	C	G	G	C	C	C	C	C	T	T	C	G	T	A	G	C	A	T	C	T	G	G
CLA1	T	A	A	A	C	T	T	T	C	G	G	C	C	C	C	C	T	T	C	G	T	A	G	C	A	T	C	T	G	G
CLA2	T	A	A	A	C	T	T	T	T	G	A	C	C	C	C	C	T	T	C	G	T	A	G	C	A	T	C	T	A	G

	120										130											140									
MYS1	C	A	C	C	G	A	C	A	A	C	G	C	T	G	T	T	A	T	C	A	A	T	G	C	C	A	T	T	T	C
MYS3	C	A	C	C	A	A	C	A	A	C	G	C	T	G	T	T	A	T	C	A	A	T	G	C	C	A	T	T	T	C
MYS12	C	C	C	C	G	A	C	A	A	C	G	C	T	G	T	T	A	T	C	A	A	T	G	C	C	A	T	T	T	C
MYS21	C	A	C	C	G	A	C	A	A	C	G	C	T	G	T	T	A	T	C	A	A	T	G	C	C	A	T	T	T	C
SCS57	C	A	C	C	G	A	C	A	A	C	A	C	T	G	A	T	A	T	T	A	A	T	A	C	C	A	T	T	T	C
CLA1	C	A	C	C	G	A	C	A	A	C	A	C	T	G	A	T	A	T	C	A	A	T	A	C	C	A	T	T	T	C
CLA2	C	A	C	C	G	A	C	A	A	C	A	C	T	G	A	T	A	T	C	A	A	T	A	C	C	A	T	T	T	C

	150										160											170									
MYS1	C	A	C	G	C	A	C	A	G	C	C	C	G	C	C	G	C	T	G	A	C	G	T	A	G	C	T	T	A	A
MYS3	C	A	C	G	C	A	C	A	G	C	C	C	G	C	C	G	C	T	G	A	C	G	T	A	G	C	T	T	A	A
MYS12	C	A	C	G	C	A	C	A	G	C	C	C	G	C	C	G	C	T	G	A	C	G	T	A	G	C	T	T	A	A
MYS21	C	A	C	G	C	A	C	A	G	C	C	C	G	C	C	G	C	T	G	A	C	G	T	A	G	C	T	T	A	A
SCS57	C	A	C	G	C	A	C	A	A	C	C	C	A	C	T	G	C	T	A	G	C	G	T	A	G	C	T	T	A	A
CLA1	C	A	C	G	C	A	C	A	A	C	C	C	A	C	T	G	C	T	A	G	C	G	T	A	G	C	T	T	A	A
CLA2	C	A	C	G	C	A	C	A	A	C	C	T	G	C	T	G	C	T	A	G	C	G	T	A	G	C	T	T	A	A

	180					
MYS1	C	T	A	A	A	G
MYS3	C	T	A	A	A	G
MYS1	C	T	A	A	A	G
MYS2	C	T	A	A	A	G
SCS5	C	T	A	A	A	G
CLA1	C	T	A	A	A	G
CLA2	C	T	A	A	A	G

Variable nucleotide locations are outlined in comparison with MYS1.

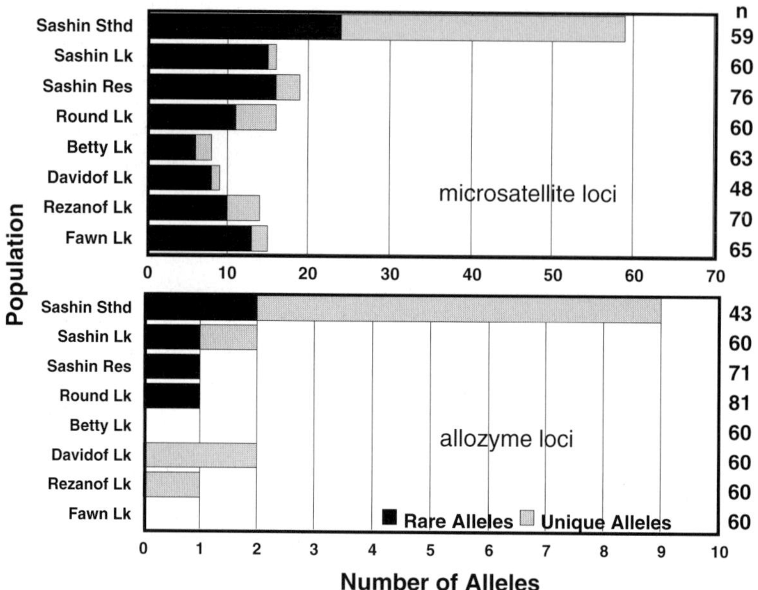

Figure 2. Incidence of unique and rare microsatellite and allozyme alleles in Sashin Creek steelhead and seven derived freshwater populations.

unique or rare allele per four fish. The sample from the primary transplant population, Sashin Lake, had two unique and one rare alleles which, when adjusted for sample size, implies only one unique or rare allele per 40 fish or about one order of magnitude fewer than in the anadromous steelhead (Figure 2). For the secondary transplants, the range of unique and rare alleles varied from zero (Betty and Fawn lakes and Sashin Creek residents) to one each in Rezanof and Round lakes and two in Davidof Lake. Since the unique alleles found in the secondary populations were at low frequencies, and because relatively little time (60 years; 15 generations) has passed from transplantation, it is likely these alleles were present in the source population(s) at low frequency and are not new mutations.

The anadromous steelhead sample contained all of the 14 common alleles. The Sashin Lake sample was lacking only one common allele as were Betty, Davidof and Round lakes. All the other study population samples had the full complement of common alleles. In contrast, the Deer Lake sample had variation in two additional loci (mAAT-2 and ADA-1) both of which were fixed in the study samples. The Deer Lake sample also had three unique alleles in the variable loci and were fixed at two of the polymorphic loci. The 'Bottleneck' analysis did not indicate a heterozygosity excess in any of the populations.

Discussion

The genetic variability of the Sashin Lake population is relatively low when compared to the ancestral, anadromous steelhead population of lower Sashin Creek. The mtDNA evidence indicates perhaps as few as three and probably no more than 7 or 8 females successfully reproduced to start the new population in Sashin Lake, unless the ancestral haplotype proportions were dramatically different from the recent samples. Even with this restriction, the population expanded rapidly in the new habitat and has remained at a relatively large size since at least 1934 when first inventoried. The genetic evidence provided by all three techniques used supports the transplant records of single transplants of Sashin Lake fish into the five other study lakes. The unique mtDNA haplotypes, microsatellite and allozyme alleles found in the Deer Lake population, that was known to have had at least one additional transplant from a source other than Sashin Lake, also supports the lack of additional successful transplants to the study lakes. Virtually all of the lake populations showed some reductions in genetic variability when compared to the donor Sashin Lake population. These reductions range from a loss of unique and rare alleles of approximately 50% (Betty Lake) to virtually no change in frequency (Round Lake).

All five lakes currently have robust populations of naturally reproducing fish. These transplants of 50 and 85 fish appear to have been much more successful at transferring genetic variation than the original transplant into Sashin Lake from lower Sashin Creek. No heterozygosity excess (Cornuet & Luikart 1996) was found in any of the study populations which indicates that recent population bottlenecks have not occurred.

While survival in a quality habitat without competitors or piscine predators cannot be considered the same as that in the original habitat, the establishment of a large, free breeding population is perhaps the most essential element in the preservation of endangered species. This has not been possible with many species of large endangered mammals and some species of fish (Baugh & Deacon 1988, Flagg et al. 1995). However in the case of endangered steelhead, it does appear possible that large populations could be maintained under quasi-natural conditions in freshwater habitats for decades while their native habitat is restored and still have much of the original genetic variation of the population preserved for reintroduction efforts. In fact, in many places in California and the Pacific Northwest, important reservoirs of ancestral steelhead genetic information may still exist behind many irrigation and hydroelectric projects that were put in place in the 1800s and 1900s with no allowance for fish passage. Unfortunately, many of these populations may have been genetically compromised with introductions of stocks of fish from other areas; however, many uncontaminated populations undoubtedly still exist. The fact that these populations have not had the opportunity to express anadromous behavior for decades does not mean that the ability to reinitiate that life history type successfully under the proper conditions has been lost permanently. In fact, the Sashin Lake population and the populations of all the other study lakes, still produce fish that smolt and migrate to sea and return as mature adults to the base of waterfalls blocking access to their natal lakes. Obviously, with complete selection against anadromy in the lake populations, and no reinforcing selection in the original habitat for decades, it is likely survival of the reintroduced fish would be somewhat compromised when compared to the original endemic stock. If a large reservoir of genetic variation has been maintained, successful reinitiation of the anadromous life stage seems likely.

While the use of natural freshwater habitats for the maintenance of a normally anadromous species or stock is not preferable to the use of the original habitat, a naturally reproducing population in a wild or semi-wild state has substantial advantages over maintaining captive populations. However, the effects of freshwater sequestration for decades on the ability of a normally anadromous stock to recolonize its native habitat are unknown and should be investigated. Some evidence for reduced ability to recolonize the former native habitat does appear to exist in the case of the Sashin Lake fish. After 70 years, fish from Sashin Lake and steelhead from Sashin Creek still have substantial genetic differences despite the continued movement of fish from the upper watershed to the lower one. Downstream movement occurs through the normal smolting process of some portion of the upper watershed fish and the downstream movement of fry and juveniles through displacement and washout by floods. Using a Bayesian analysis for stock mixtures of the genotypes present in the watershed, and using the allozyme and microsatellite data independently, Pella & Masuda (2001) concluded that 25% of the anadromous adults at the Sashin Creek weir in 1996 and 1997 had originated in the upper watershed. If this proportion is typical, and the fish of upstream origin mated randomly with those of the anadromous section and offspring survival was similar, then one would expect the genetic profile of the Sashin Creek steelhead to be very similar to the three upstream populations (Sashin Lake, Round Lake and Sashin Creek residents) after 70 years of immigration (Falconer 1981, p. 22). Because differences remain (e.g. frequencies of unique and rare alleles, mitochondrial haplotypes, and PGK-2 alleles), it seems likely that non-random mating and/or differential survival of offspring could be influencing the maintenance of population differences. Research is currently underway at the Little Port Walter Research Station to determine the cause of the maintenance of these stock distinctions.

Acknowledgements

The authors would like to thank the field collection efforts of Ty Cummins at the Little Port Walter Research Station, and the laboratory assistance of Hanhvan Nguyen at the Auke Bay Laboratory and M.C. Fountain at the Hopkins Marine Laboratory of Stanford University. Thanks are also due to two anonymous reviewers whose efforts significantly improved the manuscript.

References

Anonymous. 1939. Trout planting in Alaskan lakes. Prog. Fish. Cult. 46: 31–32.

Baugh, T.M. & J.E. Decon. 1988. Evaluation of the role of refugia in conservation efforts for the Devils Hole pupfish *Cyprinodon diabolis* Wales. Zoo Biol. 7: 351–358.

Cornuet J.M. & G. Luikart. 1996. Description and power analysis of two tests for detecting recent population bottlenecks from allele frequency data. Genetics 144: 2001–2014.

Falconer, D.S. 1981. Introduction to Quantitative Genetics, 2nd edition, Longman Group Ltd. Essex, U.K. 340 pp.

Flagg, T.A., C.V. Mahnken & K.A. Johnson. 1995. Salmon recovery using captive broodstocks. pp. 81–90. *In*: H.L. Schramm, Jr., & R.G. Piper (ed.) Uses and Effects of Cultured Fishes in Aquatic Ecosystems, American Fisheries Society Symposium 15, Albuquerque, NM, 1994.

Harris, H. & D.A. Hopkinson. 1976. Handbook of Enzyme Electrophoresis in Human Genetics, American Elsevier, New York, 120 pp.

Kincaid, H.L. 1983. Inbreeding in fish populations used in aquaculture. Aquaculture 3: 215–227.

Luczynski, M., R. Bartel & A. Marczynski. 1996. Biochemical genetic characteristics of the farmed Alantic Salmon (*Salmo salar*) stock developed in Poland for restoration purposes. Inter. Counc. for the restoration of the sea, Reykjavik, Iceland. ICES-CM-1996/T:12.

Nielsen J.L., C.A. Gan & W.K. Thomas. 1994a. Differences in genetic diversity for mtDNA between hatchery and wild populations of *Oncorhynchus*. Can. J. Fish. Aquat. Sci. 51(Suppl. 1): 290–297.

Nielsen, J.L., C.A. Gan, J.M. Wright, D.B. Morris & W.K. Thomas. 1994b. Biogeographic distributions of mitochondrial and nuclear markers for southern steelhead. Mol. Mar. Biol. Biotech. 3(5): 281–293.

Nielsen, J.L., M.C. Fountain & J.M. Wright. 1997. Biogeographic analysis of Pacific trout (*Oncorhynchus mykiss*) in California and Mexico based on mitochondrial DNA and nuclear microsatellites. pp. 53–73. *In*: T. Kocher & C. Stepien (ed.) Molecular Systematics of Fishes, Academic Press, New York.

Pella, J. & M. Masuda. 2001. Bayesian methods for analysis of stock mixtures from genetic characters. Fish. Bull. 99: 151–167.

Reisenbichler, R.R. & G. Brown. 1995. Is genetic change from hatchery rearing of anadromous fish really a problem? pp. 578–579. *In*: H.L. Schramm, Jr., & R.G. Piper (ed.) Uses and Effects of Cultured Fishes in Aquatic Ecosystems, American Fisheries Society Symposium 15, Albuquerque, NM.

Reisenbichler, R.R. & S.R. Phelps. 1989. Genetic variation in steelhead (*Salmo gairdneri*) from the north coast of Washington. Can. J. Fish. Aquat. Sci. 46: 66–73.

Shaklee, J.B., C. Smith, S. Young, C. Marlowe, C. Jones & B. Sele. 1995. A captive broodstock approach to rebuilding a depleted chinook salmon stock. p. 567. *In*: H. L. Schramm, Jr. & R.G. Piper (ed.) Uses and Effects of Cultured Fishes in Aquatic Ecosystems, American Fisheries Society Symposium 15, Albuquerque, NM.

Shaklee, J., F. Allendorf, D. Morizot & G. Whitt. 1990. Genetic nomenclature for protein-coding loci in fish. Trans. Amer. Fish. Soc. 119: 2–15.

Appendix 1. Allele distributions for 10 microsatellite loci isolated in Sashin Creek steelhead and seven derived populations and Deer Lake.

Population	Base	Locus – Oneμ14											
	143	147	149	151	153	155	157	159	161	163	171	173	Total
Sashin Cr.sthd.	0	46	1	13	4	37	1	0	7	3	2	0	114
Sashin Cr. res.	0	93	0	0	0	57	0	0	0	0	0	0	150
Sashin Lake	0	68	0	2	2	43	0	0	1	0	0	0	116
Round Lake	1	74	7	0	0	28	0	0	0	0	0	0	110
Betty Lake	0	84	1	0	0	31	0	0	0	0	0	0	116
Fawn Lake	0	75	0	0	0	55	0	0	0	0	0	0	130
Rezanof Lake	0	89	0	0	0	34	0	3	0	0	0	2	128
Davidof Lake	0	78	0	0	0	18	0	0	0	0	0	0	96
Deer Lake	0	35	0	10	0	5	0	0	0	0	0	0	50
Total	1	642	9	25	6	308	1	3	8	3	2	2	1010

Locus – Ssa85

Population	97	101	105	109	117	119	121	123	125	127	129	133	135	137	145	157	159	169	Total
Sashin Cr.sthd.	0	0	23	6	0	0	0	2	0	9	5	59	4	2	0	0	0	0	110
Sashin Cr. res.	0	0	42	0	1	0	1	0	0	0	1	101	0	0	4	0	0	0	150
Sashin Lake	0	1	32	2	0	0	0	0	0	1	1	81	0	0	0	0	0	0	118
Round Lake	0	0	54	0	0	0	0	0	0	0	4	60	0	0	0	0	0	0	118
Betty Lake	1	0	56	0	0	0	0	0	0	0	0	65	0	0	0	0	0	0	122
Fawn Lake	0	0	66	0	0	6	0	0	0	0	1	55	0	2	0	0	0	0	130
Rezanof Lake	0	0	16	0	0	1	0	0	0	0	1	114	0	0	0	0	0	0	132
Davidof Lake	0	0	35	0	0	0	0	0	1	0	2	54	0	0	0	0	0	0	92
Deer Lake	0	0	39	0	0	0	0	0	0	1	5	0	1	0	0	2	1	3	52
Total	1	1	363	8	1	7	1	2	1	11	20	589	5	4	4	2	1	3	1024

Locus – Ots1

Population	157	159	161	163	165	167	169	171	173	175	177	237	239	241	Total
Sashin Cr.sthd.	0	0	0	69	2	24	0	0	18	1	1	2	1	0	118
Sashin Cr. res.	0	0	0	58	4	69	2	1	9	0	0	0	0	1	144
Sashin Lake	0	0	0	57	0	48	0	0	10	0	0	0	0	1	116
Round Lake	0	0	0	56	0	22	0	0	12	0	0	0	0	30	120
Betty Lake	0	0	0	27	2	87	0	1	5	0	0	0	0	0	122
Fawn Lake	0	0	0	24	0	95	0	0	10	1	0	0	0	0	130
Rezanof Lake	0	0	0	77	3	40	0	0	0	0	0	0	0	0	120
Davidof Lake	0	2	0	70	0	4	0	0	0	0	0	0	0	18	94
Deer Lake	4	0	7	3	2	36	0	0	0	0	0	0	0	0	52
Total	4	2	7	441	13	425	2	2	64	2	1	2	1	50	1016

Appendix 1. (*Continued*)

Population	Base	Locus – Oneμ11							Locus – Sfo8						Locus – Oneμ8				
	109	119	143	145	147	149	Total		172	174	176	178	Total		157	159	161	173	Total
Sashin Cr.sthd.	1	0	10	43	24	40	118		117	1	0	0	118		0	3	98	1	102
Sashin Cr. res.	0	0	9	52	48	41	150		151	0	1	0	152		0	4	145	1	150
Sashin Lake	0	0	0	41	33	44	118		113	0	1	0	114		0	0	114	0	114
Round Lake	0	2	3	23	17	65	110		114	0	0	0	114		0	4	116	0	120
Betty Lake	0	0	0	6	6	114	126		118	0	0	2	120		0	2	114	0	116
Fawn Lake	0	0	6	54	47	1	108		129	1	0	0	130		0	0	130	0	130
Rezanof Lake	0	0	1	57	18	58	134		130	0	0	0	130		0	2	128	0	130
Davidof Lake	0	0	0	24	2	70	96		96	0	0	0	96		0	0	94	0	94
Deer Lake	0	0	0	28	1	23	52		Not run						4	19	29	0	52
Total	1	2	29	328	196	456	1012		968	2	2	2	974		4	34	968	2	1004

Population		Locus – Omy77												
	95	99	103	107	115	117	119	121	127	129	131	137	145	Total
Sashin Cr.sthd.	0	0	24	3	17	1	38	9	0	5	15	6	0	118
Sashin Cr. res.	0	0	69	0	0	0	47	0	0	0	28	0	0	144
Sashin Lake	0	0	56	0	0	0	30	0	0	0	34	0	0	120
Round Lake	0	0	20	0	0	0	62	1	0	0	36	0	1	120
Betty Lake	0	0	97	0	0	0	29	0	0	0	0	0	0	126
Fawn Lake	1	0	15	0	0	0	39	0	0	1	58	0	0	114
Rezanof Lake	0	0	104	0	0	0	34	0	0	0	2	0	0	140
Davidof Lake	0	0	45	0	0	0	32	0	0	0	15	0	0	92
Deer Lake	0	5	2	0	1	0	22	0	8	0	14	0	0	52
Total	1	5	432	3	18	1	333	10	8	6	202	6	1	1026

Population		Locus – Oneμ2													
	206	208	212	222	226	240	242	244	246	248	250	252	254	256	258
Sashin Cr.sthd.	0	0	0	0	1	0	0	12	25	7	1	4	1	6	7
Sashin Cr. res.	0	0	0	0	0	0	0	12	36	8	20	10	3	22	5
Sashin Lake	0	0	0	0	0	0	0	0	34	11	9	25	2	9	7
Round Lake	0	0	0	0	0	4	0	0	2	5	7	7	8	28	0
Betty Lake	0	0	0	0	0	0	0	0	0	0	33	52	17	16	8
Fawn Lake	0	0	0	0	0	0	0	18	19	6	7	13	28	28	3
Rezanof Lake	0	0	0	0	0	0	3	1	24	1	15	13	4	24	0
Davidof Lake	0	0	0	0	0	0	0	0	11	1	29	6	6	13	1
Deer Lake	3	3	12	7	0	0	1	24	0	0	0	2	0	0	0
Total	3	3	12	7	1	4	4	67	151	39	121	132	69	146	31

														Total
Sashin Cr.sthd.	2	2	7	3	1	3	0	10	2	8	4	2	1	1	110
Sashin Cr. res.	0	0	0	0	0	0	2	28	0	0	0	0	0	0	146
Sashin Lake	0	0	0	0	0	0	0	21	0	0	0	0	0	0	118
Round Lake	0	0	0	0	0	0	10	1	0	0	0	0	0	0	72
Betty Lake	0	0	0	0	0	0	0	0	0	0	0	0	0	0	126
Fawn Lake	0	0	0	0	0	0	0	4	0	0	0	0	0	0	126
Rezanof Lake	0	0	0	0	0	0	3	24	0	0	0	0	0	0	112
Davidof Lake	0	0	0	0	0	0	10	17	0	0	0	0	0	0	94
Deer Lake	0	0	0	0	0	0	0	0	0	0	0	0	0	0	52
Total	2	2	7	3	1	3	25	105	2	8	4	2	1	1	956

Locus – Ssa14

	130	132	134	136	138	140	142	144	146	148	150	152	154	Total
Sashin Cr.sthd.	78	4	7	17	1	2	2	0	0	0	0	0	1	112
Sashin Cr. res.	87	0	0	55	0	10	0	0	0	0	0	0	0	152
Sashin Lake	64	0	1	43	0	6	0	0	0	0	0	0	0	114
Round Lake	43	0	39	33	0	0	0	0	1	0	0	0	0	116
Betty Lake	83	0	0	39	0	0	0	0	0	0	0	0	0	122
Fawn Lake	105	0	0	17	0	8	0	0	0	0	0	0	0	130
Rezanof Lake	71	0	0	59	0	0	0	0	0	0	0	0	0	130
Davidof Lake	32	0	0	61	0	1	0	0	0	0	0	0	0	94
Deer Lake	27	0	1	0	0	0	0	9	0	1	11	1	0	50
Total	590	4	48	324	1	27	2	9	1	1	11	1	1	1020

Locus – Omy325

	95	97	99	101	103	105	109	111	113	123	125	127	129	131	133	135	Total
Sashin Cr.sthd.	2	4	2	0	1	30	48	0	3	1	0	5	0	0	13	3	112
Sashin Cr. res.	0	0	0	0	0	52	52	0	0	0	0	1	0	0	45	0	150
Sashin Lake	0	0	0	0	0	44	44	0	0	0	0	3	0	0	25	0	116
Round Lake	0	0	0	0	0	30	51	0	0	0	0	8	2	0	27	0	118
Betty Lake	0	0	0	0	1	65	0	0	0	0	0	0	0	0	58	0	124
Fawn Lake	0	0	0	0	0	22	65	0	0	0	1	13	12	1	8	0	122
Rezanof Lake	0	6	0	0	0	96	23	0	0	0	0	6	0	2	1	0	134
Davidof Lake	0	0	0	0	0	39	37	0	0	0	0	0	0	2	18	0	96
Deer Lake	0	0	0	1	0	3	9	1	13	10	10	0	1	4	0	0	52
Total	2	10	2	1	2	381	329	1	16	11	11	36	15	9	195	3	1024

Appendix 2. Allozyme variation in Oncorhychus mykiss from southern Baranof Island, Southeast Alaska. Alleles are designated by their mobility relative to the most common allele (*100) and described in Shaklee et. al. (1990).

Location	N	mAAT-2*		N	sAAT-4*			N	ADA-1*		N	ADA-2*		N	ADH*		N	sAH*		N	BGLUA*		
		*-100	*-90		*100	*69	*80		*100	*85		*100	*90		*100	*-128		*100	*85		*100	*90	*50
Sashin steelhead	43	1.000	0.000	39	0.987	0.013	0.000	43	1.000	0.000	43	1.000	0.000	43	1.000	0.000	43	0.814	0.186	43	0.965	0.000	0.035
Sashin Lake	60	1.000	0.000	54	1.000	0.000	0.000	60	1.000	0.000	60	0.992	0.008	60	1.000	0.000	58	0.672	0.328	60	1.000	0.000	0.000
Sashin Cr. Res.	71	1.000	0.000	70	0.986	0.014	0.000	71	1.000	0.000	71	1.000	0.000	71	1.000	0.000	70	0.743	0.257	71	1.000	0.000	0.000
Round Lake	81	1.000	0.000	69	0.826	0.174	0.000	80	1.000	0.000	80	1.000	0.000	81	1.000	0.000	81	0.586	0.414	81	1.000	0.000	0.000
Betty Lake	60	1.000	0.000	59	1.000	0.000	0.000	60	1.000	0.000	60	1.000	0.000	60	1.000	0.000	60	0.992	0.008	60	1.000	0.000	0.000
Davidof Lake	60	1.000	0.000	58	1.000	0.000	0.000	60	1.000	0.000	60	1.000	0.000	60	0.992	0.008	59	0.797	0.203	60	1.000	0.000	0.000
Fawn Lake	60	1.000	0.000	51	0.951	0.049	0.000	60	1.000	0.000	60	1.000	0.000	60	1.000	0.000	57	0.868	0.132	60	1.000	0.000	0.000
Rezanof Lake	60	1.000	0.000	56	0.777	0.223	0.000	60	1.000	0.000	60	1.000	0.000	60	1.000	0.000	60	0.983	0.017	60	1.000	0.000	0.000
Deer Lake	57	0.851	0.149	56	0.607	0.116	0.277	57	0.640	0.360	57	1.000	0.000	57	1.000	0.000	52	0.990	0.010	57	0.816	0.096	0.088

Location	N	CK-C2*		N	FH*		N	G3PDH-3*		N	G3PDH-4*			N	mIDHP-2*		N	sIDHP-1,2*				N	LDH-B2*	
		*100	*104		*100	*84		*100	*121		*100	*124	*85		*100	*144		*100	*42	*121	*72		*100	*76
Sashin steelhead	43	0.988	0.012	43	1.000	0.000	43	0.977	0.023	43	0.430	0.558	0.012	43	0.860	0.140	43	0.465	0.215	0.110	0.209	43	0.942	0.058
Sashin Lake	60	0.933	0.067	60	1.000	0.000	60	1.000	0.000	60	0.192	0.808	0.000	60	1.000	0.000	60	0.446	0.192	0.171	0.192	60	1.000	0.000
Sashin Cr. Res.	71	1.000	0.000	71	1.000	0.000	71	1.000	0.000	71	0.148	0.852	0.000	70	1.000	0.000	71	0.359	0.204	0.162	0.275	71	1.000	0.000
Round Lake	80	0.994	0.006	81	1.000	0.000	79	1.000	0.000	78	0.006	0.994	0.000	81	1.000	0.000	81	0.494	0.204	0.191	0.111	81	1.000	0.000
Betty Lake	60	1.000	0.000	60	1.000	0.000	58	1.000	0.000	58	0.440	0.560	0.000	60	1.000	0.000	60	0.492	0.238	0.088	0.183	60	1.000	0.000
Davidof Lake	60	1.000	0.000	60	1.000	0.000	60	1.000	0.000	60	0.350	0.650	0.000	60	1.000	0.000	60	0.542	0.229	0.104	0.125	60	1.000	0.000
Fawn Lake	60	1.000	0.000	60	1.000	0.000	60	1.000	0.000	59	0.059	0.941	0.000	60	1.000	0.000	59	0.530	0.246	0.072	0.153	60	1.000	0.000
Rezanof Lake	60	1.000	0.000	60	0.992	0.008	60	1.000	0.000	60	0.000	1.000	0.000	60	1.000	0.000	60	0.425	0.204	0.233	0.138	60	1.000	0.000
Deer Lake	57	0.991	0.009	57	1.000	0.000	57	1.000	0.000	57	0.044	0.912	0.044	57	0.974	0.026	57	0.667	0.202	0.000	0.132	57	1.000	0.000

Location	N	sMDH-A1,2*		N	sMDH-B1,2*		N	sMEP-1*		N	PEPD-1*		N	PGK-2*				N	PGM-1*		N	sSOD*	
		*100	*155		*100	*120		*100	*83		*100	*110		*100	*115	*144	*90		*100	*Null		*100	*226
Sashin steelhead	43	1.000	0.000	43	0.942	0.058	43	0.779	0.221	42	0.988	0.012	43	0.523	0.465	0.000	0.012	43	1.000	0.000	43	0.767	0.233
Sashin Lake	60	1.000	0.000	60	1.000	0.000	60	0.533	0.467	60	1.000	0.000	60	0.000	1.000	0.000	0.000	60	1.000	0.000	60	0.817	0.183
Sashin Cr. Res.	71	1.000	0.000	71	1.000	0.000	71	0.599	0.401	68	1.000	0.000	71	0.000	0.993	0.000	0.007	71	1.000	0.000	71	0.789	0.211
Round Lake	81	1.000	0.000	81	1.000	0.000	79	0.494	0.506	67	1.000	0.000	81	0.000	1.000	0.000	0.000	81	1.000	0.000	81	0.926	0.074
Betty Lake	60	1.000	0.000	60	1.000	0.000	60	0.600	0.400	59	1.000	0.000	60	0.000	1.000	0.000	0.000	60	1.000	0.000	60	0.575	0.425
Davidof Lake	60	1.000	0.000	60	1.000	0.000	60	0.417	0.583	60	1.000	0.000	60	0.000	1.000	0.000	0.000	60	0.967	0.033	60	0.433	0.567
Fawn Lake	60	1.000	0.000	60	1.000	0.000	60	0.633	0.367	59	1.000	0.000	60	0.000	1.000	0.000	0.000	60	1.000	0.000	60	0.600	0.400
Rezanof Lake	60	1.000	0.000	60	1.000	0.000	60	0.692	0.308	55	1.000	0.000	60	0.000	1.000	0.000	0.000	60	1.000	0.000	60	0.692	0.308
Deer Lake	57	1.000	0.000	57	0.956	0.044	57	1.000	0.000	57	1.000	0.000	57	0.184	0.798	0.018	0.000	57	1.000	0.000	57	0.833	0.167

N refers to number of fish sampled.

Appendix 3. Sample sizes of monomorphic loci screened.

Location	mAAT-1	sAAT-1,2	sAAT-3	mAH-1	mAH-2	mAH-3	mAH-4	AK	ALAT	BGALA	CKA1	CKA2	CKB	CKC!
Sashin steelhead	43	43	43	40	40	40	32	43	40	43	43	43	43	43
Sashin Lake	60	60	60	60	60	60	60	60	57	60	60	60	60	60
Sashin Cr. Res.	71	71	71	71	71	71	71	71	70	71	71	71	71	71
Round Lake	60	60	60	60	60	60	60	60	59	60	60	60	60	60
Betty Lake	60	60	60	59	59	59	59	60	60	60	60	60	60	60
Davidof Lake	60	60	60	60	60	60	60	60	53	60	60	60	60	60
Fawn Lake	60	60	60	60	60	60	60	60	60	67	81	81	81	81
Rezanof Lake	81	81	81	81	81	81	81	30	67	81	81	81	81	81
Deer Lake	57	57	57	57	57	57	57	57	57	57	57	57	57	57

Location	ESTD	FDHG	GAPDH-2	GAPDH-3	GAPDH-4	GR	GPIA	GPI-B1	GPI-B2	G3PDH-1	G3PDH-2	IDDH-1	IDDH-2	mIDHP-1
Sashin steelhead	43	43	43	41	43	43	43	43	43	43	43	43	43	43
Sashin Lake	59	60	60	60	60	60	59	59	59	60	60	60	60	60
Sashin Cr. Res.	71	71	71	71	71	69	71	71	70	71	71	71	71	70
Round Lake	69	81	81	79	81	81	81	81	81	81	81	81	81	81
Betty Lake	60	60	59	59	59	59	60	60	60	60	60	60	60	60
Davidof Lake	58	60	60	60	60	60	60	60	60	60	59	60	60	60
Fawn Lake	60	60	60	60	60	60	60	60	60	60	60	60	60	60
Rezanof Lake	59	60	60	60	60	60	60	60	60	60	60	60	60	60
Deer Lake	57	57	57	57	57	57	57	57	57	57	57	57	57	57

Location	LDH-B1	LDH-C	mMDH-1	mMDH-2	mMDH-3	sMDH-A1	sMDH-B1	mMEP-1	MPI	PEPA	PEPB-1	PEPLT	PGDH	PGM-R
Sashin steelhead	43	43	43	41	43	43	43	43	43	43	42	43	43	43
Sashin Lake	60	60	60	60	60	60	60	60	60	58	60	60	60	60
Sashin Cr. Res.	71	71	71	71	71	71	71	71	71	71	70	67	69	71
Round Lake	81	81	81	81	81	81	81	81	81	74	80	81	81	81
Betty Lake	60	60	60	60	60	60	60	60	60	60	60	60	60	60
Davidof Lake	60	60	60	60	60	60	60	60	60	60	60	60	60	60
Fawn Lake	60	60	60	60	60	60	60	60	60	60	60	60	60	60
Rezanof Lake	60	60	60	60	60	60	60	60	60	60	60	60	60	60
Deer Lake	57	57	57	57	57	57	57	57	57	57	57	57	57	57

Location	PGM-2	PNP	mSOD	TPI-1	TPI-2	TPI-3	TPI-4
Sashin steelhead	43	19	43	43	43	43	43
Sashin Lake	60	60	59	60	60	60	60
Sashin Cr. Res.	71	71	71	71	71	71	71
Round Lake	81	75	50	81	81	81	81
Betty Lake	60	60	59	60	60	60	60
Davidof Lake	60	60	60	60	60	60	60
Fawn Lake	60	60	60	60	59	60	60
Rezanof Lake	60	60	60	60	60	60	60
Deer Lake	57	57	57	57	57	57	57

Environmental Biology of Fishes **69**: 127–142, 2004.
© 2004 *Kluwer Academic Publishers. Printed in the Netherlands.*

Genetic integrity and microgeographic population structure of westslope cutthroat trout, *Oncorhynchus clarki lewisi*, in the Pend Oreille Basin in Washington

Sewall F. Young[a], Jason G. McLellan[b] & James B. Shaklee[a]
[a]*Washington Department of Fish and Wildlife, 600 Capitol Way N, Olympia, WA 98501, U.S.A.*
(e-mail: youngsfy@dfw.wa.gov)
[b]*Washington Department of Fish and Wildlife, 8702 N Division Street, Spokane, WA 99218, U.S.A.*

Received 17 April 2003   Accepted 19 April 2003

Key words: microsatellite, introgression, hybrids

Synopsis

Five microsatellite DNA loci (*Ots-101**, *Ots-107**, *Oki-10**, *Ogo-3**, and *FGT-3**) were screened to evaluate the genetic characteristics and population structure for cutthroat trout from eight tributaries of the Pend Oreille River in northeastern Washington and to compare these collections with two hatchery stocks of westslope cutthroat trout, *Oncorhynchus clarki lewisi*, Yellowstone cutthroat trout, *Oncorhynchus clarki bouvieri* and a hatchery rainbow trout, *Oncorhynchus mykiss*, strain that have been stocked in northeastern Washington. Relatively high levels of variation (numbers of alleles and heterozygosity) were observed in all collections and allele frequencies were quite variable among collections. Evidence of limited introgression by rainbow and/or Yellowstone cutthroat was found at several locations. Both F_{ST} values and tests of genetic differentiation indicated the existence of numerous, reproductively isolated populations. The population in Slate Creek was very similar to the Kings Lake Hatchery strain, and we conclude that this similarity is the result of historical introductions of this hatchery strain into what was presumably a stream without a native cutthroat population. In one stream, differences in introgression and allele frequencies were found above and below a barrier falls. Because of the substantial level of population differentiation observed among the various collections, we recommend that management and conservation actions be focused at the level of individual streams in order to maintain the productivity and genetic character of the existing populations of cutthroat trout.

Introduction

Cutthroat trout, *Oncorhynchus clarki*, occur in many of the streams in the Pend Oreille River drainage in northeastern Washington and throughout the Pacific Northwest. The native cutthroat in the Pend Oreille are presumably westslope cutthroat, *O. clarki lewisi* (Behnke 1992). However, there have been numerous documented introductions of hatchery strains of cutthroat trout (both westslope and Yellowstone subspecies) into the drainage by the Washington Department of Fish and Wildlife (WDFW) and others.[1]

Furthermore, inland rainbow trout occur naturally throughout the same region and stocking of hatchery rainbow trout (of coastal California origin) has been extensive.[1] Because westslope and Yellowstone cutthroat are known to hybridize with each other and to hybridize with introduced rainbow trout, we undertook this investigation both to document the genetic character of the existing populations and to investigate population structure of the cutthroat trout in the Pend Oreille basin.

We used microsatellite DNA loci (arrays of short, tandemly repeated nucleotide sequences occurring commonly in eukaryotic organisms) because: (1) they can be studied using non-lethal tissue samples, (2) they are non-coding and therefore, assumed to be

---

[1] WDFW, unpublished stocking records.

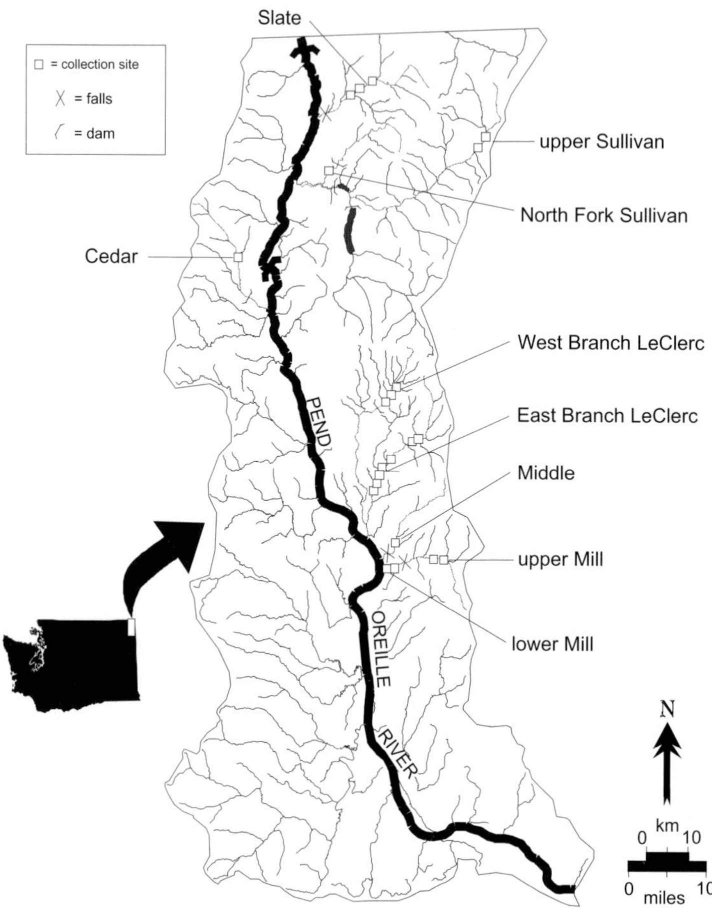

Figure 1. Map of the major tributaries of the Pend Oreille basin in Washington showing collection sites and fish passage barriers.

selectively neutral, (3) they typically exhibit biparental, Mendelian inheritance and alleles are co-dominantly expressed allowing an organism's genotype to be unambiguously inferred from its DNA phenotype, and (4) they evolve rapidly and often exhibit many alleles and high heterozygosities. These characteristics make microsatellites useful and powerful markers for investigating genetic aspects of population structure.

Materials and methods

Collections of 50 cutthroat trout were obtained from each of eight tributaries in the Pend Oreille basin in Washington (Figure 1) using a backpack electroshocker. Wherever possible, fish were collected from multiple sites within each stream to minimize the inclusion of closely related individuals. A small piece of caudal or dorsal fin was removed and placed directly into a labeled vial containing absolute ethanol. Estimates of the amount of cutthroat trout habitat, cutthroat densities, and cutthroat population sizes for each tributary sampled are given in Table 1.

Four hatchery strains are relevant to the history and current status of cutthroat populations in the Pend Oreille based on stocking records, and these were also sampled; Kings Lake Hatchery (N = 50), Twin Lakes Hatchery (N = 50), Yellowstone Hatchery (MT) cutthroat (N = 60), and Spokane Hatchery rainbow trout (N = 90). The Kings Lake Hatchery strain of westslope cutthroat was established in the early 1940s using fish obtained from Granite and Kalispell creeks; streams flowing into Priest Lake, Idaho.[2] Both

---

[2] Crawford, B. 1979. The origin and history of the trout brood stocks of the Washington Department of Game. Washington State Game Department Fishery Research Report. Olympia, WA, 76 pp.

Table 1. Collection locality information with estimated cutthroat densities and population sizes.

Collection locality/stock	N	Date(s)	Total length (mm)		Densities[b] (# 100 m^{-1})	Cutthroat habitat (km)	Estimated population sizes[a]
			Mean	Range			
West Branch LeClerc Creek	50	Oct. 5, 1999	165	90–233	93.5	7.0	6 545
East Branch LeClerc Creek	50	Oct. 6 & 12, 1999	167	66–281	37.2	20.8	7 738
Middle Creek	50	Oct. 7, 1999	144	62–241	40.1	12.2	4 892
Cedar Creek	50	Oct. 12, 1999	138	95–232	24.7	13.4	3 310
Mill Creek (upper & lower)	50	Oct. 14 & 19, 1999	171	93–237	6.8	14.3	972
Slate Creek	50	Oct. 19 & 20, 1999	155	78–231	21	10.0	2 100
Upper Sullivan Creek	50	Oct. 14, 1999	177	78–250	17	28.0	4 760
North Fork Sullivan Creek	50	Oct. 21, 1999	134	81–224	Unknown	15.0	Unknown

[a]Based on electroshocking or snorkel surveys.
[b]Estimated density × estimated cutthroat habitat.

Table 2. Loci screened, alleles observed, and overall variability.

Locus[a]	Reported repeat unit(s)	# Alleles		Overall size range	Heterozygosity (H_0)	Gene diversity (H_T)
		Cutthroat	Rainbow			
Ogo-3	GT	30	4	181–265	0.68	0.90
Oki-10	CTGT	20	3	92–160	0.67	0.87
Ots-101[b,c]	GACA, AAC	30	8[c]	143–484[c]	0.66	0.92
FGT-3	CA	22	7	143–244	0.66	0.85
Ots-107[b]	GATA, GT, GTCT	33	8	133–233	0.78	0.94

[a]*Ogo-3* (Olsen et al. 1998); *Oki-10* (Smith et al. 1998); *Ots-101* (Small et al. 1998); *FGT-3* (Sakamoto et al. 1994); *Ots-107* (Nelson & Beacham 1999).
[b]Alleles in Yellowstone offset by 2 bp.
[c]Rainbow with: dinucleotide repeat pattern, allele(s) >500 bp, and null allele(s) present.

of these streams originate in northeastern Washington, just east of the divide that separates the LeClerc, Mill, Middle, and Sullivan creek drainages in Washington from those flowing east into Idaho. The Kings Lake Hatchery strain has been used by WDFW since 1943 as a source of cutthroat trout for stocking in northeastern Washington. The Twin Lakes Hatchery strain of westslope cutthroat was probably established with fish from Wenatchee Lake, the Wenatchee River, or the Chelan Hatchery at Stehekin Creek, all in Chelan County.[2] The Twin Lakes Hatchery strain has been used since 1915 by WDFW as a source of cutthroat for numerous stocking programs. Behnke's (1992) morphological analysis of both of these westslope hatchery strains suggested they were unaffected by hybridization with Yellowstone cutthroat. The Yellowstone Hatchery strain is reared at the Yellowstone River Trout Hatchery in Big Timber (MT). This strain was founded with fish from McBride Lake in Yellowstone National Park,[3] and is

representative of the Yellowstone subspecies of cutthroat. The Spokane Hatchery strain of rainbow trout, which has been referred to as the 'Cape Cod strain', was founded with fish from the McCloud River (near Mt. Shasta, CA[2]) and has been extensively planted throughout the region by WDFW since approximately 1943.

Genomic DNA was extracted from a small piece of fin sample using a 5% chelex (BioRad Chelex 100 resin) suspension containing Proteinase K. Following extraction at 65°C for 30–180 min, the samples were heated for 5 min at 95°C to denature proteins. The microsatellite DNA loci were amplified and fluorescently labeled using the polymerase chain reaction (PCR; Saiki et al. 1988) with specific primers. The loci analyzed and their characteristics are listed in Table 2. Three loci were multiplexed together in one PCR: *Ots-101** (0.2 μM; labeled with 6-Fam), *Oki-10** (0.09 μM; labeled with Hex), and *Ogo-3** (0.2 μM; labeled with Ned). The remaining two loci – *FGT-3** (0.09 μM; labeled with 6-Fam) and *Ots-107** (0.2 μM; labeled with Hex) – were amplified in a second PCR. The PCR reactions for both multiplexes consisted of 40

---

[3] Daryl Hodges, Yellowstone River Trout Hatchery manager, pers. comm.

cycles with a 50°C annealing temperature and a 30 min final extension at 72°C. PCR products were mixed with Genescan-500 Rox size standard and analyzed on an ABI-377 automated sequencer (PE Biosystems) running in Genescan mode. Ned, Hex, 6-Fam, and Rox are fluorescent amidite dyes commonly used to label DNA fragments. The resulting data were processed using Genescan, version 2.1 and Genotyper, version 2.0 software (PE Biosystems). The raw size calls from Genotyper were imported into MS Excel where allele calling was accomplished using bins defined based on the primary classes observed in plotted distributions of the Genotyper size calls for each locus and the presumed minimum repeat motif of each microsatellite. Allele bins were centered on median values, with bin boundaries defined to leave interbin spaces equal to one-half the interval between adjacent bins (see Figure 2). This was done to ensure that all DNA

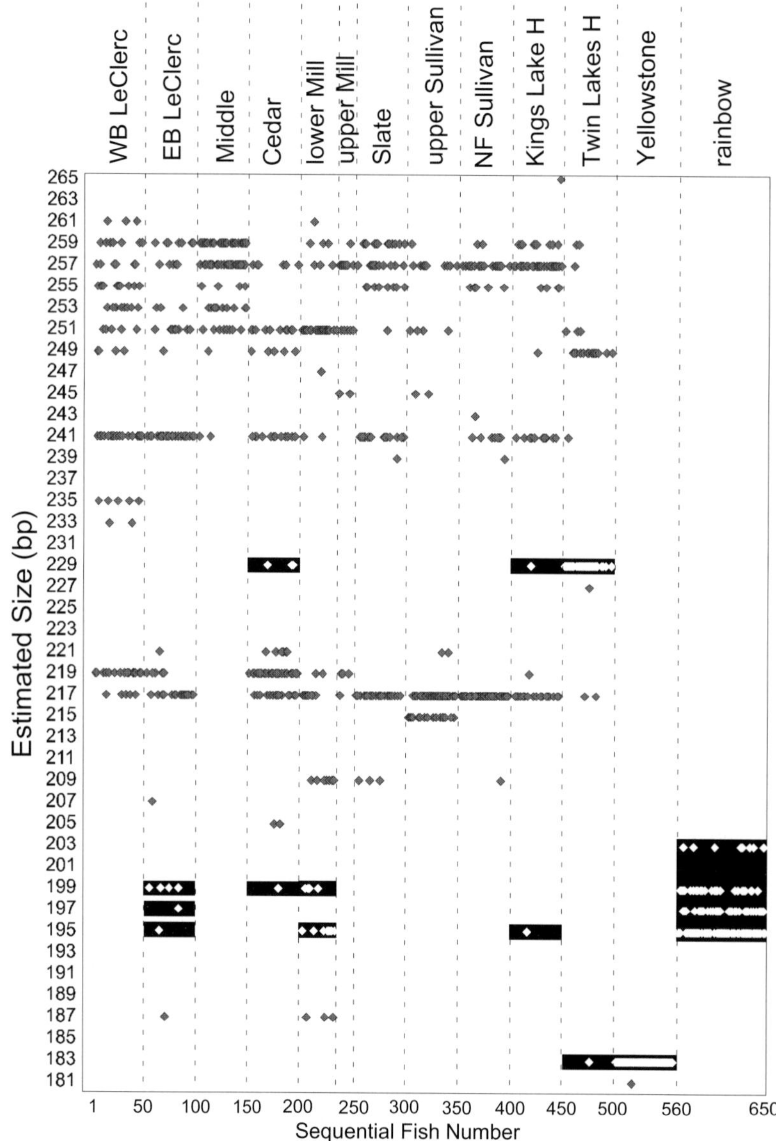

Figure 2. Plot of *Ogo-3** alleles (by bin) observed in the 650 fish screened in this study. Each vertical column shows the two alleles found in a fish. Presumed diagnostic alleles for hatchery rainbow, Yellowstone cutthroat, and the Twin Lakes Hatchery stain of westslope cutthroat are identified by open symbols included in shaded boxes.

fragments falling within an allele bin were more similar in size to each other than any was to a fragment in another recognized allele bin. Single-locus genotypes that included any raw size calls falling into an interbin space were zeroed to remove ambiguous allele scores from the data set. Given the 650 fish screened at five loci in this study, there were a total of 6 500 possible allele scores. The allele binning procedure we used resulted in a total of 52 ambiguous calls; 0.8% of all possible allele calls.

Statistical analyses of the data were accomplished using several computer programs. GENEPOP, version 3.3 (Raymond & Rousset 1995a) was used to calculate allele frequencies and observed heterozygosities, to test for Hardy–Weinberg equilibrium, and to perform explicit tests for population subdivision by assessing genotypic differentiation (Raymond & Rousset 1995b). We used FSTAT, version 2.9.3.2 (Goudet 1995) to generate descriptive statistics and to test for genotypic disequilibrium. We calculated Nei (1978) genetic distances and Weir & Cockerham (1984) coancestry values (theta) with GDA, version 1.0 (Lewis & Zaykin 2001[4]). We calculated F_{ST} values using Arlequin, version 1.1 (Schneider et al. 1997[5]). We used MICROSAT (version 2; available from: http://hpgl.stanford.edu/projects/microsat/programs) to calculate Cavalli-Sforza and Edwards (1967) chord distances and to bootstrap (by individuals) these values. We used PHYLIP (Felsenstein 1989) version 3.5c to create a consensus neighbor-joining tree of the Cavalli-Sforza and Edwards distances (using NEIGHBOR and CONSENSE). NTSYS-pc, version 2.0 (Rohlf 2000[6]) was used to conduct non-metric multidimensional scaling analyses of F_{ST} values and of genetic distances, to create minimum spanning trees, and to generate plots of the resulting information. We used STRUCTURE (Pritchard et al. 2000) to evaluate population structure and the proportional ancestry of individual fish to identified groups.

---

[4] Lewis, P.O. & D. Zaykin. 2001. Genetic data analysis: Computer program for the analysis of allelic data, Version 1.0 (d16c). Free program distributed by the authors over the internet from http://lewis.eeb.uconn.edu/lewis/lewishome/software.html.

[5] Schneider, S., J.-M. Kueffer, D. Roessli & L. Excoffier. 1997. Arlequin ver. 1.1: A software for population genetic data analysis, Genetics and Biometry Laboratory, University of Geneva, Switzerland. 000 pp.

[6] Rohlf, F.J. 2000. NTSYS-pc. Numerical Taxonomy and Multivariate Analysis System. Version 2.0. Exeter Software, Setauket, NY. 000 pp.

Results

Despite exhibiting tetranucleotide patterns of allelic variation in nearly all the cutthroat screened, *Ots-107** exhibited a dinucleotide pattern of variation in the rainbow (only one collection analyzed). Furthermore, the five alleles of *Ots-107** observed in the Yellowstone cutthroat were all 2-bp out of phase with the predominant pattern of alleles seen in the westslope cutthroat collections. Similarly, although *Ots-101** exhibited tetranucleotide patterns of allelic variation in all collections (cutthroat and rainbow), four high frequency alleles at *Ots-101** in Yellowstone (**237*, **241*, **253*, and **257*; combined frequency = 0.90) were 2-bp out of phase with the distribution of alleles seen in all westslope cutthroat collections. For these reasons, we binned alleles at all five loci assuming a 2-bp repeat motif. This approach had a pronounced effect on the between-collection resolution obtained with *Ots-101** and *Ots-107** in the current data set. All five loci screened exhibited high numbers of alleles and relatively high values of observed heterozygosity (Table 2). Additionally, large differences in allele frequencies were observed among collections at all loci (Appendix 1).

Tests of Hardy–Weinberg equilibrium revealed that genotypes were in expected proportions in all but four cases. The four heterozygote deficiencies that were significant after correction for multiple tests occurred at: *Oki-10** in Twin Lakes, *Ots-107** in rainbow, and *Ots-101** in both the total collection from Mill (upper and lower Mill combined) and in rainbow. We believe that these results indicate that one or more null alleles occur at both *Ots-101** and *Ots-107** in rainbow. Furthermore, we believe that it is likely that the significant heterozygote deficiency at *Ots-101** for the combined Mill Creek collection was due to population admixture in this combined sample and the presence of rainbow-cutthroat hybrids in lower Mill (see below).

Nearly all analyses of genetic data assume independence of loci (i.e., that the loci are not genetically linked and, therefore, that alleles at different loci recombine and assort randomly at reproduction). This assumption was examined for the loci screened in this study by conducting tests of genotypic disequilibrium. A total of 128 pairwise tests were done and only one test (*Ots-101** × *Ots-107** in lower Mill) yielded a significant outcome. We believe that the results of the Hardy–Weinberg tests and of the genotypic disequilibrium tests indicate that the five loci screened are suitable markers for the analysis of population structure and genetic interrelationships in cutthroat trout.

Thirty-one alleles occurred at frequencies of 0.05 or greater in either the rainbow or the Yellowstone cutthroat collections and were either absent or at very low frequencies in the various westslope cutthroat collections (e.g., Figure 2). We considered these alleles to be diagnostic of hatchery rainbow and Yellowstone cutthroat and interpreted their presence in Pend Oreille cutthroat as an indication of probable introgression (via past hybridization). Using this approach, we found one fish that appeared to be a pure rainbow and five others that each had two or three of these diagnostic alleles (we concluded that these five fish were introgressed individuals). Interestingly, three of these five fish had both 'rainbow' and 'Yellowstone' alleles. These six fish were removed from the data set for subsequent analyses. Additionally, we found a total of 22 other fish that had one diagnostic allele each. We considered these individuals as probably introgressed, but left them in the data set because we did not believe that the presence of a single, presumed diagnostic allele was definitive, given the limited scope of our data. Presumed introgression was most pronounced in the lower Mill collection (15/36), followed by the East Branch LeClerc (6/50), and upper Sullivan (3/50) collections. Both the West Branch LeClerc and the Cedar collections had one fish with a single diagnostic allele. No evidence of introgression was observed in the Middle, upper Mill, North Fork Sullivan, or Slate collections. We treated the lower and upper Mill Creek fish as two separate groups in the analyses described below because of the strong signal of introgression in the lower Mill fish, the significant Hardy–Weinberg equilibrium test at *Ots-101** in the combined Mill collection (upper and lower Mill), and substantial differences in allele frequencies at all five loci between lower and upper Mill (see below).

There was substantial allelic variation among the 13 collections at all loci. A plot of *Ogo-3** alleles exemplifies this variation (Figure 2). The Yellowstone collection had only **183* and **181* alleles and these alleles were absent in all other collections (except for presumed introgressed individuals). The rainbow trout population only had alleles between 195 and 203 bp at this locus. We believe that the rare occurrence of these *Ogo-3** alleles in a few of the cutthroat was due to hybridization. Similarly, the Twin Lakes collection had a high frequency of the **229* allele and its rare presence in two other cutthroat collections was attributed to introgression.

Variation among the Pend Oreille cutthroat collections was considerable. Ignoring alleles presumed to have come from introgression, the number of alleles per collection of Pend Oreille cutthroat averaged 43.4 and ranged from 33 in upper Mill (N = 14) to 52 in West Branch LeClerc (Appendix 1). Each locus had multiple alleles that exhibited large frequency differences among the eight Pend Oreille tributaries. Examples included: *Ots-101*195*, p = 0.810 in upper Sullivan but absent in Middle and p = 0.014 in East Branch LeClerc; *Ots-107*145*, p = 0.490 in Middle but absent in North Fork Sullivan; *Oki-10*104*, p = 0.772 in North Fork Sullivan and p = 0.094 in Middle; *FGT-3*171*, p = 0.511 in Middle Creek but absent in upper Sullivan and p = 0.031 in North Fork Sullivan; and *Ogo-3*217*, p = 0.633 in North Fork Sullivan but absent in Middle and p = 0.054 in West Branch LeClerc. There were three cases of apparent 'private alleles' (alleles occurring at a frequency of over 5% in one collection but absent from all others): *Ogo-3*215*, p = 0.202 in upper Sullivan and *Ogo-3*235*, p = 0.054 in West Branch LeClerc (Figure 2); and *FGT-3*208*, p = 0.054 in lower Mill. Figure 2 also shows large frequency differences among the 10 Pend Oreille cutthroat collections for several other *Ogo-3** alleles (e.g., **219, *241, *251*, and **259*).

Population subdivision was formally investigated in several ways. Fixation indices (F_{ST}; see Wright 1969) were calculated using Arlequin version 1.1. The resulting values of F_{ST} indicated considerable genetic differentiation among the collections (Table 3). Pairwise F_{ST} values for the five loci ranged from 0.014 (Slate vs. Kings Lake) to 0.436 (Twin Lakes vs. Yellowstone) and all values were significantly larger than zero according to Arlequin. The average pairwise F_{ST} among the nine Pend Oreille tributaries was 0.128. Collectively, these F_{ST} values indicated substantial population substructure among the collections, presumably resulting from reproductive isolation. Explicit tests of population differentiation (Raymond & Rousset 1995b) were conducted using contingency table exact probability tests of the genotypic data to examine this issue further. Out of a total of 390 tests (78 per locus), only 20 were non-significant, even after Bonferonni correction for multiple tests. All collection pairs except Slate *versus* Kings Lake yielded significant outcomes for at least two loci and the vast majority yielded significant outcomes at all five loci. When all five loci were considered together, these tests indicated that all collection pairs were significantly different. These results (and the F_{ST} values discussed above) suggested that virtually all the Pend Oreille cutthroat collections represented distinct populations (separate gene pools). At the same time, it was clear that there was a relatively

Table 3. Pairwise Cavalli-Sforza & Edwards (1967) chord distances above diagonal and pairwise F_{ST} values below diagonal (all values significantly larger than zero; C–S & E distances calculated using MICROSAT; F_{ST} values calculated using Arlequin).

	WB LeClerc	EB LeClerc	Middle	Cedar	lower Mill	upper Mill	Slate	upper Sullivan	NF Sullivan	Kings Lake	Twin Lakes	Yellow-stone	Rainbow
WB LeClerc	—	0.390	0.488	0.469	0.571	0.539	0.484	0.610	0.569	0.479	0.707	0.986	0.975
EB LeClerc	0.036	—	0.472	0.404	0.485	0.520	0.425	0.571	0.526	0.374	0.703	0.975	0.918
Middle	0.138	0.125	—	0.599	0.583	0.546	0.514	0.707	0.653	0.501	0.714	0.992	0.939
Cedar	0.058	0.039	0.137	—	0.610	0.623	0.576	0.644	0.656	0.537	0.705	0.993	0.984
lower Mill	0.086	0.065	0.131	0.075	—	0.535	0.502	0.674	0.587	0.490	0.703	0.992	0.881
upper Mill	0.097	0.101	0.149	0.134	0.073	—	0.550	0.620	0.512	0.497	0.781	0.990	0.979
Slate	0.093	0.049	0.138	0.079	0.058	0.114	—	0.556	0.410	0.256	0.664	0.995	0.962
upper Sullivan	0.204	0.172	0.273	0.187	0.188	0.201	0.149	—	0.477	0.525	0.740	0.972	0.992
NF Sullivan	0.172	0.137	0.261	0.179	0.158	0.131	0.122	0.090	—	0.405	0.776	0.991	0.997
Kings Lake H	0.095	0.051	0.132	0.095	0.068	0.085	0.014	0.149	0.114	—	0.656	0.993	0.956
Twin Lakes H	0.236	0.258	0.267	0.234	0.191	0.319	0.204	0.318	0.338	0.222	—	0.973	0.887
Yellow-stone	0.345	0.365	0.393	0.347	0.329	0.421	0.326	0.418	0.426	0.335	0.436	—	0.921
Rainbow	0.266	0.257	0.284	0.249	0.199	0.296	0.240	0.336	0.341	0.250	0.245	0.363	—

close affinity between the Slate Creek population and the Kings Lake Hatchery strain from both the pairwise F_{ST} value (0.014) and the results of the population differentiation tests.

The interrelationships among the collections were examined further by calculating several different genetic distance measures (see Table 3 for Cavalli-Sforza & Edwards (1967) chord distance values). While the precise values varied among distance measures, most measures showed the same general relationships among populations. The Yellowstone subspecies, *O. clarki bouvieri*, and the rainbow trout yielded the largest distance values in comparisons with all other collections, followed by the Twin Lakes Hatchery strain of westslope cutthroat. The pairwise distances among the remaining 10 collections were smaller, but were themselves somewhat heterogeneous. For all measures, the Slate–Kings Lake Hatchery comparison showed the smallest distance.

A neighbor-joining tree of Cavalli-Sforza and Edwards chord distances indicated that the upper Sullivan and North Fork Sullivan populations formed a distinct cluster as did the Kings Lake and Slate populations (Figure 3). Likewise, there was support for the branch to Mill (combined upper and lower Mill). The

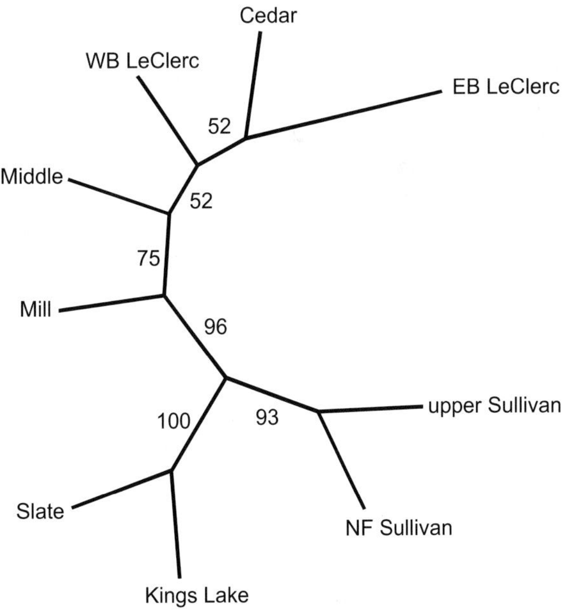

Figure 3. Consensus neighbor-joining tree of pairwise Cavalli-Sforza & Edwards (1967) chord genetic distances for the nine Pend Oreille collections (upper and lower Mill combined). Values on branches indicate the number of times the more distal branching patterns were observed in 100 bootstraps.

Table 4. Proportional membership of each collection in each of eight clusters identified using the program STRUCTURE.

Collection	Cluster							
	1	2	3	4	5	6	7	8
WB LeClerc	0.79	0.07	0.02	0.02	0.08	0.01	0.00	0.00
EB LeClerc	0.55	0.10	0.03	0.06	0.24	0.01	0.00	0.00
Middle	0.02	0.93	0.01	0.01	0.02	0.01	0.00	0.00
Cedar	0.87	0.03	0.01	0.02	0.05	0.02	0.00	0.00
lower Mill	0.07	0.06	0.60	0.04	0.20	0.02	0.00	0.01
upper Mill (N = 14)	0.18	0.17	0.06	0.14	0.43	0.01	0.00	0.00
Slate	0.05	0.04	0.02	0.08	0.79	0.01	0.00	0.00
upper Sullivan	0.01	0.01	0.01	0.95	0.02	0.01	0.00	0.00
NF Sullivan	0.01	0.01	0.01	0.76	0.20	0.01	0.00	0.00
Kings Lake H	0.06	0.06	0.02	0.11	0.73	0.02	0.00	0.01
Twin Lakes H	0.01	0.01	0.01	0.01	0.01	0.94	0.00	0.00
Yellowstone	0.00	0.00	0.00	0.00	0.00	0.00	0.98	0.00
Rainbow	0.00	0.00	0.00	0.00	0.00	0.00	0.00	0.98

Boxes identify estimated memberships of greater than 0.50.

relationships among Middle, the two LeClerc populations (East Branch and West Branch) and Cedar were ambiguous, with low bootstrap support. We also conducted multidimensional scaling (MDS) analysis with minimum spanning trees using several of the genetic distance matrixes to provide additional graphic comparisons. The resulting MDS diagrams were fundamentally similar for all distance measures and showed the Yellowstone cutthroat subspecies, *O. clarki bouvieri*, as the most distant outlier and that the Spokane Hatchery rainbow strain and the Twin Lakes Hatchery westslope cutthroat strain were also quite divergent from the remaining 10 Pend Oreille populations (including the Kings Lake Hatchery strain) (data not shown). MDS plots using only the 10 Pend Oreille populations emphasized the close similarity between the Slate and Kings Lake Hatchery populations and the relative divergence of the Middle, North Fork and upper Sullivan, and upper Mill populations (data not shown).

We also explored the data set without *a priori* regard for the collection locality of each fish using the model-based clustering program STRUCTURE (Pritchard et al. 2000). This program identifies clusters of individuals that minimize Hardy–Weinberg and genotypic disequilibrium in the overall data set and then probabilistically assigns individuals to these clusters. Repeated runs with the possible number of clusters (k) ranging from 6 to 11 were done to identify the most likely number of clusters. Maximum probabilities were obtained when the number of clusters was assumed to be eight. Table 4 summarizes the proportional membership of each of the 13 collections in the eight clusters identified by STRUCTURE. This analysis yielded proportional memberships in unique clusters exceeding 0.90 for four collections (Yellowstone, rainbow, Twin Lakes, and Middle). Additionally, it yielded proportional membership in a unique cluster that exceeded 0.70 for one other collection (Slate). The analysis assigned the majority of fish in the West Branch LeClerc, East Branch LeClerc, and Cedar collections to the same cluster, it assigned the majority of North Fork Sullivan to the same cluster as the upper Sullivan fish, and it assigned most of the Slate and Kings Lake Hatchery fish to another cluster. The upper Mill fish were not clearly affiliated with any single cluster, although the largest proportion of these fish were associated with the cluster containing most of the Slate and Kings Lake fish. However, any conclusions regarding the upper Mill population are tentative given the extremely small sample size (N = 14) for this population.

Discussion and conclusions

Throughout the past century, WDFW and many other agencies and individuals have introduced non-native trout (including cutthroat and rainbow) into the Pend Oreille River basin and other areas throughout Washington[1,7]. Because cutthroat subspecies and

[7] Trotter, P.C., B. McMillan, N. Gayeski, P. Spruell & A. Whiteley. 2001. Genetic and phenotypic catalog of native

rainbow trout can successfully interbreed and produce hybrid offspring and because other investigators have documented the existence of such hybrids in the region,[7] we felt it necessary to investigate the extent of such introgression in the Pend Oreille cutthroat collections before analyzing them for population divergence. Both this study and a previous investigation[7] found introgressed individuals in several Pend Oreille tributaries with the largest percentage in both studies occurring in lower Mill. Both studies also found introgressed individuals in East Branch LeClerc and Cedar but did not detect them in North Fork Sullivan or Slate. We believe it is noteworthy that, of the four populations in the present study that exhibited no evidence of introgression, three (Middle, upper Mill, and Slate) occur above falls that likely act as absolute barriers to upstream migration. Our investigation indicated that 9 of the 10 Pend Oreille collections had either a low frequency of introgressed individuals or none at all and that the vast majority of presumed introgressed individuals had only a single non-native allele (out of the 10 at the five loci screened). While even this low level of introgression may be of concern with regard to the genetic health and integrity of the cutthroat populations, it is unlikely to have had a substantial influence on population divergence and interrelationships within the basin.

The question of whether or not stocking of Twin Lakes and/or Kings Lake hatchery strains has significantly affected Pend Oreille populations is more difficult to resolve unambiguously because of the greater similarity of these populations to those in the Washington tributaries of the Pend Oreille. The Twin Lakes Hatchery collection had five alleles at *Ots-101** that were not observed in any of the collections from the Pend Oreille (**243, *247, *249, *251*, and **255*) and these had a cumulative frequency of 0.744. Additionally, *Ogo-3*229* occurred at a frequency of 0.635 in Twin Lakes but was only seen in two Pend Oreille collections (Cedar; p = 0.032 and Kings Lake, p = 0.010). These very large frequency differences, together with frequency differences for *Oki-10*108* and *FGT-3*177* suggest that there has been relatively little introgression of Twin Lakes Hatchery strain genes into Pend Oreille populations (including Kings Lake).

Rigorous exclusion of genetic effects from the Kings Lake Hatchery strain is even more difficult because this strain – which was founded using fish from tributaries to the Pend Oreille River in Idaho – would be expected to be, and is, genetically similar to the collections from the Washington tributaries of the Pend Oreille. Nevertheless, the allele frequency data, genetic distances, F_{ST} values, and the population differentiation tests and STRUCTURE analyses make it very unlikely that the genetic character of the native Washington Pend Oreille westslope cutthroat stocks has been dramatically altered by the Kings Lake strain.

These same data and analyses, however, clearly demonstrate a close genetic affinity between the Kings Lake Hatchery stock and the cutthroat population in Slate Creek. It seems likely that this genetic similarity is the result of past introductions of Kings Lake Hatchery fish into Slate Creek (in 1945, 1948, 1954, and 1981)[1] and the subsequent survival and successful reproduction of these fish. In fact, Slate Creek is inaccessible to trout due to an impassable, 6.0 m vertical falls with a relatively smooth bedrock face ~1 200 m upstream of the mouth that presumably acts as a complete barrier to upstream fish migration.[8] Assuming that this barrier is complete, the genetic data are consistent with the conclusion that past introductions of fish from the Kings Lake Hatchery into Slate Creek created a self-sustaining population in the creek. The fact that allele frequencies in Slate are not identical to those in the Kings Lake Hatchery stock could be due to genetic drift in one or both populations since the last stocking, and/or founder (population bottleneck) effects associated with the stocking of hatchery fish. If the Slate Creek population represents a population of non-local origin because the Kings Lake Hatchery stock was founded using fish from tributaries to Priest Lake (Idaho) it should probably be managed differently from the other, presumably native cutthroat stocks in the Pend Oreille River drainage.

The among-collection variation in allele frequencies at all five loci screened suggests limited or no

____

resident trout in the interior Columbia River basin. FY-99 report: Populations in the Pend Oreille, Kettle, and Sanpoil River Basins of Colville National Forest. BPA project 1998-026-00 report. 188 pp and references therein.

____

[8] McLellan, J.G. 2001. 2000 WDFW annual report for the project resident fish stock status above Chief Joseph and Grand Coulee Dams, Part I: Baseline assessment of Boundary Reservoir, Pend Oreille River, and its tributaries. pp. 63–218. *In*: Lockwood, N., J. McLellan & B. Crossley. 2001. Resident fish stock status above Chief Joseph and Grand Coulee Dams, 2000 annual report. Report to Bonneville Power Administration, Contract No. 00004819, Project No. 199700400, BPA Report DOE/BP-00004619–1. 291 electronic pp.

136

interbreeding among any of the 13 populations characterized in this study. This inference is strongly supported by the relatively large F_{ST} values observed (mean $F_{ST} = 0.215$ for all 13 collections and mean $F_{ST} = 0.128$ for the nine Pend Oreille tributary samples). The direct genotypic tests of population differentiation statistically confirm this inference and establish that the collections represent independent, reproductively isolated gene pools ($=$'stocks'). The F_{ST}, genetic distance, MDS, and STRUCTURE analyses all provide insights into the interrelationships among the cutthroat populations in the Pend Oreille basin. These all indicate that the Middle and Sullivan (upper and North Fork) populations are the most divergent and that the West Branch and East Branch LeClerc populations are the most similar. Whether or not some of the divergence of the Middle and the Sullivan populations is due to the fact that each was represented by fish from only one or two sampling locations is unknown at present but can be tested by analyzing additional collections from these streams.

The existence of multiple, genetically differentiated subpopulations of cutthroat trout is, in itself, not surprising given that both allozyme and microsatellite DNA studies have documented subpopulation structure in cutthroat trout in many locations (e.g., Phelps and Allendorf 1982, Leary et al. 1985, Campton and Utter 1987, Allendorf and Leary 1988, Wenburg et al. 1998).[9] Wenburg et al. (1998) investigated microsatellite DNA variation in 13 anadromous populations of coastal cutthroat trout, *O. clarki clarki*, in western Washington. They reported relatively high values of among-population F_{ST} (0.121) and R_{ST} (0.093) and concluded that the pattern of interpopulation differences was consistent with isolation by distance. Our results for westslope cutthroat trout populations within the Pend Oreille River drainage indicate substantial population substructure, just as those of Wenburg et al. (1998) did for coastal cutthroat in western Washington. However, while the study of Wenburg et al. (1998) examined 13 populations spanning a range of ~800 km of shoreline (from Willapa Bay in the South, through Hood Canal and south Puget Sound, to the Nooksack River in the North), the present study examined populations in eight tributary streams spanning a distance of only about 113 km of the Pend Oreille River. Furthermore, the present study included three pairs of collections from different portions of the same stream (East and West branches of LeClerc Creek, upper and

---

[9] Also Baker, B.M. WDFW; unpubl. data.

North Fork Sullivan Creek, and lower and upper Mill Creek) that were each significantly divergent. Thus, the westslope cutthroat trout populations in the present study showed considerable genetic differentiation even within tributaries on this fine scale (<32 km). Possible explanations for the observed microgeographic genetic differences include: (1) very limited gene flow between populations sampled due to physical and/or other barriers, (2) high levels of genetic drift due to very small population sizes in some or all streams in the study, and (3) demographic/sampling artifacts whereby collections were dominated by full or half sibs (or otherwise closely related individuals). We believe that limited gene flow is the most likely explanation for the observed genetic divergence among populations because the allelic diversity was relatively high in all collections (Appendix 1), the estimated census population sizes exceeded 2 000 for all streams except upper and lower Mill Creek (Table 1), and sampling was often spread among multiple locations (Figure 1) and included a range of size classes of fish in most collections (Table 1).

We conclude that westslope cutthroat trout in the Pend Oreille drainage in northeastern Washington are subdivided into distinct genetic populations at least at the level of major tributaries (if not at the level of individual creeks). We recommend that management and conservation activities for these fish be focused at this fine geographic scale in order to maintain natural levels of productivity and genetic diversity for cutthroat trout in this region.

Acknowledgements

We wish to thank Cherril Bowman, Alice Pichahchy, and Kari Schilling for laboratory assistance and Ken Warheit for advice and assistance with data analysis. We also acknowledge personnel from the Kalispel Tribe Natural Resources Department, Eastern Washington University, and WDFW for their assistance with field collections. We thank Jim Ebel (WDFW Colville Hatchery), Mike Lewis (WDFW Spokane Hatchery), and Daryl Hodges (Yellowstone River Trout Hatchery) for providing the hatchery samples. This study was part of the project entitled: 'Resident Fish Stock Assessment Above Chief Joseph and Grand Coulee Dams' that was funded by the Bonneville Power Administration via a subcontract with the Kalispel Tribe.

References

Allendorf, F.W. & R.F. Leary. 1988. Conservation and distribution of genetic variation in a polytypic species, the cutthroat trout. Conserv. Biol. 2: 170–184.

Behnke, R.J. 1992. Native trout of Western North America. Amer. Fish. Soc. Monogr. 6. 275 pp.

Campton, D.E. & F.M. Utter. 1987. Genetic structure of anadromous cutthroat trout (*Salmo clarki clarki*) populations in the Puget Sound area: Evidence for restricted gene flow. Can. J. Fish. Aquat. Sci. 44: 573–582.

Cavalli-Sforza, L.L. & A.W.F. Edwards. 1967. Phylogenetic analysis, models and estimation procedures. Evolution 21: 550–570.

Felsenstein, J. 1989. PHYLIP – Phylogeny Inference Package (version 3.2). Cladistics 5: 164–166.

Goudet, J. 1995. FSTAT (version 1.2): A computer program to calculate F-statistics. J. Hered. 86: 485–486.

Leary, R.F., F.W. Allendorf & S.R. Phelps. 1985. Population genetic structure of westslope cutthroat trout: Genetic variation within and among population. Proc. Mont. Acad. Sci. 45: 37–45.

Nei, M. 1978. Estimation of average heterozygosity and genetic distance from a small number of individuals. Genetics 89: 583–590.

Nelson, R.J. & T.D. Beacham. 1999. Isolation and cross species amplification of microsatellite loci useful for study of Pacific salmon. Anim. Genet. 30: 228–229.

Olsen, J.B., P.B. Bentzen & J.E. Seeb. 1998. Characterization of seven microsatellite loci derived from pink salmon. Mol. Ecol. 7: 1083–1090.

Phelps, S.R. & F.W. Allendorf. 1982. Genetic comparison of upper Missouri cutthroat trout to other *Salmo clarki lewisi* populations. Proc. Mont. Acad. Sci. 41: 14–22.

Pritchard, J.K., M. Stephens & P. Donnelly. 2000. Inference of population structure using multilocus genotype data. Genetics 155: 945–959.

Raymond, M. & F. Rousset. 1995a. GENEPOP (ver. 1.2): A population genetics software for exact test and ecumenicism. J. Hered. 86: 248–249.

Raymond, M. & F. Rousset. 1995b. An exact test for population differentiation. Evolution 49: 1280–1283.

Saiki, R.K., D.H. Gelfand, S. Stoffel, S.J. Scharf, R. Higuchi, G.T. Horn, K.B. Mullis & H.A. Erlich. 1988. Primer-directed enzymatic amplification of DNA with a thermostable DNA polymerase. Science 239: 487–490.

Sakamoto, T., N. Okamoto & Y. Ikeda. 1994. Rapid communication: Dinucleotide repeat polymorphism of rainbow trout, FGT3. J. Anim. Sci. 72: 2766.

Small, M.P., T.D. Beacham, R.E. Withler & R.J. Nelson. 1998. Discriminating coho salmon (*Oncorhynchus kisutch*) populations within the Fraser River, British Columbia. Mol. Ecol. 7: 141–155.

Smith, C.T., B.F. Koop & R.J. Nelson. 1998. Isolation and characterization of coho salmon (*Oncorhynchus kisutch*) microsatellites and their use in other salmonids. Mol. Ecol. 7: 1613–1621.

Weir, B.S. & C.C. Cockerham. 1984. Estimating F-statistics for the analysis of population structure. Evolution 38: 1358–1370.

Wenburg, J.K., P. Bentzen & C.J. Foote. 1998. Microsatellite analysis of genetic population structure in an endangered salmonid: The coastal cutthroat trout (*Oncorhynchus clarki clarki*). J. Mol. Evol. 7: 733–749.

Wright, S. 1969. Evolution and the Genetics of Populations. Vol. 2. The Theory of Gene Frequencies, University of Chicago Press, Chicago. 511 pp.

Appendix 1. Allele frequencies at five microsatellite DNA loci in 12 collections of cutthroat trout and one collection of rainbow trout.

Locus allele	WB LeClerc	EB LeClerc	Middle	Cedar	lower Mill	upper Mill	Slate	upper Sullivan	NF Sullivan	Kings Lake	Twin Lakes	Yellowstone	Rainbow
Ogo-3													
181	—	—	—	—	—	—	—	—	—	—	—	0.009	—
183	—	—	—	—	—	—	—	—	—	—	0.014	0.991	—
187	—	0.012	—	—	0.043	—	—	—	—	—	—	—	—
195	—	—	—	—	0.143	—	—	—	—	0.010	—	—	0.528
197	—	—	—	—	—	—	—	—	—	—	—	—	0.244
199	—	0.012	—	0.011	0.057	—	—	—	—	—	—	—	0.178
203	—	—	—	—	—	—	—	—	—	—	—	—	0.050
205	—	—	—	0.021	—	—	—	—	—	—	—	—	—
207	—	0.012	—	—	—	—	—	—	—	—	—	—	—
209	—	—	—	—	0.100	—	0.033	—	0.010	—	—	—	—
215	—	—	—	—	—	—	—	0.202	—	—	—	—	—
217	0.054	0.209	—	0.181	0.114	0.042	0.356	0.585	0.633	0.292	0.027	—	—
219	0.185	0.058	—	0.309	0.029	0.125	—	—	—	0.010	—	—	—
221	—	0.012	—	0.064	—	—	—	0.021	—	—	—	—	—
227	—	—	—	—	—	—	—	—	—	—	0.014	—	—
229	—	—	—	0.032	—	—	—	—	—	0.010	0.635	—	—
233	0.022	—	—	—	—	—	—	—	—	—	—	—	—
235	0.054	—	—	—	—	—	—	—	—	—	—	—	—
239	—	—	—	—	—	—	0.011	—	0.010	—	—	—	—
241	0.196	0.372	0.020	0.149	0.043	—	0.222	—	0.082	0.146	0.014	—	—
243	—	—	—	—	—	—	—	—	0.010	—	—	—	—
245	—	—	—	—	—	0.083	—	0.021	—	—	—	—	—
247	—	—	—	—	0.014	—	—	—	—	—	—	—	—
249	0.043	0.012	0.010	0.064	—	—	—	—	—	0.010	0.216	—	—
251	0.054	0.093	0.090	0.106	0.343	0.208	0.011	0.043	—	—	0.041	—	—
253	0.098	0.035	0.120	—	—	—	—	—	—	—	—	—	—
255	0.120	—	0.040	—	—	—	0.089	—	0.051	0.042	—	—	—
257	0.065	0.058	0.410	0.064	0.043	0.500	0.133	0.117	0.184	0.365	0.014	—	—
259	0.076	0.116	0.310	—	0.043	0.042	0.144	0.011	0.020	0.104	0.027	—	—
261	0.033	—	—	—	0.029	—	—	—	—	—	—	—	—
265	—	—	—	—	—	—	—	—	—	0.010	—	—	—
nall	12	12	7	10	12	6	8	7	8	10	9	2	4
(N)	(46)	(43)	(50)	(47)	(35)	(12)	(45)	(47)	(49)	(48)	(37)	(58)	(90)

FGT-3													
143	—	—	—	—	0.036	—	—	—	—	—	—	—	—
157	0.033	—	—	0.095	—	—	—	—	—	—	—	—	—
159	—	—	—	0.012	—	—	—	—	—	—	—	—	—
161	0.022	—	—	—	0.036	0.115	0.177	—	0.146	0.181	—	—	—
163	0.467	0.559	0.370	0.298	0.268	0.654	0.146	0.446	0.563	0.266	0.011	0.010	—
165	—	—	—	—	—	—	—	—	0.021	—	—	—	—
167	—	—	—	—	0.018	—	—	0.011	—	—	—	—	—
169	—	—	—	—	—	—	—	—	—	—	—	0.010	—
171	0.300	0.294	0.511	0.321	0.250	0.038	0.156	—	0.031	0.074	0.189	—	—
173	—	—	0.022	0.012	—	—	—	—	—	—	—	—	—
175	0.011	0.118	0.065	0.060	0.071	—	0.260	0.043	0.125	0.287	0.078	—	—
177	0.056	—	0.011	—	0.054	—	0.125	0.141	—	0.106	0.689	—	—
179	—	0.015	—	—	—	—	—	0.315	—	0.011	0.033	—	—
181	0.011	0.015	0.011	—	—	—	0.073	0.011	0.021	0.021	—	—	—
183	0.011	—	0.011	—	—	0.192	0.063	0.011	0.094	0.032	—	—	—
187	—	—	—	0.012	—	—	—	—	—	—	—	—	—
191	0.089	—	—	0.190	0.161	—	—	0.011	—	0.021	—	—	—
193	—	—	—	—	0.018	—	—	—	—	—	—	—	—
202	—	—	—	—	—	—	—	—	—	—	—	0.031	—
204	—	—	—	—	—	—	—	—	—	—	—	0.337	0.086
206	—	—	—	—	—	—	—	—	—	—	—	0.010	—
208	—	—	—	—	0.054	—	—	—	—	—	—	—	—
214	—	—	—	—	—	—	—	—	—	—	—	0.592	0.270
220	—	—	—	—	0.036	—	—	—	—	—	—	0.010	0.023
222	—	—	—	—	—	—	—	—	—	—	—	—	0.069
224	—	—	—	—	—	—	—	0.011	—	—	—	—	0.218
234	—	—	—	—	—	—	—	—	—	—	—	—	0.270
244	—	—	—	—	—	—	—	—	—	—	—	—	0.063
nall	9	5	7	8	11	4	7	9	7	9	5	7	7
(N)	(45)	(34)	(46)	(42)	(28)	(13)	(48)	(46)	(48)	(47)	(45)	(49)	(87)
Oki-10													
92	—	—	—	0.021	0.061	—	—	—	—	—	—	—	—
96	—	—	—	—	0.015	—	—	—	—	—	—	—	—
100	0.163	0.039	0.063	0.042	0.030	—	0.022	0.020	0.043	0.032	—	—	0.011

Appendix 1. (Continued)

Locus allele	WB LeClerc	EB LeClerc	Middle	Cedar	lower Mill	upper Mill	Slate	upper Sullivan	NF Sullivan	Kings Lake	Twin Lakes	Yellowstone	Rainbow
102	—	—	—	—	0.076	—	—	—	—	—	—	—	—
104	0.296	0.303	0.094	0.177	0.273	0.536	0.233	0.408	0.772	0.277	0.095	—	—
106	—	—	—	—	—	—	—	—	—	—	0.071	—	—
108	0.020	—	0.260	—	0.212	0.036	0.056	—	—	0.032	0.762	—	0.697
112	0.224	0.289	—	0.417	—	—	0.078	—	—	0.021	0.036	—	—
114	0.010	—	—	—	—	—	—	—	—	—	—	—	—
116	—	—	—	—	—	—	—	—	—	—	—	—	0.292
124	—	0.013	—	0.083	—	—	—	0.020	—	—	—	—	—
128	—	0.053	0.385	0.104	0.045	0.143	0.056	0.031	0.011	0.128	0.012	—	—
132	0.020	0.184	0.042	0.010	0.076	0.107	0.156	0.092	0.109	0.298	0.024	—	—
136	0.214	0.092	0.094	0.094	0.061	—	0.044	0.102	—	0.085	—	—	—
140	0.031	0.026	0.063	0.052	0.152	0.143	0.333	0.020	0.065	0.128	—	—	—
144	0.020	—	—	—	—	0.036	0.022	0.245	—	—	—	—	—
148	—	—	—	—	—	—	—	0.031	—	—	—	0.029	—
152	—	—	—	—	—	—	—	0.031	—	—	—	0.471	—
156	—	—	—	—	—	—	—	—	—	—	—	0.471	—
160	—	—	—	—	—	—	—	—	—	—	—	0.029	—
nall	9	8	7	9	10	6	9	10	5	8	6	4	3
(N)	(49)	(38)	(48)	(48)	(33)	(14)	(45)	(49)	(46)	(47)	(42)	(52)	(89)
Ots-107													
133	—	—	—	—	—	—	0.011	—	0.071	—	—	—	—
137	—	—	0.020	—	—	—	0.032	—	—	0.020	0.010	—	—
139	0.010	—	—	—	—	—	—	—	—	—	—	—	—
141	0.327	0.150	0.110	0.087	0.031	0.036	0.128	0.051	0.082	0.120	0.060	—	—
143	0.010	—	—	—	—	—	—	—	—	—	—	—	—
145	0.112	0.088	0.490	0.228	0.109	0.143	0.117	0.020	—	0.040	0.040	—	—
147	—	—	0.020	—	—	—	—	—	—	—	—	—	—
149	0.071	0.088	0.020	0.109	0.063	0.036	0.234	0.122	0.020	0.280	0.030	—	—
151	0.010	—	—	—	—	—	—	—	—	—	—	—	—
153	0.194	0.088	0.030	—	0.125	0.107	0.053	0.031	0.224	0.090	—	—	—
155	—	—	—	—	—	—	—	—	—	—	0.010	—	—
157	0.031	0.025	0.020	0.011	0.078	—	0.096	0.112	0.082	0.060	0.040	—	—
161	0.031	0.175	0.010	0.217	0.047	—	—	0.214	0.082	0.070	0.060	—	—
163	0.010	—	—	—	—	—	—	—	—	—	—	0.167	—
165	0.112	0.213	0.280	0.054	—	0.071	0.106	0.306	0.102	0.110	0.030	—	—
167	—	—	—	—	—	—	—	—	—	—	—	0.200	—

169	0.071	—	—	0.022	—	—	0.096	0.143	0.092	0.030	0.410	—	—
171	—	—	—	—	—	—	—	—	—	—	—	0.250	—
173	0.010	0.038	—	0.043	—	—	0.106	—	—	0.050	0.050	—	0.067
175	—	—	—	—	—	—	—	—	—	—	—	0.258	—
177	—	0.063	—	0.109	0.047	0.179	0.021	—	—	0.010	0.040	—	—
179	—	0.013	—	—	—	—	—	—	—	—	—	—	0.124
181	—	0.038	—	0.022	0.125	0.071	—	—	—	0.050	0.010	—	—
183	—	—	—	—	—	—	—	—	—	—	—	0.125	—
185	—	0.013	—	—	0.141	0.286	—	—	—	0.020	0.020	—	—
187	—	—	—	—	0.016	—	—	—	—	0.010	—	—	0.112
189	—	—	—	—	—	0.071	—	—	0.224	0.030	0.010	—	—
193	—	—	—	0.098	—	—	—	—	0.010	0.010	0.130	—	—
197	—	—	—	—	0.016	—	—	—	0.010	—	0.050	—	—
199	—	0.013	—	—	0.172	—	—	—	—	—	—	—	0.006
201	—	—	—	—	—	—	—	—	—	—	—	—	0.101
203	—	—	—	—	—	—	—	—	—	—	—	—	0.191
207	—	—	—	—	—	—	—	—	—	—	—	—	0.343
227	—	—	—	—	0.016	—	—	—	—	—	—	—	—
229	—	—	—	—	0.016	—	—	—	—	—	—	—	—
233	—	—	—	—	—	—	—	—	—	—	—	—	0.056
nall	13	13	9	11	14	9	11	8	11	16	16	5	8
(N)	(49)	(40)	(50)	(46)	(32)	(14)	(47)	(49)	(49)	(50)	(50)	(60)	(89)
Ots-101													
143	—	—	—	—	—	—	—	—	—	—	—	0.038	—
187	—	—	—	0.050	—	—	0.034	—	0.051	—	—	—	—
191	0.415	0.086	0.044	0.083	—	0.125	0.011	0.030	—	0.038	0.027	—	—
195	0.032	0.014	—	0.050	0.103	0.125	0.182	0.810	0.551	0.103	0.014	—	—
199	—	—	—	—	—	—	0.034	0.090	0.010	0.051	0.014	—	—
203	0.011	—	—	—	—	—	—	—	—	—	—	—	—
207	—	0.057	0.011	—	0.147	—	0.205	—	0.061	0.269	—	—	—
211	0.032	0.086	0.433	0.050	—	—	0.057	—	0.020	0.090	—	—	—
215	0.319	0.200	0.211	—	0.338	0.458	0.205	—	0.092	0.192	0.081	—	—
217	—	—	—	—	—	0.042	—	—	—	—	—	—	—

Appendix 1. (*Continued*)

Locus allele	WB LeClerc	EB LeClerc	Middle	Cedar	lower Mill	upper Mill	Slate	upper Sullivan	NF Sullivan	Kings Lake	Twin Lakes	Yellowstone	Rainbow
219	0.106	0.257	0.233	0.167	0.235	0.083	0.170	0.050	0.173	0.115	0.041	—	—
223	—	0.071	—	0.067	0.118	—	0.045	—	—	0.051	0.041	—	—
233	—	—	—	—	—	—	—	—	—	—	—	0.013	—
237	—	—	—	—	—	—	—	—	—	—	0.014	0.238	—
241	—	—	—	—	—	—	—	—	—	—	—	0.063	—
243	—	—	—	—	—	—	—	—	—	—	0.041	—	—
247	—	—	—	—	—	—	—	—	—	—	0.162	—	—
249	—	—	—	—	—	—	—	—	—	—	0.054	0.025	—
251	—	—	—	—	—	—	—	—	—	—	0.311	—	—
253	—	—	—	—	—	—	—	—	—	—	—	0.500	—
255	—	—	—	—	—	—	—	—	—	—	0.176	—	—
257	—	—	—	—	—	—	—	—	—	—	—	0.100	—
259	—	0.014	—	0.183	—	—	—	—	—	0.026	0.014	—	—
263	0.011	0.071	0.022	0.117	—	0.083	0.034	—	—	0.038	—	—	—
267	0.053	0.129	—	0.133	—	0.042	—	0.020	—	0.013	—	—	—
271	0.021	0.014	0.044	0.100	—	0.042	—	—	—	—	—	—	—
275	—	—	—	—	0.015	—	—	—	0.010	0.013	0.014	—	—
279	—	—	—	—	0.029	—	0.023	—	0.031	—	—	—	—
283	—	—	—	—	0.015	—	—	—	—	—	—	—	—
287	—	—	—	—	—	—	—	—	—	—	—	0.025	—
318	—	—	—	—	—	—	—	—	—	—	—	—	0.115
330	—	—	—	—	—	—	—	—	—	—	—	—	0.020
382	—	—	—	—	—	—	—	—	—	—	—	—	0.304
386	—	—	—	—	—	—	—	—	—	—	—	—	0.189
390	—	—	—	—	—	—	—	—	—	—	—	—	0.061
440	—	—	—	—	—	—	—	—	—	—	—	—	0.014
448	—	—	—	—	—	—	—	—	—	—	—	—	0.027
484	—	—	—	—	—	—	—	—	—	—	—	—	0.270
nall	9	11	7	10	8	8	11	5	9	12	14	8	8
(N)	(47)	(35)	(45)	(30)	(34)	(12)	(44)	(50)	(49)	(39)	(37)	(40)	(74)

'—' = allele not observed; nall = number of alleles observed; N = number of fish successfully scored.

Environmental Biology of Fishes **69**: 143–152, 2004.

Microsatellite variation in populations of Atlantic salmon from North Europe

Vidar Wennevik[a], Øystein Skaala[a], Sergej F. Titov[b], Igor Studyonov[c] & Gunnar Nævdal[d]

[a]*Institute of Marine Research, P.O. Box 1870 Nordnes, N-5817 Bergen, Norway*
(e-mail: vidar.wennevik@imr.no)
[b]*State Research Institute on Lake and River Fisheries, Makarova Emb. 26, 199053 St. Petersburg, Russia*
[c]*Northern Branch of Polar Research Institute of Marine Fisheries and Oceanography, SevPINRO,*
Uritskovo ul. 17, 163002 Arkhangelsk, Russia
[d]*Institute of Fisheries and Marine Biology, University of Bergen , P.O. Box 7800, N-5020 Bergen, Norway*

Received 17 April 2003   Accepted 19 April 2003

Key words: genetics, distribution, population, divergence

Synopsis

Our aim was to investigate the level of genetic differentiation in northern European populations of Atlantic salmon, to establish the genetic relationship among major salmon populations in Russia and North Norway, and to compare these to populations from the western Atlantic lineage. Samples were collected along an east–west axis, from Pechora River in Russia to Restigouche River in Quebec, Canada. A total of 439 individual salmon were collected from seven rivers (sample sizes from 50 to 84 individuals). The samples were analysed for variation at four microsatellite loci; Ssa13.37, Ssa14, Ssa171 and Ssa171. Significant differences were found between most of the European populations, and the populations from the Tana and Pechora Rivers were most distinct. The samples from the Rivers Mezenskaya Pizhma and Emtsa in Arkhangelsk oblast in Russia were not significantly different from each other in an exact test of population differences. All other river pairs were significantly different. These results confirmed the deep genetic divergence between American and European salmon populations demonstrated in earlier studies, with alleles specific to continent found in three of the microsatellites.

Introduction

Genetic differentiation in Atlantic salmon is maintained by high homing fidelity and correspondingly low levels of gene flow among populations. Early studies of genetic population structure based on polymorphisms at protein coding loci indicated multiple genetically differentiated and more or less reproductively isolated populations in individual rivers (Møller 1970, Payne et al. 1971, Ståhl 1987). From his studies of data from 23 major river systems draining into Atlantic Ocean and the Baltic Sea, Ståhl (1987) concluded that three major groups of populations could be identified; Western Atlantic, Eastern Atlantic and Baltic. Studies using minisatellites and microsatellites

also demonstrate a significant differentiation between European and American populations (Taggart et al. 1995, McConnell et al. 1995a,b, Norris et al. 1999, Sanchez et al. 2000, King et al. 2001). Relatively few comparative studies have been conducted on North Norwegian and Russian populations (Heggberget et al. 1986, Kazakov & Titov 1991, 1993, Skaala et al. 1998), and none so far using DNA markers. The aim of the present study was to obtain a better understanding of the genetic relationships among major salmon rivers in Russia and northern Norway through an examination of allelic variation at four microsatellite loci, and previously published molecular genetic data, and to compare the levels of variation within this region to the differentiation to the samples from the western Atlantic Ocean.

Methods

Populations and sampling

Samples of fin or muscle tissue from Atlantic salmon were collected from seven different rivers in northern Norway, northwestern Russia and Canada (Figure 1). The samples from the different rivers were obtained by several methods including electrofishing, beach seine and smolt trap (Table 1). Samples were frozen or ethanol preserved until analysis. The frozen samples were kept at $-80°$C.

We selected Russian rivers that have some of the largest Atlantic salmon populations in Russia. Pechora River was, and to some extent still is, one of the most productive Atlantic salmon rivers in the world. In earlier years the average weight of returning spawners was about 7 kg and the annual catch was 350 metric tonnes (V. Antonova, SevPINRO, Arkhangelsk, Russia, pers. comm.). The sample from Pechora River was collected from returning spawners, and fin tissue from 87 fish of the ca. 1000 caught during 1.5 month of operation of a salmon trap was collected. The dispersion in time over which the sample was collected increases the probability that the sample is representative for the river as a whole. The sample from Mezen River system consisted of parr and was collected from only one of several tributaries where salmon spawns and may not be representative for the river system a whole. It was collected in the main spawning stretch of the tributary Pizhma, over a period of 2 days and on several locations separated by some hundred meters. The biological data for the individuals in the sample verified that several generations and both sexes were represented. The sample from Emtsa River, a tributary of the Northern Dvina River system, was collected by a smolt trap over a period of 16 days and only two generations are represented.

Two collections were obtained from Norwegian rivers. The sample from Tana River was collected from spawners in three different tributaries. No age data is available for these fish. The parr sample from Ånes River was collected by electrofishing. The Ånes River is quite small with a length navigable to salmon of only 5 km (Halvorsen & Svenning 2000).

Finally, for comparison, collections representing the western Atlantic Ocean lineage described by Ståhl (1987) were obtained from two Canadian rivers. These samples were from Restigouche River and Dartmouth River and were obtained from Professor Dag Møller at the University of Bergen. The samples, which were of parr, were collected in 1998 near Gaspé, Quebec, as part of a study on the genetic variation in serum transferrin in salmon.

Genetic analysis

The extraction of DNA was carried out mainly as outlined in Dahle et al. (1997), but with some modifications. Approximately 50–100 mg of muscle or fin tissue was subsampled and transferred to 2.0 ml Microtubes with screw cap. Before extraction of DNA, samples stored in 96% EtOH (Pechora and Mezenskaya Pizhma) were washed in deionised water at $4°$C overnight to remove the EtOH. The tissue was digested overnight at $37°$C with $380\,\mu$l 0.2 mol/l EDTA (pH 8.0), 0.5% sacrocyl and $5\,\mu$l 20 mg/ml proteinase K with constant stirring. The following morning $2\,\mu$l 10 mg/ml DNase-free Rnase was added to each tube and they were stirred for 1 h at $37°$C. The contents of the tubes were transferred to Vacutainer SST gel tubes and $800\,\mu$l of phenol (equilibrated at pH 8.0) was added to each tube. After mixing for 20 min, the tubes were centrifuged at 3000 × g for 10 min. This step was repeated with an equal volume of chloroform/isoamyl alcohol (24 : 1). The aqueous fraction was transferred from the gel tubes to Microcon 50 Centrifugal Filter Devices (Millipore) for desalting and concentration of the DNA. The procedure given by the manufacturer was followed. After the cleaning procedure, the DNA extracted was dissolved in $40\,\mu$l of TE-buffer. The Tana samples were extracted by the use of a Qiagen Dneasy$^{®}$96 Tissue Kit. Approximately 20–30 mg of tissue was used, and the DNA was extracted according to the manufacturer's procedure. The extracted DNA was measured spectrophotometrically and diluted to a standard concentration of $25\,\mu$g/ml.

Four microsatellite loci were amplified; Ssa14 (McConnell et al. 1995a), Ssa171 (O'Reilly et al. 1996), Ssa20.19 (Sanchez et al. 1996) and Ssa13.37 (Sanchez et al. 2000). The PCR reactions were performed in a Gene AMP 9700 thermocycler (PE Biosystems), using $2\,\mu$l of DNA template (25 ng/μl) and $8\,\mu$l of PCR-mix in 0.2 ml PCR-tubes. The reaction was run using heated lid to avoid evaporation of samples. The general pattern for the PCR cycle used for amplification of all microsatellite loci was as follows; initial denaturation at $95°$C for 5 min, 30/35 cycles of: $95°$C for 50 s, 40 s at locus specific annealing temperature, 50 s at $72°$C, followed by a final extension at $72°$C for 5 min. The locus specific annealing temperatures were as follows; Ssa14: $57°$C, Ssa13.37 and Ssa171: $60°$C

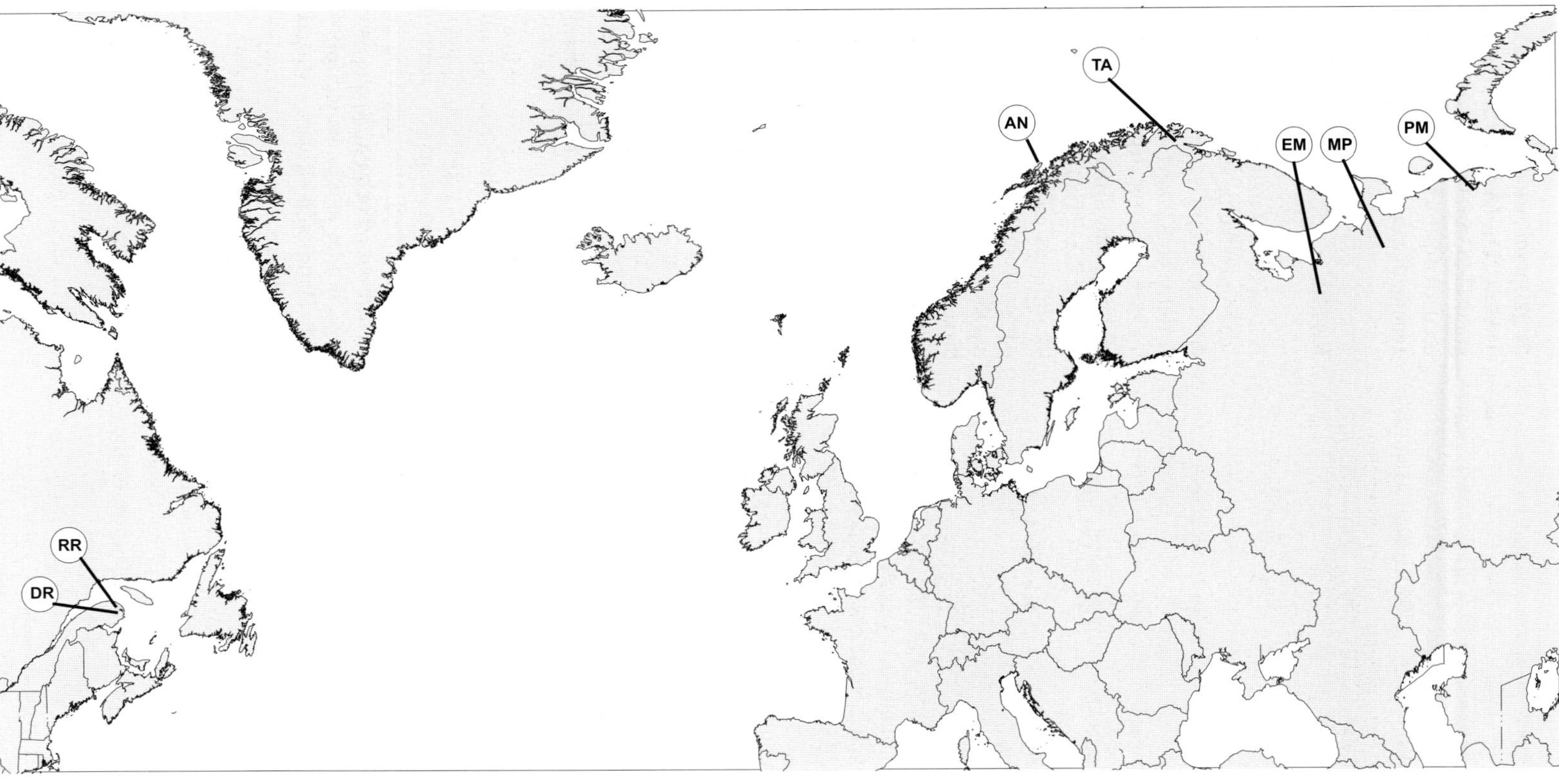

Figure 1. Sampling localities in Canada, Norway and Russia (RR = Restigouche River, DR = Dartmouth River, AN = Åneselva River, TA = Tana River, EM = Emtsa River, MP = Mezenskaya Pizhma River, PM = Pechora River).

146

Table 1. Account of samples of salmon collected for microsatellite analysis.

Sample	Area	Type	# ind.	Date collected
PM	Pechora River, Komi region, Russia	A	83	Sept. 2000
MP	Mezenskaya Pizhma River, Arkhangelsk region, Russia	P	50	July 2000
EM	Emtsa River, Arkhangelsk region, Russia	S	50	June 1999
TA	Tana River, Finnmark, Norway	A	68	Oct. 1997
AN	Ånes River, Nordland, Norway	P	60	Oct. 1998
RR	Restigouche River, Quebec, Canada	P	63	Oct. 1998
DR	Dartmouth River, Quebec, Canada	P	61	Oct. 1998

A: adult returning spawners, P: parr, S: smolt.

and Ssa20.19: 62°C. Usually 30 cycles were used, but for some samples which yielded low amounts of PCR product, the number of cycles was increased to 35 for Ssa171 and Ssa14.

The sizes of the fragments were determined on an ALFexpress II fragment analyser (Amersham Pharmacia Biotech, Bucks, UK). The results of the runs on the ALFexpress II were interpreted and alleles were scored using the program ALFwin Fragment Analyser 1.0 (Amersham Pharmacia Biotech).

Statistical analyses

The number of alleles, observed and expected heterozygosities, and conformity with Hardy–Weinberg equilibrium was calculated separately for all loci in all samples. The analysis was conducted by an exact test employing a Markov–Chain procedure. Loci were tested for linkage disequilibrium across all pairs of loci in all samples. For evaluating differentiation between samples, several test were conducted. An exact test of allelic frequencies (Raymond & Rousset 1995a) was used to test for differences between all pairs of samples at each locus, and an overall P-value was estimated between all pairs of samples across loci by Fisher's method. All these test were conducted by use of the program GenePop 3.3 (Raymond & Rousset 1995b). As a further numerical measure of differentiation, F_{ST} (Wright 1951) values with associated estimates of P-values, were calculated for all population pairs, using

the program Arlequin (Schneider et al. 2000). Genetic distance, D_A, (Nei 1972) was calculated with the program TFPGA1.3 (Miller 1997). This program was also used to construct an UPGMA dendrogram based on the genetic distance values calculated.

Results

A total of 439 individual samples of fin or muscle tissue from Atlantic salmon was collected from the seven different rivers.

Between 429 and 439 fish were successfully scored for each locus (98–100%). The number of alleles across samples at the different loci varied from three in Ssa14 to 28 in Ssa171. (Table 2). Observed heterozygosity at the different microsatellite loci averaged over samples ranged from 0.46 in Ssa13.37 to 0.86 at Ssa171, reflecting the great difference in number of alleles between these loci. The variation in observed heterozygosity between samples for some microsatellites was also substantial, ranging from 0.33 in Tana to 0.82 in Dartmouth River at Ssa20.19. There were no significant deviations from Hardy–Weinberg equilibrium at any locus in any sample after sequential Bonferroni correction of the significance level (Rice 1989). A test of linkage disequilibrium showed no significant deviations for any locus after sequential Bonferroni correction.

The Canadian and the European samples had very different allele distributions at Ssa13.37, Ssa171 and Ssa20.19 (Table 2). The two largest alleles at Ssa13.37, 122 and 126, were only found in the Canadian populations, while the two shortest were found in all populations.

For Ssa13.37, the frequency differences among populations within regions were smaller; and some apparent differences were not significant in the exact test at the 5% level after applying sequential Bonferroni corrections. Pechora River and Ånes River were most similar at this locus, followed by Restigouche and Dartmouth Rivers, Mezenskaya Pizhma and Emtsa Rivers, Pechora and Emtsa Rivers, Mezenskaya Pizhma and Ånes Rivers, and Mezenskaya Pizhma River and Pechora River.

At the Ssa14 locus only three alleles were found (Table 2), and the largest allele, 149, was only found in low frequency in Ånes River and Restigouche River. The differences in allele distributions did not follow a clear geographic pattern, as the most similar distributions were found in Mezen & Emtsa Rivers followed by Restigouche & Emtsa Rivers, and

Table 2. Allele frequencies at four microsatellite loci investigated in the seven rivers.

Locus	Allele	Pechora	M. Pizhma	Emtsa	Tana	Ånes	Restig.	Dartm.
Ssa13.37	114	0.8049	0.6875	0.7400	0.4559	0.8250	0.0714	0.0328
	118	0.1951	0.3125	0.2600	0.5441	0.1750	0.4365	0.4426
	122	0.0000	0.0000	0.0000	0.0000	0.0000	0.4841	0.5164
	126	0.0000	0.0000	0.0000	0.0000	0.0000	0.0079	0.0082
Ssa14	143	0.1818	0.3061	0.3100	0.5441	0.5000	0.3333	0.4590
	147	0.8182	0.6939	0.6900	0.4559	0.4831	0.6667	0.5328
	149	0.0000	0.0000	0.0000	0.0000	0.0169	0.0000	0.0082
Ssa171	197	0.0000	0.0000	0.0000	0.0250	0.0075	0.0000	0.0000
	201	0.0824	0.0000	0.0000	0.0000	0.0522	0.0000	0.0000
	203	0.0000	0.0000	0.0000	0.0083	0.0000	0.0000	0.0000
	209	0.0059	0.1875	0.1600	0.1167	0.0522	0.0000	0.0000
	213	0.0235	0.0729	0.2000	0.1000	0.1343	0.0159	0.0000
	217	0.2412	0.3125	0.3100	0.2333	0.3284	0.0159	0.0164
	219	0.0000	0.0000	0.0000	0.0000	0.0000	0.0000	0.0082
	221	0.0588	0.0000	0.0700	0.0583	0.0224	0.0238	0.0164
	223	0.0000	0.0000	0.0000	0.0000	0.0000	0.0079	0.0000
	225	0.1118	0.0417	0.1100	0.0333	0.1418	0.0159	0.0000
	227	0.0000	0.0000	0.0000	0.0000	0.0075	0.0397	0.0164
	229	0.2765	0.1250	0.0200	0.2833	0.0821	0.0635	0.0738
	231	0.0059	0.0000	0.0000	0.0000	0.0000	0.2222	0.2377
	233	0.1588	0.1146	0.0500	0.0417	0.0299	0.0397	0.0656
	235	0.0000	0.0000	0.0000	0.0000	0.0000	0.0714	0.1721
	237	0.0118	0.1250	0.0500	0.0667	0.0746	0.0317	0.1066
	239	0.0059	0.0000	0.0000	0.0000	0.0000	0.0873	0.0164
	241	0.0059	0.0208	0.0100	0.0000	0.0299	0.0635	0.0410
	243	0.0059	0.0000	0.0000	0.0083	0.0149	0.0794	0.0246
	245	0.0000	0.0000	0.0100	0.0083	0.0000	0.0556	0.0410
	247	0.0000	0.0000	0.0000	0.0000	0.0000	0.0476	0.0000
	249	0.0000	0.0000	0.0000	0.0000	0.0224	0.0317	0.0246
	251	0.0000	0.0000	0.0000	0.0167	0.0000	0.0317	0.0000
	253	0.0059	0.0000	0.0000	0.0000	0.0000	0.0238	0.0246
	255	0.0000	0.0000	0.0000	0.0000	0.0000	0.0159	0.0246
	257	0.0000	0.0000	0.0000	0.0000	0.0000	0.0079	0.0738
	259	0.0000	0.0000	0.0100	0.0000	0.0000	0.0079	0.0000
	261	0.0000	0.0000	0.0000	0.0000	0.0000	0.0000	0.0164
Ssa20.19	78	0.0000	0.0000	0.0000	0.0000	0.0000	0.2302	0.3033
	84	0.0000	0.0000	0.0102	0.0299	0.0083	0.5397	0.4016
	96	0.0750	0.2600	0.2245	0.0672	0.0000	0.0397	0.1311
	98	0.5500	0.6500	0.7041	0.8284	0.6083	0.1032	0.1311
	100	0.0063	0.0000	0.0000	0.0075	0.1333	0.0556	0.0246
	102	0.0000	0.0600	0.0306	0.0075	0.1750	0.0317	0.0082
	104	0.3688	0.0300	0.0306	0.0597	0.0750	0.0000	0.0000

Restigouche & Mezen Rivers. These, and a number of other population pairs were not significantly different at this locus. The two Norwegian populations were also quite similar, while the most distinct sample was from Pechora River.

Ssa171 had a much higher number of alleles than the other loci. This is also reflected in the consistent high observed heterozygosity values at Ssa171, ranging from 0.79 to 0.90. The highest number of alleles was found in the Canadian populations, with 22 alleles in Restigouche River and 18 alleles in Dartmouth River. The lowest number of alleles was found in the Mezen River population, which had only eight alleles. Alleles were found not only at 4 bp intervals, but also intermittently at 2 bp intervals. The frequency of the latter was much higher in both Canadian populations.

Table 3. Pair-wise genetic differentiation between seven populations detected by four microsatellite loci.

Population	Pechora	M. Pizhma	Emtsa	Tana	Ånes	Dartmouth	Restigouche
Pechora	—	0.075	0.084	0.215	0.100	0.792	0.704
M. Pizhma	0.063	—	0.004	0.086	0.067	0.632	0.625
Emtsa	0.071	0.003	—	0.085	0.070	0.701	0.675
Tana	0.151	0.064	0.068	—	0.120	0.560	0.607
Åneselva	0.080	0.051	0.054	0.086	—	0.769	0.768
Dartmouth	0.281	0.232	0.255	0.225	0.257	—	0.023
Restigouche	0.270	0.237	0.256	0.241	0.263	0.012	—

F_{ST} values below diagonal and D_A (Nei 1972) values above.

Only six scorings of these 'in-between' alleles were made in the European populations, while 68 were made in the Canadian populations. The allele distributions in the different populations were also quite variable, and almost all population pairs were significantly different from each other. The most similar pair was Emtsa & Tana River populations. In general, the distribution in the two Canadian populations was shifted towards larger alleles than in the European populations, with several large alleles appearing only in these two populations. Similarly, the shortest alleles found in the European populations were not present in the Canadian populations.

At Ssa20.19, the shortest allele, 78, was only found in the Canadian populations; and the allele 84, which was found in high frequency in the Canadian populations, was only found in low frequency in three of the five European populations. This allele was not present in the samples from the two populations furthest east (Pechora and Mezenskaya Pizhma). The largest allele, 104, was not found in the two Canadian populations. This allele is possibly the allele 106 reported by Sanchez et al. (2000). They found this allele in only one sample from the populations investigated; the Skogsfjord sample from North Norway. The differences between populations were in general greater than for Ssa13.37. The observed heterozygosity ranged from 0.33 in Tana to 0.82 in Dartmouth river, reflecting the larger number of alleles in the Canadian populations. A possible new allele with size 94 bp was found at this locus in the sample from the Emtsa River population. Since the existence of this allele needs to be verified with further analysis, it was not included in the statistical analysis, and the individual was scored as having the allele 96.

An exact test combined over loci (Fisher's method) gave significant differences between all population pairs, except Mezen *versus* Emtsa where the difference was not significant after sequential Bonferroni correction of the significance level.

A test of linkage disequilibrium showed no significant deviations for any locus after sequential Bonferroni correction, though the locus pair Ssa13.37 and Ssa14 was close to significance.

The differences are also reflected in the matrix of F_{ST} and D_A values estimated for all population pairs (Table 3).

The UPGMA phenogram shows a major differentiation between European and Canadian populations (Figure 2). Mezenskaya Pizhma River and Emtsa River, which were not significantly different from each other, form the closest cluster, while the two Canadian populations also cluster closely. The most surprising feature of the dendrogram is Pechora being more closely related to both Norwegian populations than to Mezenskaya Pizhma River and Emtsa River.

Discussion

All four microsatellite loci examined in the seven populations were polymorphic to varying extent: from the low variability in Ssa13.37 and Ssa14, via medium variability in Ssa20.19 to high variability in Ssa171. The number of alleles found in this study is comparable to that found by others examining the same loci.

At the Ssa13.37 locus however, four alleles were found, one more than previously reported by Sanchez et al. (2000). Sanchez et al. (2000) found only three alleles at this locus: 112, 116 and 120. From the allele frequencies they reported for their samples from Canada and Europe, it seems reasonable to assume that these are the same as the alleles 114, 118 and 122 reported in this study. The largest allele found here, 126, was not found by Sanchez et al. (2000), but this is probably due to the limited number of individuals in

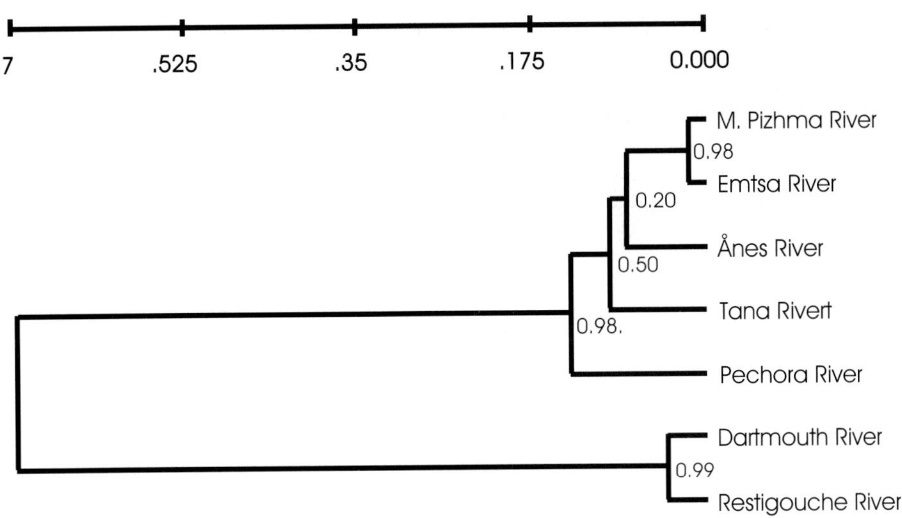

Figure 2. UPGMA dendrogram showing clustering and genetic distances (D$_A$) between the different populations of salmon. The numbers at the node indicate proportion of similar nodes in the bootstrapping.

the two populations that they analysed. The frequency of this allele was very low in both populations (0.08).

Three alleles were found at the Ssa14 locus. This is comparable to the 3–5 alleles reported in other studies (McConnell et al. 1995a,b, 1997, Beacham & Dempson 1998, King et al. 2001).

The total count of alleles at the Ssa171 locus was high (28) and this was also reflected in the heterozygosity values ranging from 0.79 to 0.90. Other studies using this microsatellite have reported allele counts of 13–37, reflecting differences in spatial scale of the studies (O'Reilly et al. 1996, McConnell et al. 1997, Tessier et al. 1997, Norris et al. 1999, 2000, Primmer et al. 2000, King et al. 2001). The number of alleles was quite different in the different samples, ranging from eight in Mezenskaya Pizhma to 22 in Restigouche River (Table 2). Average allele counts in Canadian and European samples were 20 and 12, respectively. The low number of alleles found in the rivers Mezenskaya Pizhma and Emtsa relative to the other European populations may be explained by differences in sample size, but in allozyme studies with larger sample sizes, a lower variability in these rivers compared to Pechora River was found (S. Titov, unpubl. data).

At the Ssa20.19 locus a total of seven alleles was found. Published studies using this marker have reported from four to seven alleles (Sanchez et al. 1996, 2000, Reilly et al. 1999, Norris et al. 1999, 2000). The number of alleles in the different populations ranged from four to six, with the lowest number in the samples

from rivers Pechora and Mezenskaya Pizhma. Allele 78 was specific to Canadian populations and allele 104 was specific to European populations. The high frequency of the 104 allele in the Pechora sample is interesting, as this allele is only found in low frequencies in the other populations. The absence of the 96 allele in the sample from Ånes River is also noteworthy, as this allele was found in all other samples. A possible new allele, 94, not reported previously was found in one individual in the Emtsa River population.

The results obtained in this study showed that alleles at the Ssa171 locus did not only differ in size by 4 bp, but alleles at 2 bp intervals were also found. The compound repeat motif of Ssa171 is (TGTA)$_{14}$(TG)$_7$ (O'Reilly et al. 1996). The different interval lengths observed between fragments indicate that variation in fragment length is not only due to variation in number of tetranucleotide repeats (TGTA), but that variation in the dinucleotide motif may also occur. This needs to be verified by sequencing of fragments from homozygote individuals. Primmer et al. (2000) also reported alleles at 2 bp intervals and assumed that this was due to variation in number of (TG) repeats (Craig Primmer, Department of Ecology and Systematics, University of Helsinki, pers. comm.). Interestingly, the frequency of 2 bp intervals was much higher in the Canadian populations.

The results from the exact test of allele frequencies showed that all population pairs were significantly different from each other, with the exception of the rivers

150

Mezenskaya Pizhma and Emtsa. Though the data set is limited, with only two Canadian populations included, the magnitude of the measures of genetic distance observed here indicates a major genetic divergence between the American and European continents. This is in accordance with what has been reported previously by authors using a variety of markers like blood proteins (Nyman 1966, Payne et al. 1971, Nyman & Pippy 1972), allozymes (Verspoor 1986, Ståhl 1987, Bourke et al. 1997, Makhrov et al. 1998), mitochondrial DNA (Bermingham et al. 1991, Birt et al. 1991, McVeigh & Davidson 1991, King et al. 2000, Nilsson et al. 2001), ribosomal RNA markers (Cutler et al. 1991), minisatellites (Taggart et al. 1995) and microsatellites (McConnell et al. 1995a,b, Sanchez et al. 1996, 2000, Norris et al. 1999, King et al. 2001, Koljonen et al. 2002). In the material available for the present study, alleles specific to continent were found in three of the four microsatellite loci investigated, and alleles specific to region were also possibly detected. At the locus Ssa20.19, the allele 104 in this study may be specific to northern European populations of salmon, assuming that this is the same allele that Sanchez et al. (2000) classified as 106. In their study, this allele was only found in the most northern population, from Skogsfjord in Troms County, and not in any of the southern European or Canadian populations investigated. The possible existence of a distinct northern European population group was also discussed in the allozyme study by Blanco et al. (1992). Skaala et al. (1998) investigated two populations from south-west Norway, one from North Norway, one from the Kola Peninsula and one population from the Karelian side of the White Sea using allozymes. They found genetic structuring that suggested a northern population group, distinct from the Karelian White Sea and south-western Norwegian populations. Bourke et al. (1997), in a study of allozyme variation throughout Europe, also found an allele at the ESTD locus only present in the northern populations (Tana River, Norway and Pecha River, Kola peninsula, Russia), which is also was present in Canadian populations, supporting the idea of a possible northern group of populations. Makhrov et al. (1998) found that populations from North America and Europe up to the north of Norway are fixed for alternative alleles at the ESTD locus. Thus the locus distinguishes between populations from North America and Europe. However, in populations in the north of Norway and on the Kola peninsula, also the slow 'North American' allele has been found, most likely reflecting an early

partial colonization of this area from North American salmon.

The low differentiation observed between Mezenskaya Pizhma River and Emtsa River, a tributary of Northern Dvina, was surprising as the river outlets are 390 km apart. The Mezen River system flows into the sea at the opening of the White Sea, while the Northern Dvina River system enters the ocean via a wide river delta on the eastern shore of the White Sea. Salmon returning from the Barents Sea to spawn in the Northern Dvina River system, pass the freshwater flow of the Mezen system on its migration route, and straying of Northern Dvina salmon into the Mezen River could therefore be expected to reduce differentiation between these populations. However, comprehensive tagging studies conducted by SevPINRO during the 1980s did not detect any such straying (I. Studyonov, unpubl. data). Pechora River, which is located at the absolute north-eastern limit of Atlantic salmon is ranked as more different from Emtsa River and Mezenskaya Pizhma River than Ånes River by several of the estimators of genetic distance, which is surprising. However, this may be a result of the differing colonization patterns of salmon populations in north-west Russia after the last glaciation discussed by Kazakov & Titov (1991). They hypothesized that Pechora River was recolonized from one or more glacial ice lakes, while the rivers on the Kola coast, and possibly Eastern shore of the White Sea, were colonized from the Atlantic Ocean. They also found similarities in allele frequencies in ESTD, PX2 and IDH between the Pechora and Onega Rivers and rivers of the Baltic. This led them to further hypothesize that there at times had been connections between the glacial refugia holding the salmon populations that would colonize these rivers and refugia that held salmon that would colonize the Baltic rivers. Nilsson et al. (2001) in their study of phylogeography of Baltic salmon, also found strong evidence for the existence of a glacial lakes as a refugia for salmon during the last glacial period and that the Baltic and possibly rivers along the southern coast of the White Sea were colonized from such lakes. The differences observed by Kazakov & Titov (1991), and also by Skaala et al. (1998), between populations from the Kola peninsula and populations from the southern coast of the White Sea, support the theory of different colonization histories for these rivers, with the Kola rivers being colonized from the Atlantic.

The genetic difference between the two Norwegian rivers was more pronounced than the difference between Mezenskaya Pizhma River and Emtsa River, and also than the difference between the two Canadian rivers. The Tana River sample is quite distinct from the sample from Ånes River at the locus Ssa13.37. In all three Russian rivers, and in Ånes River, 114 is the most common allele, while in Tana River the frequency of the 118 allele is higher. This is also the case for the Canadian samples, where the frequency of the 114 allele is very low. A similar distribution of alleles has also been found in the Tana's neighboring River Neiden, (Øystein Skaala pers. comm.).

The two Canadian populations grouped closely together in the cluster analysis and the genetic distance between them was small (0.023). Fontaine et al. (1997) investigated seven Canadian salmon populations, of which three were in rivers located around the Gaspé peninsula, and found relatively small differences between populations. Two of the rivers located within 40 km of each other were not significantly different. The distance between the two rivers in the present study is 260 km. King et al. (2001) observed less genetic differentiation at microsatellite loci within North American populations than within European populations. They also observed fewer alleles and lower heterozygosity at the loci studied, in contrast to the observations in our study. However, their study included a relatively large number of European samples, from Spain to West Norway. The pair-wise intercontinental genetic distances (D_A) of 0.46–0.75 they found are comparable to those found in our study, which ranged from 0.56 to 0.79. Koljonen et al. (2002) found intercontinental F_{ST} values between 0.63–0.84 in a comparison of Baltic and Barents Sea salmon with populations from Maine and Labrador.

A number of independent studies have demonstrated significant genetic differences between Atlantic salmon from the European and North American continents. Our study confirmed and expanded the knowledge about differences between the two groups of Atlantic salmon and underlines the need for further studies to clarify the phylogeography and colonization routes of Atlantic salmon in the northern part of Europe following the last glaciation.

Acknowledgements

This study was financed by the Institute of Marine Research, Bergen Norway. We are deeply grateful to all who contributed to this work, especially to Dag Møller who provided the Canadian samples and Morten Halvorsen and Øyvind Walsø for the Norwegian samples. We also thank Valentina Antonova at SevPINRO for making available samples from Pechora. Thanks also to Anne Grete Eriksen and Ingunn Haaverstad for assistance with analysing the samples. We also thank two anonymous referees for thoughtful and helpful comments on the manuscript.

References

Beacham, T.D. & J.B. Dempson. 1998. Population structure of Atlantic salmon from the Conne River, Newfoundland as determined from microsatellite DNA. J. Fish Biol. 52: 665–676.

Bermingham, E., S.H. Forbes, K. Friedland & C. Pla. 1991. Discrimination between Atlantic Salmon (*Salmo salar*) of North American and European origin using restriction analyses of mitochondrial-Dna. Can. J. Fish. Aquat. Sci. 48: 884–893.

Birt, T.P., J.M. Green & W.S. Davidson. 1991. Mitochondrial-Dna variation reveals genetically distinct sympatric populations of anadromous and nonanadromous Atlantic Salmon, *Salmo Salar*. Can. J. Fish. Aquat. Sci. 48: 577–582.

Blanco, G., J.A. Sanchez, J.R. Vazquez & F.M. Utter. 1992. Genetic differentiation among natural European populations of Atlantic salmon, *Salmo salar* L., from drainages of the Atlantic Ocean. Anim. Genet. 23: 11–18.

Bourke, A.A., J. Coughlan, H. Jansson, P. Galvin & T.F. Cross. 1997. Allozyme variation in populations of Atlantic salmon located throughout Europe: Diversity that could be compromised by introductions of reared fish. ICES J. Mar. Sci. 54: 974–985.

Cutler, M.G., S.E. Bartlett, S.E. Hartley & W.S. Davidson. 1991. A polymorphism in the ribosomal-Rna genes distinguishes Atlantic salmon (*Salmo salar*) from North-America and Europe. Can. J. Fish. Aquat. Sci. 48: 1655–1661.

Dahle, G., M. Rahman & A.G. Eriksen. 1997. Rapid fingerprinting used for discriminating among three populations of Hilsa Shad (*Tenualosa ilisha*). Fish. Res. 32: 263–269.

Fontaine, P.-M., J. Dodson, L. Bernatchez & A. Slettan. 1997. A genetic test of metapopulation structure in Atlantic salmon (*Salmo salar*) using microsatellites. Can. J. Fish. Aquat. Sci. 54: 2423–2442.

Halvorsen, M. & M.A. Svenning. 2000. Growth of Atlantic Salmon parr in fluvial and lacustrine habitats. J. Fish Biol. 57: 145–160.

Heggberget, T.G., R.A. Lund, N. Ryman & G. Ståhl. 1986. Growth and genetic variation of Atlantic salmon (*Salmo salar*) from different sections of the River Alta, North Norway. Can. J. Fish. Aquat. Sci. 43: 1828–1835.

Kazakov, R. & S. Titov. 1991. Geographical patterns in the population genetics of Atlantic salmon, *Salmo salar* L., on U.S.S.R. territory, as evidence for colonization routes. J. Fish Biol. 39: 1–6.

Kazakov, R. & S. Titov. 1993. Population genetics of salmon, *Salmo salar* L., in northern Russia. Aquacult. Fish. Manage. 24: 495–506.

King, T.L., S.T. Kalinowski, W.B. Schill, A.P. Spidle & B.A. Lubinski. 2001. Population structure of Atlantic salmon (*Salmo salar* L.): A range-wide perspective from microsatellite DNA variation. Mol. Ecol. 10: 807–821.

King, T.L., A.P. Spidle, M.S. Eackles, B.A. Lubinski & W.B. Schill. 2000. Mitochondrial DNA diversity in North American and European Atlantic salmon with emphasis on the Downeast rivers of Maine. J. Fish Biol. 57: 614–630.

Koljonen, M.-L., J. Tähtinen, M. Säisä & J. Koskiniemi. 2002. Maintenance of genetic diversity of Atlantic salmon by captive breeding programmes and the geographic distribution of microsatellite variation. Aquaculture 212: 69–92.

Makhrov, A.A., Ø. Skaala, Yu.P. Altukhov & R.L. Saunders. 1998. The allozyme ESTD*locus as a marker of genetic differentiation between European and North American populations of Atlantic salmon (*Salmo salar* L.). Doklady Biol. Sci. 360: 281–283.

McConnell, S., L. Hamilton, D. Morris, D. Cook, D. Paquet, P. Bentzen & J. Wright. 1995a. Isolation of salmonid microsatellite loci and their application to the population genetics of Canadian east coast stocks of Atlantic salmon. Aquaculture 137: 19–30.

McConnell, S.K.J., D.E. Ruzzante, P.T. O'Reilly, L. Hamilton & J. Wright. 1997. Microsatellite loci reveal highly significant genetic differentiation among Atlantic salmon (*Salmo salar* L.) stocks from the east coast of Canada. Mol. Ecol. 6: 1075–1089.

McConnell, S.K., P. O'Reilly, L. Hamilton, J.M. Wright & P. Bentzen. 1995b. Polymorphic microsatellite loci from Atlantic salmon (*Salmo salar*): Genetic differentiation of North American and European populations. Can. J. Fish. Aquat. Sci. 52: 1863–1872.

McVeigh, H.P. & W.S. Davidson. 1991. A salmonid phylogeny inferred from mitochondrial Cytochrome-B gene-sequences. J. Fish Biol. 39: 277–282.

Miller, M.P. 1997. Tools for population genetic analyses (TFPGA)1.3.

Møller, D. 1970. Transferrin polymorphism in Atlantic salmon (*Salmo salar*). J. Fish. Res. Board Can. 27: 1617–1625.

Nei, M. 1972. Genetic distance between populations. Am. Nat. 106: 283–292.

Nilsson, J., R. Gross, T. Asplund, O. Dove, H. Jansson, J. Kelloniemi, K. Kohlmann, A. Loytynoja, E.E. Nielsen, T. Paaver, C.R. Primmer, S.F. Titov, A. Vasemagi, A. Veselov, T. Ost & J. Lumme. 2001. Matrilinear phylogeography of Atlantic salmon (*Salmo salar* L.) in Europe and postglacial colonization of the Baltic Sea area. Mol. Ecol. 10: 89–102.

Norris, A.T., D.G. Bradley & E.P. Cunningham. 1999. Microsatellite genetic variation between and within farmed and wild Atlantic salmon (*Salmo salar*) populations. Aquaculture 180: 247–264.

Norris, A.T., D.G. Bradley & E.P. Cunningham. 2000. Parentage and relatedness determination in farmed Atlantic salmon (*Salmo salar*) using microsatellite markers. Aquaculture 182: 73–83.

Nyman, L. 1966. Geographic variation in Atlantic salmon (*Salmo salar* L.). LFI Medd. 3: 1–6.

Nyman, O.L. & J.H.C. Pippy. 1972. Differences in Atlantic salmon, *Salmo salar*, from North America and Europe. J. Fish. Res. Board Can. 29: 179–185.

O'Reilly, P., L. Hamilton, S. Mcconnell & J. Wright. 1996. Rapid analysis of genetic variation in Atlantic salmon (*Salmo salar*) by PCR multiplexing of dinucleotide and tetranucleotide microsatellites. Can. J. Fish. Aquat. Sci. 53: 2292–2298.

Payne, R.H., A.R. Child & A. Forrest. 1971. Geographical variation in the Atlantic Salmon. Nature 231: 250–252.

Primmer, C.R., M.T. Koskinen & J. Piironen. 2000. The one who did not get away: Individual assignment detects a case of fishing competition fraud. Proc. R. Soc. Lond. B 267: 1699–1704.

Raymond, M. & F. Rousset. 1995a. An exact test for population differentiation. Evolution 49: 1280–1283.

Raymond, M. & F. Rousset. 1995b. Genepop (Version-1.2) – population-genetics software for exact tests and ecumenicism. J. Hered. 86: 248–249.

Reilly, A., N.G. Elliot, P.M. Grewe, C. Clabby, R. Powell & R.D. Ward. 1999. Genetic differentiation between Tasmanian cultured Atlantic salmon (*Salmo salar* L.) and their ancestral Canadian population: Comparison of microsatellite DNA and allozyme and mitochondrial DNA variation . Aquaculture 173: 459–469.

Rice, W.R. 1989. Analyzing tables of statistic tests. Evolution 43: 223–225.

Sanchez, J.A., C. Clabby, D. Ramos, G. Blanco, F. Flavin, E. Vazquez & R. Powell. 1996. Protein and microsatellite single locus variability in *Salmo salar* L (Atlantic salmon). Heredity 77: 423–432.

Sanchez, J.A., M.D. Ramos, H. Pineda, Y. Borrel, E. Vasquez & G. Blanco. 2000. The application of genetic variation at microsatellite loci in Atlantic salmon (*Salmo salar* L.) stock identification. ICES Council Meeting Y: 06.

Schneider, S., D. Roessli & L. Excoffier. 2000. Arlequin: A software for population genetics data analysis, version 2.000. Genetics and Biometry Laboratory, Department of Anthropology, University of Geneva, Switzerland.

Skaala, Ø., A.A. Makhrov, T. Karlsen, K.E. Jørstad, Y.P. Altukhov, D.V. Politov, K.V. Kuzishin & G.G. Novikov. 1998. Genetic comparison of salmon from the White Sea and north-western Atlantic Ocean. J. Fish Biol. 53: 569–580.

Ståhl, G. 1987. Genetic Population Structure of Atlantic Salmon. pp. 121–140. *In*: N. Ryman & F. Utter (ed.) University of Washington Press, Seattle.

Taggart, J., E. Verspoor, P. Galvin, P. Moran & A. Ferguson. 1995. A minisatellite DNA marker for discriminating between European and North American Atlantic salmon (*Salmo salar*). Can. J. Fish. Aquat. Sci. 52: 2305–2311.

Tessier, N., L. Bernatchez & J.M. Wright. 1997. Population structure and impact of supportive breeding inferred from mitochondrial and microsatellite DNA analyses in land-locked Atlantic salmon *Salmo salar* L. Mol. Ecol. 6: 735–750.

Verspoor, E. 1986. Spatial correlations of transferrin allele frequencies in Atlantic salmon, *Salmo salar*, populations from North America. Can. J. Fish. Aquat. Sci. 43: 1074–1078.

Wright, S. 1951. The genetical structure of populations. Ann. Eugenics 15: 323–354.

Environmental Biology of Fishes **69**: 153–166, 2004.
© 2004 *Kluwer Academic Publishers. Printed in the Netherlands.*

The rainbow smelt, *Osmerus mordax,* complex of Lake Utopia: threatened or misunderstood?

R. Allen Curry[a], Steve L. Currie[a,c], Louis Bernatchez[b] & Robert Saint-Laurent[b]
[a]*New Brunswick Cooperative Fish and Wildlife Research Unit, Canadian Rivers Institute, Biology and Forestry and Environmental Management, University of New Brunswick, Fredericton, New Brunswick, Canada, E3B 6E1 (e-mail: racurry@unb.ca)*
[b]*Département de biologie, Université Laval, Sainte-Foy, Quebec, Canada G1K 7P4*
[c]*Present address: New Brunswick Department of Natural Resources and Energy, Region 3, Islandview, New Brunswick, Canada E3E IG3*

Received 21 April 2003   Accepted 16 June 2003

Key words: genetics, predation, spawning, threatened

Synopsis

We report on the spawning ecology, genetic characteristics, and predation threats to spawning groups of rainbow smelt, *Osmerus mordax,* in Lake Utopia, New Brunswick where a dwarf morpho-type has been listed as a threatened species. Two spawning groups in three inlet streams had been previously identified; we observed three groups using four inlet streams. The earliest group was the largest in body size (12–29 cm fork length (FL)), lowest in numbers (~1 000), and completed spawning approximately two weeks before the second group. The early spawners were previously identified as the normal morpho-type, but we now classify these as a giant morpho-type. The second group spawned in three different streams. They were intermediate in body size (10–15 cm FL) and numbers (~10 000). The dwarf group began spawning as the intermediate group completed spawning and within the same three streams. The dwarfs were numerous (~1 000 000), small in size (<12 cm), and with higher gill raker counts. Microsatellite analyses suggested that gene flow among groups occurred, but genetic divergence was high and genetic separation among populations of the same group among streams and within a stream occurred. Stable isotopes and stomach contents indicated the dwarf group were likely consumed by a variety of fishes, but they were not the sole food resource of any predator including a population of landlocked salmon. These are some of the complexities of smelt ecology, but there are clearly life history tactics that we do not yet understand.

Introduction

The rainbow smelt, *Osmerus mordax*, is one of the several species of north temperate, freshwater fishes that display a complex of morpho-types occurring in sympatry (e.g., Ryman et al. 1979, Skulason et al. 1996, Bernatchez 1997). Normal and dwarf morpho-types have been described for lake-dwelling rainbow smelt of North America (Kendall 1927, Nellbring 1989). Additionally, a larger body-sized, anadromous form exists (Frechet et al. 1983), although these populations have not been reported to occur in sympatry with other forms. Within anadromous populations, dwarf forms have been suggested (McKenzie 1964) and sub-populations separated by distinct spawning sites/times occur (Baby et al. 1991, Bernatchez & Martin 1996).

The lacustrine populations have provided strong evidence of distinct morpho-types existing in sympatry. Coexisting normal and dwarf forms were described in the early literature (see review by Lanteigne & McAllister 1983). Most recent studies describe morphological differences in body form

154

(Bridges & Delisle 1974, Copeman 1977, Copeman & McAllister 1978). Rupp & Redmond (1966) using experimental transplants concluded that body morphology was controlled by physical and biological characteristics of the environment more than genotypic differences between normal and dwarf forms. Copeman & McAllister (1978) re-examined the transplanted populations after several more generations and observed that the two forms were again appearing. A search for genetic differences found mixed levels of genetic separation between forms and concluded that smelt were monophyletic (Taylor & Bentzen 1993a,b).

Both morpho-types have been described in Lake Utopia of southwestern New Brunswick. First described in the early 1900s, a normal form was observed to be larger, spawning earlier, and using different tributary streams for spawning and incubation. The normal form begins to spawn as the lake becomes devoid of ice in spring (Lanteigne & McAllister 1983) with local fishers' accounts of spawning when the lake is still ice bound in some years. Spawning was reported in the two largest inlet tributaries located in the northeast of the lake. Dwarf smelt spawn in late April and May in two smaller streams located in the northwest.

The existing data suggest normal smelt range in body length from about 150 to 250 mm with 31–33 gill rakers. Dwarf smelt are 80–150 mm long with 33–37 gill rakers (Lanteigne & McAllister 1983, Taylor & Bentzen 1993a,b). An examination of mtDNA and mini-satellite DNA of normal and dwarf forms as well as allopatric, sympatric, and anadromous populations from across the region concluded the sympatric forms originated repeatedly within individual systems (Taylor & Bentzen 1993a,b). In Lake Utopia, the normal and dwarf forms were genetically more similar than differences between all dwarf and normal forms from across the region.

The few documented occurrences of the dwarf form combined with its genetic distinctness, apparent limited number of spawners, and apparent threats of predation and fishing pressure, were sufficient evidence for the Lake Utopia dwarf smelt to be classified as a 'threatened' species in Canada[1] (Taylor & Bentzen 1993b). However, there are ecological gaps in the knowledge base for the smelt of Lake Utopia and lacustrine populations in general. In Lake Utopia for example, there has been no complete description of the spawning ecology or habitats of the smelts. Collections in the lake have focused on four tributaries, yet there are at least two more inlet streams in the northern portion of the lake and a total of 17 inlet streams that represent potential spawning and incubation habitats. In addition, sampling of smelt has been of short duration (1 or 2 nights), small sample sizes (<50 smelt), and targeted the known spawners. This sampling may have biased results if the spawning period is longer or other streams support spawning and successful incubation. In addition, the lake is connected to the Magaguadavic River, which has a series of lakes where smelt exist and a fishway to the Bay of Fundy where anadromous smelt are common.

A dipnet fishery for smelt has occurred over many years in Lake Utopia and the northeast streams, Trout and Mill Streams, are the primary locations of fishing (i.e., the apparent normal smelt are targeted; Cronin, P. New Brunswick Department of Natural Resources and Energy, pers. comm.). A population of landlocked Atlantic salmon, *Salmo salar*, also inhabits the lake and supports an important recreational fishery. The population has received supplemental stocking of salmon 12 times since 1984 and presently is stocked at a rate of 3 400 salmon every other year (Collet et al. 1999). Landlocked salmon can target smelt as their principle prey in lakes and fisheries managers specifically use lakes with smelt populations for developing salmon fisheries, or they introduce smelt to enhance the forage base for salmon (Sayers et al. 1989, Kirn & Labar 1996). In Lake Utopia, there is no information on the diet of salmon or the trophic interactions of the fish community in general.

It is impossible to assess the threats accurately to the dwarf smelt of Lake Utopia without a more complete understanding of the spawning ecology and trophic status of the smelt complex. Moreover, such information may lead to resolution of questions concerning the ecological adaptability and evolutionary history of the species. In this context, our first objective was to describe the spawning ecology and genetic characteristics of the spawning groups of smelt in Lake Utopia. The second objective was to examine the trophic relationships within the fish community and assess the potential threats of predation for dwarf smelt in Lake Utopia. A third objective was to test the null hypothesis of no genetic differentiation among dwarf smelt spawning in different tributaries and assess their relationships with smelt of other forms in the lake.

---

[1] COSEWIC 2001. The Lake Utopia Dwarf Smelt. http://www.speciesatrisk.gc.ca.

Methods

Study area

Lake Utopia is located in southwestern New Brunswick, Canada (45°10′, 66°47′, Figure 1). It is a 1 400 ha oligotrophic lake averaging 11 m in depth. It is connected by its outflow to the Magaguadavic River that flows into the Bay of Fundy at St. George, New Brunswick. An unsurpassable waterfall exists at St. George where a fishway exists although there is no evidence that anadromous smelt can traverse the fishway (Carr, J. Atlantic Salmon Federation, St. Andrew's, N.B., pers. comm.). The river also drains Magaguadavic and Digideguash Lakes with known populations of smelt, but dwarf forms have not been reported.

In Lake Utopia, there are four inlet streams that have been identified as spawning areas for smelt (Figure 1). Mill (Lake) Stream and Trout (Lake) Stream are outflows from smaller lakes. A dam at Mill Lake prohibits upstream migration into the lake. Smelt can be captured moving upstream near the mouth of Trout Stream, which does not have apparent suitable habitat for smelt spawning and incubation. An inlet stream to Trout Lake may provide appropriate habitat (Spear Brook). Mill Stream averages 4 m wide and <1 m in depth.

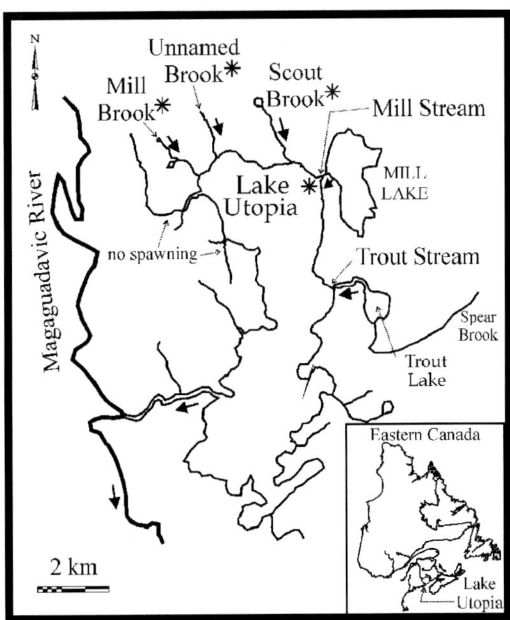

Figure 1. Lake Utopia and its tributaries. Spawning and incubation habitats are identified by an *.

Trout Stream averages 10 m wide with slow moving water and deeper pools. These streams were sampled in previous studies for what were described as 'normal' smelt in the system.

We surveyed a number of additional streams located around the north half of the lake (Figure 1). Two of these were previously described as Mill Brook (Lanteigne & McAllister 1983 – or Smelt Brook, Taylor & Bentzen 1993b) and Unnamed 'A' Brook (Taylor & Bentzen 1993b). A fourth stream, Scout (Second) Brook, with later spawning smelt was also discovered (Figure 1). All three of these inlet streams are small, ≤1 m wide with <500 m of accessible habitats. They drain upland, mostly forested areas that are undeveloped.

Field sampling

In 1998, sampling of spawning smelt began in mid-April and continued until mid-May. The objective was to identify streams used for spawning and examine basic morphometrics and spawning timing to match spawner groups to earlier studies. Sampling occurred weekly, but no smelt were encountered in Mill and Trout Streams and Scout Brook was not discovered until the completion of the spawning run.

In 1999, sampling began in late March while the lake was still ice covered. Some smelt were initially collected in Mill Stream, but none were encountered during the spring in Trout Stream. Bi-weekly sampling continued into mid-May. Approximately 30 females and 30 males roughly sorted into large and small smelt (four groups of 15 smelt) were collected by dip net and seine. Length (nearest 1 mm) and live wet weight (nearest 0.1 g) were recorded. After euthanization, individual fish were stored in 95% EtOH for later counts of lateral line scales (left side) and gill rakers (first gill arch, left side).

In the laboratory, otoliths were removed and stored dry. They were eventually mounted in epoxy, cut, ground, and aged (years) by three readers. Age was assigned only when there was a consensus among readers (n = 208 smelt in total). Tissue samples were taken and stored in 95% EtOH for later analyses of stable isotopes of carbon and nitrogen (epaxial muscle – Stable Isotopes in Nature Laboratory, Canadian Rivers Institute, University of New Brunswick) and microsatellite DNA (caudal fin – Laval University).

Estimates of spawner abundance were attempted for each stream and evening of sampling in 1999. A barrier net was placed across the stream mouth and seine net sweeps (1–3 over 1.5 h) of the stream mouth and

156

lake shoreline were made. Smelt were marked with fin clips (unique to each stream and evening) and returned immediately to the water. Schnabel estimates of abundances were calculated (Ricker 1975). These estimates represented the numbers of smelt entering the stream during the period of actual sampling (typically 1.5 h). From the 1998 and 1999 surveys, it was observed that spawners entered the streams from 21:30 to 04:30 h with an approximate normal distribution of numbers over time each evening, i.e., the peak was 00:00 to 01:30 h. From the distribution, evening and stream specific estimates for five time periods were summed to provide the total evening abundance for each stream. It is expected that this method underestimated the actual abundance. If too few smelt were encountered for a mark-recapture estimate, then numbers were estimated from actual counts of smelt observed in the stream. There were too many smelt in Unnamed Brook to enumerate with mark-recapture techniques on 3 May 1999. The estimate that evening was made by quickly isolating 1 m sections of the stream, counting all smelt within the section, and extrapolating across the entire length of accessible stream habitat.

In 2000, the earlier spawners detected in previous years in Mill Stream were targeted for tissue sampling for stable isotope and genetic analyses. No other analyses were undertaken.

The lake environment from March to May was generally similar among years based on lake levels: water levels were lowest in March in 1998; peak levels occurred in early April in each year; and levels declined during the remainder of smelt spawning except in 2000 when levels remained high into early May (J.D. Irving Limited, unpubl. data).

Comparisons of FL ($\log_{10}$ transformed) were made by an analysis of covariance (ANCOVA, $\alpha = 0.05$; SAS Institute Inc. 1990) with age as a covariate. Gill raker and scale counts ($\log_{10}$ transformed) were compared with an ANCOVA with FL as a covariate ($\alpha = 0.05$). Age was compared among sites by a one-way analysis of variance (ANOVA, $\alpha = 0.05$). A Tukey pairwise comparison of streams and dates followed analyses when appropriate ($\alpha = 0.05$). Analyses of stable isotopes of carbon and nitrogen were made graphically and with a cluster analysis of the Bray–Curtis similarity matrix (Primer version 6.0). Because of the potential of age-biased measures of morphological characteristics, examinations of age-3 smelt were also undertaken.

For isotope analyses, tissue samples were dried at 50°C for 48 h and ground into a fine powder with a mortar and pestle. Approximately 0.2 mg aliquots were packed into 3 mm × 5 mm tin cups. Samples were combusted to gas using a Thermoquest NC 2500 elemental analyzer, and gases were submitted via a continuous flow of helium to a Finnigan MAT Delta Plus isotope-ratio mass spectrometer. Results of ^{13}C : ^{12}C and ^{15}N : ^{14}N isotope ratios are reported as:

$$\delta X = [(R_{\text{sample}}/R_{\text{standard}}) - 1]*1000$$

where $X = {}^{15}$N or ^{13}C and $R = {}^{15}$N/^{14}N or ^{13}C/^{12}C. Values are said to be enriched when values are more positive than a comparison sample and depleted when more negative. Replicates of commercially available isotope standards yielded results that were both accurate and precise (International Atomic Energy Agency (IAEA)). Ten percent of the samples were analysed in duplicate.

Genetic analysis

Total DNA was extracted from the muscle tissues using a standard proteinase K phenol–chloroform protocol (Sambrook et al. 1989). The extent of genetic differentiation and gene flow between ecotypes on each spawning ground was assessed using five microsatellite loci: four developed from rainbow smelt (Saint-Laurent & Bernatchez, unpubl. data) and one for the eulachon, *Thaleichthys pacificus* (McLean & Taylor 2001; Table 1; Appendix 1). Rainbow smelt specific loci were developed as detailed in Wirth et al. (1999). Microsatellite polymorphism was analysed by the fluorescent dye detection method. One of the primers of each locus was 5′-labelled with three different dye (colours); HEX (yellow) for *OSMO-Lav 12* and *OSMO-Lav 16*, TET (green) for *OSMO-Lav 45* and 6-FAM (blue) for *OSMO-Lav 16* and *TPA 26*. PCR was carried out in a 10 μl reaction volume containing 1 U of *Taq* polymerase, 1.0 μl of reaction buffer (10 mM Tris–HCl (pH 9.0), 1.5 mM MgCl$_2$, 0.1% Triton X-100, 50 mM KCl), 750 μmol of dNTPs, between 25 and 50 ng template DNA, and 0.03 pmol of each primer. Simplex and duplex (for *OSMO-Lav 12–OSMO-Lav 16*) PCR was used and performed in a Perkin-Elmer 9600 thermocycler (version 2.01) with the following profile: an initial denaturing step of 5 min at 95°C, followed by 30 cycles of 30 s at 94°C, 30 s at annealing temperature and 30 s at 72°C. Samples were heated to 95°C for 5 min and chilled on ice prior gel loading.

Electrophoresis procedures were conducted on a denaturing 5% polyacrylamide gel with an ABI 377 automated sequencer/gene scanner and analysis

Table 1. Allelic variability at five microsatellite loci from sympatric rainbow smelt ecotype of Lake Utopia. Number of samples successfully used for genetic analysis (N), number of alleles at each locus (A), mean number of alleles at six loci (Am) most common allele (Ac; in base pairs), frequencies of the most commun alleles (Fc), range of allele size (Ar), observed heterozygosity (Ho), gene diversity (He) at each locus and mean allelic diversity (Am) and heterozygosity (Hm) within each sample. d indicates significant heterozygote deficit following the sequential bonferoni correction ($\alpha = 0.005$, k $= 30$), whereas D indicates significant heterozygote deficit across sample (*OSMO-Lav45*) and across loci within sample (Unnamed Brook, dwarf) (global test, Fisher's method).

Sample		*OSMO-Lav16*	*OSMO-Lav157*	*OSMO-Lav12*	*OSMO-Lav45* (D)	*Tpa 26*	Am	Hm
Mill Stream	N	30	30	30	30	30		
giant (00)	A	3	6	10	11	9	8	
	Ac	80	225	168	233	217		
	Fc	0.9	0.6	0.36	0.34	0.5		
	Ar	80–84	221–239	152–180	195–289	215–231		
	He	0.10	0.56	0.78	0.81	0.65		0.58
	Ho	0.10	0.57	0.84	0.80	0.66		0.59
Mill Stream	N	12	12	12	11	12		
giant (99)	A	3	3	8	6	3	5	
	Ac	80	217	162	215	217		
	Fc	0.58	0.371	0.18	0.33	0.59		
	Ar	80–84	217–225	152–184	205–265	217–227		
	He	0.57	0.66	0.85	0.75	0.55		0.67
	Ho	0.59	0.85	0.81	0.50d	0.58		0.69
Mill Brook	N	30	30	30	30	30		
dwarf	A	6	6	16	18	6	10	
	Ac	80	221	170	215	217		
	Fc	0.6	0.41	0.12	0.18	0.56		
	Ar	74–88	217–231	150–184	189–277	213–221		
	He	0.56	0.66	0.91	0.91	0.61		0.73
	Ho	0.6	0.67	0.9	0.40d	0.62		0.74
Mill Brook	N	30	30	30	30	30		
intermediate	A	6	7	17	21	5	11	
	Ac	80	225	164	209	217		
	Fc	0.65	0.36	0.14	0.1	0.5		
	Ar	74–84	217–233	152–188	137–285	213–221		
	He	0.53	0.72	0.9	0.93	0.6		0.74
	He	0.55	0.72	0.92	0.53d	0.8		0.72
Scout Brook	N	30	30	30	30	30		
dwarf	A	8	11	17	18	6	12	
	Ac	80	227	176	211	217		
	Fc	0.45	0.48	0.16	0.18	0.75		
	Ar	72–88	215–239	156–190	173–285	209–225		
	He	0.65	0.72	0.9	0.91	0.4		0.71
	He	0.66	0.73	0.93	0.62d	0.4		0.73
Scout Brook	N	30	30	30	30	30		
intermediate	A	4	6	12	14	2	7	
	Ac	80	223	168	217	217		
	Fc	0.8	0.44	0.19	0.17	0.92		
	Ar	74–82	221–231	138–188	205–279	217–225		
	He	0.31	0.73	0.84	0.87	0.14		0.59
	He	0.30	0.75	0.88	0.69d	0.16		0.60

Table 1. (*Continued*)

Sample		OSM O-Lav16	OSMO-Lav157	OSMO-Lav12	OSMO-Lav45 (D)	Tpa 26	Am	Hm
Unnamed	N	30	30	30	30	30		
Brook dwarf	A	6	5	17	20	6	10	
	Ac	80	225	164	219	217		
	Fc	0.62	0.36	0.2	0.14	0.78		
	Ar	74–86	217–229	148–190	183–305	213–221		
	He	0.56	0.73	0.9	0.92	0.37		0.70
	He	0.67	0.60d	0.89	0.53d	0.34d		0.61D
Unnamed	N	30	30	30	30	30		
Brook	A	6	6	15	14	4	9	
intermediate	Ac	72	227	180	235	217		
	Fc	0.55	0.46	0.13	0.17	0.56		
	Ar	70–92	211–231	148–188	191–285	215–227		
	He	0.63	0.68	0.91	0.89	0.53		0.73
	He	0.64	0.8	0.93	0.85	0.53		0.74

software. Loading product consisted of 0.2 μl internal sizing standard (TAMRA 350 bp red colour) and 2 μl deionized formamide that were combined to 0.8 μl (0.3 μl for *TPA 14–TPA 26*) of the PCR reaction. Gels were run for 2 h at 3 000 V. Allelic size was determined (using GENESCAN software version 2.1; ABI 1996a) by reference to the internal sizing standard and by comparison with the same standard sample of known allelic size that was run on each gel. The final scoring of allelic size and tabulation of data for each locus were conducted with the GENOTYPER software, version 2.0 (ABI 1996b).

Genetic polymorphism was quantified by the number of alleles per locus (A), observed heterozygosity (Ho), and unbiased gene diversity (He) with the GENETIX software, version 4.02 (Belkhir et al. 2000). Samples were tested for conformity with Hardy–Weinberg equilibrium under the alternative hypotheses of heterozygote excess or heterozygote deficiency with the score test (U-test) described in Rousset & Raymond (1995). Global tests across loci and across samples were also made with GENEPOP version 3.1d (Rousset & Raymond 1995).

The null hypothesis of no difference in allelic frequency distribution at each locus between all sample pairs was tested with the Markov chain method to obtain unbiased estimates of Fisher's exact test through 1 000 iterations (Guo & Thompson 1992) available in GENEPOP 3.1d. Probability values over all loci were obtained by the Fisher method (Sokal & Rohlf 1995). The extent of genetic differentiation between each population unit (defined as group of samples for which the above null hypothesis was not rejected) was

estimated from the θ estimator (Weir & Cockerham 1984) of F_{st} with F-stat 1.2 (Goudet & Raymond 1996). Statistical significance levels in all the above tests were adjusted for multiple comparisons with a sequential Bonferroni adjustment (Rice 1989) to minimize Type 1 errors.

The hierarchical partitioning of genetic diversity among populations was quantified with an analysis of variance framework (AMOVA) using variance in allelic frequency (Weir & Cockerham 1984) in the program ARLEQUIN, version 1.1 (Schneider et al. 1997). By this procedure, we performed two hierarchical analyses of gene diversity in order to assess the component of genetic variance corresponding to: (i) variance among individuals within sample; (ii) among samples within groups (F_{SC}); and (iii) among groups (F_{CT}). In the first analysis, tributary of origin was considered the first grouping level, whereas the date of sampling was the grouping level in the second analysis.

We then assessed relationships among all populations analysed by chord distances (D_{CE}; Cavalli-Sforza & Edwards 1967), which assume pure genetic drift. Pairwise distances were computed with the GENEDIST program included in the PHYLIP computer package, version 3.57c (Felsenstein 1993). The matrix of pairwise distances obtained for D_{CE} was used to construct a population phenogram (Figure 2) with the neighbour-joining algorithm available in PHYLIP. Confidence levels on tree topology were estimated by the percentage of 2 000 bootstraps performed from re-sampling loci, and compiled with the CONSENSE program of PHYLIP.

Results

Spawning chronology

The most complete year of sampling took place in 1999. Spawning smelt were regularly encountered in four of six inlet streams in the northern portion of Lake Utopia (Figure 1). The earliest spawners were encountered in Mill Stream. They began spawning when stream temperatures were $\leqslant 6°C$ and the lake was ice covered ($6°C$) in the years they were observed (1999 and 2000). Their period of spawning ended by 2–5 April 1999 and may have lasted a total of 7–14 days, but numbers were sufficiently low that accurate estimates of abundance were not possible (i.e., no recaptures were made and estimates of removals by fishers was impossible).

After the completion of this early run, no smelt were observed in any stream until mid- to late-April, and no smelt were observed in Mill Stream for the remainder of the spawning period in any year. By 21 April 1999, spawning smelt were observed in Scout and Smelt Brooks (Table 2). Stream water temperatures were $\leqslant 6°C$. Smelt continued to move into these streams until mid-May. Mill Brook was used from 26 April to 13 May at least. Stream temperatures were $\leqslant 9°C$ during the spawning period. There were regular occurrences of the mid- to late-spawning smelt moving between streams in one evening and between successive sampling dates. This occurred most often between Smelt Brook and Mill Brook.

In 1998, spawning smelt were captured in Mill Brook from 20 April until 5 May. Spawners were subsequently discovered in Unnamed and Scout Brooks on 5 May.

Body morphology

The early spawning, Mill Stream smelt were the largest in body size ranging from 12 to 26 cm FL in 1999 (Tukey test, $p < 0.05$; Figure 3 – smelt as large as 29.0 cm FL were captured in 2000). Body sizes declined on average over the duration of the spawning period, although there was substantial overlap among sample sites and dates. By early May, the average size of spawning, female smelt had declined from 17.6 ± 2.7 (Mill Stream; average ± 1SD) to 11.3 ± 1.2 cm. Mill Stream smelt had the fewest gill rakers ($F = 0.64$, $p < 0.0$; Figure 4). Gill raker counts suggested Mill Brook smelt spawning on 26 April were more similar to later spawners than concurrently spawning smelt in Unnamed and Scout Brooks. Mill Stream smelt (earliest spawning) were the oldest on average (3.1 ± 0.7 years), although the difference was not statistically

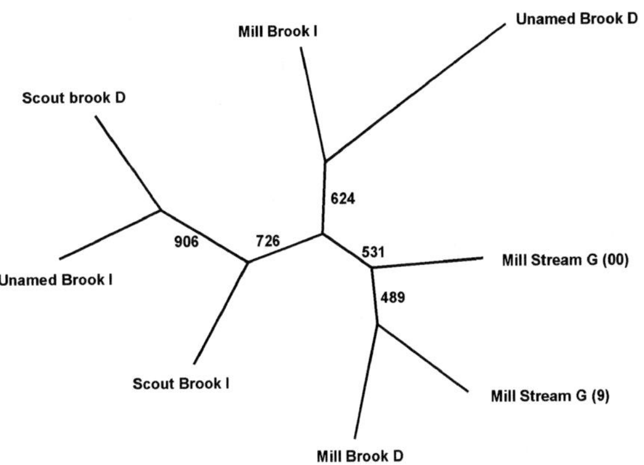

Figure 2. Population phenogram developed using the neighbour-joining algorithm (PHYLIP).

Table 2. Single evening estimates of abundance and 95% confidence limits of spawning smelt in four stream of Lake Utopia in 1999.

Date	Abundance			
	Mill Stream	Scout Brook	Unnamed Brook	Mill Brook
Apr. 2	<500[1]	0	0	0
Apr. 15	0	0	0	0
Apr. 21	0	43 000	64 000	
		39 000–48 000	52 000–82 000	0
Apr. 26	0	< 500[1]	100 000[2]	45 000
				31 000–61 000
May 3	0	<100[1]	169 000	17 000
			130 000–237 000	15 000–19 000
May 13	0	0	<100[1]	<100[1]

[1]Too few smelt to estimate.
[2]Density-based estimate.

160

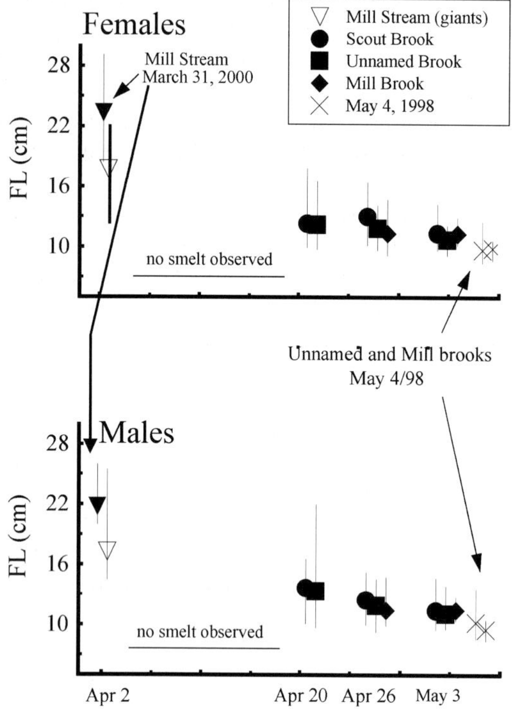

Figure 3. Size distribution of spawning smelt in Lake Utopia in 1998, 1999, and 2000.

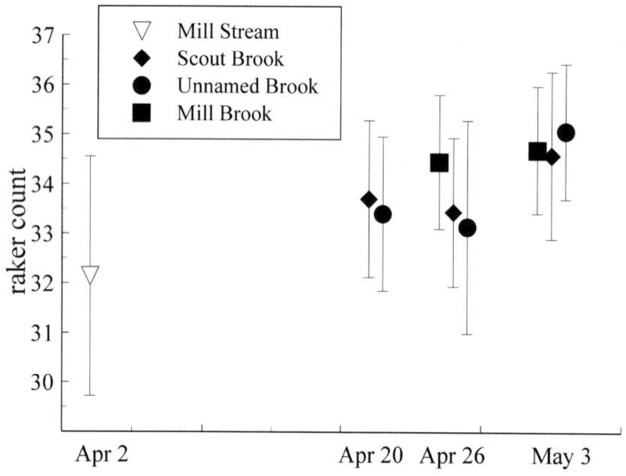

Figure 4. Gill raker counts for the smelt of Lake Utopia, 1999.

significant (overall average $= 2.8 \pm 0.6$ years). Lateral line scale counts were consistent among sites and sampling dates (overall average $= 64 \pm 2$). The overall trends remained the same for age-3 smelt only, i.e., early spawners were the largest and late spawners were smallest in size and gill raker counts increased from early to late spawners (Figure 5).

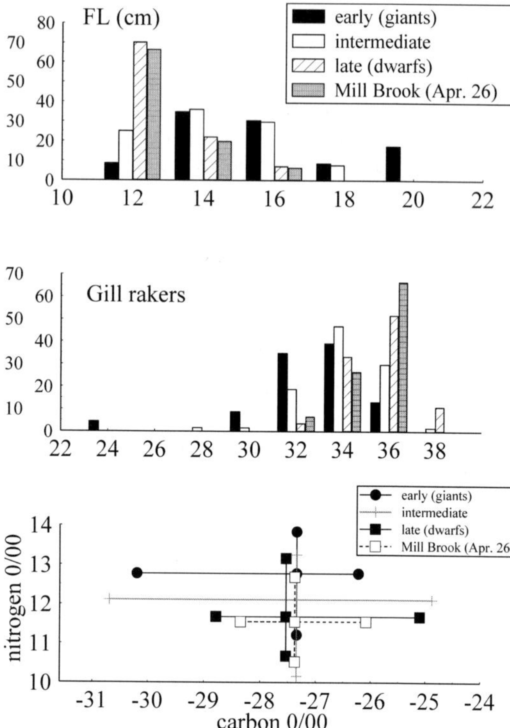

Figure 5. Size, gill raker counts, and stable isotopes of C and N of age-3 smelt of Lake Utopia, 1999.

Population abundance

The latest spawning group had the greatest abundance (Table 2). Single evening estimates were in the order of 100 000 suggesting a total population of 1–1.5 million spawners. The intermediate-spawning group had single evening estimates in the order of 50 000 suggesting a total population of 250 000–500 000. The early spawning Mill Stream group was substantially smaller and most probably numbered <5 000 spawners in total.

Trophic relationships

The fish community was relatively complex in Lake Utopia with 12 species encountered. Atlantic salmon (a landlocked population), brook trout, and smallmouth bass displayed enriched nitrogen and carbon signatures identifying them as potential predators of the late spawning, small smelt (Figure 6). Larger, predatory smelt were the only species with gut contents composed of smelt. The single gaspereau captured had a carbon signature of -18.5 0/00 and would be considered to have migrated to the lake from the marine environment.

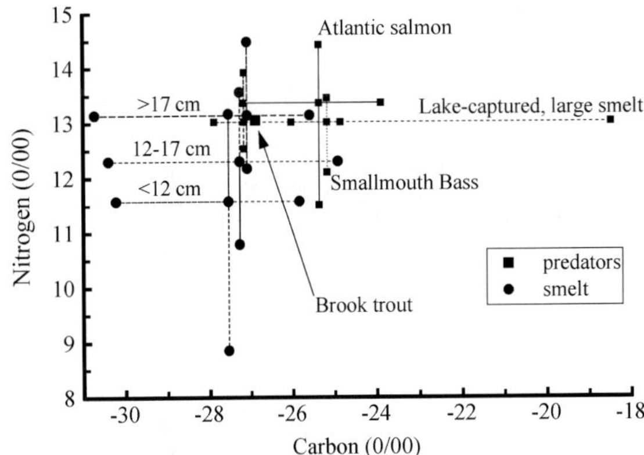

Figure 6. Trophic relationships among major predatory species (solid lines and squares) and spawning smelt of three size classes (dashed lines and solid circles) in Lake Utopia, 1999.

One large smelt captured offshore displayed a similar carbon signature (−18.4 0/00).

Genetic relationships

All microsatellite loci generally displayed high levels of polymorphism, with the total number of alleles per locus per sample varying between 3 and 21 (average = 10), and the mean gene diversity (He) per locus per sample varying between 0.10 and 0.93 (average = 0.68; Table 1). No significant departure from Hardy–Weinberg equilibrium was detected by multi-locus probability tests within each sample. However, a significant heterozygote deficiency was detected across sample for locus *Osmo-45*. This locus is the most variable, and had the widest allelic size range. It is therefore possible that this deficit was caused either by null alleles or miss-amplification of the largest alleles. Levels of intra-sample genetic diversity, expressed both in terms of number of alleles and gene diversity was generally very similar among samples, except for Mill Stream 'giant' and Scout Brook 'intermediate' smelts for which diversity was lower (Table 1). A t-test comparing mean number of alleles (A) and mean gene diversity (He) among each pair of parks for each locus revealed no significant differences in either mean values of number of alleles or in gene diversity.

All samples differed in their allelic composition at one or more loci (Figure 7). Consequently, homogeneity tests of allele frequency distributions revealed significant differences following Bonferroni sequential corrections in all pairwise sample comparisons (Table 3). This provided a first indication that each of the sample analysed represented a genetically distinct population. This was generally corroborated by θ estimates of population differentiation (Table 3). The average pairwise θ value was relatively high (average = 0.091), but varied substantially among comparisons (range = −0.010 and 0.233). The extent of differentiation between dwarf and intermediate smelt varied among tributaries. The lowest value was observed in Mill Brook. The extent of differentiation was intermediate in Scout Brook and highest in Unnamed Brook. Clearly, the giant smelt population from Mill Stream was the most genetically distinct from all others.

The AMOVAs did not reveal any significant pattern of hierarchical clustering (Table 4). The first AMOVA revealed that grouping by tributary of origin did not explain any significant component of total genetic variance. In contrast, the extent of genetic variance imputable to among populations within tributary was substantial and highly significant. Similarly, the component of genetic variance imputable to grouping by form (intermediate vs. dwarf) was null whereas genetic variance due to inter-population within form was important and highly significant (Table 3). Therefore, these results do not support either a more common origin of all smelt populations of a given form or more genetic exchange between populations within a given tributary relative to populations from other tributaries.

The topology of the population phenogram constructed from the pairwise D_{CE} matrix corroborated this pattern of population relationships; no apparent clustering either by form of tributary was observed (Figure 2).

Discussion

The evidence from the timing of spawning, body sizes, and gill raker counts, with some corroboration from trophic analyses has demonstrated that at least three identifiable groups compose the spawning population of the Lake Utopia smelt complex. Overall measures of genetic divergence among the groups were relatively high exhibiting greater differences than commonly reported among anadromous populations (e.g., Garant et al. 2000). In addition, genetic analyses indicated that two of these groups (dwarf and intermediate) each comprise several genetically distinct populations.

The first group is an early spawning, giant morphotype that used one inlet stream (Mill Stream) from a time when the lake was still ice covered in late March until the first week of April. This group appears to have been sampled in all earlier studies as 'normal' smelt.

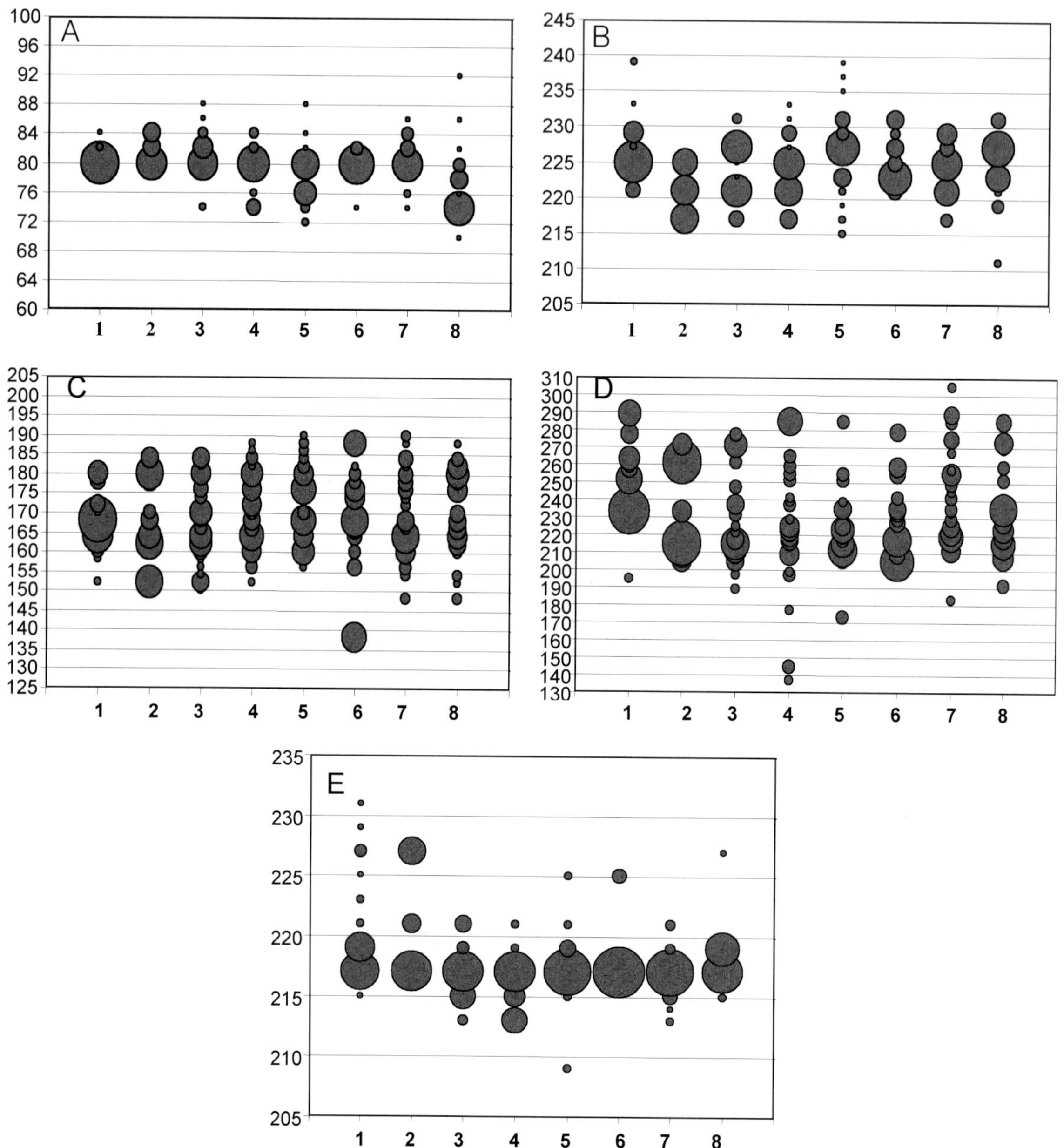

Figure 7. Allele frequency histogram of loci (A) *OSMO-Lav16*, (B) *OSMO-Lav157*, (C) *OSMO-Lav12*, (D) *OSMO-Lav45*, and (E) *Tpa-26*. Allele designation is in base pairs and size of circles is proportional to allele frequency within each sample. Populations: (1) Mill Stream giant (00); (2) Mill Stream giant (99); (3) Mill Brook dwarf; (4) Mill Brook intermediate; (5) Scout Brook dwarf; (6) Scout Brook intermediate; (7) Unnamed Brook dwarf; and (8) Unnamed Brook intermediate.

They were also previously sampled from a second stream, Trout Brook, but it produced very few smelt and not in all years of the present study. Interestingly, spawning in a tributary of Trout Stream's source lake was confirmed in 2002 (Currie, S. unpubl. data). An intermediate-sized group of smelt began spawning in three small, northwestern streams 1–5 km from the early spawning streams in mid-April. These smelt were most similar in body size to normal populations of smelt in other lakes (e.g., Lanteigne & McAllister 1983,

Table 3. Test of heterogeneity of allele frequency between samples (above main diagonal) over all loci with the Fisher's method and the level of significance adjusted after sequential Bonferroni correction (S = significant, α = 0.05, k = 28). Below the main diagonal are the pairwise estimates of population differentiation (θ values, *statistically significant following sequential Bonferroni). G = giant, D = dwarf, and I = intermediate morph-types.

	1	2	3	4	5	6	7	8
1. Mill Stream G (1999)		S	S	S	S	S	S	S
2. Mill Stream G (2000)	0.070*		S	S	S	S	S	S
3. Mill Brook D	−0.010	0.159*		S	S	S	S	S
4. Mill Brook I	−0.010	0.089*	0.051*		S	S	S	S
5. Scout Brook D	0.058*	0.166*	0.057*	0.084*		S	S	S
6. Scout Brook I	0.061*	0.150*	0.100*	0.092*	0.079*		S	S
7. Unnamed Brook D	0.001	0.091*	0.052*	0.014*	0.070*	0.072*		S
8. Unnamed Brook I	0.118*	0.234*	0.120*	0.145*	0.078*	0.164*	0.155*	

Table 4. Hierarchical analyses of molecular variance (AMOVA) as shown by (A) grouping by tributary and (B) grouping by form. Mill Stream giant smelt samples from 1999 were not included in this analysis.

Source of variation	d.f.	Sum of squares	Variance components	Percentage of variation
(A) Grouping by tributary				
Among groups	2	20.940	0.00376	0.22[ns]
Among populations within groups	4	36.701	0.14516	8.31*
Within populations	375	599.455	1.59855	91.48*
(B) Grouping by form				
Among groups	1	7.507	−0.02665	−1.50[ns]
Among populations within groups	4	49.023	0.17825	10.01*
Within populations	352	573.372	1.62890	91.49*

ns = non-significant; *p < 0.0001.

Taylor & Bentzen 1993a). Two of the streams were previously identified as habitats for spawning 'dwarf' smelt (Mill Brook and Unnamed 'A' in Taylor & Bentzen 1993b) and the third stream, Scout (Second) Brook, was previously not known to support spawning. Within 7 days of the appearance of intermediate spawners, a third group of smaller smelt began spawning in the same streams on the same evenings. Their small size and increased number of gill rakers were previously used to characterize these late spawners as 'dwarf' smelt. The dwarf smelt were clearly distinct from the giant smelt in terms of size, gill raker counts, trophic status, and genetic characteristics.

It was previously reported that the giant smelt were abundant and dwarf smelt few in numbers (Taylor & Bentzen 1993b). Based on a more complete sampling of the spawning period and streams, we demonstrated that the actual abundances of spawning smelt were in the order of a few thousand for giants, hundreds of thousands for the intermediates, and over 1 000 000 for the dwarfs in 1999 with similar abundances observed in 2000. It was clear that the dwarf smelt abundance was limited by habitats. In both Mill and Unnamed Brooks, all suitable substrates, i.e., secure surfaces susceptible to egg attachment, were densely packed with eggs in some instances creating 5 cm deep mats of eggs covering the entire width of the streams for distances of 5 m. Smaller mats have also been observed in Scout Brook in 2001 and 2002 (Currie, S.L. unpubl. data). At the peak of their spawning, dwarf smelt completely clogged the streams after dark and significant numbers remained in the streams during the day such that they could be easily captured by hand in 1999 and 2000. There was no indication that they were spawning during the day, but there would be some reproductive advantages to being at a spawning area as opposed to attempting to enter the stream and access spawning areas once the observed mass of smelt entered at sunset. In 2001, we observed drifting smelt larvae at rates ranging from 1 to 44 larvae cm^{-3} h^{-1} from these streams and therefore we know that embryos successfully develop to at least the larval interval (Curry, R.A. unpubl. data).

164

It would appear that one major limitation to dwarf smelt production in the lake is limited availability of spawning and incubation habitats. In addition, the over-saturation of available spawning habitats suggest the population size of small, dwarf smelt is near its maximum level.

The important role smelt can play as forage for land-locked Atlantic salmon populations is well known (e.g., Sayers et al. 1989). In Lake Utopia, there was evidence from stable isotope analysis that landlocked salmon consumed smaller bodied smelt. It was most probable that other predatory fishes also consumed dwarf smelt, including larger, predatory smelt which were the only species with smelt found in the stomach contents. How important the smaller smelt were to the landlocked salmon was not discernable, but smelt were clearly not the sole food of salmon and numerous other potential forage species displayed stable isotope signatures in the predicted region for prey of salmon, e.g., reduced carbon (1 0/00) and nitrogen (3.4 0/00) signatures (Vander Zanden et al. 1999). Salmon may target immature smelt which may have different isotope signatures than the mature smelt of the present study. Additional studies will be required to determine a more accurate description of the salmon and other predatory fishes' diets and the importance of the various life history stages of smelt as prey. Given the apparent abundance of the small, dwarf smelt, it does not appear that predatory species of the fish community are affecting the survival of the dwarf smelt.

Despite a significant increase in spatial and temporal sampling of the spawning smelt, there is evidence that more groups exist within the smelt complex of Lake Utopia. First, at the time of spawning for early, giant smelt, some smelt enter Trout Stream and these smelt spawn in a tributary to Trout Lake. This is some distance from Mill Stream where giant smelt spawn annually. Although straying among stream occurs for later spawners at least, the distance between Mill Stream and Trout Lake may represent a spatially isolating mechanism for an additional group of smelt. Second, one larger, lake-captured smelt displayed a stable isotope signature most similar to the one anadromous gaspereau captured. Both had signatures suggesting a marine origin, i.e., carbon = −18 0/00 compared with the lake's freshwater clams at carbon = −27 0/00 and marine-dwelling gaspereau can access Lake Utopia via the fishway on the Magaguadavic River (Carr, J. Atlantic Salmon Federation, pers. comm.). The possibility that anadromous smelt enter the lake and their contribution to reproduction and trophic dynamics requires further exploration. Third, shore or shoal spawning by smelt is not uncommon (Scott & Crossman 1973). We did not search for such spawning and incubation habitats; neither did we survey all the inlet streams representing potential habitats. Fourth, we observed intermediate spawners in Mill Brook at the onset of the second spawning on 20 April 1998 and some mix of intermediates and dwarf spawners a week after the second spawning began on 26 April 1999. An earlier study captured giant smelt, 19–27 cm in length from Mill Stream on 11 April 1922 (MacLeod 1922). Such observations suggest that the timing of spawning and patterns of stream use may vary annually. Indeed, in 2002 early spawning giants were not observed until 24 April, their abundance was >5 000, and normal smelt began spawning 1 May (Currie, S.L. unpubl. data). We have demonstrated that the smelt of Lake Utopia present a more complex population structure, as well as behavioural and phenotypic variability than previously believed, but we may have only touched on the complexity that could exist.

The levels of genetic divergence among these three groups of smelt in Lake Utopia were higher than commonly observed among populations of anadromous salmon in geographically larger systems (e.g., Garant et al. 2000). In addition, there was clear evidence of genetic separation among populations of the same group among streams and within a stream over time (Scout, Unnamed, and Mill Brooks). This indicates that during the intermediate and late spawning, normal and dwarf smelt maintain some degree of assortive mating or suffer reduced hybrid survivability as embryos or a later life history stage. There is good evidence that all three groups mix in the lake because a distinct trophic separation was not apparent (although some giant smelt were top predators). Our examination of the spawning ecology begins to define the complexity of the smelt's intra- and inter-specific relationships in this lake and lacustrine environments in general, but there are clearly life history tactics that we do not yet understand.

Acknowledgements

Numerous individuals assisted with collections and we are particularly indebted to R. Hawkins and M. Gautreau. P. Cronin and P. Seymour provided historical data and logistic support. K. Gosse undertook the morphometrics and age analyses. This project was funded in part by the New Brunswick Wildlife Trust Fund.

References

Baby, M.-C., L. Bernatchez & J.J. Dodson. 1991. Genetic structure and relationships among anadromous and landlocked populations of rainbow smelt (*Osmerus mordax* Mitchill) as revealed by mtDNA restriction analysis. J. Fish Biol. 39(Suppl. A): 61–68.

Belkhir, K., P. Borsa, L. Chikhi, N. Raufaste & F. Bonhomme. 2000. GENETIX 4.02, logiciel sous Windows™ pour la génétique des populations. Laboratoire Génome, Populations, Interactions, CNRS UMR 5000, Université de Montpellier II, Montpellier (France). 000 pp.

Bernatchez, L. 1997. Mitochondrial DNA analysis confirms the existence of two glacial races of rainbow smelt *Osmerus mordax* and their reproductive isolation in the St Lawrence River estuary (Quebec, Canada). Mol. Ecol. 6: 73–83.

Bernatchez, L. & S. Martin. 1996. Mitochondrial DNA diversity in anadromous rainbow smelt, *Osmerus mordax* Mitchill: A genetic assessment of the member-vagrant hypothesis. Can. J. Fish. Aquat. Sci. 53: 424–433.

Bridges, C.D.B. & C.E. Delisle. 1974. Postglacial evolution of the visual pigments of the smelt, *Osmerus eperlanus mordax*. Vision Res. 14: 345–356.

Cavalli-Sforza, L.L. & A.W.F. Edwards. 1967. Phylogenetic analysis: Models and estimation procedures. Evolution 32: 550–570.

Collet, K.A., T.K. Vickers & P.D. Seymour. 1999. The contribution of stocking to the recreational Landlocked salmon fishery in six New Brunswick lakes, 1996–1997. Management Report. New Brunswick Department of Natural Resources and Energy Fisheries Program.

Copeman, D.G. 1977. Population differences in rainbow smelt, *Osmerus mordax*: Multivariate analysis of mensural and meristic data. J. Fish. Res. Board Can. 34: 1220–1229.

Copeman, D.G. & D.E McAllister. 1978. Analysis of the effect of transplantation on morphometric and meristic characters in lake populations of the rainbow smelt *Osmerus mordax* (Mitchill). Env. Biol. Fish. 3: 253–259.

Felsenstein, J. 1993. PHYLIP (Phylogeny Inference Package) Version 3.5c. Department of Genetics, University of Washington, Seattle, WA.

Frechet, A., J.J. Dodson & H. Powles. 1983. Use of variation in biological characters for the classification of anadromous rainbow smelt (*Osmerus mordax*) groups. Can. J. Fish. Aquat. Sci. 40: 718–727.

Garant, D., J.J. Dodson & L. Bernatchez. 2000. Ecological determinants and temporal stability of the within-river population structure in Atlantic salmon (*Salmo salar* L.). Mol. Ecol. 9: 615–628.

Goudet, J. & M. Raymond. 1996. Testing differentiation in diploid populations. Genetics 144: 1933–1940.

Guo, S.W. & E.A. Thompson. 1992. Performing the exact test of Hardy–Weinberg proportion for multiple alleles. Biometrics 48: 361–372.

Kendall, W.C. 1927. The smelts. Fish. Bull. Fish & Wildlife Serv. 42: 217–225.

Kirn, R.A. & G.W. Labar. 1996. Growth and survival of rainbow smelt, and their role as prey for stocked salmonids in Lake Champlain. Trans. Amer. Fish. Soc. 125: 87–96.

Lanteigne, J. & D.E. McAllister. 1983. The pygmy smelt, *Osmerus spectrum* Cope, 1870, a forgotten sibling species of Eastern North American Fish. Ichthyology Section, National Museum of Natural Sciences, Canada.

MacLeod, N. 1922. An investigation of the Lake Utopia smelt. Biological Board of Canada, Atlantic Biological Station, St. Andrew's, N.B.

McKenzie, R.A. 1964. Smelt life history and fishery in the Miramichi River, New Brunswick. Fisheries Research Board of Canada, Bulletin 144.

McLean, J. & E.B. Taylor. 2001. Resolution of population structure in a species with high gene flow: Microsatellite variation in the eulachon (Osmeridae: *Thaleichthys pacificus*). Mar. Biol. 139: 411–420.

Nellbring, S. 1989. The ecology of smelts (Genus *Osmerus*): A literature review. Nordic J. Freshw. Res. 65: 116–145

Rice, W.R. 1989. Analysing tables of statistical tests. Evolution 43: 223–225.

Rousset, F. & M. Raymond. 1995. Testing heterozygote excess and deficiency. Genetics 140: 1413–1419.

Rupp, R.S. & M.A. Redmond. 1966. Transfer studies of ecologic and genetic variation in American smelt. Ecology 47: 253–259.

Ryman, N., F.W. Allendorf & G. Stahl. 1979. Reproductive isolation with little genetic divergence in sympatric populations of brown trout (*Salmo trutta*). Genetics 92: 247–262.

Sambrook, J., E.F. Fritsch & T. Maniatis. 1989. Molecular cloning: A laboratory manual. Cold Spring Harbor Laboratory Press, New York. 1659 pp.

SAS Institute Inc. 1990. SAS User's guide: Statistics, version 6, 4th edition SAS Institute Inc., Cary, NC.

Sayers, R.E. Jr., J.R. Moring, P.R. Johnson & S.A. Roy. 1989. Importance of rainbow smelt in the winter diet of landlocked Atlantic salmon in four Maine lakes. N. Amer. J. Fish. Mgt. 9: 298–302.

Schneider S., J.M. Kueffer, D. Roessli & L. Excoffier. 1997. ARLEQUIN version 1: An exploratory population genetics software environment. Genetics and Biometry Laboratory, University Geneva, Geneva, Switzerland.

Scott, W.B. & E.J. Crossman. 1973. Freshwater fishes of Canada. Fisheries Research Board of Canada, Ottawa, Ontario.

Skulason, S., S.S. Snorrason, D.L.G. Noakes & M.M. Ferguson. 1996. Genetic basis of life history variations among sympatric morphs of Arctic char, *Salvelinus alpinus*. Can. J. Fish. Aquat. Sci. 53: 1807–1813.

Taylor, E.B. & P. Bentzen. 1993a. Evidence for multiple origins and sympatric divergence of trophic ecotypes of smelt (*Osmerus*) in northeastern North America. Evolution 47: 813–832.

Taylor, E.B. & P. Bentzen. 1993b. Molecular genetic evidence for reproductive isolation between sympatric populations of smelt *Osmerus* in Lake Utopia, south-western New Brunswick, Canada. Mol. Ecol. 2: 345–357.

Vander Zanden, M.J., B.J. Shuter, N. Lester & J.B. Rasmussen. 1999. Patterns of food chain length in lakes: A stable isotope study. Amer. Nat. 154: 406–416.

Weir, B.S. & C.C. Cockerham. 1984. Estimating F-statistics for the analysis of population structure. Evolution 38: 1358–1370.

Wirth, T., R. Saint-Laurent & L. Bernatchez. 1999. Isolation and characterization of microsatellite loci in the walleye (*Stizostedion vitreum*), and cross-species amplification within the family Percidae. Mol. Ecol. 8: 1961–1963.

Appendix 1. Primer sequences for the five loci used in the study of Lake Utopia's rainbow smelt.

Name	Motif	Primer sequence (5′–3′)	Cloned allele (pb)	Ta (°C)
Osmo-Lav 12	$(GT)_{37}$	1: CTG TAA TAT TCC ACTGCT GC	164	55
		2: CAA GTA GAC AGT AGO GAGA		
Osmo-Lav 16	$(GT)_{15}$	1: GGA TCT TGG ATG AGA AC A T	92	55
		2: GGC TCT TTC ATT ACA CAG G		
Osmo-Lav 45	$(CT)_{33}$	1: CTG TTG ATA GAT TGG CAT C	201	55
		2: CCC ATT CAA TTA GAC AGT G		
Osmo-Lav 157	$(GT)_{33}AAGAGA(GT)_{7}$	1: CTT GCT TAT GTA AAG GTG GG	246	55
		2: GAT CCA CCA GTT CTC ACA		
Tpa-26		1 :AGG ACT GGC GTG GGA AAT		57
		2: CTG CAC TGC TGT CTG GAG AA		

Environmental Biology of Fishes **69**: 167–175, 2004.
© 2004 *Kluwer Academic Publishers. Printed in the Netherlands.*

Mitochondrial DNA variation in northwestern Bering Sea walleye pollock, *Theragra chalcogramma* (Pallas)

Vladimir A. Brykov[a], N.E. Polyakova[a], T.F. Priima[a] & O.N. Katugin[b]
[a]*Institute of Marine Biology, Vladivostok, Russia (e-mail: vladbrykov@hotmail.com)*
[b]*Pacific Research Fisheries Centre (TINRO-Centre), Vladivostok, Russia*

Received 21 April 2003   Accepted 16 June 2003

Key words: population genetics stock structure

Synopsis

Genetic variability of the highly valuable gadid species, walleye pollock, *Theragra chalcogramma* (Pallas 1811), from five spatially separated northwestern Bering Sea areas (Ozernoi Bay, Olutorskyi Bay, Koryak shelf, Navarin region, and Anadyr Gulf) was investigated. Haplotype diversity within the samples ranged from 0.8788 ± 0.0393 to 0.9436 ± 0.0162. Nucleotide diversity within the samples ranged from 0.0108 to 0.0127. Nucleotide diversity among the regional collections ranged from 0% to 0.18%. Walleye pollock from the Anadyr Gulf appeared genetically separate from the other four samples in a clustering of genetic distances. Total heterogeneity among all five samples was significant, while there was no heterogeneity among the four samples excluding that from the Anadyr Gulf. Pair-wise comparisons using tests for heterogeneity and F_{st} supported the dendrogram of genetic distances in that only the collection from the Anadyr Gulf was significantly different from the others. The observed pattern of genetic differentiation among walleye pollock from the northwestern Bering Sea presumably emerged as a result of a population of the species subdividing in the northernmost part of its geographic range, though further analysis is needed to verify this supposition.

Introduction

The walleye Pollock, *Theragra chalcogramma*, is one of the major fish resources in the subarctic Pacific Ocean region, often contributing to more than 1 million metric tons (t) of annual harvest. It is widely distributed over continental slopes of the western coast of North America, of the Aleutian and Kuril islands, and of the Bering, Okhotsk, and Japan seas. A number of spatially distinct spawning areas of the species have been recognized, and they presumably reflect general population differentiation of the species over its geographical range (Shuntov et al. 1993). Based on the spatial distribution of spawners and free-floating developing eggs, three main reproductive centers of walleye pollock have been recognized in the Bering Sea (Fadeev 1991): the eastern Bering Sea stock with spawning sites concentrated over the southeastern shelf, the Bogoslof/Aleutian Basin stock,

and the western Bering Sea stock with a spawning center in Olutorskyi Bay (Figure 1). The eastern Bering Sea stock of walleye pollock is the most abundant (Karp & Traynor 1989, Stepanenko 2001); its estimated biomass was assessed at 10.3 million t in 1993, which declined almost by half by 1998 (Ianelli et al. 1998). A substantial, almost five-fold decrease of pollock abundance was indicated for the Bogoslof area as well, where the estimated biomass dropped to around one half million t in 1998 (Honkaletho & Williamson 1999). The western Bering Sea stock is much less abundant than eastern Bering Sea stocks (Balykin 1996), and currently it is also considerably depleted, with an estimated biomass barely reaching 300,000 t. Besides the mass spawning events within the reproductive centers, there are minor spawning grounds of walleye pollock on the Bering Sea continental shelf. For example, regular spawning occurs in bays along the eastern coast of the Kamchatka Peninsula (Antonov & Zolotov 1987). Some data from

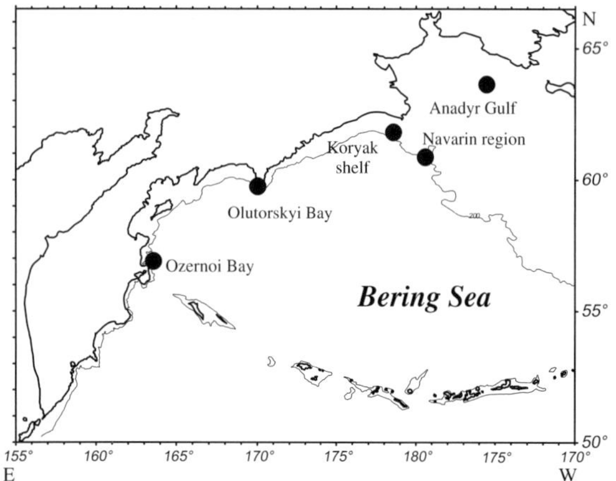

Figure 1. Map of the northwestern Bering Sea showing sampling locations of walleye pollock.

trawl surveys also indicate that spawning of walleye pollock may occur in the area adjacent to Cape Navarin in the northern Bering Sea (Datskyi 2002) though the relationship of the Navarin fish with either western or eastern stocks is poorly documented. Besides distribution patterns, size structure, and biological traits, considerable differentiation among walleye pollock from the northwestern Bering Sea was revealed by morphometric analyses (Datskyi & Makoyedov 2002, O.N. Katugin & Berkutova, unpublished data). In order to sustain commercial fishing enterprises, and considering large fluctuations in abundance, walleye pollock resources should be managed rationally, based on a composite analysis of stock structure particularly using data measuring genetic variability and differentiation of stocks.

Several researchers have applied genetic approaches to the differentiation of stocks of walleye pollock in the Bering Sea, and some of them used data obtained from mitochondrial DNA (mtDNA) analyses (e.g., Mulligan et al. 1992, Shields & Gust 1995, Numachi & Kobayashi 1994, Okazaki 1996, 1997, Yanagimoto et al. 1997). On the whole, they found a low variability of mtDNA structure within the surveyed areas, a result concordant with the generally small genetic differentiation of the species revealed by other molecular approaches such as protein electrophoresis and nuclear DNA analysis (Bailey 1998, Bailey et al. 1999). In spite of the low intraspecific genetic diversity, a major subdivision between Asian and North American stocks of walleye pollock has been observed from analysis of variation of allozyme

frequencies and frequencies of restriction fragment length polymorphisms (RFLP) of mtDNA, a result not confirmed by analysis of microsatellite gene markers (Olsen et al. 2002). Similar east–west differentiation between walleye pollock from the southeastern Bering Sea shelf and the western Bering Sea was revealed by the sequencing of 300 DNA base pairs (bp) of two mitochondrial genes (Powers 1998). Here we report on the use of mtDNA as a potential gene marker for fine stock structure analysis of walleye pollock from the northwestern Bering Sea, including geographic sites not covered by other genetic studies of the species.

Materials and methods

Mature walleye pollock were collected from geographically distinct localities along the western and northern Bering Sea shelf, across a vast area from Ozernoi Bay to Anadyr Gulf (Figure 1; Table 1). In total, five collections were taken for genetic analysis. Small pieces of heart tissue were dissected and fixed in absolute ethanol until they were processed in the laboratory. Total genomic DNA was isolated using standard procedures (Sambrook et al. 1989), and mtDNA segments were amplified using primers developed for two salmonid species, *Oncorhynchus mykiss* and *Oncorhynchus kisutch* (Zardoya et al. 1995, Gharrett et al. 2001). The primer pairs targeted three regions of the walleye pollock mtDNA in pieces ranging from 2330 to 2600 bp. Three regions were amplified by polymerase chain reaction (PCR): ND3/ND4 (including genes for NADH dehydrogenase-3 subunit and the NADH dehydrogenase-4L and -4 subunit), ND5/ND6 (including genes for the NADH dehydrogenase-5 and -6 subunits), and Cyt*b*/D-loop (including almost the entire cytochrome *b* gene and the control region) (Figure 2). The three mtDNA fragments were amplified in a Perkin-Elmer 9600 thermocycler. The process consisted of denaturation at 94°C for 5 min, followed by 30 cycles of the PCR of 1 min at 94°C, 1 min at 55°C, and 3 min at 72°C.

Aliquots of the PCR amplified mtDNA fragments were digested by an array of restriction endonucleases, and then the reaction mixtures were subjected to electrophoresis in 1.5% agarose gels (a mixture of 1/3 common agarose, Ultra Pure™, BRL Gibco, Grand, NY, and 2/3 Synergel agarose, Diversified Biotech Inc., Boston, MA). The electrophoretic buffer was 90 mM tris-borate, 2 mM EDTA, pH 7.5. Gels were stained in ethidium bromide and photographed in transmitted

Table 1. Sampling locations, dates, and size of walleye pollock from the northwestern Bering Sea, taken for the RFLP analysis of mitochondrial DNA (s.e. = standard error).

Region	Coordinates	Date	Average fork length mm ± s.e
Ozernoi Bay	56°59′N, 163°18′E	22 Aug 1998	503.66 ± 5.88
Olutorskyi Bay	59°51′N, 170°45′E	19 May 2000	444.75 ± 3.74
Koryak shelf	61°47′N, 177°37′E	10 Sep 1999	490.00 ± 8.36
Navarin region	60°41′N, 179°01′W	18 May 2000	453.75 ± 8.56
Anadyr Gulf	63°40′N, 175°36′W	4 Nov 1999	652.95 ± 7.68

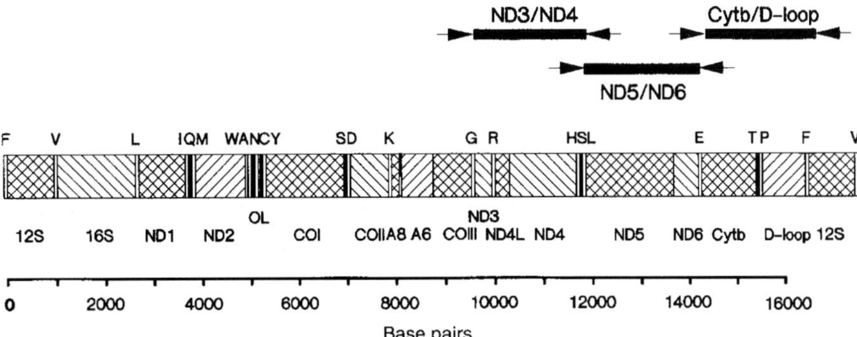

Figure 2. Scheme of linearized mitochondrial DNA and PCR-amplified segments of the molecule in walleye pollock. The map is based on the sequence of *Oncorhynchus mykiss* (Zardoya et al. 1995).

UV light. A 100 bp ladder of synthetic oligonucleotides served as a molecular weight reference marker.

PCR products of each mtDNA fragment were initially digested with each of the following 29 restriction endonucleases: *Bsp19* I, *Bgl* I, *Hind* III, *Pst* I, *Bst*HP I, *Acc*B I, *Pvu* II, *Bam*H I, *Sty* I, *Sec* I, *Sau96* I, *Asp*LE I, *Bsu*R I, *Msp* I, *Hpa* I, *Hinf* I, *Eco*RV, *Dra* I, *Bme18* I, *Mva* I, *Ama87* I, *Rsa* I, *Mbo* I, *Dde* I, *Fnu* II, *Sse9* I, *Taq* I, Alu I, *Vsp* I. Variable restriction sites of walleye pollock mtDNA were revealed using *Hinf* I in the ND3/ND4 region, *Rsa* I and *Hinf* I in the Cyt*b*/D-loop region, and *Ama87* I,*Vsp* I, *Msp* I, and *Rsa* I in the ND5/ND6 region. Each restriction variant received a letter, and mtDNA fragments of every individual over all restriction sites for the six informative enzymes were used to obtain composite haplotypes. A site matrix for each of the composite haplotypes resolved was constructed using the software package REAP (McElroy et al. 1992). A network of minimum restriction site changes between haplotypes was made with the use of the software package Arlequin 2.000 (Schneider et al. 2000).

Heterogeneity between samples of walleye pollock was assessed with contingency χ^2-tests for differences in haplotype frequencies among populations (Roff & Bentzen 1989) by using Monte-Carlo simulation (10,000 iterations) performed using the REAP software package (McElroy et al. 1992). Genetic divergence among samples was assessed with the F-statistic, F_{st} and its significance was estimated from distributions of the statistics generated by 20,000 permutations at the probability level of 0.05% using the Arlequin 2.000 software package (Schneider et al. 2000). The number of nucleotide changes per site, d, and their standard deviation among haplotypes were estimated using REAP (McElroy et al. 1992). The amount of gene divergence was calculated according to Nei & Li (1979). Nucleotide (π), and haplotype (h), diversities were assessed from Nei (1987), and Nei & Tajima (1981) using REAP (McElroy et al. 1992). As a measure of genetic diversity among populations, we computed the average numbers of nucleotide substitutions per site between populations (nucleotide divergence) (Nei 1987) using the DA program in REAP. Cluster analysis of genetic distances between sampled populations was conducted using the unweighted pair-group method with arithmetic averages (UPGMA) (Sneath & Sokal 1973) in the NTSYS computer program (Rohlf 1990).

Results

PCR-analyzed segments of the walleye pollock mitochondrial genome are localized one after another almost without overlap. The total length of the amplified mtDNA constituted $\sim$7 kb, thus accounting for a little bit less than one half of 16.8 ± 0.078 kb, the estimated size of the entire mtDNA molecule of the species (Mulligan et al. 1992). Digestion patterns obtained from six restriction enzymes generated 48 composite mtDNA haplotypes among 194 specimens of walleye pollock (Table 2). Most of the neighboring haplotypes in the table differed by a single nucleotide substitution. Thirty of the 48 haplotypes were only observed in single individuals. Haplotypes 3 and 18 were the most frequent, appearing in 29 (14.9%) and 33 (17%) of the individuals.

Haplotype frequencies differed among walleye pollock from various geographical sites. In the Anadyr Gulf, haplotype 12 was the most common, and was found in 10 (25.6%) of the fish; haplotype 18 was dominant in the Ozernoi Bay and Navarin region, where it was found in 10 (29.4%) and 10 (25%) of the fish, respectively. Haplotype 3 was almost equally represented in all the surveyed regions. Haplotype 1 was common in four of the five collections, ranging in frequency from 5.8% to 12.5% , but was not observed in the Navarin region. Haplotype 22 was found in four collections at high frequency (5–11.8%), but was not observed in the Anadyr Gulf.

Estimated percent nucleotide divergence between sampled aggregations of walleye pollock, d, ranged from 0.000 to 0.176 (Table 3), mean value 0.051 ± 0.020 (standard error of the mean). The highest values were characteristic of pairs of samples involving fish from the Anadyr Gulf, ranging from 0.032 to 0.176, mean 0.106 ± 0.036. Percent divergence among the four samples excluding that from the Anadyr Gulf appeared much lower, with a range from 0.000 to 0.035, and mean 0.014 ± 0.005. A UPGMA tree produced from those distances revealed a pattern of relationships among walleye pollock collections from different northwestern Bering Sea areas shows the Anadyr Gulf collection differing from the others most of all (Figure 3). The remaining four samples of walleye pollock formed a compact group on the graph of genetic distances, with a considerably lower level of divergence, not exceeding 0.03%. The divergence of Anadyr Gulf fish from the others ranged from 0.03% to 0.18%.

A χ^2-test for genetic heterogeneity computed from raw data for all five samples of walleye pollock was marginally significant ($\chi^2 = 206.76$, P $= 0.04$). Of the ten pair-wise comparisons between the five samples, all of those involving the sample from the Anadyr Gulf were significant except for the pair Anadyr Gulf *versus* Olutorskyi Bay (p $= 0.058$) (Table 4). The other six pair-wise comparisons between four samples excluding that from the Anadyr Gulf indicated no genetic differences of walleye pollock over the area from the Ozernoi Bay to Navarin region. The total χ^2-test for these four samples also did not reveal any significant heterogeneity ($\chi^2 = 125.32$, p $= 0.70$). The use of F_{st} as an indicator of inter-sample genetic divergence showed six of possible ten significant outcomes at the 5% probability of rejecting the null hypothesis of no differentiation (Table 4). All the four pair-wise comparisons involving the Anadyr Gulf walleye pollock sample appeared significant, thus confirming genetic distinctiveness of this population. Two $F_{st\,t}$ values for sample pairs: Olutorskyi Bay–Ozernoi Bay and Olutorskyi Bay–Navarin region, were also significant.

Certain geographic differences emerged when estimated levels of haplotype, π, and nucleotide, h, diversity were taken into consideration (Table 5). The lowest levels of genetic diversity were observed in samples from the Ozernoi Bay and Anadyr Gulf, while the highest were in the Olutorskyi Bay and Koryak shelf samples. This may indicate that the effective population size of walleye pollock from the central part of the studied area, where most of the western Bering Sea pollock spawn, is greater than that from the distant areas, where a much smaller part of the Asian stock reproduces.

Discussion

The use of mtDNA as a genetic marker in population studies has certain advantages over other biochemical genetic methodologies, and they partly stem from the mode of its inheritance (Lansman et al. 1981). Maternal transmission of mtDNA from parents to progeny results in a four-fold decrease of population effective number as assessed from nuclear genes, and therefore in comparatively quick accumulation of genetic differences resulting from random genetic drift processes (Avise & Lansman 1983, Avise et al. 1984, Nei 1987, Billington & Hebert 1991, Avise 1994, 2000). Moreover, it appears that mtDNA evolves quickly, at an average rate of 1–2% nucleotide substitutions per 1 million years, i.e., 5–10 times higher than in unique nuclear genes (Brown et al. 1979), but slower than microsatellites. It has also been shown that

Table 2. Number of fish observed with each of 48 mitochondrial DNA haplotypes in five samples of walleye pollock from the northwestern Bering Sea.

Haplotype number	Haplotype profile	Ozernoi Bay	Olutorskyi Bay	Koryak shelf	Navarin region	Anadyr Gulf	Total
1	AAAAAAA	2	3	5	0	2	12
2	AAAAABA	0	0	1	0	1	2
3	AABAAAA	6	5	7	4	7	29
4	AABAAAC	1	0	1	10	31	03
5	AABAAAG	0	1	0	0	0	1
6	AABAABA	1	5	2	3	1	12
7	AABAACA	0	0	0	42	2	24
8	AABAAEA	0	0	0	0	1	1
9	AABAAFA	0	0	0	1	0	1
10	AABAAGA	0	1	0	0	0	1
11	AABABAA	1	2	1	1	2	7
12	AABABAB	0	2	2	1	10	15
13	AABABBA	0	0	1	0	0	1
14	AABADAA	1	0	0	1	0	2
15	AABAFAA	0	1	0	0	1	2
16	AABDAAA	0	1	0	0	0	1
17	AAEAAAA	0	1	0	0	0	1
18	ABAAAAA	10	5	5	10	3	33
19	ABAAABA	0	0	1	0	0	1
20	ABAAACA	1	0	0	0	0	1
21	ABAAADA	1	0	0	0	0	1
22	ABBAAAA	4	4	4	2	0	14
23	ABBAABA	0	0	0	0	1	1
24	ABBAACA	1	3	1	4	2	11
25	ABBAACB	0	0	0	1	0	1
26	ABBAACE	0	0	0	1	0	1
27	ABBAADA	0	1	0	1	0	2
28	ABBAADE	0	0	0	1	0	1
29	ABBAAFA	0	0	0	1	0	1
30	ABBABAA	0	0	0	0	1	1
31	ABBABCA	0	41	12	21	10	04
32	ABBACCA	1	0	0	0	0	1
33	ABBAFCA	1	0	0	0	0	1
34	ABBAIAA	0	0	1	0	0	1
35	ABBAKBA	0	0	0	1	0	1
36	ABBBAAA	0	0	0	1	1	2
37	ABBBACA	2	1	3	0	0	6
38	ABBBACF	0	0	0	1	0	1
39	ABBBBCA	0	1	0	0	0	1
40	ABBCACA	0	0	0	1	0	1
41	ABBEICA	0	0	1	0	0	1
42	ACBAAAA	1	0	0	0	0	1
43	BABAAAA	0	0	0	0	2	2
44	BABAABA	0	1	0	0	0	1
45	BBAAACA	0	0	0	1	0	1
46	CAAAAAA	0	1	0	0	0	1
47	CABAAAA	0	0	1	0	0	1
48	CBBAACA	0	0	1	0	0	1

Digestion profiles of haplotypes are given in letters.

Table 3. Percent nucleotide divergence between regional samples of the northwestern Bering Sea walleye pollock based on RFLP of PCR amplified mitochondrial DNA fragments.

Region	Ozernoi Bay	Olutorskyi Bay	Koryak shelf	Navarin region	Anadyr Gulf
Ozernoi Bay	0	0.000278	0.000019	0.000076	0.001598
Olutorskyi Bay		0	−0.000183	0.000347	0.000324
Koryak shelf			0	0.000139	0.000582
Navarin region				0	0.001761
Anadyr Gulf					0

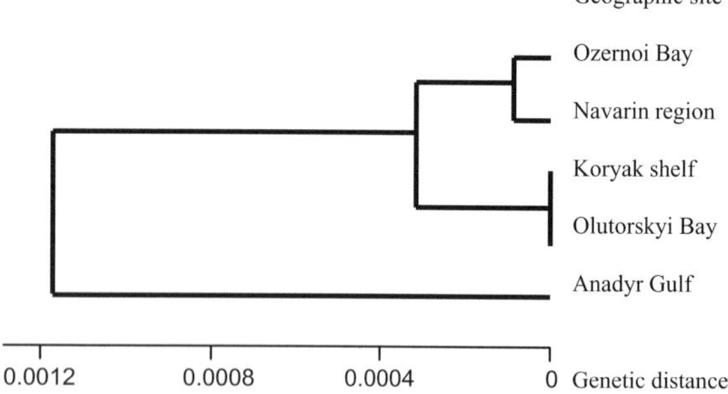

Figure 3. UPGMA clustering among walleye pollock geographical samples constructed from genetic distances.

various mtDNA segments accumulate mutations at different rates, and this may vary among species (Avise 1994, 2000).

Within-sample mtDNA haplotype distribution patterns for walleye pollock from the northwestern Bering Sea apparently are concordant with such patterns in the species' entire mtDNA (Mulligan et al. 1992), and are generally similar to patterns in other anadromous and marine fish species in which the greater part of haplotypes are either unique or infrequent (Avise 1994, 2000).

Our data also are similar to recent findings that some parts of the species' mitochondrial genome are more variable (i.e., have more nucleotide substitutions per site) and informative for population studies than others. Analysis of RFLPs generated by six endonucleases (*Hae* III, *Hinf* I, *Aci* I, *Alu* I, *Bfa* I, *Msp* I) in the control and ND5/ND6 regions of mtDNA of walleye pollock from the northwestern, southeastern, and central Bering Sea suggested the possibility of differences among collections, though they appeared insignificant due to small samples of only 10 individuals per sample (Okazaki 1996). Expansion of his initial study to include RFLPs generated by 10 endonucleases in the ND1 and 16sRNA regions also failed to reveal significant differentiation among six Bering Sea areas in collections consisting of 19–27 individuals

(Okazaki 1997). Within the eastern Bering Sea shelf, geographic variability in mtDNA digestion patterns was observed after applying *Hinf* I to the control region, and *Msp* I to ND5/ND6 region (Yanagimoto et al. 1997). Like those authors, we have found that the ND5/ND6 mtDNA fragment provides not only relatively high polymorphism, but also shows differences in pattern of haplotype distribution over a considerable part of the species' range. Digestion profiles of only three sectors of the molecule, about 7 kb long, with six endonucleases revealed significant regional divergence among samples.

Genetic structuring among geographical collections of walleye pollock could be suggested based on the observed differences in levels of haplotype and nucleotide diversities. Genetic variability as revealed from putatively neutral polymorphism in mtDNA can reflect population histories, including bottlenecks and expansions. Generally, the lower the level of intrapopulation diversity, the lower the effective population number, and vice versa (Avise 2000). Estimates of average haplotype, π, and nucleotide, h, diversities were high, 0.9139 ± 0.0001 and 0.0126 ± 0.0000, respectively, among the five samples of walleye pollock analyzed, characteristic of other highly abundant marine fish species (Billington & Hebert 1991, Avise 2000). At the same time, the highest π and h values were characteristic of Olutorskyi Bay and adjacent Koryak shelf samples, within the reproduction center of the western Bering Sea stock.

Interpretation of our data regarding population structure should be cautious for two reasons. First, only a small portion of the species' reproductive range was included by our samples, and walleye pollock from the eastern Bering Sea and Aleutians were not present in this study. As a result, we are not aware of the extent of genetic influence of those pollock populations on the fish sampled in the northwestern part of the sea. Second, some samples of walleye pollock were taken from postspawning aggregations, and exact reproduction sites of those fish could only be suggested from

Table 4. χ^2-tests for genetic heterogeneity (above diagonal) and f_{st} values (below diagonal) between regional samples of walleye pollock from the northwestern Bering Sea.

Region	Ozernoi Bay	Olutorskyi Bay	Koryak shelf	Navarin region	Anadyr Gulf
Ozernoi Bay	0	25.0	20.2	28.6	36.4*
Olutorskyi Bay	0.05*	0	21.8	28.8	33.2
Koryak shelf	0.03	0.00	0	35.8	32.1*
Navarin region	0.02	0.03*	0.02	0	37.9*
Anadyr Gulf	0.18*	0.05*	0.07*	0.16*	0

* = significant values at 0.05 level of probability.

Table 5. Levels of genetic diversities within population samples of walleye pollock from five regions of the northwestern Bering Sea.

Region	Haplotype diversity ± s.e.	Nucleotide diversity
Ozernoi Bay	0.879 ± 0.039	0.011
Olutorskyi Bay	0.944 ± 0.016	0.013
Koryak shelf	0.932 ± 0.020	0.012
Navarin region	0.921 ± 0.030	0.014
Anadyr Gulf	0.895 ± 0.033	0.011

data on the species biology and migrations. For example, walleye pollock caught on the Koryak shelf in the fall must have migrated there from Olutorskyi Bay, which is also evidenced by the absence of significant genetic differences between samples from these areas.

The Anadyr Gulf walleye pollock population, the one that appeared to be the most unique genetically, was sampled from a putatively resident near-bottom aggregation of fish characterized by low migration, high growth rate, and large size-at-age (they attain 85 cm fork length at 18 years old, while actively migrating pelagic pollock of the same age usually grow only to 57 cm (Karp & Traynor 1989, Ilyinskyi et al. 1990)).

It is also questionable whether the apparent genetic structuring of walleye pollock revealed in our study is a result of local population differentiation in the northernmost part of the species range, or is due to the influence of the large eastern Bering Sea population, or whether both factors have accounted for the observed pattern. The latter seems more likely when we take into account the observed genetic distance clustering and the observed pattern of the haplotype network constructed from our data (Figure 4). The UPGMA dendrogram (Figure 3) unequivocally separates large presumably resident fish from the northern Anadyr Gulf, while the 'gene tree' (Figure 4) for haplotypes

suggests an independent pattern. The haplotype network revealed two well-differentiated phylogroups: A and B. Each group is of putatively separate origin, and consists of a central haplotype, haplotypes 3 and 24, respectively, with star-like structures emerging from the two cores, and a single bond between the groups. Multiple radii from a central core may suggest that rapid rises in abundance were characteristic for both phylogroups during population history (Rogers & Harpening 1992, Nee et al. 1995), supporting the general idea of considerable fluctuations of stock abundance in walleye pollock. Phylogroup A apparently consists of the 'Asian' haplotypes, while phylogroup B is of doubtful origin, because it incorporates six of eight haplotypes unique to the Navarin region sample. Rare haplotypes unique to the Anadyr Gulf fish appeared equally distributed in the two phylogroups thus supporting the hypothesis that two genetically different sources influenced this population's modern composition. The A-group influence seems stronger in the Anadyr Gulf pollock because most of its haplotypes, including the most frequent haplotype 12, are found within this group. Phylogroup A is complex, with homoplasmic elements present as a result of repeat and reverse mutations. It is likely that the phylogroups could have originated in different geographic areas, though their connection to the eastern Bering Sea populations should be verified. In any case, our data on mtDNA RFLP unequivocally suggest that there exists low but significant genetic differentiation of walleye pollock in the northwestern Bering Sea. The approach proved to be effective in revealing potential heritable markers for further analysis of the genetic structure of walleye pollock populations.

Whether any of the mtDNA haplotypes revealed in our study of walleye pollock from different Bering Sea regions can serve as genetic markers of the fish from different regions of origin might be verified during further investigations with an expanded sampling program particularly involving much larger samples, additional

174

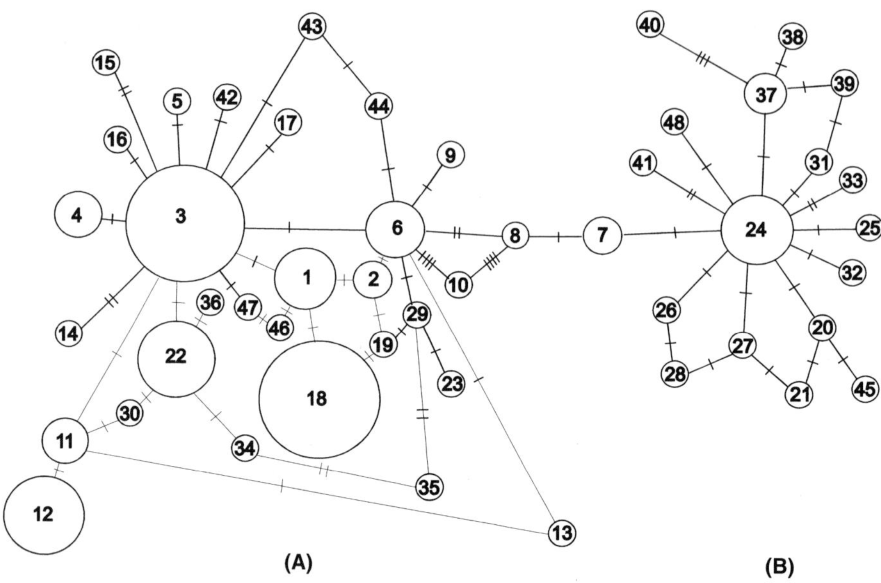

Figure 4. Minimum spanning haplotype network of walleye pollock mtDNA. Tic-marks on the branches indicate mutational changes.

reproductive areas over the species' reproductive range, and yearly sampling.

References

Antonov, N.P. & O.G. Zolotov. 1987. Peculiarities of the eastern Kamchatka walleye pollock reproduction. pp. 123–132. *In:* N.S. Fadeev (ed.) Population Structure, Dynamics of Abundance and Ecology of Walleye Pollock, TINRO, Vladivostok (in Russian).

Avise, J.C. 1994. Molecular Markers, Natural History and Evolution, Chapman & Hall, New York. 511 pp.

Avise, J.C. 2000. Phylogeography. The History and Formation of Species, Harvard University Press, Cambridge. 447 pp.

Avise, J.C. & R.A. Lansman. 1983. Polymorphism of mitochondrial DNA in populations of higher animals. pp. 147–184. *In:* M. Nei & R.K. Koehn (ed.) Evolution of Genes and Proteins, Sinauer Associates, Inc., Sunderland, MA.

Avise, J.C., J.E. Neigel & J. Arnold. 1984. Demographic influences on mitochondrial DNA lineage survivorship in animal populations. J. Mol. Evol. 20: 90–105.

Bailey, K.M. 1998. Population ecology and structural dynamics of walleye pollock, *Theragra chalcogramma*. pp. 3–54. *In:* S.A. Macklin (ed.) Bering Sea FOCI Final Report (Bering Sea Fisheries-Oceanography Coordinated Investigations), NOAA Coastal Oceans Program, Alaska Fisheries Science Center, Seattle.

Bailey, K.M., T.J. Quinn, P. Bentzen & W.S. Grant. 1999. Population structure and dynamics of walleye pollock, *Theragra chalcogramma*. Adv. Mar. Biol. 37: 179–255.

Balykin, P.A. 1996. Dynamics and abundance of western Bering Sea walleye pollock. pp. 177–182. *In:* O.A. Mathisen & K.O. Coyle (ed.) Ecology of the Bering Sea: A Review of Russian Literature. Alaska Sea Grant College Program, University of Alaska Fairbanks, Fairbanks.

Billington, N. & P.D.N. Hebert. 1991. Mitochondrial DNA diversity in fish and its implications for introductions. Can. J. Fish. Aquat. Sci. 48: 80–94.

Brown, W.M., M. George & A.C. Wilson. 1979. Rapid evolution of animal mitochondrial DNA. Proc. Natl Acad. Sci. U.S.A. 76: 1967–1971.

Datskyi, A.V. 2002. On the spawning of walleye pollock in the Anadyr–Navarin region of the Bering Sea. pp. 64–65. *In:* G.I. Nesvetova (ed.) Abstracts from Reports on the All-Russian Conference of Young Scientists, Dedicated to the 140th Anniversary of N.M. Knipovitch, PINRO Press, Murmansk (in Russian).

Datskyi, A.V. & A.N. Makoyedov. 2002. On the population status of walleye pollock in the northwestern Bering Sea. pp. 67–69. *In:* G.I. Nesvetova (ed.) Abstracts from Reports on the All-Russian Conference of Young Scientists, Dedicated to the 140th Anniversary of N.M. Knipovitch, PINRO Press, Murmansk (in Russian).

Fadeev, N.S. 1991. Distribution and migrations of walleye pollock in the Bering Sea, VNIRO Press, Moscow, 54 pp. (in Russian).

Gharrett, A.J., A.K. Gray & V.A. Brykov. 2001. Phylogeographic analysis of mitochondrial DNA variation in Alaskan coho salmon, *Oncorhynchus kisutch*. Fish. Bull., U.S. 99: 528–544.

Honkaletho, T. & N. Williamson. 1999. Walleye pollock (*Theragra chalcogramma*) abundance in the southeastern Aleutian Basin near Bogoslof Island during March 1998. *In:* Stock Assessment and Fishery Evaluation Report for Groundfish Resources in the Bering Sea/Aleutian Islands Regions.

North Pacific Fishery Management Council, 605 West 4th Ave., Anchorage, AK 99501, U.S.A.

Ianelli, J., L. Fritz, T. Honkalehto, N. Williamson & G. Walters. 1998. Bering Sea–Aleutian Islands walleye pollock assessment for 1999. *In*: Stock Assessment and Fishery Evaluation Report for the Groundfish Resources of the Bering Sea/Aleutian Islands Regions. North Pacific Fishery Management Council, Section 1: 1–79. NPFMC, 605 West 4th Ave., Anchorage, AK 99501, U.S.A.

Ilyinskyi, E.N., V.I. Radchenko & K.M. Gorbatenko. 1990. A hypothesis on the existence of near-bottom and pelagic forms of walleye pollock. pp. 41–42. *In:* Yu.I. Zuenko (ed.) Ecology, Migrations, and Distribution Patterns of Marine Commercial Species. Functioning of Marine Ecosystems and Anthropogenic Influence Upon Them. Abstracts of Reports at the Conference of Junior Scientists of TINRO, Vladivostok (in Russian).

Karp, W.A. & J.J. Traynor. 1989. Assessments of the abundance of eastern Bering Sea walleye pollock stocks. pp. 433–456. *In*: Proceedings of the International Symposium on Biology and Management of Walleye Pollock. Alaska Sea Grant College Program, University of Alaska Fairbanks, Fairbanks.

Lansman, R.A., R.O. Shade, J.F. Shapira & J.C. Avise. 1981. The use of restriction endonucleases to measure mitochondrial DNA sequence relatedness in natural populations. J. Mol. Evol. 17: 214–226.

McElroy D., P. Moran, E. Birmingham & I. Kornfield. 1992. REAP: The restriction enzyme package. J. Hered. 83: 157–158.

Mulligan, T.J., R.W. Chapman & B.L. Brown. 1992. Mitochondrial analysis of walleye pollock, *Theragra chalcogramma*, from the eastern Bering sea and Shelikov Strait, Gulf of Alaska. Can. J. Fish. Aquat. Sci. 49: 319–326.

Nee, S., E.C. Holmes & P.H. Harvey. 1995. Inferring population history from molecular phylogenies. Philos. Trans. R. Soc. London B. 349: 25–31.

Nei, M. 1987. Molecular Evolutionary Genetics, Columbia University Press, New York, NY. 512 pp.

Nei, M. & W. Li. 1979. Mathematical model for studying genetic variation in terms of restriction endonucleases. Proc. Natl Acad. Sci. U.S.A. 76: 5259–5273.

Nei, M. & T. Tajima. 1981. DNA polymorphism detectable by restriction endonucleases. Genetics 97: 145–163.

Numachi, K. & T. Kobayashi. 1994. MtDNA analysis: General research of groundfish in the northern Pacific high seas in 1993. Fisheries Agency of Japan, pp. 60–76 (in Japanese).

Okazaki, T. 1996. MtDNA analysis: General research of groundfish in the northern Pacific high seas in 1995. Fisheries Agency of Japan, pp. 60–68 (in Japanese).

Okazaki, T. 1997. MtDNA analysis: General research of groundfish in the northern Pacific high seas in 1996. Fisheries Agency of Japan, pp. 50–56 (in Japanese).

Olsen, J.B., S.E. Merkouris & J.E. Seeb. 2002. An examination of spatial and temporal genetic variation in walleye pollock (*Theragra chalcogramma*) using allozyme, mitochondrial DNA, and microsatellite data. Fish. Bull., U.S. 100: 752–764.

Powers, D.A. 1998. The use of molecular techniques to dissect the genetic architecture of pollock populations. pp. 55–63. *In*: S.A. Macklin (ed.) Bering Sea FOCI Final Report. Bering Sea Fisheries-Oceanography Coordinated Investigations (FOCI), NOAA Coastal Oceans Program, Alaska Fisheries Science Center, Seattle.

Roff, D.A. & P. Bentzen. 1989. The statistical analysis of mitochondrial DNA polymorphisms: χ^2 and the problem of small samples. Mol. Biol. Evol. 6: 539–545.

Rogers, A.R. & H. Harpending. 1992. Population growth makes waves in the distribtuion of pairwise genetic differences. Mol. Biol. Evol. 9: 552–569.

Rohlf, F.J. 1990. NTSYS-pc: Numerical Taxonomy and Multivariate Analysis System. Version 1.60, Exeter Publishing, Ltd., Setauket, New York, 117 pp.

Sambrook, J., E.F. Fritsch & T. Maniatis. 1989. Molecular Cloning: A Laboratory Manual, Cold Spring Harbor Lab. Press, New York, 1659 pp.

Schneider, S., D. Roessli & L. Excoffier. 2000. Arlequin Version 2.000: A Software for Population Genetic Data Analysis, Genetics & Biometry Laboratory, University of Geneva, Geneva, Switzerland, 111 pp.

Shields, G.F. & J.R. Gust. 1995. Lack of geographical structure in mitochondrial DNA sequences of Bering Sea walleye pollock, *Theragra chalcogramma*. Mol. Mar. Biol. Biotechnol. 4: 69–82.

Shuntov, V.P., A.F. Volkov, O.S. Temnykh & E.P. Dulepova. 1993. Walleye pollock in the ecosystems of the Far Eastern Seas, TINRO, Vladivostok, 426 pp (in Russian).

Sneath, P.A.H. & R.R. Sokal. 1973. Numerical Taxonomy, W.H. Freeman & Co., San Francisco, CA, p. 214.

Stepanenko, M.A. 2001. Stock condition, interannual variability in number of recruits, and commercial utilization of the eastern Bering Sea walleye pollock population in 1980–1990s. Izvestiya TINRO 128: 145–152 (in Russian with English summary).

Yanagimoto, T., K. Kidokoro & T. Kobayashi. 1997. PCR-RFLP analysis: General research of groundfish in the northern Pacific high seas in 1996. Fisheries Agency of Japan, pp. 55–77 (in Japanese).

Zardoya, R., A. Garrido-Pertierra & J.M. Bautista. 1995. The complete nucleotide sequence of the mitochondrial DNA genome of the rainbow trout, *Oncorhynchus mykiss*. J. Mol. Evol. 41: 942–951.

Environmental Biology of Fishes **69**: 177–185, 2004.
© 2004 *Kluwer Academic Publishers. Printed in the Netherlands.*

Analysis of the genetic structure of northwestern Bering Sea walleye pollock, *Theragra chalcogramma*

Elena A. Shubina[a], Marina N. Mel'nikova[a], Aleksandr I. Glubokov[b] & Boris M. Mednikov[a]
[a]*A.N. Belozersky Institute of Physico-Chemical Biology, Moscow State University, Vorobyovi gori, Moscow 119899, Russia (e-mail: shubina@genebee.msu.su)*
[b]*Russian Federal Research Institute of Fisheries & Oceanography, 17, V. Krasnoselskaya, Moscow 107140, Russia*

Received 21 April 2003   Accepted 16 June 2003

Key words: microsatellite analysis, loci Tch12, Tch14, Tch15, Tch18, heterozygote deficiency, GENEPOP, F-statistics, stock differentiation

Synopsis

The intraspecific structure of major populations of the walleye pollock *Theragra chalcogramma* from the northwestern part of the Bering Sea was studied. Specimens from Navarin, Olyutor and Shirshov shoals were sampled in the 2001 spawning season. Preliminary cluster analysis (TREECON) of PCR-RAPD data revealed the existence of cluster with low level of bootstrap support, which generally corresponds to geographic localization of the shoals. Processing of microsatellite sequence data (loci Tch12, Tch14, Tch15 and Tch18) with GENEPOP demonstrated the deviation from HWE and differentiation between the samples. The value of the inter-population variance ($F_{st} = 0.02$) corresponded to published data on marine stocks, which were subject to high levels of gene flow. The Shirshov group was found to be equidistant from the Navarin and Olyutor groups, with the genetic distance between the latter two being significantly less. Analysis of the microsatellite loci inferred disparities in their selective capacities. Loci Tch12 and Tch15 are found to be highly homozygous with low levels of polymorphism. Tch12 clearly follows the '3 band' pattern. Tch15 shows sign of linkage disequilibrium. Comparison of F_{st} values for the loci with the mean standardized variance f_0 suggests that Tch12 is under pressure of disruptive selection, while Tch15 is rather neutral and Tch14 and Tch18 are under the influence of balance selection. Also a discussion is conducted in the context of undeterminate group migration.

Introduction

The walleye pollock *Theragra chalcogramma* (Pallas), family Gadidae, is one of the most widespread species of fish in the northern part of the Pacific Ocean and is economically important to several countries, including Russia and the U.S.A. The problem of structural organization of its populations is a part of the task to define structure of major marine fish shoals in order to optimize fishery strategies and control allocation of wild populations. It is well known that conventional morphological and genetic methods of research, which were successfully applied in numerous studies of freshwater and anadromous fishes, appeared to be of limited use in studies of marine species. Impaired 'homing', passive migration of eggs and larvae with oceanic currents, and other factors result in lower levels of genetic differentiation in marine fishes. Nevertheless, it is commonly accepted that the association of bulk spawning pollock to particular locations of the continental shelf has resulted in intraspecific differentiation of *T. chalcogramma* into discrete populations. The subspecies status of walleye pollock in the Bering Sea has been well documented (Iwata 1975, Serobaba 1977, Grant & Utter 1980, Glubokov & Kotenev 1999, Datsky 2000; for review, see Bailey et al. 1999). There are several large population groups of walleye pollock in the Bering Sea. Analysis of morphometric features distinguished three relatively isolated populations: the western Bering Sea, the eastern Bering Sea and the Navarin population (Datsky 2000). Genetic analysis on the basis of three isoenzyme systems permitted the division of the major Bering Sea population into six subpopulations, with two of them – Koryak-Olyutor and Navarin – within the fishery zone of Russia (Flusova & Bogdanov 1987). Iwata (1975) on the basis of analysis of tetrazole oxidase polymorphism data found that Olyutor and Navarin populations constitute one integral population. Analogous results were obtained with analysis of mitochondrial DNA restriction data (Katugin, unpublished data).

A cursory examination of the level of polymorphism of genomic DNA in walleye pollock was preliminarily assessed with the PCR-RAPD technique (E. Shubina, unpublished data). Any correlation between cluster topology with the Neighbor-Joining method and geographic localization of the populations was detected, though values of bootstrap support for clusters did not exceed 30%. Both statistic and distance analyses of PCR-RAPD data revealed disparity between the level of RAPD marker polymorphism and the extent of genetic differentiation of the Bering Sea walleye pollock populations, and demonstrated the necessity of using higher polymorphic markers, such as di-, tri- and tetranucleotide microsatellite sequences.

O'Reilly et al. (2000) conducted population studies using microsatellite DNA cloned from a library of *T. chalcogramma*. These markers were able to show the genetics for determining the genetic population structure in North Pacific pollock, including the western Bering Sea subpopulation. Microsatellite loci were characterized by polymorphism (number and size of alleles). Primary population genetic parameters were estimated for several large Pacific populations. We conducted analogous research with some of the markers defined by O'Reilly et al. (2000) on samples taken from spawning stocks from the northwestern part of the Bering Sea in 2001. Discrimination of stocks in mixed populations depends on the choice of appropriate genetic markers. We studied the genetic structure of two major populations – Navarin and Olyutor and a group located to the east of the Gulf of Olyutor, within the area of the underwater Shirshov ridge.

Materials and methods

Samples were taken during the 2001 spawning period. Tissue samples, mainly milt at II–IV stages of development, were examined in 81 specimens. Samples were collected and fixed in alcohol. The map of sampling sites is depicted in Figure 1.

Two procedures were used for DNA extraction from alcohol-fixed tissue: (i) the technique modified from Arrighi et al. (1968), lysis was done with pronase E or proteinase K with subsequent deproteinization with chloroform and phenol-chloroform, (ii) the technique using DIA*tom*™ DNA Prep kit, which utilizes guanidinetiocyanate as a lytic agent. In the presence of this reagent DNA is selectively retained by *NucleoS*TM ion-exchange resin, salts and proteins are washed away with alcohol-containing buffer, and DNA is eluted with

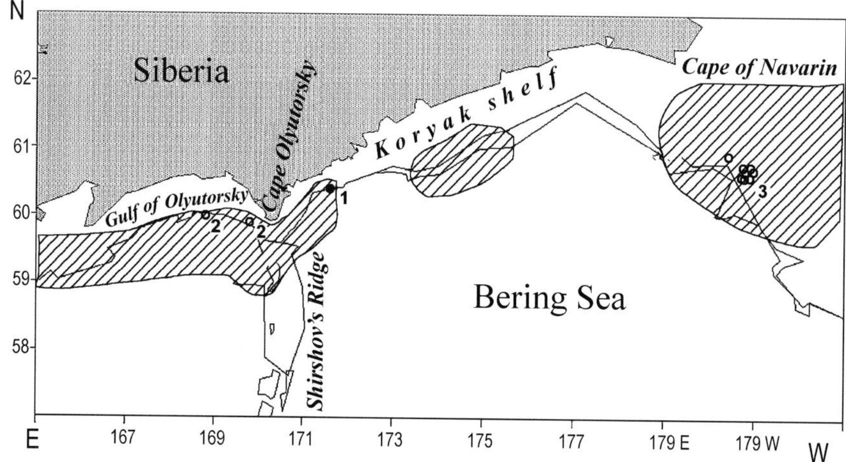

Figure 1. Localities of walleye pollock genetic samples in the northwestern Bering Sea, 2001. Sample localities: 1 = Shirshov; 2 = Olyutor; 3 = Navarin. The area of the major spawning stocks are marked by hatches.

a balanced buffer or water. We followed the protocol recommended by the manufacturer (Institute of General Genetic RAS, G. Shaikhaev). The average DNA yield was 0.5–1 g per 5 µl solution. The polymeric nature of native DNA was checked selectively on 0.6% agarose gel (agarose Type II, Sigma) for about 30% of the samples. Purity and size (40–50 kb) of the samples satisfied the PCR reaction standards.

Microsatellite analysis. A 20-clone library of microsatellite DNA of walleye pollock was constructed and submitted to GenBank (O'Reilly et al. 2000). Preliminary analysis conducted by the authors showed that five clones exhibited the highest polymorphism in walleye pollock from the northern part of the Bering Sea, and these clones were selected as markers in the present study. We failed to apply one of the clones, Tch19, due to a poor quality of Taq polymerase. For the analysis of genetic structure of the walleye pollock populations four loci were used: Tch12, Tch14, Tch15 and Tch18.

PCR reactions were set in 25 µ volume, the reaction mixture contained ~100 ng DNA, 10 mM Tris–HCl, pH 8.3, 50 mM KCl, 2 mM $MgCl_2$, 200 µM each dNTP, 0.5 µM forward and reverse primers and 2 u Taq polymerase. Cycling conditions: 94°C, 3 min (preheat); 94°C, 1 min; 54°C, 0.5 min (5 cycles); 90°C, 3 min; 54°C, 0.5 min; 72°C, 0.5 min (40 cycles). The total of 81 DNA samples was used in PCR. Cycling was done with 'Tertsik' multichannel thermal cycler with the capacity of handling 40 reactions at the same time (DNA-Technology, Russia).

Electrophoresis of microsatellite PCR products was done in non-denaturing vertical 6–8% polyacrylamide gel ($200 \times 200 \times 0.4$ mm^3), TBE buffer, 400 V, 10 mA, 3.5–4 h. The inward side of one of the glass plates was covered with molecular glue (Bind-silan, Sigma), the other one – with silicone. Up to 10 µl of samples were applied to the gel in a buffer containing bromphenol dye with 50% glycerine. One gel accommodated up to 33 samples. Thus, each reaction was divided among three gels. For the purposes of compiling the image, each gel was supplemented with cross-reference samples. The 20 bp Molecular Ruler (BioRad, U.S.A.) was applied to three lanes in each gel as a size marker. We also used an allele marker (pooled PCR-products from 4–5 samples). After the run was complete, the silicone glass plate was removed and the gel attached to the other plate was dyed with ethidium bromide, visualized in transmitted UV light and the colored image was registered with the aid of Kodak Electrophoresis Documentation and Analysis System (Edas) 290.

For the purposes of capturing black-and-white prints of photographs we used photofilm 'Mikrat' (Rus). Adobe PhotoShop application was used to edit and process graphics. Results were tabulated manually.

Computer analysis. We used TFPGA 1.3[1] and GENEPOP 3.1d[2] (Raimond & Rousset 1995) to estimate the level of genetic divergence between populations on the basis of microsatellite sequence polymorphism. Primary population genetic analysis conducted with TFPGA produced estimates of allele and heterozygote frequencies, polymorphism level of studied loci, values of observed and expected heterozygosity under HWE (genetic diversity within populations) as well as the unbiased estimate of population differentiation calculated according to Nei (1978). The non-parametric Wilconson's signed rank test is appropriate to estimate the differences in average unbiased observed heterozygosities between population samples. But it was impossible to fulfill this estimator with only four loci data. Therefore, Student's t-criterion for the small sample number for the comparisons of heterozygosities was calculated by the following formula:

$$t = \frac{D}{m_D}, \quad \text{where } D = \overline{H}_1 - \overline{H}_2 \text{ and } m_D = \sqrt{m_1^2 + m_2^2},$$

in which H is the mean value of observed heterozygosity in populations compared, and m_D stands for squariance for each sample.

The goodness of fit of the data to Hardy–Weinberg proportion was estimated by means of conventional c2 and G-tests and Haldane's exact test (Haldane 1954) (TFPGA), as well as by Guo & Thompson (1992), provided polynomial distribution derived from the Markov chain analysis (Raimond & Rousset 1995a,b, GENEPOP). The number of batches, number of reiterations per sample, and a dememorization step are predefined, and the probability of the null hypothesis to be correct is estimated with an exact probability test. We set the number of dememorizations to 1 000, the number of batches to 150 and the number of reiterations to 2 000 per sample. In case of high error values (>0.05), the number of iterations during the chain's run (batches × iterations) was

---

[1] Miller, M.P. 1997. Tools for population genetic analysis (TFPGA) 1.3: A Windows program for the analysis of allozyme and molecular population genetic data. Computer software distributed by author.

[2] Raimond, M. & F. Rousset. 1999. GENEPOP (version 3.1d). Computer software, distributed by authors.

increased to $1\,000 \times 1\,000$. The system equilibrium in these algorithms was estimated with several tests. Summarized probability tests were used to defining null-hypothesis rejection zone, i.e., Haldane's exact test (1954), Weir (1990) and Robertson & Hill (1984). Significant levels of disequilibrium were detected for two loci in all populations and for all loci and populations, when Fisher's combined probability method was employed. The same analysis was provided with pairs of pooled data. Genotypic linkage disequilibrium was estimated pairwise between the loci (GENEPOP). Fisher's $R \times C$ test was used to detect differentiation of populations (with contingency tables). Allele frequencies of each locus were compared between pairs of populations and all populations simultaneously (Raimond & Rousset 1995a,b).

F_{st} value (q in Weir & Cockerham data) was estimated with 95% significance level ($10\,000$ bootstrap replicates). Genetic distances between populations were obtained according to Nei (1972, 1978) with TFPGA program. Despite the fact that the tests are valid only under the assumption of discrete populations without any exchange of genetic material, we give them here to illustrate the relative level of divergence between walleye pollock aggregations.

Results

As evident from Table 1, two out of four loci – Tch12 and Tch15 – clearly exhibited heterozygote deficiency in all three aggregations studied. Both loci showed low polymorphism and narrow size interval of alleles. In a PCR reaction with an appropriate pair of primers an extra sequence is synthesized for a particular locus, of approximately the same size. This fragment is present in all DNA samples and has the same electrophoretic mobility in all samples studied. The origin of this

sequence remains unknown. It is quite possible that this is another locus relative in sequence to Tch12 (see in O'Reilly et al. 2000, Figure 1). In our analysis of population genetic structure this sequence was disregarded. Both microsatellite loci distinguished by observed heterozygosity possess at the minimum a 4-nucleotide long repetitive sequence, which renders unlikely the possibility of the systematic error in determination of the allele number. Of course, it is not justified to completely rule out the null-allele situation, when null-allele heterozygotes would be seen as homozygotes. However, the fact of such a common occurrence of null-alleles in the present case required separate discussion (see below). As follows from Table 1, the average number of alleles per locus varies between 6.8 in Shirshov and 8.8 in Navarin. Under the aligned to the minimum sample size the reduction in the average number of alleles was observed as far as 6.5 in Navarin and 7.8 in Olyutor population groups. Thus, reduction in polymorphism became noticeable under reduction of the sample size from 46 to 14 specimens in the Navarin population and was small when the Olyutor specimens were reduced from 21 to 14. Percentage of polymorphic loci (a measure of population genetic heterogeneity) gave 100% under the 95% and 99% confidence intervals. Sharp distinctions between loci resulted in clear overevaluations of standard deviation and coefficient of variation (Table 2). Consequently, pairwise comparisons of all population samples over four loci with respect to average unbiased heterozygosity were evidently meaningless. To somehow estimate the probability of their genetic differentiation under this criterion, computations were conducted separately for two loci of low heterozygosity and for two loci of relatively equal values of expected and observed heterozygosity. Data on average estimates of heterozygosity and variance are shown in Table 3. As evident from the table, the error amount has clearly diminished.

Table 1. Standard analysis of genetic variants of four microsatellite loci in three large groups of Bering Sea walleye pollock: N = sample size, NA = number of alleles, H_e = expected, unbiased heterozygosity (under Hardy–Weinberg equilibrium), H_o = observed heterozygosity.

Loci description	NA	Size bp	Stock 1 (Shirshov)				Stock 2 (Olyutor)				Stock 3 (Navarin)			
			N	NA	H_e	H_o	N	NA	H_e	H_o	N	NA	H_e	H_o
			14				21				46			
Tch12	7	100–120		4	0.7275	0.3571		6	0.8258	0.3333		7	0.7157	0.3696
Tch18	14	80–110		8	0.7011	0.6429		11	0.8374	0.8571		12	0.8058	0.6522
Tch15	6	70–80		4	0.5053	0.2857		6	0.7596	0.2381		6	0.6211	0.4565
Tch14	11	120–200		11	0.9233	0.7857		9	0.8641	0.7619		10	0.8715	0.7391
Mean	9.5			6.8	0.7143	0.5179		8	0.8217	0.5476		8.8	0.7535	0.5543

However, in this case differences among populations estimated with Student's criterion did not exceed the 5% significance level. Hence, there were no significant differences between the populations in average heterozygosity.

The goodness-of-fit of the sampling data to Hardy–Weinberg equilibrium (HWE) was estimated for separate loci in different populations (TFPGA), as well as for total populations with respect to each locus and total loci in each population (GENEPOP). Deviations from HWE were analyzed with intrapopulation inbreeding coefficients Fis (Weir 1990, Robertson & Hill 1984) with exact probability tests. Total tests of loci and populations were based on Fisher Combined Probability of HWE with deviation from the null-hypothesis set at the 5% significance level. This analysis shows that loci Tch12 and Tch15 with low Hz demonstrate disequilibrium in all populations. Locus Tch15 in the

Olyutor population did not significantly exceed the 5% significance level (p = 0.0579). Robertson & Hill's F_{is} estimations were two times less than Weir & Cockerham's estimations in the Shirshov samples. And in pairwise analysis, when Shirshov and Navarin samples were computed together, the summarized variance under Weir & Cockerham estimations were two times less than in other pools. Locus Tch18 in the Navarin population was characterized with the probability of 0.05. Total sample analysis of the Tch18 locus was in equilibrium (p = 0.24 > 0.05). Locus Tch14 detected no deviation from HWE in separate populations, nor in the total sample. In total analysis of populations and all four loci, HWE was met in all of population groups, as well as across the sample of 81 specimens. In Table 4 the summarized pairwise equilibrium tests for each loci are shown.

Studies of genotypic disequilibrium with the aid of Fisher's contingency tables revealed linkage between loci Tch15 and Tch18 (p = 0.057), and Tch15 and Tch14 (p = 0.028) in the total sample, while linkage disequilibrium between Tch18 and Tch14 was absent. Linkage disequilibrium analysis conducted separately for pairs of loci in each population was significant only for the Navarin stock.

Analysis for population differentiation with respect to gene and genotypic frequencies was conducted for each locus in each population and separately between

Table 2. The mean observed (H_o) and expected (H_e) heterozygosities with their standard deviation across four loci for three walleye pollock stocks.

Locations	$H_o \pm$ S.D.	CV%	$H_e \pm$ S.D.	CV%
Shirshov	0.5179 ± 0.2043	39	0.7143 ± 0.15	21
Olyutor	0.5476 ± 0.2662	48.6	0.8217 ± 0.039	4.7
Navarin	0.5543 ± 0.1522	27.4	0.7535 ± 0.094	12.5

CV = coefficient of variability.

Table 3. Mean pairwise heterozygosities with associated errors for the loci Tch12–Tch15 and Tch14–Tch18.

	H_e	S.D.	CV%	H_o	S.D.	CV%	H_e	S.D.	CV%	H_o	S.D.	CV%
Shirshov	0.6164	0.11	18	0.3214	0.036	11	0.8122	0.11	13.5	0.7143	0.07	10
Olyutor	0.7927	0.03	4.2	0.2857	0.048	16.6	0.8118	0.25	3.15	0.8095	0.05	5.9
Navarin	0.6684	0.04	7	0.4080	0.038	8	0.6956	0.03	3.9	0.6956	0.04	6.2

Abbreviations the same as in the Table 2.

Table 4. The pairwise test for a goodness-of-fit to HWE was performed by GENEPOP. F_{is} was calculated under Weir & Cockerham. P = the 'exact probability' of Hardy–Weinberg expectation (see text).

Population	Loci							
	Tch12		Tch14		Tch15		Tch18	
	1	2	1	2	1	2	1	2
2								
F_{is}	0.588		0.133		0.628		0.13	
P	0.0000		0.0087		0.0000		0.5124	
3								
F_{is}	0.523	0.533	0.151	0.154	0.299	0.430	0.174	0.119
P	0.0000	0.0000	0.825	0.1332	0.022	0.0000	0.0104	0.2116

Populations are: 1 = Shirshov; 2 = Olyutor; 3 = Navarin.

182

pairs of populations. For loci Tch12 and Tch15 allelic differentiation among all populations was detected, confirmed by the Combined Fisher's test. However, in pairwise comparisons, locus Tch15 exhibited no differentiation between the Olyutor and Navarin groups. In estimations of genotype distribution among all populations differences were detected with respect to Tch12 only, but Fisher's combined probabilistic method detected differentiation of samples in all loci studied ($c2 = 20.127$, $df = 8$: null hypothesis of identically distributed genotypes is rejected at 1% significance level). Pairwise comparisons of populations with respect to separate loci again detected differences only in Tch12 (between Navarin and Shirshov and between Olyutor and Shirshov). There was no difference detected between the Navarin and Olyutor stocks.

Differentiation level of walleye pollock populations with 'weighted' variances of gene frequencies (Wright's F-statistic) in Weir's equations (1990) were determined with respect to separate loci in the total sample as well as in pairwise comparison of populations. Results on the differentiated population are given in Table 5. Unbiased estimates of corresponding values of F, q and f obtained with the jacknife approach are enclosed in brackets. Threshold values of F-statistic within 95% confidence interval were determined by bootstrapping. As seen in Table 5, sharp disparities between biased and unbiased values of standard deviation were common for all four loci studied.

Overlap of the F-statistic values with the chosen level of confidence interval suggests that statistic confidence of differentiation is not sufficient. Nevertheless, the exact probability test for differentiation by Raimond & Rousset (1995a,b) confirmed differentiation between populations. Estimation of the pairwise intergroup differentiation in F_{st}-terms is shown in Table 6. With respect to all loci, the total value of interpopulation differentiation, F_{st}, between Navarin and Shirshov aggregations gives 0.0489; between Olyutor and Shirshov $= 0.0417$; between Olyutor and Navarin $= 0.0173$. Estimation of genetic distances with Nei's method ($F_{st} = 2D^2$) yields the values of 0.156, 0.144 and 0.093, respectively (Table 7). Thus, the Shirshov population was equidistant from Navarin and Olyutor aggregations, and the distance between the latter two is much shorter.

As another useful characteristic, Table 5 contains normalized (mean) values of inter-population variances with respect to separate loci, and the index of population genetic differentiation f_0 obtained

Table 5. Biased and unbiased values of the F-statistics (Weir & Cockerham 1984) for three aggregations of northwestern walleye pollock.

Loci	F	q	f	f_{0i}
Tch12	0.5622	0.0785	0.5249	0.0610
	(0.2168)	(0.0052)	(0.2128)	
Tch18	0.1154	−0.0030	0.1181	−0.00248
	(0.3710)	(0.0334)	(0.3493)	
Tch15	0.4384	0.0261	0.4234	0.02145
	(0.2715)	(0.0237)	(0.2539)	
Tch14	0.1423	−0.0031	0.1450	0.0037
	(0.3699)	(0.0348)	(0.3472)	
All	0.3065	0.0242	0.2548	0.021
	(0.3039)	(0.0240)	(0.2848)	
S.D.	0.1144	0.0205	0.1027	
95% C.I.	$0.1227 > F < 0.5069$	$−0.031 < q < 0.0585$	$0.1253 < f < 0.4782$	

$F = F_{IT}$, $q = F_{st}$ & $f = F_{is}$; f_{0i} = the mean interpopulation variance for each locus and overall loci. Explanations are in the text.

Table 6. Estimation of pairwise intergroup differentiation in F_{st} terms.

	TCh12			Tch14			Tch15			Tch18			All loci		
	1	2	3	1	2	3	1	2	3	1	2	3	1	2	3
1		0.0752	0.1419	X	−0.021	0.0048	X	0.0580	−0.0149	X	−0.0031	0.0117	X	0.0417	0.0489
2	X		0.0342	X	X	−0.0418	X	X	0.0390	X	X	−0.0101	X	X	0.0173

Stocks of walleye pollock: 1 = Shirshov, 2 = Olyutor, 3 = Navarin.

Table 7. Pairwise Nei's distances in accordance with microsatellite data by TFPGA and GENEPOP.

Populations	TFPGA	GENEPOP
Shirshov–Navarin	0.128	0.156
Shirshov–Olyutor	0.132	0.144
Olyutor–Navarin	0.058	0.093

on their basis (Rychkov & Sheremetiyeva 1977, 1980).

It is calculated as a mean $\bar{f}_0 = (1/I) \sum_{i=1}^{I} f_{0i}$, for I gene loci and can be considered a selectively neutral evaluation of micro-differentiation.

Discussion

The ultimate task of the present work was to estimate genetic distance between several groups of Bering Sea walleye pollock. Using four microsatellite loci we revealed an interesting phenomenon, which is rather worth discussing in more detail. There is a striking difference in behavior between different microsatellite loci. In studies of major population groupings of walleye pollock interconnected by migration routes it has been shown that different loci differ among groups with respect to allele frequencies (Table 5). Because of the nature of microsatellite DNA, one does not associate genotypes of different loci with fluctuations in migration patterns, and disparities in normalized values of f_0 between different loci may be accounted for by different pressures of natural selection (Lewontin & Krakauer 1973, cited from Altukhov 1989).

Deviation of F_{st} values for each locus from the mean variance indicated the direction and strength of selection. In case of neutral evolution, the F_{st} value approximately equals the mean variance. This is observed for locus Tch15. Locus Tch12 falls under the pressure of disruptive selection, because the F_{st} value exceeds the mean variance. Both loci exhibit sharp heterozygote deficiency, which cannot be accounted for by inbreeding. Loci Tch14 and Tch18 fall under the pressure of balanced selection. As long as adaptive capacities of microsatellite DNA itself are doubtful, one can suggest that locus Tch12 is associated with functionally important parts of chromosomes. The gel profile for locus Tch12 contains a third, monomorphic band across the entire sample set (O'Reilly et al. 2000). It has been known that the large microsatellite areas and/or difficulties in typing alleles are generally accepted to explain the abundance of the null-alleles in fishes (O'Connel & Wright 1997).

However, there is one more possible supposition for the locus Tch12. This is competition for the primers due the '3-d band' synthesis. Against a background of the preferential and successful amplification of the similar-in-size but nevertheless shorter additional product, the missing of the long alleles of the microsatellite loci can take place. Since the nature of this accompanying fragment is still unknown, we do not see a way to verify this assumption except by alteration of the reaction conditions and maybe more careful selection of the primer sequence to reduce the effect of the unwanted synthesis.

A limited quantity of highly divergent, in terms of genetic nature, markers may have affected some of the results. Thus, locus Tch15 reveals genotype linkage disequilibrium with loci Tch14 and Tch18 in the Navarin stock only, but Fisher's probabilistic criterion detects linkage in all populations. Insufficient loci resulted in a difference between biased and unbiased estimates of the F-statistic. In the present study we also faced problems associated with sample size. Recently, reduction of the average polymorphism level was reported under random alignment of sample sizes to a minimal size, in our case to 14 specimens (Table 3). This reflects an insufficient sample size that should be increased up to a minimum of 50 for each population. Apart from that, in a traditionally used protocol, DNA was extracted from milt, thus confining the sample to males only. In case of selectively neutral autosome loci, sex ratio in the sample usually does not affect confidence in the result. Here, however, a low heterozygous locus Tch12 possesses an unusual third band. Analysis of this locus on a sample consisting entirely of females, as well as a mixed representative sample, is necessary and will be conducted in the future.

Two large walleye pollock spawning stocks are distributed in the western and northwestern Bering Sea. The Olyutor-Karagin stock is distributed on the area about 400 n.m.sq. and the Navarin stock 300 n.m.sq. In this study the genetic heterogeneity of northwestern walleye pollock has been shown. The microsatellite marker results indicate that genetic differentiation between Olyutor and Navarin stocks is about two times less than between either of the two stocks and the Shirshov group. The latter sampling was done in the eastern area of the bank of the underwater Shirshov Ridge. This geographic locality is between the Gulf of Olyutor and Cape of Navarin, not far from a morphologically different Koryak spawning group (Datsky 2000).

The last overlap of the Olyutor and Navarin stocks, the midpoints are 500 n.m. apart, was observed before

184

1995. During the last 7–8 years the Olyutor and Navarin stocks essentially did not mix with one another. Therefore it would be possible to expect a high level of genetic distance between the Olyutor and Navarin stocks and some intermediate value for the Shirshov sample. Genetic isolation of the Shirshov ridge area walleye pollock leads one to consider the possibility of the fish migrating to this region from remote reproduction areas. Shirshov Ridge separates the Aleutian and Kommandor basins. In this connection both specimens from the Olyutor-Koryak and migrants from the Kommandor or Aleutian Islands waters may be found here. This supposition is supported by small number of prespawning and spawning adults in this area. In contrast, a considerable part of the catches in the neighboring areas of the Gulf of Olyutor and Koryak shelf is composed of spawning males and females , and there are developing eggs in icthyoplankton samples. In periods of large walleye pollock populations in Bering Sea deep waters, fall migrations were observed from the central area to the Shirshov Ridge. These migration routes still likely exist. However, documented survey data for these issues and hydrological surveys over the past few years are absent. Therefore one only may suppose the passive transfer of eggs and larvae by the West Bering Slope Current.

Data obtained in this present study do not contradict the results published by O'Reilly et al. (2000). A smaller number of alleles per locus is probably accounted for by narrower geographic bounds for the groupings studied here. It was shown that the three stocks differ according to which genetic markers were used, and the extent of differentiation under the value of intergroup variance ($F_{st} = 0.02$) suggests a high level of gene flow. The results of this study are very preliminary. Examining more loci, making sampling more representative, and supplementing comparative analysis with samples from local populations with a stronger genetic isolation will allow us to better determine population genetic structure of northwestern Bering Sea walleye pollock.

Acknowledgements

We thank the Russian Federal Research Institute of Fisheries & Oceanography for sampling and financial support, Tagir Samigullin for computer help, and Valentina A. Sheremetiyeva for valuable consultations and remarks. The constructive remarks of two anonymous referees and referee for English corrections are acknowledged. This study partially was supported by the Russian Foundation Basic Research, grants 01-04-48613 and 00-15-97905.

References

Altukhov, Yu.P. 1989. Genetic Processes in Populations, Nauka, Moscow, pp. 328 (Russian).

Arrighi, F.E., G. Bergendahl & M. Mandel. 1968. Isolation and characterization of DNA from fixed cells and tissues. Expt. Cell. Res. 50: 40–47.

Bailey, K.M., T.J. Quinn, P. Bentzen & W.S. Grant. 1999. Population structure and dynamics of walleye pollock, *Theragra chalcogramma*. Adv. Mar. Biol. 37: 179–255.

Datsky, A.V. 2000. On the population heterogeneity of walleye pollock *Theragra chalcogramma* in the Anadyr–Navarin area. Problems Fish. 1: 74–90 (Russian).

Flusova, G.D. & L.V. Bogdanov. 1987. On the population structure of pollock according to genetic works. pp. 79–88. *In*: The Gadidae of the Far-Eastern Sea, TINRO, Vladivostok (Russian).

Glubokov, A. & B. Kotenev. 1999. Alaska pollock in the Navarin Island area. Ribnoe hozáystvo (The Fisheries) 5: 36–37 (Russian).

Grant, W.S. & F.M. Utter. 1980. Biochemical genetic variation in walleye pollock, *Theragra chalcogramma*: Population structure in the southeastern Bering Sea and Gulf of Alaska. Can. J. Fish. Aquat. Sci. 37: 1093–1100.

Guo, S.W. & E.A. Thompson. 1992. Performing the exact test of Hardy–Weinberg proportion for multiple alleles. Biometry 48: 361–372.

Haldane, J.B.S. 1954. An exact test for randomness of mating. J. of Genetics 52: 631–635.

Iwata, M. 1975. Population genetics of the breeding groups of walleye pollock *Theragra chalcogramma* based on tetrazolium oxidase polymorphism. Sci. Rep. Hokkaido Fish. Exp. Stat. 17: 1–14.

Lewontin, R.C & J. Krakauer. 1973. Distribution of gene frequency as a test of the theory of the selective neutrality of polymorphisms. Genetics 74: 175–195.

Nei, M. 1972. Genetic distance between populations. Amer. Nat. 106(949): 283–293.

Nei, M. 1978. Estimation of average heterozygosity and genetic distance from a small number of individuals. Genetics 89: 583–590.

O'Connel, M. & J.M. Wright. 1997. Microsatellite DNA in fishes. Rev. Fish Biol. Fish. 7: 331–363.

O'Reilly, P., M. Canino, K. Bailey & P. Bentzen. 2000. Isolation of twenty low stutter di- and tetranucleotide microsatellites for population analyses of walleye pollock and other gadoids. J. Fish Biol. 56: 1074–1086.

Raimond, M. & F. Rousset. 1995a. An exact test for population differentiation. Evolution 49: 1280–1283.

Raimond, M. & F. Rousset. 1995b. GENEPOP (version 1.2): Population genetics software for exact tests and ecumenism. J. Hered. 86: 248–249.

Robertson, A. & W.G. Hill. 1984. Deviations from Hardy–Weinberg proportions: Sampling variances and use in estimation of inbreeding coefficients. Genetics 107: 713–718.

Rychkov, Yu.G. & V.A. Sheremetiyeva. 1977. The genetic process in the system of ancient human isolates in North Asia. pp. 47–108. *In*: G.A. Harrison (ed.) Population Structure and Human Variations, Cambridge University Press, New York.

Rychkov, Yu.G. & V.A. Sheremetiyeva. 1980. The genetics of circumpolar populations of Eurasia related to the problem of human adaptation. pp. 37–80. *In*: F. Milan (ed.) The Human Biology of Circumpolar Populations, Cambridge University Press, New York.

Serobaba, I.I. 1977. Data of the population structure of the Bering Sea Alaska pollock *Theragra chalcogramma* (Pallas). Voprosi ikhtiologii (J. Ichthyol.) 17: 247–260 (Russian).

Weir, B.S. 1990. Genetic Data Analysis: Methods for Discrete Population Genetic Data, Sinauer Associates, Inc., Sunderland, MA, 445 pp.

Weir, B.S. & C.C. Cockerham. 1984. Estimating F-statistics for the analysis of population structure. Evolution 38: 1358–1370.

Environmental Biology of Fishes **69**: 187–199, 2004.
© 2004 *Kluwer Academic Publishers. Printed in the Netherlands.*

Genetic population structure of Pacific hake, *Merluccius productus*, in the Pacific Northwest

Eric Iwamoto, Michael J. Ford & Richard G. Gustafson
National Marine Fisheries Service, Northwest Fisheries Science Center, Conservation Biology Division, 2725 Montlake Blvd. E., Seattle, WA 98112, U.S.A. (e-mail: eric.iwamoto@noaa.gov)

Received 17 April 2003   Accepted 19 April 2003

Key words: stock structure, marine fish genetics, allozyme electrophoresis, genetic differentiation, distinct population segment, Pacific whiting, heterozygosity

Synopsis

This study presents the first protein electrophoretic study of population structure within the Georgia Basin Pacific hake Distinct Population Segment, as defined under the U.S. Endangered Species Act. Forty-one allozyme loci (29 polymorphic) were analyzed in samples from three Pacific hake spawning populations on the west coast of North America: (1) Port Susan, Puget Sound, Washington (three temporal samples); (2) south-central Strait of Georgia, British Columbia, Canada (two temporal samples); and (3) offshore of southern California (two temporal samples) (total n = 664). Mean heterozygosity over all loci was 12–13% for all populations. Within-population temporal samples were not significantly different from one another, but statistically significant differences were detected at 15 of the 29 polymorphic loci (p < 0.05) among the three populations. Differences at eight of these loci were highly significant (p < 0.001): *ADA**, *ALAT**, *bGALA**, *GPI-A**, *sIDHP**, *LDH-A**, *MPI**, and *PEP-B**. The two Georgia Basin populations were significantly different at six loci: *bGALA**, *sIDHP**, *LDH-A**, *MPI**, *PGK**, and *PGM-2** (p < 0.05). Nei's genetic distance (D) was 0.0006 between Port Susan and Strait of Georgia pooled temporal samples, and 0.005 between these populations and offshore Pacific hake. F_{ST} was 0.02 and 0.0046 among all three populations and among the Georgia Basin populations, respectively. Both F_{ST} estimates were significantly greater than zero, and the results suggest a high degree of demographic isolation among all three populations.

Introduction

Pacific hake, *Merluccius productus* (Ayres 1855), also known as Pacific whiting, were one of 18 species of marine fish in Puget Sound, Washington petitioned for protection under the U.S. Endangered Species Act (ESA) in 1999 (NMFS 1999, Gustafson et al.[1]). Pacific hake is a gadiform groundfish that is currently the most abundant commercial fish species on the U.S. West Coast (Methot & Dorn 1995). There are three

recognized stocks of Pacific hake: a highly migratory offshore (or coastal) stock that ranges from southern California to Queen Charlotte Sound, a central-south Puget Sound stock, and a Strait of Georgia (SOG) stock (Bailey et al. 1982, Methot & Dorn 1995) (Figure 1). A previously recognized 'dwarf' hake stock, limited to waters off the west coast of Baja California, is morphologically and genetically distinct from *M. productus* (Vrooman & Paloma 1976, Grant & Leslie 2001), and has been described as a separate species, *Merluccius hernandezi* Mathews (Mathews 1985). Pacific hake in the offshore stock are primarily caught with midwater trawls over bottom depths of 100–500 m off the coasts of northern California, Oregon, Washington, and British Columbia in the months of April to October. The

---

[1] Gustafson, R.G., W.H. Lenarz, B.B. McCain, C.C. Schmitt, W.S. Grant, T.L. Builder & R.D. Methot. 2000. Status review of Pacific hake, Pacific cod, and walleye pollock from Puget Sound, Washington. U.S. Department Commerce, NOAA Techical Memorandum NMFS-NWFSC-44. 275 pp.

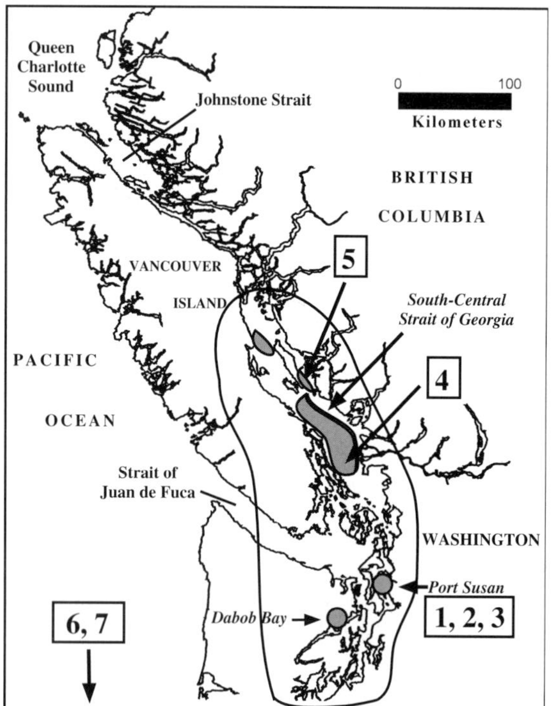

Figure 1. Known or suspected spawning locations (shaded) and collection locations for Pacific hake in Puget Sound and Strait of Georgia. Sample numbers correspond to collection sites in Table 1. The solid line encompasses the approximate boundaries of the 'Georgia Basin Pacific hake Distinct Population Segment'.

combined foreign and U.S. trawl fishery catch for this stock averaged over 206,000 metric tonnes annually between 1966 and 2000 (Helser et al.[2]). Pacific hake in Puget Sound and the SOG have also been subject to commercial exploitation (McFarlane & Beamish 1985, Pedersen 1985), although recent declines in abundance resulted in closure of the Puget Sound fishery in 1990 (Palsson et al.[3]).

Offshore Pacific hake spawn off south-central California to Baja California in the winter months of January and February (Dorn 1995, Methot & Dorn 1995). In spring and summer, adults migrate northward to feed to as far as central Vancouver Island (and as far as Queen Charlotte Sound in some years). In the fall, adults migrate southward toward spawning grounds (Dorn 1995). Resident stocks in the inside waters of the SOG and Puget Sound undergo a similar migration, but on a greatly reduced scale (McFarlane & Beamish 1986, Shaw et al. 1990). Within Puget Sound, Pacific hake are known to spawn in Port Susan (Pedersen 1985) and in Dabob Bay (Bailey & Yen 1983, Fox 1997) from February through April (Figure 1). In the SOG, the main stock aggregates to spawn in the deep basins of the south-central SOG (Figure 1) where peak spawning occurs from March to May (Shaw et al. 1990). Previously, the offshore, SOG, and Puget Sound Pacific hake stocks have been considered to be discrete from one another for management purposes on the basis of: (1) allozyme frequencies, (2) spawning localities, (3) size- and age-at-maturity, (4) growth, (5) year-class strength, (6) effective fecundity, (7) otolith morphology and annuli formation, or (8) the degree of infestation with a protozoan parasite (Gustafson et al.[1]).

In response to the ESA petition, a Biological Review Team (BRT), formed by the National Marine Fisheries Service (NMFS), used a combination of the above stock-specific information to conclude that Pacific hake in Puget Sound are part of a larger entity that qualifies as a 'species' under the ESA and that this grouping consists of inshore resident Pacific hake from Puget Sound and the SOG. This ESA 'species' was designated as the 'Georgia Basin Pacific hake Distinct Population Segment' (DPS) (Gustafson et al.,[1] NMFS 2000). The Georgia Basin encompasses Puget Sound and the transboundary marine waters of the SOG and the eastern Strait of Juan de Fuca (Figure 1).

Delineation of the Georgia Basin Pacific hake DPS relied heavily upon limited allozyme genetic data that showed significant frequency differences between Puget Sound and offshore Pacific hake at four loci (lactate dehydrogenase, transferrin, muscle protein, and esterase) (Utter 1969, Utter & Hodgins 1969, 1971, Utter et al. 1970). However, no genetic studies have adequately addressed the relationship between Pacific hake in Puget Sound and the SOG, although demographic differences indicate that significant population structure may exist (Gustafson et al.[1]). Uncertainties regarding the relationship of Pacific hake stocks within the Georgia Basin DPS, and their status, resulted in NMFS placing this DPS on the agency's list of ESA candidate species pending additional genetic data that

[2] Helser, T.E., M.W. Dorn & M.W. Saunders. 2000. Pacific whiting assessment update for 2000. Pacific Fishery Management Council. Portland, Oregon. 13 pp. (Available online at: http://www.pcouncil.org/groundfish/gfother/gfstocks.htm.)

[3] Palsson, W.A., J.C. Hoeman, G.G. Bargmann & D.E. Day. 1997. 1995 status of Puget Sound bottomfish stocks (revised). Report No. MRD97-03, Washington Department of Fish and Wildlife, Olympia, Washington. 98 pp.

might resolve the DPS's population structure (NMFS 2000). A more accurate definition of the DPS structure is vital to estimating extinction risk, since Puget Sound Pacific hake are severely depressed while SOG Pacific hake are estimated to be 10 times as abundant, and are not believed to be at risk of extinction (Gustafson et al.,[1] NMFS 2000).

Previous studies have failed to find significant genetic differentiation within the migratory offshore Pacific hake stock (Utter 1969, Utter & Hodgins 1969, 1971, Utter et al. 1970, Grant & Leslie 2001). However, the complex physiography of the Georgia Basin may present sufficient physical barriers to larval drift and adult migration to provide the necessary isolating mechanism to allow for genetic differentiation (Grant & Leslie 2001). The present contribution is the first in a series of anticipated molecular genetic analyses of Pacific hake population structure and examines allozyme variation among spawning aggregations of Pacific hake in Puget Sound, the SOG, and the offshore stock.

Methods

Samples

With one exception, samples were collected from ripe adults on the spawning grounds to avoid mixed stock sampling, a problem common to all previous genetic studies of Pacific hake (Table 1). Spawning Pacific hake were obtained in Port Susan in Puget Sound in 2000, 2001, and 2002 and in the Canadian portion of the SOG in 2000 and 2002 in association with hydro-acoustic/trawl surveys of the Washington Department of Fish and Wildlife and Canadian Department of Fisheries and Oceans (DFO), respectively (Table 1, Figure 1). Offshore Pacific hake were collected from a spawning aggregation off southern California during the joint NMFS/DFO January 2001 offshore Pacific hake survey (Table 1, Figure 1). A sample of offshore hake outside the spawning season and area was also obtained during the NMFS West Coast Slope Survey in the summer of 2000 off south-central California for comparison with a sample of spawning offshore Pacific hake (Table 1, Figure 1).

As Pacific hake tissues were samples of opportunity, methods varied with each survey. Each fish was sexed and measured for length and samples of liver (L), muscle (M), eye (E), and heart (H) were frozen on dry ice or placed in a $-20°C$ freezer at sea until transfer to and storage at $-80°C$ in the laboratory. Differences in sampling technique did not affect the quality of the resulting allozyme data.

Electrophoretic techniques

Unpublished allozyme data for Pacific hake showed the following loci to be polymorphic: *CK-A**, *PGM-2**, *GAPDH**, *GPDH-3**, *GPI-1**, *IDHP-2**, *LDH-A**,

Table 1. Collection data for seven Pacific hake populations from Puget Sound, Strait of Georgia (SOG), and offshore U.S. West Coast. Sample locations are shown in Figure 1.

Sample number and name	Collection location	Latitude; longitude	Collection date	Vessel	Condition	Sample size
1 Port Susan 2000	Puget Sound, Washington	48°08′N; 122°20′W	7 Mar. 2000	*FV Chasina*	Spawning	100
2 Port Susan 2001	Puget Sound, Washington	48°08′N; 122°20′W	13 Mar. 2001	*FV Chasina*	Spawning	100
3 Port Susan 2002	Puget Sound, Washington	48°08′N; 122°20′W	5 Mar. 2002	*FV Chasina*	Spawning	100
4 SOG 2000	Strait of Georgia, British Columbia	49°33′N; 124°07′W	9 Apr. 2000	*CCGS W.E. Ricker*	Spawning	100
5 SOG 2002	Strait of Georgia, British Columbia	49°02′N; 123°24′W	22 Mar. 2002	*CCGS W.E. Ricker*	Spawning	100
6 California 2000	Offshore south-central California	35°20′N; 121°10′W	30 Jul. 2000	*FV Coast Pride*	Non-spawning	80
7a California 2001	Offshore southern California	32°35.1′N; 120°44.1′W	27 Jan. 2001	*CCGS W.E. Ricker*	Spawning	40
7b California 2001	Offshore southern California	32°37.3′N; 120°45.6′W	27 Jan. 2001	*CCGS W.E. Ricker*	Spawning	10

190

*LDH-B**, *PEP-B**, and *SOD** (S. Grant[4]). We attempted histochemical staining for enzymes encoded by these and other loci for a total of 35 enzymes. Extracts from the four tissues were prepared and analyzed for enzymatic activity as described in Aebersold et al.[5]. Electrophoretic patterns of 33 enzymes encoded by 41 presumptive gene loci could be scored in all samples of Pacific hake examined in this study (Table 2). Alleles for each locus were designated by their relative electrophoretic mobility relative to the most common allele. The most common allele was arbitrarily designated as 100. Laboratory conditions for enzymes representing the 41 loci that showed sufficiently clear activity are presented in Table 2. Nomenclature for loci and alleles follows Shaklee et al. (1990).

Statistical techniques

Population genetic and statistical analyses were performed using the statistical programs BIOSYS-1.7 (Swofford & Selander 1981) and GDA (Lewis & Zaykin[6]). A locus was considered to be polymorphic if more than one allele was observed in at least one sample. Genotypic frequencies of polymorphic loci for each sample were examined for departures from Hardy–Weinberg expectations using Chi-square goodness-of-fit tests.

Nei's (1978) genetic distances (D) were calculated from multi-locus allelic frequencies for all pairs of samples, and D values were used to construct a Unweighted Pair Group Method using Arithmetic averages (UPGMA) tree (Sneath & Sokal 1973). Cavalli-Sforza & Edwards (1967) chord distances were calculated but are not herein reported as they yielded similar results to Nei's D and produced a tree with identical topology. Nei's D was used to compare with previous estimates reported in the hake genetic literature. F_{ST} (a measure of the proportion of total gene diversity within a species that is due to

subdivision among populations – Wright (1978)) was also used to estimate the magnitude of genetic differentiation, and was estimated using Weir's method (Weir 1990) as implemented in GDA. Confidence intervals for F_{ST} were determined by bootstrapping over loci (1000 replicates).

Population models

The F_{ST} estimates were used to fit two simple population models. The first model, called the complete isolation model, assumes that each population has diverged by drift alone in complete isolation for t generations. Under this model, $-\ln(1-F_{ST}) \approx t/2N$, where N is the effective size of each population (which are assumed to be equal) (Reynolds et al. 1983, Weir 1990). The second model, called the migration model, assumes that the samples are drawn from a group of populations (not all of which have necessarily been sampled) which are connected by geneflow such that each generation a fraction m of each population consists of migrants drawn randomly from the other populations (Crow & Aoki 1984). The populations are assumed to be at equilibrium with respect to drift and migration. Under the migration model, $F_{ST} \approx 1/(4Nm\alpha + 1)$, where $\alpha = [n/(n-1)]^2$ and n is the actual number of subpopulations. For purposes of model fitting, temporal samples from the same populations were combined into single population samples.

Results

We identified the gene products of 41 protein-coding loci. Twenty-nine loci (71%) were polymorphic and 13 loci (32%) were variable at more than two alleles (Appendix 1). Mean heterozygosity over all loci ranged from 12.1% to 13.3% for all populations. Most of the variant alleles observed in Pacific hake populations occurred at low frequencies. Of the 45 variant alleles seen, only eight occurred greater than 25% of the time. Significant (p < 0.05) deviations from expected Hardy–Weinberg proportions were detected at two loci in a single sample, Port Susan 2002. In both cases an excess of heterozygotes was observed.

Within-population variability

Private alleles (an allele found in only one sample) were observed at low frequencies (generally <0.01) in

[4] Grant, S. Alaska Department of Fish and Game, 333 Raspberry Road, Anchorage, Alaska 99518-1599. Pers. comm.

[5] Aebersold, P.B., G.A. Winans, D.J. Teel, G.B. Milner & F.M. Utter. 1987. Manual for starch gel electrophoresis: A method for the detection of genetic variation. U.S. Department of Commerce, NOAA Technical Report NMFS 61. 19 pp.

[6] Lewis, P.O. & D. Zaykin. 2001. Genetic Data Analysis: Computer program for the analysis of allelic data. Version 1.1. Free program available online at http://lewis.eeb.uconn.edu/lewishome/software.html.

Table 2. Enzymes and electrophoretic conditions for 41 enzyme loci detected in *M. productus.*

Enzyme name	EC number	Locus abbreviation	Tissue	Buffer	Polymorphic populations
Aspartate aminotransferase	2.6.1.1	*sAAT-2**	M, H	ACE7, TBE	1
		*sAAT-4**	L, E	ACE7	7
Acid phosphatase	3.1.3.2	*ACP**	L	TBE	0
Adenosine deaminase	3.5.4.4	*ADA**	L, M, E	TBE	7
Adenylate kinase	2.7.4.3	*AK**	M	ACE7	0
Alanine aminotransferase	2.6.1.2	*ALAT**	M	TBE	7
Creatine kinase	2.7.3.2	*CK-A**	M	TBCLE	0
Esterase-D	3.1.1	*ESTD**	M	TBCLE	0
Fructose-bisphosphate aldolase	4.2.1.13	*FBALD-3**	E	TBE	0
		*FBALD-4**	E	TBE	0
Fumarate hydratase	4.2.1.2	*FH**	E	ACE7	7
β-N-acetylgalactosaminidase	3.2.1.53	*bGALA**	L	ACE7, TBE	7
Glyceraldehyde-3-phosphate dehydrogenase	1.2.1.12	*GAPDH-4**	E	ACE7	0
		*GAPDH-5**	E	ACE7	1
N-acetyl-β-glucosaminidase	3.2.1.30	*bGLUA**	L	TC4, TBE	2
Glucose-6-phosphate isomerase	5.3.1.9	*GPI-B**	M	TBCLE	7
		*GPI-A**	M	TBCLE	7
Glutathione reductase	1.6.4.2	*GR**	H	ACE7, TC4	7
L-iditol dehydrogenase	1.1.1.14	*IDDH**	L	TBCL	7
Isocitrate dehydrogenase	1.1.1.42	*mIDHP**	L, E	ACE7	4
		*sIDHP**	M, H	ACE7	7
L-lactate dehydrogenase	1.1.1.27	*LDH-A**	M, E	ACE7	7
		*LDH-B**	M, E, H	ACE7	1
		*LDH-C**	L, M	ACE7	2
α-Mannosidase	3.2.1.24	*aMAN**	L	TBE	0
Malate dehydrogenase (NAD+)	1.1.1.37	*sMDH-A**	M, H	ACE7, TC4	7
		*sMDH-B**	M, H	ACE7, TC4	7
Malate dehydrogenase (NADP+)	1.1.1.40	*sMEP**	L, M, E, H	ACE7, TC4	0
Mannose-6-phosphate isomerase	5.3.1.8	*MPI**	L, E, H	TBE	3
Nucleoside-triphosphate pyrophosphatase	3.6.1.19	*NTP**	M	TBCLE	1
Dipeptidase	3.4.-.-	*PEP-A**	L, M, E	TBE	1
Tripeptide aminopeptidase	3.4.-.-	*PEP-B**	L, M	TC4, TBE	7
Peptidase-C	3.4.-.-	*PEP-C**	E	TBE	0
Proline dipeptidase	3.4.-.-	*PEP-D**	M	TBCLE	7
Phosphoglycerate kinase	2.7.2.3	*PGK**	E, 0H	ACE7	7
Phosphoglucomutase	5.4.2.2	*PGM-1**	M, E, H	ACE7, TBCLE	5
		*PGM-2**	M, E, H	ACE7, TBCLE	7
Pyruvate kinase	2.7.1.40	*PK-2**	L, H	ACE7	0
Superoxide dismutase	1.15.1.1	*sSOD**	L	TBCL	7
Triose-phosphate isomerase	5.3.1.1	*TPI**	E, M	ACE7, TC4	2
Xanthine oxidase	1.2.3.2	*XO**	M	ACE7	0

Enzyme names and enzyme commission (EC) numbers follow recommendations of IUBMBNC (1992). Locus abbreviations follow Shaklee et al. (1990). Tissues are muscle (M), liver (L), heart (H), and retinal tissue/eye fluid (E). Polymorphic populations are the number of populations with allelic variance.

the California 2000 trawl (*GAPDH-5*110, GPI-B*40, IDDH*120, PEP-A*85, NTP*135*), SOG 2002 (*sAAT-2*150, ALAT*65, GPI-A*140, PGM-2*150*), Port Susan 2001 (*LDH-C*65, TPI*50*), Port Susan 2002 (*TPI*180*), and the California 2001 sample of the spawning offshore stock (*PEP-B*75*). The *bGLUA*90* allele was seen only in the Port Susan 2001 (0.005 frequency) and the offshore California 2000 (0.022 frequency) trawl samples. Chi-square probabilities of comparisons of multiple samples across all loci within the three Port Susan samples (2000, 2001, and 2002) and the two SOG samples (2000, 2002) revealed no significant temporal differences in overall allele frequencies (Table 3). Likewise, the two samples of

Table 3. Tests for differences in allele frequencies of all variable loci for different groupings of Pacific hake. Probabilities based on contingency Chi-squared analyses.

Locus	Comparison				
	All samples	Temporal Port Susan samples (n = 3)	Temporal SOG samples (n = 2)	Georgia Basin (n = 5)	Offshore (n = 2)
sAAT2*	0.506	1.000	0.314	0.457	1.000
sAAT4*	0.148	0.205	0.707	0.084	0.284
ADA*	0.000***	0.350	0.190	0.323	0.226
ALAT*	0.000***	0.095	0.539	0.346	0.936
FH*	0.002**	0.785	0.787	0.968	0.134
bGALA*	0.000***	0.203	0.184	0.000***	0.135
bGLUA*	0.003**	0.367	1.000	0.405	0.136
GAPDH-5*	0.614	1.000	1.000	1.000	0.507
PEP-A*	0.566	1.000	1.000	1.000	0.507
GPI-A*	0.000***	0.680	0.516	0.739	0.531
GPI-B*	0.001**	0.242	0.791	0.646	0.234
GR*	0.251	0.084	0.450	0.238	0.981
IDDH*	0.463	0.743	0.480	0.261	0.834
mIDHP*	0.846	0.904	0.562	0.927	1.000
sIDHP*	0.000***	0.235	0.221	0.000***	0.195
LDH-A*	0.000***	0.385	0.065	0.047*	0.663
LDH-B*	0.003**	0.367	1.000	0.405	0.136
LDH-C*	0.464	0.367	1.000	0.405	1.000
MDH-A*	0.003**	0.862	0.077	0.295	0.028*
MDH-B*	0.359	0.053	0.563	0.167	0.737
MPI*	0.000***	0.044	0.401	0.000***	1.000
NTP*	0.612	1.000	1.000	1.000	0.507
PEP-B*	0.000***	0.196	0.298	0.062	0.065
PEP-D*	0.058	0.363	0.200	0.052	1.000
PGK*	0.039*	0.117	0.077	0.045*	0.977
PGM-1*	0.054	0.049*	0.845	0.367	0.643
PGM-2*	0.001**	0.060	0.004**	0.000***	0.260
sSOD*	0.061	0.643	0.703	0.063	0.911
TPI*	0.505	0.406	1.000	0.433	1.000
Totals	0.000***	0.067	0.101	0.000***	0.122

Asterisks associated with numerical data (not loci names) are meant to indicate level of significance: * indicates $p < 0.05$, ** indicates $p < 0.01$, *** indicates $p < 0.001$.

the offshore stock (California 2000, 2001) were not significantly different from one another (Table 3).

Among-population variability

We found no fixed differences in allele frequency between samples that would allow for easy differentiation of populations. However, there were significant frequency differences between the three populations at 15 of 29 polymorphic loci ($p < 0.05$). Differences at 8 of these 15 loci were highly significant ($p < 0.001$): *ADA1*, *ALAT*, *bGALA*, *GPI-A*, *sIDHP*, *LDH-A*, *MPI*, and *PEP-B* (Table 3). Both Georgia Basin populations expressed the alleles *mIDHP*75* and *sIDHP*75* that were absent in the offshore stock. The variant allele *LDH-A*0* occurred at much lower frequencies in the offshore stock samples (0.020 and 0.030) than in the Georgia Basin samples (0.140–0.258). Conversely, the variant alleles *GPI-A*125* and *PEP-B*110* occurred at higher frequency levels in the offshore stock samples (0.48–0.52 and 0.10–0.14, respectively) than in the Georgia Basin samples (0.32–0.35 and 0.01–0.05, respectively). Nei's genetic distances (D) between the offshore samples and the Georgia Basin samples ranged from 0.003 to 0.006 with an overall D of 0.005 between pooled offshore and pooled Georgia Basin data (Figure 2).

Samples within Georgia Basin (Puget Sound and SOG) were significantly different at five loci: *bGALA*, *sIDHP*, *MPI*, *PGM-2* ($p < 0.001$), and *LDH-A*

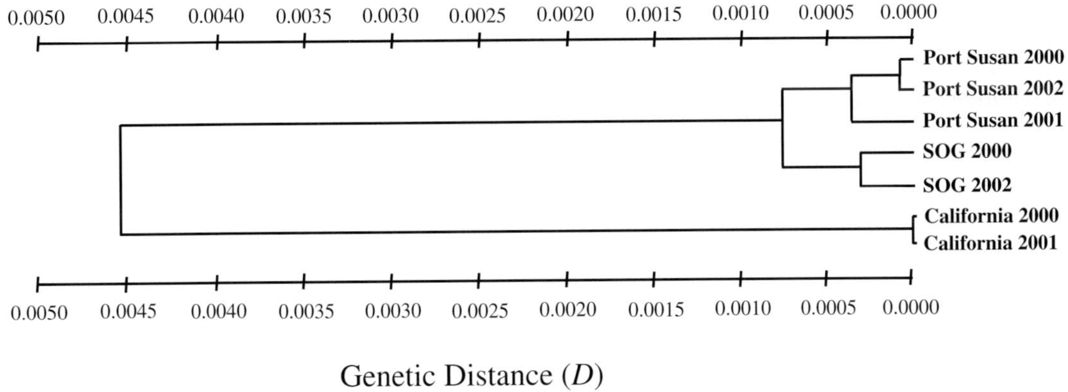

Genetic Distance (*D*)

Figure 2. UPGMA tree of relationships among Pacific hake samples, based on Nei's (1978) genetic distances (D).

Table 4. Pairwise estimates of F_{ST}, t/2N, and number of effective migrants per generation (Nm) for Pacific hake populations.

Populations	F_{ST} (95% CI)	t/2N (95% CI)	Nm (95% CI)
Puget Sound/SOG	0.0046 (0.00155–0.00906)	0.0046 (0.00155–0.00911)	24.0 (12.1–71.7)
SOG/Offshore	0.0261 (0.0155–0.0376)	0.0265 (0.0156–0.0383)	4.1 (2.8–7.1)
Puget Sound/Offshore	0.0347 (0.0197–0.0518)	0.035 (0.0199–0.0532)	3.1 (2.0–5.5)

(p < 0.05) (Table 3). Totals of the contingency Chi-square analyses for differences in allele frequencies for all polymorphic loci (Table 3) indicate that these two populations are significantly different from one another (p < 0.001). The variant allele *MPI*90* occurred in both SOG samples at low frequencies (0.06 and 0.08) but was not observed in the offshore samples, and over the three years of sampling in Port Susan this allele occurred in only the 2002 sample, at a frequency of 0.005 (Appendix 1). The *PEP-D*75* allele was present in both SOG samples (frequencies of 0.008 and 0.015) but not in Port Susan and at a frequency of only 0.010 in the California 2001 sample. Nei's genetic distances (D) between Port Susan and SOG samples ranged from 0.0004 to 0.001 with an overall D of 0.0006 between pooled Port Susan and pooled SOG data (Figure 2). The estimates of F_{ST} and their 95% confidence intervals are provided in Table 4. The confidence intervals for each pairwise comparison do not include zero.

Discussion

When Pacific hake in Puget Sound were petitioned for protection under the ESA, only limited protein electrophoretic (Utter 1969, Utter & Hodgins 1969, 1971, Utter et al. 1970) and very limited mtDNA data

(Goñi 1988) were available to address the question of genetic population structure. Essentially no genetic data were available that could be used to infer relationships among spawning aggregations of Pacific hake within the Puget Sound and the SOG. In this study, we have extended the work by Utter et al. by greatly expanding the number of loci, sampling over multiple years, and by adding samples from the major spawning aggregate within the central SOG.

In earlier studies of Pacific hake, Utter et al. (Utter 1969, Utter & Hodgins 1969, 1971, Utter et al. 1970) found significant allelic variation at four polymorphic allozyme loci between offshore and Puget Sound Pacific hake, and concluded that these populations were reproductively isolated. The results of the present study confirm these earlier findings. In particular, the estimate of F_{ST} between the inshore and coastal populations is ~0.03, a relatively high value for a marine fish species. For example, Ward et al. (1994) reviewed 47 population genetic studies of marine fish, and found the mean value of F_{ST} to be 0.02. The level of differentiation among the two inshore populations was considerably lower than between the inshore and coastal populations, but was still significantly greater than zero (F_{ST} = 0.0046; Table 4). Nei's D (Nei 1978) among all three species was very small, ranging from 0.0006 between the two inshore populations to 0.005 between the inshore and coastal populations (Figure 2).

194

These low values indicate that essentially no mutations have accumulated in the populations since the time they diverged from each other and that the differentiation that has occurred has been primarily due to the drift of alleles that mostly exist in all three populations.

It is of some interest to compare the patterns of variation we observed in *M. productus* to those observed in other *Merluccius* species. In addition to studies on Pacific hake, allozyme variation among populations has been studied in the following species of *Merluccius*: *M. merluccius* (European hake) (Mangaly & Jamieson 1978, Pla et al. 1991, Lo Brutto et al. 1998, Roldan et al. 1998, Imsiridou & Triantaphyllidis 2001), *M. australis* (New Zealand hake) (Smith et al. 1979), *M. capensis* (shallow-water Cape hake), *M. paradoxus* (deep-water Cape hake) (Grant et al. 1987), *M. hubbsi* (Argentinian hake) (Roldan 1991), and *M. gayi* (Chilean hake) (Galleguillos et al. 2000). In most cases, Nei's D between populations has been shown to be less than 0.005 (Grant & Leslie 2001); however, Roldan et al. (1998) found a Nei's genetic distance of 0.014 between Atlantic and Mediterranean samples of European hake. In general, species of *Merluccius* that have been investigated tend to show subdivided population structure around geographically complex coastlines (Roldan et al. 1998, Grant & Leslie 2001), but not along linear coastlines (Smith et al. 1979, Grant et al. 1987). For example, the relatively high levels of population differentiation evident among European hake populations in the Atlantic and Mediterranean (D = 0.014) are likely the result of periodic isolation due to sea level changes across the Straits of Gibraltar (Roldan et al. 1998, Grant & Leslie 2001). Our results also fit this pattern, in that we found much higher levels of differentiation between the coastal and combined inshore stocks than between the two inshore stocks (Figure 2, Table 4).

Our results clearly lend support to the current management policy of treating offshore and Georgia Basin Pacific hake as separate stocks (Methot & Dorn 1995) and to the BRT's conclusion that at least one DPS of Pacific hake resides in the inshore waters of the Georgia Basin (NMFS 2000, Gustafson et al.[1]). In order to use directly our results to make inferences about Pacific hake population structure, however, it is more useful to estimate explicitly parameters of interest, such as the time of divergence or rate of migration among populations. There are two commonly used models that can be used to estimate population parameters from estimates of F_{ST}: the isolation model and the island migration model. The isolation model assumes that two populations of equal effective size N have diverged

from a common ancestral population t generations ago. The populations are assumed to exchange no migrants after the time of divergence, and all evolution is assumed to occur through random genetic drift. Under this model, F_{ST} can be used to estimate the quantity t/2N (Weir 1990). The migration model, in contrast, assumes that a group of populations of equal size N exchange migrants such that a proportion m of each population consists of migrants from all other populations each generation. The model assumes that current patterns of variation reflect an equilibrium between migration and random genetic drift, and under this model F_{ST} can be used to estimate the quantity Nm (Crow & Aoki 1984).

Under the isolation model, estimates of t/2N vary among population pairs and range from 0.0046 to 0.035 (Table 4). Assuming an average generation time of 6 years, and post-glacial divergence time of ~15,000 years (Pielou 1991, Porter & Swanson 1998), the effective population sizes for each of the three populations would need to be on the order of 10^4–10^5 to explain the level of observed divergence. The lower level of divergence between the two inshore populations compared to the level of divergence between the coastal and inshore populations could in principle be due to either a more recent divergence time for the inshore populations or to a larger effective population size for the inshore populations than the coastal population. The latter alternative seems unlikely, because all three populations are roughly equally variable (Table 5) suggesting that they have roughly equal long-term effective population sizes. Under the migration model, estimates of Nm range from 3 to 24 (Table 4). Under the reasonable assumption that N is relatively large ($> 10^3$), this suggests that genetically effective migrants make up a small proportion of the populations.

Simple population genetic models, such as the island migration model, have come under some criticism for being unrealistic and subject to being used in situations where their assumptions clearly do not apply

Table 5. Summary of allozyme genetic variation within Pacific hake populations (pooled temporal samples).

Population	Sample size	Proportion loci variable	Alleles per variable locus	Heterozygosity
Port Susan	287	0.44	2.39	0.122
SOG	197	0.49	2.55	0.129
Offshore	159	0.51	2.57	0.130
Mean	214	0.48	2.50	0.127

(Whitlock & McCauley 1999). In the case of Pacific hake, however, we believe both models are potentially appropriate as an approximation of reality. In particular, all three populations are very large and have probably been relatively stable for long periods of time (at least until relatively recently). All three populations are roughly equally variable (Table 5), suggesting that long-term effective sizes are similar. The assumption of strict selective neutrality is questionable for many allozyme loci (Gillespie 1991, Ford 2002), but by examining a large number of different loci the effects of selection on individual loci will probably not result in severely biased parameter estimates. Nonetheless, because neither model is likely to be strictly correct and because we have no good way of evaluating which model is most appropriate, the parameter estimates in Table 4 are best viewed as approximations under two 'what if' scenarios, rather than as truly accurate estimates. Importantly, under either model it appears reasonable to conclude that all three populations are demographically quite isolated from each other, and this conclusion is unlikely to be affected by minor violations of model assumptions. It is possible that other factors, such as non-random sampling, may have imparted an upwards bias to our estimates of F_{ST} (Waples 1998); however, sampling in multiple years, sampling multiple age classes, and the use of large sample sizes should have reduced some of this bias.

A conservative upper range for the time of divergence of Pacific hake populations in the eastern North Pacific dates to the last deglaciation event in the Georgia Basin at the end of the Vashon Stade of the Fraser Glaciation, ~15,000 years BP (Pielou 1991, Porter & Swanson 1998, Tunnicliffe et al. 2001). Following deglaciation, colonization of the inshore waters of the Georgia Basin may have occurred when southward migrating adult hake entered the inside waters of the Georgia Basin from the north, through Queen Charlotte Sound and Johnstone Strait, or from the west, through the Strait of Juan de Fuca, and became entrained in the basin and remained to spawn in suitable deepwater areas such as Port Susan (Figure 1). The complex physiography of the basin and the north–south migratory pattern of Pacific hake may have facilitated this colonization event. The recent northward expansion of the feeding migration of Pacific hake, attributed to El Niño events (Hollowed 1992) and changes in climate regime (McFarlane et al. 2000), suggests a mechanism that may have allowed this colonization event to occur. Since 1990, adult offshore Pacific hake have been migrating and spawning further north each

year (McFarlane et al. 2000). In 1998, they were distributed north of the Queen Charlotte Islands and have been observed spawning off the west coast of Vancouver Island since 1994 (McFarlane et al. 2000). If a similar climate regime and northward expansion of Pacific hake distribution obtained around the end of the last ice-age, colonization of the Georgia Basin may have resulted through a comparable northward spread of Pacific hake. The genetic data in the present study are consistent with this scenario under the assumption that effective population sizes have been large (10^4–10^5; Table 4). Assuming that the differences in our estimates of t/N between the populations are due primarily to differences in divergence time rather than differences in effective population size, our data suggest that the inshore population initially split from the coastal population. Much later, the inshore population then split into the SOG population and the Puget Sound population. Assuming the initial split between the coastal and inshore stocks occurred immediately after the glacial retreat ~15,000 years ago, the divergence between the two inshore populations would date to ~2,300 years ago, based on the relative estimates of $t/2N$ (Table 4).

In summary, this study has confirmed previous work that suggested that inshore and coastal Pacific hake populations are genetically differentiated and suggests that the division of North Eastern Pacific hake into at least two management units (Methot & Dorn 1995, Gustafson et al.[1]) is biologically appropriate. This study also found lower, but statistically significant, genetic differentiation between the Puget Sound and SOG populations. These two inshore populations are only ~15% as diverged from each other as each is from the coastal population, but the observed level of differentiation between them is nonetheless consistent with a high degree of demographic isolation (Table 4). In their status review of Pacific hake for the ESA, Gustafson et al.[1] hypothesized that significant population structure may exist within the Georgia Basin DPS (which includes both the SOG and Puget Sound populations), and the results from this study support this hypothesis. By definition, a DPS must be both 'discrete' from other conspecifics and biologically or ecologically significant (USFWS & NMFS 1996). The current study primarily addresses the criteria of discreteness, and the results suggest that the Puget Sound and SOG populations are reasonably considered discrete from each other. The question of biological or ecological significance was not directly addressed by the present study, and it is therefore premature to speculate about whether or not the Puget Sound and SOG populations might reasonably be considered to be separate DPSs under the ESA.

Acknowledgements

We thank the crew of the *FV Chasina*, Wayne Palsson of Washington Department of Fish and Wildlife, and Anna Elz of the Northwest Fisheries Science Center (NWFSC) for assistance in sampling in Port Susan; Gail Jewsbury, Mark Saunders, and Gordon (Sandy) McFarlane of the Pacific Biological Station, Department of Fisheries and Oceans Canada (DFO) for providing samples from the SOG; and Mark Saunders of DFO, and Larry Hufnagle, Tonya Builder, and Herb Sanborn of the NWFSC for providing samples of offshore hake. We gratefully acknowledge the assistance of Delia Patterson, Stephanie Holtz, and Ben Wright of the NWFSC Protein Laboratory with protein electrophoresis, and Paul Aebersold of the NWFSC for interpretation of allozyme gels. We thank Robin Waples, Stew Grant, and one anonymous reviewer for valuable comments on the manuscript.

References

Bailey, K.M., R.C. Francis & P.R. Stevens. 1982. The life history and fishery of Pacific whiting, *Merluccius productus*. Calif. Coop. Ocean. Fish. Invest. Rep. 23: 81–98.

Bailey, K.M. & J. Yen. 1983. Predation by a carnivorous marine copepod, *Euchaeta elongata* Esterly, on eggs and larvae of the Pacific hake, *Merluccius productus*. J. Plankton Res. 5: 71–82.

Cavalli-Sforza, L.L. & A.W.F. Edwards. 1967. Phylogenetic analysis: Models and estimation procedures. Evolution 21: 550–570.

Crow, J.F. & K. Aoki. 1984. Group selection for a polygenic behavioral trait: Estimating the degree of population subdivision. Proc. Natl. Acad. Sci. U.S.A. 81: 6073–6077.

Dorn, M.W. 1995. Effects of age composition and oceanographic conditions on the annual migration of Pacific whiting, *Merluccius productus*. Calif. Coop. Ocean. Fish. Invest. Rep. 36: 97–105.

Ford, M.J. 2002. Applications of selective neutrality tests to molecular ecology. Mol. Ecol. 11: 1245–1262.

Fox, D.A. 1997. Otolith increment analysis and the application toward understanding recruitment variation in Pacific hake (*Merluccius productus*) within Dabob Bay, WA. Master's Thesis, University of Washington. 73 pp.

Galleguillos, R., L. Troncoso, C. Oyarzun, M. Astorga & M. Penaloza. 2000. Genetic differentiation in Chilean hake *Merluccius gayi gayi* (Pisces: Merlucciidae). Hydrobiologia 420: 49–54.

Gillespie, J.H. 1991. The Causes of Molecular Evolution, Oxford University Press, New York. 336 pp.

Goñi, R. 1988. Comparison of Pacific hake (*Merluccius productus* Ayres, 1855) stocks in inshore waters of the Pacific Ocean: Puget Sound and Strait of Georgia. Master's Thesis, University of Washington. 104 pp.

Grant, W.S. & R.W. Leslie. 2001. Inter-ocean dispersal is an important mechanism in the zoogeography of hakes (Pisces: *Merluccius* spp.). J. Biogeogr. 28: 699–721.

Grant, W.S., R.W. Leslie & I.I. Becker. 1987. Genetic stock structure of the southern African hakes *Merluccius capensis* and *M. paradoxus*. Mar. Ecol. Prog. Ser. 41: 9–20.

Hollowed, A.B. 1992. Spatial and temporal distributions of Pacific hake, *Merluccius productus*, larvae and estimates of survival during early life stages. Calif. Coop. Ocean. Fish. Invest. Rep. 33: 100–123.

Imsiridou, A. & C. Triantaphyllidis. 2001. Allozyme electrophoretic studies in European hake *Merluccius merluccius* (Gadidae) populations. Biologia 56: 545–547.

International Union of Biochemistry and Molecular Biology, Nomenclature Committee (IUBMBNC). 1992. Enzyme nomenclature 1992: Recommendations of the Nomenclature Committee of the International Union of Biochemistry and Molecular Biology on the nomenclature and classification of enzymes. Academic Press, San Diego. 862 pp.

Lo Brutto, S., M. Arculeo, A. Mauro, M. Scalisi, M. Cammarata & N. Parrinello. 1998. Allozymic variation in Mediterranean hake *Merluccius merluccius* (Gadidae). Ital. J. Zool. 65: 49–52.

Mangaly, G. & A. Jamieson. 1978. Genetic tags applied to the European hake, *Merluccius merluccius* (L.). Anim. Blood Grps Biochem. Genet. 9: 39–48.

McFarlane, G.A. & R.J. Beamish. 1985. Biology and fishery of Pacific whiting, *Merluccius productus*, in the Strait of Georgia. Mar. Fish. Rev. 47: 23–34.

McFarlane, G.A. & R.J. Beamish. 1986. Biology and fishery of Pacific hake *Merluccius productus* in the Strait of Georgia. Int. N. Pac. Fish. Comm. Bull. 50: 365–392.

McFarlane, G.A., J.R. King & R.J. Beamish. 2000. Have there been recent changes in climate? Ask the fish. Progr. Oceanogr. 47: 147–169.

Mathews, C.P. 1985. Meristic studies of the Gulf of California species of *Merluccius*, with a description of a new species. J. Nat. Hist. 19: 697–718.

Methot, R.D. & M.W. Dorn. 1995. Biology and fisheries of North Pacific hake (*Merluccius productus*). pp. 389–414. *In*: J. Alheit & T.J. Pitcher (ed.) Hake: Biology, Fisheries and Markets, Chapman and Hall, London.

National Marine Fisheries Service (NMFS). 1999. Listing endangered and threatened species and designating critical habitat: Petition to list eighteen species of marine fishes in Puget Sound, Washington. Federal Register, 21 June 1999. 64: 33037–33040.

National Marine Fisheries Service (NMFS). 2000. Notice of determination: Endangered and threatened species: Puget Sound populations of Pacific hake, Pacific cod, and walleye pollock. Federal Register, 24 November 2000. 65: 70514–70521.

Nei, M. 1978. Estimation of average heterozygosity and genetic distance from a small number of individuals. Genetics 89: 583–590.

Pedersen, M. 1985. Puget Sound Pacific whiting, *Merluccius productus*, resource and industry: An overview. Mar. Fish. Rev. 47: 35–38.

Pielou, E.C. 1991. After the Ice Age: The Return of Life to Glaciated North America, University of Chicago Press, Chicago. 366 pp.

Pla, C., A. Vila & J.L. Garcia-Marin. 1991. Differentiation de stocks de merlu (*Merluccius merluccius*) par l'analyse genetique: Comparition de plusiers populations mediterraneenes et atlantiques du littoral espagnol. FAO, Rapp. sur Pesches 447: 87–93.

Porter S.C. & T.W. Swanson. 1998. Radiocarbon age constraints on rates of advance and retreat of the Puget Lobe of the Cordilleran ice sheet during the last glaciation. Quaternary Res. 50: 205–213.

Reynolds, J., B.S. Weir & C.C. Cockerham. 1983. Estimation of the coancestry coefficient: Basis for a short-term genetic distance. Genetics 105: 767–779.

Roldan, M.I. 1991. Enzymatic polymorphisms in the Argentinian hake, *Merluccius hubbsi* Marini, of the Argentinian continental shelf. J. Fish Biol. 39(Suppl. A): 53–59.

Roldan, M.I., J.L. Garcia-Marin, F.M. Utter & C. Pla. 1998. Population genetic structure of European hake, *Merluccius merluccius*. Heredity 81: 327–334.

Shaklee, J.B., F.W. Allendorf, D.C. Morizot & G.S. Whitt. 1990. Gene nomenclature for protein-coding loci in fish. Trans. Am. Fish. Soc. 119: 2–15.

Shaw, W., G.A. McFarlane & R. Kieser. 1990. Distribution and abundance of the Pacific hake (*Merluccius productus*) spawning stocks in the Strait of Georgia, British Columbia, based on trawl and acoustic surveys in 1981 and 1988. Int. N. Pac. Fish. Comm. Bull. 50: 121–134.

Smith, P.J., G.J. Patchell & P.G. Benson. 1979. Glucosephosphate isomerase and isocitrate dehydrogenase polymorphisms in the hake, *Merluccius australis*. NZ J. Mar. Freshw. Res. 13: 545–547.

Sneath, P.H.A. & R.R. Sokal. 1973. Numerical Taxonomy, W.H. Freeman, San Francisco. 573 pp.

Swofford, D.L. & R.B. Selander. 1981. BIOSYS-1: A FORTRAN program for the comprehensive analysis of electrophoretic data in population genetics and systematics. J. Heredity 72: 281–283.

Tunnicliffe, V., J.M. O'Connell & M.R. McQuoid. 2001. A Holocene record of marine fish remains from the Northeastern Pacific. Mar. Geol. 174: 197–210.

United States Fish and Wildlife Service & National Marine Fisheries Service (USFWS & NMFS). 1996. Policy regarding the recognition of distinct vertebrate population segments under the Endangered Species Act. Federal Register, 7 February 1996. 61: 4722–4725.

Utter, F.M. 1969. Transferrin variants in Pacific hake (*Merluccius productus*). J. Fish. Res. Board Can. 26: 3268–3271.

Utter, F.M. & H.O. Hodgins. 1969. Lactate dehydrogenase isozymes of Pacific hake (*Merluccius productus*). J. Exp. Zool. 172: 59–68.

Utter, F.M. & H.O. Hodgins. 1971. Biochemical polymorphisms in the Pacific hake (*Merluccius productus*). Rapp. P.-V. Reun. Cons. Perm. Int. Explor. Mer. 161: 87–89.

Utter, F.M., C.J. Stormont & H.O. Hodgins. 1970. Esterase polymorphism in vitreous fluid of Pacific hake, *Merluccius productus*. Anim. Blood Grps Biochem. Genet. 1: 69–82.

Vrooman, A.M. & P.A. Paloma. 1977. Dwarf hake off the coast of Baja California, Mexico. Calif. Coop. Ocean. Fish. Invest. Rep. 19: 67: 72.

Waples, R.S. 1998. Separating the wheat from the chaff: Patterns of genetic differentiation in high gene flow species. J. Heredity 89: 438–450.

Ward, R.D., M. Woodward & D.O.F. Skibinski. 1994. A comparison of genetic diversity levels in marine, freshwater, and anadromous fishes. J. Fish Biol. 44: 213–232.

Weir, B.S. 1990. Genetic Data Analysis, Sinauer, Sunderland, Massachusetts. 370 pp.

Whitlock, M.C. & D.E. McCauley. 1999. Indirect measures of gene flow and migration: F_{ST} doesn't equal $1/(4Nm + 1)$. Heredity 82: 117–125.

Wright, S. 1978. Evolution and the Genetics of Populations, Vol. 4. Variability Within and Among Populations, University of Chicago Press, Chicago. 580 pp.

Appendix 1. Allele frequencies for 29 polymorphic loci found in the seven Pacific hake samples. Sample numbers and locations as in Table 1. Loci monomorphic in all populations were ACP^*, AK^*, CK-A^*, EST-D^*, $FBALD$-3^*, $FBALD$-4^*, $GAPDH$-4^*, aMAN*, sMEP*, PEP-C^*, PK^*, and XO^*.

Locus	Allele	Sample number						
		1	2	3	4	5	6	7
sAAT-2*	N	100	60	100	100	99	114	50
	*100	1.000	1.000	1.000	1.000	0.995	1.000	1.000
	*150	0.000	0.000	0.000	0.000	0.005	0.000	0.000
sAAT-4*	N	100	99	100	100	99	114	50
	*100	0.835	0.904	0.850	0.905	0.919	0.882	0.870
	*130	0.130	0.061	0.105	0.075	0.071	0.105	0.090
	*80	0.035	0.035	0.045	0.020	0.010	0.013	0.040

198

Appendix 1. (Continued)

Locus	Allele	Sample number						
		1	2	3	4	5	6	7
ADA*	N	100	99	100	99	100	75	50
	*100	0.385	0.389	0.460	0.480	0.405	0.673	0.550
	*95	0.265	0.232	0.250	0.182	0.260	0.153	0.240
	*90	0.140	0.152	0.150	0.152	0.175	0.060	0.080
	*110	0.210	0.227	0.140	0.187	0.160	0.113	0.130
ALAT*	N	80	100	100	80	100	114	50
	*100	0.425	0.500	0.395	0.438	0.470	0.605	0.610
	*90	0.575	0.500	0.605	0.563	0.525	0.395	0.390
	*65	0.000	0.000	0.000	0.000	0.005	0.000	0.000
FH*	N	100	100	99	100	99	114	50
	*100	0.705	0.735	0.712	0.710	0.722	0.618	0.530
	*150	0.295	0.265	0.288	0.290	0.278	0.382	0.470
bGALA*	N	100	100	88	100	99	114	50
	*100	0.935	0.915	0.960	0.880	0.833	0.974	0.940
	*95	0.065	0.085	0.040	0.120	0.167	0.026	0.060
GAPDH-5*	N	100	60	99	100	100	114	50
	*100	1.000	1.000	1.000	1.000	1.000	0.996	1.000
	*110	0.000	0.000	0.000	0.000	0.000	0.004	0.000
bGLUA*	N	100	100	100	100	100	114	50
	*100	1.000	0.995	1.000	1.000	1.000	0.978	1.000
	*90	0.000	0.005	0.000	0.000	0.000	0.022	0.000
GPI-A*	N	100	100	100	100	100	114	50
	*100	0.665	0.675	0.635	0.680	0.650	0.482	0.520
	*125	0.335	0.325	0.365	0.320	0.345	0.518	0.480
	*140	0.000	0.000	0.000	0.000	0.005	0.000	0.000
GPI-B*	N	100	100	99	99	99	114	50
	*100	0.680	0.760	0.682	0.682	0.667	0.680	0.640
	*20	0.295	0.230	0.308	0.308	0.313	0.272	0.330
	*180	0.010	0.000	0.000	0.005	0.005	0.004	0.010
	*−100	0.015	0.010	0.010	0.005	0.015	0.009	0.020
	*40	0.000	0.000	0.000	0.000	0.000	0.035	0.000
GR*	N	80	96	100	80	96	114	50
	*100	0.850	0.755	0.805	0.819	0.786	0.759	0.760
	*135	0.150	0.245	0.195	0.181	0.214	0.241	0.240
IDDH*	N	99	80	98	100	100	114	40
	*100	0.636	0.644	0.607	0.580	0.545	0.640	0.650
	*75	0.364	0.356	0.393	0.420	0.455	0.355	0.350
	*120	0.000	0.000	0.000	0.000	0.000	0.005	0.000
mIDHP*	N	100	100	20	100	100	80	49
	*100	0.995	0.995	1.000	0.990	0.995	1.000	1.000
	*75	0.005	0.005	0.000	0.010	0.005	0.000	0.000
sIDHP*	N	100	99	100	100	100	114	50
	*100	0.940	0.934	0.960	0.885	0.905	0.846	0.900
	*85	0.045	0.066	0.035	0.065	0.030	0.154	0.100
	*75	0.015	0.000	0.005	0.050	0.065	0.000	0.000
LDH-A*	N	100	99	99	100	100	114	50
	*100	0.800	0.763	0.742	0.860	0.790	0.978	0.970
	*0	0.200	0.237	0.258	0.140	0.210	0.022	0.030
LDH-B*	N	100	100	100	100	100	114	50
	*100	1.000	0.995	1.000	1.000	1.000	1.000	1.000
	*65	0.000	0.005	0.000	0.000	0.000	0.000	0.000

Appendix 1. (Continued)

Locus	Allele	Sample number						
		1	2	3	4	5	6	7
*LDH-C**	N	100	100	100	100	100	114	50
	*100	0.995	1.000	1.000	1.000	1.000	0.978	1.000
	*−100	0.005	0.000	0.000	0.000	0.000	0.022	0.000
*sMDH-A**	N	100	100	98	100	100	114	50
	*100	0.855	0.825	0.827	0.845	0.905	0.908	0.970
	*150	0.140	0.170	0.163	0.140	0.095	0.083	0.010
	*200	0.005	0.005	0.010	0.015	0.000	0.009	0.020
*sMDH-B**	N	100	100	100	100	100	114	50
	*100	0.920	0.890	0.955	0.920	0.935	0.921	0.910
	*135	0.080	0.110	0.045	0.080	0.065	0.079	0.090
*MPI**	N	52	98	91	97	96	31	50
	*100	1.000	1.000	0.995	0.943	0.922	1.000	1.000
	*90	0.000	0.000	0.005	0.057	0.078	0.000	0.000
*NTP**	N	80	100	100	80	100	114	50
	*100	1.000	1.000	1.000	1.000	1.000	0.996	1.000
	*135	0.000	0.000	0.000	0.000	0.000	0.004	0.000
*PEP-A**	N	100	100	100	100	100	114	50
	*100	1.000	1.000	1.000	1.000	1.000	0.996	1.000
	*85	0.000	0.000	0.000	0.000	0.000	0.004	0.000
*PEP-B**	N	100	100	100	100	99	114	50
	*100	0.990	0.970	0.990	0.970	0.949	0.860	0.880
	*110	0.010	0.030	0.010	0.030	0.051	0.140	0.100
	*75	0.000	0.000	0.000	0.000	0.000	0.000	0.020
*PEP-D**	N	99	100	100	60	97	114	50
	*100	0.960	0.965	0.955	0.983	0.943	0.965	0.990
	*90	0.030	0.035	0.025	0.008	0.042	0.022	0.000
	*110	0.010	0.000	0.020	0.000	0.000	0.013	0.000
	*75	0.000	0.000	0.000	0.008	0.015	0.000	0.010
*PGK**	N	100	99	99	100	100	114	50
	*100	0.535	0.596	0.636	0.595	0.680	0.548	0.550
	*165	0.465	0.404	0.364	0.405	0.320	0.452	0.450
*PGM-1**	N	100	100	100	100	100	114	50
	*100	1.000	1.000	0.985	0.985	0.990	0.961	0.970
	*90	0.000	0.000	0.015	0.010	0.005	0.035	0.020
	*50	0.000	0.000	0.000	0.005	0.005	0.004	0.010
*PGM-2**	N	100	100	100	100	100	74	50
	*100	0.955	0.950	0.990	0.880	0.960	0.912	0.950
	*125	0.045	0.050	0.010	0.120	0.035	0.088	0.050
	*150	0.000	0.000	0.000	0.000	0.005	0.000	0.000
*sSOD**	N	100	100	100	100	100	114	50
	*100	0.945	0.960	0.940	0.980	0.985	0.978	0.980
	*800	0.055	0.040	0.060	0.020	0.015	0.022	0.020
*TPI**	N	100	100	100	100	100	114	50
	*100	1.000	0.995	0.995	1.000	1.000	1.000	1.000
	*180	0.000	0.000	0.005	0.000	0.000	0.000	0.000
	*50	0.000	0.005	0.000	0.000	0.000	0.000	0.000

Environmental Biology of Fishes **69**: 201–210, 2004.
© 2004 *Kluwer Academic Publishers. Printed in the Netherlands.*

Population structure of Alaskan shortraker rockfish, *Sebastes borealis*, inferred from microsatellite variation

Andrew P. Matala[a], Andrew K. Gray[a,b], Jonathan Heifetz[b] & Anthony J. Gharrett[a,c]
[a]*Fisheries Division, School of Fisheries and Ocean Sciences, University of Alaska, Fairbanks, 11120 Glacier Highway, Juneau, AK 99801, U.S.A.*
[b]*National Marine Fisheries Service, Auke Bay Laboratory, 11305 Glacier Highway, Juneau, AK 99801, U.S.A.*
[c] *Corresponding author (e-mail: ffajg@uaf.edu)*

Received 17 April 2003   Accepted 19 April 2003

Key words: population genetics, marine fisheries, Wakefield Symposium

Synopsis

Alaskan shortraker rockfish population structure was analyzed by examining allelic variation at eight microsatellite loci. Samples were collected along the continental shelf and upper slope from the south end of Baranof Island to the western Aleutian Islands, and collections were pooled into eight geographically distinct groups. An exact test of homogeneity indicated population structure ($p < 0.0006$) among groups. The proportion of the total variation that was attributable to divergence among populations ($\theta = 0.0014$) was not statistically significant, and no evidence of a geographic cline of structure was detected. Finer scale analyses that compared adjacent collections indicated that the collection from the southern end of the range differed from all remaining collections at three loci. Structure related to geographic location was detected by partitioning the variation among populations. The size distributions of shortraker rockfish varied among collections from east to west. The size differences may reflect divergent oceanographic and biological factors acting on populations that have restricted migration and movement. Alternatively, if there is substantial movement accompanied by lengthy reverse migration to natal grounds, the size differences may be related to ages of cohorts that are differentially distributed along the Pacific Rim. Further biological information including size, age composition, and age of maturity data, as well as information on other life history characteristics will be required to explain shortraker rockfish population structure.

Introduction

Shortraker rockfish *Sebastes borealis* are common throughout Alaskan waters, and range from California, along the northern Pacific Rim, to Japan (McDermott 1994, Schnute & Haigh 1999[1], Orlov 2001). They are often found between 300 and 600 m along the continental slope, but occur at depths as great as 875 m (Allen & Smith 1988[2], Tokranov & Davydov 1997).

Shortraker rockfish are long-lived (157 years) and slow growing, attaining lengths up to 100 cm (Munk 2001). The estimated length and age at 50% maturity are 44.9 cm and 20 years (McDermott 1994); however, little is known about the life history of shortraker rockfish. Information on the recruitment of deeper dwelling rockfish, such as shortraker rockfish, is limited because of difficulties involved in identification of larvae and juveniles, and because barotrauma hampers tag-recapture experiments (Kendall 1991, Love et al. 1991, Moles et al. 1998). Shortraker rockfish, like all members of the *Sebastes* genus, are ovoviviparous and

____

[1] Schnute, J. & R. Haigh. 1999. Shortraker rockfish British Columbia Coast. Canadian DFO Science, Stock Status Report A6–14. 3 pp.
[2] Allen, M.J. & G.B. Smith. 1988. Atlas and zoogeography of common fishes in the Bering Sea and Northeastern Pacific. U.S.

Department of Commerce, NOAA Technical Report NMFS 66, 151 pp.

extrude live planktonic larvae. Early life stages of many species of rockfish are frequently distributed in mid-water and near-surface habitats often at great distances from adult habitat (Moser & Boehlert 1991); however, data on the extent of dispersal of larval shortraker rockfish are lacking.

In Alaska, shortraker rockfish are more valuable than more plentiful species such as Pacific ocean perch, *Sebastes alutus* (Ackley & Heifetz 2001). In 1991, the North Pacific Fisheries Management Council established three subgroups within the slope rockfish assemblage in an effort to protect the most sought-after rockfish species from overfishing. One of the subgroups combined shortraker and rougheye *Sebastes aleutianus* rockfish (Heifetz et al. 2000[3]). The shortraker rockfish is one of the highest valued commercial rockfish species. It is popular in the Asian market because of its large size and red coloration. By regulation, there is no targeted fishery for shortraker rockfish in Alaska, but it is a major bycatch component of several target species within Alaskan fisheries, including Pacific ocean perch *S. alutus*, Pacific halibut (*Hippoglossus stenolepis*), and sablefish (*Anoplopoma fimbria*). In 2001, the shortraker rockfish exploitable biomass in the Gulf of Alaska (GOA) was estimated to be 25 473 MT, from which managers defined an acceptable biological catch of 586 MT (Heifetz et al. 2001[4]). Geographical allocation of shortraker rockfish harvests is based on the distribution of biomass, with little or no emphasis on population structure or biological influences (Heifetz et al. 2000[3]).

Generally, marine species have potential for high gene flow through larval transport, juvenile dispersal, and adult migratory behavior (Moser & Boehlert 1991, Stanley et al. 1994, Seeb 1998). Heavy localized fishing effort, or 'hot spots' can diminish the resource without proper delineation of productivity units to guide managers. Populations of slow growing marine species such as shortraker rockfish, that may have a restricted adult migratory range and whose larvae are dispersed by planktonic transport, can be expected to recover slowly from local depletion (Mason 1998, Yoklavich 1998[5]), and may require many years for recolonization.

Genetic analyses are fundamental in characterizing patterns of population structure or productivity units, which result from restrictions in dispersal and gene flow, and may allow inference on the extent of transport and migration of larvae and juveniles (Withler et al. 2001). Several recent studies using mtDNA, allozymes, and microsatellites reported genetic structure and restricted gene flow within rockfish species (Hawkins et al. 1997[6], Seeb 1998, Rocha-Olivares & Vetter 1999, Roques et al. 1999, Withler et al. 2001).

Shortraker rockfish were sampled in eight geographic areas ranging from Southeast Alaska to the western Aleutian Islands (ALEU). Population genetic analysis was conducted using a suite of eight microsatellite loci. Three specific questions were addressed: (1) Is population structure for shortraker rockfish detectable? (2) If divergence or structure exists, what is the geographic basis of that structure? and (3) Is the population structure compatible with current management?

Materials and methods

Tissue preparation and extraction

Shortraker rockfish were collected in Alaska in a cooperative effort by the National Marine Fisheries Service (NMFS) Auke Bay Laboratory and the University of Alaska Fairbanks, School of Fisheries and Ocean Science. Sample sites ranged south of Baranof Island to the ALEU. Samples were grouped geographically for analysis, which resulted in eight groups hereafter referred to as populations: southern Southeast Alaska (SSE), northern Southeast Alaska (NSE), Yakutat (YAK), Cordova (CORD), Seward (SEW), Kodiak (KOD), Shumagin Islands (SHU), and ALEU, the last of which included collections from the same area in both 1994 and 1996. Sample sizes and locations are given in Figure 1.

[3] Heifetz, J., J.N. Ianelli, D.M. Clausen, D.L. Courtney & J.T. Fujioka. 2000. Slope rockfish. Stock assessment and fishery evaluation report for the groundfish resources of the Gulf of Alaska, Section 6. Compiled by: The plan team for the groundfish fisheries of the Gulf of Alaska. North Pacific Fishery Management Council, 605 W 4th Ave., Suite 306 Anchorage, AK 99501. 63 pp.

[4] Heifetz, J., D.L. Courtney, D.M. Clausen, J.T. Fujioka & J.N. Ianelli. 2001. Slope rockfish. Stock assessment and fishery evaluation report for the groundfish resources of the Gulf of Alaska, Section 6. Compiled by: The plan team for the groundfish fisheries of the Gulf of Alaska. North Pacific Fishery Management Council, 605 W 4th Ave, Suite 306 Anchorage, AK 99501. 72 pp.

[5] Yoklavich, M.M. (ed.) 1998. Marine Harvest Refugia For West Coast Rockfish: A Workshop. NOAA/NMFS Technical Memorandum. NOAA-TM-NMFS-SWFSC-255. 159 pp.

[6] Hawkins, S., J. Heifetz, J. Pohl & R. Wilmot. 1997. Genetic population structure of rougheye rockfish *Sebastes aleutianus* inferred from allozyme variation. Alaska Fishery Science Quarterly Report, July–August–September 1977. NMFS, USDOC Alaska Fisheries Science Center, Seattle, WA. pp. 1–10.

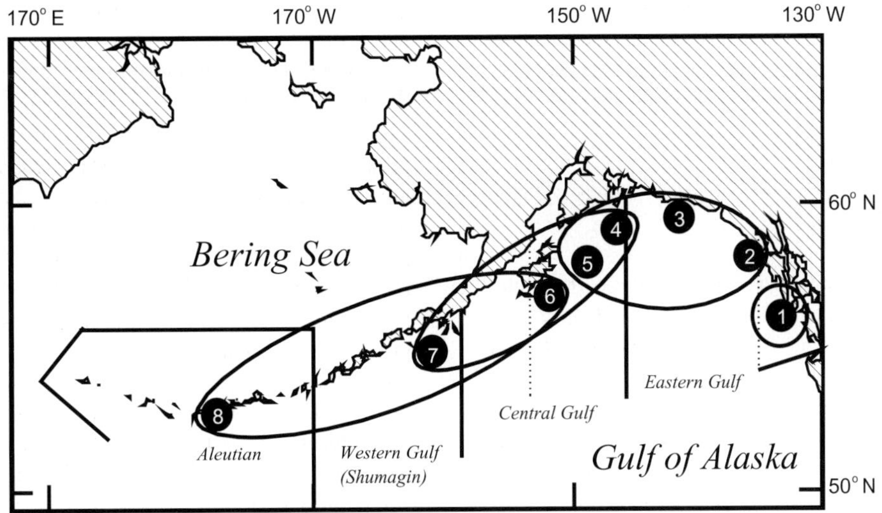

Figure 1. The eight groups of Alaskan shortraker rockfish collections sampled from geographically defined fishery regulatory areas are designated as the following: (1) SSE – south of Baranof Island (n = 42) and (2) NSE – northwest of Chichagof Island from the Southeast Outside District (n = 55); (3) YAK – northwest of Yakutat from the West Yakutat District (n = 74); (4) CORD – south of Prince William Sound (n = 38), (5) SEW – south of Seward (n = 48), and (6) KOD – south of Kodiak Island (n = 47) from the Kodiak District; (7) SHU – Shumagin Islands (n = 38) from the Shumagin District; and (8) ALEU – western Aleutian Islands from the Western Aleutian District. The ALEU site consists of collections from two years: 1994 (n = 89) and 1996 (n = 97). The location of each collection on the map is the mean location of the contributing hauls. Ovals designate homogeneous population groupings. Solid lines on the map distinguish regulatory area boundaries in the GOA and Aleutian Islands. For reference, the Chirikof (west) and KOD (east) INPFC statistical areas are shown in the Central GOA regulatory area (dashed line), and the YAK and Southeast INPFC areas of the Eastern GOA regulatory area are shown (dashed line).

Heart tissue was dissected from specimens and pre-served in either a cell lysis solution [50 mM Tris–HCl, 20 mM ethylenediaminetetraacetic acid (EDTA), 1% sodium dodecyl sulfate pH 8.0], or in a preserva-tion solution [20% dimethyl sulfoxide, 0.25 M EDTA pH 8.0, saturated with NaCl (Seutin et al. 1991)], and stored at $-20°C$. Total cellular DNA was extracted from heart tissue using Puregene DNA^{TM} isolation kits (Gentra Systems, Inc., Minneapolis, MN) and rehy-drated in 1X TE (TE is 0.01 M Tris–HCl, 0.001 M EDTA, pH 8.0).

Microsatellite amplification and allele scoring

In a preliminary survey, 12 microsatellite loci were polymerase-chain reaction (PCR)-amplified and eval-uated: μSma-1, μSma-2, μSma-3, μSma-4, μSma-5, μSma-6, μSma-7, μSma-10, and μSma-11 (Wim-berger et al. 1999), and μSr7-2, μSr7-7, μSr7-25[7].

Eight loci were identified: μSma-2, μSma-3, μSma-5, μSma-7, μSma-11, μSr7-2, μSr7-7, and μSr7-25 that had variability, amplified reliably, and could be easily interpreted (minimal band stutter). These eight loci were used in the analysis of population structure. Another locus, which was nearly fixed at a single allele (μSma-6; 99.6%), was omitted from the analysis.

PCR amplification of loci was carried out in 10 μl reactions containing: 5.5 μl of ddH$_2$O, 1 μl of 10X buffer (10X buffer is 50 mM KCl$_2$, 10 mM Tris–HCl pH 9.0, 0.1% Triton × 100; Promega^{TM}), 0.75 μl of 25 mM MgCl, 0.5 μl of 25 mM dNTPs, 0.35 μl of 10 μM forward primer (Operon Technologies^{TM}), 0.4 μl of 10 μM reverse primer, 0.4 μl of 1 μM forward label (LICOR^{TM}), 0.05 μl of Taq polymerase (Perkin Elmer), and 1.0 μl of DNA template (50–100 ng). Target sequences were amplified in a Strategene Robocyler^{TM} with the following regime: 95°C for 3.75 min; one cycle of 94°C for 1.5 min, 58°C for 0.5 min, and 72°C for 0.75 min; 30 cycles of 94°C for 0.5 min, 58°C for 0.5 min, 72°C for 0.75 min; one cycle of elongation at 68°C for 1 min. Following PCR-amplification, 0.5–1.0 μl of sample for each

<hr>

[7] Westerman, M., V.P. Buonaccorsi & R.D. Vetter. unpubl. data. GenBank accession numbers AF269054, AF269056 and AF269055.

of 45–60 samples and a LICOR DNA standard (50–300 bp) were loaded on a 6% polyacrylamide gel and analyzed in a LICOR LongReadIR[TM] sequencing system. Alleles were scored on the same system using LICOR ImageIR[TM] software.

Data analysis

Allele frequencies at each locus in each population were estimated using GENEPOP v. 3.2a (Raymond & Rousset 1995a). Tests for departure from Hardy–Weinberg equilibrium (HWE) were performed for all combinations of loci and populations with an exact test in GENEPOP v. 3.2a (Raymond & Rousset 1995a), using the GENEPOP default settings. Probabilities of linkage disequilibrium for each pair of loci within each population were estimated using the default settings in GENEPOP v. 3.2a (Raymond & Rousset 1995a). Ohta's two-locus analysis of population subdivision (D-statistics) was calculated in POPGENE v. 1.31 (Yeh et al. 1999[8]). Ohta's (1982) analysis partitions linkage disequilibrium into within-subpopulation and among-subpopulation components, and resolves the two main causes of non-random association of alleles (natural selection involving epistatic interactions of loci, and random genetic drift). Simultaneous tests in this study were adjusted using a Bonferroni correction with an initial α level of 0.05 (Rice 1989, Sankoh 1997).

Wright's (1978) inbreeding coefficient (F_{is}) and both the observed heterozygosity (H_o) and Nei's unbiased expected heterozygosity (H_e; Nei 1978), were estimated for each combination of locus and population using POPGENE v. 1.31 (Yeh et al. 1999[8]). The heterozygosities and the index of inbreeding (F_{is}) show the direction of HWE departures; values of F_{is} less than zero indicate an excess of heterozygous types, whereas values greater than zero indicate a deficit of heterozygotes.

A pseudo-exact test performed in GENEPOP v. 3.2a (Raymond & Rousset 1995b) was used to test for overall population heterogeneity (H_o: allele frequencies among populations are homogeneous). The analyses were run using the default settings, except that 1 000 batches were run to reduce the standard errors of the estimates. The combined probability over populations

at each locus and an overall probability were estimated using Fisher's method (Winer 1971).

Collections were distributed along the arc of the north Pacific Rim. Adjacent populations would be expected to be more similar than those separated by greater distance, whether or not allele frequencies among them form a geographic cline (detected by isolation by distance tests). To obtain information about the geographic basis of genetic structure, we partitioned the total genetic variation observed among populations by grouping adjacent populations into homogeneous groups. The total variation observed among populations can be quantified by a log-likelihood ratio (G-statistic), which measures heterogeneity among the populations. This total can be partitioned into two components, (1) the total of the heterogeneities measured among populations within groups and (2) the heterogeneity among groups. The procedure followed identifies all groups of populations that include the maximum number of adjacent populations that are not heterogeneous (p > 0.05). From those homogeneous groups, the sets of partitioning schemes were constructed that included at least one, but as many as possible of those groups. There may be multiple ways to partition the total variation because some homogeneous groups might be overlapping (e.g. if one group includes populations 1, 2, and 3 (but not population 4) and another group includes populations 3, 4, and 5 (but not 2)). The most efficient partitioning, however, removes the largest amount of variation from the total as the sum of within-group heterogeneities (note that the original non-partitioned set of collections has the maximum among-group heterogeneity). An approximate F-ratio quantifies the relative amounts of divergence and provides an index for the efficiency of the partitioning schemes. The index used was the ratio of standardized within-group heterogeneity to the total standardized heterogeneity:

$$F_{\text{df within, df total}} = \frac{G_{\text{within}}/\text{df}_{\text{within}}}{G_{\text{total}}/\text{df}_{\text{total}}}$$

The goal was to examine the magnitude of the geographic area included by each of the homogeneous groupings as opposed to identifying specific geographic boundaries, which may be ambiguous if there are overlapping homogeneous groups. This method also provides information on suboptimal partitions, which provide alternative perspectives on the geographic coverage of genetically homogeneous units. Probabilities for the homogeneous groups

---

[8] Yeh, F.C., R-c. Yang & T. Boyle. 1999. POPGENE version 1.31: Microsoft Window-based Freeware for Population Genetic Analysis, University of Alberta, and Centre for International Forestry Research.

(the significance of the G-statistic) were determined in these analyses using the exact test in GENEPOP v. 3.2a (Raymond and Rousset 1995a).

Population substructure was also examined by tests that partition genetic variation based on allele identity (F_{ST} or θ; Weir & Cockerham 1984) using the analysis of molecular variance option in Arlequin v. 2.0 (Schneider et al. 2000), or based on allele size under the stepwise mutational model (R_{ST}; Slatkin 1995) using GENEPOP v. 3.2a (Raymond & Rousset 1995a). Tests were performed for each allele, for each locus, and overall.

Neighbor-joining (NJ) and unweighted pair-group method with arithmetic mean (UPGMA) trees were constructed using chord distances (Cavalli-Sforza and Edwards 1967) in the software package PHYLIP v. 3.57c (Felsenstein 1995[10]). The Ewens–Watterson neutrality test, a test for selection at each locus within each population, was performed in POPGENE v. 1.31 (Yeh et al. 1999[8]) using the algorithm of Manly (1985). Isolation by distance was examined with a Mantel test in GENEPOP v. 3.2a (Raymond & Rousset 1995a), which permutes the lines of both a half-matrix of pairwise F_{ST} values and a geographic distance half-matrix constructed from Mercator coordinates (Mantel 1967).

Descriptive statistics

The size distribution of shortraker rockfish used in the population structure analysis was evaluated to determine if there was a relationship between observed genetic structure and regionally based biological variation. The International North Pacific Fisheries Commission (INPFC) statistical areas were used to group samples for the comparison (Figure 1). The fork lengths of all specimens used in the genetic analysis were obtained from measurements taken at the time of collection by the NMFS. These data were compared to NMFS survey data, including trawl surveys (1978–2000) and longline surveys (1988–2000). The relationship between fork lengths and geographic locations for both genetic samples and survey data was evaluated

using regression analysis (Microsoft Excel[TM]), and Kendall's coefficient of rank correlation (Sokal & Rohlf 1995).

Results

Microsatellite variability

A wide range of polymorphism was observed among the loci. Allele frequency data were obtained at eight loci for each of eight populations (Table 1). The number of alleles per locus across all populations ranged from seven alleles at μSma-5, to 34 alleles at μSr7–25. The mean number of alleles for all loci was 16. The minimum average expected heterozygosity was 31% at μSma-11, and the maximum was 92.3% at μSr7–2, both in the CORD collection. The mean heterozygosity over all loci and populations was 59.6%. Tests for HWE revealed departure from expectations (F_{is} significantly greater than 0; Table 1) in 5 of 81 individual tests on loci and populations, and at μSma-2 over all populations (p = 0.0302; Fisher's combined probability). No test for HWE was significant after Bonferroni adjustment for simultaneous testing ($\alpha \sim 0.05/k$, k = 81).

The microsatellite allele frequency data were tested for linkage disequilibrium and neutrality. After Bonferroni adjustment, none of the populations had pairs of loci that exhibited disequilibrium, although Fisher's combined probability over all populations for the locus pair μSr7–2 and μSma-2 remained significant (p < 0.01). Ohta's analysis of the distribution of gametic disequilibrium among populations was consistent with random genetic drift in 31 of the 36 locus pair comparisons. The remaining five comparisons, including the μSr7–2 and μSma-2 locus pair, were consistent with the existence of epistasis between alleles in some but not all populations. The results of linkage disequilibrium and Ohta's partitioning may be explained several ways; (1) the loci are partially linked, (2) the samples include an admixture of populations, (3) the microsatellite loci are linked to loci under selective pressure, or (4) epistatic interactions maintain allele associations. The Ewens–Watterson test indicates that the eight microsatellite loci are selectively neutral (results not shown).

Genetic structure and population divergence

Population structure was observed among the eight shortraker rockfish populations by tests of

---

[9] Schneider, S., J.M. Kueffer, D. Roessli & L. Excoffier. 2000. Arlequin ver. 2.0: A software for population genetics data analysis. Genetics and Biometry Laboratory, University of Geneva, Switzerland.

[10] Felsenstein, J. 1995. PHYLIP (Phylogeny Inference Package) version 3.57c. Distributed by the author, Department of Genetics, University of Washington, Seattle, Washington.

homogeneity. Tests were significant at the μSma-3 (p = 0.048), μSma-7 (p = 0.0026), and μSr7–7 (p = 0.0029) loci, and for all loci combined (p = 0.0006; Table 1). Although a test of homogeneity between the collections from the ALEU in 1994 and 1996 was significant at μSma-2 (p = 0.029) and the overall probability between the two groups approaches significance (p = 0.056), the two collections were pooled into one group (ALEU) for all subsequent analyses that involved geographic structure.

There was no evidence for a geographic cline in genetic composition for the eight populations. Weir and

Table 1. Allelic variation and statistical analyses of shortraker rockfish population structure for eight microsatellite loci and eight populations. A is the number of alleles at each locus, bp is the range of allele sizes, H_E is the expected heterozygosity, F_{is} is the index of inbreeding for each population, F_{IS} is the overall index of inbreeding, F_{ST} is the Weir and Cockerham θ, and P_H is the exact test probability (Fisher's combined value) of homogeneity among populations (*indicates p < 0.05, **indicates p < 0.01, and do not reflect the adjustments for multiple testing).

Sample population	Statistic	Locus								Average over loci
		μSma-2	μSma-3	μSma-5	μSma-7	μSma-11	μSr-72	μSr-77	μSr-725	
SSE	A	9	9	5	5	3	16	11	19	8.6
	bp	176–194	146–164	167–175	118–132	162–166	170–208	176–200	198–270	—
	H_E	0.803	0.599	0.675	0.434	0.342	0.912	0.779	0.832	0.5973
	F_{is}	0.066	−0.033	−0.095	−0.21	0.001	−0.097	−0.028	−0.022	−0.046
NSE	A	8	10	4	8	5	20	9	19	9.2
	bp	170–186	148–168	163–175	110–138	156–166	156–206	182–198	204–244	—
	H_E	0.781	0.675	0.655	0.353	0.466	0.893	0.827	0.838	0.61
	F_{is}	−0.048	−0.16	−0.028*	0.073	−0.016	−0.008	0.011	0.032	−0.021
YAK	A	9	12	5	7	5	20	9	24	10.1
	bp	174–190	146–168	167–175	118–134	156–168	156–214	182–204	194–262	—
	H_E	0.73	0.569	0.68	0.324	0.396	0.907	0.777	0.86	0.583
	F_{is}	0.112*	0.002	0.027	0.083	0.011*	−0.028	−0.009	0.057	0.029**
CORD	A	8	8	3	7	4	16	9	20	8.3
	bp	174–192	148–170	167–175	118–136	156–166	172–212	180–200	202–258	—
	H_E	0.727	0.498	0.635	0.369	0.31	0.923	0.776	0.855	0.566
	F_{is}	0	−0.059	−0.121	0.074	−0.047	0.025	−0.069	0.082	−0.011
SEW	A	8	10	5	7	5	19	10	23	9.7
	bp	174–188	148–168	163–175	118–136	156–168	158–210	182–202	198–246	—
	H_E	0.787	0.6	0.692	0.508	0.429	0.886	0.849	0.892	0.627
	F_{is}	−0.087	−0.042	−0.054	−0.108	0.175	−0.011	0.019	−0.004	−0.019
KOD	A	9	9	5	6	6	17	12	18	9.1
	bp	174–190	146–164	163–177	116–132	156–166	158–206	180–204	198–250	—
	H_E	0.772	0.632	0.669	0.46	0.458	0.906	0.801	0.869	0.619
	F_{is}	−0.076	−0.146	0.079	−0.111	0.024	−0.01	0.018	−0.029	−0.028
SHU	A	8	9	4	7	5	15	9	19	8.4
	bp	174–188	148–168	167–177	118–134	156–170	172–202	180–204	196–248	—
	H_E	0.759	0.526	0.656	0.33	0.353	0.895	0.816	0.82	0.573
	F_{is}	−0.005*	0.151	−0.248	0.044	0.158	0.007*	0.033	0.006	0.004**
94 and 96AL	A	12	11	4	11	5	21	13	25	11.3
	bp	168–192	146–168	167–175	114–136	156–168	158–210	180–204	198–264	—
	H_E	0.748	0.6285	0.633	0.348	0.4395	0.8915	0.827	0.838	0.5955
	F_{is}	−0.1305	−0.039	−0.0705	0.075	−0.1745	−0.0105	−0.034	−0.0325	−0.052
Summary Over all populations	A	13	13	7	14	8	28	14	34	16.4
	H_E	0.762	0.595	0.659	0.386	0.404	0.901	0.809	0.849	0.596
	F_{IS}	−0.0302*	−0.0408	−0.0373	−0.0041	−0.0227	−0.0154	−0.0058	0.0046	−0.0182
	F_{ST}	0	0.0026	−0.0016	0.0037	0.0019	0.0018	0.0047*	−0.001	0.0014
	P_H	0.2049	0.0480*	0.2555	0.0026**	0.2618	0.175	0.0029**	0.4123	0.0006**

Cockerham's (1984) θ, which is an analog of Wright's F_{ST}, exceeded zero only at locus μSr7–7 (Table 1), but was not significant after Bonferroni adjustment. A Mantel test provided no evidence of isolation by distance, and NJ and UPGMA trees showed no obvious geographical clustering (results not shown).

Finer scale tests of homogeneity provided evidence of underlying structure. Tests of homogeneity on adjacent populations and groups of adjacent populations reveal that population SSE differs from the adjacent NSE population over all loci (p = 0.019), and differs from all seven other populations at μSma-2 (p = 0.002), μSr7–2 (p = 0.014), μSr7–7 (p = 0.013), and over all loci (p = 0.004). The remaining seven populations were partitioned into three overlapping, homogeneous groups (Figure 1) using exact test probabilities to determine significance of the log-likelihood (*G*-statistic). From those overlapping groups, six partitioning schemes were possible, which assigned populations to non-overlapping sets of populations that were not heterogeneous (Figure 2). The most efficient partitioning of overall genetic structure among the eight populations (in which the within-group component of overall heterogeneity is maximized) included a Southeast Alaska group (SSE), a NSE group and Central GOA group, and a group extending from KOD Island to the Western Aleutians (Figure 2a).

Shortraker size distribution

Regional variation in the size of shortraker rockfish was observed within the range of the study, which may suggest localized oceanographic and/or biological

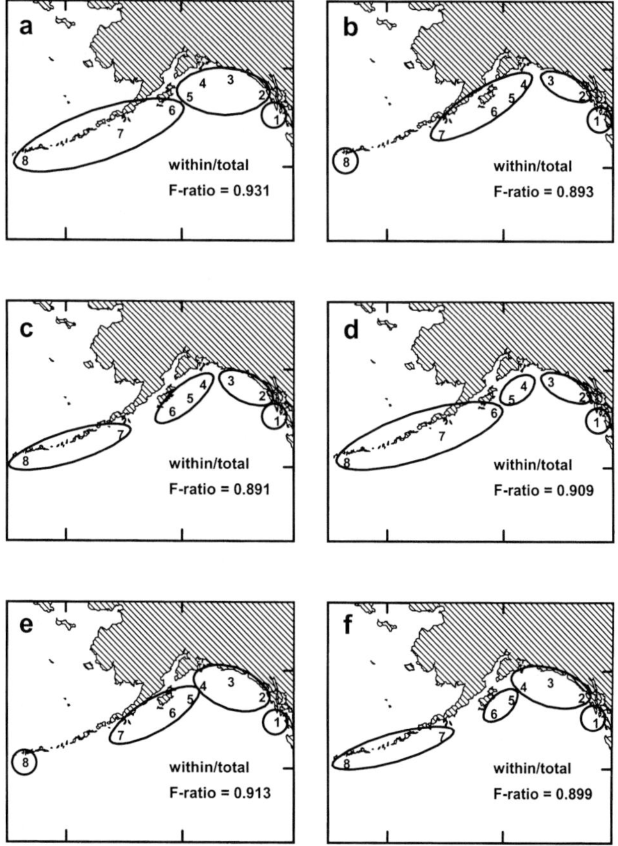

Figure 2. Partitioning schemes derived from the four homogeneous groups of populations (Figure 1) that eliminated overlap between groups. For each scheme, a population occurring in an area of overlap may be assigned to one or the other group, but not both. Shown are the six possible optimal schemes for partitioning variation among populations. The maximum efficiency of partitioning corresponds to the maximum *F*-ratio of standardized within to total log-likelihood statistics, which is in panel (a).

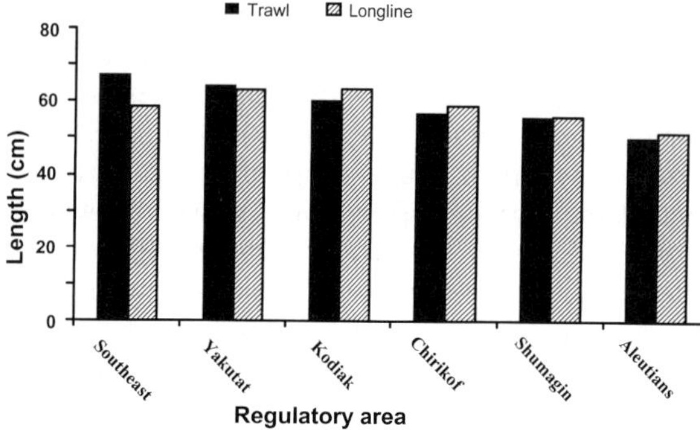

Figure 3. Mean fork length of shortraker rockfish caught in trawl surveys between 1978 and 2000 and from longline surveys between 1988 and 2000. Not all regulatory units were surveyed in each year of the trawl surveys. The management units are ordered from east to west. The decrease in fork length from east to west (trawl data) is nearly identical to that seen in the samples used for genetic analysis ($r^2 = 0.98$; p < 0.01).

influences on population structure. Fork length generally decreased from east to west for both the microsatellite data and NMFS trawl survey data (Figure 3). The trend was nearly identical between microsatellite and NMFS data sets ($r^2 = 0.98$), and the correlation was significant (Kendall's coefficient of rank correlation; p < 0.01). Although the largest mean fork lengths in the longline surveys were found in the West YAK and KOD Districts, correlation between the length data from the longline surveys and of fish analyzed for microsatellite variation was also strong ($r^2 = 0.80$). The shortest mean fork length from both trawl and longline data was observed in the far ALEU.

Discussion

This genetic analysis of shortraker rockfish provides evidence for genetic structure among collections sampled along the Pacific Rim from Southeast Alaska to the Aleutian Islands. Although the divergence did not produce a geographic pattern of clusters or isolation by distance, analyses of adjacent populations and groups of populations indicated that homogeneity among allele frequencies was restricted geographically, with the most distinct separation occurring in the Southeast Alaska region.

The interpretation of these observations is not unequivocal. There is little information about the biology and life history of shortraker rockfish. The geographic restrictions (structure) observed among shortraker rockfish may be related to the size of the

basic units of productivity if dispersal distance is limited. On the other hand heterogeneity must be explained in some other way if adult migration and larval drift are prevalent.

If neither shortraker adults nor their offspring move far from their natal sites, the wide distribution of the species suggests that there are numerous spawning areas throughout the range. These data are consistent with the assumption of limited movement, and may conform to the neighborhood model of migration, in which structure occurs within the continuum of a species if the average movement of individuals between birth and reproduction is small compared to the species range (Malecot 1950). Stanley et al. (1994) suggest limited latitudinal movement of deeper demersal rockfish along the continental shelf, although some species (including Pacific ocean perch) make seasonal bathymetric migrations. Withler et al. (2001) described a Pacific ocean perch population near Vancouver Island, B.C., which differed genetically from two Queen Charlotte Sound populations, suggesting that the divergence may have resulted from limited adult dispersal and larval retention within local currents.

Planktonic shortraker rockfish larvae may, however, have the potential to drift in oceanic currents and disperse widely. If spawning occurs at numerous sites along the Pacific Rim with high adult migration and high larval dispersal, one would expect the species to be panmictic. This is inconsistent with these data, which indicate a geographically based structure. Orlov (2001) speculated that shortraker juveniles from the eastern GOA are subject to transport in the Alaska Gyre and westward currents along the Alaskan coastline, causing them to drift and settle in the western Aleutians, and then undergo lengthy reverse migrations seeking natal reproductive grounds of the eastern GOA. If shortraker juveniles return to natal sites to mature and spawn as adults, population genetic structure may result. This structure might reflect cohorts (not geographic isolation) derived from a sweepstakes-chance effect in which a few adult spawners successfully contribute offspring each season (Hedgecock 1994, Geiger et al. 1999).

Data on shortraker rockfish fork lengths demonstrate that the average size of fish decreases significantly from east to west within the range of this study. Lunsford (1999) described a similar trend on the fork length of Pacific ocean perch at 50% maturity, showing a significant decrease of fish size from east to west through the GOA. This geographic distribution of Pacific ocean perch and shortraker rockfish sizes suggests that ecological differences may be responsible for size differences. One explanation of the geographic variation in shortraker rockfish size is restricted distribution and ecological differences among regions.

On the other hand, differences in sizes among regions could indicate that different cohorts from a single spawning area are distributed from east to west, and that the structure observed is related to the distribution of age classes. If shortraker rockfish length data reflect the presence of cohorts, then a genetic analysis based on groups arranged by size class, should reveal heterogeneity among classes. Because there were temporal differences among these collections, and length data were obtained from both trawl and longline surveys, that sample fish differently, we did not conduct this analysis.

Results of genetic analyses have management implications. Substantial harvests of shortraker rockfish are taken as incidental catch in other groundfish fisheries, and sold commercially. The observation of genetic structure among shortraker rockfish could potentially affect the size and allocation of total allowable catch among GOA and Aleutian management areas, which is currently apportioned according to the distribution of exploitable shortraker biomass. Because the sizes of genetic partitions described here reflect the scale of production units (except SSE) and are roughly as large as the GOA and the Aleutian regulatory regions, (Figures 1 and 2) it appears that present management is generally not grouping multiple productivity units. The exception in SSE should be reviewed. The structure reported here and its significance may be related to sample size and the particular loci surveyed. A broader survey including more fish, more sample sites, and additional loci may increase resolution of genetic structure.

Two divergent explanations for the observation of shortraker population genetic structure are provided, which cannot be distinguished because of a lack of biological information. However, these explanations helped identify some biological questions that need to be answered. For example, (1) What is the geographic distribution of shortraker rockfish size at age of maturity? and (2) Is there a correlation between age class and genetic structure? Answers to both questions will aid in understanding the reasons for the genetic structure detected in this study. In addition, a general life history description of shortraker rockfish (e.g. where larvae and juveniles are located in the water column, and to where juveniles disperse and ultimately settle out) is clearly important in identifying

any environmental factors that influence production, and for addressing questions of how to manage shortraker rockfish most effectively. The paucity of such information is not restricted to shortraker, and there is relatively little known about any of the northern species of *Sebastes*. Though abundance assessments can be effective in helping to understand when conservation problems arise, those data are largely ineffective in contributing to actually solving such problems. Because many rockfish species are long-lived and slow growing, it is crucial to learn about their life histories so that conservation problems can be remedied quickly, or even better avoided.

Acknowledgements

We are grateful to the NMFS scientists, survey crew members, and the University of Alaska Fairbanks personnel, who collected the samples. This work was supported by Saltonstall-Kennedy Fund project NA06FD0171 to AJG. We appreciate the constructive comments of two anonymous reviewers.

References

Ackley, D.R. & J. Heifetz. 2001. Fishing practices under maximum retainable bycatch rates in Alaska's groundfish fisheries. Alaska Fish. Res. Bull. 8: 22–44.

Cavalli-Sforza, L.L. & A.W.F. Edwards. 1967. Phylogenetic analysis: Models and estimation procedures. Evolution 21: 550–570.

Geiger, H.J., W.W. Smoker, L.A. Zhivotovsky & A.J. Gharrett. 1999. Variability of family size and marine survival in pink salmon (*Oncorhynchus gorbuscha*) has implications for conservation biology and human use. Can. J. Fish. Aquat. Sci. 54: 2684–2690.

Hedgecock, D. 1994. Does variance in reproductive success limit effective population sizes of marine organisms? pp. 122–134. *In*: A.R. Beaumont (ed.) Genetics and Evolution of Aquatic Organisms, Chapman & Hall, London.

Kendall, A.W., Jr. 1991. Systematics and identification of larvae and juveniles of the genus *Sebastes*. Environ. Biol. Fish. 30: 173–190.

Love, M.S., M.H. Carr & L.J. Haldorson. 1991. The ecology of substrate-associated juveniles of the genus *Sebastes*. Environ. Biol. Fish. 30: 225–243.

Lunsford, C.R. 1999. Distribution patterns and reproductive aspects of Pacific ocean perch (*Sebastes alutus*) in the Gulf of Alaska. M.S. Thesis, University of Alaska Fairbanks, AK. 145 pp.

Malécot, G. 1950. Quelques schémas probabilistes sur la variabilité des populations naturelles. Annales de l'Université de Lyon Sciences, Section A 13: 37–60.

Manly, B.F.J. 1985. The Statistics of Natural Selection on Animal Populations, Chapman & Hall, London & New York. 475 pp.

Mantel, N.A. 1967. The detection of disease clustering and a generalized regression approach. Cancer Res. 27: 209–220.

Mason, J.E. 1998. Declining rockfish lengths in the Monterey Bay, California, recreational fishery, 1959–1994. Mar. Fish. Rev. 60: 15–28.

McDermott, S.F. 1994. Reproductive Biology of Rougheye and Shortraker Rockfish, *Sebastes aleutianus* and *Sebastes borealis*. M.S. Thesis, University of Washington, Seattle, WA. 76 pp.

Moles, A., J. Heifetz & D.C. Love. 1998. Metazoan parasites as potential markers for selected Gulf of Alaska rockfish. Fish. Bull. 96: 912–916.

Moser, G.H. & G.W. Boehlert. 1991. Ecology of planktonic larvae and juveniles of the genus *Sebastes*. Environ. Biol. Fish. 30: 203–224.

Munk, K.M. 2001. Maximum Ages of groundfish in waters off Alaska and British Columbia and considerations of age determination. Alaska Fish. Res. Bull. 8: 12–21.

Ohta, T. 1982. Linkage disequilibrium due to random genetic drift in finite subdivided populations. Proc. Natl Acad. Sci. U.S.A. 79: 1940–1944.

Orlov, A.M. 2001. Ocean current patterns and aspects of life history of some Northwestern Pacific scorpaenids. pp. 161–184. *In*: G.H. Kruse, N. Bez, A. Booth, M.W. Dorn, S. Hills, R.N. Lipcius, D. Pelletier, C. Roy, S.J. Smith & D. Witherell (ed.) Spatial Processes and Management of Marine Populations. Pub. No. AK-SG-01–02. University of Alaska Sea Grant College Program, Fairbanks, AK.

Raymond, M. & F. Rousset. 1995a. GENEPOP (version 3.2a): Population genetics software for exact tests and ecumenicism. J. Hered. 86: 248–249.

Raymond, M. & F. Rousset. 1995b. An exact test for population differentiation. Evolution 49: 1280–1283.

Rice, W.R. 1989. Analyzing tables of statistical tests. Evolution 43: 223–225.

Rocha-Olivares, A. & R.D. Vetter. 1999. Effects of oceanographic circulation on the gene flow, genetic structure, and phylogeography of the rosethorn rockfish (*Sebastes Helvomaculatus*). Can. J. Fish. Aquat. Sci. 56: 803–813.

Roques, S., P. Duchesne & L. Bernatchez. 1999. Potential of microsatellites for individual assignment: The North Atlantic redfish (genus *Sebastes*) species complex as a case study. Mol. Ecol. 8: 1703–1717.

Sankoh, A.J., M.F. Huque & S.D. Dubey. 1997. Some comments on frequently used multiple endpoint adjustments methods in clinical trials. Stat. Med. 16: 2529–2542.

Seeb, L.W. 1998. Gene flow and introgression within and among three species of rockfishes, *Sebastes auriculatus, S. caurinus,* and *S. maliger.* J. Hered. 89: 393–404.

Seutin, G., B.N. White & P.T. Boag. 1991. Preservation of avian blood and tissue samples for DNA analysis. Can. J. Zool. 69: 82–90.

Slatkin, M. 1995. A measure of population subdivision based on microsatellite allele frequencies. Genetics 139: 457–462.

Sokal, R.R. & J.F. Rohlf. 1995. Biometry, 3rd edition, W.H. Freeman, NY. 877 pp.

Stanley, R.D., B.M. Leaman, L. Haldorson & V.M. O'Connell. 1994. Movements of tagged adult yellowtail rockfish, *Sebastes flavidus,* off the west coast of North America. Fish. Bull. 92: 655–662.

Tokranov, A.M. & L.L. Davydov. 1997. Some aspects of biology of the shortraker rockfish *Sebastes borealis* (scorpaenidae) in the Pacific waters of Kamchatka and western part of the Bering Sea: 1. spatial and bathymetric distribution. J. Ichthyol. 37: 761–768.

Weir, B.S. & C.C. Cockerham. 1984. Estimating F-statistics for the analysis of population structure. Evolution 38: 1358–1370.

Wimberger, P., J. Burr, A. Gray, A. Lopez & P. Bentzen. 1999. Isolation and characterization of twelve microsatellite loci for rockfish (*Sebastes*). Mar. Biotechnol. 1: 311–315.

Winer, B.J. 1971. Statistical Principles and Experimental Design, 2nd edition, McGraw-Hill Inc., NY. 907 pp.

Withler, R.E., T.D. Beacham, A.D. Schultze, L.J. Richards & K.M. Miller. 2001. Co-existing populations of Pacific ocean perch, *Sebastes alutus,* in Queen Charlotte Sound, British Columbia. Mar. Biol. 139: 1–12.

Wright, S. 1978. Variability within and among natural populations, Vol. 4, The University of Chicago Press, Chicago, IL. 580 pp.

Environmental Biology of Fishes **69**: 211–221, 2004.
© 2004 *Kluwer Academic Publishers. Printed in the Netherlands.*

Evidence for two highly differentiated herring groups at Goose Bank in the Barents Sea and the genetic relationship to Pacific herring, *Clupea pallasi*

Knut E. Jørstad
*Institute of Marine Research, Department of Aquaculture, Postbox 1970, Nordnes, 5817 Bergen,
Norway (e-mail: knut.joerstad@imr.no)*

Received 17 April 2003   Accepted 16 June 2003

Key words: Atlantic herring *Clupea harengus*, allozymes, intermingling of herring stocks

Synopsis

Genetic studies on Atlantic herring, *Clupea harengus*, have generally revealed a low level of genetic variation over large geographic areas. Genetically distinct herring populations in some of the Norwegian fjords are exceptions, and juvenile herring from the large oceanic herring, Norwegian Spring Spawners (NSS), are often found in mixture with local fjord populations as well as widely distributed in the Barents Sea. Research surveys in the eastern Barents Sea (Goose Bank) in 1993, 1994 and 2001 included collection of herring samples for allozyme analyses. As expected the results identified juveniles from NSS stock, but an additional unique group of herring (low vertebrae number), being almost fixed for alternative alleles at several allozyme loci, was detected. In some cases, the two groups of herring were taken in the same trawl catches as documented by highly significant departure from Hardy–Weinberg expectation with large excess of homozygotes providing evidence for population mixing. Large genetic differences (Nei's genetic distance = 1.53; F_{ST} = 0.754) were detected in pairwise comparisons based on five allozyme loci. The two herring groups were also compared with reference samples of Pacific herring, *Clupea pallasi*, including one sample from Japan Sea and three Alaskan samples. UPGMA dendrogram based on five allozyme loci revealed a close genetic relationship between the low vertebrae herring in the Barents Sea and the group of samples of Pacific herring. Although significant different in allele frequencies, one of the herring samples clustered together with the reference sample from Bering Sea with genetic distance of 0.008 and F_{ST} value of 0.032. The close genetic relationship found in this paper, suggest a re-evaluation of the taxonomic status of the Barents Sea herring populations investigated.

Introduction

Atlantic herring, *Clupea harengus*, is widely distributed over large areas at both sides of the North Atlantic Ocean. Detailed studies on life history characteristics, migration behaviour as well as morphometrics and meristics have revealed large variation between geographic areas and suggest a complicated population structure for this species (Svetovidov 1963, Parrish & Saville 1965, Blaxter 1985). This general picture has not been confirmed by a number of allozyme studies that revealed low levels of genetic differentiation (Anderson et al. 1981, Kornfield et al. 1982,

Grant 1984, Ryman et al. 1984, King et al. 1987). The genetic studies of some Norwegian fjord areas (Jørstad & Nævdal 1981, Jørstad et al. 1991), also based on polymorphic allozymes, contrast with this picture demonstrating the existence of a number of genetically distinct fjord populations of herring.

The first samples of herring from Balsfjord in northern Norway were analyses as early as 1981–1983 (Jørstad & Nævdal 1981), and a more detailed investigation was carried out in 1986 (Jørstad & Pedersen 1986). The results from the genetic analyses demonstrated large genetic differences between the Balsfjord herring and large oceanic population of herring, the Norwegian Spring Spawning (NSS) stock. For several allozyme loci the two herring groups were nearly fixed for alternative alleles at the same locus. The large NSS stock generally spawns at offshore banks along the Norwegian coast and in strong year classes the juveniles are spread and distributed in the Barents Sea as well as in Norwegian coastal and fjord areas (Dragesund et al. 1980, Røttingen 1990).

In contrast to the NSS stock the Balsfjord herring was characterised by intertidal spawning on vegetation at a particular locality in the bottom of the Balsfjord. Such intertidal spawning behaviour is more similar to the spawning strategy of Pacific herring, *Clupea pallasi*, and the herrings occurring in the White Sea and Russian coastal areas (Svetovidov 1963). Based on morphometric and meristic studies as well as similarities in spawning strategies, this herring was considered as subspecies of *C. pallasi*, but this taxonomy has been controversial (see discussion in Novikov et al. 2001). The biological information available suggested some kind of linkage with Pacific herring. For this reason a sample of herring from British Columbia was provided, transported to Norway and directly compared with samples of herring from the NSS stock and the Balsfjord. The results demonstrated a closer genetic relationship between the Balsfjord and Pacific herrings (genetic distance = 0.069), while the value between Balsfjord herring and NSS stock was 0.926, all based on five polymorphic allozyme loci (Jørstad et al. 1994). As expected, the Balsfjord herring and the sample of Pacific herring were clustered together in an UPGMA dendrogram.

Overall the genetic and biological information suggested a linkage between Balsfjord and Pacific herring occurring in Russian areas (Svetovidov 1963) and in the Pacific. Annually, the Institute of Marine Research in Bergen conducts research surveys in the Barents Sea to estimate distribution and abundance of juveniles of commercially exploited fish species. These monitoring also include herring juveniles from the oceanic NSS stock, which can be distributed in the eastern part of Barents Sea (Røttingen 1990). Occasionally, mature herring specimens of small sizes are also found in this area, especially at Goose Bank near Novaya Zemlya. Early maturation at this size (total length of about 20 cm) is not known for the NSS stock, and the observations, therefore, suggested these fish to belong to a different herring group.

The purpose of this work was to investigate the genetics of the herrings found in this area. Further,

based on the observations of Jørstad et al. (1994), it was important to estimate the genetic relationship with herrings from the Pacific Ocean. Genetic studies were initiated in 1993 in connection with the Norwegian investigations in the eastern Barents Sea. Sampling were carried out over 3 years, and the results from the genetic analyses are presented in this paper together with four reference samples of Pacific herring, *C. pallasi*.

Materials and methods

Complete mapping of geographic distribution of fish juveniles in the Barents Sea is depending of access into the Russian Economic Zone in the eastern part. In the time period in question, the Russian authorities only gave permission to Institute of Marine Research conduct research surveys in this zone in 1993, 1994 and 2001, and samples from herring were collected during winter surveys using R/V 'Johan Hjort'. The surveys were conducted in February and mapped the distribution of juvenile fish in the area, including estimates of yearclass strength of the NSS herring stock and Northeast Arctic cod, *Gadus morhua*.

In addition to measurements of abundance by echo registration, fish samples are collected with pelagic (PT) and bottom trawls (BT). Both gears caught herring, and white tissues were taken for genetic analyses, mainly from selected trawl stations in eastern part covering the Goose Bank near Novaya Zemlya (Figure 1). Where possible, samples from 96 specimens in each trawl catch, were taken for allozyme analyses. The low sample sizes in some of the cases simply reflect low catches of herring in those particular trawl hauls. Details about samples and locations are given in Table 1, and the trawl stations are shown in Figure 1. Designation of samples was recorded to research vessel, year and trawl station. Thus 'JH93ST10' is the sample collected by R/V 'Johan Hjort' in 1993 from the catch taken at trawl station number 10.

Samples were analysed for 2 years on board during surveys using a specially developed starch gel electrophoresis apparatus. Fresh samples gave excellent enzyme activities and clear banding patterns. Starch gel electrophoresis was carried out using a histidine buffer pH = 7.0. The enzymes stained for included phosphoglucomutase (PGM), glucosephosphate isomerase (GPI), lactate dehydrogenase (LDH) and maleate dehydrogenase (MDH), all of which have been described

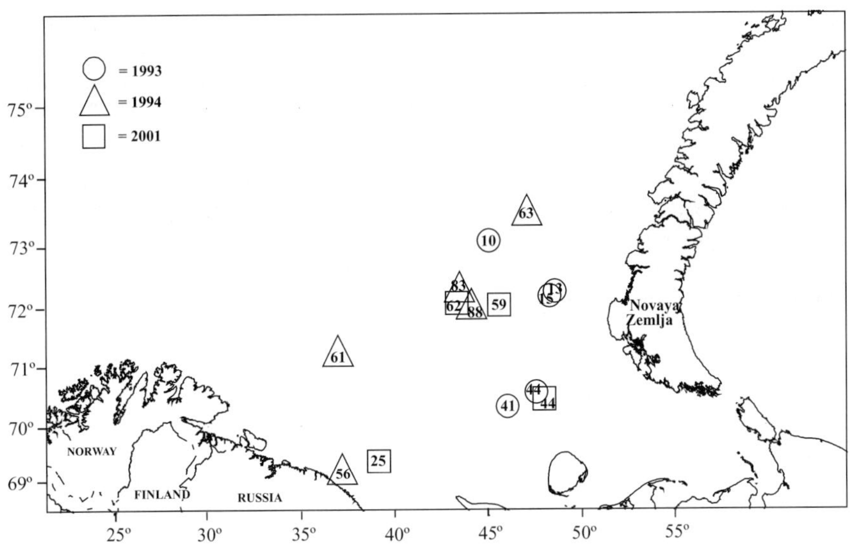

Figure 1. Sampling stations of herring in eastern Barents Sea. The surveys were conducted by R/V 'Johan Hjort' in February 1993, 1994 and 2001. The Goose Bank area is west of Novaya Zemlya. The numbers in symbols correspond to the trawl station number (see Materials and methods section; Table 1) for the three different years.

Table 1. Samples of herring collected in eastern Barents Sea and reference samples of Pacific herring (*C. pallasi*).

Sample no.	Sample code	Area/location	Sampling year	Sample size	Gear	Sample source
1	JH93ST10	Barents Sea/N 73°07′ E 45°05′	February 1993	96	PT	A
2	JH93ST13	Barents Sea/N 72°28′ E 48°59′	February 1993	96	PT	A
3	JH93ST15	Barents Sea/N 72°18′ E 48°15′	February 1993	96	BT	A
4	JH93ST41	Barents Sea/N 70°34′ E 46°02′	February 1993	50	BT	A
5	JH93ST44	Barents Sea/N 70°59′ E 47°58′	February 1993	50	BT	A
6	JH94ST56	Barents Sea/N 69°21′ E 37°20′	February 1994	96	PT	A
7	JH94ST61	Barents Sea/N 71°31′ E 37°20′	February 1994	47	BT	A
8	JH94ST63	Barents Sea/N 73°54′ E 47°11′	February 1994	96	BT	A
9	JH94ST83	Barents Sea/N 72°35′ E 43°48′	February 1994	96	BT	A
10	JH94ST88	Barents Sea/N 72°08′ E 44°12′	February 1994	96	BT	A
11	JH01ST62	Barents Sea/N 72°09′ E 43°37′	February 2001	96	BT	A
12	JH01ST59	Barents Sea/N 72°06′ E 45°59′	February 2001	84	PT	A
13	JH01ST44	Barents Sea/N 70°47′ E 48°01′	February 2001	96	BT	A
14	JH01ST25	Barents Sea/N 69°35′ E 39°15′	February 2001	86	PT	A
15	R1-HPETB	Southern Alaska/Petersburg	2000	96		B
16	R2-HPWSE	Gulf of Alaska/Prince Williams Sound	1995	96		C
17	R3-HNORT	Bering Sea/Unalakleet	1996	96		C
18	R4-JAPANSEA	Japan Sea/Sakhalin	1994	60		D

BT = bottom trawl; PT = pelagic trawl. (A) R/V 'Johan Hjort', Institute of Marine Research, Bergen. (B) Dr. Robert Larson, Alaska Department of Fish and Games, Petersburg. (C) Drs. Lisa Seeb and James Seeb, Alaska Department of Fish and Games, Gene Conservation Laboratory, Anchorage. (D) Prof. Yuri P. Altukhov, Vavilov Institute of General Genetics, Moscow.

earlier and used in population studies (Anderson et al. 1981, Jørstad & Nævdal 1981, Ryman et al. 1984), and a detailed comparison of alleles and loci is presented in Jørstad et al. (1991). All samples collected in the 3 years were screened for the enzymes mentioned following the protocol described in Jørstad et al. (1991), provided information at five polymorphic loci, *PGM-1**, *GPI-2**, *LDH-1**, *LDH-2** and *MDH-4**. After electrophoresis

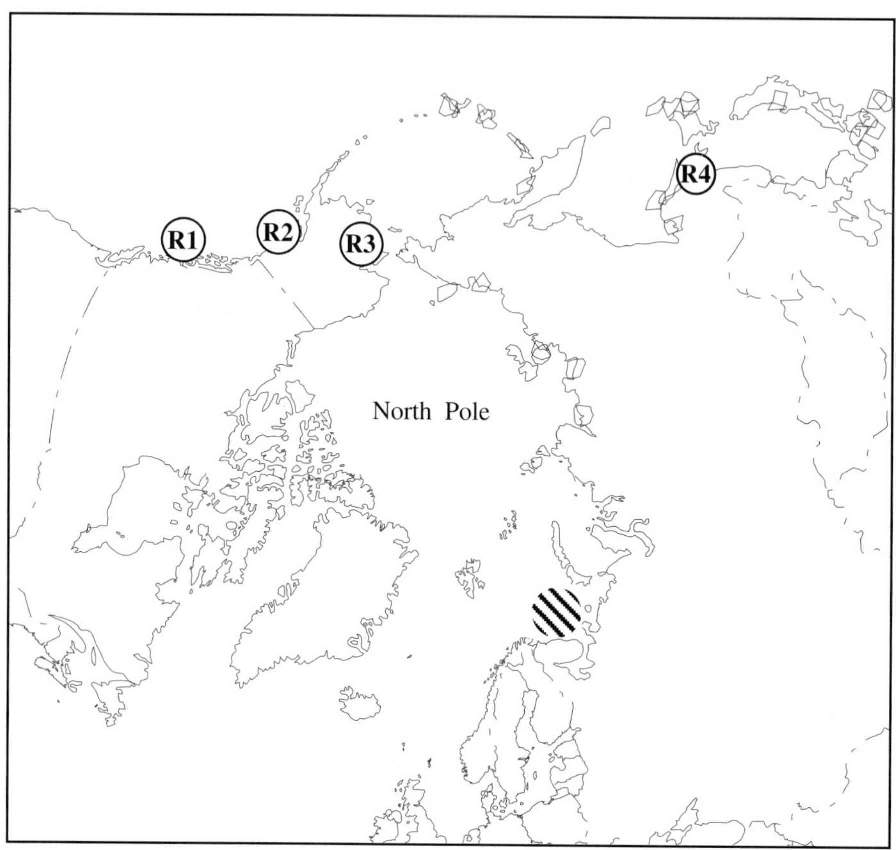

Figure 2. Location of reference samples (R1–R4) of herring from Pacific Ocean. The eastern Barents Sea region is also indicated.

and selective staining, the gels were dried for permanent storage. A picture of a dried gel obtained from analyses of LDH is shown in Figure 3. During the survey in 2001 herring samples from the same area were frozen on the research vessel and transferred to Bergen where they were analysed using the same protocols. The overall material (all years) from the Barents Sea consisted of 14 separate trawl stations and 1196 specimens.

Reference samples of Pacific herring (*C. pallasi*) were provided from several regions in the Pacific Ocean (Figure 2). Frozen samples of white muscle were provided through various contacts (see Table 1). The samples represented 340 specimens collected from four different regions. The samples were all transported frozen to Bergen and analyzed as described above.

A number of computer program packages were used to analyse the data obtained. Allele frequencies were calculated and tested for Hardy–Weinberg equilibrium by using Fisher's exact test available in the GENEPOP (version 3.3, March 2001) of Raymond & Rousset (1995). The tests were carried out with 1000 iterations per batch. The same program was used in testing for population differentiation (global and pairwise tests) as well as in pairwise F_{ST} comparisons. Nei's genetic distance (1972) between the samples was calculated using BIOSYS-2 (Swofford & Selander, 1981) modified by Black (1997[1]). The cluster analyses and the UPGMA dendrograms were obtained using TFPGA 1.3 (Miller 1997[2]).

---

[1] Black, W.C. 1997. BIOSYS-2. A computer program for the analysis of allelic variation in genetics (wcb@lamar.colostate.edu).

[2] Miller, M. 1997. Tools for population genetic analyses (TFPGA) 1.3: a window program for the analyses of allozyme and molecular population genetic data. Computer software distributed by the author.

Results

Genetics of herring in Barents Sea

The first results from the starch gel electrophoresis carried out on board the R/V 'Johan Hjort' in 1993 were very surprising. In some of the catches the herring were nearly fixed for alternative alleles compared to the common alleles found in NSS stock and elsewhere in the Atlantic (Jørstad et al. 1991). Table 2 summarize the allele frequencies found at the five polymorphic loci analysed. The herring taken in pelagic trawls (JH93ST10; JH93ST13) consisted of juveniles, which have typical allele frequencies of herring belonging to the NSS stock ($LDH-1^*100 = 0.979$; $LDH-2^*100 = 0.916$, respectively). For two of the samples collected in 1993 (JH93ST10 and JH93ST13), the allele frequencies at all loci analysed correspond to the frequencies earlier reported for the NSS stock (Jørstad et al. 1991, 1994). On the other hand, the herring taken in two of bottom trawls (JH93ST15 and JH93ST44) were almost fixed for alternative alleles at the two LDH loci mentioned above, corresponding to the common alleles detected in samples of Pacific herring (see Jørstad et al. 1994). The JH93ST44 sample was fixed for the $LDH-1^*200$ allele (1.000) and nearly fixed for the $LDH-2^*120$ allele (0.980), both alleles not found in samples of NSS stock. Banding patterns of the most discriminating loci ($LDH-1^*$ and $LDH-2^*$) are shown in Figure 3.

Even more surprising was the results obtained when analysing the herring taken at some of the bottom trawls. In Figure 3, the banding patterns observed in the samples from JH93ST41 are compared with JH93ST10 and JH93ST44. Obviously, the JH93ST41 sample consisted of a mixture of the two very different herring groups occurring in Goose Bank area. Based on these findings it was not surprising that the JH93ST41 sample departed from Hardy–Weinberg expectations. Actually there were no heterozygotes at all for the two alleles in question, providing an extreme case of population (or species) mixing, as first discussed by Wahlund (1928).

A similar situation was found in the samples collected in 1994 and 2001. In 1994 three samples (JH94ST56, JH94ST61 and JH94ST83) have alleles frequencies corresponding to the NSS herring stock, while two samples (JH94ST63 and JH94ST) have high frequencies of $LDH-1^*200$ and $LDH-2^*120$ (Table 2). Three samples from 2001 have genetic characteristic of the NSS stock, and one sample (JH01ST44) have high frequencies of $LDH-1^*200$ and $LDH-2^*120$.

Table 3 is summarizes the results from tests for HW equilibrium for all herring samples analysed for the five allozyme loci. No deviations from HW were detected at any allozyme loci for any of the samples from the NSS juveniles. Also, the JH93ST44 sample, nearly fixed for other alleles, shows no deviation from HW. In all three sample collection from the Goose Bank in the Barents Sea there were samples (JH93ST15, JH93ST41, JH94ST63, JH94ST88 and JH01ST44), which significantly deviate from HW even under the most conservative Bonferroni correction of significance level (Rice 1989). This provides evidence for physical mixing of herring specimens from two highly genetically differentiated populations in this particular area. Overall, 5 of the 14 samples demonstrated significant deviation from HW, usually for all loci investigated. Only two exceptions were seen in the in the MDH^* locus (JH93ST41; P = 0.3733 and JH94ST63; P = 0.3471; Table 3).

The allele frequencies at the five loci are summarised in Table 2. The largest variation were detected for the $LDH-1^*$ and $LDH-2^*$ loci, and the most extreme differences were found for two samples from the 1993 dataset, JH93ST10 and JH93ST44, being nearly fixed for alternative alleles. Based on the data obtained for all five allozyme loci, Nei's genetic distance (D) between the two samples was estimated to 1.532 and the overall pairwise F_{ST} value was 0.747 (Table 4). Similar results were found for the two other years, JH94ST56/JH94ST63 (D = 0.770; F_{ST} = 0.513) and JH01ST59/JH01ST44 (D = 1.053; F_{ST} = 0.597). And as expected, the pairwise genetic differences between all three samples pairs from the 3 years were all highly significant. No significant differences were found in testing the typical NSS samples (JH93ST10, JH94ST56 and JH01ST59) in different year, resulting in low values of genetic distance and F_{ST} in the different pairwise comparisons (Table 4).

Comparisons with Pacific herring

As presented above the samples of herring from the Barents Sea constituted juveniles with genetic characteristics of the NSS stock and herring taken in bottom trawls which almost fixed for alternative alleles at the $LDH-1^*$ and $LDH-2^*$ loci. In addition, some samples obviously consisted of a mixture of these groups. In

Table 2. Allele frequencies estimated for five polymorphic allozymes in herring samples from eastern Barents Sea and reference samples of Pacific herring (*C. pallasi*).

Sample no.	Sample code	N	LDH-1*			LDH-2*			PGM-1*			MDH-4*		GPI-2*					
			100	160	200	70	100	120	30	100	120	70	100	30	50	70	100	150	180
1	JH93ST10	95	0.979	0.021	0.000	0.084	0.916	0.000	0.000	0.953	0.047	0.405	0.595	0.032	0.058	0.100	0.679	0.132	0.000
2	JH93ST13	96	0.953	0.047	0.000	0.042	0.958	0.000	0.005	0.948	0.042	0.521	0.479	0.010	0.094	0.078	0.708	0.109	0.000
3	JH93ST15	96	0.021	0.005	0.974	0.000	0.073	0.927	0.000	0.224	0.776	0.005	0.995	0.000	0.005	0.005	0.021	0.589	0.380
4	JH93ST41	49	0.673	0.000	0.327	0.082	0.592	0.327	0.000	0.704	0.296	0.367	0.633	0.010	0.051	0.061	0.500	0.276	0.102
5	JH93ST44	50	0.000	0.000	1.000	0.000	0.020	0.980	0.000	0.110	0.890	0.020	0.980	0.000	0.010	0.000	0.020	0.730	0.240
6	JH94ST56	96	0.979	0.021	0.000	0.064	0.931	0.005	0.000	0.953	0.031	0.438	0.563	0.021	0.094	0.104	0.698	0.083	0.000
7	JH94ST61	47	0.979	0.021	0.000	0.085	0.915	0.000	0.000	0.948	0.052	0.500	0.500	0.000	0.073	0.042	0.781	0.104	0.000
8	JH94ST63	96	0.234	0.005	0.760	0.016	0.292	0.693	0.005	0.313	0.677	0.115	0.885	0.010	0.057	0.036	0.146	0.479	0.266
9	JH94ST83	96	0.984	0.016	0.000	0.042	0.958	0.000	0.005	0.964	0.026	0.411	0.589	0.016	0.141	0.073	0.698	0.073	0.000
10	JH94ST88	96	0.495	0.005	0.500	0.036	0.469	0.495	0.000	0.557	0.443	0.229	0.771	0.016	0.026	0.063	0.354	0.391	0.146
11	JH01ST62	96	0.963	0.032	0.005	0.083	0.917	0.000	0.000	0.984	0.016	0.438	0.563	0.005	0.094	0.099	0.688	0.115	0.000
12	JH01ST59	84	0.958	0.042	0.000	0.060	0.940	0.000	0.000	1.000	0.000	0.482	0.518	0.012	0.095	0.101	0.673	0.119	0.000
13	JH01ST44	96	0.161	0.021	0.818	0.005	0.182	0.813	0.000	0.229	0.766	0.125	0.875	0.016	0.016	0.042	0.109	0.563	0.255
14	JH01ST25	80	1.000	0.000	0.000	0.038	0.962	0.000	0.006	0.971	0.023	0.512	0.488	0.000	0.110	0.116	0.674	0.099	0.000
15	R1-HPETB	96	0.000	0.010	0.990	0.000	0.000	1.000	0.005	0.745	0.250	0.073	0.927	0.000	0.000	0.000	0.010	0.818	0.172
16	R2-HPWSE	96	0.000	0.005	0.995	0.000	0.000	1.000	0.000	0.703	0.297	0.052	0.948	0.000	0.000	0.000	0.005	0.839	0.156
17	R3-HNORT	95	0.000	0.000	1.000	0.000	0.000	1.000	0.011	0.277	0.713	0.005	0.995	0.000	0.000	0.000	0.000	0.698	0.302
18	R4-JAPANSEA	60	0.000	0.000	1.000	0.000	0.000	1.000	0.000	0.308	0.692	0.000	1.000	0.000	0.000	0.000	0.008	0.742	0.250

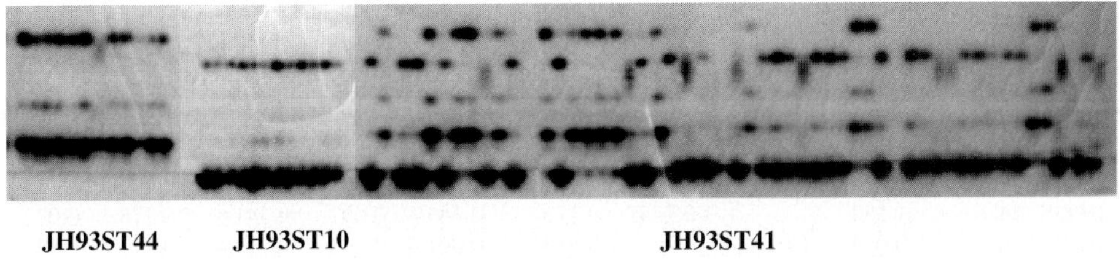

Figure 3. Banding pattern observed in herring samples after starch gel electrophoresis and staining for LDH. Sample code for the trawl stations are given (see Table 1 and Figure 1). The banding patterns observed in sample JH93ST10 are typical for the NSS herring, while the herring in sample JH93ST44 are almost fixed for alternative alleles both for the *LDH-1** and *LDH-2** loci. The sample to the right (JH93ST41) consisted of a mixture of two different herring groups, and these fish were caught in the same bottom trawl haul.

Table 3. Tests for Hardy–Weinberg equilibrium (P-values) in samples of herring collected by R/V 'Johan Hjort' in the eastern Barents Sea during February in 1993, 1994 and 2002.

Sample no.	Sample code	N	p-values				
			*LDH-1**	*LDH-2**	*PGM-1**	*MDH-4**	*GPI-2**
1	JH93ST10	95	1.0000	0.1210	1.0000	0.5303	0.5472
2	JH93ST13	96	1.0000	1.0000	1.0000	0.8378	0.3331
3	JH93ST15	96	0.0001	0.0706	0.0077	n.p.	0.0027
4	JH93ST41	49	0.0000	0.0000	0.0000	0.3733	0.0000
5	JH93ST44	50	n.p.	1.0000	1.0000	1.0000	0.6905
6	JH94ST56	96	1.0000	1.0000	1.0000	0.5351	0.8949
7	JH94ST61	47	1.0000	1.0000	1.0000	0.1472	0.6818
8	JH94ST63	96	0.0000	0.0000	0.0000	0.3471	0.0000
9	JH94ST83	96	1.0000	1.0000	1.0000	1.0000	0.9756
10	JH94ST88	96	0.0000	0.0000	0.0000	0.0071	0.0000
11	JH01ST62	96	1.0000	1.0000	1.0000	0.8363	0.3159
12	JH01ST59	84	1.0000	1.0000	n.p.	0.1887	0.0989
13	JH01ST44	96	0.0000	0.0000	0.0000	0.0005	0.0000
14	JH01ST25	80	n.p.	1.0000	1.0000	0.6661	0.3538
15	R1-HPETB	96	1.0000	n.p.	0.6944	0.3997	0.0754
16	R2-HPWSE	96	n.p.	n.p.	0.2240	1.0000	0.1996
17	R3-HNORT	95	n.p.	n.p.	0.0002	n.p.	0.4748
18	R4-JAPANSEA	60	n.p.	n.p.	1.0000	n.p.	0.1847

Reference samples from Pacific herring (*C. pallasi*) are included. The p-values are given for each of the five allozyme loci. (n.p. = no polymorphism).

the genetic comparisons with Pacific herring, two samples representing the two main groups were selected from each year of investigation (1993: JH93ST10 and JH93ST44. 1994; JH94ST56 and JH94ST63. 2001: JH01ST59 and JH01ST44).

The allele frequencies for the five loci for the reference samples of Pacific herring are given in Table 2. For the *LDH-1** and *LDH-2** loci all reference samples (R1–R4) were fixed or almost fixed for the same alleles dominating the JH93ST44 sample from the Barents Sea. When using this sample from the Barents Sea in pairwise comparisons with the reference samples, the lowest values were obtained in comparisons with the sample from the Bering Sea (R3-HNORT; $D = 0.008$; $F_{ST} = 0.032$) and the sample from the Japan Sea (R4-JAPANSEA; $D = 0.009$; $F_{ST} = 0.042$). The pairwise tests for population differentiation, however, were significant ($p = 0.00065$ and 0.00775). In contrast, comparisons of the reference samples from the Pacific with the samples of NSS stock from the

Table 4. Comparisons of Nei's (1972) genetic distance (below diagonal) and pairwise F_{ST} (above diagonal) between selected herring samples from Barents Sea and reference samples of Pacific herring (*C. pallasi*).

Sample no.	Sample code	1	5	6	8	12	13	15	16	17	18
1	JH93ST10		0.745	−0.002	0.501	0.003	0.581	0.703	0.711	0.735	0.726
5	JH93ST44	1.532		0.754	0.113	0.764	0.057	0.341	0.318	0.032	0.042
6	JH94ST56	0.001	1.627		0.513	−0.001	0.592	0.713	0.721	0.744	0.736
8	JH94ST63	0.735	0.045	0.770		0.518	0.007	0.208	0.200	0.111	0.104
12	JH01ST59	0.003	1.689	0.001	0.796		0.597	0.720	0.728	0.752	0.745
13	JH01ST44	0.973	0.016	1.020	0.009	1.053		0.215	0.202	0.063	0.062
15	R1-HPETB	1.022	0.103	1.074	0.105	1.069	0.098		−0.002	0.214	0.192
16	R2-HPWSE	1.051	0.089	1.107	0.096	1.105	0.087	0.001		0.190	0.167
17	R3-HNORT	1.387	0.008	1.468	0.041	1.510	0.019	0.059	0.049		−0.003
18	R4-JAPANSEA	1.349	0.009	1.429	0.043	1.466	0.021	0.050	0.041	0.001	

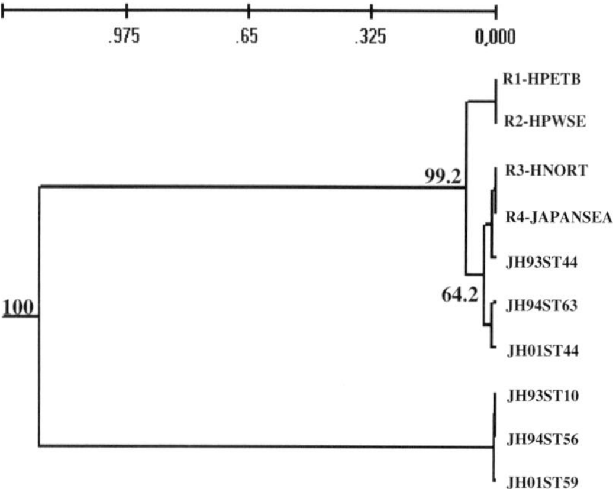

Figure 4. UPGMA dendrogram based on genetic distance (Nei 1972) for five polymorphic allozyme loci for the reference samples of Pacific herring and selected samples (see text) from the Barents Sea area. Bootstrap values (in percentage of 1 000 permutations) are given for major branches.

Figure 5. UPGMA dendrogram of genetic distance (Nei 1972) for reference samples of Pacific herring and selected samples (JH93ST44, JH94ST63 and JH01ST44) from Barents Sea, all based on five polymorphic loci. Bootstrap values (percentage of 1 000 permutations) are given.

Barents Sea (JH93ST10, JH94ST56 and JH01ST44) revealed high genetic distances as well as pairwise F_{ST} values (Table 4).

An UPGMA overall dendrogram of genetic distance (Nei 1972) based on allele frequencies at all five allozyme loci is presented in Figure 4. There are two main branches in the dendrogram, corresponding to for the Atlantic herring (*C. harengus*), represented by the samples of NSS stock in the Barents Sea ((JH93ST10, JH94ST56 and JH01ST59), and the group of reference samples of Pacific herring (*C. pallasi*). Three of the selected samples of herring from the Barents Sea

(JH93ST44; JH94ST63; JH01ST44) cluster together with the Pacific herring group. Based on the same data a more detailed dendrogram of the Pacific herring group is given on Figure 5. The two samples of herring (R1-HPETB; R2-PWSE) in the Gulf of Alaska cluster together and are also genetically different from the sample from Bering Sea (R3-HNORT) and Japan Sea (R4-JAPANSEA), as demonstrated in pairwise test for genetic differentiation at five polymorphic loci. No significant differentiation was detected between R3-HNORT and R4-JAPANSEA. Although significantly different (see above), the JH93ST44 sample clusters with R3-HNORT and R4-JAPANSEA. The two other samples from Barents Sea (JH93ST44; JH94ST63) cluster together in an intermediate branch. In contrast with the JH93ST44 sample, both these samples deviate from HW expectation at all loci presumably due to mixing of some specimens from the NSS stock.

Discussion

The data presented clearly demonstrate that two highly genetically differentiated groups of herring exist in the eastern Barents Sea in the northeast Atlantic. The groups are fixed or nearly fixed for alternative alleles at several allozymes, and the highest genetic distance observed (Nei 1972) was approximately 1.5. One group was dominated by juveniles and belongs to the NSS stock as seen by comparing the observed allele frequencies with values reported earlier (Jørstad et al. 1991, 1994). The other group of herring possibly belong to one of the 'low vertebrae' herring groups found in Russian seas including the White Sea and the north-western coastal areas. According to Svetovidov (1963) the herring populations in this region are considered as a subspecies of Pacific herring, *C. pallasi*. The taxonomy of the herring group in this region is controversial, and for a recent discussion of the problems see Novikov et al. (2001).

The results obtained by analysing herring samples from the Goose Bank in the Barents Sea, clearly demonstrate a Barents Sea herring stock with close genetic relationship to four different samples of Pacific herring (*C. pallasi*), ranging from Japan Sea, through Bering Sea and to southern Gulf of Alaska. Herring caught during February are possibly in the over wintering areas, and as demonstrated here, are often found in mixture with NSS juveniles. Considering the 14 herring samples analysed in the 3 years of investigation in the eastern Barents Sea, five samples deviated from HW expectation with strong deficiency of heterozygotes, demonstrating mixing of specimens from two highly genetically differentiated herring groups. Samples taken on the spawning site in Russian coastal areas during the actual spawning will possibly provide more correct estimates of the genetic relationship with Pacific herring. Such work has now being carried out in joint investigation between Norwegian and Russian institutions (Jørstad et al. 2001).

The mixing of two different herring populations in the same trawl catch was observed in all 3 years

of our investigation. As the two groups were nearly fixed for alternative alleles at the *LDH-1* and *LDH-2* loci, a strong departure from Hardy–Weinberg expectations were found. The use of Bonferroni corrections of significance level in multiple tests has recently been evaluated (Ryman & Jorde 2001). In herring case described here, several of the samples from Goose Bank showed significant departure even under the most conservative Bonferroni correction (Rice 1989). This is clearly an evidence that the two groups are overlapping in the eastern geographic areas investigated, but it is unclear if they are actually mixed in the same school. The 'pure' samples of the NSS juveniles were always taken in pelagic trawls, while the 'pure' catches of the other group were caught in bottom trawls. The mixed samples were also from bottom trawls, suggesting that pelagic juveniles from NSS stock could have been caught when lifting the trawl from the bottom and into the vessel. Such a situation has actually been observed in detailed investigation in Balsfjord, where juveniles from NSS stock were found pelagic and the local Balsfjord herring stock was detected at the bottom (Jørstad & Pedersen 1987).

The early genetic studies of herring based on allozymes include Pacific herring (Grant & Utter 1984) as well as Atlantic herring (Anderson et al. 1981, Jørstad & Nævdal 1981, Kornfield et al. 1982, Grant 1984, 1986). Of particular interest with regards to the work presented here, Grant (1986) compared Atlantic and Pacific herring for a large number of allozyme loci. Based on the genetic distance obtained he suggested that the two groups should be considered as species and not as subspecies as suggested by Svetovidov (1963). Both the high genetic distance obtained in comparison between Atlantic and Pacific herring from British Columbia (Jørstad et al. 1994) and the work presented here support the conclusions of Grant (1986). The differences found between the two species in allozyme comparisons are also in contrast with recent microsatellite data provided by Shaw et al. (1999). They found a high estimate (0.743) of F_{ST} (mean of five allozyme loci) which is comparable to the value (about 0.80) found in this study. The F_{ST} value obtained in the same work for five microsatellite loci was, on the other hand, only 0.032. Thus several studies, including Grant (1986), Jørstad et al. (1994), Shaw et al. (1999) and work presented here, demonstrate that allozymes, in this study especially the *LDH-1* and *LDH-2* loci, are so far the most discriminating genetic markers between the two species. The genetic differences found in this study between the sample from the Bering Sea and the

Gulf of Alaska is in agreement with earlier work on allozymes from the region (Grant & Utter 1984) as well as recent microsatellite investigation (O'Connel et al. 1998).

The genetic relationship between the Balsfjord herring (Jørstad et al. 1991, 1994, 2001) and the herring populations in the White Sea (Svetovidov 1963, Novikov et al. 2001) and Russian coastal waters has yet to be analysed in more detail. At present no direct comparison with samples have been made on the same starch gels to prove identities of alleles at the allozyme loci in question. In addition the distribution of herring in the eastern parts of Russian coastal area should be further investigated, including genetic analyses of samples from this area. A more complete data collection should be considered both for establishing detailed herring population structure for the management of herring fisheries, but also as basic for a re-evaluation of the taxonomic status of the herrings in the northeast Atlantic.

Acknowledgement

I am indebted to the staff of Institute of Marine Research for helping with sample collection during the three research vessel surveys of R/V 'Johan Hjort'. Ole Ingar Paulsen assisted in 2001 laboratory analyses, Eva Farestveit and Vidar Wennevik helped with data treatment. The samples of Pacific herring, *C. pallasi*, were kindly provided by Lisa Seeb and Jim Seeb Alaska Department of Fish and Games (ADFG); Anchorage, and Robert Larsen (ADFG), Petersburg) provided the sample of herring taken near Petersburg in southern Alaska. Professor Yuri P. Altukhov, Vavilov Institute of General Genetics in Moscow, Russia, organized the sampling of herring from Japan Sea. I am also indebted to professor Georgij G. Novikov and his colleges at Moscow State University, Moscow, for numerous discussions on herring population structure. Norwegian Ministry of Foreign Affairs supported the work financially.

References

Anderson, L., N. Ryman, R. Rosenberg & G. Stahl. 1981. Genetic variability in Atlantic herring (*Clupea harengus*): Description of protein loci and population data. Hereditas 95: 69–78.

Blaxter, J.H.S. 1985. The herring: A successful species? Can. J. Fish. Aquat. Sci. 42(Suppl. 1): 21–30.

Dragesund, O., J. Hamre & Ø. Ulltang. 1980. Biology and population dynamics of the Norwegian Spring-Spawning herring. Rapp. P.-v. Reun. Cos. Int. Explor. Mer 177: 43–71.

Grant, W.S. 1984. Biochemical population genetics of Atlantic herring *Clupea harengus*. Copeia 1984: 357–364.

Grant, W.S. 1986. Biochemical genetic divergence between Atlantic, *Clupea harengus*, and Pacific, *Clupea pallasi*, herring. Copeia 1986: 714–719.

Grant, W.S. & F.M. Utter. 1984. Biochemical genetics of Pacific herring (*Clupea pallasi*). Can. J. Fish. Aquat. Sci. 41: 856–864.

Jørstad, K.E. & G. Nævdal. 1981. Significance of population genetics on management of herring stocks. Council Meet. Cons. Int. Explor. Sea 1981/H:64.

Jørstad, K.E. & S.A. Pedersen. 1986. Discrimination of herring populations in a Northern Norwegian fjord: Genetic and biological aspects. Council Meet. Cons. Int. Expor. Sea. 1986/M:63.

Jørstad, K.E., D.P.F. King and G. Nævdal. 1991. Population structure of Atlantic herring, *Clupea harengus* L. J. Fish Biol. 39(Suppl. A): 43–52.

Jørstad, K.E., G. Dahle & O.I. Paulsen. 1994. Genetic comparison between Pacific herring *Clupea pallasi* and a Norwegian fjord stock of Atlantic herring *Clupea harengus*. Can. J. Fish. Aquat. Sci. 51(Suppl. 1): 233–239.

Jørstad, K.E., G.G. Novikov, N.J. Stasenkov, I. Røttingen, V.A. Stasenkov, V. Wennevik, A.N. Golubev, O.I. Paulsen, A.K. Karpov, L. Telitsina, A. Andreeva & A.N. Stroganov. 2001. Intermingling of herring stocks in the Barents Sea. pp. 629–633. *In*: F. Funk, J. Blackburn, D. Hay, A.J. Paul, R. Stevenson, R. Toresen & D. Witherell (ed.) Herring: Expectations for a New Millennium, University of Alaska Sea Grant, AK-SG-01-04, Fairbanks.

King, D.P.F., A. Ferguson & I.J.J. Moffett, 1987. Aspects of population genetics of herring, *Clupea harengus*, around the British Isles and in the Baltic Sea. Fish. Res. (Amst.) 6: 35–52.

Kornfield, I., B.D. Sidell & P.S. Gagnon. 1982. Stock definition in Atlantic herring (*Clupea harengus*): Genetic evidence for discrete fall and spring spawning populations. Can. J. Fish. Aquat. Sci. 39: 1610–1621.

Nei, M. 1972. Genetic distance between populations. Amer. Nat. 106: 283–292.

Novikov, G.G., A.K. Karpov, A.P. Andreeva & A.V. Semenkova. 2001. Herring of the White Sea. pp. 591–597. *In*: F. Funk, J. Blackburn, D. Hay, A.J. Paul, R. Stevenson, R. Toresen & D. Witherell (ed.) Herring: Expectations for a New Millennium, University of Alaska Sea Grant, AK-SG-01-04, Fairbanks.

O'Connell, M., M.C. Dillon, J.M. Wright, P. Benzen, S. Mercouris & J. Seeb. 1998. Genetic structuring among Alaskan Pacific herring populations identified using microsatellite variation. J. Fish Biol. 53: 150–163.

Parrish, B.B. & A. Saville. 1965. The biology of the north-east Atlantic herring populations. Oceanogr. Mar. Biol. Ann. Rev. 3: 323–373.

Raymond, M. & F. Rousset. 1995. GENEPOP (version 1.2): Population genetics software for exact tests and ecumenicism. J. Hered. 86: 248–249.

Rice, W.R. 1989. Analysing tables of statistical tests. Evolution 43: 223–225.

Røttingen, I. 1990. A review of variability in the distribution and abundance of Norwegian spring spawning herring and Barents Sea capelin. Polar Res. 8: 33–42.

Ryman N. & P.E. Jorde. 2001. Statistical power when testing genetic differentiation. Mol. Evol. 10: 2361–2371.

Ryman, N., U. Lagercrantz, L. Anderson, R. Chakraborty & R. Rosenberg. 1984. Lack of correspondence between genetic and polymorphic variability patterns in Atlantic herring *Clupea harengus*. Heredity 53: 687–704.

Shaw, P.W., C. Turan, J.M. Wright, M. O'Connell & G.R. Carvalho. 1999. Microsatellite DNA analyses of population structure in Atlantic herring (*Clupea harengus*), with direct comparison to allozyme and mtDNA RFLP analyses. Heredity 83: 490–499.

Svetovidov, A.N. 1963. Fauna of U.S.S.R. Fishes. Vol. 2. Clupeidae. 428 pp (translated from Russian by Israel Prog. Sci. Transl., Jerusalem).

Swofford, D.L. & R.B. Selander. 1981. BIOSYS-1: A FORTRAN program for the comprehensive analyses of electrophoretic data in population genetics and systematics. J. Hered. 72: 281–283.

Wahlund, S. 1928. The combination of population and the appearance of correlation examined from the stand-point of the study of heredity. Hereditas 11: 65–106.

Environmental Biology of Fishes **69**: 223–231, 2004.

Sub-arctic populations of European lobster, *Homarus gammarus*, in northern Norway

Knut E. Jørstad[a], Paulo A. Prodöhl[b], Ann-Lisbeth Agnalt[a], Maria Hughes[b], Apostolos P. Apostolidis[c,*],
Alexandros Triantafyllidis[c,*], Eva Farestveit[a], Tore S. Kristiansen[a], John Mercer[d] & Terje Svåsand[a]
[a]*Department of Aquaculture, Institute of Marine Research, Post Box 1872 Nordnes,
5817 Bergen, Norway (e-mail: knut.joerstad@imr.no)*
[b]*School of Biology and Biochemistry, Queens University of Belfast, N. Ireland*
[c]*Department of Genetics, Development and Molecular Biology School of Biology,
Aristotle University of Thessaloniki, Greece*
[d]*Shellfish Research Laboratory, National University of Ireland Galway, Galway, Ireland*

Received 17 April 2003   Accepted 16 June 2003

Key words: population structure, allozymes, microsatellites, mtDNA variation

Synopsis

The European lobster is distributed throughout the south and western regions of the Norwegian coast. A previous lobster allozyme investigation (1993) in the Tysfjord region, north of the Arctic Circle demonstrated that the lobster population from this region was genetically different from lobster samples collected in other parts of Norway. More detailed investigation including supplementary extensive sampling and additional allozyme, microsatellite and mtDNA analyses are reported here. This investigation supports the genetic distinctness of the Tysfjord population and shows that this is mainly due to a reduction (60–70%) in gene diversity (observed heterozygosities and number of alleles) compared with lobsters from more southern regions. In addition to the Tysfjord region, the comprehensive sampling also included lobsters found in the adjacent Nordfolda fjord system. Genetic analyses provided evidence for significant differences between the lobster populations of Tysfjord and Nordfolda, even though they are separated by a coastal distance of only 142 km. The two populations were also different with regards to several biological characteristics such as body size. The genetic difference between these two geographically close populations is likely to be due to the local hydrological conditions, preventing larval dispersal between the fjord systems. Assessment of lobster abundance in the north-west region suggests that the sub-arctic lobster populations are geographically isolated.

Introduction

Both the American lobster, *Homarus americanus*, and the European lobster, *Homarus gammarus*, are commercially important decapod species supporting valuable fisheries on both sides of the North Atlantic (Factor 1995, Browne et al. 2001). The European lobster is widely distributed from the Arctic Ocean in the north of Norway to Morocco in North Africa. Although to a much lesser extent nowadays, the distribution also extends throughout the Mediterranean and into the Adriatic and Aegean seas.

The general biology and distribution of the American lobster is well described (Factor 1995) and both lobster species have been evaluated of potential for aquaculture (Aiken & Waddy 1995, Nicosia & Lavalli 1999, Wickins & Lee 2002) and stock enhancement (see Gendron 1998). The first genetic studies on American lobster were carried out in the 1970s. Tracey et al. (1975) examined samples from eight localities for allozyme variation and found very small genetic differences between localities. Hedgecock et al. (1977) subsequently compared a sample of 94 individuals

---

* These authors have contributed equally to this work.

224

of American lobster with a sample of 51 individuals of European lobster, and found significant differences both in number and frequencies of common allozyme alleles. In addition, species specific alleles were found at several loci for the European lobster, and a genetic distance (Nei 1972) of 0.11 was estimated between the two species averaged over all loci investigated.

The need for more detailed information about the population structure of American lobster led to a number of other studies. Analyses of mtDNA (Kornfield & Moran 1990) suggested differences between widely separated population, but no evidence for substructuring on finer scales. No evidence for genetic differentiation was found in a study using RAPD profiling of lobsters from three different localities (Harding et al. 1997). New molecular techniques such as microsatellite DNA profiling are potentially more powerful to detect population structuring (Estoup & Angers 1998). Tam & Kornfield (1996) developed microsatellites markers specific for the American lobster, three microsatellites markers were reported, two of which were found to be highly variable.

With the exception of the work of Hedgecock et al. (1977), no detailed genetic studies have previously been undertaken on the European lobster. Recently, Jørstad & Farestveit (1999), following the methodological approach of Hedgecock et al. (1977), used allozyme markers to monitor the Norwegian Sea Ranging Program (PUSH) that started in 1991. A detailed allozyme based genetic survey of wild lobsters in Norwegian waters was also carried out. Samples from 22 different locations (2 580 individuals) were screened for four polymorphic loci (*ME**, *GPI**, *IDHP** and *PGM**). A very low level of variation was found for all loci, but significant genetic differences at all loci were detected in the sample from the most northern population collected in Tysfjord region (Jørstad & Farestveit 1999).

A fundamental requirement for the rational management of exploited species, and for the development of aquaculture, is knowledge of the genetic variability within and among populations of the species. Recent interest in cultivation (Nicosia & Lavalli 1999, Wickins & Lee 2002) and stock enhancement/ranching of lobster (Addison & Bannister 1994, Bannister & Addison 1998, Agnalt et al. 1999) in Europe, has emphasized the need for more detailed knowledge of genetic structure in European lobster. The establishment of the EU funded project 'Genetics of European

Lobster' (Ferguson 2002[1]) provided a more detailed study of the lobsters in Norway found north of the Arctic Circle (Jørstad & Farestveit 1999), including more extensive geographic sampling. In addition to allozyme analyses, both microsatellite and mtDNA characterizations were conducted, and the results are here discussed in relation to the local hydrographic systems in the area.

Materials and methods

Samples were collected during fieldwork organized by Institute of Marine Research in Bergen, Norway, from 1995 to 2001 (Figure 1). In the northern region, a small research vessel (R/V 'Fangst') was used, and lobster/crab pots and eel traps to collect lobster specimens. In the Tysfjord region, individual specimens were collected from the smaller Stefjord in 1995 and 1999, and the adjacent fjord system Nordfolda in 1999. In addition, two samples, one comprised of individual lobsters from western Norway (Bjørnefjord) and the other comprised of individuals from the northern part of Scotland (Orkney, ScotlandN), were obtained for comparison (see Figure 1 for details).

For each lobster specimen sampled, measures as carapace length (CL), sex, shell condition and female egg condition were recorded. For allozyme analysis, the more proximal segment of the walking leg was removed, frozen on dry ice for transport to the laboratory at the Institute of Marine Research in Bergen where tissues were stored at $-80°C$. For DNA analyses, the distal segment of the walking leg was removed and placed immediately in ethanol (99%). These samples were transported to Queen's University in Belfast where DNA was extracted following methodology described by Taggart et al. (1992) for microsatellite DNA analysis. Purified samples of DNA were subsequently aliquoted and sent to the Aristotle University of Thessaloniki, Greece, for mtDNA screening.

The allozyme screening analyses were based on starch gel electrophoresis and selective staining of enzymes present in white muscle, and were carried out as described in Jørstad & Farestveit (1999) for four enzymes (glucosephosphate isomerase, GPI; isocitrate dehydrogenase, mIDHP; phosphoglucomutase, PGM

---

[1] Ferguson, A. (coordinator) 2002. Genetic diversity in the European lobster (*Homarus gammarus*): Population structure and impacts of stock enhancement. [Online] Available, www.qub.ac.uk/bb/prodohl/GEL/gel.html.

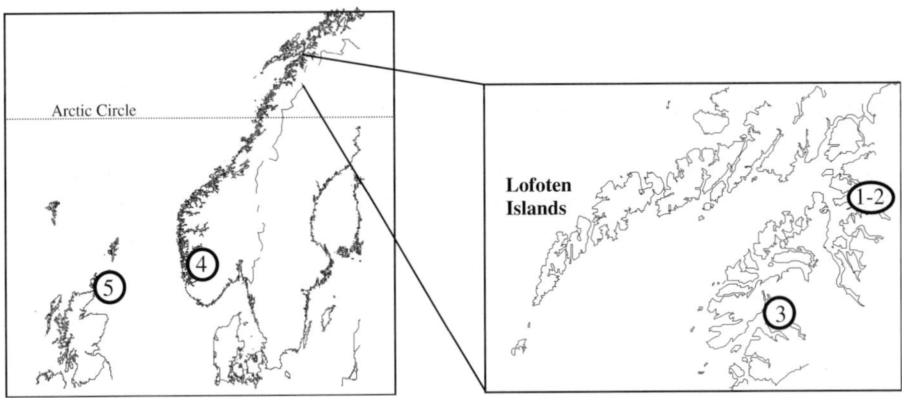

Figure 1. Sampling localities of sub-arctic lobster populations in northern Norway and two reference samples. 1, 2: Tysfjord/Stefjord 1995 and 1999; 3: Nordfolda; 4: Bjørnefjord; 5: ScotlandN.

and malic enzyme, sMEP). Two loci were expressed for GPI and PGM, giving a total of six polymorphic loci surveyed in this study.

European lobster specific microsatellite primers were developed at Queens University, and a number of such loci were fully optimized in different multiplex set for high throughput automated screening (Ferguson 2002,[1] Prodöhl et al. unpubl.). The samples described here were screened for one microsatellite multiplex set comprising six loci (*Hgam*-106, *Hgam*-21, *Hgam*-47b, *Hgam*-153, *Hgam*-105 and *Hgam*-197b).

Statistical analyses of allozyme and microsatellite data was carried out using BIOSYS-2 (Swofford & Selander 1981, Black[2]) and GENEPOP (Raymond & Rousset 1995; version 3.3). Summary population sample statistics was described in terms of the number of alleles per locus (na) and the observed heterozygosity (Ho). Deviations from Hardy–Weinberg equilibrium (HWE) were tested by the Markov chain method as implemented in GENEPOP. Genetic differentiation between all pairs of population samples, were tested by means of exact tests for allelic frequency distribution using the same program. For allozyme data, Nei's genetic distance (Nei 1972) between samples were calculated using BIOSYS-2, while subsequent cluster analyses and UPGMA trees were obtained with the TFPGA 1.3 (Miller[3]) program package. For the

microsatellite data, UPGMA topologies among samples were constructed based on Nei's (1972) genetic distance using the programs MICROSAT (Minch et al. 1995) and PHYLIP (Felsenstein 1995[4]).

The mtDNA screening was based on amplification of a 3.0 kb segment by specific designed primers developed by Aristotle University of Thessaloniki (Ferguson 2002,[1] Katsares et al. 2003). This segment in the *Homarus* genus encompasses part of the cytochrome oxidase I (CoI) gene, the entire CoII, CoIII, subunits 6 and 8 of ATPase, subunit 3 of the NAD dehydrogenase genes and several tRNAs. The primers for this region are 3F-GCCCGCTATATTCCCCAATG, 3R-CCTTAGGCGAGACTAAGTGCAA, Katsares et al. 2003. Purified DNA samples were provided from Queens University and restriction fragment length polymorphism screening was conducted in Thessaloniki using five of the most polymorphic, restriction enzymes (i.e. *Ase*I, *Ava*II, *Hinc*II, *Hinf*I and *Tac*I). Restriction patterns were analysed based on presence–absence of bands. A specific letter identified distinct endonuclease patterns. Each individual lobster was assigned a multi letter code that described its composite mtDNA genotype/haplotype. Composite haplotype frequencies were calculated for each population sample. Haplotype frequencies of samples were used to calculate Nei's (1972) genetic distances among samples with the GENDIST option of the PHYLIP package (Felsenstein[4]). Based on these distances, a UPGMA

---

[2] Black, W.C. 1997. BIOSYS-2. A computer program for the analysis of allelic variation in genetics (wcb@lamar.colostate.edu).

[3] Miller, M. 1997. Tools for population genetic analyses (TFPGA) 1.3: A window program for the analyses of allozyme and molecular population genetic data. Computer software distributed by the author.

[4] Felsenstein, J. 1995. PHYLIP, Phylogeny Inference Package, Version 3.57c. Department of Genetics, SK-50, University of Washington, Seattle. http://evolution.genetics.washington.edu/phylip/software.html.

dendrogram was constructed with the PHYLIP package depicting relationships among samples.

In autumn 2001 a lobster survey was carried out in order to access the lobster abundance along the coast from Tysfjord in the north to Bessaker in mid Norway. Sampling locations were chosen based on information gathered from local fishermen. A large number of lobster/crab pots and eel traps were used in each case, and the abundance of lobster at each locality is given as catch per unit effort (CPUE), i.e. the number of lobsters caught per pot lift. Similar data from two other samples (Bjørnefjord and Kvitsøy; Agnalt pers. comm.) were included as comparisons with south-western Norway.

Results

The lobster populations north of the Arctic Circle, Tysfjord and Nordfolda, have reduced gene diversities compared to other regions in the Atlantic coastal areas for all three molecular approaches used (allozyme, microsatellite and mtDNA; Table 1). Compared to the two reference samples from western Norway (Bjørnefjord) and Scotland, the latter has one of the highest allozyme gene diversities observed in the recent study of European lobster genetics. Only one allozyme locus (mIDHP*) was polymorphic in the northern lobster samples, resulting in low estimate of observed heterozygosities (0.004–0.028), as well as an average low number of alleles per locus (1.17–1.19). A similar result was observed from the microsatellite analyses where the average number of alleles per locus was found to be 7.8 in the reference samples, but only about 4.2 in the samples from northern Norway (Table 1). Not entirely unexpected, given its proportionally small genome size, an even more striking difference is observed during the comparison of the number of mtDNA haplotypes among samples. In the sample from Scotland, 31 haplotypes were found, but in Tysfjord and Nordfolda the values were down to only 4 and 5, and the estimates of haplotype diversities were reduced to about 40–50%. No deviation from HWE were found for both allozyme and microsatellite loci surveyed during this investigation.

Significant allele frequency differences were also detected between the lobsters in Nordfolda compared with the samples from Tysfjord. Indeed, all pairwise genetic comparisons between samples from the two fjord systems were significant. Similar results were obtained from all three molecular approaches, allozymes, microsatellites and mtDNA (Table 2), thus providing strong evidence for genetic differentiation

Table 1. Estimates of gene diversities in northern Norway lobster samples and comparison with two selected reference samples (Bjornefjord and ScotlandN). N = number of individuals screened. Observed heterozygosity (Ho) and the number of alleles per locus (Na/locus) are given for allozymes and microsatellites. For mtDNA analyses, n = number of haplotypes observed, while h = haplotype diversity.

	Allozymes			Microsatellites			mtDNA		
	N	Ho	Na/locus	N	Ho	Na/locus	N	h	n
1 Tysfjord/Stefjord 1995	100	0.023	42736	84	0.519	37656	70	0.562	5
2 Tysfjord/Stefjord 1999	94	0.028	43466	86	0.595	37656	73	0.543	4
3 Nordfolda	93	0.004	42736	86	0.577	37684	89	0.446	4
4 Bjørnefjord	96	0.055	24473	86	0.730	37840	81	0.911	22
5 ScotlandN	100	0.084	2.00	87	0.716	37840	96	0.944	31

Table 2. Pairwise tests for genetic differences between samples of European lobster based on allozyme allele frequencies, microsatellite allele frequencies and mtDNA haplotype frequencies. * p < 0.05; ** p < 0.01; *** p < 0.001; — : not significant.

	Allozymes				Microsatellites				mtDNA			
	1	2	3	4	1	2	3	4	1	2	3	4
1 Tysfjord/Stefjord 1995												
2 Tysfjord/Stefjord 1999	—				—				—			
3 Nordfolda	***	**			***	**			**	*		
4 Bjørnefjord	***	***	***		***	***	***		***	***	***	
5 ScotlandN	***	***	***	—	***	***	***	—	***	***	**	—

between the lobster populations separated by only 142 km (shore line). No significant allele frequency differences were found between the two different samples from Tysford/Stefjord collected in 1995 and 1999, using allozymes, microsatellites and mtDNA, indicating no genetic changes at least during the period investigated.

The resulting UPGMA tree topologies, based on genetic distance (Nei 1972) for the three different data sets, are shown in Figure 2. The samples from northern Norway are clustered together in all three systems (allozymes, microsatellites and mtDNA), and the reference samples (Bjørnefjord and ScotlandN) constitute a separate branch. The differences between Nordfolda

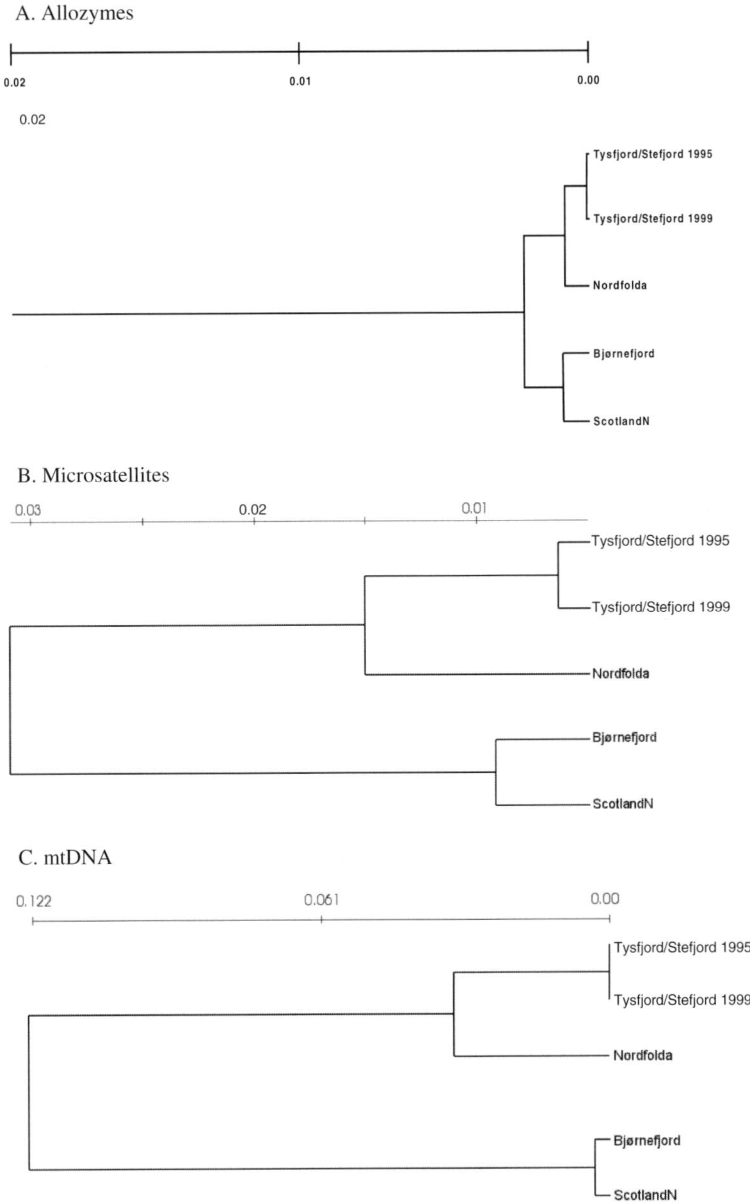

Figure 2. UPGMA dendrograms of genetic distance (Nei 1972), based on (A) Allele frequencies at six allozyme loci; (B) Allele frequencies at six microsatellite loci; (C) mtDNA haplotype frequencies of a 3 KB fragment with five restriction enzymes.

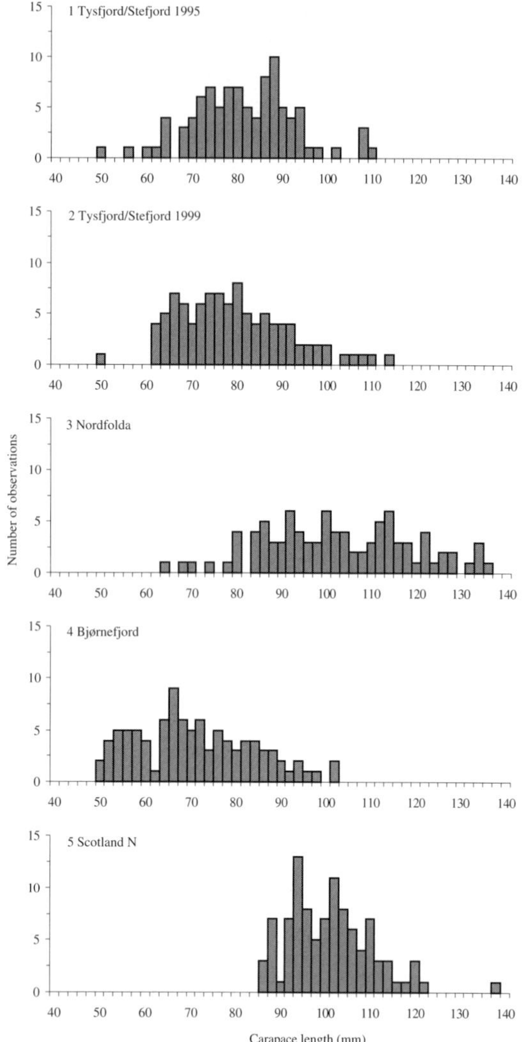

Figure 3. Size distribution (CL in mm) in the lobster samples.

and the two temporal samples from Tysfjord/Stefjord are also indicated in the distinct trees.

The genetic differences between the lobster populations in Nordfolda and Tysfjord/Stefjord, is further supported by important biological differences. Most striking was the difference in size. This difference is summarized in terms of size distributions for the two population samples in Figure 3. The mean CL in the 1999 sample of lobsters from Tysfjord/Stefjord was only 81 mm, while the corresponding value in the Nordfolda sample was 102 mm. Comparatively small lobsters also dominated the 1995 sample from Tysfjord/Stefjord. Other relevant differences such as size of egg-bearing females and size at maturation were

Table 3. Catch per unit of effort estimated as number of lobster caught per pot lift in different localities along the Norwegian coast.

Locality	# lobsters	# pots	CPUE (lobster/pot)
Tysfjord/Stefjord	31	43	0.7209
Nordfolda	72	172	0.4186
Terra/Bodø	0	43	0.0000
Træna	0	114	0.0000
Vikna	0	43	0.0000
Bessaker	3	146	0.0205
Bjørnefjord*			0.1900
Kvitsøy*			0.0600

*Unpublished data provided by A.-L. Agnalt.

also observed between the two populations (Agnalt, pers. comm.).

The genetic differences observed between the two sub-arctic lobster populations in Nordfolda and Tysfjord, and the lobsters found in the more southern regions of Norway, indicate that the populations are reproductively isolated. According to the first written reports (Gundersen 1966) on lobsters from this area, there were only occasional captures of single large lobsters outside the fjord systems, possibly on migration. No regular commercial fishery on lobsters was carried out in the northern region, and lobster abundance was very low in the coastal area between Tysfjord and southward to Bessaker in North Trøndelag. This earlier observation was subsequently confirmed through the fishing information obtained during lobster survey carried out with R/V 'Fangst' in 2001. The CPUE data (Table 3; Figure 4) demonstrated high abundance of lobsters at the locations investigated in Nordfolda (0.42) and Tysfjord/Stefjord (0.72). Interestingly, these CPUE values were also comparable to the previous earlier reports (Gundersen 1966) and, thus, suggest population stability. Only one lobster was caught on the long coastal areas south to Bessaker. This area was a good lobster fishing location in the past, but over the past 5 years, stocks have decline considerably. The CPUE values for the two reference localities Bjørnefjord and Kvitsøy were 0.19 and 0.06, respectively. The latter value does not include catches of cultured lobsters.

Discussion

The results of the present investigation clearly confirm the existence of a genetically distinct lobster stock

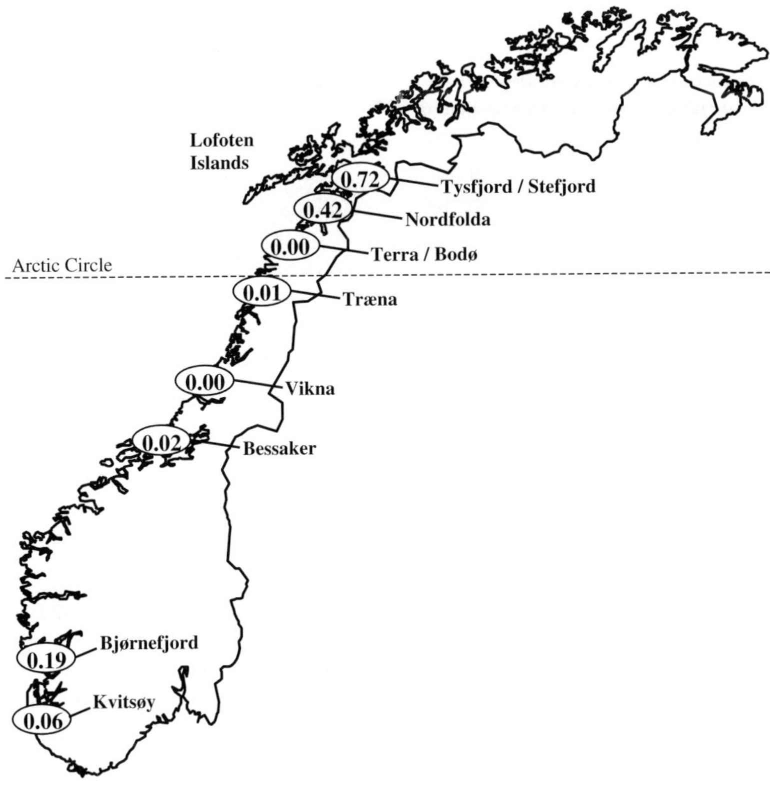

Figure 4. CPUE of lobsters along the Norwegian coast from Tysfjord to Bessaker. CPUE is estimated as the number of lobsters caught per pot lift. Data from lobster research survey carried out with R/V 'Fangst', September 2001. Values estimated for two reference localities in southern Norway (Bjørnefjord and Kvitsøy) are also included.

in the Tysfjord/Stefjord region, as previously reported (Jørstad & Farestveit 1999). Samples collected 5 years apart from this area were genetically identical as confirmed by three independent molecular approaches (allozymes, mtDNA and microsatellite) indicating that this population is temporally stable. The extended sampling, carried out during this study, throughout the northern region, revealed the existence of further levels of genetic structure in the area. Indeed, the samples collected in a adjacent fjord Nordfolda were shown to be significantly different from Tysfjord/Stefjord by all three molecular systems, even though there is only 142 km shore line between the two fjords. Biological differences were also found between the lobsters from the two fjord systems.

All the samples collected in northern Norway have low gene diversities in comparison with the southern reference samples. As well as having lowered gene diversity, the northern Norway populations were significantly different in allele frequencies from all samples taken from the rest of the European distributions, including the Mediterranean (Ferguson 2002[1]). This reduced allelic diversity is likely to be associated with one or more bottleneck events. Thus, the northern Norway groups are differentiated from the rest not due to the existence of unique alleles, but rather due to a reduction in variability resulting in changes in frequency of the remaining alleles. These populations are living north of the Arctic Circle at the extreme of the distribution range, and presumably environmental tolerance, of the species (Figure 4). Over time it is not unlikely that they experience periodic reductions in population size as the result of extreme environmental conditions. On the other hand, recent estimates of CPUE in Tysfjord and Nordfolda correspond to the values reported in the 1960s by Gundersen (1966), suggesting population stability at least during the last 40 years. This contrasts with the general decline of the Norwegian lobster harvest in southern regions of Norway (Agnalt et al. 1999), and indicates no apparent problems with over-exploitation in the area.

As far as adult lobsters are concerned, tagging studies have demonstrated that they undergo very limited migrations. Greater opportunity for gene flow, however, is likely to occur in the pelagic part of the larval phase at 14–20 days, depending on temperature. The appropriate settling requirements (such as optimal bottom substrate for the larvae) have been described for American lobsters (Wahle & Steneck 1991). Although detailed information is so far not available for the European lobster (Mercer et al. 2001), high survival and growth have recently been observed for juveniles in experiments using natural bottom substrate supplemented by shelters (Jørstad et al. 2001). Recently, it has been suggested that larval dispersal of marine organisms could be much less limited than originally thought (Bailey et al. 1997, Benzie 2000). The oceans are much more structured than was previously considered, with various subsurface oceanographic fronts and eddies, which are likely to limit larval dispersal. Also larvae may have behavioural mechanisms such as vertical movements that actually prevent dispersal. This particular behaviour has recently been described for hake (*Merluccius* spp.) in the Benguela upwelling ecosystem in Africa (Sundby et al. 2001). This upwelling system is characterized by very strong offshore transportation, where pelagic egg and larvae are likely to be lost from coastal waters. The hake, however, spawn offshore at depths between 100 and 400 m. The spawned eggs very slowly ascend through in the water masses due to the specific buoyancy of the eggs. After hatching, the larvae display a downward swimming activity, which possibly explains why the hake juveniles are concentrating near shore, opposite to what might be expected based on offshore advection.

Indeed it might be expected that, for lobsters, given the patchiness and limited depth-distribution of suitable habitat, there would be strong selection for any mechanism that limited dispersal, which could easily carry the larvae to areas unsuitable for benthic settlement. In the case of Nordfolda and Tysfjord, both fjord systems drain into the larger circulation system in Vestfjorden. The present description of this circulation system demonstrates a complex hydrographic regime consisting of a number of eddies limiting water exchange between the two adjacent fjord systems (Mitchelson-Jacob & Sundby 2001). Thus if any lobster pelagic larvae are transported out from the two fjords by surface currents, they are likely to be trapped in the large eddies outside the fjords and will be unable to settle on benthic environments suitable for lobsters. In addition,

measurements of sea surface temperature from the region (Mitchelson-Jacob & Sundby 2001) indicate that the highest and most optimal temperatures for hatching of lobster eggs ($>13°C$) is in June, and are actually found within the two fjords.

The two sub-arctic populations of European lobster in the northern Norway provide an example of population differentiation over short distances in a marine species with pelagic larval stage. The differentiation also emphasizes the importance of the local hydrographical conditions for explaining barriers for gene flow through larval dispersal. Even for a pelagic species like *Meganyctiphanes norwegica*, genetic structuring has been observed (Zane et al. 2000). The most striking example is from the Isle of Man in the Irish Sea were two different species of spider crabs have been studied (Weber et al. 2000). Both species have a relatively long larval period (40–80 days). Allozyme analyses of samples collected from localities separated by only 40 km revealed significant genetic differences in several loci for both species.

The Tysfjord and Nordfolda regions in Norway are located north of the Arctic Circle, and the genetically distinct populations are possibly adapted to the local environmental conditions. At present, lobster is harvested within the general management in Norway based on limited fishing season and minimum legal size. The genetic and biological differences between the two populations, however, suggest the importance for the development of new management approaches for future conservation of these unique lobster populations. Unless special precautions are taken, development of future lobster stock enhancement and farming could represent a potential threat. From a scientific view, these two populations offer exciting opportunities for a number of interesting comparative studies focused on genetics as well as life-history characters.

Acknowledgements

This work was supported by the EC FAIR programme (CT98 4266; 'Genetics of European Lobster'; Coordinator: Prof. A. Ferguson, Queens University, Belfast). Odd Jakobsen, Haukøy, gave valuable information about the lobsters in Stefjord. Johan R. Lillehaug, Bergen, first informed about the lobsters in Nordfolda, while Ludvik Larsen, Nordfold, provided detailed information about the lobster fishing localities in the same area.

References

Addison, J.T. & R.C.A. Bannister. 1994. Re-stocking and enhancement of clawed lobster stocks: a review. Crustaceana 67: 131–155.

Agnalt, A.-L., G.I. van der Meeren, K.E. Jørstad, H. Næss, E. Farestveit, E. Nøstvold, T. Svåsand, E. Korsøen & L. Ydstebø. 1999. Stock enhancement of European lobster (*Homarus gammarus*); A large scale experiment off southwestern Norway (Kvitsøy). *In*: B. Howell, E. Moksness & T. Svåsand (ed.) Stock Enhancement and Sea Ranching, Fishing News Books, Blackwell Science, Oxford, U.K. 606 pp.

Aiken, D.E. & S.L. Waddy. 1995. Aquaculture. pp. 153–175. *In*: J.R. Factor (ed.) Biology of the Lobster *Homarus americanus*, Academic Press, New York. 528 pp.

Bannister, R.C.A. & J.T. Addison. 1998. Enhancing lobster stocks: A review of recent European methods, results and future prospects. Bull. Mar. Sci. 62: 369–387.

Bailey, K.M., P.J. Stabeno & D.A. Powers. 1997. The role of larval retention and transport features in mortality and potential gene flow of walleye pollock. J. Fish Biol. 51(Suppl. A): 135–154.

Benzie, J.A.H. 2000. The detection of spatial variation in widespread marine species: Methods and bias in the analyses of population structure in the thorns of starfish (Echinodermata: Asteroidea). Hydrobiologia 420: 1–14.

Browne, R., J.P. Mercer & M.J. Duncan. 2001. An historical overview of the Republic of Ireland's lobster (*Homarus gammarus* Linnaeus) fishery, with reference to European and North American (*Homarus americanus* Milne Edwards) Lobster Landings. Hydrobiologia 465: 49–62.

Estoup, A. & B. Angers. 1998. Microsatellites and mini-satellites for molecular ecology: Theoretical and empirical considerations. pp. 55–86. *In*: G.R. Carvalho (ed.) Advances in Molecular Ecology, IOS Press, Amsterdam.

Factor, J.R. (ed). 1995. Biology of the Lobster *Homarus americanus*, Academic Press, New York. 528 pp.

Gendron, L. (ed.) 1998. Proceedings of a workshop on lobster stock enhancement held in the Magdalen Islands (Québec) from October 29 to 31, 1997. Can. Ind. Rep. Fish. Aquat. Sci. 244: 135 pp.

Gundersen, K. 1966. Rapport om prøvefiske etter hummer i Nordland fylke i 1964 og 1965 (Report of an exploratory fishery for lobster in Norland county in 1964 and 1965). Fiskets Gang, nr. 29, 1966.

Harding, G.C., E.L. Kenchington, C.J. Bird, D.S. Pezzack & D.C. Landry. 1997. Genetic relationship among subpopulations of the American lobster (*Homarus americanus*) as revealed by random amplified polymorphic DNA. Can. J. Fish. Aquat. Sci. 54: 1762–1771

Hedgecock, D., K. Nelson, J. Simons & R. Shleser. 1977. Genic similarity of American and European species of the lobster Homarus. Biol. Bull. 152: 41–50.

Jørstad, K.E. & E. Farestveit. 1999. Population genetic structure of lobster (*Homarus gammarus*) in Norway, and implications for enhancement and sea-ranching operation. Aquaculture 173: 447–457.

Jørstad, K.E., A-L. Agnalt, T.S. Kristiansen & E. Nøstvold. 2001. High survival and growth of European lobster juveniles (*Homarus gammarus*), reared communally with natural bottom substrate. Mar. Freshw. Res. 52: 1431–1438.

Katsares, V., A. Apostolidis, A. Triantafyllidis, A. Kouvatsi & C. Triantaphyllidis. 2003. Development of mitochondrial DNA primers for use with homarid lobsters. Mar. Biotechnol. 5: 469–479.

Kornfield, I. & P. Moran. 1990. Genetics of population differentiation in lobsters. pp. 23–24. *In*: I. Kornfield. (ed.) Life History of the American Lobsters, Lobster Institute, Orono, Maine.

Mercer, J.P., R.C.A Bannister, G.I. van der Meeren, V. Debuse, D. Mazzoni, S. Lovewell, R. Browne, A. Linnane & B. Ball. 2001. An overview of the LEAR (Lobster Ecology and Recruitment) project: The results of field and experimental studies on the juvenile ecology of *Homarus gammarus* in cobble. Mar. Freshw. Res. 52: 1291–1302.

Minch, E., A. Ruiz-Linares, D. Goldstein, D. Feldman & L.L. Cavalli-Sforsa. 1995. MICROSAT (Version 1.5b): A Computer Program for Calculating Various Statistics on Microsatellite Allele Data, Standford University, Standford, California.

Mitchelson-Jacob, G. & S. Sundby. 2001. Eddies of Vestfjorden, Norway. Cont. Shelf Res. 21: 1901–1918.

Nei, M. 1972. Genetic distance between populations. Am. Nat. 106, 283–292.

Nicosia, F. & K. Lavalli. 1999. Homarid lobster hatcheries: Their history and role in research, management, and aquaculture. Mar. Fish. Rev. 61: 1–57.

Raymond, M. & F. Rousset. 1995. GENEPOP: Population genetics software for exact tests and ecumenicism. J. Hered. 86: 248–249.

Sundby, S., A.J. Boyed, L. Hutchings, M.J. O'Toole, K. Thorisson & A.Thorsen. 2001. Interaction between cape hake spawning and the circulation in the northern Benguela upwelling ecosystem. S. Afr. J. Mar. Sci. 23: 317–336.

Swofford, D.L. & R.B. Selander.1981. BIOSYS-1: A FORTRAN program for the comprehensive analyses of electrophoretic data in population genetics and systematics. J. Hered. 72: 281–283.

Taggart, J.B., R.A. Hynes, P.A. Prodohl & A. Ferguson. 1992. A simplified protocol for routine total DNA isolation from salmonid fishes. J. Fish Biol. 40: 963–965.

Tam, Y.K & I. Kornfield. 1996. Characterisation of microsatellite markers in *Homarus* (Crustacea, Deacapoda). Mol. Marine Biol. Biotechnol. 5: 230–238.

Tracey, M.L., K. Nelson, D. Hedgecock & R.A. Shleser. 1975. Biochemical genetics of lobsters: Genetic variation and the structure of American lobster (*Homarus americanus*) populations. J. Fish. Res. Board Can. 32: 2091–2101.

Wahle, R.A. & R. Steneck. 1991. Recruitment habitats and nursery grounds of the American lobster *Homarus americanus*: A demographic bottleneck? Mar. Ecol. Prog. Ser. 69: 231–243.

Weber, L.I., R.G. Hartnoll & J.P. Thorpe. 2000. Genetic divergence and larval dispersal in two spider crabs (Crustacea: Decapoda). Hydrobiology 420: 211–219.

Wickins, J.F. & D.O.C. Lee. 2002. Crustacean Farming. Ranching and Culture, Blackwell Science Ltd., London. 446 pp.

Zane, L., L. Ostellari, L. Maccatrozzo, L. Bargelloni, J. Cuzin-Roudy, F. Buchholz & T. Paternello. 2000. Genetic differentiation in a pelagic crustacean (*Meganuctiphanes norvegica*: Euphausiaca) from the north east Atlantic and the Mediterranean Sea. Mar. Biol. 136: 191–199.

Environmental Biology of Fishes **69**: 233–243, 2004.
© 2004 *Kluwer Academic Publishers. Printed in the Netherlands.*

Detecting specific populations in mixtures

Joel Howard Reynolds[a] & William David Templin
*Gene Conservation Laboratory, Alaska Department of Fish and Game, 333 Raspberry Road, Anchorage,
AK 99518-1599, U.S.A.*
[a]*Current address: Division of Natural Resources, U.S. Fish and Wildlife Service, 1011 E. Tudor Road,
MS 221, Anchorage, AK 99503, U.S.A. (e-mail: joel_reynolds@fws.gov)*

Received 17 April 2003   Accepted 16 June 2003

Key words: component contribution, genetic stock identification, mixed stock analysis, mixture models, SPAM,
species conservation

Synopsis

Mixed stock analysis (MSA) estimates the relative contributions of distinct populations in a mixture of organisms.
Increasingly, MSA is used to judge the presence or absence of specific populations in specific mixture samples.
This is commonly done by inspecting the bootstrap confidence interval of the contribution of interest. This method
has a number of statistical deficiencies, including almost zero power to detect small contributions even if the
population has perfect identifiability. We introduce a more powerful method based on the likelihood ratio test
and compare both methods in a simulation demonstration using a 17 population baseline of sockeye salmon,
Oncorhynchus nerka, from the Kenai River, Alaska, watershed. Power to detect a nonzero contribution will vary
with the population(s) identifiability relative to the rest of the baseline, the contribution size, mixture sample size,
and analysis method. The demonstration shows that the likelihood ratio method is always more powerful than the
bootstrap method, the two methods only being equal when both display 100% power. Power declines for both
methods as contribution declines, but it declines faster and goes to zero for the bootstrap method. Power declines
quickly for both methods as population identifiability declines, though the likelihood ratio test is able to capitalize on
the presence of 'perfect identification' characteristics, such as private alleles. Given the baseline-specific nature of
detection power, researchers are encouraged to conduct *a priori* power analyses similar to the current demonstration
when planning their applications.

Introduction

Mixed stock analysis (MSA) is used to estimate the relative contributions of distinct populations in a mixture
of organisms. This is an important tool in fisheries management and research, with genotypes commonly used
as natural markers to distinguish major populations or
stocks (e.g. genetic stock identification) (Begg et al.
1999, Shaklee et al. 1999, Pearce et al. 2000). Other
characteristics commonly used in fisheries include parasite assemblages (Urawa et al. 1998, Moles & Jensen
2000), scale patterns (Marshall et al. 1987), morphometrics and meristics (Fournier et al. 1984), artificial
tags such as thermal marks, coded wire tags, or fin clips
(Ihssen et al. 1981).

Increasingly, MSA is used to judge the presence
or absence of specific populations in specific mixture
samples. For example, management of an interception
fishery may be heavily influenced by the presence or
absence, in the harvest or bycatch, of a specific population that is threatened, weakened, or otherwise of
special interest.

MSA can overestimate the contributions of populations that actually contribute little, or nothing, to
a mixture (Pella & Milner 1987). Managers and
researchers therefore face two questions when using
MSA to judge the absence of a specific population in
a mixture sample. Q1: What method should be used
to test if a specific population is absent from the mixture and just receiving a nonzero estimate due to bias?

Q2: What is the method's power to detect a contribution of x% of the population from a mixture sample of size N?

Testing Population A's absence is equivalent to testing for a nonzero contribution: $H_0: \theta = 0$ *versus* $H_A: \theta > 0$, where θ is Population A's contribution to the mixture. The most common test assesses the one-sided lower 95% bootstrap confidence interval for θ: if the interval's limit is >0, the test rejects H_0 at a significance level of 0.05 and the population is deemed 'present'. If the interval's limit is 0, the observations present insufficient evidence to reject H_0 at 0.05 and the population contribution is deemed 'statistically indistinguishable from zero' (Seeb & Crane 1999).

This method has a number of statistical flaws, some subtle (Reynolds & Templin 2003), some more obvious. Most glaring is the method's low statistical power for detecting small contributions in application. Consider an ideal marker and an ideal population: a gene for which Population A is fixed for an allele that is unique among the other populations in the baseline, that is, a private allele. For illustration, we briefly ignore the impact of sampling the mixture and speak directly of the contribution in the mixture sample. Population A is perfectly identifiable, so a mixture sample of size N containing N times θ individuals from Population A will produce a nonzero contribution estimate, $\hat{\theta} > 0$. Even with reasonably sized mixture samples, if θ is small, say $\theta < 0.05$, then there is a positive probability that a bootstrap resample will not have any individuals from Population A; for that resample $\hat{\theta}^{resample} = \theta^{resample} = 0.0$ (Appendix 1, Part 1). The probability of this occurring increases as θ decreases to 0. Considering the roughly 1 000 resamples required for an adequate bootstrap confidence interval (Davison & Hinkley 1997, p. 156), this small probability of a resample 'without Population A'

can lead to a moderate probability that the lower confidence interval will have a limit of 0 and therefore fail to detect the nonzero contribution (Appendix 1, Part 2). While this probability of a 0 lower limit is ameliorated somewhat by considering the full process – random sampling from the mixture followed by bootstrap resampling from the random sample (Appendix 1, Part 3), yet even with perfect identifiability the bootstrap confidence interval method has only little to moderate power to detect small contributions at common sample sizes (Table 1). Most importantly, as demonstrated below, the method's power is drastically reduced in application, where perfect identifiability is the rare exception. A more powerful alternative, using a likelihood ratio test, is introduced below.

We briefly review the standard MSA model and estimation method, conditional maximum likelihood estimation (Millar 1987, Pella & Milner 1987), develop the likelihood ratio test, and describe how to estimate P values using Monte Carlo simulation. The method is demonstrated in a simulation study of mixtures of sockeye salmon (*Oncorhynchus nerka*) from the Kenai River, Alaska. The likelihood ratio test and the confidence interval approach are compared in terms of their power to detect a nonzero contribution from a specific population or group of populations. The comparison demonstrates how to conduct *a priori* power analyses for a given baseline, a specific stock of interest, and a range of stock contributions and mixture sample sizes. Alternative approaches using individual assignment methods are discussed.

The likelihood ratio test is more powerful than the bootstrap confidence interval, though both methods display lower power than desired. Both methods lose power as population identifiability and contribution decline, but the bootstrap method loses power faster

Table 1. Power to detect nonzero, perfectly identifiable, contributions in MSA using the bootstrap confidence interval method, as a function of true population contribution to the original mixture (θ) and mixture sample size. In applications with less than perfect identifiability, power will be much lower (see Figure 3).

θ	Mixture sample size						
	50	75	100	150	200	300	350
0.10	0.85	0.97	1.00	1.00	1.00	1.00	1.00
0.05	0.40	0.66	0.83	0.97	0.99	1.00	1.00
0.04	0.28	0.50	0.69	0.90	0.97	1.00	1.00
0.03	0.16	0.32	0.49	0.76	0.90	0.99	1.00
0.02	0.06	0.15	0.25	0.48	0.68	0.90	0.95
0.01	0.01	0.03	0.06	0.14	0.24	0.47	0.58

Calculations are based on 1000 nonparametric bootstrap resamples and the one-sided lower 95% percentile bootstrap confidence interval (Davison & Hinkley 1997). Calculation details are in Appendix 1.

and has zero power in some scenarios. The likelihood ratio method retains a nonzero power in all scenarios. In the absence of perfect identifiability, the likelihood ratio method's low power limits application to those problems involving small to moderate-sized baselines of potentially contributing populations.

Methods

The finite mixture model

The following model describes mixtures of contributions from finitely many source populations (see, e.g. Millar 1987 or Pella & Milner 1987). Although the presentation assumes discrete characteristics are observed on each individual, such as a genotype, this is not essential; the model holds for continuous characteristics as well.

Let N individuals be randomly sampled from a mixture of J populations. Let the jth population contribute an unknown proportion $\theta_j \geq 0$ to the mixture, $\Sigma \theta_j = 1$; $\Theta = (\theta_1, \ldots, \theta_J)$. If the characteristic measured on the ith sample individual is denoted by x_i, then the probability of observing the sample $\mathbf{X} = \{x_1, x_2, \ldots, x_n\}$ is

$$\Pr(\mathbf{X} \mid \Theta, \Pi) = \prod_{i=1}^{n} \Pr(x_i \mid \Theta, \Pi)$$

$$= \prod_{i=1}^{n} \left\{ \sum_{j=1}^{J} \theta_j \Pr(x_i \mid \boldsymbol{\pi}_j) \right\} \qquad (1)$$

where $\boldsymbol{\pi}_j$ is a column vector of parameters specifying the probability density function for the characteristic in population j, and Π is the matrix $[\boldsymbol{\pi}_1 | \cdots | \boldsymbol{\pi}_J]$. For a discrete characteristic with k possible outcomes, $\boldsymbol{\pi}_j = (\pi_1^j, \ldots, \pi_k^j), \pi_i^j \geq 0, \sum_i \pi_i^j = 1$, the vector of multinomial probabilities. This assumes that the set $\{\text{Pop.}1, \ldots, \text{Pop. J}\}$ includes all potentially contributing populations (see Smouse et al. 1990). Expanding the $\Pr(x_i \mid \boldsymbol{\pi}_j)$ terms allows for multivariate characteristics.

Estimation

Estimating the mixture proportions, Θ, requires information regarding the (possibly multivariate) characteristic probability density function, $\boldsymbol{\pi}_j$, for each contributing population. This is generally available in the form of a sample from each baseline population. In most fisheries applications researchers fix the nuisance parameters, $\boldsymbol{\pi}_j$, at their estimates from the baseline samples, $\hat{\boldsymbol{\pi}}_j$ (Millar 1987). Maximum likelihood is then used to estimate the unknown Θ conditional on $\boldsymbol{\pi}_j = \hat{\boldsymbol{\pi}}_j$. This conditioning is justified by the small amount of information on $\boldsymbol{\pi}_j$ in the mixture sample, relative to the baseline sample (Milner et al. 1981). Recently developed Bayesian methods utilize this information, which may be an important consideration when analyzing related mixture samples collected through space or time (Pella & Masuda 2001).

Identifiability of a stock in a mixture requires, among other things (Pella & Milner 1987), that the probability density functions of the characteristics, $\boldsymbol{\pi}_j$, differ across the contributing populations (Redner & Walker 1984). Characteristics commonly used in fisheries include parasite assemblages (Urawa et al. 1998, Moles & Jensen 2000), scale patterns (Marshall et al. 1987), morphometrics and meristics (Fournier et al. 1984), artificial tags such as thermal marks, coded wire tags, or fin clips (Ihssen et al. 1981), and increasingly, genetic markers (Seeb & Crane 1999, Ruzzante et al. 2000).

Uncertainty in the mixture proportion estimates, $\hat{\Theta}$, arises from sampling uncertainty in both the mixture and the population baselines samples. In practice, these sampling uncertainties can be accounted for by nonparametric bootstrap resampling from the mixture sample and parametric bootstrap resampling from the baseline characteristic distributions, $\hat{\boldsymbol{\pi}}_j$. Bootstrap resampling of the baseline samples increases the width of the resulting confidence intervals, reducing the power to detect nonzero contributions. The following demonstration only resamples the mixture sample.

Testing population absence

Assume a sample is taken from a mixture consisting of contributions from a known set of baseline populations, with specific interest in testing the absence of Population A, $H_0: \theta_A = 0$. The likelihood ratio test compares the likelihood of the observed sample under the general model, in which Population A contributes, to the likelihood under the null model, in which Population A does not contribute (that is, $\theta_A = 0$). The likelihood ratio test statistic, conditional on $\boldsymbol{\pi}_j = \hat{\boldsymbol{\pi}}_j$, is

$$
\begin{aligned}
\text{LR} &= \frac{L(\{\theta_1, \theta_2, \ldots, \theta_J\} \mid \mathbf{X}, \hat{\Pi})}{L(\{\theta_1', \theta_2', \ldots, \theta_A' = 0, \ldots, \theta_J'\} \mid \mathbf{X}, \hat{\Pi})} \\
&= \frac{\prod_{i=1}^{n} \left\{ \sum_{j=1}^{J} \theta_j \Pr(x_i \mid \hat{\boldsymbol{\pi}}_j) \right\}}{\prod_{i=1}^{n} \left\{ \sum_{\substack{j=1 \\ j \neq A}}^{J} \theta_j' \Pr(x_i \mid \hat{\boldsymbol{\pi}}_j) \right\}} \qquad (2)
\end{aligned}
$$

with $\{\theta_1, \ldots, \theta_J\}$ and $\{\theta_1', \ldots, \theta_J'\}$ replaced by their conditional maximum likelihood contribution estimates under their respective models. The observed ratio, $\mathrm{LR}^{\mathrm{obs}}$, is calculated by fitting the mixture sample using the full baseline (the general model, the numerator), then fitting the mixture sample using the reduced baseline with Population A dropped to force $\theta_A = 0$, (the null model, the denominator). The test can be extended to the joint contribution of a specific group of populations, H_0: $\theta_{A1} = \theta_{A2} = \cdots \theta_{Av} = 0$ *versus* H_a: one or more of $\{\theta_{A1}, \ldots, \theta_{Av}\} > 0$.

If Population A has a unique characteristic, relative to the other baseline populations, that also occurs in the mixture sample, the likelihood under the reduced baseline will be zero, giving a likelihood ratio of ∞. Commonly used MSA software (Debevec et al. 2000) assigns such individuals to an 'unknown' baseline component, clearly identifying the nonzero contribution of Population A.

Completing the test requires comparing the observed ratio, $\mathrm{LR}^{\mathrm{obs}}$, to its expected distribution under H_0. There are two ways to estimate this distribution.

Null reference distribution method 1: asymptotic theory

In theory, the null hypothesis can be tested by comparing -2 times $\ln(\mathrm{LR}^{\mathrm{obs}})$ to its asymptotic distribution under the null model, a χ^2 with degree of freedom equal to the number of populations being simultaneously tested for zero contribution (Stuart et al. 1999). The conditions underlying this asymptotic result do not hold when other populations in the baseline fail to contribute to the mixture (Stuart et al. 1999). As this is often the case in fisheries genetic stock identification problems (Millar 1987), the asymptotic results are frequently unreliable. Even when all populations are expected to have nonzero estimates, experience has shown that the asymptotic results may remain unreliable.

Null reference distribution method 2: Monte Carlo simulation

The null reference distribution can be approximated by Monte Carlo simulation under H_0, conditional on Θ_0 (Davison & Hinkley 1997, p. 138). The value of Θ_0 will generally not be known prior to analysis and must be estimated from fitting the null model. Estimate Θ_0 from the observed mixture sample by fitting model (1) using the reduced baseline, giving $\hat{\Theta}_0$. Then simulate R sets of

N observations from model (1) using the estimated null mixture proportions, $\hat{\Theta}_0$, and the baseline population characteristic densities $\hat{\Pi}$. Take each set of simulated observations and fit model (1) using the full baseline (Population A included), giving an estimate $\hat{\Theta}^{*r}$. Take each set of simulated observations and fit model (1) using the reduced baseline (no Population A included), giving an estimate $\hat{\Theta}_0^{*r}$. For each r, calculate and record the likelihood ratio (Equation (2)), LR^{*r}.

This process gives a sample of size R, $\{\mathrm{LR}^{*r}: r = 1, \ldots, R\}$, from the unknown null reference distribution. Calculate the observed likelihood ratio, $\mathrm{LR}^{\mathrm{obs}}$, by fitting the general and restricted models to the actual mixture sample and plugging the estimates into Equation (2). An approximate P value for the test is given by $\left(1 + \sum_r I(\mathrm{LR}^{*r} \geq \mathrm{LR}^{\mathrm{obs}})\right)/(1 + R)$ (Davison & Hinkley 1997, p. 141), where the indicator function $I(\,)$ takes the value one when the argument is true and zero otherwise. Generally, R in the range 1000–5000 provides sufficient precision (Davison & Hinkley 1997, sec. 4.2.5).

Uncertainty in the conditional values of the nuisance parameters, $\hat{\pi}_j$, can be incorporated into the Monte Carlo simulation approach by parametric bootstrap resampling from each $\hat{\pi}_j$ before constructing the null mixture during each of the R simulation rounds. The resulting null reference distribution will actually be a mixture of null reference distributions, one for each resampled baseline. This resampling will increase the dispersion in the null reference distribution and hence potentially overestimate the tail areas of interest. For ease of comparison, the following demonstration does not include resampling of the baseline estimates.

Demonstration

The two methods were compared in terms of their power to detect a nonzero population contribution. The simulation study used an allozyme baseline of 19 markers for the sockeye salmon populations of Kenai River, Alaska (collection and analysis details in Seeb et al. 2000, nomenclature following Shaklee et al. 1990): *mAAT-1**, *mAAT-2**, *mAH-1,2**, *mAH-4**, *sAH**, *ALAT**, *G3PDH-1,2**, *GPI-A**, *GPI-B1,2**, *sIDHP-1**, *LDH-B2**, *sMDH-A1,2**, *PEPA**, *PEPC**, *PEPB-1**, *PEPD-1**, *PEPLT**, *PGM-1**, *PGM-2**. To investigate how population identifiability influences detection power, we explored three population sets of declining, though relatively high, identifiability.

Kenai River sockeye baseline

The Kenai River is the major producer of sockeye salmon in Cook Inlet, Alaska, supporting a commercial fishery in the inlet, a personal use fishery at the river mouth, and a recreational fishery within the river itself (Figure 1). The river fisheries are managed to allow a set range of individuals to reach the spawning grounds. Resource managers are interested in detecting the presence of specific populations at different times during the fishing season.

The Kenai River baseline consists of seventeen populations. Geography (Figure 1) and genetic diversity (Figure 2), followed by extensive simulation analyses, were used to define five regions, or population aggregates, that were reliably identified in mixture estimation (Seeb et al. 2000). A region had to demonstrate a contribution estimate of 90% or more, averaged over 500 bootstrap resamples, from simulated mixture samples of 400 genotypes, where each mixture sample was drawn uniformly from the populations within the region of interest, and each genotype was parametrically bootstrapped from its population of origin's allele frequency estimates. Five regions were identified (Figures 1 and 2): Upper Russian River (two populations), Hidden Creek, Trail Lakes (three populations), Tern Lake, and Kenai/Skilak (10 populations).

Three scenarios were investigated: detecting the highly identifiable Upper Russian River region, the moderately identifiable Trail Lakes region, and the somewhat less identifiable Tern Lake region.

The Upper Russian River drainage occurs above a waterfall, which acts as a partial barrier to upstream movement. The populations spawning above the falls are relatively genetically distinct (Seeb et al. 2000), though they do not exhibit any private alleles at the allozyme markers considered. The Railroad Creek population in the Trail Lakes region exhibits a private allele at $mAH\text{-}1,2^*$ (relative frequency 0.013); the Tern Lake population exhibits a private allele at $mAH\text{-}4^*$ (relative frequency 0.01).

Simulated mixtures

For each reporting region of interest, mixture samples of genotypes from 200 individuals were simulated over a range of population contributions, $\theta_{\text{Region\_of\_interest}} = \{10\%, 5\%, 4\%, 3\%, 2\%, 1\%\}$. For Upper Russian River or Trail Lakes, $\theta_{\text{Region\_of\_interest}}$ was evenly split among the region's populations. The remaining populations in the baseline evenly contributed the rest of the mixture. A contribution from a given population was simulated by randomly generating a genotype from that population's allele frequencies

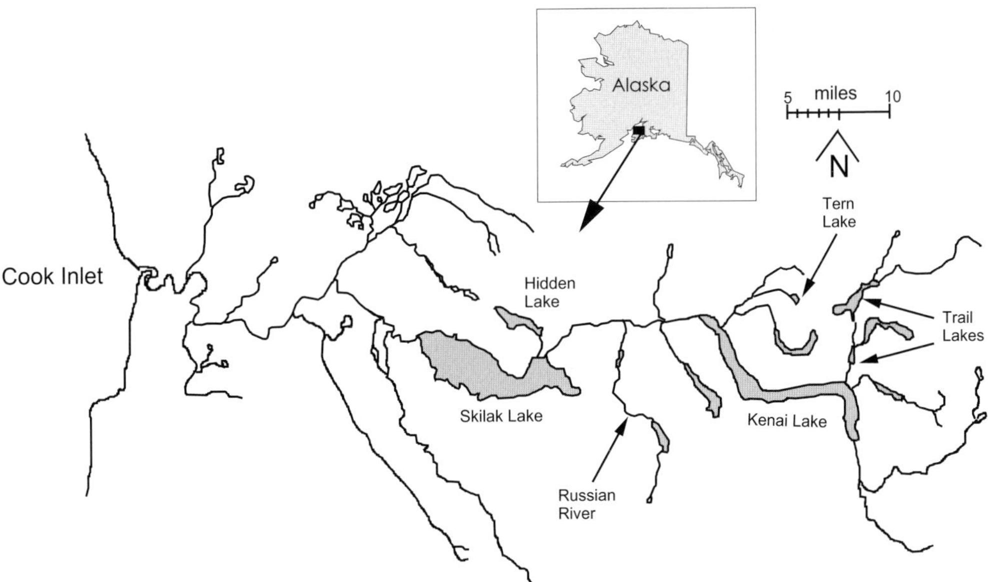

Figure 1. Major sockeye salmon producing lake systems of the Kenai River watershed, Cook Inlet, Alaska. Each labeled lake system constitutes a reporting region in the baseline (Figure 2) with the exception of Skilak and Kenai Lakes, which are combined into one region.

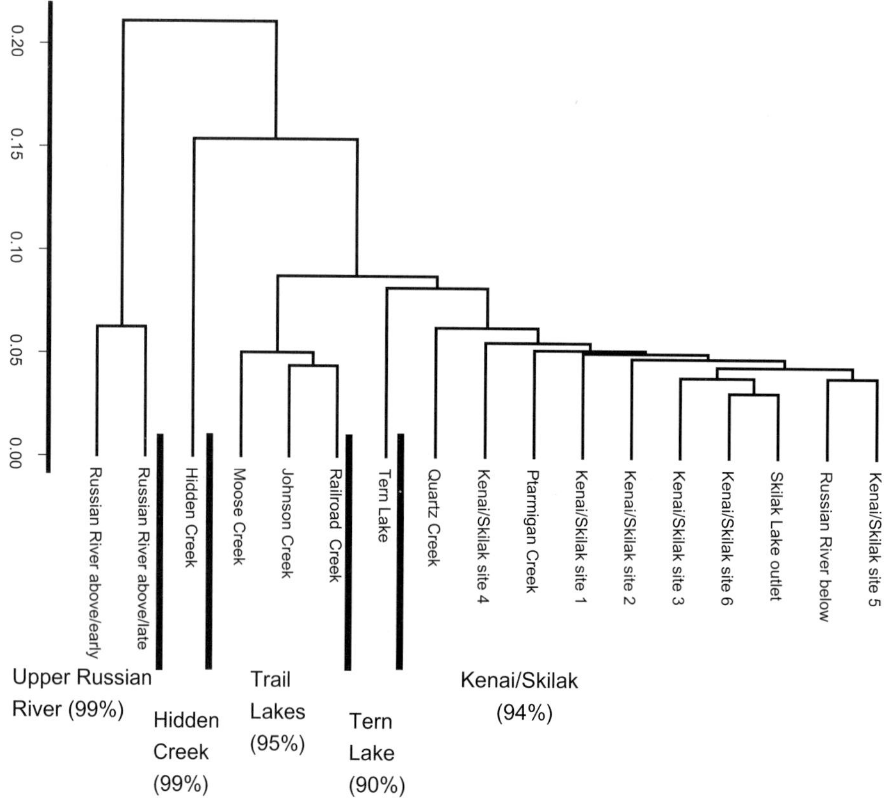

Figure 2. UPGMA tree of populations in the Kenai River sockeye salmon baseline, using Cavalli-Sforza and Edwards genetic distance (described in Weir 1996) on 19 allozyme markers (see text for details). The populations are aggregated into five reporting regions for MSA, where a reporting region is the smallest set of populations that achieves, on average, a 90% or greater contribution estimate for simulated mixtures generated strictly of individuals from the populations themselves ('100% simulations') (see Seeb et al. 2000 for details). The region labels also give the mean contribution estimate for each region's 100% simulations. The Kenai/Skilak sites are located along the reach between the two lakes (Figure 1).

for each of the 19 allozyme markers. Baseline allele frequencies are available in Seeb et al. (2000). Fifty mixture samples were simulated for each combination of region of interest and contribution level.

Analyses

Each mixture sample was analyzed to estimate: the reporting region contributions under both the full and reduced baseline models, the one-sided 95% lower percentile bootstrap confidence interval for $\theta_{Region\_of\_interest}$ under the full baseline model, and all quantities required to conduct the likelihood ratio test of H_0: $\theta_{Region\_of\_interest} = 0$ *versus* H_A: $\theta_{Region\_of\_interest} > 0$. The bootstrap confidence interval used $B = 1000$ resamples; for equal numerical accuracy in tail estimation,

the Monte Carlo approximation to the null reference distribution used $R = 1000$ simulations.

The bootstrap and likelihood ratio test were compared in terms of their power to detect the nonzero contribution of the region of interest. For a given scenario – (method and region of interest and contribution), power was estimated as the percentage of the 50 simulated mixture samples for which the method detected a nonzero contribution from the region of interest. Detection was defined as: bootstrap method – nonzero limit lower limit on the one-sided 95% confidence interval when rounded to two significant digits; likelihood ratio – p value ≤ 0.05 or nonzero contribution assignment to the 'unknown' category when fitting the mixture using the reduced baseline model. Baseline allele frequencies were not resampled.

Mixture samples were generated using S-Plus 2000 (Insightful, Inc., Seattle, WA, U.S.A.) and locally written programs. Mixture analyses were conducted using the freeware package SPAM 3.5 (Reynolds 2001).

Demonstration results

The likelihood ratio test was as or more powerful than the bootstrap confidence interval method in testing population absence (Figure 3), detecting at least every contribution the bootstrap method detected. Equality occurred only when both methods displayed 100% power. Both methods displayed less than 'ideal' power (Figure 3 vs. Table 1 for the bootstrap confidence interval method; Figure 3 vs. 100% for the likelihood ratio method). The likelihood ratio method always displayed positive power.

Even under the most optimistic scenario (perfect identifiability), the discrete nature of bootstrap resampling drives the confidence interval method's detection power to near zero for very small contributions,

$\theta \leq 2\%$, and common sample sizes (Table 1). In practice, the method displays a drastically reduced power that quickly drops to very low levels for even relatively sizeable contributions, $\theta \leq 5\%$, for any populations with less than extremely high identifiability (Figure 3).

Both methods lost power as the region of interest's identifiability declined or the true contribution declined (Figure 3). The likelihood ratio method's power did not decline as quickly as that of the bootstrap confidence interval method (Figure 3).

Discussion

Increased usage of MSA has increased demand for methods of detecting small nonzero contributions from a specific population, and hence for distinguishing, to the extent possible, such contributions from mere biased estimates for an absent population. The likelihood ratio test is a more powerful method than the current bootstrap confidence interval approach. It detected every contribution the bootstrap method detected, and

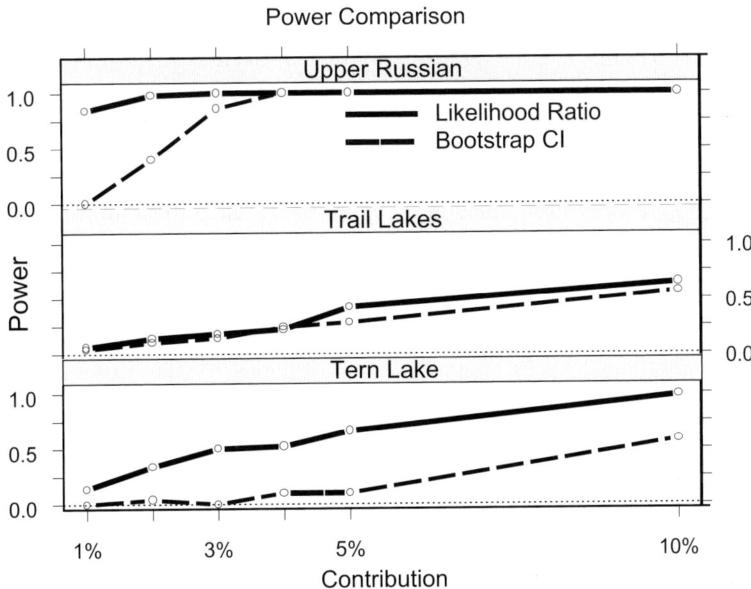

Figure 3. Power to detect nonzero contributions, by method, for each investigated region of interest. The likelihood ratio test displays higher power than the bootstrap confidence interval method for detecting a specific region's contribution, regardless of which specific region (panel), or contribution level (%) (x axis). The likelihood ratio is able to maintain ideal power to detect even two individuals out of 200 (Russian River panel). While both methods lose power with decreasing identifiability of the population(s) of interest (refer to Figure 2), the likelihood ratio method is able to retain a positive, albeit small, power at every scenario; the power of the bootstrap method quickly goes to zero in all but the high identifiability situations. Fifty mixtures of 200 individuals were simulated for each contribution level by region of interest combination; see text for details. Power estimates are the proportion of each scenario's 50 simulated mixtures in which the method detected the contribution.

more (Figure 3), while maintaining a positive detection power in every scenario.

The increase in power of the likelihood ratio test over the bootstrap confidence interval is not surprising. The bootstrap's usefulness for generating confidence intervals stems from its almost complete absence of parametric assumptions. In the case of MSA, however, the parametric assumptions associated with the mixture model and Hardy–Weinburg equilibrium are generally not contentious. Accepting these assumptions, which is already implicit in using standard methods of MSA estimation, lets one employ traditional parametric methods of parameter testing and their associated gain in power.

Further, using the bootstrap confidence interval for a parameter test actually entails more subtle statistical assumptions that, in this application, are known to be false. One must assume the contribution estimate is a pivotal statistic (see Lunneborg 2000), and one must accept the arithmetic difference as a useful measure of distance between compositional data vectors (see Aitchison 1992).

The likelihood ratio test's increase in power, relative to the bootstrap confidence interval method, varies with both the identifiability of the population(s) of interest relative to the rest of the baseline and the number of observations the population actually contributes to the mixture sample. When the populations in a region of interest are highly identifiable, both methods are capable of 100% power even for moderately small contributions (Russian River, Figure 3). In these situations, the likelihood ratio method retains 100% power at every contribution level.

As the population(s) of interest becomes increasingly similar to other members of the baseline, the few individuals actually contributed by the population(s) of interest may be adequately explained as having originated from the similar populations. When these similar populations are themselves contributing to the mixture, such as during the Tern Lake scenarios where the Kenai/Skilak region contributes a relatively large portion of the mixture, the principle of parsimony underlying the likelihood ratio test will lead to absorption of the individuals from the population(s) of interest into that component contributed by the similar populations. For example, at low-contribution levels, the few observations from Tern Lake appear to be easily absorbed into the already substantial Kenai/Skilak region contribution.

This absorption is avoided if the population(s) of interest has either sufficiently distinct characteristics that allow clear detection of even a single contribution (e.g. private alleles or the Russian River scenario – Figures 2 and 3), or has a large enough contribution to the mixture sample such that the likelihood ratio test is able to detect the signal in the sample's joint distribution of characteristics (Figure 3). It is sobering to see how easily a contribution can be absorbed, that is, how low the power to detect a nonzero contribution can be (Figure 3, Trail Lakes). The power demonstrated here, with a baseline of only 17 populations, may decrease even more as baseline size increases.

Normal standards of mixture analysis identifiability, that is, the performance on 100% simulations, do not appear to directly relate to expected power in detecting small contributions. For example, relative to Trail Lakes, Tern Lake has a lower correct contribution estimate from its 100% simulations yet it appears to exhibit a higher power of detectability when testing for population absence. This may be an artifact of the simulation method: the Trail Lakes contribution is simulated as equal portions from all three component populations. Relative to the single population contribution from Tern Lake, this partitioning of the Trail Lakes contribution lessens any population-specific contribution, weakening the regional signal.

The likelihood ratio test is very sensitive to private alleles due to the conditional likelihood estimation method. If one is confident that the private allele is, indeed, a population-specific marker, then this makes for a very sensitive test. However, if the private allele is likely an artifact of limited sampling of highly polymorphic markers, this sensitivity may mislead. A substantial portion of the Trail Lakes and Tern Lake contribution detections involved such rare alleles (Table 2). The current demonstration does not allow us to judge whether or not a contribution would have

Table 2. Detections of contribution as a result of private alleles (left) compared with the total detections of contribution using the likelihood ratio method (right), out of 50 simulations per scenario.

Region	Contribution (%)					
	10	5	4	3	2	1
Trail lakes	13/32	11/21	4/11	5/9	0/7	1/3
Tern lake	22/49	6/33	8/26	9/25	6/17	0/7

The likelihood ratio method detects the presence of even a single individual, regardless of mixture sample size, if that individual displays a characteristic unique to its source population. The likelihood ratio method detected every contribution from the Upper Russian River, even though it has no private alleles.

been detected in the absence of a rare allele, that is, strictly by the signal inherent in the joint distribution of characteristics contributed by the population(s) of interest. The Russian River results clearly demonstrate that the likelihood ratio method does not require private alleles to detect contributions (Figure 3). The impact on detection of a given 'private allele' would be revealed by repeating the analysis with the allele in question recoded as a common allele, that is, by binning alleles.

One may consider abandoning MSA for this question and using an analysis that directly assigns each observation in the mixture sample to the 'most probable' population of origin. This can lead to a less powerful test. Conditional maximum likelihood-based individual identification methods (e.g. Cornuet et al. 1999, Banks & Eichert 2000) analyze each observation independently. Since they ignore the information in the mixture sample's joint distribution of characteristics, the methods cannot provide a more powerful test of absence than the MSA likelihood ratio method, and generally will be less powerful. In contrast, Bayesian methods of MSA that estimate, for each observation and each population, the probability the observation came from that population, do utilize the information in the mixture sample's joint distribution of characteristics (Pella & Masuda 2001). For the question of interest, such Bayesian methods will likely have comparable power to the MSA likelihood ratio.

The Bayesian method provides two other appealing features for the question under consideration. First, Bayesian MSA will provide a direct estimate of the distribution of any specific population's mixture contribution, θ. This allows the researcher to directly assess the Prob($\theta > 0$) for that population. Further, one can explore whether a large Prob($\theta > 0$) arises because of a couple of highly distinguishable observations, or simple due to the collective weight of a large number of observations each of which displays some possibility of having originated from the population of interest. Explorations of this application of Bayesian methods are currently in progress.

Regardless of which method one utilizes to assess population absence, one should repeat the demonstration process illustrated here to assess the method's power. Such *a priori* analyses allow one to determine the sample size required to detect a given contribution with a given power, as well as compare methods in a specific context. Note that posterior power analyses, while unfortunately rather common, are generally uninformative and, therefore, misleading (Hoenig &

Heisey 2001). *A priori* power analysis methods in the general context of MSA are discussed elsewhere (Reynolds 2001).

Acknowledgements

We are indebted to P.A. Crane for motivating discussions, L.W. Seeb for support and encouragement, and Sandra Sonnichsen for Figure 1. The manuscript, and the authors' thinking on this topic, were greatly enhanced by the contributions of two anonymous reviewers. Contribution PP-227 of the Alaska Department of Fish and Game, Commercial Fisheries Division, Juneau.

References

Aitchison, J.A. 1992. On criteria for measures of compositional difference. Math. Geol. 24: 365–379.

Banks, M.A. & W. Eichert. 2000. WHICHRUN (Version 3.2) a computer program for population assignment of individuals based on multilocus genotype data. J. Hered. 91: 87–89.

Begg, G.A., K.D. Friedland & J.B. Pearce. 1999. Stock identification – its role in stock assessment and fisheries management. Fish. Res. 43: 1–3.

Cornuet, J.M., S. Piry, G. Luikart, A. Estoup & M. Solignac. 1999. New methods employing multilocus genotypes to select or exclude populations as origins of individuals. Genetics 153: 1989–2000.

Davison, A.C. & D.V. Hinkley. 1997. Bootstrap Methods and their Application, Cambridge University Press, Cambridge, UK. 582 pp.

Debevec, E.M., R.B. Gates, M. Masuda, J.J. Pella, J.H. Reynolds & L.W. Seeb. 2000. SPAM (version 3.2): Statistics Program for Analyzing Mixtures. J. Hered. 91: 509–511.

Fournier, D.A., T.D. Beacham, B.E. Ridell & C.A. Busack. 1984. Estimating stock composition in mixed stock fisheries using morphometric, meristic, and electrophoretic characteristics. Can. J. Fish. Aquat. Sci. 41: 400–408.

Hoenig, J.M. & D.M. Heisey. 2001. The abuse of power: The pervasive fallacy of power calculations of data analysis. Amer. Stat. 55: 19–25.

Ihssen, P.E., H.E. Booke, J.M. Casselman, J.M. McGlade, N.R. Payne & F.M. Utter. 1981. Stock identification: Materials and methods. Can. J. Fish. Aquat. Sci. 38: 1838–1855.

Lunneborg, C.E. 2000. Data Analysis by Resampling: Concepts and Applications, Duxbury Press, Pacific Grove, CA, U.S.A. 568 pp.

Marshall, S., D. Bernard, R. Conrad, B. Cross, D. McBride, A. McGregor, S. McPherson, G. Oliver, S. Sharr & B. Van Allen. 1987. Application of scale patterns analysis to the management of Alaska's sockeye salmon (*Onchorhynchus nerka*) fisheries. pp. 207–326. *In*: H.D. Smith, L. Margolis & C.C. Wood (ed.) Sockeye Salmon (*Oncorhynchus nerka*) Population Biology and Future Management, Canadian Special Pub. on Fish. and Aquat. Sci. 96.

Millar, R.B. 1987. Maximum likelihood estimation of mixed stock fishery composition. Can. J. Fish. Aquat. Sci. 44: 583–590.

Milner, G.B., D.J. Teel, F.M. Utter & C.L. Burley. 1981. Columbia River stock identification study: Validation of genetic method. Final report of research (FY80) financed by Bonneville Power Administration Contract DE-A179-80BP18488, National Marine Fisheries Service, Northwest and Alaska Fisheries Center, Seattle, WA. 35 pp.

Moles, A. & K. Jensen. 2000. Prevalence of the sockeye salmon brain parasite *Myxobolus arcticus* in selected Alaska streams. Alaska Fish. Res. Bull. 6: 85–93.

Pearce, J.M., B.J. Pierson, S.L. Talbot, D.V. Derksen, D. Kraege & K.T. Scribner. 2000. A genetic evaluation of morphology used to identify harvested Canada geese. J. Wild. Manag. 64: 863–874.

Pella, J.J. & M. Masuda. 2001. Bayesian methods for stock-mixture analysis from genetic characters. Fish. Bull. 99: 151–167.

Pella, J.J. & G.B. Milner. 1987. Use of genetic marks in stock composition analysis. pp. 247–276. *In*: N. Ryman & F. Utter (ed.) Population Genetics and Fishery Management, Washington Sea Grant Program, Seattle, WA.

Redner, R.A. & H.F. Walker. 1984. Mixture densities, maximum likelihood and the EM algorithm. Soc. Indust. Appl. Math. Rev. 26: 195–239.

Reynolds, J.H. 2001. SPAM version 3.5: User's guide addendum. Addendum to special publication No. 15, Alaska Dept. of Fish and Game, Commercial Fisheries Division, Gene Conservation Lab, 333 Raspberry Rd., Anchorage, AK, 99518–1599 (available: www.cf.adfg.state.ak.us/geninfo/research/genetics/Software/SpamPage.htm).

Reynolds, J.H. & W.D. Templin. 2003. Testing component contributions in finite discrete mixtures: detecting specific populations in mixed stock fisheries. pp. 2873–2878. *In*: Proceedings of the Joint Statistical Meetings, New York, New York, Aug. 11–15, 2002. Am. Stat. Assoc., Alexandria, VA.

Ruzzante, D.E., C.T. Taggart, S. Lang & D. Cook. 2000. Mixed-stock analysis of Atlantic cod near the Gulf of St. Lawrence based on microsatellite DNA. Ecol. Appl. 10: 1090–1109.

Seeb, L.W. & P.A. Crane. 1999. Allozymes and mitochondrial NDA discriminate Asian and North American populations of chum salmon in mixed-stock fisheries along the south coast of the Alaska Peninsula. Trans. Amer. Fish. Soc. 128: 88–103.

Seeb, L.W., C. Habicht, W.D. Templin, K.E. Tarbox, R.Z. Davis, L.K. Brannian & J.E. Seeb 2000. Genetic diversity of sockeye salmon of Cook Inlet, Alaska, and its application to management of populations affected by the *Exxon Valdez* Oil Spill. Trans. Amer. Fish. Soc. 129: 1223–1249.

Shaklee, J.B., F.W. Allendorf, D.C. Morizot & G.S. Whitt. 1990. Gene nomenclature for protein-coding loci in fish. Trans. Amer. Fish. Soc. 119: 2–15.

Shaklee, J.B., T.D. Beacham, L. Seeb & B.A. White. 1999. Managing fisheries using genetic data: Case studies from four species of Pacific salmon. Fish. Res. 43: 45–78.

Smouse, P.E., R.S. Waples & J.A. Tworek. 1990. A genetic mixture analysis for use with incomplete source population data. Can. J. Fish. Aquat. Sci. 47: 620–634.

Stuart, A., J.K. Ord & S. Arnold. 1999. Kendall's Advanced Theory of Statistics Vol 2A: Classical Inference and the Linear Model, 6th edition, Oxford University Press, New York, NY, U.S.A. 885 pp.

Urawa, S., K. Nagasawa, L. Margolis & A. Moles. 1998. Stock identification of chinook salmon (*Onchorhynchus tshawytscha*) in the North Pacific Ocean and Bering Sea by parasite tags. Nor. Pacific Anadro. Fish. Com. Bull. 1: 199–204.

Weir, B.S. 1996. Genetic data analysis II, Sinauer Associates, Sunderland, MA, U.S.A. 445 pp.

Appendix 1

Ideal power to detect nonzero contributions using bootstrap confidence intervals.

Part 1. Assume that the population of interest, Population A, contributes a proportion $0 < \theta < 1$ of uniquely identifiable individuals to the mixture and that N individuals are randomly and independently sampled from the mixture. For example, consider red balls (Population A) randomly mixed with blue balls (Population B), in proportion θ to $1 - \theta$, in an infinite barrel. N balls are randomly selected as the mixture sample. A one-sided lower 95% confidence interval is constructed from 1000 bootstrap resamples using the percentile method (Lunneborg 2000).

Let k be the number of individuals from Population A in the original mixture sample, $k \sim$ binomial(N, $p = \theta$). Then Prob(no Population A individuals in one resample $|$ k) $=$ Prob(no red balls in random sample, with replacement, of original N balls) $= (1 - k/N)^N = v$.

Part 2. Let X be the number of times out of 1000 resamples in which no red balls are found. X is a binomial random variable with N $=$ 1000 trials and 'success' probability $= v$. Then the probability a confidence interval calculated using the one-sided 95% percentile method includes 0 is the probability that 50 or more of the resamples will contain no red balls, i.e. that $X \geq 50$.

$$\mathrm{Prob}(X \geq 50) = 1 - \mathrm{Prob}(X < 50)$$

$$= 1 - \sum_{X=0}^{49} \binom{1000}{X} v^X (1 - v)^{1000-X}$$

So Prob(CI calculated using one-sided 95% percentile method does not include 0)

$$\text{Prob}(X < 50)$$

$$= 1 - \left[1 - \sum_{X=0}^{49} \binom{1000}{X} v^X (1-v)^{1000-X} \right]$$

$$= \sum_{X=0}^{49} \binom{1000}{X} v^X (1-v)^{1000-X}$$

Part 3. The power to detect Population As nonzero contribution θ to the mixture, using the bootstrap confidence interval method, is

$$\sum_{k=0}^{N} (\text{Power} \mid k, N) \text{Prob}(K = k \mid \theta, N)$$

$$= \sum_{k=0}^{N} \left(\sum_{X=0}^{49} \binom{1000}{X} v^X (1-v)^{1000-X} \right)$$

$$\times \text{Prob}(K = k \mid \theta, N)$$

$$= \sum_{k=0}^{N} \left(\sum_{X=0}^{49} \binom{1000}{X} \left(\left(1 - \frac{k}{N}\right)^N \right)^X \middle/ \left(1 - \left(1 - \frac{k}{N}\right)^N \right)^{1000-X} \right)$$

$$\times \binom{N}{k} \theta^k (1-\theta)^{N-k}$$

Environmental Biology of Fishes **69**: 245–259, 2004.

Sampling issues affecting accuracy of likelihood-based classification using genetical data

B. Guinand[a,d], K.T. Scribner[a], A. Topchy[b], K.S. Page[a,e], W. Punch[b] & M.K. Burnham-Curtis[c,f]
[a]*Department of Fisheries and Wildlife,* [b]*Department of Computer Science and Engineering, Michigan State University, East Lansing, MI 48824, U.S.A. (e-mail: guinand@univ-montp2.fr)*
[c]*U.S.G.S/BRD, Great Lakes Science Center, 1451 Green Rd., Ann Arbor, MI 48105, U.S.A.*
[d]*Current address: UMR CNRS 5000 Génome, Populations et Interactions, Université des Sciences et Techniques du Languedoc, Montpellier, France.*
[e]*Current address: Minnesota Department of Natural Resources, Fisheries Research Unit, 1601 Minnesota Dr., Brainerd, MN 56401, U.S.A.*
[f]*Current address: U.S. Fish and Wildlife Service National Forensic Laboratory, Ashland, OR, U.S.A.*

Received 21 April 2003   Accepted 24 April 2003

Key words: assignment test, genetic algorithm, locus selection, genetic differentiation, microsatellite, lake trout

Synopsis

We demonstrate the effectiveness of a genetic algorithm for discovering multi-locus combinations that provide accurate individual assignment decisions and estimates of mixture composition based on likelihood classification. Using simulated data representing different levels of inter-population differentiation ($F_{st} \sim 0.01$ and 0.10), genetic diversities (four or eight alleles per locus), and population sizes (20, 40, 100 individuals in baseline populations), we show that subsets of loci can be identified that provide comparable levels of accuracy in classification decisions relative to entire multi-locus data sets, where 5, 10, or 20 loci were considered. Microsatellite data sets from hatchery strains of lake trout, *Salvelinus namaycush*, representing a comparable range of inter-population levels of differentiation in allele frequencies confirmed simulation results. For both simulated and empirical data sets, assignment accuracy was achieved using fewer loci (e.g., three or four loci out of eight for empirical lake trout studies). Simulation results were used to investigate properties of the 'leave-one-out' (L1O) method for estimating assignment error rates. Accuracy of population assignments based on L1O methods should be viewed with caution under certain conditions, particularly when baseline population sample sizes are low (<50).

Introduction

In recent years, new molecular markers and novel methods of analysis have enabled researchers to extend investigations of ecological and evolutionary processes below the population level to the level of individuals. Examples include estimates of coefficients of relationship and parentage analysis (e.g., Queller & Goodnight 1989), assignments of individuals to putative breeding populations or to different subspecies, and detection of sex-specific patterns of dispersal (review in Hansen et al. 2001). Fisheries science has recently emphasized methods and applications of assigning individuals to populations of origin (e.g., Potvin & Bernatchez 2001, Nielsen et al. 2001). Population assignment methods can be separated into two groups (Cornuet et al. 1999): 'general methods' such as discriminant analysis (e.g., Cornuet et al. 1996) and neural networks (e.g., Aurelle et al. 1999) that can be used for any data set, and 'population genetics methods' including likelihood (Paetkau et al. 1995) and Bayesian approaches (Rannala & Mountain 1997, Cornuet et al. 1999, Pritchard et al. 2000) that are specific to genetic data. Most current empirical research in the area of individual assignment uses likelihood-based approaches (review in Waser & Strobeck 1998) in which assignment is based on features (e.g., loci) and parameters (e.g., allele frequencies) conditioned on samples of baseline populations. Cornuet et al. (1999), expanding upon earlier work by Paetkau et al. (1995) and Rannala & Mountain (1997), have shown in simulations that four factors are important for accuracy in assignment of individuals to populations of origin: the number of loci, the number of alleles per locus, the population size, and degree of population differentiation as measured by F_{st} (Weir & Cockerham 1984). Greater numbers of loci of high discriminatory power (i.e., larger F_{st}) result in greater accuracy (e.g., Smouse et al. 1982, Cornuet et al. 1999). Inclusion of additional loci is generally considered to lead to greater improvements in assignment accuracy over that achieved by increased sample sizes or selection of loci with high allelic diversity (e.g., Smouse & Chevillon 1998). Bernatchez & Duchesne (2000) compared, in simulations, assignment of parentage and assignment to populations with varying numbers of loci and with loci of different allelic diversity. They found that selection of loci with high allelic diversity is most important for assigning parentage (see also Letcher & King 1999), whereas increasing the number of loci has the greatest effect on the accuracy of assignment to population of origin. Other measures have also been shown to predict assignment accuracy (see Shriver et al. 1997), but have been less thoroughly evaluated. Recent empirical applications of population assignment methods often contradict the general guidelines offered above. For instance Roques et al. (1999) assigned individuals of the genus *Sebastes* to different populations or subspecies, showing that the probability of correct assignment was highest when four of eight polymorphic loci were used. Similar results have been demonstrated in simulations (Bernatchez & Duchesne 2000), and empirical studies (e.g., Girman et al. 1997, Norris et al. 2000).

Many current methods of assignments based on genetic data assume that allele frequencies are known accurately. However, in most 'practical' applications, allele frequencies are estimated from a finite set of samples, often from the same data used for classification (details in Hansen et al. 2001). When states (e.g., allele frequencies) of large numbers of features (e.g., loci) are estimated with error, the inclusion of additional features beyond a certain number can lead to higher assignment error. This phenomenon is known as the 'curse of dimensionality' in applications of pattern recognition research (Trunk 1979). For many empirical population genetics applications, allele frequencies used in assignment tests are estimated from small or moderate samples of baseline populations. For instance, assignment tests are often used for conservation (e.g., Martinez et al. 2001) or management of populations (e.g., Hansen et al. 2000, Taylor et al. 2000, Neraas & Spruell 2001, Nielsen et al. 2001), and small samples (<50 individuals per population) in which few loci ($\leq$10) are observed are generally the rule. Because the variance of estimates of allele frequencies is generally ignored, true probabilities of assignment error can increase beyond the estimated probabilities as dimensionality (number of loci) increases, particularly if the loci are characterized by high allelic diversity. Moreover, because most assessments of accuracy of assignment tests are based on the 'leave-one-out' (L1O) method of resubstitution (Efron 1983), and although L1O may provide unbiased estimates of generalization error, it may perform poorly for discontinuous error functions such as the number of misclassified cases.

Under circumstances routinely encountered in fisheries, it would be prudent to ascertain what errors and biases may exist in classification decisions, evaluation of both locus selection methods and error estimation methods (e.g., L1O) are crucial for applications to management and conservation. Using locus selection techniques, it is important to know if comparable assignment accuracies are possible when fewer loci are employed.

In this study, we contrast estimates of assignment accuracies based on all loci with estimates from 'exhaustive' searches in which assignment accuracies of all possible multi-locus combinations, including subsets of loci, are evaluated. We use a genetic algorithm (GA; Holland 1994, Goldberg 1989) to show that under many situations commonly realized in studies of natural fish populations, optimal subsets of loci exist that provide more cost-effective means of attaining desired assignment accuracy than examination of all loci and with comparable accuracy. We simulate assignment with ranges of values of number of loci, allelic diversity, inter-population variance in allele frequency, and population size (influencing reliability estimates of allele frequency), that are routinely observed in natural populations and that have been shown to influence accuracy of assignment decisions. We show that L1O assessments of assignment error rate may not reveal how well the loci (or subsets of loci) perform 'generally' when used to assign an entirely new group of individuals. As an empirical example, we use a GA to select loci for an assignment of individual lake trout (*Salvelinus namaycush*; Pisces, Salmonidae) to their respective hatchery stocks. We extend analyses using GA-reduced data sets to analyses of mixtures (mixed-stock analyses).

Methods

Classification (assignment) involves the evaluation of measured attributes that occur in different states in different individuals or observations (Duda et al. 2000, Jain et al. 2000). In population genetics, most classification methods are 'supervised' in the sense that assessment of classifier accuracy is based on baseline (training) data collected from each of several known or predefined populations (e.g., Paetkau et al. 1995, Rannala & Mountain 1997, but see Pritchard et al. 2000). One implicit assumption is that accuracy demonstrated for individuals from known baseline populations will translate to comparable classification accuracy for unknown individuals that are the subjects of study, and form the basis on which ecological and evolutionary inferences will be drawn (e.g., Hansen et al. 2001). Good inferences about unknown individuals thus require maximum accuracy of the baseline assignment based on individuals of known origin. Further, knowledge of properties of data that translate to improved assignment accuracy may lead to greater predictability of the utility of attributes (loci) in different contexts (e.g., Shriver et al. 1997, Bernatchez & Duchesne 2000). Here we describe methods emphasizing likelihood-based methods of assignment of individuals to population of origin. We show that these methods are directly applicable to mixed-stock analyses routinely employed in fisheries research and management (Pella & Milner 1987).

Assignment accuracy

Given estimates of allele frequencies $\hat{\theta}_A$ and $\hat{\theta}_B$ determined from training samples of size n′ in two populations (A and B), the goal is to use the genotype of an individual of unknown origin to assign it to one of the two populations. For all simulated and empirical case studies described below, classification accuracy is estimated by the deterministic likelihood method defined

by Paetkau et al. (1995). An individual is assigned to population A if the ratio

$$\frac{L_A}{L_B} = \frac{\text{Prob(genotype)}|\hat{\theta}_A}{\text{Prob(genotype)}|\hat{\theta}_B} \tag{1}$$

is greater than one, and to population B if the ratio is less than one. In Equation (1) L_A is the probability of the individual's multi-locus genotype (assuming both Hardy–Weinberg and linkage equilibriums within the source population), conditional on its origin being in a population with allele frequencies $\hat{\theta}_A$. A complication occurs when an allele in an unknown individual does not occur in the baseline sample of a candidate population so that the estimate of that allele frequency in the candidate population is zero (e.g., Cornuet et al. 1999). In such a case the deterministic likelihood method leads to a likelihood of zero and eliminates *a priori* the candidate population from consideration even though the allele may actually occur in the candidate population and the likelihood is not actually zero. In our simulated and empirical data sets null frequencies have been replaced by $[1/(2n + 1)]$ where n is the number of sampled individuals in each population baseline, and no candidate populations are eliminated *a priori*.

Several measures are available that provide estimates of true or predictive accuracy: cross-validation accuracy, bootstrap accuracy, and L1O accuracy as a specific case of cross-validation (see review in Shao 1993). L1O is most commonly used, since it has been shown to be unbiased (i.e., on average, the true accuracy is correctly estimated). An estimate of L1O accuracy represents a mean of accuracies over a number of classifiers, where each classifier is obtained from the original sample from which a single individual has been removed. However, L1O estimates of assignment accuracy often have large variances, since estimated accuracy can change greatly across different baseline samples (e.g., for baselines comprised of individuals with genotypes made up of common alleles compared with baselines comprised of individuals characterized by rare alleles). In our simulation studies we evaluate a range of parameter values generated from classifiers identified for naïve (all loci used) or for reduced data sets. We use a GA to identify, based on L1O criteria, subsets of reduced data that can be used to classify other untested individuals (i.e., reduced data sets that have high predictive accuracy).

Genetic algorithms and their operations

Holland (1994) described GAs as a part of a computational methodology for studying natural adaptive systems. They are based on analogies to processes of natural selection in biology. Evolutionary, phylogenetic, and genetic applications of GAs have increased in recent years (e.g., Lewis 1998, Mitchell & Taylor 1999, Stefanini & Camussi 2000). A GA has the advantage that it can explore the space of different locus groupings as combinations of loci, rather than treating each locus separately (Shriver et al. 1997). This is advantageous because the information available for classifying individuals resides not merely in which particular alleles appear within the individuals but also in which single-locus and multi-locus genotypes appear in individuals. The search process in GA eliminates uninformative features (i.e., loci), and retains optimal subsets of all features. In the context of assignment testing in population genetics, a GA provides a search process for locus selection. A GA is not a 'classifier' as are likelihood-based approaches commonly used in molecular ecology (Equation (1)). A GA is simply a tool that can be used to select combinations of loci that provide comparable or better accuracy of estimates (benefits) at reduced cost.

Genetic algorithm terminology defines a classification solution as a point or 'individual' in the search space as a 'string'. Or, in a population genetics context, an individual is a subset of possible loci. A GA requires a population of 'individuals' (locus combinations) that each represents a subset of the total set of available locus combinations (Figure 1). An 'evaluation function' [F(n, s)] is used to calculate the 'fitness' of the multi-locus combinations, in our present case determined by likelihood-based estimates of individual classification accuracy. The GA manipulates the population of 'individuals' in a manner analogous to evolutionary processes and natural selection (details in, e.g., Mitchell & Taylor 1999). The optimization process in a GA is performed in cycles called 'generations' (Figure 1). During each generation, operators create a set of new individuals (Figure 1). These operators structurally change initial locus combinations by processes analogous to crossover, inversion, and mutation. For example, a mutation within a string produces a nearly identical string with only local alterations (e.g., use [1] or disuse [0] of a locus, Figure 1). The cycle repeats until the population 'converges' to an optimal value, where the GA is unable to find locus

combinations that cannot be outperformed by others in additional generations. Optimal solutions retain only a fraction of the original number of loci. As such, GAs offer an attractive means of locus selection before attempting to make assignment decisions for unknown individuals.

We use a GA in this study to manipulate locus combinations and use accuracy of likelihood classification (the proportion of correctly assigned individuals, Figure 1) to define optimality for each GA cycle or generation. The following steps define the use of the GA for both simulated and empirical data (Figure 1):

Step 1 – Creation of 'individuals' or subsets of features. We are interested in the effects of inclusion or exclusion of loci in the assignment procedure on classification accuracy. A convenient representation of a locus subset is shown as 'individuals' whose strings contain a number of variables equal to the number of loci studied. A locus included in the analysis is encoded by '1' in a string and if unused, by a '0' (Figure 1).

Step 2 – The evaluation function F(n, s). We use a likelihood function (i.e., the assignment score or the accuracy as defined in Equation (1)) that assigns a numerical value (accuracy of assignment) to each subset of loci (Figure 1).

Step 3 – The operators. Operators include selection, crossover, and mutation. We use one-point crossover, one-bit mutation, and proportional selection sequentially (Figure 1; details below).

Other important parameters include GA population size, number of GA generations, and number of GA runs. The population size (the number of individuals to be evaluated before operations are performed) is a critical value. Too few 'individuals' in the population may lead to poor searches of multi-feature space, while too many locus combinations unnecessarily slow the search process. Further, the nature of the initial population is important. Typically the population is created with 'random' feature combinations to ensure that the search is not biased, pushing the GA to a particular portion of the search space. Similarly, a sufficient number of GA generations (Figure 1) must be performed to allow operators to investigate new strings, and then to optimize the GA process as described above. Finally, because a GA starts with a random population of different locus combinations (Figure 1), the results of various experiments can vary and 'convergence' times across generations may vary. Several runs (for a fixed population size and a fixed number of GA generations) should be conducted to evaluate GA performance.

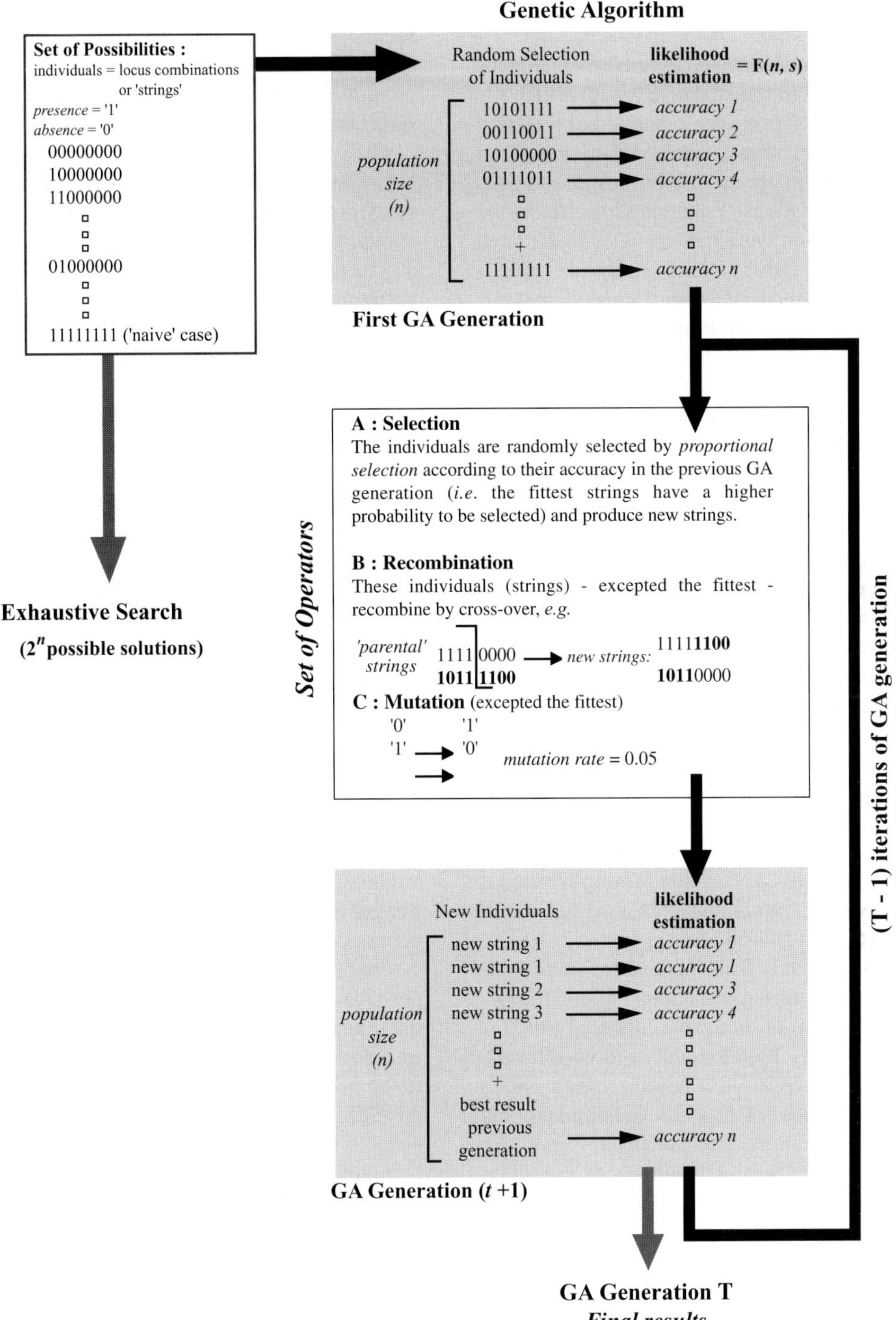

Figure 1. Diagrammatic representation of two analyses that search for locus combinations maximizing the accuracy of assignment: an exhaustive search (i.e., test of all-locus combinations) and an exploratory search based on a GA. Randomly selected 'individuals' (or 'strings', or combinations of loci) are analyzed in groupings of user specified population size in each run of the GA search. Runs consist of a succession of GA generations where suits of locus combinations are manipulated using operators that mimic evolutionary processes and that facilitate exploration of multi-locus parameter space.

Design of GA experiments

Considerable variation exists in the implementation of the GA with respect to the stringency with which the 'survival of the fittest' principle is applied to selection of multi-locus combinations. 'Fitness', here measured as the accuracy of assignment with respect to particular locus combinations (Equation (1)), affects the probability that locus combinations will be retained in subsequent generations. In this study, we used a crossover rate of 80% and a mutation rate of 5% per GA generation. Those values are typical in pattern recognition research (Duda et al. 2000). The selection of locus combinations between successive GA generations was proportional (i.e., locus combinations with higher assignment accuracy had a higher probability of being selected in the next GA generation). We initially seeded each population of locus combinations with an individual string of loci representing the all-locus combination (Figure 1). This procedure forces an immediate comparison with the all-locus case and drives the GA to an early search for locus combinations with greater classification accuracy. Seeding the population avoids a totally random GA in which the likelihood estimated for the all-locus case might not be evaluated.

GA population sizes, number of GA generations and number of runs were determined by successive tests using the protocol illustrated in Figure 1. For both simulated and empirical data sets we used 20 GA generations and 10 GA runs (experiments), numbers based on preliminary tests. For simulated data the GA population sizes were 20, 40, and 50 strings for the 5, 10, and 20 loci cases, respectively. This increase in GA population size with increasing numbers of loci allows the GA to explore a sufficient subset of solutions (Figure 1). A GA population size of 20 individuals was used for the empirical lake trout data set. Since the entire search space (i.e., total number of multi-locus combinations) for the empirical example was relatively small (for eight loci; 2^8 or 256 possible multi-locus combinations), an exhaustive search of all possible multi-locus combinations was also performed.

Evaluation of GA using simulated data

We conducted a series of experiments using an idealized simulation setting, whereby the underlying data distribution of baseline population allele frequencies was known. Full knowledge allows using new individuals (independent of training samples) for the search process as well as for testing of true classification accuracy with high precision. The results provide upper bounds on evaluations of GA locus selection improvement.

The effects on classification accuracy of two values of average F_{st} (0.01 'low' and 0.1 'high'; Wright 1965), two values of allelic diversity (four and eight alleles per locus), three numbers of loci (5, 10, and 20), three baseline sample sizes (20, 40, 100 individuals per population) were evaluated (36 different parameter combinations) using simulated data for each of two baseline populations. Data distributions defining allelic probabilities at each locus for each baseline population were prepared corresponding to each parameter combination. All evaluations were performed with populations of multi-locus individuals in Hardy–Weinberg and linkage equilibrium. Multi-locus distributions with 'low' average F_{st} had actual F_{st} values of loci uniformly distributed between 0.02 (twice the mean) and 0.0 (e.g., in simulations employing five loci F_{st} values were 0.020, 0.015, 0.010, 0.005, and 0.00). Distributions with 'high' F_{st} used loci uniformly distributed over the range of F_{st} [0.20–0.00]. Thus different loci within each multi-locus suite for all simulations have different discriminatory power. For each distribution we performed 100 experiments with different baseline samples drawn from the same distribution. We report the mean (± 1 standard deviation) assignment accuracy for each simulation.

Once frequencies were estimated, a likelihood-based assignment test was performed (Equation (1)). To evaluate performance of any subset of loci we always used a separate 'holdout' sample of 400 individuals (200 per population) that was independently drawn from distributions associated with each baseline population. Samples were drawn randomly and independently in each experiment. Holdout sets of such large size provided reliable estimates of true accuracy and reasonable trade off between precision and computation complexity.

Subsequently, the GA search for loci subsets was conducted for each of the 100 simulated data sets for all 36 parameter combinations. The search used GA for 20 generations with 24 strings (subsets). The accuracy of the best subset on the 400 holdout individuals at the end of search is reported and averaged over 100 experiments for all the distributions. The naïve classifier (using all loci) was also tested against the holdout sample. To estimate predictive accuracy of each classifier, we generated 20 000 (10 000 per population) new individuals and evaluated the best classifier over all simulations.

True accuracy of each naïve classifier (all loci included) was also obtained for the same 20 000 individuals using the GA-reduced subsets across all experiments. The later analysis allowed us to answer to our main question of whether comparable predictive accuracy was achievable with GA-reduced subsets of loci and under what plausible range of parameter combinations. The values of accuracy were also compared using a significance test, which assumed a binomial distribution for classification errors. If the 95% confidence interval for assignment accuracy did not overlap between the naïve classifier and that found by the GA, then we reported this difference in accuracies as significant.

Evaluation of GA using empirical data

Three lake trout, *S. namaycush* (Salmonidae) strains [Marquette (SMD), Isle Royale (SIW) and Seneca Lake (SLW)] were considered in this study. The SMD and SIW strains originated from native Lake Superior lake trout populations. The SLW strain was derived from a native population from Seneca Lake in the state of New York. U.S. Fish and Wildlife Service hatchery personnel sampled adult lake trout (153 individuals; 57 SIW, 45 SMD, 51 SLW) during 1998 spawning. Portions of the caudal fin (1 cm^2) were collected, and kept at $-80°C$ until analysis. DNA was extracted by proteinase K digestion and a modification of the Puregene protocol (Gentra Inc.). DNA was resuspended in TE buffer (10 mM Tris–Hcl, pH 8.0, 1 mM EDTA). Concentrations were determined by fluorimetry.

Eight polymorphic microsatellite loci were screened: *Sfo1*, *Sfo12*, and *Sfo18* (Angers et al. 1995), *Oneμ9*, *Oneμ10* (Scribner et al. 1996), *Ogo1a* (Olsen et al. 1998), *Scoμ19* (Taylor et al. 2001), and *Ssa85* (O'Reilly et al. 1996). PCR reactions using 100 ng DNA were performed in 25 μl volumes according to specifications outlined by each author. Microsatellite polymorphisms were screened on 6% acrylamide denaturing sequencing gels and visualized with a Hitachi FMBIO II$^®$ Multi-View scanner and appropriate software (Hitachi Software).

Tests of Hardy–Weinberg equilibrium were calculated for each locus and population (GENETIX; Belkhir et al. 1996). Between-population levels of genetic differentiation (F_{st}) and within-population variation (F_{is}) were estimated as in Weir & Cockerham (1984) for each microsatellite locus. Statistical significances of observed F_{st} and F_{is} values for each population and for each locus were assessed in each case using 1000 bootstrap replicates. Linkage disequilibrium was examined as a factor affecting locus selection in assignment tests. We refer to 'linkage disequilibrium' as a composite measure of degree of association between loci. Exact χ^2 tests of linkage disequilibrium were calculated for each pair of alleles and for each pair of loci following Weir (1979) using GENETIX (Belkhir et al. 1996). The step-down Holm–Šidák adjustment for multiple tests (Sokal & Rohlf 1995) was applied for each test to evaluate significance.

Mixture analyses

Identical GA analyses were performed using likelihood-based mixture analysis (Pella & Milner 1987) instead of likelihood assignment test as the evaluation function in Figure 1. Further evaluation of classifiers with holdout data sets was similarly used to test predictive accuracy. For seek of simplicity, we report results for only one case using this approach.

Results

GAs improves accuracy of assignment

Simulated data sets for individual assignment
Accuracy of assignment based on all loci increased with increasing sample size in the simulated baseline populations, presumably related to decreasing error in allele frequency estimation in population samples of larger size (Table 1). The increase of assignment accuracy with increasing sample size for cases in which all loci were included was always greater for populations exhibiting greater variance in allele frequency (high F_{st} cases). Accuracy generally increased with loci characterized by high allelic diversity (eight vs. four alleles per locus), and when increasing numbers of loci were evaluated.

Simulations were initially conducted by drawing a sample of 20, 40, or 100 individuals from distributions to represent a sample baseline from which allele frequencies at each locus were estimated. We drew an additional 10 000 individuals from the same distributions simulating a situation of 'total knowledge' of the population from which baseline samples were drawn. Using the likelihood estimator (Equation (1))

Table 1. Results of simulation analyses assessing potential bias in estimates of classification accuracy for different methods of cross-validation and of the accuracy and precision by which GA are able to elucidate subsets of loci that optimize classification accuracy. Average number of loci used by GA were computed over 100 replicates.

Baseline size		Four alleles per locus							Eight alleles per locus						
		Validation using						Average number of loci in GA	Validation using						Average number of loci in GA
		Large set of individuals			Leave-one-out		Holdout (SD)		Large set of individuals			Leave-one-out		Holdout (SD)	
		All loci (SD)	GA-reduced (SD)	GA sign. better (%)	All loci (SD)	GA-reduced (SD)			All loci (SD)	GA-reduced (SD)	GA sign. better (%)	All loci (SD)	GA-reduced (SD)		
Low F_{st}															
5 loci	20	0.645	0.654	33	0.627	0.634	0.694	3.41	0.678	0.697	60	0.663	0.659	0.734	3.21
		0.024	0.023		0.095	0.096	0.024		0.032	0.028		0.089	0.083	0.031	
	40	0.671	0.672	16	0.675	0.672	0.710	3.61	0.728	0.733	23	0.732	0.729	0.771	3.82
		0.014	0.016		0.061	0.059	0.020		0.018	0.018		0.050	0.043	0.018	
	100	0.693	0.689	3	0.694	0.690	0.726	4.06	0.766	0.766	3	0.767	0.763	0.798	4.00
		0.007	0.012		0.033	0.030	0.016		0.007	0.010		0.031	0.031	0.016	
10 loci	20	0.690	0.691	26	0.677	0.673	0.727	6.96	0.734	0.748	47	0.710	0.703	0.782	6.84
		0.024	0.023		0.096	0.096	0.024		0.028	0.028		0.077	0.074	0.025	
	40	0.727	0.720	1	0.725	0.713	0.758	7.46	0.800	0.797	9	0.799	0.795	0.831	7.45
		0.164	0.019		0.054	0.059	0.021		0.015	0.016		0.051	0.050	0.018	
	100	0.757	0.744	1	0.757	0.744	0.780	7.72	0.841	0.833	0	0.840	0.833	0.862	7.96
		0.007	0.014		0.035	0.034	0.015		0.006	0.012		0.031	0.029	0.014	
20 loci	20	0.754	0.744	4	0.738	0.735	0.785	13.74	0.799	0.806	34	0.779	0.750	0.834	13.7
		0.024	0.025		0.083	0.082	0.026		0.028	0.024		0.066	0.080	0.024	
	40	0.798	0.783	2	0.809	0.777	0.818	14.82	0.883	0.868	2	0.890	0.872	0.897	15.00
		0.014	0.018		0.048	0.058	0.018		0.011	0.018		0.038	0.037	0.017	
	100	0.833	0.816	0	0.835	0.816	0.848	15.46	0.920	0.907	0	0.918	0.905	0.928	17.7
		0.006	0.014		0.028	0.027	0.014		0.005	0.011		0.017	0.025	0.011	
High F_{st}															
5 loci	20	0.949	0.954	20	0.942	0.933	0.969	3.83	0.889	0.900	37	0.875	0.881	0.924	3.45
		0.014	0.014		0.034	0.044	0.013		0.022	0.021		0.061	0.060	0.020	
	40	0.966	0.966	4	0.965	0.962	0.980	4.07	0.938	0.941	11	0.933	0.931	0.956	3.98
		0.010	0.010		0.019	0.020	0.010		0.013	0.013		0.028	0.028	0.013	
	100	0.974	0.973	0	0.975	0.975	0.984	4.40	0.958	0.957	0	0.958	0.956	0.972	4.24
		0.002	0.003		0.010	0.010	0.004		0.002	0.006		0.015	0.016	0.006	
10 loci	20	0.987	0.985	1	0.985	0.975	0.993	7.47	0.950	0.951	8	0.937	0.928	0.965	7.35
		0.007	0.007		0.021	0.023	0.006		0.014	0.012		0.041	0.046	0.012	
	40	0.994	0.993	0	0.995	0.992	0.998	7.84	0.982	0.981	0	0.981	0.975	0.990	7.97
		0.003	0.003		0.009	0.010	0.003		0.006	0.007		0.015	0.018	0.006	
	100	0.997	0.995	0	0.996	0.994	1.000	8.06	0.993	0.990	0	0.993	0.990	0.997	8.01
		0.000	0.002		0.004	0.006	0.001		0.001	0.003		0.005	0.007	0.003	
20 loci	20	0.999	0.997	0	0.998	0.996	1.000	14.66	0.989	0.987	0	0.987	0.976	0.996	14.96
		0.001	0.002		0.007	0.009	0.001		0.007	0.007		0.018	0.025	0.004	
	40	1.000	0.999	0	1.000	1.000	1.000	16.24	0.998	0.996	0	0.997	0.996	1.000	14.30
		0.000	0.001		0.000	0.002	0.000		0.001	0.003		0.005	0.007	0.001	
	100	1.000	1.000	0	1.000	1.000	1.000	18.18	1.000	0.999	0	1.000	0.999	1.000	16.62
		0.000	0.000		0.000	0.001	0.000		0.001	0.001		0.001	0.002	0.000	

we used the parameter estimates (allele frequencies associated with $N = 20, 40, 100$ baselines) to assign the entire set of 20 000 individuals. When large numbers of independent individuals were available for verification of assignment accuracy, the GA-reduced set of loci consistently provided comparable (or even greater) assignment accuracy (Table 1). Estimates of variance in estimates of accuracy for the naïve and GA-reduced sets of loci were comparable suggesting a high degree of predictive accuracy with classifiers developed using GA. In situations with low baseline sample size and few numbers of loci, the improvement is assignment accuracy of GA-reduced *versus* naïve sets were statistically significant in a large number of replicate trials (Table 1).

Estimates of accuracy were based on L1O resampling of the baseline samples. Results (naïve L1O vs GA-L1O; Table 1) revealed that accuracy estimated using L1O were consistently downwardly biased and imprecise as seen by much higher inter-replicate variance in assignment accuracy across the 100 replicates (Table 1). GA-reduced data sets typically reduce the number of loci 15–40% below the naïve cases. However, general predictive estimates of accuracy (i.e., to unknown samples) may be misleading. If a hold-out method of cross-validation whereby an additional 200 individuals per population was used to conduct GA experiments, the estimates of accuracy were high (perhaps artificially so). However, lower standard deviations suggest that GA-reduced locus sets are likely to be more generally able to classify unknown individuals with levels of accuracy consistent with estimates offered solely from the baseline samples. Improvements in accuracy using GAs were most notable when populations were poorly differentiated (e.g., low F_{st}) and in situations were sampling error is most likely to occur (e.g., with low baseline sample size and high allelic diversity; Table 1). The mean F_{st} of the loci retained in the GA search using the holdout method were consistently higher than the mean (results not shown), indicating that loci with greater discriminatory power (F_{st}) are preferentially selected by GA.

Simulated data sets for analyses of mixtures

Population allocations to simulated mixtures were more accurately estimated when baseline populations were sampled using larger sample sizes and based on greater numbers of loci with higher allelic diversity, and inter-population variance in allele frequency (Table 2). When one of 100 simulated data sets was randomly selected from each of the 36 individual assignment analyses presented in Table 1, GA-reduced sets of loci representing 15–40% fewer loci consistently provided levels of accuracy and precision comparable to those observed when all loci were employed. Results were then similar to results reported for individual assignment testing (Table 1).

Empirical data sets

Each of the three lake trout strains was in Hardy–Weinberg equilibrium for each locus, with the exception of locus *Ssa85* in the SLW strain ($p < 0.01$). No significant population substructure suggesting potential familial relationships within strains was detected except for two loci (*Sfo18* and *Ssa85*) suggesting that coancestry within those strains was not a significant factor affecting assignment accuracy. Evidence for linkage disequilibrium is weak with one pairwise locus combination exhibiting significant disequilibrium in each strain (*Scoμ19–Sfo12* for strains SLW and SIW, *Oneμ9–Sfo12* for strain SMD). Mean allelic diversities for each strain ranged from 3.5 to 4.3 alleles per locus for strain SLW and SIW, respectively (mean allelic diversity over the three strains: 3.96 alleles per locus). In each strain, loci *Sfo1* and *Oneμ10* (two alleles), and locus *Scoμ19* (7–9 alleles depending on strain) exhibited lowest and highest allelic diversity, respectively. The degree of genetic differentiation varied by as much as 10-fold among the pairwise comparisons (Table 3). Average allelic diversities and the range in F_{st} values were comparable to levels of population differentiation used in the simulations.

Four separate GA analyses were considered for the lake trout empirical data set: (i) 'low F_{st}' [SMD–SIW] case; (ii, iii) two 'high F_{st}' cases [SLW–SMD, case 1; SLW–SIW, case 2]; (iv) all three strains considered simultaneously. Results from each empirical data set confirmed findings from simulated data (Table 1). The GA effectively and rapidly searched multi-locus space and always found combination(s) of loci maximizing assignment (Table 3). Accuracy levels associated with the all-locus case based on all eight polymorphic loci were never the optimal solutions (Table 3). Using relatively large baseline sample sizes (~50 individuals), predictive accuracy of GA-reduced sets was believed to be high (i.e., as per the simulated data set; Table 1). The magnitude of accuracy gains differed among case studies. Improvements in accuracy up to 4.9% were observed (three strain case; Table 3). Importantly, the subsets of loci found to maximize accuracy for each comparison were not identical across all strain

Table 2. Results of mixed stock analyses* using sets of loci selected by the GA on the basis of accuracy of individual-based likelihood assignment. Comparisons are made of the mean proportional contribution of one of two populations to a simulated mixture of 400 individuals across a range of variables combinations of numbers of loci, allelic diversity, and sample size.

| Number of loci | Number of individuals | Low F_{st} | | | | | | High F_{st} | | | | | |
| | | Four alleles per locus | | | Eight alleles per locus | | | Four alleles per locus | | | Eight alleles per locus | | |
		All loci (SD)	GA-reduced (SD)	Average number of loci used	All loci (SD)	GA-reduced (SD)	Average number of loci used	All loci (SD)	GA-reduced (SD)	Average number of loci used	All loci (SD)	GA-reduced (SD)	Average number of loci used
5 loci	20	0.524	0.461	4	0.683	0.693	4	0.478	0.477	4	0.499	0.515	3
		0.063	0.067		0.046	0.048		0.037	0.037		0.037	0.038	
	40	0.437	0.442	4	0.484	0.482	4	0.503	0.509	4	0.498	0.504	4
		0.059	0.059		0.052	0.052		0.037	0.037		0.038	0.038	
	100	0.656	0.643	4	0.601	0.606	4	0.493	0.494	4	0.506	0.503	4
		0.067	0.069		0.049	0.05		0.037	0.036		0.038	0.038	
10 loci	20	0.479	0.438	7	0.493	0.509	8	0.498	0.498	8	0.503	0.510	5
		0.036	0.05		0.035	0.036		0.035	0.036		0.035	0.036	
	40	0.561	0.567	8	0.547	0.548	8	0.501	0.503	9	0.484	0.482	9
		0.042	0.055		0.038	0.038		0.035	0.036		0.035	0.035	
	100	0.445	0.453	9	0.5	0.49	8	0.498	0.495	8	0.503	0.503	9
		0.043	0.055		0.035	0.035		0.036	0.036		0.036	0.035	
20 loci	20	0.531	0.553	15	0.506	0.485	12	0.500	0.500	17	0.500	0.500	15
		0.038	0.039		0.036	0.034		0.035	0.035		0.035	0.035	
	40	0.433	0.439	15	0.538	0.543	17	0.497	0.501	14	0.507	0.500	14
		0.031	0.031		0.038	0.038		0.035	0.035		0.036	0.035	
	100	0.473	0.469	16	0.507	0.505	16	0.500	0.500	16	0.500	0.500	17
		0.033	0.033		0.036	0.036		0.035	0.035		0.035	0.035	

*Baseline population proportions were equal (0.500) for both populations considered.

255

Table 3. Summary of 'naïve' likelihood estimates and GA assignment accuracies (absolute numbers and probabilities P) for each of four F_{st} cases from the lake trout data. Performances of GA are expressed as the average number of evaluation functions F(n, s) (see Figure 1) used (20 per GA generation) to find a solution equal or better than the 'naïve' case, or to find the best solution(s) reported by the exhaustive search algorithm (range indicated in brackets).

Strain comparison	SMD–SIW Low F_{st}	SLW–SMD High F_{st}	SLW–SIW High F_{st}	Three strains
Total number of individuals	102	96	108	153
'naïve' likelihood score	72	90	98	111
p ['naïve' likelihood]	0.706	0.938	0.907	0.725
Best GA solution(s) scores	75	92	103	118
p [best solution(s)]	0.735	0.958	0.954	0.771
Average number of functions evaluated to reach a better accuracy level	34	38	24	42
Range	[20–60]	[20–60]	[20–40]	[20–140]
Average number of functions evaluated to reach the optimal accuracy level[1]	60	80	102	124
Range[1]	[20–120]	[40–140]	[20–260]	[20–380]
Numbers of GA runs where the optimal solution was not found	0	7	0	1
Optimal solutions found[2] (number of locus implied)	10000111 (**4**) 10001111 (**5**) 10110111 (**6**)	11000010 (**3**) 11100010 (**4**)	10001110 (**4**) 11001001 (**4**) 10001111 (**5**) 10101111 (**6**)	10001111 (**5**)
All optimal solutions of the exhaustive search found?	Yes	Yes	Yes	Yes
Loci deleted in the optimal solution(s)[3]	***Sfo1*** *Oneμ9* *Oneμ10* *Ogo1a*	***Oneμ10*** ***Ogo1a*** ***Scoμ19*** ***Sfo12*** *Oneμ9*	***Oneμ10*** *Oneμ9* *Sfo1* *Scoμ19* *Sfo12* *Ssa85*	*Sfo1* *Oneμ9* *Oneμ10*
Loci also deleted in solutions better or equal to 'naïve' likelihood estimate	None	*Ssa85* *Sfo1*	*Ogo1a*	*Scoμ19* *Ssa85* *Sfo12*

[1]Calculated only on the number of GA runs out of ten where the optimal solution(s) were found. A best or equal solution to 'naïve' likelihood estimates was always found in each of the ten runs for each of the four F_{st} cases studies examined. Hence, range and averages are not biased in such a case.
[2]Order of the 8 loci in strings: *Sfo18*, *Sfo1*, *Oneμ9*, *Oneμ10*, *Ogo1a*, *Scoμ19*, *Ssa85*, and *Sfo12*.
[3]When more than one optimal solutions is found, the locus or loci deleted in all optimal solutions were indicated in bold.

comparisons (i.e., the same subset of loci is not selected in each comparison, Table 3). This suggests that the 'information content' conferred by individual loci can vary depending on the population/strain comparison made; a conclusion made previously based on other methods (e.g., Shriver et al. 1997).

GA and characteristics of loci used

In the design of study, prior knowledge of locus characteristics (i.e., levels of inter-population variation, heterozygosity, allelic diversity) that would confer greater classification accuracy will be advantageous. In our simulations (Table 1) we consistently observed that GA-reduced data sets were characterized by higher mean F_{st} values than the mean (either 0.01 or 0.10 for low and high F_{st} cases, respectively). GA analyses show the same trend for the empirical lake trout data in that loci with higher than average F_{st} values are generally retained in optimal GA-solutions (Table 3). However, a few exceptions occurred such as locus *Ogo1a*; it presents the highest F_{st} value in the SLW–SMD comparison, but was never selected by

GA (Table 3). Locus *Sfo1* (highest F_{st} value for the SLW–SIW comparison) was also left out of three over four optimal GA-solutions (Table 3). Analysis of other variables identified in earlier studies as influencing assignment accuracy (e.g., allelic diversity) also lead to identical conclusions. Overall, the optimal multi-locus combination depended on the strain comparison made (Table 3). Thus, results from simulated and empirical case studies suggest there are few universal locus selection criteria. Rather, locus selection would most efficiently be conducted on a case-by-case basis.

Discussion

Herein, we present simulation and empirical data that pose the question 'is it possible to elucidate multi-locus combinations that afford classification accuracies in individual assignment (and mixture analyses) comparable to those based on all loci?' Methods that improve assignment accuracy while using smaller subsets of available loci can be of great value (e.g., Shriver et al. 1997), particularly in situations where few data are available, as is routinely the case for rare or endangered fishes or for fishes whose habits preclude extensive sampling. Further, selection of loci provides insights on the relative importance of factors (e.g., levels of allelic diversity, heterozygosity, level of inter-population variance) that may affect accuracy of assignments. We extend earlier investigations of assignment tests (e.g., Cornuet et al. 1999, Bernatchez & Duchesne 2000) by inclusion of validation methodologies (e.g., L1O vs. holdout methods) in simulation analyses.

Assignment accuracy and efficiency with GA

GAs provide a systematic method to explore subsets of all possible multi-locus combinations (Figure 1). For most simulated data sets, the GA seeks to identify multi-locus solutions using fewer loci of comparable or greater assignment accuracy than would be realized using a conventional assignment test as proposed by Paetkau et al. (1995) (Table 1). Simulated data of varying numbers of loci (5, 10, 20) were constructed for populations where allele frequencies (and thus expected genotypic frequencies) were characterized with different sample sizes (20, 40, 100 individuals), and for loci having different levels of both allelic diversity and levels of inter-population variance in allele frequency. Complementary empirical examples included hatchery strains of fish that exhibit non-zero levels of coancestry within-strains and a wide range of intra- and inter-strain variation. Given the range of conditions examined, concordance of results across the data sets surveyed with GA methodology suggests that comparable levels of assignment accuracy and computational efficiency are likely to be realized when GAs are applied to natural population data (Table 3) given that population baselines are adequately sampled and additional methods of validation (other than L1O) are employed. For example, we show that the GA finds all the optimal solutions for the empirical lake trout data sets (Table 3). Population surveys, even of modest size (e.g., 8–10 loci; Tables 1 and 3) can incur substantial savings in time and resources by analyzing fewer loci without compromising assignment accuracy.

Previous studies (e.g., Smouse & Chevillon 1998, Cornuet et al. 1999) have shown that the accuracy of assignment improves with degree of population differentiation (i.e., the mean F_{st} value). This was observed both for simulated and empirical data (Tables 1 and 3). However, empirical data also revealed that the locus-specific levels of population differentiation were not sufficient to predict which loci would confer optimal assignment accuracy and be retained in 'maximally informative' multi-locus combinations. This suggests that parameters other than F_{st} (see Gomulkiewicz et al. 1990; Shriver et al. 1997, Olsen et al. 2000) should be investigated further. Allelic diversity is another important factor affecting individual assignment (Bernatchez & Duchesne 2000). Results confirmed that higher allelic diversity conferred greater assignment accuracy (Table 1), even though this fact should be investigated on a case-by-case basis for empirical data.

Magnitude of improvement in assignment accuracy using GA

For simulated case studies, increase in individual assignment accuracy were typically lower for inter-population comparisons when variance in allele frequency was high (generally <5% increase in assignment accuracy; Table 1), and larger for inter-population comparisons where levels were lower (generally >10% increase in accuracy; Table 1). Results associated with empirical data are slightly different from those using simulated data. For empirical data sets, the highest gains in assignment accuracy with GA over all-locus estimates were realized for highly differentiated strains (SLW–SIW) and for the three strain case (>4.5% increase in accuracy; Table 3).

Results from analyses of mixtures (Table 2) suggest that individual likelihood-based assignment accuracies estimated using GAs may prove useful in screening baselines for multi-locus combinations for use in analyses of mixtures. When the sample size of unknown individuals is large, analyses of mixtures (Pella & Milner 1987) is likely to prove far superior in terms or estimation of accuracy and precision of mixture composition because of added information content available in the genotypic composition within the mixture. Individual assignment tests have been used in lieu of or concurrent with mixture analyses (e.g., Potvin and Bernatchez 2001). While simulations were employed on a single and randomly selected GA-reduced data set and mixtures were based solely on two populations and a single mixture scenario, GAs may have to prove useful and warrant further evaluation.

Leave-one-out

Estimates of accuracy offered on the basis of any validation methodology are often imprecise, particularly when a finite number of samples are drawn from populations. Estimates of accuracy based on L1O methods of validation appear to be particularly prone to higher variance in situations of low baseline sample size. Bias is reduced when baselines are developed with higher sample size. Generalizations of assignment accuracies for GA-reduced data sets to samples of unknown origin is possible across the range of variables evaluated. Even using a holdout method, variance in classifier accuracy was often lower than L1O estimates though estimates were optimistically biased (e.g., mean accuracies of estimates were higher for holdout validation that when 20 000 unknown samples were used). Results only reporting L1O error rates in assignment studies should then be cautiously considered for management purposes if baseline population sample sizes are low.

Relative merits of GA for locus selection

GAs represent one of several approaches for searching optimal multi-locus combinations. Shriver et al. (1997) proposed a method based on a likelihood test to select the most informative loci used for estimation of ancestral origin. Their method has been criticized,

and should be reassessed more clearly (Brenner 1998). The GA has the advantage that it can explore the space of different locus groupings as combinations of loci, rather than treating each locus separately as proposed by Shriver et al. (1997). This is an advantage because the information available for classifying individuals resides not merely in which particular alleles appear within the individuals but also in which single-locus and multi-locus genotypes appear in individuals. Pairwise discriminant analysis may also be used to select loci maximizing assignment scores. However, population genetics data are categorical, and discriminant analysis is more appropriately applied to continuous data. Roques et al. (1999) used a permutation procedure in a study of *Sebastes* fishes to find subsets of loci that performed comparably or better than the set of all loci. However, they could not precisely determine which loci were represented in optimal subsets of loci as GA does for our empirical data set (Table 3).

No single algorithm search approach including GA is capable of handling all problems associated with all possible data sets (e.g., Duda et al. 2000). GAs have been proved resistant to 'deceptive' problems where noisy data can deceive other learning algorithms (Goldberg 1989). Furthermore, GAs tend to resist training ordering problems (i.e., the order in which populations in the training data set are introduced do not affect the resulting model). Finally, GAs are less prone to finding non-optimal solutions in a 'rugged' search landscapes (e.g., Jain & Zongker 1997).

Conclusions

When populations parameters (allele frequencies) upon which classification decisions are based are estimated with error, the use of additional loci will not always confer greater accuracy of assignment. We have shown with simulated data (Tables 1 and 2) and with empirical examples (Table 3) that GAs are useful locus selection tools that can identify subsets of loci that maximize assignment accuracy of individuals to populations. GAs have advantages over methods previously reported in the literature for locus selection. Loci are not uniformly informative across all case studies (Table 3). Information contributed by each locus must be evaluated in the context of each applied study. Reported estimates of accuracies based on L1O validation should be viewed with caution under certain conditions, particularly where baseline populations have been inadequately sampled. Hence, management issues solely based on such error rates estimates should be decided with caution. Use of unknown individuals (Hansen et al. 2001) and additional methods of cross-validation (i.e., holdout methods) should be employed in empirical studies to better estimate assignment error rates.

Acknowledgements

This study was supported by the Great Lakes Fishery Trust, the Great Lakes Protection Fund, The Michigan University Sea Grant Program, and by Partnership for Ecosystem Research and Management between the Michigan Department of Natural Resources and Michigan State University. Samples of hatchery lake trout were provided by D. Bast, F. Bland, D. Johnson, C. LeGault and D. Radloff of the U.S. Fish and Wildlife Service. J. Warrillow conducted analyses of simulated mixtures. We thank two anonymous reviewers and B. Smoker for comments that improved the presentation.

References

Angers, B.A., L. Bernatchez, A. Angers & L. Desgroseillers. 1995. Specific microsatellite loci for brook charr reveal strong population subdivision on a microgeographic scale. J. Fish Biol. 47(Suppl. A): 177–185.

Aurelle, D., S. Lek, J.L. Giraudel & P. Berrebi. 1999. Microsatellites and artificial neural networks: Tools for the discrimination between natural and hatchery brown trout (*Salmo trutta*, L.) in Atlantic populations. Ecol. Model. 120: 313–324.

Belkhir, K., P. Borsa, J. Goudet, L. Chikhi & F. Bonhomme. 1996. GENETIX v. 3.0, logiciel sous Windows™ pour la génétique des populations. Montpellier, Laboratoire Génome et Populations, Université Montpellier 2, France.

Bernatchez, L. & P. Duchesne. 2000. Individual-based genotype analysis in studies of parentage and population assignment: How many loci, how many alleles? Can. J. Fish. Aquat. Sci. 57: 1–12.

Brenner, C.H. 1998. Difficulties in the estimation of ethnic affiliation. Am. J. Hum. Genet. 62: 1558–1560.

Cornuet, J.M., S. Aulagnier, S. Lek, P. Franck & M. Solignac. 1996. Classifying individuals among infra-specific taxa using microsatellite data and neural networks. C. R. Acad. Sci. Paris, Life Sci. 319: 1167–1177.

Cornuet, J.M., S. Piry, G. Luikart, A. Estoup & M. Solignac. 1999. New methods employing multi-locus genotypes to select or exclude populations as origins of individuals. Genetics 153: 1989–2000.

Duda, R.O., P.E. Hart & D.G. Stork. 2000. Pattern Classification. 2nd edition, John Wiley and Sons, New York. 654 pp.

Girman, D.J., M.L.G. Mills, E. Geffen & R.K. Wayne. 1997. A molecular genetic analysis of social structure, dispersal, and interpack relationships of the African wild dog (*Lycaon pictus*). Behav. Ecol. Sociobiol. 40: 187–198.

Goldberg, D. 1989. Genetic Algorithms in Search, Optimization and Machine Learning, Addison-Wesley, Reading. 452 pp.

Gomulkiewicz, R., J.K.T. Brodziak & M. Mangel. 1990. Ranking loci for genetic stock identification by curvature methods. Can. J. Fish. Aquat. Sci. 47: 611–619.

Hansen, M.M., E. Kenchington & E.E. Nielsen. 2001. Assigning individual fish to populations using microsatellite DNA markers. Fish Fish. 2: 93–112.

Hansen, M.M., D.E. Ruzzante, E.E. Nielsen & K.L.D. Mensberg. 2000. Microsatellite and mitochondrial DNA polymorphism reveals life-history dependent interbreeding between hatchery and wild brown trout (*Salmo trutta* L.). Mol. Ecol. 9: 583–594.

Holland, J. 1994. Adaptation in Natural and Artificial Systems, MIT Press, Cambridge. 221 pp.

Jain, A.K. & D. Zongker. 1997. Feature selection: Evaluation, application and small sample performance. IEEE Trans. Patt. Anal. Mach. Intell. 19: 153–158.

Jain, A.K, R.P.W. Duin & J. Mao. 2000. Statistical pattern recognition: A review. IEEE Trans. Patt. Anal. Mach. Intell. 22: 4–37.

Letcher, B.H. & T.L. King. 1999. Targeted stock identification using multi-locus genotype 'familyprinting'. Fish. Res. 43: 99–111.

Lewis, P.O. 1998. A genetic algorithm for maximum-likelihood inference using nucleotide sequence data. Mol. Biol. Evol. 15: 277–283.

Martinez, J.L., J. Dumas, E. Beall & E. Garcia-Vazquez. 2001. Assessing introgression of foreign strains in wild Atlantic salmon populations: Variation in microsatellites assessed in historic scale collections. Freshw. Biol. 46: 835–844.

Mitchell, M. & C.E. Taylor. 1999. Evolutionary computation: An overview. Annu. Rev. Ecol. Syst. 30: 593–616.

Neraas, L.P. & P. Spruell. 2001. Fragmentation of riverine systems: The genetic effects of dams on bull trout (*Salvelinus confluentus*) in the Clark Fork River system. Mol. Ecol. 10: 1153–1164.

Nielsen, E.E., M.M. Hansen, C. Schmidt, D. Meldrup & P. Grønkjaer. 2001. Population of origin of Atlantic cod. Nature 413: 272.

Norris, A.T., D.G. Bradley & E.P. Cunningham. 2000. Parentage and relatedness determination in farmed Atlantic salmon (*Salmo salar*) using microsatellite markers. Aquaculture 182: 73–83.

Olsen, J.B., P. Bentzen & J.E. Seeb. 1998. Characterization of seven microsatellite loci derived from pink salmon. Mol. Ecol. 7: 1087–1089.

Olsen, J.B., P. Bentzen, M.A. Banks, J.B. Shaklee & S. Young. 2000. Microsatellites reveal population identity of individual pink salmon to allow supportive breeding of a population at risk of extinction. Trans. Amer. Fish. Soc. 129: 232–242.

O'Reilly, P.T., L.C. Hamilton, S.K. McConnell & J.W. Wright. 1996. Rapid analysis of genetic variation in Atlantic salmon (*Salmo salar*) by PCR multiplexing of dinucleotide and tetranucleotide microsatellite. Can. J. Fish. Aquat. Sci. 53: 2292–2298.

Paetkau, D., W. Calvert, I. Stirling & C. Strobeck. 1995. Microsatellite analysis of population structure in Canadian polar bears. Mol. Ecol. 4: 347–354.

Pella, J.J. & G.B. Milner. 1987. Use of genetic marks in stock composition analysis. pp. 247–276. *In*: N. Ryman & F. Utter (ed.) Population Genetics and Fisheries Management. Univeristy of Washington Press, Seattle, WA.

Potvin, C. & L. Bernatchez. 2001. Lacustrine spatial distribution of landlocked Atlantic salmon populations assessed across generations by multi-locus individual assignment and mixed-stock analysis. Mol. Ecol. 10: 22375–22388.

Pritchard, J.K., M. Stephens & P. Donnelly. 2000. Inference of population structure using multi-locus genotype data. Genetics 155: 945–959.

Queller, D.C. & K.F. Goodnight. 1989. Estimating relatedness using genetic markers. Evolution 43: 258–275.

Rannala, B. & J.L. Mountain. 1997. Detecting immigration using multi-locus genotypes. Proc. Natl. Acad. Sci. USA 94: 9197–9202.

Roques, S., P. Duchesne & L. Bernatchez. 1999. Potential of microsatellites for individual assignment: The North Atlantic redfish (genus *Sebastes*) species complex as a case study. Mol. Ecol. 8: 1703–1717.

Scribner, K.T., J.R. Gust & R.L. Fields. 1999. Isolation and characterization of novel salmon microsatellite loci: Cross-species amplification and population genetics applications. Can. J. Fish. Aquat. Sci. 53: 833–841.

Shao, J. 1993. Linear model selection by cross-validation. J. Amer. Stat. Assoc. 88: 486–494.

Shriver, M.D., M.W. Smith, L. Jin, A. Marcini, J.M. Akey et al. 1997 Ethnic-affiliation estimation by use of population-specific DNA markers. Amer. J. Hum. Genet. 60: 957–964.

Smouse, P. E. & C. Chevillon. 1998. Analytical aspects of population-specific DNA fingerprinting for individuals. J. Hered. 89: 143–150.

Smouse, P.E., R.S. Spielman & M.H. Park. 1982. Multiple-locus allocation of individuals to groups as a function of the genetic variation within and differences among human populations. Amer. Nat. 119: 445–463.

Smouse, P.E., R.S. Waples & J.A. Tworek. 1990 A genetic mixture analysis for use with incomplete source population data. Can. J. Fish. Aquat. Sci. 47: 20–634.

Sokal, R.R. & J.F. Rohlf. 1995. Biometry. 2nd edition, Freeman, USA. 887 pp.

Stefanini, M.F. & A. Camussi. 2000. The reduction of large molecular profiles to informative components using a genetic algorithm. Bioinformatics 16: 923–931.

Taylor, E.B., A. Kuiper, P.M. Troffe, D.J. Hoysak & S. Pollard. 2000. Variation in developmental biology and microsatellite DNA in reproductive ecotypes of kokanee, *Oncorhynchus nerka*: Implications for declining populations in a large British Columbia lake. Conserv. Genet. 1: 213–249.

Taylor, E.B., Z. Redenbach, A.B. Costello, S.J. Pollard & C.J. Pacas. 2001. Nested analysis of genetic variation in northwestern North American char, Dolly Varden (*Salvelinus malma*) and bull trout (*S. confluentus*). Can. J. Fish. Aquat. Sci. 58: 406–420.

Trunk, G.V. 1979. A problem of dimensionality: A simple example. IEEE Trans. Patt. Anal. Mach. Intell. 1: 306–307.

Waser, P.M. & C. Strobeck. 1998. Genetic signatures of interpopulation dispersal. Trends Ecol. Evol. 13: 43–44.

Weir, B.S. 1979. Inferences about linkage disequilibrium. Biometrics 25: 235–254.

Weir, B.S. & C.C. Cockerham. 1984. Estimating F-statistics for the analysis of population structure. Evolution 43: 1358–1370.

Wright, S. 1965. The interpretation of population structure by F-statistics with special regards to system of mating. Evolution 19: 395–420.

Environmental Biology of Fishes **69**: 261–273, 2004.
© 2004 *Kluwer Academic Publishers. Printed in the Netherlands.*

Moderately and highly polymorphic microsatellites provide discordant estimates of population divergence in sockeye salmon, *Oncorhynchus nerka*

Jeffrey B. Olsen[a,c], Chris Habicht[a], Joel Reynolds[a,b] & James E. Seeb[a]
[a]*Gene Conservation Laboratory, Alaska Department of Fish and Game, 333 Raspberry Road, Anchorage, AK 99518-1599, U.S.A.*
[b]*Division of Natural Resources, U.S. Fish & Wildlife Service, 1011 East Tudor Road, Anchorage, AK 99503, U.S.A.*
[c]*Corresponding author: Conservation Genetics Laboratory, U.S. Fish & Wildlife Service, 1011 East Tudor Road, Anchorage, AK 99503, U.S.A. (e-mail: jeffrey_olsen@fws.gov)*

Received 17 April 2003   Accepted 16 June 2003

Key words: mutation, migration, population genetics, population structure, F_{ST}, Pacific salmon

Synopsis

Mutation rate can vary widely among microsatellite loci. This variation may cause discordant single-locus and multi-locus estimates of F_{ST}, the commonly used measure of population divergence. We use 16 microsatellite and five allozyme loci from 14 sockeye salmon populations to address two questions about the affect of mutation rate on estimates of F_{ST}: (1) does mutation rate influence F_{ST} estimates from all microsatellites to a similar degree relative to allozymes?; (2) does the influence of mutation rate on F_{ST} estimates from microsatellites vary with geographic scale in spatially structured populations? For question one we find that discordant estimates of F_{ST} among microsatellites as well as between the two marker classes are correlated with mean within-population heterozygosity (H_S) and thus are likely due to differences in mutation rate. Highly polymorphic microsatellites ($H_S > 0.84$) provide significantly lower estimates of F_{ST} than moderately polymorphic microsatellites and allozymes ($H_S < 0.60$). Estimates of F_{ST} from binned allele frequency data and R_{ST} provide more accurate measures of population divergence for highly polymorphic but not for moderately polymorphic microsatellites. We conclude it is more important to pool loci of like H_S rather than marker class when estimating F_{ST}. For question two we find the F_{ST} values for moderately and highly polymorphic loci, while significantly different, are positively correlated for geographically proximate but not geographically distant population pairs. These results are consistent with expectations from the equilibrium approximation of Wright's infinite island model and confirm that the influence of mutation on estimates of F_{ST} can vary in spatially structured populations presumably because the rate of migration varies inversely with geographic scale.

Introduction

The use of microsatellite loci for estimating the amount of genetic variation between natural populations has received much scrutiny. Of particular appeal and concern is the high level of polymorphism often associated with these loci. High polymorphism provides high statistical power to detect genetic divergence, especially between weakly differentiated populations (Estoup et al. 1998). In contrast, highly polymorphic loci may underestimate the degree of genetic divergence between populations when using popular measures of differentiation such as F-statistics (Hedrick 1999). This is because mutation rate, a locus-specific force that is often high in microsatellites, counteracts the loss of heterozygosity within populations due to genetic drift. The equilibrium approximation $F_{ST} \approx 1/(1 + 4N_e(m + \mu))$ from the infinite island model of Wright (1969) shows that F_{ST}, the commonly used index of genetic population structure, is inversely related to both the rate of migration (m) and mutation (μ). It is generally assumed that μ is much smaller than m and that the estimate of

F_{ST} reflects population differentiation due primarily to the opposing forces of migration and genetic drift. If this assumption is not valid and μ is similar to or greater than m, then F_{ST} will be misinterpreted and inferences regarding the influence of effective population size (N_e) and migration on the degree of genetic population structure will be biased. This mutation bias may result in an over estimate of gene flow (N_em) which may in turn lead to inappropriate inferences regarding the forces influencing population structure.

Three empirical methods are commonly used to reveal mutation bias in estimates of F_{ST} from microsatellites. First, F_{ST} estimates from microsatellites are compared with those from another codominant nuclear marker that has a low mutation rate, such as allozymes. Allozyme loci provide a good standard from which to test for mutation bias in microsatellites, assuming the allozymes are not influenced by selection. Selection will confound the marker comparison by biasing the F_{ST} estimate from the locus under selection. Therefore, it is preferable to use allozymes with empirical evidence of selective neutrality. If only migration and genetic drift are influencing population divergence then both marker classes are expected to provide similar estimates of F_{ST}. If mutation is seriously biasing the F_{ST} estimate from microsatellites then the estimate of F_{ST} from allozymes should be significantly larger (Balloux et al. 2000). Second, alternative estimators of genetic divergence are computed for microsatellites and compared to the original F_{ST} values. Two examples are estimates of the parameter R_{ST} (Slatkin 1995) and estimates of F_{ST} based on binned allele frequency data. R_{ST} is commonly estimated because it assumes mutation properties that presumably reflect those exhibited by microsatellites (Balloux & Goudet 2002). In contrast, binning allele frequencies will reduce mean within-population heterozygosity for highly polymorphic loci and this will increase estimates of F_{ST} if the rate of mutation is as large or larger than migration (Buonaccorsi et al. 2001). Finally, estimates of an absolute measure of genetic divergence such as D'_{ST} (Nei 1987) are computed for locus groups that do and do not exhibit high polymorphism. If these values do not differ among the locus groups but the values of F_{ST} do, then mutation bias may be inferred if the lowest F_{ST} values belong to the highly polymorphic locus group (Charlesworth 1998).

Microsatellites may exhibit a wide range in mutation rate and therefore heterozygosity (Jarne & Lagoda 1996). Most studies, however, treat microsatellites as a single marker class and assume that estimates of F_{ST} from each locus are affected similarly by mutation rate bias relative to other marker classes (Balloux et al. 2000). As long as the variance in single-locus mutation rate is low then this assumption is probably valid and microsatellites can be pooled for estimating measures of population differentiation (e.g., Buonaccorsi et al. 2001). For studies in which the variance in single-locus heterozygosity is large, the mutation rate could vary significantly among microsatellites, which could contribute to discordant single-locus and multi-locus estimates of population differentiation (e.g., Bowcock et al. 1994, Freville et al. 2001). These studies imply that the assumption above may not be valid and that groups of microsatellites with different ranges of heterozygosity may provide discordant estimates of F_{ST} and should be analyzed separately rather than pooled as a single marker class.

Spatially structured populations are often organized in a hierarchical pattern of relatedness in which the rate of gene flow varies within and between geographically defined population groups. For these populations the effect of mutation bias may vary with spatial scale and should be examined both at different spatial scales and over all populations (e.g., Shaw et al. 1999). This is particularly relevant for broad scale studies because the ratio of migration rate to mutation rate may vary substantially between geographically nearby and geographically distant populations (e.g., Shaw et al. 1999).

In this case study we use data from 16 microsatellite and five allozyme loci in 14 sockeye salmon populations to address two questions about the influence of mutation rate on estimates of F_{ST} derived from frequencies of microsatellite alleles: (1) do microsatellite loci with different mutation rates provide similar estimates of F_{ST}?; (2) does the influence of mutation rate on F_{ST} estimates from microsatellites vary with geographic scale in spatially structured populations? Sockeye salmon, *Oncorhynchus nerka*, is an excellent organism with which to address these questions. Loci and methods for both allozyme and microsatellite analysis are well developed for sockeye salmon (Seeb et al. 2000, Olsen et al. 2000). Also, studies of allozyme variation show that genetic differentiation among populations is spatially structured (Wood 1995). Finally, sockeye salmon exhibit behavioral traits (e.g., anadromy, natal philopatry) and ecological adaptations that are strong reproductive isolating barriers, assuring gene flow among populations is limited (Hendry et al. 2000).

Methods

Sample collection

Tissue samples for allozyme and microsatellite analysis were obtained from 14 spawning populations of sockeye salmon representing seven river systems from two geographic regions: Bristol Bay in Western Alaska and the Kamchatka Peninsula in Russia (Figure 1, Table 1). Tissue samples included heart, liver, muscle, and eye, and all samples were stored at $-80°C$ until analyzed.

Allozyme and microsatellite genotyping

Twenty five allozyme loci were screened for polymorphism: *sAAT-3*, mAAT-1*, mAAT-2*, ADA-1*, mAH-3*, mAH-4*, ALAT*, CK-A1*, CK-A2*, FDHG*, FH*, GAPDH-2*, sIDHP-1*, sIDHP-2*, LDH-A1*, LDH-B2*, mMEP-1*, MPI*, PEPA*, PEPB-1*, PEPD1*, PEPLT*, PGDH*, PGM-2*,* and *TPI-4**. Allozyme alleles were resolved using horizontal starch–gel electrophoresis and enzyme-specific histochemical staining procedures (Seeb et al. 2000 and references therein). Five loci showed a common allele frequency no greater than 0.95 in at least 1 of the 14 populations.

Allele frequencies at these loci (*mAAT-1*, ALAT*, LDHB-2*, PEPLT*, PGM-2**) were used to estimate genetic variation in sockeye salmon.

Allele frequencies at 16 microsatellite loci were also used to estimate genetic variation in sockeye salmon (Table 2). DNA isolation and amplification by PCR was carried out by methods given in Olsen et al. (2000). Microsatellite alleles were size fractionated using an Applied Biosystems Inc. (ABI) 377-96 automated DNA sequencer operated in GeneScan™ mode (ABI 1996a). Alleles for each locus were scored and data were tabulated for importing into statistical software with Genotyper software, Version 2.1 (ABI 1996b).

Genetic variation within and among populations

Estimates of allele frequency, number of alleles, observed and expected heterozygosity (H_O, H_E) per locus and population, total heterozygosity (H_T), and mean within-population heterozygosity (H_S), were calculated using the computer program FSTAT version

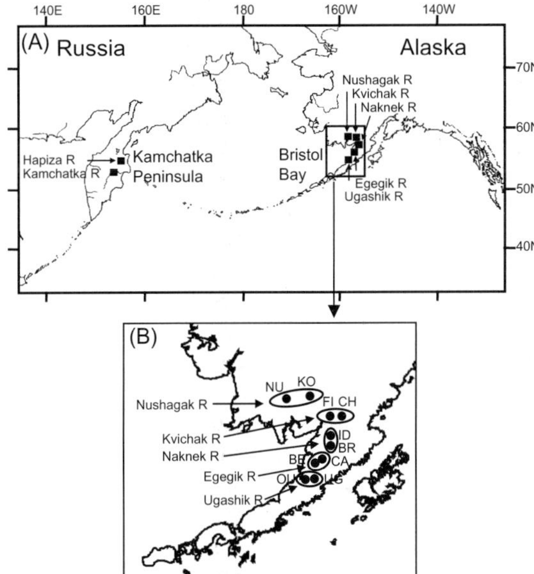

Figure 1. (A) Location of seven major river systems sampled in Bristol Bay, Alaska, and on the Kamchatka Peninsula, Russia. (B) The location of populations sampled in Bristol Bay. The population abbreviations are as indicated in Table 1.

Table 1. Region and major river system, site abbreviation (Abbr.), sample date, and sample size (n) for 14 sockeye salmon populations from Bristol Bay, Alaska, and the Kamchatka Peninsula, Russia.

Region/ river system	Population	Abbr.	Sample date	n
Bristol Bay				
Nushagak River	Koktuli River	KO	08/13/00	100
Nushagak River	Nuyakuk Lake	NU	08/16/00	100
Kvichak River	Chinkelyes Creek	CH	08/23/00	100
Kvichak River	Finger Beach	FI	08/24/00	85
Naknek River	Brooks Lake	BR	08/22/00	100
Naknek River	Idavain Creek	ID	08/23/00	100
Egegik River	Becharof Creek	BE	08/11/00	100
Egegik River	Cabin Creek	CA	08/14/00	100
Ugashik River	Outlet Stream	OU	08/26/00	100
Ugashik River	Ugashik Narrows	UG	08/24/00	100
Kamchatka Peninsula				
Kamchatka River	Kamchatka Late Run	KL	07/21/98	119
Kamchatka River	Kamchatka Early Run	KE	06/01/98	80
Hapiza River	Hapiza Early Run	HE	07/17/98	100
Hapiza River	Hapiza Late Run	HL	09/02/98	81

Table 2. Sample size (n), number of alleles (A), expected and observed heterozygosity (H_E, H_O) for 14 sockeye salmon populations from seven river systems.

	Nushagak		Kvichak		Naknek		Egegik		Ugashik		Kamchatka		Hapiza	
	KO	NU	FI	CH	BR	ID	BE	CA	OU	UG	KL	KE	HE	HL
Microsatellite														
Omy77														
n	100	100	83	93	98	94	96	98	100	94	93	68	89	79
A	8	7	7	6	5	7	10	8	7	8	4	5	5	4
H_E	0.64	0.61	0.55	0.56	0.59	0.57	0.64	0.62	0.68	0.65	0.52	0.54	0.45	0.37
H_O	0.68	0.59	0.61	0.49	0.66	0.62	0.60	0.60	0.58	0.60	0.43	0.51	0.44	0.38
One102														
n	100	98	82	98	99	98	98	100	100	97	97	70	99	79
A	15	12	16	14	13	16	18	14	15	15	12	16	11	12
H_E	0.85	0.82	0.87	0.83	0.79	0.87	0.87	0.83	0.81	0.84	0.85	0.87	0.85	0.83
H_O	0.85	0.83	0.90	0.83	0.76	0.89	0.85	0.83	0.78	0.82	0.80	0.86	0.88	0.92
One103														
n	100	100	83	98	100	100	99	98	100	97	95	70	99	79
A	31	34	31	29	25	31	39	33	39	34	22	26	31	30
H_E	0.95	0.94	0.95	0.95	0.94	0.95	0.95	0.95	0.96	0.95	0.91	0.93	0.95	0.94
H_O	0.98	0.94	0.98	0.92	0.91	0.93	0.95	0.94	0.91	0.94	0.88	0.93	0.90	0.90
One104														
n	100	100	83	98	100	100	97	99	100	96	94	70	98	79
A	19	15	17	19	18	18	18	20	18	20	20	21	17	14
H_E	0.86	0.89	0.91	0.91	0.93	0.91	0.92	0.91	0.92	0.91	0.91	0.92	0.90	0.88
H_O	0.83	0.94	0.88	0.93	0.93	0.94	0.91	0.95	0.90	0.95	0.87	0.90	0.92	0.87
One105														
n	100	99	85	98	100	100	99	99	100	96	95	70	95	79
A	6	6	4	4	3	4	5	5	6	5	7	6	5	4
H_E	0.30	0.36	0.37	0.37	0.40	0.35	0.44	0.38	0.36	0.41	0.62	0.57	0.50	0.59
H_O	0.27	0.33	0.35	0.39	0.38	0.40	0.42	0.38	0.34	0.41	0.53	0.53	0.43	0.61
One108														
n	100	100	84	98	100	98	98	100	98	96	97	70	98	79
A	15	16	17	14	15	19	20	18	17	19	17	16	19	16
H_E	0.89	0.85	0.90	0.90	0.81	0.90	0.91	0.90	0.88	0.88	0.92	0.89	0.92	0.91
H_O	0.88	0.80	0.93	0.89	0.81	0.88	0.90	0.82	0.84	0.92	0.94	0.90	0.90	0.95
One109														
n	99	100	84	98	100	98	98	100	100	97	97	70	99	79
A	13	10	13	13	13	13	14	14	15	13	13	11	13	12
H_E	0.87	0.86	0.84	0.87	0.88	0.89	0.89	0.90	0.88	0.88	0.88	0.86	0.88	0.85
H_O	0.94	0.86	0.85	0.91	0.92	0.86	0.89	0.86	0.84	0.84	0.91	0.87	0.87	0.91
One110														
n	100	100	84	98	100	100	96	99	99	97	96	69	99	79
A	15	13	18	19	17	16	16	19	16	15	12	10	15	14
H_E	0.89	0.90	0.88	0.90	0.89	0.89	0.91	0.92	0.91	0.91	0.88	0.88	0.89	0.89
H_O	0.94	0.90	0.89	0.87	0.91	0.90	0.94	0.91	0.94	0.92	0.83	0.88	0.84	0.89
One111														
n	100	100	80	98	100	100	98	100	99	96	96	70	95	79
A	25	25	28	26	25	28	29	29	26	28	21	22	23	23
H_E	0.80	0.82	0.92	0.92	0.90	0.89	0.92	0.93	0.89	0.91	0.87	0.86	0.92	0.92
H_O	0.74	0.84	0.90	0.84	0.87	0.90	0.94	0.94	0.89	0.92	0.88	0.86	0.96	0.90
One112														
n	100	100	84	98	100	99	99	100	100	96	96	70	99	79
A	21	20	28	27	21	27	26	30	29	24	26	24	23	23
H_E	0.85	0.88	0.87	0.88	0.91	0.90	0.93	0.94	0.90	0.90	0.90	0.92	0.91	0.91
H_O	0.84	0.87	0.83	0.85	0.92	0.88	0.95	0.93	0.89	0.88	0.89	0.94	0.88	0.95

Table 2. (*Continued*)

	Nushagak		Kvichak		Naknek		Egegik		Ugashik		Kamchatka		Hapiza	
	KO	NU	FI	CH	BR	ID	BE	CA	OU	UG	KL	KE	HE	HL
One114														
n	100	100	84	98	100	100	99	100	100	97	96	70	78	79
A	19	19	21	23	17	19	20	20	20	20	17	16	18	15
H_E	0.91	0.92	0.94	0.94	0.92	0.94	0.94	0.94	0.93	0.92	0.91	0.91	0.91	0.90
H_O	0.94	0.93	0.93	0.93	0.94	0.96	0.94	0.97	0.91	0.90	0.86	0.96	0.96	0.89
One115														
n	98	100	84	98	100	100	97	100	100	97	95	70	99	79
A	15	14	18	17	16	19	18	18	15	16	16	13	15	16
H_E	0.91	0.90	0.91	0.91	0.90	0.92	0.92	0.92	0.92	0.91	0.91	0.92	0.91	0.92
H_O	0.83	0.84	0.92	0.85	0.89	0.94	0.90	0.91	0.90	0.93	0.91	0.93	0.87	0.92
Ots103														
n	100	99	84	98	100	98	97	99	99	97	92	70	94	79
A	14	16	17	17	17	20	17	18	15	17	19	21	19	18
H_E	0.90	0.90	0.92	0.93	0.91	0.91	0.92	0.92	0.92	0.92	0.93	0.94	0.93	0.93
H_O	0.85	0.90	0.93	0.92	0.87	0.91	0.94	0.90	0.93	0.92	0.92	0.93	0.95	0.96
Ots107														
n	100	94	84	94	100	99	96	100	100	96	93	69	94	79
A	4	5	4	5	5	6	8	7	7	7	5	5	4	3
H_E	0.27	0.39	0.26	0.20	0.26	0.25	0.30	0.24	0.30	0.32	0.49	0.49	0.36	0.43
H_O	0.24	0.39	0.25	0.22	0.29	0.24	0.29	0.25	0.32	0.32	0.49	0.43	0.34	0.47
Ots3														
n	97	100	84	98	99	98	99	99	98	97	92	70	92	77
A	5	6	6	6	6	7	5	4	5	5	6	5	7	8
H_E	0.21	0.34	0.17	0.20	0.53	0.42	0.12	0.13	0.28	0.18	0.42	0.33	0.75	0.71
H_O	0.21	0.33	0.18	0.20	0.57	0.47	0.12	0.11	0.32	0.18	0.39	0.31	0.70	0.68
μSat60														
n	100	100	84	98	100	95	98	99	100	97	97	70	99	79
A	5	5	7	8	5	9	12	9	9	6	6	5	5	3
H_E	0.59	0.53	0.55	0.51	0.51	0.55	0.55	0.52	0.58	0.54	0.58	0.57	0.53	0.56
H_O	0.55	0.63	0.57	0.50	0.55	0.51	0.60	0.52	0.61	0.48	0.53	0.54	0.56	0.57
Mean														
n	99.6	99.4	83.5	97.4	99.8	98.6	97.8	99.4	99.6	96.4	95.1	69.8	95.4	78.9
A	14.4	13.9	15.8	15.4	13.8	16.2	17.2	16.6	16.2	15.8	13.9	13.9	14.4	13.4
H_E	0.73	0.74	0.74	0.74	0.75	0.76	0.76	0.75	0.76	0.75	0.78	0.77	0.78	0.78
H_O	0.72	0.75	0.74	0.72	0.76	0.76	0.76	0.74	0.74	0.74	0.75	0.77	0.77	0.80
Allozyme														
mAAT-1 *														
n	99	100	85	98	99	99	99	100	96	100	96	69	96	79
A	2	2	2	2	1	2	2	2	2	2	2	2	2	2
H_E	0.19	0.23	0.21	0.12	0.00	0.21	0.22	0.23	0.31	0.17	0.17	0.10	0.16	0.20
H_O	0.17	0.22	0.19	0.13	0.00	0.17	0.19	0.24	0.29	0.17	0.17	0.10	0.18	0.23
ALAT *														
n	100	100	84	98	100	100	99	100	100	99	97	69	99	79
A	4	4	4	3	4	3	3	3	3	4	3	4	3	4
H_E	0.54	0.52	0.55	0.55	0.55	0.54	0.61	0.57	0.63	0.61	0.41	0.35	0.46	0.44
H_O	0.67	0.56	0.55	0.52	0.58	0.59	0.59	0.55	0.66	0.64	0.38	0.38	0.45	0.48
LDHB-2 *														
n	100	100	85	98	100	100	100	100	100	100	98	71	99	79
A	2	2	2	2	2	2	2	2	2	2	2	2	2	2
H_E	0.20	0.23	0.14	0.15	0.13	0.19	0.24	0.17	0.08	0.11	0.41	0.49	0.40	0.44
H_O	0.18	0.22	0.11	0.16	0.12	0.21	0.27	0.19	0.08	0.09	0.36	0.51	0.42	0.51

Table 2. (*Continued*)

	Nushagak		Kvichak		Naknek		Egegik		Ugashik		Kamchatka		Hapiza	
	KO	NU	FI	CH	BR	ID	BE	CA	OU	UG	KL	KE	HE	HL
*PEPLT**														
n	100	100	83	98	100	100	99	100	100	100	93	70	99	79
A	1	1	1	1	3	1	1	1	1	1	1	1	1	1
H_E	0.00	0.00	0.00	0.00	0.30	0.00	0.00	0.00	0.00	0.00	0.00	0.00	0.00	0.00
H_O	0.00	0.00	0.00	0.00	0.28	0.00	0.00	0.00	0.00	0.00	0.00	0.00	0.00	0.00
*PGM-2**														
n	100	100	85	98	99	100	100	99	100	100	98	72	99	79
A	2	2	2	2	2	2	2	2	2	2	2	2	2	2
H_E	0.50	0.44	0.32	0.40	0.26	0.42	0.36	0.43	0.45	0.43	0.43	0.38	0.36	0.39
H_O	0.57	0.42	0.29	0.51	0.28	0.36	0.29	0.43	0.36	0.35	0.36	0.43	0.31	0.35
Mean														
n	99.8	100	84.4	98.0	99.6	99.8	99.4	99.8	99.2	99.8	96.4	70.2	98.4	79.0
A	2.2	2.2	2.2	2.0	2.4	2.0	2.0	2.0	2.0	2.2	2.0	2.2	2.0	2.2
H_E	0.29	0.28	0.24	0.25	0.25	0.27	0.29	0.28	0.29	0.26	0.28	0.26	0.28	0.29
H_O	0.32	0.28	0.23	0.27	0.25	0.27	0.27	0.28	0.28	0.25	0.25	0.28	0.27	0.31

Population abbreviations are as indicated in Table 1.

2.9.3 (Goudet 2001[1]) Data randomization methods (Goudet 2001) were used to test conformity to Hardy–Weinberg equilibrium for each locus and population combination and to test for genotypic disequilibrium among locus pairs. The data set was typically permuted 5,000 times and the threshold for statistical significance ($\alpha = 0.05$) was corrected for simultaneous tests using the sequential Bonferroni method (Rice 1989).

Estimates of the relative measures of population divergence, F_{ST} (allozymes and microsatellites) and R_{ST} (microsatellites only), were computed for each locus and locus group over all populations and for each population pair. Estimates of the absolute measure D'_{ST} (Nei 1987) was calculated for each locus and locus group over all populations. Estimates of F_{ST} and D'_{ST} were computed using FSTAT version 2.9.3 (Goudet 2001[1]) and estimates of R_{ST} were calculated according to Goodman (1997) using the computer program S-PLUS (Statistical Sciences 1995). Confidence intervals (95%) for each multilocus F_{ST} and R_{ST} estimate were generated by bootstrap sampling over loci.

Allele binning was used to reduce within-population genetic variation for microsatellites. A method similar to that described by Buonaccorsi et al. (2001) was used to group alleles from each locus into two or three size frequency clusters. A size frequency cluster was defined as any group of contiguous alleles with a composite allele frequency >10% in at least one population but absent or of much lower frequency in one or more populations. For loci that exhibit a single normally distributed mode that overlaps significantly in all populations, the distinguishing allele clusters were defined as the size extremes, and alleles at the center of the distribution were grouped to form the third cluster. Estimates of 'binned' F_{ST} ($F_{ST}b$) with confidence intervals were computed for each microsatellite.

Comparison of F_{ST} estimates from microsatellites and allozymes

Four methods were used to address question one. First, the Spearman rank order correlation test was used to test for a negative correlation between mutation rate and F_{ST} over all loci and for both marker classes. As in other studies (e.g., Hedrick 1999, Balloux et al. 2000) values of H_S were used as a surrogate for mutation rate because mutation rate is generally not known and is difficult to estimate. This approach assumes selection does not influence the loci and that large differences in H_S are primarily due to differences in mutation rate and not the stochastic influence of genetic drift. Second, four locus groups were examined for non-overlapping confidence intervals around F_{ST} estimates. The locus groups include: all five allozymes, all 16 microsatellites, five moderately polymorphic microsatellites ($H_S < 0.60$), and 11 highly polymorphic microsatellites ($H_S > 0.84$). Third, each of the three microsatellite groups were

---

[1]Goudet, J. 2001. FSTAT, a program to estimate and test gene diversities and fixation indices, Version 2.9.3. Available from http://www.unil.ch/izea/softwares/fstat.html.

examined for non-overlapping confidence intervals between estimates of F_{ST} and R_{ST}, and F_{ST} and $F_{ST}b$. Finally, estimates of absolute (D'_{ST}) and relative (F_{ST}) genetic divergence for pairs of locus groups were compared as suggested by Charlesworth (1998). The Mann–Whitney U-test was used to test for differences in estimates of both D'_{ST} and F_{ST}. Significant differences in estimates of F_{ST} but not D'_{ST} were interpreted as evidence of mutation bias. The program STATISTICA (StatSoft, Inc., Tulsa, Oklahoma) was used to perform the Spearman rank order correlation and Mann–Whitney U-tests.

Tests for spatial variation in mutation bias of F_{ST}

Graphical and statistical methods were used to address question two. Estimates of F_{ST} were computed for all population pairs using two locus groups: the five allozymes and the five moderately polymorphic microsatellites called locus group M; 11 highly polymorphic microsatellites called locus group H. Locus group H contained all microsatellites that exhibited evidence of mutation bias. The estimates of F_{ST} from the two locus groups were compared at two spatial scales: (1) population pairs within the same region; (2) population pairs from different regions (Bristol Bay and Kamchatka Peninsula). Scatter plots were used to graphically examine the relationship of F_{ST} estimates from the two locus groups for all population pairs and for population pairs at each spatial scale. This graphical examination included comparing the scatter plots of actual F_{ST} values with a scatter plot of F_{ST} values generated from the equilibrium approximation $F_{ST} \approx 1/(1 + 4N_e(m + \mu))$ for two hypothetical loci having mutation rates (μ) of 10^{-3} and 10^{-6}, effective population size (N_e) of 10,000, and migration rates (m) between 1.5×10^{-4} and 4.0×10^{-3}. Two statistical tests were performed to compare the estimates of genetic differentiation from locus groups M and H and to evaluate the influence and pattern of mutation bias on F_{ST} within and among each spatial scale. First, a paired sample Wilcoxon signed-ranks test was used to determine if estimates of F_{ST} from locus group M were significantly greater than estimates of F_{ST}, $F_{ST}b$, and R_{ST} from locus group H. Second, the Spearman rank order correlation test was used to test for correlation between the estimates of genetic differentiation from the two locus groups. The program STATISTICA (StatSoft, Inc., Tulsa, Oklahoma) was used to perform both tests.

Results

Genetic variation within and among populations

Estimates of mean expected heterozygosity (H_E) and mean number of alleles (A) over all loci for each population were higher for microsatellites (H_E range = 0.73–0.78, A range = 13.4–17.2) than for allozymes (H_E range = 0.24–0.29, A range = 2.0–2.4, Table 2). Nevertheless, five microsatellite loci (*Omy77, One105, Ots3, Ots107, μSat60*) had similar or lower H_E in most populations than the two most polymorphic allozymes (*ALAT*, PGM-2**). Single-locus estimates of mean within-population heterozygosity (H_S) for both marker classes varied widely and overlapped but were significantly higher for microsatellites than for allozymes (Mann–Whitney U-test, p < 0.004, Table 3). Estimates of absolute and relative measures of genetic divergence between populations also varied among loci and were generally low with the exception of the F_{ST} estimates from microsatellite *Ots3* and allozyme *PEPLT** (Table 3).

Randomization tests of conformity to Hardy–Weinberg equilibrium for each locus and population combination revealed 13 instances where the P-value for the test statistic F was below 0.05. These instances were not common to any one locus or population and were not judged significant when the α-level was adjusted for 14 simultaneous tests (each locus across all populations) and 21 simultaneous tests (each population across all loci). Randomization tests for genotypic disequilibrium revealed 13 locus pairs with a p-value for the G-statistic of <0.05. These tests were not judged significant when the α-level was adjusted for 210 simultaneous tests.

Comparison of F_{ST} estimates from microsatellites and allozymes

Estimates of F_{ST} *versus* H_S are shown for each locus in Figure 2. The Spearman rank order correlation test revealed a strong and significant inverse relationship between F_{ST} and H_S for all 21 loci (R_S = -0.805, p < 0.001), all microsatellites (R_S = -0.832, p < 0.001), and highly polymorphic microsatellites (R_S = -0.771, p < 0.006). In contrast, the inverse relationship between F_{ST} and H_S was weaker and not significant for moderately polymorphic microsatellites (R_S = -0.200, p = 0.747) and allozymes (R_S = -0.300, p = 0.624). This latter result probably

Table 3. Single-locus statistics from 14 sockeye salmon populations: sample size (n), total heterozygosity (H_T), mean within-population heterozygosity (H_S), estimates of absolute (D'_{ST}) and relative genetic differentiation (F_{ST}, $F_{ST}b$, R_{ST}), number of alleles (A), and frequency of the most common allele over all populations (P_a).

Locus	n	H_T	H_S	D'_{ST}	F_{ST}	$F_{ST}b$	R_{ST}	A	P_a
Microsatellite									
One103	1318	0.952	0.945	0.007	0.008	0.013	0.052	63	0.074
One114	1301	0.931	0.924	0.007	0.008	0.019	0.031	28	0.100
Ots103	1306	0.932	0.919	0.013	0.014	0.076	0.067	24	0.101
One115	1317	0.919	0.913	0.006	0.007	0.012	0.023	21	0.109
One104	1314	0.915	0.905	0.011	0.012	0.018	0.022	23	0.166
One112	1320	0.908	0.901	0.008	0.009	0.033	0.019	36	0.232
One110	1316	0.902	0.895	0.007	0.008	0.012	0.020	27	0.137
One108	1316	0.904	0.890	0.015	0.016	0.075	0.051	27	0.158
One111	1311	0.902	0.889	0.015	0.017	0.032	0.031	40	0.197
One109	1319	0.894	0.873	0.022	0.023	0.028	0.015	17	0.168
One102	1315	0.860	0.843	0.019	0.021	0.054	0.039	19	0.250
Omy77	1285	0.598	0.570	0.030	0.048	0.059	0.058	12	0.533
μSat60	1316	0.554	0.547	0.007	0.013	0.021	0.024	16	0.539
One105	1315	0.441	0.430	0.012	0.025	0.035	0.056	10	0.732
Ots3	1300	0.392	0.341	0.055	0.140	0.138	0.030	12	0.774
Ots107	1298	0.338	0.326	0.013	0.034	0.046	0.033	9	0.806
Mean	1310	0.771	0.757	0.016	0.020	0.040	0.036	24	0.317
Allozyme									
*ALAT**	1324	0.546	0.523	0.025	0.043	—	—	5	0.537
*PGM-2**	1329	0.406	0.398	0.009	0.023	—	—	2	0.726
*LDHB-2**	1330	0.263	0.239	0.025	0.088	—	—	2	0.607
*mAAT-1**	1315	0.183	0.180	0.003	0.015	—	—	2	0.899
*PEPLT**	1321	0.025	0.021	0.004	0.170	—	—	3	0.990
Mean	1324	0.285	0.272	0.013	0.044	—	—	3	0.752

Estimates of $F_{ST}b$ were derived using the allele binning method described in the text.

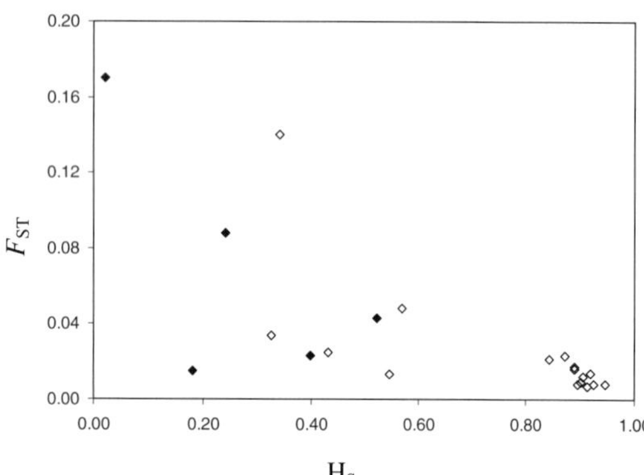

Figure 2. Scatter plot of F_{ST} estimates *versus* mean within-population genetic diversity (H_S) for five allozyme (♦) and 16 microsatellite (◊) loci from 14 sockeye salmon populations.

Table 4. Estimates of genetic divergence among 14 sockeye salmon populations from four locus groups with n loci. The abbreviations are moderately polymorphic (MP) and highly polymorphic (HP).

Locus group	n	Parameter	Value	95% CI
Allozymes	5	F_{ST}	0.044	0.024–0.076
Microsatellites	16	F_{ST}	0.020	0.013–0.032
		$F_{ST}b$	0.040	0.028–0.055
		R_{ST}	0.036	0.029–0.044
MP microsatellites	5	F_{ST}	0.049	0.021–0.096
		$F_{ST}b$	0.055	0.029–0.094
		R_{ST}	0.040	0.028–0.052
HP microsatellites	11	F_{ST}	0.013	0.010–0.016
		$F_{ST}b$	0.034	0.022–0.048
		R_{ST}	0.034	0.025–0.044

Estimates of $F_{ST}b$ were derived using the allele binning method described in the text.

reflects the high variance in F_{ST} values for moderately polymorphic microsatellites and allozymes. Two groups of loci are evident based on estimates of H_S: (1) 11 microsatellites ($H_S > 0.84$); (2) five allozymes and five microsatellites ($H_S < 0.60$).

The estimate of F_{ST} from allozymes was twice as large as the estimate of F_{ST} from microsatellites (Table 4). Regardless, the 95% confidence intervals overlapped so the two estimates were not judged significantly different. This was not the case when the microsatellites were divided into two groups; moderately polymorphic ($H_S < 0.60$) and highly polymorphic ($H_S > 0.84$). The estimate of F_{ST} from the highly polymorphic microsatellites was significantly lower than the F_{ST} estimates from both the allozymes and moderately polymorphic microsatellites. The latter two locus groups provided similar estimates of F_{ST} and a similar degree of variance around the estimate as indicated by the confidence intervals. The original estimate of F_{ST} from the highly polymorphic microsatellites was significantly lower than the estimates of R_{ST} and $F_{ST}b$. For all microsatellites and for moderately polymorphic microsatellites R_{ST} and $F_{ST}b$ provided very similar values that were larger, but not significantly different, than the original F_{ST} estimate (Table 4).

The tests for differences in F_{ST} estimates among locus groups using the Mann–Whitney U-test provided results similar to those described above with the exception of the comparison of allozymes and all microsatellites ($p < 0.043$). In contrast, no significant differences in absolute genetic divergence (D'_{ST}) between populations were revealed among any pair of locus groups, regardless of locus polymorphism.

Test for spatial variation in mutation bias of F_{ST}

The relationship between estimates of F_{ST} from locus groups H ($F_{ST}H$) and M ($F_{ST}M$) for all population pairs and for population pairs at two spatial scales is depicted graphically in Figure 3. It is clear from these scatter plots that the F_{ST} values, while generally low overall, tend to be largest for population pairs from scale 2. It is also evident that, regardless of spatial scale, the estimates of $F_{ST}M$ are nearly always larger than the estimates of $F_{ST}H$. Indeed, the Wilcoxon sign-rank tests revealed significant differences between F_{ST} estimates from the two locus groups for all population pairs ($p < 0.001$) and for population pairs from scale 1 ($p < 0.001$) and scale 2 ($p < 0.001$, Table 5). The estimates of $F_{ST}H$ increased for most population pairs when alleles for each locus were binned into two or three size categories (Figure 3). In fact, the scatter plot of $F_{ST}Hb$ *versus* $F_{ST}M$ for scale 1 population pairs clearly clusters around the 1 : 1 line. A second series of Wilcoxon sign-rank tests showed the differences between estimates of $F_{ST}M$ and $F_{ST}Hb$ were not significant for scale 1 population pairs but were significant for all population pairs ($p < 0.001$) and scale 2 population pairs ($p < 0.001$, Table 5). Finally, estimates of R_{ST} were generally larger than estimates of $F_{ST}H$.

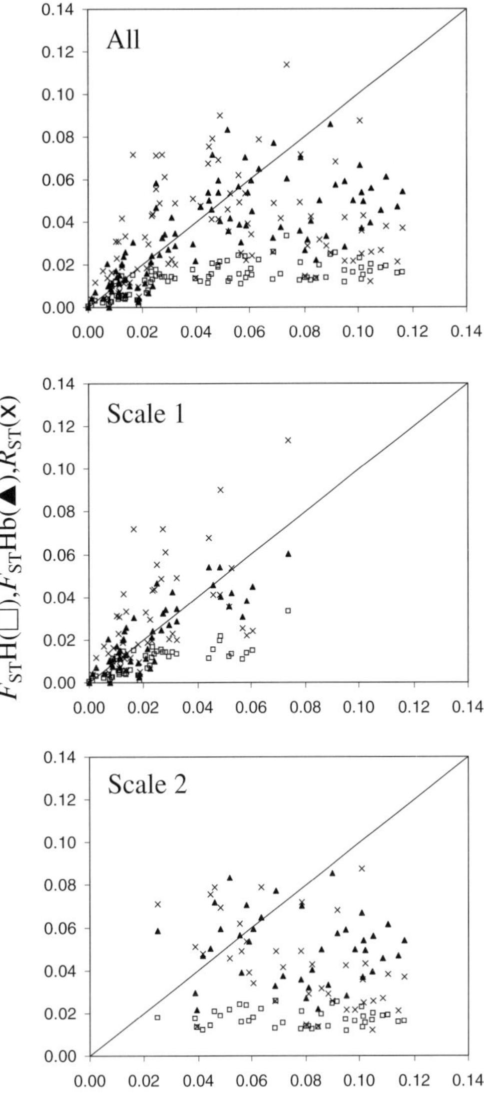

Figure 3. Scatter plot of population divergence estimates from 11 highly polymorphic microsatellites (y-axis) and 10 moderately polymorphic loci (x-axis) for all population pairs and for population pairs at two spatial scales: scale 1, within regions; scale 2, between regions (Bristol Bay and Russia). Estimates of F_{ST} for moderately polymorphic loci ($F_{ST}M$) are compared to three estimates of genetic diversity for highly polymorphic microsatellites: $F_{ST}M \times F_{ST}H$ ($\square$), $F_{ST}M \times F_{ST}Hb$ ($\blacktriangle$), and $F_{ST}M \times R_{ST}$ ($\times$).

Interestingly, a final series of Wilcoxon sign-rank tests showed the estimates of R_{ST} were significantly larger than estimates of $F_{ST}M$ for scale 1 population pairs and significantly smaller than estimates of $F_{ST}M$ for scale 2 population pairs.

Table 5. Tests of the relationship between estimates of genetic divergence from the moderately polymorphic locus group (LGM) and highly polymorphic locus group (LGH) for sockeye salmon population pairs at two spatial scales.

Test	LGM	LGH	All	Scale 1	Scale 2
WSR – H_O: Estimates are equal					
$F_{ST}M$	$F_{ST}H$		<0.001	<0.001	<0.001
$F_{ST}M$	$F_{ST}Hb$		<0.001	NS	<0.001
$F_{ST}M$	R_{ST}		<0.014	<0.046	<0.001
SRO – H_O: Estimates not correlated					
$F_{ST}M$	$F_{ST}H$		<0.001	<0.001	NS
$F_{ST}M$	$F_{ST}Hb$		<0.001	<0.001	NS
$F_{ST}M$	R_{ST}		<0.001	<0.001	<0.002

The Wilcoxon sign-rank (WSR) and Spearman rank order correlation (SRO) tests were used to compare genetic divergence estimates. The two spatial scales are: scale 1, within regions (n = 51); scale 2, between regions (n = 40). Estimates of $F_{ST}Hb$ were derived using the allele binning method described in the text.

Although the estimates of $F_{ST}H$ are nearly always smaller, they appear to be positively correlated with estimates of $F_{ST}M$ for scale 1 population pairs (Figure 3). The Spearman rank order correlation tests confirmed a positive correlation for scale 1 population pairs (R_S = 0.748, p < 0.001). There is clearly no correlation between estimates of F_{ST} from the two locus groups for scale 2 population pairs. Subsequent Spearman rank order correlation tests supported a positive correlation between estimates of $F_{ST}M$ and estimates of $F_{ST}Hb$ and R_{ST} for scale 1 population pairs (R_S = 0.773, p < 0.001; R_S = 0.624, p < 0.001; Table 5). Interestingly, estimates of R_{ST} and $F_{ST}M$ for scale 2 populations pairs were negatively correlated (R_S = −0.491, p < 0.002).

Discussion

Mutation bias and microsatellite F_{ST} estimates

We found a negative correlation between single-locus estimates of F_{ST} and mean within-population heterozygosity (H_S) for all microsatellites and for highly polymorphic microsatellites, but not for moderately polymorphic microsatellites and allozymes. This suggests that F_{ST} estimates are constrained in microsatellites but only in extremely polymorphic loci. We assume the large differences in locus polymorphism (H_S) are due to differences in locus mutation rate and not the result of random genetic drift. This assumption

is supported by the fact that the maximum value of F_{ST} ($1 - H_S$, Hedrick 1999) for 10 of the 11 highly polymorphic loci fall well below the F_{ST} estimates for two of the moderately polymorphic loci, microsatellite *Ots3* ($F_{ST} = 0.140$) and allozyme *PEPLT* ($F_{ST} = 0.170$).

The initial comparison of F_{ST} estimates from the two marker classes, allozymes and microsatellites, is equivocal. Although the F_{ST} value for allozymes (0.044) is twice as large as the F_{ST} value for microsatellites (0.020), the difference between the two estimates is not significant based on bootstrap sampling (95% confidence intervals overlap) and only marginally significant using the Mann–Whitney U-test (p < 0.043). Subdividing the microsatellites, however, into moderately and highly polymorphic groups shows that when all microsatellites are pooled, the moderately polymorphic loci appear to conceal evidence of mutation bias in estimates of F_{ST} for the highly polymorphic loci. For example, estimates of F_{ST} for allozymes and moderately polymorphic microsatellites are nearly identical and significantly larger (non-overlapping 95% confidence intervals) than the F_{ST} estimate for highly polymorphic microsatellites. Indeed, an important finding of this study is that mutation bias can lead to discordant estimates of F_{ST} among moderately and highly polymorphic microsatellites.

Two other tests provide evidence that mutation is downwardly biasing the F_{ST} estimate from highly polymorphic (but not moderately polymorphic) microsatellite loci. First, the estimates of absolute divergence (D'_{ST}) for highly polymorphic microsatellites are not significantly different from the D'_{ST} estimates for allozymes and moderately polymorphic microsatellites. The fact that estimates of D'_{ST} do not differ among the three locus groups, but estimates of F_{ST} for highly polymorphic microsatellites differ from F_{ST} estimates for both moderately polymorphic microsatellites and allozymes, is consistent with mutation bias in F_{ST} estimates for highly polymorphic microsatellites (Charlesworth 1998). Second, the estimate of F_{ST} for the highly polymorphic microsatellites is significantly lower than the estimates of $F_{ST}b$ and R_{ST} whereas for moderately polymorphic microsatellites these three statistics are similar. Estimates either of F_{ST} from binned allele frequencies or of R_{ST} will only be greater than estimates of F_{ST} if mutation is a significant force influencing microsatellite variation (Slatkin 1995). The fact that $F_{ST}b$ and R_{ST} are greater than F_{ST} for only the highly polymorphic microsatellites implies that mutation is not a significant force influencing genetic variation of the moderately polymorphic microsatellites.

Our conclusions rest on the assumption that estimates of F_{ST} from both allozymes and microsatellites are not disproportionately influenced by exceptional loci. Allendorf & Seeb (2000) showed that differences in estimates of F_{ST} from different marker classes could be explained by one or two exceptional loci with unusually high F_{ST} values. None of the loci in this study exhibited values of F_{ST} that are as uniquely large as the loci reported in Allendorf & Seeb (2000). Nevertheless, in our study two moderately polymorphic loci stand out: microsatellite *Ots3* and allozyme *PEPLT**. To examine the influence of these loci on our conclusions, they were removed from the data base and the F_{ST} estimate from moderately polymorphic loci was recalculated. The remaining eight loci provided an F_{ST} value of 0.035 with a 95% confidence interval of 0.023–0.049 that does not overlap with the confidence interval for the F_{ST} estimate from highly polymorphic microsatellites.

Spatial population structure and mutation bias in F_{ST}

The Wilcoxon sign-rank test results demonstrate that mutation bias is influencing estimates of $F_{ST}H$ for most population pairs. Estimates of genetic divergence for locus group H increase for almost all population pairs from each spatial scale when $F_{ST}Hb$ or R_{ST} are used in place of $F_{ST}H$. Overall, these results show that mutation bias can influence estimates of F_{ST} from highly polymorphic loci even when the degree of genetic divergence is very low ($F_{ST} < 0.02$).

Two patterns in the estimates of genetic divergence clearly distinguish population pairs in scale 1 from those in scale 2. First, the Wilcoxon sign-rank test results show that $F_{ST}Hb$ and R_{ST} reveal the same degree of differentiation as the $F_{ST}M$ estimates for population pairs from scale 1 but not scale 2. This relationship is also evident from the scatter plots in which the genetic divergence estimates from the two locus groups tend to cluster evenly around the 1 : 1 line for scale 1. This is the expectation for estimators that are not biased by locus-specific forces. The results for scale 1 corroborate those reported by Buonaccorsi et al. (2001) who showed that underestimates of genetic divergence due to mutation bias in F_{ST} could be reconciled by using binned allele frequencies to compute F_{ST} or by using R_{ST}. The results for scale 2 are equivocal, however. For these population

pairs, estimates of $F_{ST}Hb$ and R_{ST} are generally greater than estimates of $F_{ST}H$, but appear to underestimate genetic divergence with respect to $F_{ST}M$. More estimates of $F_{ST}Hb$ and R_{ST} lie below the 1 : 1 line than above it. The cause of this pattern is unclear, however, a likely explanation is that it simply reflects stochastic variation among loci influenced predominantly by genetic drift. The rate of migration should be extremely low for population pairs at scale 2 relative to those at scale 1. Under this relatively high degree of isolation the single-locus variation in estimates of F_{ST} will be greatest (Hutchison & Templeton 1999). It is therefore possible that, due to locus sampling error, the majority of estimates of $F_{ST}M$ are greater than $F_{ST}Hb$ and R_{ST}.

Second, the Spearman rank order correlation test reveals a positive correlation between the three estimates of genetic divergence for locus group H ($F_{ST}H$, $F_{ST}Hb$, R_{ST}) and estimates of $F_{ST}M$ for spatial scale 1 but not 2. Again, these relationships are apparent in the scatter plots (Figure 3). The positive correlation between estimates of $F_{ST}H$ and $F_{ST}M$ for scale 1 population pairs suggests that the migration rate is sufficiently large to influence estimates of $F_{ST}H$. In fact, the equilibrium approximation $F_{ST} \approx 1/(1 + 4N_e(m + \mu))$ from the infinite island model of Wright (1969) provides support for this conclusion and may explain the lack of correlation between $F_{ST}H$ and $F_{ST}M$ for scale 2 population pairs. The example in Figure 4 shows that, as the rate of migration (m) decreases, the value of F_{ST} for the high mutation rate loci ($\mu = 10^{-3}$) approaches an asymptote

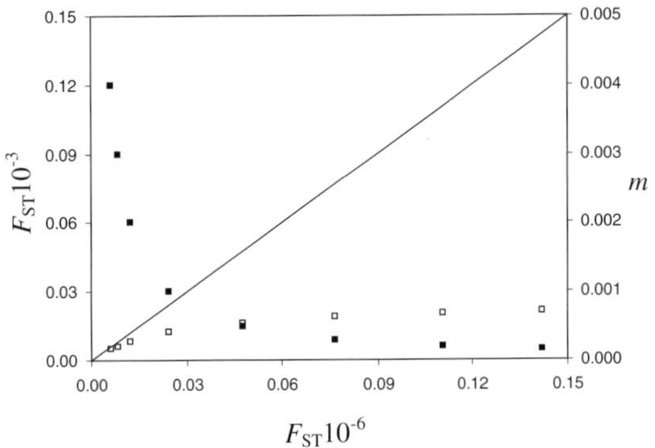

Figure 4. Expected values of F_{ST} ($\square$) for loci with a mutation rate (μ) of 10^{-3} (y-axis) and 10^{-6} (x-axis). Each value was derived for a different migration rate (m, $\blacksquare$) using the equilibrium approximation $F_{ST} \approx 1/(1 + 4N_e(m + \mu))$ with an effective population size (N_e) = 10, 000 (selected to provide values of F_{ST} observed in this study).

while the value of F_{ST} for the low mutation rate loci ($\mu = 10^{-6}$) continues to increase. The relationship between $F_{ST}10^{-3}$ and $F_{ST}10^{-6}$ in Figure 4 is very similar to the relationship between $F_{ST}H$ and $F_{ST}M$ in Figure 3. From the two figures we infer that m is nearly as large (if not larger than) μ for scale 1 population pairs, resulting in a positive correlation between $F_{ST}H$ and $F_{ST}M$. When the alternative estimates of genetic divergence ($F_{ST}Hb$, R_{ST}) are used, the two locus groups are positively correlated around the 1 : 1 line – the expectation if genetic drift and migration are the primary forces influencing both locus groups (Landry & Bernatchez 2001). In contrast, we infer that μ is much larger than m for scale 2 population pairs and consequently the estimates of $F_{ST}H$ show little correlation to $F_{ST}M$, similar to the relationship between $F_{ST}10^{-3}$ and $F_{ST}10^{-6}$ on the right side of Figure 4. The fact that $F_{ST}Hb$, and R_{ST} show no correlation to $F_{ST}M$ for scale 2 population pairs is interesting since both $F_{ST}Hb$, and R_{ST} account for the mutation that is biasing $F_{ST}H$, and should therefore reveal a similar degree of genetic divergence for each population pair. Because the rate of migration is low for these population pairs, the lack of correlation here is likely an outcome of locus sampling error due to the high variance in single locus estimates of genetic divergence when population structure is influenced mostly by genetic drift.

Conclusions

Our results do not contradict the arguments of Neigel (2002) that support the continued use of F_{ST} to describe population structure. Rather, we demonstrate that caution should be used when comparing estimates of F_{ST} from loci with different levels of heterozygosity (H_S). Large differences in H_S can result in discordant estimates of F_{ST} among microsatellite loci, which may result in an underestimation of the degree of genetic divergence among populations if all loci are pooled for analysis. To avoid this error, we suggest using a combination of single-locus evaluation of microsatellites as suggested by Freville et al. (2001) and locus-group evaluation. For those loci that appear to underestimate F_{ST}, alternative estimators of genetic divergence can be computed, such as R_{ST} or estimates of F_{ST} from binned allele frequency data. This approach provides an alternative to excluding loci simply because they are highly polymorphic.

The influence of mutation rate on estimates of F_{ST} from microsatellites may vary substantially among pairs of populations in broad-scale studies of populations that are spatially structured. This is because the ratio of migration to mutation is not constant and the influence of mutation will be greatest at the largest spatial scale. To avoid incorrect conclusions regarding the migration rate among populations in a broad-scale study, we suggest comparing pairwise estimates of F_{ST} from highly polymorphic microsatellites, stratified by spatial scale, to theoretical expectations and, if available, F_{ST} estimates from moderately polymorphic loci.

Acknowledgements

Funding was provided by the State of Alaska and National Marine Fisheries Service through the Western Alaska research project. Genetic data were collected by Andy Barclay, Judy Berger, Zac Grauvogel, Sheri Wilson, and Bruce Whelan of the Alaska Department of Fish and Game. Substantial assistance in organizing sample collections was provided by Steve Morstad, Keith Weiland, Jim Browning (Alaska Department of Fish and Game), Tom Quinn (University of Washington), and Nataly Varnavskaya (Kamchatka Research Institute of Fisheries Oceanography). Daniel Ruzzante (Dalhousie University) provided the S-PLUS program that computes R_{ST}. This is contribution PP-228 of the Alaska Department of Fish and Game, Division of Commercial Fisheries, Juneau.

References

ABI (Applied Biosystems Inc.) 1996a. GeneScan 672 Users Manual Rev. A. Perkin-Elmer Corp., Foster City, CA. 337 pp.

ABI (Applied Biosystems Inc.) 1996b. Genotyper 2.0 Users Manual. Perkin-Elmer Corp., Foster City, CA. 75 pp.

Allendorf, F.W. & L.W. Seeb. 2000. Concordance of genetic divergence among sockeye salmon populations at allozyme, nuclear DNA, and mitochondrial DNA markers. Evolution 54: 640–651.

Balloux, F., H. Brünner, N. Lugon-Moulin, J. Hausser & J. Goudet. 2000. Microsatellites can be misleading: An empirical and simulation study. Evolution 54: 1414–1422.

Balloux, F. & J. Goudet. 2002. Statistical properties of population differentiation estimators under stepwise mutation in a finite island model. Mol. Ecol. 11: 771–783.

Bowcock, A.M., A. Ruiz-Linares, J. Tomfohrde, E. Minch, J.R. Kidd & L.L. Cavalli-Sforza. 1994. High resolution of human evolutionary trees with polymorphic microsatellites. Nature 368: 455–457.

Buonaccorsi, V.P., J.R. McDowell & J.E. Graves. 2001. Reconciling patterns of inter-ocean molecular variance from four classes of molecular markers in blue marlin (*Makaira nigricans*). Mol. Ecol. 10: 1179–1196.

Charlesworth, B. 1998. Measures of divergence between populations and the effect of forces that reduce variability. Mol. Biol. Evol. 15: 538–543.

Estoup, A., F. Rousset, Y. Michalakis, J.-M. Cornuet, M. Adriamanga & R. Guyomard. 1998. Comparative analysis of microsatellite and allozyme markers: A case study investigating microgeographic differentiation in brown trout (*Salmo trutta*). Mol. Ecol. 7: 339–353.

Freville, H., F. Justy & I. Olvieri. 2001. Comparative allozyme and microsatellite population structure in a narrow endemic plant species, *Centaurea corymbosa* Pourret (Asteraceae). Mol. Ecol. 10: 879–889.

Goodman, S.J. 1997. RST Calc: A collection of computer programs for calculating estimates of genetic differentiation from microsatellite data and determining their significance. Mol. Ecol. 6: 881–886.

Hedrick, P.W. 1999. Perspectives: Highly variable loci and their interpretation in evolution and conservation. Evolution 53: 313–318.

Hendry, A.P., J.K. Wenburg, P. Bentzen, E.C. Volk & T.P. Quinn. 2000. Rapid evolution of reproductive isolation in the wild: Evidence from introduced salmon. Science 290: 516–518.

Hutchison, D.W. & A.R. Templeton. 1999. Correlation of pairwise genetic and geographic distance measures: Inferring the relative influences of gene flow and drift on the distribution of genetic variability. Evolution 53: 1898–1914.

Jarne, P. & J.L. Lagoda. 1996. Microsatellites, from molecules to populations and back. TREE 11: 424–429.

Landry, C. & L. Bernatchez. 2001. Comparative analysis of population structure across environments and geographical scales at major histocompatibility complex and microsatellite loci in Atlantic salmon (*Salmo salar*). Mol. Ecol. 10: 2525–2539.

Nei, M. 1987. Molecular Evolutionary Genetics, Columbia University Press, New York. 512 pp.

Neigel, J.E. 2002. Is F_{ST} obsolete? Conserv. Genet. 3: 167–173.

Olsen, J.B., S.L. Wilson, E.J. Kretschmer, K.C. Jones & J.E. Seeb. 2000. Characterization of 14 tetranucleotide microsatellite loci derived from sockeye salmon. Mol. Ecol. 9: 2155–2234.

Rice, W.R. 1989. Analyzing tables of statistical tests. Evolution 43: 223–225.

Seeb, L.W., C. Habicht, W.D. Templin, K.E. Tarbox, R.Z. Davis, L.K. Brannian & J.E. Seeb. 2000. Genetic diversity of sockeye salmon of Cook Inlet, Alaska, and its application to management of populations affected by the *Exxon Valdez* oil spill. Trans. Amer. Fish. Soc. 129: 1223–1249.

Shaw, P.W., C. Turan, J.M. Wright, M. O'Connell & G.R. Carvalho. 1999. Microsatellite DNA analysis of population structure in Atlantic herring (*Clupea harengus*), with direct comparison to allozyme and mtDNA RFLP analysis. Heredity 83: 4990–4499.

Slatkin, M. 1995. A measure of population subdivision based on microsatellite allele frequencies. Genetics 139: 457–462.

Wood, C.C. 1995. Life history variation and population structure in sockeye salmon. pp. 195–216. *In*: J.L. Nielsen (ed.) Evolution and the Aquatic Ecosystem: Defining Unique Units of Population Conservation, American Fisheries Society Symposium 17, Bethesda, Maryland.

Wright, S. 1969. Evolution and the Genetics of Populations, The Theory of Gene Frequencies, The University of Chicago Press, Chicago, Illinois. 511 pp.

Environmental Biology of Fishes **69**: 275–285, 2004.
© 2004 *Kluwer Academic Publishers. Printed in the Netherlands.*

Forensic DNA analysis of Pacific salmonid samples for species and stock identification

Ruth E. Withler, John R. Candy, Terry D. Beacham & Kristina M. Miller
Department of Fisheries and Oceans Canada, Pacific Biological Station, Nanaimo, BC, V9T 6N7 Canada
(e-mail: withlerr@pac.dfo-mpo.gc.ca)

Received 17 April 2003   Accepted 19 April 2003

Key words: genetic stock identification, PCR, illegal harvest, conviction

Synopsis

Identification of salmonid tissue samples to species or population of origin has been conducted for over 20 forensic cases in British Columbia. Species identification is based on published sequence variation in exon and intron regions of coding genes. Identification of source populations or regions is carried out using microsatellite and major histocompatibility complex allele frequency data collected from populations throughout the species range and with standard genetic stock identification (GSI) methods. Regional contributions to mixture samples are estimated using maximum likelihood mixture analysis and classification of individual genotypes is carried out with Bayesian methods. DNA has been obtained successfully from salmon scale samples, fresh, frozen and canned tissue samples and bloodstains in clothing. Results from DNA analyses have been instrumental in a number of convictions. A major benefit has been cost savings resulting from the number of guilty pleas entered after disclosure to the defendant of results from genetic testing. In two cases, GSI analysis resulted in exoneration of suspects under investigation for possible illegal sales of Fraser River sockeye salmon by substantiating their claim that the fish originated from the Skeena River watershed. DNA analysis has generally corroborated the species and stock identification carried out by fishery officers, but has revealed that species identification of samples from sources such as restaurants and fish plants can be erroneous. Forensic DNA analysis has facilitated the conviction of those who purchase fish not caught under the authority of licence, thus bringing those who buy fish illegally as well as those involved in illegal harvest and sales within the scope of law enforcement.

Introduction

Conservation management of harvested species is frequently complicated by the misrepresentation of product acquired by illegal methods as legal commercial product in the marketplace or the illegal sale of fish legally harvested from recreational and aboriginal fisheries. In Canada, Pacific salmonids are harvested in separately managed aboriginal, commercial and sport fisheries subject to gear-, time- and location-specific openings. Regional fishery closures occur commonly when legal harvest is underway elsewhere within Canadian or American waters, and when fresh cultured salmon are present in the marketplace.

Moreover, the first listing of a Canadian salmon population (Interior Fraser coho salmon) as 'endangered' ushers in a new management era that will include increased demand for stock-specific fisheries management, selective harvest regimes and forensic identification of confiscated samples. Illegal harvest and sales occur in both marine and freshwater environments, often in remote locations or under the cover of legal fisheries. Therefore, prosecution of individuals in possession of fresh, frozen or canned illegal salmon products often involves not only the identification of product to species, but also to region of origin as a means of establishing the probable location and/or time of capture.

DNA methodologies of animal identification are becoming commonplace and have gained acceptance in legal proceedings (Zehner et al. 1998, Hsieh et al. 2001). Methods to identify tissue samples to species based on nuclear or mitochondrial DNA sequences have been developed for a wide variety of organisms, including commercially important fish (Chow et al. 1993, DeSalle & Birstein 1996, McKay et al. 1997, Parson et al. 2000, Calvo et al. 2001, Wetton et al. 2002, Shivji et al. 2002). For domestic and wild animal species, microsatellite loci have been used to match an individual with tissue collected from an accident or kill site putatively involving the animal (Pádár et al. 2001, Poetsch et al. 2001). Both species identification and the identification of multiple tissues from a single individual are based on concepts of a complete 'match' of genetic profiles. Thus, the genetic profile of the unknown must meet criteria that enable it to be considered identical to the profile of the known tissue. This match enables exclusion of other species or individuals as highly probable sources of the unknown tissue. Genetic species identification methods have been developed for Pacific and Atlantic salmon based on a variety of nuclear (Withler et al. 1997, McKay et al. 1997) and mitochondrial (McKay et al. 1997, Hold et al. 2001) DNA sequences and applied to fresh, frozen and smoked tissues.

The genetic structure of Pacific salmonid species brought about by their propensity to return to their natal freshwater habitat for spawning is evident in the differentiation observed among populations (stocks) at allozyme, minisatellite, microsatellite and major histocompatibility complex (MHC) loci (Shaklee et al. 1999, Miller et al. 1996, 2001, Withler et al. 2001). We have compiled large microsatellite and MHC databases covering much of the species range for coho, chinook and sockeye salmon for use in genetic stock identification (GSI) of fishery samples by means of mixture analysis and classification of individual multilocus genotypes (Beacham et al. 2001a,b, 2003). The demonstrated advantages of microsatellite data for salmonid GSI include high levels of polymorphism with strong interpopulational differentiation in allele frequencies, temporal stability of allele frequencies within populations, and the regional nature of population structure (Beacham et al. 2001a, Withler et al. 2000). Allele frequencies at MHC class I and II loci are also highly variable among salmonid populations and, influenced by both balancing and directional selection, display different patterns of population subdivision than the neutral microsatellite loci (Miller et al. 2001). The combined information from the neutral and adaptive loci provides powerful discrimination among populations (Beacham et al. 2001a). In addition, the ubiquitous presence and relative stability of DNA in body tissues enables non-lethal sample collection from almost any tissue, including dried, frozen and ethanol-preserved tissues and processed (canned, smoked) food products.

The fact that alleles are not population-specific but instead merely vary in frequency among salmonid populations precludes the possibility of obtaining with GSI methodology the population-specific match that can be achieved at the levels of species and individuals. Instead, mathematical methods are employed to determine the most probable origin of unknowns (i.e. the population or region in which the observed multilocus genotype has the highest probability of occurring) but are generally incapable of excluding other populations/regions as potential sources. Thus, empirical determination of the level of correct assignment of unknowns to the source population/region is required.

In this study, we validate the use of microsatellite and MHC databases developed for coho, chinook and sockeye salmon for identification of tissue samples to regional sources by reviewing the accuracy and precision of allocation of samples of known origin by maximum likelihood mixture analysis and Bayesian classification of individual multilocus genotypes. We then review the results obtained for forensic analyses conducted for fishery officers since 1995 that have been used to support species identification and regional origins of fish tissues gathered as evidence in cases of alleged fishery violations. Tissues most commonly used for DNA analysis include scales or fin clips (from whole fish) or muscle tissue (from processed fish) that have been frozen and/or preserved in ethanol for storage and transport to the laboratory. However, a slurry of fish tissue collected from a cooler used to transport fish, dried fish blood extracted from clothing and canned salmon have also been examined successfully.

Material and methods

Species identification

Species identification of salmonid tissue samples was carried out primarily by PCR-RFLP of 72 codons and adjoining intron sequence of the salmonid MHC class II DAB gene (Withler et al. 1997). The amplified sequence ranges from 800 to 1050 bp in length among eight Pacific salmonid species, *Oncorhynchus* spp., and between 1 000 and 3 000 bp in length in Atlantic salmon, *Salmo salar* and brown trout, *S. trutta*. Restriction with between one and three enzymes enables separation of all 10 species, except for rainbow and cutthroat trout. These two species were distinguished by PCR-RFLP analysis of the growth hormone II gene (McKay et al. 1997). Differentiation of Atlantic salmon and brown trout was carried out using the method of Pendas et al. (1995).

Validation and application of forensic GSI

Genetic stock identification of coho, chinook and sockeye salmon samples for forensic purposes is conducted using large microsatellite and MHC databases accumulated in studies of population structure (Beacham et al. 2001a,b, 2003). The software program SPAM 3.2 (Debevec et al. 2000) is used for maximum likelihood allocation of proportions of mixture samples to populations of origin. The Bayesian routine of the GeneClass program (Cornuet et al. 1999) is used to classify individual samples to population of origin. Allocations to individual populations within a 'reporting region' are summed to provide the estimated contribution of fish from each region in the sample of 'unknowns' provided by the fishery officer. Reporting regions are geographically defined areas of interest to fishery regulation enforcement for which the accuracy of fish classification is 90% or higher. In this study we examine the accuracy and precision of the allocation to reporting region of origin by both the maximum likelihood and Bayesian analyses for sockeye, coho and chinook salmon.

Accuracy of classification of fish to reporting region using SPAM was investigated by analysis of test samples of 50 fish originating from populations within that reporting region. For the analysis, some microsatellite alleles were binned to reduce the number of genotypic frequencies to be estimated. Binning consisted of combining low-frequency alleles with adjacent higher-frequency alleles and has been shown to result in little loss of discrimination among populations. Observed allele frequencies were used to generate the genotypic frequencies expected for each population under conditions of Hardy Weinberg equilibrium for model inputs. Details of these procedures are available (Beacham et al. 2001a,b, in press). For coho salmon, the baseline data consisted of ~33 000 fish scored at eight microsatellite loci and two exons of the MHC class I UBA gene. For chinook salmon, 37 000 fish were scored at 13 microsatellite loci, and for sockeye

salmon 39 000 fish were scored at 14 microsatellite loci and the MHC class II DAB gene. Each of the simulated samples of 'unknowns' was generated 100 times by re-sampling of the baseline data. Each baseline population and 'unknown' mixture was sampled with replacement to simulate random variation associated with the collection of the baseline and unknown samples. The estimated contribution of populations within the true region of origin to the unknown mixture sample is reported as the bootstrap mean, with the standard deviation of the 100 estimates also shown.

Accuracy of classification of individual chinook and sockeye salmon to reporting region of origin was examined using the Bayesian leave-one-out 'self-classification' option of GeneClass. In this procedure each fish in the baseline data is removed from the baseline, treated as an 'unknown', and classified to population of origin. The percentage of fish classified to their true population or another population within the reporting region of origin was calculated. The smaller number of loci scored in coho salmon precludes classification of individual genotypes with sufficient accuracy to use in forensic analyses. Because of limitations on the number of fish that can be 'self-classified' by GeneClass, the baseline data used for self-classification was a subset of the existing baseline set for each species, consisting of only fish for which the full suite of genetic markers had been scored. The numbers of fish were 13 589 and 13 147 for sockeye and chinook salmon, respectively.

Verification of the accuracy and precision of salmon stock composition analysis for fishery management purposes has also been conducted through analysis of fishery samples of fish carrying coded wire tags that identify their stream of origin and analysis of samples caught within large river systems that are considered to consist entirely of fish originating from that watershed. These samples may contain fish from populations not included in the baseline and thus are more challenging GSI samples. We outline the results of genetic stock identification of coho and sockeye salmon carrying coded wire tags to reporting regions of interest in forensic analyses.

GSI analysis of 20 sets of sockeye and chinook salmon unknown samples was conducted for 17 cases investigated by fishery officers between 1998 and 2002. In all cases, fishery officers suspected that the fish may have originated from the Fraser River watershed. Illegal harvest and sales of Fraser River sockeye and chinook salmon are common and occur in conjunction with legal aboriginal, sports and commercial fisheries in the heavily populated drainage. Tissues were preserved by drying (scales and bloodstained clothing) or by freezing and/or preservation in 95% undenatured alcohol. Partially degraded DNA obtained from canned salmon was also used successfully for GSI analysis. Sixteen GSI analyses were performed on sockeye salmon (with sample sizes ranging from 1 to 144) and four on chinook salmon (with sample sizes from 15 to 76). Eleven of the cases involved charges or potential charges against harvesters, often apprehended after delivery of fish to processors without proper documentation. In the remaining cases charges were contemplated or laid against commercial establishments suspected of purchasing fish caught without the authority of a licence, generally as the result of information provided by employees, health inspectors or the public.

The stock composition of samples of fish harvested from within the Fraser River drainage varies according to the time and location of harvest. Sockeye salmon enter the drainage from June to October, with those returning to many of the interior Fraser and Thompson tributaries upstream of the Fraser Canyon preceding those returning to tributaries of the lower Fraser River (Gable & Cox-Rogers 1993). Tributaries of the upper or interior Fraser River, those of the Thompson River and those of the lower Fraser river comprise three genetically distinctive groups (Withler et al. 2000) and can be easily identified in GSI analysis. Catches of sockeye salmon made anywhere in the drainage prior to August, and from within or upstream of the Fraser River Canyon after July, are expected to consist almost entirely of fish originating from interior Fraser and Thompson River tributaries, whereas catches made downstream from the canyon in or after August should also contain sockeye returning to lower Fraser River tributaries. Much of the harvest of sockeye salmon that is illegally purchased and sold is believed to occur within the Fraser Canyon region, leading to the expectation that only fish returning to interior Fraser and Thompson river tributaries should be identified in confiscated samples. We estimated the contributions of sockeye salmon populations in the interior Fraser and Thompson drainage regions in the forensic sockeye samples.

Results and discussion

Species identification

Species identification is straightforward and based on achievement of a match between the unknown(s) and standard species-specific RFLP profiles. The Pacific/Atlantic salmon species identification method most commonly used in our analyses has been validated over a wide geographic area, including the North American and Asian Pacific coastlines, and is dependent on digestion of the PCR product with three restriction enzymes (Withler et al. 1997). Species identification results based on a single mutational difference between species or novel genetic profiles resulting from mutation can be verified using other salmonid species identification methods (McKay et al. 1997, Hold et al. 2001).

Species identification was the primary objective of six analyses conducted for fishery officers between 1995 and 2000 (Table 1). In case 1, charges of illegal sale of coho salmon were supported by analysis of DNA extracted from all 24 scales removed from fish and from four out of five frozen liquid samples collected from two plastic coolers and wooden barrels allegedly used to transport the fish. The liquid was primarily water but also contained blood, scales and slime from the transported fish. No amplification of DNA with primers for the 840 bp MHC sequence used in the RFLP identification of coho salmon was obtained from the fifth liquid sample.

For cases 2–5, charges were contemplated or actually laid as a result of suspected violations of the restrictions imposed on coho salmon fisheries in British Columbia since 1998. In case 2, following a dispute between the owner/operator of a small restaurant and his cook over working conditions, the cook abandoned her position. She left with a fish concealed under her jacket and subsequently delivered the fish to the local fishery officer as evidence of the illegal coho salmon she had been cooking. In the process of conducting a thorough investigation, the fishery officer confiscated additional salmon fillets from the restaurant freezer, some of which were subsequently determined to be coho salmon. This was fortunate, as the fish delivered to the fishery officer by the cook was a chum salmon. The resulting DNA identification of coho salmon, combined with testimony from a food inspection expert that the coho salmon had not been frozen since the previous year's legal harvest, was sufficient for conviction of the restaurant owner (Table 1). Case 3 involved a recreational fisherman charged by a provincial conservation officer for illegal retention of coho salmon. In case 4, alleged dealers of illicit drugs were under investigation in Newfoundland and Labrador, and fish samples were discovered in a freezer search for illegal drugs. However, no charges were laid for illegal

Table 1. Species identification of salmonid tissue samples collected by fishery officers.

Case (year)	Tissues	Result	Defendant	Legal outcome	Fine ($)
1 (1995)	24 scales; 4 blood/scales/ slime from containers	Coho	Harvester	Conviction	1 500
2 (1998)	35 muscle	Chum Chinook Coho	Harvester	Conviction	1 800
3 (1998)	4 muscle	Coho	Harvester	Conviction	?
4 (1999)	20 muscle	Atlantic Chinook Coho	Harvester	No charges	—
5 (2000)	1 muscle	Coho	Harvester	Guilty Plea	7 500
6 (2000)	1 muscle	Sockeye	Restaurant	Conviction	1 000

In cases 1–5, the DNA analysis confirmed that the fish examined included coho salmon that had been illegally harvested or sold. In case 6, analysis confirmed that a restaurant was serving illegally purchased sockeye salmon rather than farmed Atlantic salmon.

Table 2. Accuracy and precision of maximum likelihood analysis of sample mixtures and Bayesian classification of individual genotypes of coho, chinook and sockeye salmon for reporting regions used in forensic GSI analysis of 'unknowns'.

Regions	Sockeye salmon		Chinook salmon		Coho salmon
	ML % correct	Bayes % correct	ML % correct	Bayes % correct	ML % correct
Columbia R	90.8 (5.5)	97.4	90.0 (4.6)	90.1	92.2 (4.0)
Washington	90.9 (4.7)	92.0	92.7 (3.7)	91.9	90.3 (4.7)
Vancouver Isl/South Coast	96.8 (2.6)	99.0	97.2 (2.1)	94.9	95.7 (2.8)
L. Fraser R	96.3 (2.5)	94.0	91.4 (4.2)	90.3	92.3 (4.1)
Int. Fraser/Thompson R	99.1 (1.3)	98.9	99.0 (1.5)	97.7	96.2 (2.7)
Central/North	94.4 (3.5)	91.2	96.3 (2.9)	90.1	92.1 (4.0)
Skeena R	93.4 (3.6)	94.7	94.9 (3.3)	—	91.7 (4.3)
Nass R	98.5 (1.6)	93.8	93.8 (4.2)	—	90.2 (5.2)
Skeena/Nass R	—	—	—	91.1	—
Queen Charlotte Isl	91.3 (4.3)	99.2	91.1 (4.6)	100.0	91.0 (5.4)
Transboundary	95.7 (2.6)	93.8	92.8 (4.3)	—	—
Alaska/Yukon	97.3 (2.4)	96.6	95.4 (3.0)	—	93.1 (4.0)
Trans/AK/Yukon	—	—	—	91.3	—

Bayesian classification of individual coho salmon genotypes with the current 10-locus database is insufficiently accurate to use in forensic analyses.

possession of coho salmon because it was not possible at the time to disprove the suspect's claim that the coho salmon (four of 20 tissue samples) were of farmed origin. In case 5, a commercial fisherman was accused of illegally retaining large amounts of coho salmon and halibut aboard his vessel. The fish had been filleted, so external characteristics for species identification were not available. In the most recent case (6), a restaurant owner was fined for illegal possession of sockeye salmon that he claimed was farmed Atlantic salmon.

Genetic stock identification of sockeye, chinook and coho salmon

Both the maximum likelihood analysis of mixtures and the Bayesian classification of individual genotypes provided accurate estimates of the proportions of fish originating from reporting regions comprised of major watersheds and large geographic areas (Table 2). Maximum likelihood analysis differentiated Nass and Skeena River chinook salmon with sufficient accuracy to treat each of the watersheds as a separate reporting region, whereas for a Bayesian analysis the two watersheds would be considered a single reporting region. Similarly, the transboundary rivers and those to the north in Alaska and the Yukon Territory would be considered two reporting regions for chinook salmon in a maximum likelihood analysis and a single region in a Bayesian classification of individual genotypes. The Bayesian self-classification routine provided a minimal estimate of the accuracy of classification since only a third of the baseline data for any species could be 'self-classified' with the GeneClass program. Bayesian classification of unknown samples in forensic cases is carried out with the entire baseline.

Accurate classification of fish to many smaller regions is possible (Beacham et al. 2001a,b, unpublished data) and can be conducted should the information be required. For example, classification of fish to the large Babine River tributary within the Skeena River system is highly accurate and could be used in forensic analysis if required. Similarly, identification of sockeye salmon to some of the individual rivers within the transboundary region (consisting of the Stikine, Alsek, Taku and Unuk rivers) is possible.

Analysis of fishery samples of known origin has shown that accurate classification of individual genotypes to geographic regions is achieved even when missing or inadequate baseline data precludes correct assignment to population or smaller geographic area (Beacham et al. 2001a, in press). This results from the regional structure of genetic differentiation in Pacific salmonids. Allocations of sockeye salmon caught in fisheries within the Fraser and Skeena River watersheds were greater than 95% to populations within the correct watershed, even though the fishery samples may have contained fish from unsampled populations within each watershed (Beacham et al. 2001b). Similarly, classification of 264 coded-wire-tagged coho salmon from fishery samples to region of origin was accurate and precise, with the largest discrepancies between the true and estimated regional contributions never exceeding 4% (Beacham et al. 2001a).

The use of regions for which the accuracy of identification of an unknown is at least 90% means that misclassification of up to 10% of single fish samples is expected, a value that might be considered unacceptably high in a forensic case where no independent evaluation of probable source is available. However, increasing the sample size can increase the confidence in determination of probable source. For a sample of 10 fish subjected to individual classification under these conditions, seven or more of the 10 fish should be classified to the correct source in 98.7% of analyses.

Between 1998 and 2002, both the number of fish (and populations) in the baseline data sets and the number of loci surveyed for each species increased (Table 3). All of the sockeye GSI analyses involved fish sampled in southern British Columbia or Washington State that were possibly harvested or sold illegally from the Fraser River (Table 4). The Fraser River drainage was identified as the origin of all or the majority of the fish sampled for all but two of the cases (15 and 17) (Table 4).

In eight analyses for which the Fraser River was identified as the probable source of sockeye salmon, 100% of the fish were allocated to the Fraser River and in the remaining six analyses the estimated contributions from Fraser River populations ranged from 96% to 99.5% (Table 4). Results for samples analyzed by both methods were similar, as demonstrated for the samples examined in case 12, on which both maximum likelihood mixture analysis and the Bayesian classification of individuals were carried out (Table 5). These results support the contention that the sockeye salmon sampled in each case originated from Fraser River populations.

DNA analysis provides direct evidence of where the salmonid fish originate from, and where they are returning to, but no direct evidence of where or when the fish were harvested. Since fishery openings are generally site- and time-specific, the stock identification results are often examined for consistency with the postulated location and/or time of capture of the fish. GSI analysis has generally supported the assumption that fish harvested within a watershed or its estuary all originate from populations within the watershed and that fish harvested in marine locations generally are mixtures of fish from more than one watershed (Shaklee et al. 1999). The mixture of populations within the harvest is more diverse in more northern harvest locations (Alaska and northern BC) than in more southern locations (southern BC and Washington) because migrating adult Pacific salmonids tend to travel from north to south along the North American coastline and 'turn off' into their natal watersheds as they are reached. However, even commercial and sport chinook and coho salmon harvests in southern BC consist of seasonally variable mixtures of fish originating from Vancouver Island, Fraser River, the southern BC mainland, Washington and Oregon (Beacham et al. 2001a, Candy et al. 2002). Similarly, sockeye salmon harvested in the Strait of Juan de Fuca generally

Table 3. Loci and sample locations for GSI baseline data used in the forensic analysis of confiscated sockeye and chinook salmon tissues.

Species	Baseline populations					Number of fish	Loci	Primer source
	BC (Fraser)	WA	Columbia	AK	Total			
Sockeye								
Baseline 1	64 (28)	0	1	0	65	15 000	Omy77 Ots3 Ots100 Ots103 Ots107, Ots108	Morris et al. 1996 Banks et al. 1999 Nelson et al. 1998 Beacham et al. 1998 Nelson & Beacham 1999
Baseline 2	84 (30)	0	1	5	90	27 000	Omy77, Ots3, Ots100, Ots103, Ots107, Ots108 Ots2 One8	As above Banks et al. 1999 Scribner et al. 1996

Baseline 3	124 (38)	3	2	15	143	35 000	Oki1a, Oki1b, Oki6,
							  Oki10, Oki16, Oki29
							Same microsatellite loci
							  as Baseline 2
							MHC class II DAB
								Smith et al. 1998,
								  unpub.

Chinook								

Baseline 1	97 (55)	12	10	1	120	15 000	Ots100, Ots101, Ots102,
							  Ots104, Ots107
							Ssa197
								Nelson &
								  Beacham 1999
								O'Reilly et al. 1996

Baseline 2	97 (55)	12	10	1	120	27 000	Ots100, Ots101, Ots102, Ots104, Ots107, Ssa197 Ogo2, Ogo4 Oke4 Omy325 Oki100 Ots2, Ots9	As above Olsen et al. 1998 Buchholz et al. 1999 O'Connell et al. 1997 Unpub. Banks et al. 1999
Baseline 3	121 (56)	12	23	10	166	35 000	Same 13 loci as in Baseline 2	As above

Samples are from populations in British Columbia (BC), Washington (WA), the Columbia River drainage, and Alaska (AK). The Fraser populations are a subset of the BC populations. For both species, the number of fish sampled constituting the baseline, and the number of microsatellite loci surveyed in the baseline fish, increased over time.

contain high proportions of Fraser River fish, but also include fish originating from Washington populations (Beacham et al. 2001b).

For every sample of Fraser River sockeye salmon, except one consisting of 95 sockeye examined in case 6(a) (Table 4), at least 95% of the fish in the sample were allocated to interior Fraser and Thompson tributaries and less than 3% were allocated to lower Fraser tributaries, indicating that the fish were caught prior to August and/or upstream of the lower Fraser River tributaries. Sockeye salmon belonging to Thompson River tributaries were identified in all of the samples. These results supported fishery officer suspicions that much of the illegal harvest was taking place in or near the Fraser River canyon, upstream of the lower Fraser tributaries but downstream of the confluence of the Fraser and Thompson rivers. The fish were recovered from locales as varied as highway diners in Hope and Lillooet, BC, small towns located on the Fraser River in the vicinity of the Fraser Canyon (cases 9, 10, 14), an upscale restaurant near Vancouver, BC (case 8) and fish plants and cold storage facilities in North Vancouver (case 4) and Washington State (case 6(a)).

In case 16, fisheries officers patrolling the Fraser River by boat gave chase to a boat apparently being used for illegal harvest, and pursued the boat to the mouth of the Fraser River and across the Strait of Georgia at night. The suspect was apprehended in waters close

Table 4. GSI analysis of sockeye and chinook samples collected by fishery officers. N_{SAMP} gives the number of tissue samples provided by the fishery officer and N is the number of distinct genotypes observed among the samples.

Case (year)	N_{SAMP}(N)	Species	Analysis	Base-line	Results	Defendant	Outcome	Fine ($)
1 (1998)	144 fin clips	Sockeye	ML mixture	1	100% Fraser; 97.5% IF&T	Harvester	No charges; fish seized in U.S.A	—
2 (1998)	20 fin clip	Sockeye	ML mixture	1	96.5% Fraser; 96.5% IF&T	Harvester	Guilty Plea	2 000
3 (1999)	5 muscle	Sockeye	ML mixture	1	100% Fraser	Harvester	No charges laid	—
4 (1999)	90 muscle	Sockeye	ML mixture	1	96% Fraser; 95% IF&T	Harvester	No conviction, under appeal	
5 (1999)	50 cans	Sockeye	ML mixture	2	100% Fraser; 100% IF&T	Harvester	Conviction	15 000
6 (1999)	69 muscle	Chinook	ML mixture	1	97% Fraser; 97% IF&T	Harvester	No conviction, under appeal	
	(a) 95 muscle	Sockeye	ML mixture	2	100% Fraser; 89% IF&T			
	(b) 95 muscle	Sockeye	ML mixture	2	99.5% Fraser; 97.8% IF&T			
7 (1999)	75(71) muscle	Chinook	ML mixture	1	91.4% Fraser	Harvester	No conviction, under appeal	
8 (2000)	95 muscle	Sockeye	ML mixture	2	100% Fraser; 100% IF&T	Restaurant	Guilty plea	8 000
	93(76) muscle	Chinook	ML mixture	2	97.8% Fraser; 97.8% IF&T			
9 (2000)	26 muscle	Sockeye	ML mixture	2	100% Fraser; 100% IF&T	Restaurant	Guilty plea	500
10 (2000)	67 muscle	Sockeye	ML mixture	2	100% Fraser; 100% IF&T	Restaurant	Guilty plea	500
11 (2000)	92 muscle (5 samples of 1–58)	Sockeye	ML mixture	2	100% Fraser; 100% IF&T	Harvester	Ongoing	—
12 (2000)	61 muscle (6 samples of 10–11)	Sockeye	ML mixture Bayes individual	3	97% Fraser; 95% IF&T	Harvester	Ongoing	—
13 (2001)	15 liver and fin clip	Chinook	Bayes individual	3	93.3% Fraser; 93.3% IF&T	Store	Ongoing	—
14 (2001)	45 muscle	Sockeye	Bayes individual	3	97.8% Fraser; 97.8% IF&T	Restaurant	Guilty plea	3 000
15 (2001)	56 muscle	Sockeye	Bayes individual	3	92.9% Skeena	Restaurant	No charges laid	—
16 (2001)	17(10) scales; 1 blood-stain	Sockeye	Bayes individual	3	100% Fraser; 100% IF&T	Harvester	Guilty plea	—
17 (2001)	6 fin clips	Sockeye	Bayes individual	3	100% Skeena	Harvester	No charges laid	—

The loci used in the baseline data sets are shown in Table 3. Analysis of unknowns was carried out by maximum likelihood (ML) analysis or Bayesian (Bayes) classification of individual genotypes. In all cases, the fishery officer suspected that the source of fish was the Fraser River and, more specifically, tributaries of the Interior Fraser and Thompson River (IF&T) drainages.

to Vancouver Island after his boat ran out of gas. The samples collected by the fishery officer for DNA analysis included scales from 'fish in the net' and from the interior of the vessel, as well as a shirt and pair of pants stained with fish blood. Extraction of DNA from a small piece of the shirt yielded high quality sockeye salmon DNA that was included in the microsatellite GSI analysis to determine the populations of origin.

Table 5. Maximum likelihood (ML) and Bayesian classification of 61 sockeye salmon 'unknown' genotypes to region of origin.

Region	ML (%) 61 samples	Bayesian (individuals)					
		Truck 1		Truck 2		Truck 3	
		Tote 1	Tote 2	Tote 3	Tote 4	Tote 5	Tote 6
Interior Fraser/Thompson R	95.0	9	10	9	9	10	9
Lower Fraser River	1.6	1			1		1
Fraser River total	96.6	10/11	10/10	9/10	10/10	10/10	10/10
Columbia River	0						
Washington	0						
Vancouver Island	0						
BC Central Coast	1.7						
Skeena River	1.6			1			
Nass River	0						
Queen Charlotte Islands	0						
Transboundary rivers	0	1					
Alaska	0						

The unknown fish and the baseline fish representing 143 sockeye salmon populations were screened at 14 microsatellite loci and one major histocompatibility (MHC) locus. The genotypes were obtained from muscle tissue sampled from 10 or 11 whole sockeye salmon confiscated from each of six totes of fish being transported in three trucks. The ML analysis was carried out on all 61 genotypes whereas the Bayesian analysis was conducted on each genotype independently.

Outcomes of completed cases in which charges were laid have ranged from conviction (with or without a guilty plea) to failure to convict due to case procedural or technical errors, with some cases under appeal by the Crown. Charges were not laid in case 1 because the fish were seized in the U.S.A. and prosecutors in that country were unwilling to deal with Canadian case law. In case 3, no charges were laid because of the small number of fish available for DNA analysis (5) and a lack of supporting evidence. Convictions were not obtained in case 5 because a key witness 'forgot' who delivered the confiscated fish and in cases 6 and 7 because critical non-DNA evidence was ruled inadmissible in court. When convictions occurred, the penalty was variable with the small-town diners levied fines ranging between $500 and $3 000 and the upscale urban restaurant paying a penalty of $8 000. The highest fine ($15 000) was levied in the most recent conviction in which Fraser River sockeye salmon were identified in the analysis of 50 tins of canned salmon (case 5). The defendant, charged with attempting to sell salmon not caught under a licence authorizing sale, was in possession of over 100 000 cans of salmon with an estimated 'street value' of between $300 000 and $400 000 and did not enter a guilty plea.

Sockeye salmon sampled from a restaurant in the lower Fraser River Valley (case 15) and from a fish plant on Vancouver Island (case 17) were claimed to originate from a legal fishery in the Skeena River drainage. In both cases, there was inadequate documentation to support the claim of Skeena River origin and the investigating fishery officers wanted to confirm that the fish were not of Fraser River origin. GSI analyses supported the contention that the fish sampled in both cases did, in fact, originate from the Skeena River drainage (Table 5). In Bayesian classification of individuals, 52 of the 56 restaurant fish were classified to tributaries of the Skeena River, as were all six fish sampled from the fish plant. In both cases, no charges were laid. Thus, DNA evidence has facilitated exoneration of the innocent and minimized time and money spent on unnecessary investigation, as well as assisting in conviction of the guilty. In addition, the fishery officer who investigated case 17 has indicated that word is getting out among his 'clientele' about his ability to establish the origins of sockeye and he believes that this has had a strong deterrent effect.

All four chinook salmon GSI analyses also identified the Fraser River drainage as the sole or primary source of the sampled fish (Table 4). The estimated Fraser River contribution ranged from 91.4% to 97.8%, and was accounted for entirely by interior Fraser and Thompson populations in all cases. Chinook salmon returning to lower Fraser tributaries enter the drainage later in the year than the corresponding sockeye salmon populations. Thus, chinook salmon harvested anytime prior to September are likely to consist entirely or primarily of fish from interior populations. It is therefore not surprising that none of 15 chinook salmon sampled from a fish store during May 2001 (case 13), believed to have been caught in the lower Fraser River, were allocated to lower Fraser tributaries.

Forensic PCR of salmonid tissue for species and stock identification tends not to be affected by some of the complications involved in the forensic analysis of human tissue. In general, scales, muscle or fin tissues collected from fresh or frozen fish destined for human consumption yield high quality, undegraded DNA template for the PCR analyses. Thus, a lack of amplified product, or loci mis-scored because of 'upper allele dropout' are seldom problematic. In addition, because the evidentiary tissue is not human, there is no concern that even an inexperienced sampler will contaminate the samples with tissue from himself or another person. However, occasional mis-identification of the species under examination can occur in the field.

Mis-identified Pacific salmon in forensic samples are quickly apparent in GSI analysis because they result in amplification failure and/or amplification of alleles outside expected size ranges at some microsatellite loci. Three steelhead trout and a coho salmon contained in a sample of 75 putative chinook salmon sampled in a Washington State processing plant by a local enforcement officer at the request of a Canadian officer (case 7) were all identified as non-chinook and removed from the GSI analysis even before the species identification test was conducted for definitive identification.

When processed fish samples, rather than entire fish are confiscated, microsatellite analysis can be used to determine how many fish produced the fillets or steaks in question. This is important when fish pieces sampled from retail outlets are subjected to GSI analysis, as any duplicate genotypes must be removed in order to avoid 'double counting' of fish. The 92 chinook salmon fillets sampled from the upscale restaurant (case 8) provided only 76 unique multi-locus profiles, revealing that in some cases two or three of the fillets sampled had come from the same fish. Moreover, the nine sockeye salmon scales from 'fish in the net' submitted by the fishery officer in case 16 were found after analysis to consist of three scales from each of three fish, reducing a limited sample size even further. The number of fish harvested may also be pertinent when quotas for a particular species are in effect and large amounts of processed fish are found in the possession of sports fishermen.

The results of this study indicate that the genetic methodologies developed for fishery management species and stock identification of Pacific salmon are sufficiently accurate and precise to use in classification of samples collected as evidence in court proceedings. We have found that fishery officers are generally correct in their assessment of the origin of fish based on their own investigations. However, since the onus is on the prosecutors to establish that fish have been harvested illegally, rather than on the defendant to establish that fish have been harvested or sold legally, the DNA results provide strong supporting evidence for conviction or exoneration of the defendant. This is especially true when the investigation is based on information from the public and the fishery officer has no independent indication of the source of the fish, such as is often the case in fish confiscated from stores and restaurants. The use of DNA evidence in the conviction of the upscale restaurant found guilty of purchasing fish from a vendor without a commercial licence was highly publicized in the local media and picked up as an interest item in a number of popular science magazines (e.g. Anonymous 2001). It is possible that the most important outcome of the foregoing forensic analyses will be a greatly diminished demand for illegally purchased or sold fish and an increase in the number of guilty pleas in cases supported with GSI analyses.

References

Anonymous. 2001. News in brief: Fish restaurant fingered by DNA data. Nature 411: 407.

Banks, M.A., M.S. Blouin, B.A. Baldwin, V.K. Rashbrook, H.A. Fitzgerald, S.M. Blankenship & D. Hedgecock. 1999. Isolation and inheritance of novel microsatellites in chinook salmon (*Oncorhynchus tshawytscha*). J. Hered. 90: 281–288.

Beacham, T.D., J.R. Candy, K.J. Supernault, T.J. Ming, B. Deagle, A. Schultz, D. Tuck, K. Kaukinen, J.R. Irvine, K.M. Miller & R.E. Withler. 2001a. Evaluation and application of microsatellite and major histocompatibility complex variation for stock identification of coho salmon in British Columbia. Trans. Am. Fish. Soc. 130: 1116–1155.

Beacham, T.D., J.R. Candy, B. McIntosh, C. MacConnachie, A. Tabata, K. Kaukinen, K.M. Miller & R.E. Withler. 2001b. Estimation of stock composition of sockeye salmon in the North Pacific Ocean. (NPAFC Doc. 552). Department of Fisheries and Oceans, Pacific Biological Station, Nanaimo, BC, Canada V9R 5N7.

Beacham, T.D., L. Margolis & R.J. Nelson. 1998. A comparison of methods of stock identification for sockeye salmon (*Oncorhynchus nerka*) in Barkley Sound, British Columbia. N. Pac. Andr. Fish Comm. Bull. 1: 227–239.

Buchholz, W., S.J. Miller & W.J. Spearman. 1999. Summary of PCR primers for salmonid genetic studies. U.S. Fish. Wild. Serv. Alaska Fish. Prog. Rep. 99–1.

Calvo, J.H., P. Zaragoza & R. Osta. 2001. A quick and more sensitive method to identify pork in processed and unprocessed food by PCR amplification of a new specific DNA fragment. J. Animal Sci. 79: 2108–2112.

Candy, J.R., J.R. Irvine, C.K. Parker, S.L. Lemke, R.E. Bailey, M. Wetklo & K. Jonsen. 2002. A discussion paper on proposed stock groupings (conservation units) for Fraser River chinook salmon. Can. Stock Assess. Secret. Res. Doc. 52002–08.

Cornuet, J.M., S. Piry, G. Luikart, A. Estoup & M. Solignac. 1999. Comparison of methods employing multilocus genotypes to select or exclude populations as origins of individuals. Genetics 153: 1989–2000.

Debevec, E.M., R.B. Gates, M. Masuda, J. Pella, J. Reynolds & L.W. Seeb. 2000. SPAM (version 3.2): Statistics program for analyzing mixtures. J. Hered. 91: 509–510.

DeSalle, R. & V. Birstein. 1996. PCR identification of black caviar. Nature 381: 197.

Gable, J. & S. Cox-Rogers. 1993. Stock identification of Fraser River sockeye salmon: Methodology and management application. Pac. Sal. Comm. Tech. Rep. 5, 36 p.

Hold, G.L., V.J. Russell, S.E. Pryde, H. Rehbain, J. Quinteiro, M. Rey-Mendez, C.G. Sotelo, R.I. Perez-Martin, A.T. Santos & C. Rosa. 2001. Validation of a PCR-RFLP based method for the identification of salmon species in food products. Eur. Food Res. Technol. 212: 385–389.

Hsieh, H.-M., H.-L. Chiang, L.-C. Tsai, S.-Y. Lai, N.-E. Huang, A. Linacre & J. Chun-I Lee. 2001. Cytochrome *b* gene for species identification of the conservation animals. Forensic Sci. Int. 122: 7–18.

McKay, S.J., M.J. Smith & R.H. Devlin. 1997. Polymerase chain reaction-based species identification of salmon and coastal trout in British Columbia. Mol. Mar. Biol. Biotechnol. 6: 131–140.

Miller, K.M., K.H. Kaukinen, T.D. Beacham & R.E. Withler. 2001. Geographic heterogeneity in natural selection on an MHC locus in sockeye salmon. Genetica 111: 237–257.

Miller, K.M., R.E. Withler & T.D. Beacham. 1996. Stock identification of coho salmon (*Oncorhynchus kisutch*) using minisatellite DNA variation. Can. J. Fish. Aquat. Sci. 53: 181–195.

Morris, D.B., K.R. Richard & J.M. Wright. 1996. Microsatellites from rainbow trout (*Oncorhynchus mykiss*) and their use for genetic study of salmonids. Can. J. Fish. Aquat. Sci. 53: 120–126.

Nelson, R.J. & T.D. Beacham. 1999. Isolation and cross species amplification of microsatellite loci useful for study of Pacific salmon. Animal Genet. 30: 228–229.

Nelson, R.J., T.D. Beacham & M.P. Small. 1998. Microsatellite analysis of the population structure of a Vancouver Island sockeye salmon (*Oncorhynchus nerka*) stock complex using non-denaturing gel electrophoresis. Mol. Mar. Biotech. 7: 312–319.

Olsen, J.B., P. Bentzen & J.E. Seeb. 1998. Characterization of seven microsatellite loci derived from pink salmon. Mol. Ecol. 7: 1083–1090.

O'Connell, M., R.G. Danzmann, J.M. Cornuet, J.M. Wright & M.M. Ferguson. 1997. Differentiation of rainbow trout populations in Lake Ontario and the evaluation of the stepwise mutation and infinite allele mutation models using microsatellite variability. Can. J. Fish. Aquat. Sci. 54: 1391–1399.

O'Reilly, P.T., L.C. Hamilton, S.K. McConnell & J.M. Wright. 1996. Rapid analysis of genetic variation in Atlantic salmon (*Salmo salar*) by PCR multiplexing of dinucleotide and tetranucleotide microsatellites. Can. J. Fish. Aquat. Sci. 53: 2292–2298.

Pádár, Z., M. Angyal, B. Egyed, S. Füredi, J. Woller, L. Zöldág & S. Fekete. 2001. Canine microsatellite polymorphisms as the resolution of an illegal animal death case in a Hungarian Zoological Gardens. Int. J. Legal Med. 115: 79–81.

Pendas, A.M., P. Moran, J.L. Martinez & E. Garcia-Vazquez. 1995. Applications of 5S rDNA in Atlantic salmon, brown trout, and in Atlantic salmon × brown trout hybrid identification. Mol. Ecol. 4: 275–276.

Poetsch, M., S. Seefeldt, M. Maschke & E. Lignitz. 2001. Analysis of microsatellite polymorphism in red deer, roe deer, and fallow deer – possible employment in forensic applications. Forensic Sci. Int. 116: 1–8.

Shaklee, J.B., T.D. Beacham, L.W. Seeb & B.A. White. 1999. Managing fisheries using genetic data: Case studies from four species of Pacific salmon. Fish. Res. 43: 45–78.

Smith, C.T., B.F. Koop & R.J. Nelson. 1998. Isolation and characterization of coho salmon (*Oncorhynchus kisutch*) microsatellites and their use in other salmonids. Mol. Ecol. 7: 1614–1616.

Shivji, M., S. Clarke, M. Pank, L. Natanson, N. Kohler & M. Stanhope. 2002. Genetic identification of pelagic shark body parts for conservation and trade monitoring. Conserv. Biol. 16: 1036–1047.

Wetton, J.H., C.S. Tsang, C.A. Roney and A.C. Spriggs. 2002. An extremely sensitive species-specific ARMS PCR test for the presence of tiger bone DNA. Forensic Sci. Int. 126: 137–144.

Withler, R.E., K.D. Le, R.J. Nelson, K.M. Miller & T.D. Beacham. 2000. Intact genetic structure and high levels of genetic diversity in bottlenecked sockeye salmon, *Oncorhynchus nerka*, populations of the Fraser River, British Columbia, Canada. Can. J. Fish. Aquat. Sci. 57: 1985–1998.

Withler, R.E., T.D. Beacham, T.J. Ming & K.M. Miller. 1997. Species identification of Pacific salmon by means of a major histocompatibility complex gene. N. Am. J. Fish. Management 17: 929–938.

Zehner, R., S. Zimmermann & D. Mebs. 1998. RFLP and sequence analysis of the cytochrome *b* gene of selected animals and man: Methodology and forensic application. Int. J. Legal Med. 111: 323–327.

Environmental Biology of Fishes **69**: 287–297, 2004.
© 2004 *Kluwer Academic Publishers. Printed in the Netherlands.*

Outbreeding depression in hybrids between spatially separated pink salmon, *Oncorhynchus gorbuscha*, populations: marine survival, homing ability, and variability in family size

Sara E. Gilk[a], Ivan A. Wang[a], Carrie L. Hoover[a], William W. Smoker[a], S.G. Taylor[b], Andrew K. Gray[a,b] & A.J. Gharrett[a,c]
[a]*Fisheries Division, School of Fisheries and Ocean Sciences, University of Alaska Fairbanks, 11120 Glacier Highway, Juneau, AK 99801, U.S.A.*
[b]*Auke Bay Laboratory, Alaska Fisheries Science Center, National Marine Fisheries Service, NOAA, 11305 Glacier Highway, Juneau, AK 99801, U.S.A.*
[c]*Corresponding author (e-mail: ffajg@uaf.edu)*

Received 17 April 2003   Accepted 24 April 2003

Key words: conservation biology, local adaptation, microsatellites, salmonid

Synopsis

Hybridization between distinct populations and introgression of nonnative genes can erode fitness of native populations through outbreeding depression, either by producing a phenotype intermediate to that of both contributing genomes (and maladapted in either population's environment) or by disrupting distinct coadapted complexes of epistatic genes. In salmon, fitness-related traits such as homing ability or family-size distribution may be eroded. We investigated geographically separated pink salmon populations in repeated trials in independent broodyears (odd and even). Hybrids were made between female Auke Creek (Southeast Alaska) pink salmon and Pillar Creek (Kodiak Island, $\sim$1 000 km away) males; hybrids and their offspring were compared to offspring of control crosses of the same females with Auke Creek males. Parentage assignment from microsatellite analysis was used to improve estimates of survival and straying and to examine variation of family size. Hybridization reduced return rates of adults (a proxy for survival at sea) in the F_1 generation in the odd-year broodline (p < 0.0001) but not in the even-year broodline (p = 0.678). Hybridization reduced survival in both the odd- and even-broodyear F_2 (p < 0.005 and p < 0.0001). Hybridization did not appear to impair homing ability; weekly surveys revealed similar straying rates ($\sim$2%) by both hybrid and control fish into nearby ($\sim$1 km) Waydelich Creek in both generations in both trials. Hybridization did not increase the index of variability (σ^2/μ) in family size. Decreased survival in the hybrid F_2 generation supports an epistatic model of outbreeding depression; nonepistatic effects may have contributed to reduced survival in the odd-broodyear F_1 hybrid fish. Outbreeding depression in hybrids of geographically separated populations demonstrates that introgression of nonnative fish can erode fitness, and should be recognized as a potential detriment of both aquaculture and management practices.

Introduction

Pacific salmon, *Oncorhynchus* spp., have been transplanted between zoogeographic regions and between watersheds within regions and have been artificially produced and released in large numbers for over a century, generally for harvest enhancement and species restoration or mitigation for lost habitat (reviewed by Lichatowich 1999). Pacific salmon generally home to natal sites for spawning, thereby promoting reproductive isolation and exposing populations to diversifying selective pressures (Carvalho 1993). Releases of cultured or transplanted fish into environments inhabited by indigenous salmonid populations remove natural barriers to gene flow between populations. Recurrent introgression from transplanted fish or a hatchery stock derived by transplantation may alter the structure of locally evolved genomes, thereby reducing the fitness of the local populations (Hindar et al. 1991, Campton 1995). Although introductions of hatchery-raised fish may aid in restoration of populations at risk of extinction or may increase the productivity of populations in degraded habitat, hybridization between salmon populations adapted to different local environments can also reduce the average fitness of wild salmon populations, known as outbreeding depression (Shields 1982, Hard 1995).

Outbreeding depression can result from an additive combination of average genetic values in the parental populations, disruption of an epistatic interaction among alleles at different loci, or both. A fitness loss expected from nonepistatic effects would occur in the first (F_1) and subsequent generations. Examples include intermediate resistance to *Ceratomyxa shasta* in hybrids between resistant and nonresistant coho salmon, *O. kisutch*, stocks (Hemmingson et al. 1986) and reduced tolerance to DDT in F_1 mosquitoes (King 1955).

Reproductively isolated populations may also evolve different genomes comprising coadapted, epistatic genes (Wallace 1981, Lynch 1991). Hybridization would disrupt the coadapted genomes (Dobzhansky 1950), but disruption may not appear in F_1 hybrids because each offspring possesses a complete coadapted genome from each parent. In fact, hybrid individuals may demonstrate heterosis (hybrid vigor; Emlen 1991), but reduction in fitness would be expected in the second (F_2) and later generations (Dobzhansky 1950). Outbreeding depression from a disrupted coadapted genome has been suggested in several species (e.g. Dobzhansky 1948, 1950, Brncic 1954, Gordon & Gordon 1957, Leberg 1993, Philipp & Claussen 1995), including salmonids (Gharrett & Smoker 1991, Danzmann et al. 1999, Gharrett et al. 1999).

Outbreeding depression may be directly observed as reduced survival. It may also be observed in the degradation of individual traits related to fitness such as impaired homing ability or of population traits such as a change in the distribution of family size distribution. Homing ability has been suggested as an important adaptive trait that responds to local selective pressures such as stream stability (Quinn 1993). Obvious changes in homing ability may reflect changes in population fitness. In a previous transplantation and hybridization study on pink salmon, both hybrids and transplants had similar survivals; however, many fewer transplants than hybrid native/transplanted salmon negotiated their way up the stream to the release site, suggesting that terminal homing may have a genetic basis (Bams 1976). However, transplantation/hybridization studies have not followed hybrids into the F_2 generation where outbreeding depression would be expected (reviewed by Quinn 1993).

Changes in family size distribution may also reflect fitness loss. Genetic differences between families may result in different responses (survival) to variable marine conditions. Consequently, some families of a particular generation are better adapted to particular environmental conditions, but the most favored genotypes may change from generation to generation (Geiger et al. 1997) as conditions fluctuate. Hybridization might reduce the number of contributing families such that a small number of parents contribute disproportionately to subsequent generations. Coupled with genetic drift, this may increase inbreeding and decrease the effective population size. A smaller effective population size would tend to reduce genetic variability and most likely the average fitness of a population (Falconer & Mackay 1996) and could exacerbate outbreeding depression due to stochastic losses of essential alleles (Nevo et al. 1986, Emlen 1991). Observations of nonrandom variation of marine survival between families of Pacific salmon populations (Simon et al. 1986, Geiger et al. 1997, Hard et al. 2000) underscore the importance to population fitness of maintaining genetic variation in a population.

Molecular markers make it possible to reconstruct pedigrees. The parental sources of returning fish can be confirmed, and family size distributions can be determined from these data. Polymorphic microsatellite loci provide information on parent–offspring relationships and have been used to study parentage in many species, including fishes (e.g. Estoup et al. 1998, Waldbieser & Wolters 1999).

Previous work at Auke Creek near Juneau, Alaska, studied the survivals of hybrids between odd- and even-broodyear pink salmon. Gharrett & Smoker (1991) observed that F_1 hybrid and control survival rates were similar, and observations of F_2 hybrids suggested a decrease in survival, although controls were not possible. When these experiments were repeated to include controls, returns of F_2 controls exceeded returns of F_2 hybrids (Gharrett et al. 1999). Hybrids between broodyears do not represent crosses that are likely to occur naturally or as the result of management practices; therefore, this study was begun in

1996 to examine the outbreeding depressive effects of hybridization of spatially separated pink salmon from the same broodyear.

Our objective was to examine the effects of hybridization between pink salmon from widely separated populations. To accomplish this objective, hybrids were produced between pink salmon from Auke Creek and pink salmon from Pillar Creek, a population from Kodiak Island, Alaska. Controls had only Auke Creek parents. Marine survivals were observed in the F_1 and F_2 generations for both hybrids and controls. The effects of hybridization on two fitness related traits, homing ability, and the distribution of family size, are examined here. Parentage determined from microsatellite inheritance was used to improve estimates of survival and straying and to examine family size distributions. Evidence of outbreeding depressive effects on survival suggests additive effects on one of the F_1 hybrids and epistatic effects in both F_2 hybrids. There was no evidence of an effect on either homing ability or family size distribution, which suggests that these mechanisms may not be important contributors to decreased survival.

Materials and methods

Crosses and incubation procedures

The experiments evaluated hybrids between pink salmon from two Alaskan populations: Auke Creek near Juneau, Alaska (58°23′N, 134°37′W) and Pillar Creek on Kodiak Island (57°47′N, 152°28′W). Auke Creek, a lake-fed stream of moderate grade, is about 350 m long and has spawning populations that vary between 2 000 and 20 000.[1] Pillar Creek is a reservoir-fed stream of shallow gradient, runs about 1 800 m, and has spawning populations that vary between 1 000 and 40 000.[2] The distance between these streams (great circle distance 1 048 km; the continental shelf distance is considerably greater) prevents direct gene flow between the populations, even though they are at the same latitude.

---

[1] Taylor, S.G. & J.L. Lum. 2002. Annual Report Auke Creek Weir 2001: Operations, Fish Counts, and Historical Summaries. Unpublished Report 37 pp. National Marine Fisheries Service. Auke Bay Laboratory, 11305 Glacier Highway, Juneau, Alaska 99801-8626.

[2] Alaska Department of Fish and Game, Kodiak. 2000. Stream survey data of Pillar Creek, Kodiak, AK from 1968 to 2000. Unpublished data.

The first generation hybrids were made at Auke Creek in 1996 and 1997 in an experiment that modeled Auke Creek as the recipient of a transplantation and Pillar Creek as the donor population. The mating scheme, a blocked incomplete-factorial design, was intended for quantitative genetic analyses not reported here. Eggs of each of 40 Auke Creek females were divided between hybrid and control lines. Two Auke Creek males and two Pillar Creek males were crossed with each of two females (hereafter referred to as a 'block') to make control and hybrid crosses. Twenty different blocks were produced by mating 40 Auke Creek males, 40 Auke Creek females, and 40 Pillar Creek males.

Each family was subdivided into two portions that were randomly assigned to incubation cells in vertical FAL™ incubators (MariSource, Milton, WA) partitioned with dividers. Hybrid families were incubated in different cabinets than control families. However, high flow from a common water source maintained very similar environmental conditions in the two incubators; temperature records of each incubator document virtually identical temperature regimes. Embryos were reared until about 5% of their yolk remained, when they were differentially marked by mirror image pelvic fin excisions and an experiment-identifying adipose fin excision. The polarity of the marks was alternated between hybrid and control groups over the years of the experiment. About 20 000 hybrid and 20 000 control fish were released each spring, and equal numbers of each family were released when possible. The young fish were released at or near the peak of emigration of the wild fish in Auke Creek.

Returning F_1 individuals were bred in 1998 and 1999. To produce each of 20 blocks of F_2 control and 20 blocks of hybrid progeny, two control males were crossed with each of two control females, and two hybrid males were crossed with each of two hybrid females. The control blocks included 40 F_1 Auke Creek males and 40 F_1 Auke Creek females; the hybrid blocks had 40 F_1 hybrid males and 40 F_1 hybrid females. About 20 000 F_2 hybrid and 20 000 F_2 control fish were differentially clipped and released at the peak of the emigration of the wild fish in Auke Creek in mid-April of 1999 and 2000.

Survival and straying in even- and odd-broodyears

Returning F_1 and F_2 adults were recovered in August and September 1998–2001 at a permanent weir located just above the high tideline at Auke Creek. During the return, intertidal Auke Creek and all of nearby Waydelich Creek were surveyed weekly with beach seines and dipnets during the spawning period in order to collect straying experimental fish. Survival was defined as return to the weir or recovery in a survey. Fish recovered at the weir were held in pens until they could be used for crosses or processed for data. The carcass of each returning fish was tagged, its fin mark (designating hybrid or control) and sex noted, and heart or muscle tissue samples were taken for genetic analysis. The equality of rates of survival and straying between groups was tested using log-likelihood ratios (G-tests; Sokal & Rohlf 1995). Parentage information from microsatellite analysis was used to improve estimates of survival and straying made from observation of marks by confirming the parental origin (control or hybrid) of returning fish, including those that had indistinct fin excisions.

Microsatellite and parentage analysis

Tissues taken for DNA analysis were placed in preservative (Seutin et al. 1991) and stored at $-20°C$ prior to DNA extraction. Total DNA was isolated with Puregene DNA™ isolation kits (Gentra Systems, Inc., Minneapolis, MN). Polymerase chain reaction (PCR) amplification was done in 96-well microtitre plates in a Stratagene 96 Robocycler. Reactions were performed in $10\,\mu l$ volumes [10 mM Tris–HCl at pH 8.3, 50 mM KCl, 25 mM MgCl$_2$, 2.5 mM each deoxyribonucleotide triphosphate (dNTP), 0.5 units *Taq* polymerase, $0.1–0.5\,\mu M$ each primer, and $50–100\,ng$ DNA template] overlaid with mineral oil. In addition to unlabeled forward and reverse primers for each locus, each mixture included a forward primer labeled with an infrared-sensitive dye, IRDye™ (LI-COR, Inc., Lincoln, NE). After evaluating several microsatellite loci, six loci were chosen: *Ssa*197 (O'Reilly et al. 1996), *Ots*1 (Banks et al. 1999), *Ots*103 (Small et al. 1998), *Ogo*1a (Olsen et al. 1998), and *Oki*10 and *Oki*11 (Smith et al. 1998). In general, PCR conditions were as follows: 1 cycle at 95°C for 3 min; 30 cycles at 95°C for 1 min, x°C for 1 min, and 72°C for 1 min; and finally 1 cycle at 72°C for 5 min; where x is the annealing temperature (52°C for *Ssa*197, 49°C for *Ots*1, 47°C for *Ots*103, 59°C for *Ogo*1a, 55°C for *Oki*10, and 53°C for *Oki*11). After amplification, DNA products were denatured by adding an equal volume of stop buffer (95% formamide, 0.1% Bromophenol Blue) and heating for 3 min at 95°C. Finally, $1–1.5\,\mu l$ of

PCR product was loaded onto polyacrylamide denaturing gels composed of 6% PAGE-PLUS™ 40% concentrate (AMRESCO, Inc., Solon, OH), 7 M urea, and 5X TBE (TBE is 90 mM tris-boric acid and 2 mM EDTA, pH 7.5) in a reaction catalyzed by ammonium persulfate and TEMED (N,N,N′,N′-tetramethylethylenediamine). Electrophoresis and detection were performed on a LI-COR automated sequencer (LongReadIR 4200™, LI-COR, Inc., Lincoln, NE) in 1X TBE running buffer, with running parameters 31.5 W, 1500 V, 35 mA, and 50°C plate temperature. GeneImagIR RFLP scan software (Version 3.52, Scanalytics, LI-COR, Inc., Lincoln, NE) was used to estimate the sizes of microsatellite alleles. Allele sizes were determined by comparing allele band patterns with IRD700™ or IRD800™ standard ladders (LI-COR, Inc., Lincoln, NE) and standardized using reference alleles.

Allele frequencies were estimated for the Auke and Pillar Creek populations for both even- and odd-broodline crosses. GenePop version 3.2a (Raymond & Rousset 1995) was used to test conformance to Hardy–Weinberg expectation using the following parameters: 10 000 dememorization steps, 1 000 batches, and 1 000 iterations per batch. Excess homozygosity can be an indication of null alleles, which complicate parentage analysis.

To demonstrate genetic differences between the Auke and Pillar Creek populations, allele frequencies were estimated for both the even- and odd-broodline parents. Homogeneity of allele frequencies between populations at all six loci was tested using a G-test for which the probability levels were estimated by 20 000 iterations in a Monte-Carlo simulation (e.g. Roff & Bentzen 1989). Probabilities were combined to examine overall heterogeneity for each broodyear using Fisher's method (Winer 1971).

Parentage of each returning fish was deduced using PROBMAX with default tolerance levels and allowing mis-scored loci (version 1.2, Danzmann 1997). Parentage information was used to confirm parental origins of well-marked fish, to determine parental origins (control, hybrid, or not experimental) of fish returning with indistinct marks, and to estimate family size of the even-broodyear returns. The parentage information also made it possible to estimate inbreeding coefficients (F; defined as the probability that alleles at a locus in an individual are identical by descent) in each segment of the experiment to determine if inbreeding had been imposed as a result of the limited sample of parents in the F$_2$ crosses

(Falconer & Mackay 1996). The expected inbreeding coefficient was calculated assuming that 1/2N is the probability that uniting gametes have identical genes. The realized inbreeding coefficients were estimated from pedigrees obtained from parentage analysis.

Variance in family size of even-broodyears

The underlying model for the estimation of effective population size is that the family size is approximately distributed as a Poisson. An increase in variance in a population (for a given mean) reduces the effective population size (Crow & Kimura 1970), a possible consequence of outbreeding depression that could occur if only a few families produced most of the offspring.

The observed family sizes in the even-broodyear F_1 and F_2 hybrid and control returns were compared to Poisson distributions expected from the average family size with a goodness of fit G-test (Sokal & Rohlf 1995). (The larger number of odd-broodyear returns will be analyzed and reported elsewhere.) Then, to avoid effects of correlations between families, the data blocks were randomly reduced to single full-sib families and bootstrapped. The bootstrap analysis randomly reduced the data so that each female in the analysis was assigned to a single control male and a single hybrid male. Consequently, the program randomly chose one control male and one hybrid male for the first female in each block and assigned the other males to the second female in the block. These bootstrap values were used to compare the means and variances of the distribution of family sizes of F_1 and F_2 returns, and to estimate the index of variability in family size. The index of variability contrasts the variance and mean in family size (σ^2/μ), and tends to 1 under random survival (Crow & Morton 1955), which is the Poisson expectation. A larger index of variability in hybrids than controls may indicate that increased variability in family size contributed to a reduction in survival.

Results

Microsatellite variation in Auke and Pillar Creek, and parentage analysis

Homogeneity tests showed significant divergence between 40 Pillar Creek pink salmon and 80 Auke Creek pink salmon used as parents in each F_1 experiment. Two loci in the even-broodyear (1996) samples showed significant heterogeneity between the populations (Ogo1a, p < 0.001; Oki11, p < 0.001) and four loci in the odd-broodyear (1997) samples showed significant heterogeneity (Oki10, p < 0.0001; Ogo1a, p < 0.01; Ots103, p < 0.01; Oki11, p < 0.01). The probabilities combined across loci demonstrated overall heterogeneity between the Auke Creek and Pillar Creek pink salmon populations used in breeding experiments in both the even- and odd-broodyears (both p < 0.0001). Homogeneity tests based on genotype frequencies had similar results.

Not all loci conformed to expected Hardy–Weinberg frequencies. Tests of Hardy–Weinberg suggested the presence of null alleles at two loci: Ssa197 and Ots103. Pedigree information from progeny was used to reconstruct Ssa197 genotypes and identify null alleles. There was insufficient pedigree information available to reconstruct all of the Ots103 genotypes. However, the high variability at most of the loci made all of the loci useful in contributing to parentage analyses.

Microsatellite markers were used to determine parentage of all even-broodyear fish, the odd-broodyear fish used for breeding experiments, and the returning odd-broodyear fish that had indistinct marks. Of 355 fish carrying indistinct fin marks, microsatellites revealed the parentage of 327 fish. The remaining 28 fish to which parents could not be assigned were included in hybrid counts to provide conservative hybrid survival estimates. Of the 1 266 individual progeny with distinct fin excisions that were genotyped, 59 did not correspond to combinations of parents that were recorded when the matings were conducted; these fish were excluded from subsequent analyses. The seven fish with undetermined parentage that were used as parents in the odd-broodyear F_1 broodstock probably resulted from misreading fin marks when matings were conducted. These seven fish and their progeny were not included in estimates of the inbreeding coefficient of odd-broodyear fish, and we did not analyze the odd-broodyear fish for family size variation. In both the even- and odd-broodyear F_1 breeding experiments, one individual was bred as a hybrid even though subsequent microsatellite genotyping indicated it was a control fish. Blocks including these two fish were removed from analyses.

Survival and straying in even- and odd-broodyears

The F_1 generation fish from even-broodyear experiments returned to Auke Creek in 1998 (Table 1); returns

Table 1. Number of young and returning adults for F_1 and F_2 pink salmon controls (Auke Creek parentage) and hybrids (Auke Creek by Pillar Creek). If their origin could not be determined with microsatellite allele analysis, returning fish with indistinct marks were counted as hybrids. p-values for testing differences in return rates (survival at sea) between hybrids and controls are from log-likelihood ratio tests (Sokal & Rohlf 1995). Estimates of wild releases and returns (Auke Creek) are from annual counts of fry and adults[d].

Brood-year		Control			Hybrid			p	Wild		
		Young released	Adult returns	% returns	Young released	Adult returns	% returns	(G-test)	Young released	Adult returns	% returns
1996	F_1	19 912	284	1.43	20 852	294[a]	1.41	0.6774	31 092	2 267	7.29
1997	F_1	19 918	1023	5.14	19 916	731[b]	3.67	<0.0001	60 785	28 128	46.3
1998	F_2	20 050	176	0.88	20 024	108	0.54	<0.0001	53 535	2 181	4.07
1999	F_2	20 000	254	1.27	20 000	192[c]	0.96	0.0033	1 32 075	7 859	5.95

[a]Includes 2 fish with indistinct marks.
[b]Includes 25 fish with indistinct marks.
[c]Includes 18 fish with indistinct marks.
[d]Taylor, S.G. & J.L. Lum. 2002. Annual report Auke Creek weir 2001: Operations, fish counts, and historical summaries. Unpublished Report 37 pp. National Marine Fisheries Service. Auke Bay Laboratory, 11305 Glacier Highway, Juneau, Alaska 99801–8626.

Table 2. Number of adult pink salmon recovered at Waydelich Creek. Returning fish with incomplete fin excisions were counted as hybrids if their origin could not be determined with microsatellite parentage analysis.

Brood-year		Control			Hybrid		
		Total returns	Number captured	% captured	Total returns	Number captured	% Captured
1996	F_1	284	1	0.35	292	3	1.03
1997	F_1	1023	0	0	706	1	0.14
1998	F_2	172	1	0.58	104	2	1.92
1999	F_2	254	1	0.39	192	4[a]	2.08

[a]Includes 2 fish with indistinct marks.

(survival rates to adulthood) of F_1 control (1.43%) and hybrid (1.41%) fish were nearly identical. F_1 odd-broodyear fish returned to Auke Creek in 1999; the F_1 control return (5.14%) exceeded that of F_1 hybrid fish (3.67%). In 2000, returns of even-broodyear F_2 fish (0.88%) were higher than F_2 hybrid returns (0.54%). Similarly, odd-broodyear F_2 control returns (1.27%) exceeded F_2 hybrid returns (0.96%) in 2001.

Survival rates[3] of cultured fish in these experiments were lower than those of wild fish returning to Auke Creek in all years. In the even-broodyears, returns of wild fish (7.29% and 4.07%, respectively) exceeded those of cultured fish. In odd-broodyears, survival of wild fish (46.3% and 5.95%, respectively) was also higher than that of cultured fish.

Weekly recovery efforts in nearby (about 1 km) Waydelich Creek revealed similar straying rates (about 2% or less) of both hybrid and control fish in all years (Table 2). Although the straying rates of hybrid fish were higher than those of the control fish, the straying rates were too low to detect statistical differences. The levels of straying do not compensate for the much larger differences in return rate.

Inbreeding in even- and odd-broodyears

Inbreeding in the hybrid F_2 generations did not result from the choice of parents in either the even- or

---

[3] Survival rates were estimated from counts of fry passing the weir in spring and adults passing the weir in summer of the following year. Data can be found in: Taylor, S.G. & J.L. Lum. 2002. Annual Report Auke Creek Weir 2001: Operations, Fish Counts, and Historical Summaries. Unpublished Report 37 pp. National Marine Fisheries Service. Auke Bay Laboratory, 11305 Glacier Highway, Juneau, Alaska 99801-8626.

odd-broodyear F_1 generations. The even-broodline F_2 inbreeding coefficient for controls (F = 0.034) exceeded that of F_2 hybrids (F = 0.003). Similarly, in the odd-broodline, the inbreeding coefficient for F_2 controls (F = 0.013) was higher than that of F_2 hybrids (F = 0.006). Both hybrid broodlines had low inbreeding coefficients that were similar to those expected theoretically (F = 0.003).

Variance in family size of even-broodyears

Parentage data of the F_2 generation were available only for the even-broodyear returns. Distributions of the F_1 control and hybrid offspring were not Poisson (Figure 1; p < 0.001 and p < 0.01 respectively, G-test). Removing an outlier in the F_1 control group did not alter the results. The distribution of F_2 control family size was not Poisson (p < 0.001). However, the distribution observed for the F_2 hybrid family size was weakly consistent with a Poisson distribution (p = 0.051).

The index of variability did not increase in either the controls (2.91 and 1.78 for the F_1 and F_2 generations, respectively) or the hybrids (1.76 and 1.30 for the F_1 and F_2 generations, respectively). In addition, the index of variability in the F_2 hybrids did not exceed that of controls in either the F_1 or F_2 generation. Removing an outlier in the F_1 control group did not change these results.

Discussion

Microsatellite variability in Auke and Pillar Creek, and parentage analysis

Outbreeding depression is most likely to occur in hybrids of geographically separated populations if there is no direct gene flow and they have evolved separately for a considerable time. By choosing populations from approximately the same latitude, we attempted to match environments, though the match was not perfect. The two populations would be expected to have distinct coadapted gene complexes if they have adapted independently to their environments and if epistasis was involved in adaptation. The microsatellite analyses confirmed that the Auke Creek and Pillar Creek pink salmon populations differ genetically. These differences are consistent with previous observations that populations of pink salmon diverge over large geographic distances (e.g. Gharrett et al. 1988, Noll et al. 2001, Hawkins et al. 2002).

Survival and straying in even- and odd-broodyears

A reduction in survival is the most direct indication of outbreeding depressive effects of hybridization. By examining the returns in both the F_1 and F_2 generations, it is possible to distinguish between epistatic or nonepistatic effects on outbreeding depression.

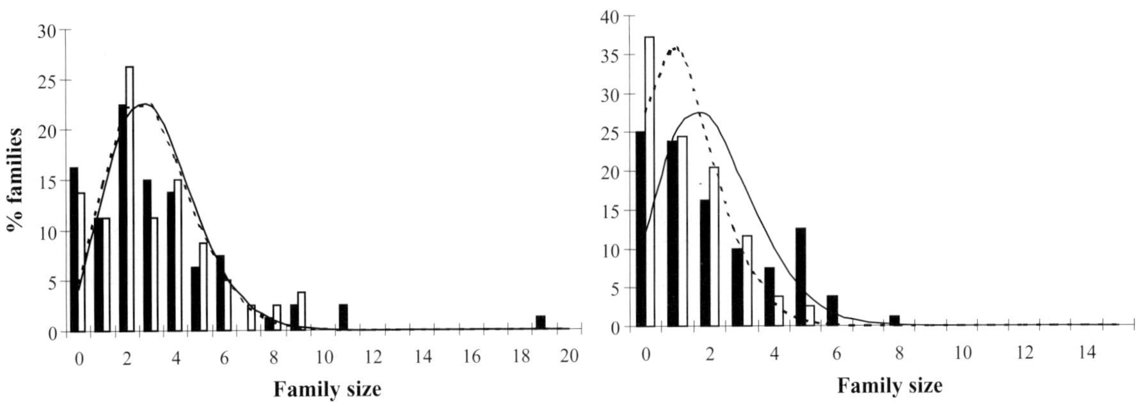

Figure 1. Individual family sizes for even-broodyear F_1 and F_2 pink salmon returns to Auke Creek. Vertical bars show the observed distribution of offspring; filled bars are control distributions and open bars are hybrid distributions. The curves show the Poisson distribution with the same mean as the observed family size; solid lines are control distributions and dashed lines are hybrid distributions. (a) 1998 F_1 returns. Neither the distribution of the control (p < 0.001) nor the hybrid (p < 0.01) samples fit a Poisson distribution (William's corrected G-test; Sokal and Rohlf 1995). (b) 2000 F_2 returns. The distribution of the control samples did not fit a Poisson distribution (p < 0.0001). The hybrid fit to a Poisson distribution was nearly significant (p = 0.0505).

The return rates for hybrids and controls were similar for the even-broodyear F_1 fish, consistent with the survival reported for F_1 hybrids between even- and odd-broodyear pink salmon (Gharrett & Smoker 1991, Gharrett et al. 1999). In a somewhat similar experiment, Bams (1976) released first-generation hybrid and transplanted (from the Kakweieken River in BC, Canada) pink salmon fry into Tsolum River/Headquarters Creek, about 146 km away; no control release was possible. Both the hybrids and transplants had similar survivals, as measured by recoveries in the commercial fishery. However, many fewer transplants than hybrid native/transplanted salmon negotiated their way about 24 km up the Tsolum River to the release site, suggesting that terminal homing may have a genetic basis.

In contrast to the even-broodyear, the return rate of F_1 control fish exceeded that of the hybrids in the odd-broodyear. This is the first of the Auke Creek pink salmon outbreeding depression experiments that demonstrated a reduced survival in F_1 hybrids and suggests that nonepistatic effects can contribute to overall survival. For the crosses reported here, developmental characters were also examined (Wang,[4] in preparation). In those studies, timing of mid-hatch was influenced by nonepistatic effects in the first generation. The idea that different genetic mechanisms may have evolved in the two broodyears is consistent with large genetic differences observed between odd- and even-broodlines of pink salmon (e.g. Aspinwall 1974, McGregor 1982). In fact, little similarity in the distribution of mtDNA variable sites was observed between the even- and odd-broodlines of pink salmon (Churikov et al. 2001), which may reflect differences in postglacial recolonizations of the two broodyears (Churikov & Gharrett 2002).

Both the even- and odd-broodline F_2 hybrids between Pillar Creek and Auke Creek pink salmon had reduced survivals relative to control fish, suggesting that outbreeding depression results from the disruption of coadapted, epistatically acting gene complexes. These results are consistent with the reduced survival of F_2 pink salmon hybrids between even- and odd-broodyear pink salmon at Auke Creek (Gharrett & Smoker 1991, Gharrett et al. 1999).

Other factors that could have influenced these results include interannual variation and incomplete control over culture conditions. Natural populations experience year-to-year environmental variation. These experiments sampled only two environmental sequences and it is possible that the environmental conditions sampled in our experiments fortuitously favored one group over the other. Fluctuating environmental differences are difficult to assess because only one set of environmental experiences can be observed at a time. In addition, it is possible that subtle incubation differences or marking effects may have influenced the differences in survival. It is unlikely, however, that either of these factors produced the primary effects observed as outbreeding depression because every pink salmon hybridization experiment performed at Auke Creek has yielded similar results (Gharrett & Smoker 1991, Gharrett et al. 1999).

The differences observed in the survivals of wild and cultured fish are not uncommon (e.g. Miller 1953, Bachman 1984). The reasons for the differences may include the effects of fin excisions, the contribution of uncounted fry derived from intertidal spawning to returns of wild salmon, or delayed mortality in cultured fish resulting from culture conditions.

The small number of experimental fish recovered at Waydelich Creek suggests that, at least in some instances, the straying rate of pink salmon is low. The number of hybrid strays was slightly higher than the number of control strays, but the total number of strays each year was less than 5 and is not statistically significant. Local straying does not account for survival differences. The straying rates observed in experimental fish were all less than a point estimate (2.8%) of straying of native Auke Creek fish within 10 km made from observations of otolith-marked fish,[5] but were similar to estimates (less than 1%) made using a genetic marker (Gharrett et al. 2001).

Failure to observe differences in straying rates of F_1 hybrids and controls is consistent with Smoker & Thrower (1995), who did not observe increased straying for either transplanted or hybrid chum salmon compared to native fish. Those results contrast with the study of Bams (1976), who found that hybridization affected homing ability. It is difficult to compare

---

[4] Wang, I.A., W.W. Smoker & A.J. Gharrett. Evidence of outbreeding depression through differences in development rates of pink salmon hybrids. *In*: Proceedings of the Northeast Pacific Pink and Chum Salmon Workshop, 3–5 March, 1999, Juneau, Alaska. National Marine Fisheries Service, 11305 Glacier Hwy., Juneau, AK 99802. Cited by permission.

[5] Mortensen, D., A. Wertheimer, J. Maselko & S. Taylor. 2000. Survival and straying of pink salmon using recoveries of coded-wire tags and thermally induced otolith marks. Feature article, AFSC Research Activities quarterly report for July–Aug–Sept 2000. Alaska Fisheries Science Center, National Marine Fisheries Service, NOAA.

these experiments because they used different kinds of crosses and evaluated homing ability differently. Moreover, neither Bams (1976) nor Smoker & Thrower (1995) followed the fish into the F_2 generation where outbreeding depression might be expected.

Inbreeding in even- and odd-broodyears

Because the broodstock for the F_2 generation derived from a limited number of parents, the possibility existed that a disproportionate number of full- or half-sib matings might have been included in the experiment, which would increase inbreeding. A loss of genetic diversity due to inbreeding can decrease fitness if essential alleles are lost and deleterious recessive alleles are expressed (Nevo et al. 1986, Emlen 1991) and might mask depressive effects due to outbreeding. The inbreeding coefficients for the F_2 experiments were of the same order of magnitude as would be expected from random breeding; and fortuitously, the hybrid broodstocks had slightly lower inbreeding coefficients than the control broodstocks. Inbreeding was not increased by our choice of broodstock for this study, and it is unlikely that the reduced survival in the F_2 hybrids resulted from inbreeding.

Variance in family size of even-broodyears

Observations of variability in family size and nonrandom marine survival in Pacific salmon populations (Simon et al. 1986, Geiger et al. 1997, Hard et al. 2000) suggest that certain combinations of genotypes resulting from epistatic relationships among genes may cause some individuals to have higher fitness in some environments, but these same combinations may not be favorable as conditions change. If, by chance, an F_2 family receives a favorable set of alleles and has a greater return, its offspring may contribute disproportionately to the return and subsequent production. If family size variability increases, the few successful families may not be the most fit in subsequent generations, thereby extending the effects of outbreeding depression to additional generations. In this experiment, more families completely failed and a few families produced more offspring than predicted by the Poisson frequency distribution for both hybrids and controls in the even-broodyear returns; in aggregate, the overall distribution was overdispersed as described by Geiger et al. (1997). However, our results suggest that an increase in variability of family size does not account for the decrease in survival seen in the hybrid fish. The relatively small number of F_2 returns in the even-broodyears of this study may have provided insufficient power to detect small differences in the variability of family size.

Conclusion

Hybrids between two spatially separated Pacific salmon populations exhibited reduced survival due to outbreeding depression, and evidence of both epistatic and nonepistatic effects was observed. It does not appear that a decreased local homing ability or increased variability in family size accounted for the decrease in hybrid survival. Whether due to nonepistatic or epistatic effects, the decrease in hybrid survival could jeopardize biodiversity of wild stocks if repeated introductions are made. Small, one-time introductions may be of minor concern because natural selection might be expected to remove maladapted genotypes rapidly from a population; however, repeated introductions or development of nonlocal hatchery brood stocks have the potential to decrease productivity in local wild streams. The possibility of outbreeding depression suggests that broodstocks should be developed from locally adapted natural populations and that salmon managers should consider the potentially detrimental effects of transplanting stocks from one hatchery to another.

Acknowledgements

Thanks to M. Adkison, S. Hall, Z. Li, J. Lum, A. Matala, E. Peterson, and many others who assisted in seining efforts, tissue collection, and breeding experiments. Thanks also to J. Seeb and the reviewers whose valuable comments contributed to this manuscript. The U.S. National Marine Fisheries Service Auke Bay Fisheries Laboratory, the Alaska Department of Fish and Game, and the University of Alaska Fairbanks cooperated in operation of the Auke Creek weir. This publication is the result of research sponsored by Alaska Sea Grant with funds from the National Oceanic and Atmospheric Administration Office of Sea Grant, Department of Commerce, under grant no. NA 16RG2321 (projects nos. R/31-06 and R/31-02), and from the University of Alaska with funds appropriated by the state.

References

Aspinwall, N. 1974. Genetic analysis of North American populations of the pink salmon, *Oncorhynchus gorbuscha*, possible evidence for the neutral mutation-random drift hypothesis. Evolution 28: 295–305.

Bachman, R.A. 1984. Foraging behavior of free-ranging wild and hatchery brown trout in a stream. Trans. Am. Fish. Soc. 113: 1–32.

Bams, R.A. 1976. Survival and propensity for homing as affected by presence or absence of locally adapted paternal genes in two transplanted populations of pink salmon (*Oncorhynchus gorbuscha*). J. Fish. Res. Board Can. 33: 2716–2725.

Banks, M.A., M.S. Blouin, B.B. Baldwin, V.K. Rashbrook, H.A. Fitzgerald, S.M. Blankenship & D. Hedgecock. 1999. Isolation and inheritance of novel microsatellites in chinook salmon (*Oncorhynchus tshawytscha*). J. Hered. 90: 281–288.

Brncic, D. 1954. Heterosis and the integration of the genotype in geographic populations of *Drosophila pseudoobscura*. Genetics 39: 77–88.

Campton, D.E. 1995. Genetic effects of hatchery fish on wild populations of Pacific salmon and steelhead: What do we really know? Am. Fish. Soc. Symp. 15: 337–353.

Carvalho, G.R. 1993. Evolutionary aspects of fish distribution: Genetic variability and adaptation. J. Fish Biol. 43(Suppl. A): 53–73.

Churikov, D. & A.J. Gharrett. 2002. Comparative phylogeography of the two pink salmon broodlines: An analysis based on mitochondrial DNA genealogy. Mol. Ecol. 11: 1077–1101.

Churikov, D., M. Matsuoka, X. Luan, A.K. Gray, V.L.A. Brykov & A.J. Gharrett. 2001. Assessment of concordance among genealogical reconstructions from various mtDNA segments in three species of Pacific salmon (genus *Oncorhynchus*). Mol. Ecol. 10: 2329–2339.

Crow, J.F. & M. Kimura. 1970. An Introduction to Population Genetics Theory. Harper and Row Publishers, New York, 591 pp.

Crow, J. & N.E. Morton. 1955. Measurement of gene frequency drift in small populations. Evolution 9: 202–214.

Danzmann, R.G. 1997. PROBMAX: A computer program for assigning unknown parentage in pedigree analysis from known genotypic pools of parents and progeny. J. Hered. 88: 333.

Danzmann, R.G., T.R. Jackson & M.M. Ferguson. 1999. Epistasis in allelic expression at upper temperature tolerance QTL in rainbow trout. Aquaculture 173: 45–58.

Dobzhansky, T. 1948. Genetics of natural populations XVII. Experiments on chromosomes of *Drosophila pseudoobscura* from different geographic regions. Genetics 33: 588–602.

Dobzhansky, T. 1950. Genetics of natural populations XIX. Origin of heterosis through natural selection in populations of *Drosophila pseudoobscura*. Genetics 35: 288–302.

Emlen, J.M. 1991. Heterosis and outbreeding depression: A multilocus model and an application to salmon production. Fish. Res. 12: 187–212.

Estoup, A., K. Gharbi, M. SanCristobal, C. Chevalet, P. Haffray & R. Guyomard. 1998. Parentage assignment using microsatellites in turbot (*Scophthalmus maximus*) and rainbow trout (*Oncorhynchus mykiss*) hatchery populations. Can. J. Fish. Aquat. Sci. 55: 715–725.

Falconer, D.S. & T.F.C. Mackay. 1996. Introduction to Quantitative Genetics, 4th edn. Longman Group Ltd., Essex, England, 464 pp.

Geiger, H.H., W.W. Smoker, L.A. Zhivotovsky & A.J. Gharrett. 1997. Variability of family size and marine survival in pink salmon (*Oncorhynchus gorbuscha*) has implications for conservation biology and human use. Can. J. Fish. Aquat. Sci. 54: 2684–2690.

Gharrett, A.J. & W.W. Smoker. 1991. Two generations of hybrids between even- and odd-year pink salmon (*Oncorhynchus gorbuscha*): a test for outbreeding depression? Can. J. Fish. Aquat. Sci. 48: 1744–1749.

Gharrett, A.J., S. Lane, A.J. McGregor & S.G. Taylor. 2001. Use of a genetic marker to examine genetic interaction among subpopulations of pink salmon (*Oncorhynchus gorbuscha*). Genetica 111: 259–267.

Gharrett, A.J., W.W. Smoker, R.R. Reisenbichler & S.G. Taylor. 1999. Outbreeding depression in hybrids between odd- and even-broodyear pink salmon. Aquaculture 173: 117–129.

Gharrett, A.J., C. Smoot, A.J. McGregor & P.B. Holmes. 1988. Genetic relationships of even-year northwestern Alaskan pink salmon. Trans. Am. Fish. Soc. 117: 536–545.

Gordon, H. & M. Gordon. 1957. Maintenance of polymorphism by potentially injurious genes in eight natural populations of the platyfish, *Xiphophorus maculatus*. J. Genet. 55: 1–44.

Hard, J.J. 1995. Genetic monitoring of life-history characters in salmon supplementation: Problems and opportunities. Am. Fish. Soc. Symp. 15: 212–225.

Hard, J.J., L. Connell, W.K. Hershberger & L.W. Harrell. 2000. Genetic variation in mortality of Chinook salmon during a bloom of the marine alga *Heterosigma akashiwo*. J. Fish Biol. 56: 1387–1397.

Hawkins, S.L., N.V. Varnavskaya, E.A. Matzak, V.V. Efremov, C.M. Guthrie III, R.L. Wilmot, H. Mayama, F. Yamazaki & A.J. Gharrett. 2002. Population structure of odd-broodline Asian pink salmon and its contrast to the even-broodline structure. J. Fish Biol. 60: 370–388.

Hemmingsen, A.R., R.A. Holt, R.D. Ewing & J.D. McIntyre. 1986. Susceptibility of progeny from crosses among three stocks of coho salmon to infection by *Ceratomyxa shasta*. Trans. Am. Fish. Soc. 115: 492–495.

Hindar, K., N. Ryman & F. Utter. 1991. Genetic effects of cultured fish on natural fish populations. Can. J. Fish. Aquat. Sci. 48: 945–957.

King, J.C. 1955. Evidence for the integration of the gene pool from studies of DDT resistance in *Drosophila*. pp. 311–317. *In*: Cold Spring Harbor Symposium on Quantitative Biology 20.

Lane, S., A.J. McGregor, S.G. Taylor & A.J. Gharrett. 1990. Genetic marking of an Alaskan pink salmon population, with an evaluation of the mark and the marking process. Am. Fish. Soc. Symp. 7: 395–406.

Leberg, P.L. 1993. Strategies for population reintroduction: Effects of variability. Cons. Biol. 7: 194–199.

Lichatowich, J. 1999. Salmon Without Rivers: A History of the Pacific Salmon Crisis. Island Press, Washington, DC, 317 pp.

Lynch, M. 1991. The genetic interpretation of inbreeding depression and outbreeding depression. Evolution 45: 622–629.

McGregor, A.J. 1982. A biochemical genetic analysis of pink salmon (*Oncorhynchus gorbuscha*) from selected streams in northern Southeast Alaska. M.S. Thesis, University of Alaska Juneau, Juneau, AK.

McGregor, A.J., S. Lane, M.A. Thomason, L.A. Zhivotovsky, W.W. Smoker & A.J. Gharrett. 1998. Migration timing, a life history trait important in the genetic structure of pink salmon. N. Pac. Anadr. Fish Comm. Bull. 1: 262–273.

Miller, R.B. 1953. Comparative survival of wild and hatchery-reared cutthroat trout in a stream. Trans. Am. Fish. Soc. 83: 120–130.

Nevo, E., R. Noy, B. Lavie, B. Beiles & S. Muchtar. 1986. Genetic diversity and resistance to marine pollution. Biol. J. Linn. Soc. 29: 139–144.

Noll, C., N.V. Varnavskaya, E.A. Matzak, S.L. Hawkins, V.V. Midanaya, O.N. Katugin, C. Russell, N.M. Kinas, C.M. Guthrie III, H. Mayama, F. Yamazaki, B.P. Finney & A.J. Gharrett. 2001. Analysis of contemporary genetic structure of even-broodyear populations of Asian and western Alaskan pink salmon, *Oncorhynchus gorbuscha*. Fish. Bull. 99: 123–138.

Olsen, J.B., P. Bentzen & J.E. Seeb. 1998. Characterization of seven microsatellite loci derived from pink salmon. Mol. Ecol. 7: 1087–1089.

O'Reilly, P., L.C. Hamilton, S.K. McConnell & J.M. Wright. 1996. Rapid detection of genetic variation in Atlantic salmon (*Salmo salar*) by PCR multiplexing of dinucleotide and tetranucleotide microsatellites. Can. J. Fish. Aquat. Sci. 53: 2292–2298.

Philipp, D.P. & J.E. Claussen. 1995. Fitness and performance differences between two stocks of largemouth bass from different river drainages within Illinois. Am. Fish. Soc. Symp. 15: 236–243.

Quinn, T.P. 1993. A review of homing and straying of wild and hatchery-produced salmon. Fish. Res. 18: 29–44.

Raymond, M. & F. Rousset. 1995. An exact test for population differentiation. Evolution 49: 1280–1283.

Roff, D.A. & P. Bentzen. 1989. The statistical analysis of mitochondrial DNA polymorphisms: χ^2 and the problem of small samples. Mol. Biol. Evol. 6: 539–545.

Seutin, G., B.N. White & P.T. Boag. 1991. Preservation of avian blood and tissue samples for DNA analyses. Can. J. Zool. 69: 82–90.

Shields, W.M. 1982. Philopatry, Inbreeding, and the Evolution of Sex. State University of New York Press, Albany, NY, 245 pp.

Simon, R.C., J.D. McIntyre & A.R. Hemmingsen. 1986. Family size and effective population size in a hatchery stock of coho salmon (*Oncorynchus kisutch*). Can. J. Fish. Aquat. Sci. 43: 2434–2442.

Small, M.P., T.D. Beacham, R.E. Withler & R.J. Nelson. 1998. Discriminating coho salmon (*Oncorhynchus kisutch*) populations withing the Fraser River, British Columbia, using microsatellite DNA markers. Mol. Ecol. 7: 141–155.

Smith, C.T., B.F. Koop & R.J. Nelson. 1998. Isolation and characterization of coho salmon (*Oncorhynchus kisutch*) microsatellites and their use in other salmonids. Mol. Ecol. 7: 1614–1616.

Smoker, W.W. & F.P. Thrower. 1995. Homing propensity in transplanted and native chum salmon. Am. Fish. Soc. Symp. 15: 575–576.

Sokal, R.R. & F.J. Rohlf. 1995. Biometry, 3rd edn. W.H. Freeman and Company, New York, 887 pp.

Waldbieser, G.C. & W.R. Wolters. 1999. Application of polymorphic microsatellite loci in a channel catfish (*Ictalurus punctatus*) breeding program. J. World Aquacult. Soc. 30: 256–262.

Wallace, B. 1981. Coadaptation revisited. J. Hered. 82: 89–95.

Winer, B.J. 1971. Statistical Principles in Experimental Design, 2nd edn. McGraw-Hill, New York, 907 pp.

Environmental Biology of Fishes **69**: 299–306, 2004.
© 2004 *Kluwer Academic Publishers. Printed in the Netherlands.*

Effects on embryo development time and survival of intercrossing three geographically separate populations of Southeast Alaska coho salmon, *Oncorhynchus kisutch*

Karla L. Granath[a,c], William W. Smoker[a], Anthony J. Gharrett[a] & Jeffrey J. Hard[b]
[a]*University of Alaska Fairbanks, Fisheries Division, School of Fisheries and Ocean Science,*
11120 Glacier Highway, Juneau, AK 99801, U.S.A.
[b]*National Marine Fisheries Service, Northwest Fisheries Science Center, Conservation Biology Division,*
2725 Montlake Boulevard East, Seattle, WA 98112, U.S.A.
[c]*Present address: Alaska Department of Fish and Game, Commercial Fisheries Division, P.O. Box 920587,*
Dutch Harbor, AK 99692, U.S.A. (e-mail: karla_granath@fishgame.state.ak.us)

Received 17 April 2003   Accepted 24 April 2003

Key words: outbreeding depression, conservation biology, salmonid, local adaptation

Synopsis

We investigated adaptive differences among three geographically separate populations of coho salmon (*Oncorhynchus kisutch*) by forming first generation intercrosses (hybrid lines) and comparing them to parental types (control lines). Broodstock for the experiment came from the Gastineau Hatchery in Juneau, Hidden Falls Hatchery on Baranof Island, and Neets Bay Hatchery near Ketchikan, Alaska. All were isolated hatchery populations separated by 220–400 km and derived 15–20 years previously from single local wild populations. For each population, gametes were taken from 50 mature salmon of each sex and combined to form nine lines (three control and six hybrid); each line had 50 full-sibling families which were assigned to separate cells of an incubator at Gastineau Hatchery. Embryo survival and development times were measured as indicators of locally adapted fitness traits. Two of the control lines had higher survival rates than hybrid lines formed between either of their parental populations and other populations. Differences ($p < 0.05$) were found between development times for control and hybrid groups, which varied by as many as 20 days between families and as many as 30 days between control lines. The intermediate expression of development time of intercrossed lines is consistent with additive genetic variation of development time between the ancestral populations of coho salmon and indicates that important genetic divergence exists between the populations. A loss of local adaptation through a change in seasonal timing of completion of embryonic development would occur in intercrosses between the populations.

Introduction

Genetic diversity, affected within a population by differing rates or patterns of migration (gene flow), mutation, random genetic drift, recombination, and selection, increases a population's ability to adapt to environmental variation from generation to generation (Taylor 1991). Genetic diversity between populations maintains a species' ability to adapt and thrive in the face of ecological challenge. Depending on several factors, including the degree of divergence between populations, a loss of between-population diversity (local adaptation or Type I outbreeding depression) can occur when populations interbreed (Shields 1982, Templeton 1986) creating hybrid lineages.

Outbreeding depression is a threat to wild salmon populations that have the potential to interbreed with artificially cultured salmon (e.g., Hindar et al. 1991,

Waples 1992, Gharrett & Smoker 1993, Campton 1995). There are two mechanisms by which cultured salmon would be different enough genetically from wild salmon that outbreeding depression might occur. The first would occur if the hatchery broodstock were transferred into an area where wild salmon are adapted to different local conditions than are the broodstock's source population. The second would occur if cultured salmon, whether locally derived or not, were influenced by founder effect or domestication. Either mechanism can threaten wild populations through ecological or genetic mechanisms of outbreeding depression (Campbell & Waser 1986, Lynch 1991) if hatchery and wild fish hybridize in the wild. Changes due to domestication have been noted in spawning behavior (Fleming & Gross 1992, Berejikian et al. 1997), growth (Hershberger et al. 1990), body shape (Taylor 1986, Fleming & Jonsson 1994, Hard et al. 2000), life history traits (Hjort & Schreck 1982), and genetic variation (Flagg et al. 1995, Withler 1998).

This experiment is part of a larger study investigating the genetic mechanisms of outbreeding depression in Pacific salmon in which fitness traits of second-generation outbreed coho salmon are being observed. We chose Alaskan coho salmon for this study due to their high rates of smolt-to-adult survival and lack of complex age structure. Here we report observations of the F_1 hybrids from the 1998 broodyear (gametes collected in 1998) of the study. When grown in a common environment, examination of differences between F_1 hybrids and their parental population controls may allow detection of outbreeding depression through loss of local adaptation if present. First-generation progeny receive a complete genetic complement from each parent, which limits ability to characterize in these fish all the genetic interactions that can occur between distinct genomes on interbreeding. However, this study evaluates some immediate consequences of interbreeding, and focuses on evidence for genetic divergence of early life history between populations. The purpose of our study is to observe which, if any, of these adaptive differences in the population are the results of genetic differentiation.

Methods

We chose three geographically separate populations from southeast Alaska for experimental intercrossing (Figure 1). The northernmost population came from the Gastineau Hatchery (G) in Juneau. This population was established at Gastineau in 1986 and originated from Montana Creek in the Mendenhall watershed. The second population was sampled from the Hidden Falls Hatchery (H) on Baranof Island. The original broodstock was derived from farther south on Baranof Island at Sashin Creek near Little Port Walter and was established at Hidden Falls in 1992. The southernmost population was the Neets Bay Hatchery (N) north of Ketchikan and was established in 1980. The original broodstock came from Indian Creek, a tributary of the Chickamin River northeast of Ketchikan in Misty Fiords National Monument.

The three populations used in this experiment derive from geographically isolated wild populations spawning in different habitats and have been under culture for five to seven generations. Indian Creek in the south enters the Chickamin River about 32 km from the ocean at an elevation of roughly 600 m and is fed from high mountain streams. Montana Creek in the Mendenhall watershed is a low elevation system fed from mountain streams. Sashin Creek drains, Sashin Lake and salmon have access to 1 100 m of spawning ground on the low gradient 3 km creek. All three populations spawn in late fall, which allowed for the simultaneous collection of gametes from hatchery broodstocks derived from them. Conditions such as water temperature, stream flow, distance to spawning areas, and the isolation of these three locations, although modulated by domestication in similar hatchery systems for a few generations, provide an opportunity for these three populations to be very different genetically and phenotypically.

Gamete collection and development time studies

We collected gametes on November 10, 1998, from 50 pairs of coho salmon sampled at random from mature adults available at each of the three hatcheries. Individuals were identified with plastic jaw tags. Eggs (with ovarian fluid) and milt were extruded into separate plastic (Ziplock®) bags labeled with the adult tag number and placed on ice for shipping. Adult carcasses were boxed and frozen for future meristic analysis. Fertilization took place at the Gastineau Hatchery on the evening of November 10.

Approximately 500 eggs (150 ml) were used for each mating. Eggs from each female were drained of ovarian fluid and weighed to nearest centigram for an estimate of total ovary mass, then partitioned into three plastic bowls and fertilized with milt from a single male from

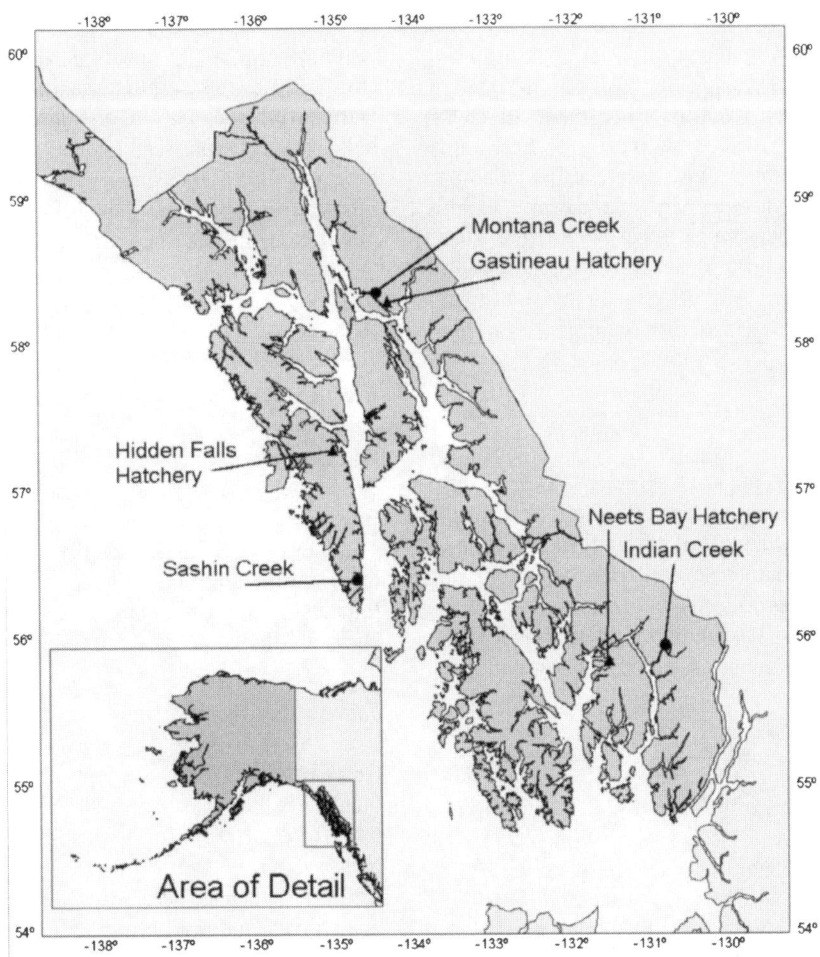

Figure 1. Map of southeast Alaska showing locations of hatchery and broodstock sites.

Table 1. Experimental design used for intercrossing three populations of southeast Alaska coho salmon (maternal population × paternal population). Control groups (on diagonal) are crosses of parents from the same population, and reciprocal intercrosses (off diagonal) represent each hybrid line.

Paternal population	Maternal population		
	Gastineau (G)	Hidden Falls (H)	Neets Bay (N)
Gastineau (G)	G × G	H × G	N × G
Hidden Falls (H)	G × H	H × H	N × H
Neets Bay (N)	G × N	H × N	N × N

each population. Milt from each male was in turn used three times to fertilize eggs from a single female in each population (Table 1). Gametes were then activated with water and placed into modified (F.A.L. HeathTecna™)

incubator trays. Plastic dividers separated each tray into 12 cells, which allowed the tracking of development and mortality in individual families. This procedure was repeated until the eggs from all 150 females were fertilized for a total of 450 families or 50 families in each of the nine lines (Table 1). Eggs were then left undisturbed in incubators until the eyed stage (when pigmentation of the eye is visible through the egg envelope). Remaining eggs, not used in the experiment, were stored on ice and individual egg masses were recorded the following day for all females (minimum of five eggs from each female).

In March of 1999, counts were made of all remaining live and dead eggs. Live eggs from each family (minimum 100, maximum 200) were then transferred to 10 cm diameter 250 ml plastic containers fitted with mesh screens and randomly placed into one of the three

302

vertical incubator stacks to randomize the possible position effects on incubation.

Hatching was observed from late March until early May. Daily visual observations were made of each family, and the dates of the beginning of hatching, mid-hatch (estimated 50% of eggs hatched), and complete hatching were recorded. Incubators were monitored four times each week post-hatch to ensure adequate water flow until transfer to rearing tanks and first feeding (ponding). Alevins were ready (less than $\sim$6% yolk mass : body mass) for ponding beginning in June.

Statistical analysis

We performed analysis of variance (ANOVA) of hatching and survival data using the SAS General Linear Model procedure. Robustness of the analysis was tested by performing ANOVA on several transformations of the data; all analyses produced similar results. Survival data for each family was transformed before it was used in the analysis using an empirical logit model (Agresti 1990):

$$\text{Logit} = \text{Ln}((D_e + 0.5)/(T_e - D_e + 0.5))$$

where D_e is the number of dead eggs and T_e is the total number of eggs (alive + dead). Approximate F-tests were constructed to test the significance of effects from the TEST option in the RANDOM statement of SAS. Maternal population and paternal population effects were analyzed with the model:

$$Y_{ijklm} = \mu + M_i + D(M)_{ij} + P_k + S(P)_{kl} + (M * P)_{ik} + e_{ijklm}$$

where Y_{ijklm} is egg survival, days to mid-hatch, or length; μ is the population mean; M is the fixed effect of maternal population ($i = 1, 3$); P is the fixed effect of paternal population ($k = 1, 3$); D is the random effect of female parent, or dam ($j = 1, 50$) within maternal population; S is the random effect of male parent, or sire ($l = 1, 50$) within paternal population k; and e_{ijklm} is the error. Because sire and dam effects are based on half sibling comparisons and there were no replicates of different sires across dams within population, there are inherent sire (paternal population) and dam (maternal population) interactions included in those terms.

Results

Body size, egg mass, and ovary size

Females originating from the Gastineau and Hidden Falls hatcheries were shorter than females from Neets Bay. Gastineau females produced the smallest eggs of all three populations (Table 2). Ovary masses differed among all three populations, with the Neets Bay females having the heaviest ovaries, followed by the Hidden Falls and Gastineau females, respectively.

Survival to eyed egg

Gastineau and Neets Bay control lines had higher survival rates than their hybrids. The Hidden Falls control line had the lowest survival, followed by intercrossed hybrid lines sired by Hidden Falls males (Table 3). An ANOVA on egg survival showed significant differences due to maternal population (M) effects ($F_{2,95} = 4.41$, $p < 0.05$) and paternal population effects ($F_{2,101} = 11.07$, $p < 0.05$), which accounted for 2.38% and 6.94% of the total variation, respectively. Dam within maternal population explained 49.77% of the variation ($F_{95,177} = 3.09$, $p < 0.05$), and 21.7% was attributed to sire within paternal population ($F_{97,177} = 2.60$, $p < 0.05$). There were no significant interaction effects detected between the maternal and paternal populations.

Table 2. Mideye–hypural length, average egg mass, and ovary mass of females from three different populations in southeast Alaska. N is the number of females or eggs samples; standard deviations are in parentheses; $p < 0.05$ for different labels denoting pairwise comparisons for which the null hypothesis of no difference could not be rejected. Length data are missing for several of the Neets Bay females.

Population	Length			Egg mass			Ovary mass		
	N	(mm)	p	N	(g)	p	N	(g)	p
Gastineau	50	537 (25)	a	685	0.16 (0.02)	a	50	59.71 (16.25)	a
Hidden Falls	48	545 (30)	a	638	0.17 (0.02)	b	50	71.20 (18.20)	b
Neets Bay	29	583 (29)	b	620	0.18 (0.02)	b	50	82.68 (13.41)	c

Table 3. Average egg survival to eyeing for the nine lines. Standard deviations are in parentheses and p < 0.05 for different labels denoting pairwise comparisons for which the null hypothesis of no difference could not be rejected.

Paternal population	Maternal population					
	G		H		N	
	Survival rate	p	Survival rate	p	Survival rate	p
G	0.927(0.057)	a	0.895(0.137)	a	0.919(0.098)	a
H	0.817(0.168)	b	0.788(0.236)	b	0.872(0.176)	a
N	0.897(0.098)	a	0.901(0.120)	a	0.923(0.135)	a

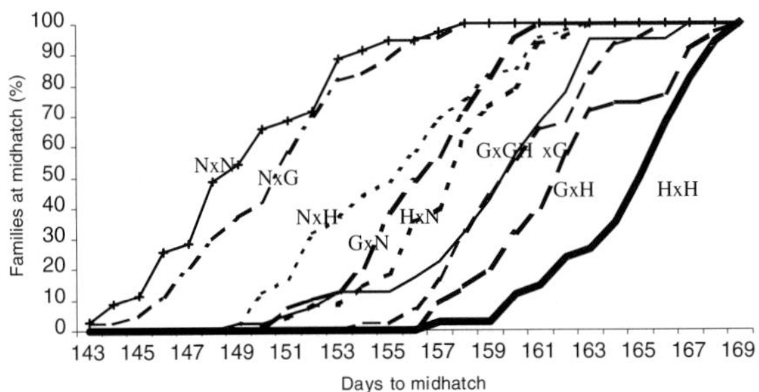

Figure 2. Development time. Cumulative percentage of families in parental and intercrossed lines to reach mid-hatch (50% hatch within families). Parental control groups are shown as solid lines (maternal × paternal): N × N, G × G, H × H. Hybrid lines are shown with broken lines and the same pattern is used for reciprocal lines: N × G, G × N; N × H, H × N; H × G, G × H.

Table 4. Development time, measured in mean days from fertilization to midhatch (50% of families) for eggs from all lines. Ranked in order by hatching time (first to last) within groups involving (A) female parents of the same population, and (B) male parents from the same population. Crosses shown as maternal population × paternal population.

Rank	Cross	Mean days	Rank	Cross	Mean days
A			*B*		
1	N × N	150	1	N × N	150
2	N × G	151	2	G × N	157*
3	N × H	155*	3	H × N	158*
1	G × N	157*	1	N × G	151*
2	G × G	160	2	G × G	160
3	G × H	162*	3	H × G	160
1	H × N	158*	1	N × H	155*
2	H × G	160*	2	G × H	162*
3	H × H	165	3	H × H	165

*Signifies a significant difference (p < 0.05) between hybrid and control line within each group of three.

Development time

The Neets Bay control line (N × N) was the first to reach mid-hatch, followed in order by: N × G, N × H, G × N, H × N, G × G, H × G, G × H, and H × H (Figure 2). Mean time to mid-hatch was significantly different (p < 0.05) among the three control groups. Time to mid-hatch also differed significantly (p < 0.05) between control and associated hybrid lines with the exception of N × N *versus* N × G and G × G *versus* H × G.

The rank order of mid-hatch was the same (Neets Bay first, Gastineau second, and Hidden Falls third) for male and female parents in all lines (Table 4). In the ANOVA on development time, maternal effect (M) ($F_{2,112}$ = 153.39, p < 0.05) and paternal effect (P) ($F_{2,125}$ = 101.06, p < 0.05) were highly significant and explained 44.9% and 19.9% of the total variation, respectively. Also significant was the dam within maternal population effect (M(P)) ($F_{91,119}$ = 1.72, p < 0.05) which accounted for 18% of the total variation. There were no significant sire within paternal population or interaction between maternal and paternal population effects.

Discussion

Geographic separation, the influence of local adaptation, and hatchery influences have produced measurable differences in length, ovary size, and egg size among the three populations observed in this experiment. Within each population, larger females produce larger eggs than smaller females as was similarly reported by Murray et al. (1990). Individual male and female parents contributed to significant variation in survival. Low survival rates detected in lines descending from Hidden Falls males, most notably in the control line, could have resulted from depressed fertility of semen (either inherent, temperature related, or caused by improper handling or storage of the semen during collection and transportation) or possible inbreeding depression in the Hidden Falls donor population. Sampling eggs 1 or 2 days after fertilization to determine which eggs were successfully fertilized would have resolved some of these questions. Both the Gastineau and Neets Bay control lines had higher survival rates than their intercrossed hybrid lines, suggesting decreased fitness as a result of outbreeding, likely due to a genotype × environment interaction (paternal genome × maternal egg environment). Discriminating genotype × environment interactions from nonadditive genetic effects is difficult because females were bred to only one male from each population. However, hybrid lines are nearly intermediate to the midpoint of parental lines, suggesting at most minor net non-additive components of population divergence. Such decreased productivity in hybrid lines would tend to limit the amount of gene flow between populations. A further reduction of hybrid productivity in subsequent generations produced from these F_1 hybrids would support evidence for a breakdown of coadapted gene complexes: the 'genetic' mechanism of outbreeding depression.

Results from this experiment suggest that local adaptation strongly influences development time under the hatchery conditions evaluated in this study. In a common environment, average time to mid-hatching was separated by as many as 15 days between control lines. These differences are probably due to variation that exists between the hatchery and source environments to which the populations have adapted. The Neets Bay population, which comes from a high mountain stream, may require fewer days to develop in its native habitat than does the Hidden Falls population, which originates in a coastal stream. Embryos of salmon from colder environments require fewer days to develop in a common environment than embryos from warm-adapted salmon (Beacham & Murray 1989, Murray et al. 1990, Konecki et al. 1995). The source of the Neets Bay population, the Chickamin River, is a cold glacially influenced mainland system. Development times of intercrossed hybrid lines were intermediate to those of control lines, suggesting that the differences between the populations reflect primarily additive genetic effects; there is no evidence for appreciable net directional dominance. Remnant 'field effects' from the source environments are probably still influencing the differences between these first generation groups. Although this experiment does not have the ability in the F_1 generation to fully address the issue of outbreeding depression, these results suggest that the ecological model for outbreeding depression, a loss of local adaptation, would manifest in progeny of these hybrid crosses if they incubated in either of the parental streams. Residual environmental effects should disappear in the F_2 generation and regenerating the F_1 groups in a subsequent generation will help answer how strongly these effects are influenced by local adaptation.

Changes in hatching times caused by outbreeding could also have repercussions later in development. Timing of hatch is important to insure that conditions will be favorable in the stream when the fry emerge (e.g. Mason & Chapman 1965, Metcalfe & Thorpe 1992). In nature, emerging too early can cause increased mortality due to unfavorable stream conditions such as high flows, lack of prey items, or increased predation. However, early emerging young also tend to be dominant and are at a competitive advantage to later emerging young, which as a result may have slower growth rates (Mason & Chapman 1965; Metcalfe & Thorpe 1992; Rhodes & Quinn 1998). Size of young at emergence can also determine when emigration occurs. Larger young tend to emigrate earlier than smaller young within the same watershed (Irvine & Ward 1989) and smaller young may require an additional 1 or 2 years of freshwater residence before they migrate to the ocean.

Our findings, that variation in development time and egg survival among populations have genetic components and can be affected by outbreeding depression within a generation of intercrossing, are consistent with the findings of Philipp and Claussen (1995), who demonstrated that local adaptations can occur among different populations of fish within a relatively small geographic area (in this case broodstocks were separated by a minimum of 220 km). Different environmental conditions within streams that are separated by geographic barriers would be expected to produce populations that are differently adapted.

The effects of outbreeding depression may be particularly important for some populations of Pacific salmon. Within natural populations, a loss of local adaptation would be stabilized by natural selection within a few generations if very few genes are involved, the affected population size is large, and the number of nonadapted migrants is low (Templeton 1986). If the local population is small (often the case for Pacific salmon) and the number of nonadapted migrants is large (perhaps the case if a large hatchery is located nearby), effects of outbreeding depression would be experienced by a larger portion of the population and the potential for an overall reduction in fitness would be high.

Salmon have a complex life history that includes life stages in both freshwater and marine environments. Timing life history events to when conditions in these dynamic environments will be favorable maximizes the fitness of individual populations. Conservation of genetic and life history diversity within and among populations is important to the survival of the species as a whole. Management practices should aim to reduce human-induced interactions between distinct populations of salmon within geographic regions. Differences between populations in this study were evaluated only in a single hatchery environment and these evaluations were limited to early life history of first generation progeny. Further studies need to focus on intercrossing effects in later life stages and generations, and in other populations.

Acknowledgements

This project was funded by the National Marine Fisheries Service, Northwest Fisheries Science Center. Field and lab support were provided by S. Hall, I. Wang, Z. Li, A. Matala, J. Pugh Jr., and hatchery staff. Special thanks to managers and staff at the Gastineau Hatchery. In 2000, the name of the Gastineau Hatchery was changed to Macaulay Hatchery after the untimely death of its founder, Ladd Macaulay.

References

Agresti, A. 1990. Categorical Data Analysis, John Wiley and Sons, New York. 558 pp.

Beacham, T.D. & C.B. Murray. 1989. Variation in developmental biology of sockeye salmon (*Oncorhynchus nerka*) and chinook salmon (*O. tshawytscha*) in British Columbia. Can. J. Zool. 67: 2081–2089.

Berejikian, B.A., E.P. Tezak, S.L. Schroder, C.M. Knudsen & J.J. Hard. 1997. Reproductive behavioral interactions between wild and captively reared coho salmon (*Oncorhynchus kisutch*). ICES J. Mar. Sci. 54: 1040–1050.

Campbell, D.R. & N.M. Waser. 1986. The evolution of plant mating systems: Multilocus simulations of pollen dispersal. Am. Nat. 129: 593–609.

Campton, D.E. 1995. Genetic effects of hatchery fish on wild populations of Pacific salmon and steelhead: What do we really know? Am. Fish. Soc. Symp. 15: 337–353.

Flagg, T.A., F.W. Waknitz, D.J. Maynard, G.B. Milner & C.V.W. Mahnken. 1995. The effect of hatcheries on native coho salmon populations in the lower Columbia River. Am. Fish. Soc. Symp. 15: 366–375.

Fleming, I.A. & B. Jonsson. 1994. Phenotypic divergence of sea-ranched, farmed, and wild salmon. Can. J. Fish. Aquat. Sci. 51: 2808–2820.

Fleming, I.A. & M.R. Gross. 1992. Reproductive behavior of hatchery and wild coho salmon (*Oncorhynchus kisutch*): Does it differ? Aquaculture 103: 101–121.

Gharrett, A.J. & W.W. Smoker. 1993. A perspective on the adaptive importance of genetic infrastructure in salmon populations to ocean ranching in Alaska. Fish. Res. 18: 45–58.

Hard, J.J., B.A. Berejikian, E.P. Tezak, S.L. Schroder, C.M. Knudsen and L.T. Parker. 2000. Evidence for morphometric differentiation of wild and captively reared adult coho salmon; a geometric analysis. Env. Biol. Fish. 59: 61–73.

Hershberger, W.K., J.M. Myers, R.N. Iwamoto, W.C. McAuley & A.M. Saxton. 1990. Genetic changes in the growth of coho salmon (*Oncorhynchus kisutch*) in marine net-pens, produced by ten years of selection. Aquaculture 85: 187–197.

Hindar, K., N. Ryman & F. Utter. 1991. Genetic effects of cultured fish on natural fish populations. Can. J. Fish. Aquat. Sci. 48: 945–956.

Hjort, R.C. & C.B. Schreck. 1982. Phenotypic differences among stocks of hatchery and wild coho salmon, *Oncorhynchus kisutch*, in Oregon, Washington, and California. Fish. Bull. 80: 105–119.

Irvine, J.R. & B.R. Ward. 1989. Patterns of timing and size of wild coho salmon (*Oncorhynchus kisutch*) smolts migrating from the Keogh River watershed on northern Vancouver Island. Can. J. Fish. Aquat. Sci. 46: 1086–1094.

Konecki, J.T., C.A. Woody & T.P. Quinn. 1995. Influence of temperature on incubation rates of coho salmon (*Oncorhynchus kisutch*) from ten Washington populations. NW Sci. 69: 126–132.

Lynch, M. 1991. The genetic interpretation of inbreeding depression and outbreeding depression. Evolution. 45: 622–629.

Mason, J.C. & D.W. Chapman. 1965. Significance of early emergence, environmental rearing capacity, and behavioral ecology of juvenile coho salmon in stream channels. J. Fish. Res. Board Can. 22: 173–190.

Metcalfe, N.B. & J.E. Thorpe. 1992. Early predictors of life-history events: the link between first feeding date, dominance and seaward migration in Atlantic salmon, *Salmo salar* L. J. Fish Bio. 41: 93–99.

Murray, C.B., T.D. Beacham & J.D. McPhail. 1990. Influence of parental stock and incubation temperature on the early development of coho *salmon (Oncorhynchus kisutch)* in British Columbia. Can. J. Zool. 68: 347–358.

Philipp, D.P. & J.E. Claussen. 1995. Fitness and performance differences between two stocks of largemouth bass from different river drainages within Illinois. Am. Fish. Soc. Symp. 15: 236–243.

Rhodes, J.S. & T.P. Quinn. 1998. Factors affecting the outcome of territorial contests between hatchery and naturally reared coho salmon parr in the laboratory. J. Fish Biol. 53: 1220–1230.

Shields, W.M. 1982. Philopatry, Inbreeding, and the Evolution of Sex, State University of New York, Albany, NY. 102 pp.

Taylor, E.B. 1986. Differences in morphology between wild and hatchery populations of juvenile coho salmon. Prog. Fish Cult. 48: 171–176.

Taylor, E.B. 1991. A review of local adaptation in Salmonidae, with particular reference to Pacific and Atlantic salmon. Aquaculture 98: 185–207.

Templeton, A. 1986. Coadaptation and outbreeding depression. pp. 105–116. *In:* M.E. Soule (ed.) Conservation Biology: The Science of Scarcity and Diversity, Sinauer, Sunderland, MA.

Waples, R.S. 1992. Genetic effects of stock transfers of fish. pp. 51–69. *In:* A.N. Other (ed.) Protection of Aquatic Biodiversity: Proceedings of the World Fisheries Congress, Oxford & IBH Pub. Co.

Withler, R. 1998. Panel conclusions: genetic effects of straying of non-native hatchery fish into natural populations. NOAA Tech. Memo NMFS NWFSC-30.

Environmental Biology of Fishes **69**: 307–316, 2004.
© 2004 *Kluwer Academic Publishers. Printed in the Netherlands.*

Major histocompatibility complex loci are associated with susceptibility of Atlantic salmon to infectious hematopoietic necrosis virus

Kristina M. Miller[a], James R. Winton[b], Angela D. Schulze[a], Maureen K. Purcell[b] & Tobi J. Ming[a]
[a]*Department of Fisheries and Oceans Canada, Pacific Biological Station, Nanaimo, BC, V9T 6N7, Canada (e-mail: millerk@pac.dfo-mpo.gc.ca)*
[b]*Western Fisheries Research Center, 6505 NE 65th St., Seattle, WA 98115, U.S.A.*

Received 17 April 2003   Accepted 24 April 2003

Key words: IHN, salmonid, MHC, viral, pathogen, association

Synopsis

Infectious hematopoietic necrosis virus (IHNV) is one of the most significant viral pathogens of salmonids and is a leading cause of death among cultured juvenile fish. Although several vaccine strategies have been developed, some of which are highly protective, the delivery systems are still too costly for general use by the aquaculture industry. More cost effective methods could come from the identification of genes associated with IHNV resistance for use in selective breeding. Further, identification of susceptibility genes may lead to an improved understanding of viral pathogenesis and may therefore aid in the development of preventive and therapeutic measures. Genes of the major histocompatibility complex (MHC), involved in the primary recognition of foreign pathogens in the acquired immune response, are associated with resistance to a variety of diseases in vertebrate organisms. We conducted a preliminary analysis of MHC disease association in which an aquaculture strain of Atlantic salmon was challenged with IHNV at three different doses and individual fish were genotyped at three MHC loci using denaturing gradient gel electrophoresis (PCR-DGGE), followed by sequencing of all differentiated alleles. Nine to fourteen alleles per exon-locus were resolved, and alleles potentially associated with resistance or susceptibility were identified. One allele (*Sasa*-B-04) from a potentially non-classical class I locus was highly associated with resistance to infectious hematopoietic necrosis (p < 0.01). This information can be used to design crosses of specific haplotypes for family analysis of disease associations.

Introduction

Genes of the major histocompatibility complex (MHC) are crucial elements of adaptive immunity in vertebrate organisms and have been linked to resistance to numerous pathogenic diseases (reviews in Apanius et al. 1997, Lamont 1998). MHC genes encode polypeptides which recognise and bind both self and foreign peptides and present them to T-cells. MHC-bound foreign peptides are recognised by the T-cells which signal the induction of a cellular or humoral immune response. For the most part, foreign peptides produced by the degradation of intracellular pathogens (e.g. viruses) are bound by MHC class I molecules and presented to cytotoxic T-cells, eliciting a cellular immune response and killing of the cell. Alternately, foreign peptides produced extracellularly (e.g. most bacteria) are bound by MHC class II molecules and presented to helper T-cells which secrete cytokine mediators required for humoral (antibody), cytotoxic and inflammatory immune responses (Brodsky et al. 1999). Because the peptide binding ability varies by the composition of the peptide binding residues (PBR) of the MHC molecules, some pathogens may escape recognition by certain MHC molecules; this can lead to increased susceptibility of individuals homozygous for those alleles (Potts & Slev 1995). Alternatively, resistance to pathogens may be derived through high affinity binding of certain peptides by specific MHC alleles (Sidney et al. 1996). However, the strong potential for MHC-based disease associations is significantly reduced in most vertebrate species by the presence of multiple polymorphic MHC loci, thereby reducing the effects of individual allele binding repertoires and creating redundancy in the binding capabilities of the MHC as a whole (Potts & Slev 1995, Kaufman et al. 1995).

The recognition and binding of foreign peptides is the primary function of classical MHC genes. Classical MHC genes can be defined by key conserved N- and C-terminal residues in their peptide binding site as well as a high rate of non-synonymous mutation in other PBRs, their ubiquitous expression, high levels of polymorphism, and the retention of old allelic lineages achieved through balancing selection (Nei & Hughes 1991). The number of classical MHC classes I and II genes varies both within and among species. Within species, haplotype variation creates different numbers and compositions of MHC molecules contained within individuals. Among species, different levels of MHC gene duplication exist as MHC genes are repeatedly gained and lost through expansion and contraction of the MHC (Parham 1994, Miller et al. 2002). Although theoretically the peptide binding repertoire should increase with the number of MHC variants carried by an individual, there appears to be a limit to the number of MHC variants tolerated. This limit may stem from restriction of the T-cell repertoire through negative selection by MHC genes (Takahata 1995). In general, most species contain two to three classical MHC classes I and II loci. Species with higher numbers of MHC class I and/or class II loci use haplotype diversity to reduce the number of MHC variants per individual. However, seven or more expressed loci have been observed in some non-mammalian species (Murray et al. 2000, Sammut et al. 1999, Malaga-Trillo et al. 1998, Timon et al. 1998), and in one teleost species, Atlantic cod, as many as 42 expressed MHC class I loci have been isolated from a single individual (Miller et al. 2002).

Non-classical MHC genes are generally derived from classical genes but carry mutations in key residues that limit or eliminate their ability to bind peptides (Parham 1994). Non-classical MHC genes thus evolve new immune system functions, often playing a supportive role for classical genes or binding only a very specific set of peptides. Highly degenerate non-classical genes can even evolve functions that are not related to immunity at all. Non-classical MHC genes

are generally not polymorphic, and their expression is limited to certain tissues and/or specific stages of development. However, the distinction between classical and non-classical MHC genes can be obscure at times, as non-classical genes are continuously evolving from classical genes, and many intermediates exist.

Most MHC-based disease associations documented in mammals are relatively weak and are greatly influenced by background genes (Apanius et al. 1997). Conversely, in chickens, which contain a 'minimal essential MHC' consisting of single highly expressed classes I and II loci, a number of strong MHC-based disease associations exist (Kaufman & Salomonsen 1997). These observations led Kaufman and Salomonsen to speculate that the lack of redundancy in the chicken MHC results in stronger levels of pathogen-driven selection, a situation that ultimately may make chickens more susceptible to disease than most mammals. Interestingly, for the most widely recognised MHC-associated disease (Marek's disease of chickens), resistance is not due to a higher affinity binding of a pathogenic peptide to a specific MHC molecule. Instead, resistance derives from the reduced expression of the MHC resistance alleles, resulting in a stronger innate immune response for individuals carrying these alleles (Kaufman et al. 1995). However, other diseases of chickens, such as Rous sarcoma virus, have more classical associations with MHC molecules, whereby the differential binding capabilities of the MHC alleles favours resistance or susceptibility to the disease (Kaufman et al. 1995).

The number of expressed MHC genes varies greatly among the teleost fishes. Most early teleost species, or euteleosts, contain 1–3 classical MHC classes I and II genes, while extensive gene duplication in the neoteleost species has greatly expanded the numbers of MHC loci (Miller & Withler 1998, Miller et al. 2002). Salmon, a somewhat intermediate euteleost, contain only a single MHC class II gene (DAB), and thus far, only one confirmed classical class I gene (UBA) has been found (UBA, Shum et al. 2001, Aoyagi et al. 2002, Miller & Withler 1996). In addition, two potentially non-classical class I genes have been discovered. The 'B' locus shares homology with the classical UBA locus and its status as a classical or non-classical gene is yet to be conclusively established (termed B by Miller & Withler 1998 or U71 by Shum et al. 2002). Alternately, the 'UAA' locus is considered a non-classical locus as it is ancient and highly differentiated from all classical loci (Shum et al. 1999).

Finally, there exists at least one psuedogene that has likely lost its function altogether (UFA – Aoyagi et al. 2002; U41 – Shum et al. 2002). Our experience in the aquaculture of salmon and trout suggests that salmonid species are quite susceptible to disease, and variation in degrees of susceptibility exist among stocks (Withler & Evelyn 1990). Given the limited numbers of classical MHC genes found in salmon, there may exist a higher potential for strong associations of MHC alleles with resistance or susceptibility to disease, as observed in chickens. Thus, in the move towards mapping QTL markers in salmon for use in selective breeding programs, targetting the MHC genes in particular may yield useful QTL markers for disease resistance.

The fish rhabdovirus, infectious hematopoietic necrosis virus (IHNV), is responsible for extensive losses among several species of salmon and trout in many areas of the world (Bootland & Leong 1999, Wolf 1988). Typical of most rhabdoviruses, IHNV has a relatively broad host range and is among the oldest and best-studied viral pathogens of fish. Historically, IHNV was endemic among many populations of wild Pacific salmon in the western portion of North America. During the past several decades, the virus has spread to stocks of cultured rainbow trout, *Oncorhynchus mykiss*, in other areas of the United States, Asia and Europe, probably as a result of the movement of infected fish or eggs (Winton 1991). More recently, IHNV has been recovered from both salmonid and non-salmonid hosts in the Pacific Ocean (Kent et al. 1998, Traxler et al. 1997) suggesting that the virus is endemic among certain marine fish species as well.

IHNV causes a disease characterised by acute viremia with resulting haemorrhage and widespread necrosis of major organ systems (Wolf 1988). Depending upon the species and size of fish, strain of virus, and environmental conditions such as temperature, an outbreak of infectious hematopoietic necrosis (IHN) may result in explosive and untreatable losses that can approach 100% (Winton 1991, Wolf 1988). Currently, the only method for control of IHN is avoidance of exposure to the causative virus through the use of virus-free stocks of fish, pathogen-free water supplies, and strict sanitation procedures (Winton 1991, Wolf, 1988). Vaccines to control IHN are in the experimental stage (Winton 1997, 1998) as are genetic approaches based on the selection of disease resistant stocks (Parsons et al. 1986, Bootland & Leong 1999, Kono et al. 2000). The purpose of this study was to determine if alleles within loci of the MHC of Atlantic salmon were associated with differences in susceptibility to IHNV.

Methods

Fish

Approximately 400 juvenile (4–5 g) Atlantic salmon were obtained from Cyprus Island Inc. in Rochester, Washington, a commercial facility with no history of IHNV. The fish originated from an outbred population of broodstock maintained at the site, and both broodstock and juveniles were reared in pathogen-free freshwater. Fish were transported to the Western Fisheries Research Center in Seattle, WA and reared in 15°C, pathogen-free freshwater until used.

Virus

IHNV isolate 4814 was provided by R. Busch (Seattle, Washington). This isolate was recovered from Atlantic salmon reared in marine netpens in British Columbia and associated with mortality in this species. The virus was propagated using monolayer cultures of the epithelioma papillosum cyprini (EPC) cell line to produce a stock containing 6.32×10^6 plaque-forming units (pfu) ml^{-1} of tissue culture fluid.

IHNV Challenge

Fish were assigned randomly to provide eight replicate groups of 50 fish for the challenge experiment. Each group of 50 fish was placed in 2 l of pathogen-free, 15°C freshwater in a 5 l aquarium provided with constant aeration. For the trial, 30 ml of minimum essential medium (MEM) was added to each of two aquaria containing groups of fish that served as negative controls, while the remaining six groups of fish were challenged by addition of 30 ml of MEM adjusted to provide IHNV concentrations in each of two aquaria of 1.05×10^2 pfu ml^{-1} (low dose), 3.16×10^3 pfu ml^{-1} (medium dose) or 9.48×10^4 pfu ml^{-1} (high dose). After 20 min in the static bath challenge, water flow was resumed at $\sim$1 l min^{-1}. The fish were observed daily for mortality. Fish were not fed during the 28-day challenge period, and once each day mortalities were collected and the dorsal or caudal fins excised and placed into 70% ethanol until assayed.

MHC genotyping

MHC class II molecules are dimers consisting of an α chain and a β chain. As these chains are encoded separately on the chromosome and both contain exons with PBR (α1 and β1), both exons of the single MHC class II gene (DA) were analysed (DAA and DAB). MHC class I genes are dimers of an α chain and a β2-macroglobulin molecule. Because the two PBR-encoding exons are linked on the α-chain (α1 and α2), we limited our survey to the α1 exon of the class I B locus. Due to the presence of numerous highly divergent allelic lineages in the two peptide encoding exons of UBA, and the inherent difficulties in geno-typing highly differentiated alleles (seven to eleven primersets per exon are utilised in our laboratory), we utilised a microsatellite from the 3' untranslated region of the UBA locus for genotyping (Grimholt et al. 2002). The three exons and microsatellite from UBA were amplified separately using the primers and PCR conditions outlined in Tables 1 and 2. Alleles derived from the amplification of exons were differentiated by denaturing gradient gel electrophoresis (DGGE) with conditions outlined in Miller et al. (1999) and summarised in Table 1. Exons were amplified using GC-clamped primer sets. Polyacrylamide gels with parallel gradients (see Table 1 for locus-specific gradi-ents) were poured using a Gradient Maker™ (Model 475 Gradient Delivery System; Bio-Rad, Hercules, Calif). Three standard sets containing all known alleles for each exon were loaded into four positions on each gel and used to directly identify individual alleles. Two microliters of PCR product were loaded into each well, and the gels were run using the conditions outlined in Table 1 with the Dcode™ electrophoresis system (Bio-Rad). Gels were stained with ethidium bromide and alleles were scored directly from the allelic ladders. Allele identifications were confirmed for a subset of individuals by direct sequencing from bands excised from the DGGE gels, with at least three confirmations

Table 1. PCR and DGGE conditions for MHC loci surveyed. PCR was carried out in a total volume of 50 μl containing 0.2–0.5 μg of DNA, 15 pmol of each primer, 200 μM of each dNTP and 2.5 U UltraTherm™ DNA polymerase. A GC-clamp, denoted[CL], with the sequence 5'-GCCCGCCCCCGCGCCCTGCCCGCGCCCCGCGCCGCCCGCCC-3', was added to 5' end of the Sense primer(s) for the B locus and the antisense (as) primer for the DAB and DAA loci.

Locus	Primer	Primers used	PCR conditions	DGGE conditions	Standard alleles
UBA		UBA-3'UTs × UBA-3'UTas	94 deg/45″	ABI 377	12 alleles
	UBA-3'UTs	5'-GGAGAGCTGCCCAGATGACTT-3'	54 deg/45″		
	UBA-3'UTas	5'-GTTTCATTACCACAAGCCCGCTC-3'	70 deg/45″		
			30 cycles		
			72 deg/10'		
B		A1B-s2[CL] × A1B-I2as1	94 deg/1'	45%–60% denat.	9 alleles
	A1B-s2[CL]	5'-GCCCTGAAGTATTTCTACACTGCG-3'	50 deg/2'	7.5% acryl.	
	A1B-I2as1	5'-TGAGGTAAAGATGAAGGAC-3'	68 deg/2'	120V/6hrs @54 deg	
			35 cycles	&	
			72 deg/10'	45%–60% denat.	
				7.0% acryl.	
				120V/6hrs @55 deg	
DAB		B1s × B1as[CL]	94 deg/1'	45%–60% denat.	13 alleles
	B1s	5'-CCGATACTCCTCAAAGGACCTGCA-3'	50 deg/2'	7.5% acryl.	
	B1as[CL]	5'-GGTCTTGACTTG[AC]TCAGTCA-3'	68 deg/2'	60V/15hrs @56 deg	
			34 cycles	&	
			72 deg/10'	60V/15hrs @54 deg	
DAA		CIIA1s × CIIA1Ias[CL]	94 deg/1'	45%–60%	9 alleles
	CIIA1s	5'-CTCTTCTGGGTTCTTGTAAGC-3'	50 deg/2'	7.5% acryl.	
	CIIA1Ias[CL]	5'-GGTTTCTTTTCTCAGTTCTGCAT	71 deg/2'	120V/6hrs @54 deg	
			35 cycles	&	
			71 deg/10'	120V/6hrs @56 deg	

Abbreviations are as follows: denat. – denaturant, acryl. – acrylamide.

Table 2. Fisher's exact test for each challenge group and exact tests for stratified RXC tests.

	Low		Mid		High		Exact estimates		
	p	OR	p	OR	p	OR	p	$p_{corrected}$	OR
UBA									
1	NA	NA	NA	NA	1.0000	∞	1.0000	1.0000	∞
2	0.7769	0.812	0.4618	2.965	1.0000	1.400	0.7500	1.0000	1.237
3	1.0000	∞	0.0077	0.000	1.0000	∞	0.1622	0.8804	0.188
4	0.2413	1.573	0.0459	0.435	0.8296	0.889	0.7547	1.0000	0.912
5	0.7873	0.777	0.7665	1.406	0.7907	0.807	0.9018	1.0000	0.920
6	0.3855	0.678	0.0494	6.922	0.0454	∞	0.1722	0.8965	1.646
7	1.0000	1.250	1.0000	1.016	0.3806	0.463	0.7792	1.0000	0.773
8	0.2616	0.3526	1.0000	0.756	1.0000	1.282	0.5457	0.9999	0.691
9	0.5246	3.173	0.495	0.500	0.1734	0.000	0.9750	1.0000	0.680
10	0.3052	0.475	0.6877	2.102	0.6287	0.613	0.6706	1.0000	0.753
11	0.6204	1.395	0.6877	2.102	0.5215	0.705	0.8117	1.0000	1.177
12	1.0000	∞	1.0000	∞	1.0000	∞	0.4411	0.9991	∞
B									
1	0.2879	∞	1	∞	0.5813	∞	0.1439	0.7885	∞
2	1	1.048	0.8032	1.324	1	1.060	0.7743	1	1.131
3	0.8803	1.091	0.4537	0.739	0.7052	1.187	1	1	0.998
4	**0.1630**	**0.522**	**0.0484**	**0.371**	**0.0071**	**0.279**	**0.0008**	**0.0080**	**0.379**
5	1	1.123	0.6903	2.473	0.3472	∞	0.3168	0.9778	1.758
6	1	1.246	0.6735	0.784	0.4934	0.630	0.7536	1	0.799
7	0.8357	1.134	0.0303	7.077	0.2554	2.571	0.0595	0.4585	1.899
8	1	1.878	1	0.788	0.5057	0.518	1	1	0.995
9	1	0.615	NA	NA	NA		1	1	0.617
10	0.1448	0	1	1.333	1	∞	1	1	0.837
DAB									
1	1	1.164	0.3401	∞	0.5828	∞	0.3297	0.9945	2.213
2	**1**	**0.953**	**0.1882**	**0.596**	**0.0392**	**0.444**	**0.0550**	**0.5207**	**0.650**
3	0.4702	0.765	0.6472	1.377	0.0918	2.383	0.4894	0.9998	1.219
4	1	1.044	0.2666	0.543	0.0873	0.348	0.1368	0.8523	0.594
5	0.2418	1.751	1	1.028	0.1907	0.523	0.9084	1	1.067
6	1	∞	1	∞	NA	NA	0.8982	1	∞
7	1	0.691	1	∞	1	∞	1	1	1.757
8	0.4472	0.561	0.5948	0.461	0.1077	0.123	0.1111	0.7837	0.370
9	**0.5303**	**1.660**	**0.0728**	**∞**	**0.0736**	**∞**	**0.0084**	**0.1039**	**4.4620**
10	0.5303	1.660	0.2212	0.371	1	1.574	1	1	0.940
11	0.7183	0.682	0.3401	∞	0.0775	∞	0.1877	0.9330	2.283
12	0.3217	0.442	0.3401	∞	1	1.850	1	1	0.986
13	NA	NA	NA	NA	1	∞	1	1	∞
DAA									
1	1	1.109	0.2920	3.814	1	1.021	0.4979	0.9980	1.489
2	1	0.909	0.0729	0.371	0.086	5.976	1	1	1.002
3	0.3020	0.691	0.2714	0.657	0.1024	0.480	0.0354	0.2770	0.620
4	0.7233	0.723	0.5743	∞	0.239	0.423	0.6701	1	0.745
5	0.7031	0.842	1	0.998	0.2622	1.899	0.8017	1	1.098
6	0.0159	3.436	1	1.048	0.1376	0.429	0.5188	0.9986	1.246
7	0.6119	1.391	0.0216	∞	0.3752	2.321	0.0367	0.2857	2.369
8	0.6513	0.479	0.6088	0.528	1	1.138	0.5669	0.9995	0.639
9	0.6513	0.479	1	∞	1	1.375	1	1	0.876

Each table shows the statistical results for one of the four loci examined: MHC class I UBA, B and MHC class II DAB and DAA. Alleles with significant p-values less than 0.01 (although not necessarily significant under $p_{corrected}$) for the exact estimates are shown in bold, with frequencies in Figure 2.

312

per allele-type to check whether all alleles were, in fact, differentiated.

Statistical analysis

Tests for genic homogeneity and fit to Hardy–Weinberg expectations were performed using GENEPOP (V.3.1) (Raymond & Rousset 1995). Survival curves were estimated by Kaplan–Meier and compared by the log-rank test using SYSTAT V.8 (SPSS Inc., Chicago, IL). Association of MHC alleles to mortality or survival for each challenge group was tested by Fisher's exact tests (Svejgaard et al. 1974). To analyse allele association across the three challenge groups combined, we first determined whether the odds ratios (OR) for each allele across all three challenges were homogeneous. If homogeneity was not rejected, the exact confidence interval for the common OR was calculated and tested against the null hypothesis of unity. All exact tests were done using STATXACT V.4 (Cytel Software Inc.). When appropriate, p-values were corrected for multiple comparisons as $p_c = 1 - (1-p)^n$, where n = number of alleles analyzed at a locus (Svejgaard and Ryder 1994).

Results

Challenge study

The challenge study used three doses of IHNV to insure a range of responses. Both mortality and mean-day-to-death during the 28-day challenge period showed a strong dose-response (Figure 1). Fish dying during the experiment showed typical signs of IHN. At the end of the experiment, 39% of the fish survived the low dose, 27% survived the mid dose, and 20% survived the high dose.

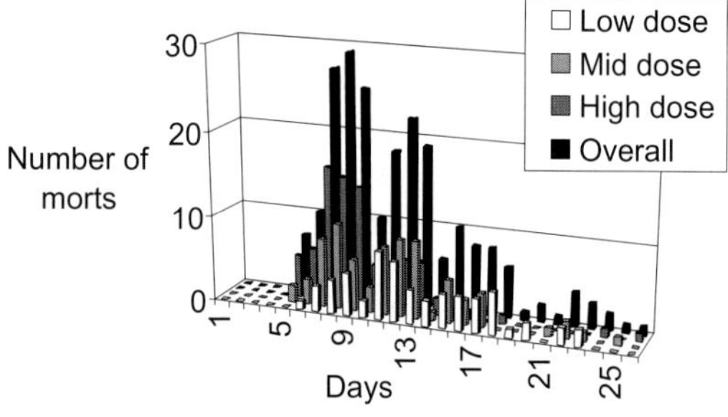

Figure 1. Challenge-dosage curve, depicting time, in days, to mortality after challenged with low, medium and high doses of IHNV. All mortality ceased after 25 days, and the experiment ended, and surviving fish sacrificed on Day 28.

Prior to the experiment, fish were randomly assigned to replicate and treatment groups. Genetic homogeneity was confirmed across all replicates and treatments (all loci combined p = 0.10). Each challenge dose (low, medium, and high titre) had two replicates. Comparisons of survival curves between the two replicates within each challenge dose detected no significant differences in survival rate (high dose, p = 0.45; medium dose, p = 0.337; low dose, p = 0.058); consequently, replicates were pooled for all further analyses.

MHC analysis

Expected heterozygosities were high for all three loci, with values of 0.81 for class I UBA, 0.76 for class II DAA, 0.79 for class I B, and 0.85 for class II DAB. Tests for Hardy–Weinberg equilibrium revealed a significant heterozygote deficiency at the class I UBA locus (p = 0.00; Ho 0.70), possibly indicative of the presence of null alleles or haplotype diversity at this locus. There was no statistically significant relationship with heterozygosity at any of the loci and susceptibility to IHNV.

Twelve class I UBA microsatellite alleles were detected, while DGGE typing of MHC exons detected, 13 alleles for class II DAB, 10 alleles for class I B, and 9 alleles for class II DAA. Each allele was tested for association to challenge outcome (survival or mortality) using Fisher's exact tests. Analysing each challenge dose separately, several loci were significant (p < 0.05) but none were highly significant and capable of withstanding p-value correction. Next, we evaluated each allele across all three challenges (strata) using a stratified RXC test. None of the alleles rejected homogeneity of OR across strata so the common odds ratio and the exact p-value were calculated for each allele. Using this approach, the class I *Sasa*-B-04 allele, which was three times more common in survivors than in the fish that died (morts) in the high dose group, was highly significant (p = 0.0008) and remained significant even after correction (p = 0.0080; Table 2 and Figure 2a). The OR for this allele was 0.38 indicating that the allele is protective. Sixty-one percent of the fish surviving the high dose of IHNV carried allele *Sasa*-B-04, whereas only 21% of fish that died in the high dose challenge group carried the allele. The class II *Sasa*-DAB-09 allele, which was only found in fish dying in the mid- and high-dose challenges, was also significant before correction (p = 0.0084) but was not significant after correction (p = 0.1039); the OR was 4.462 indicating that the allele is associated

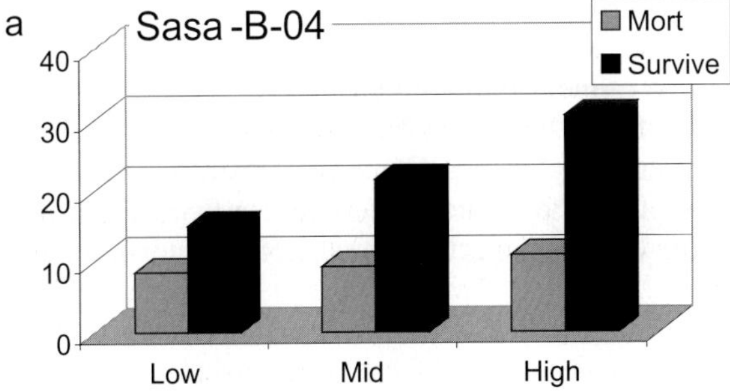

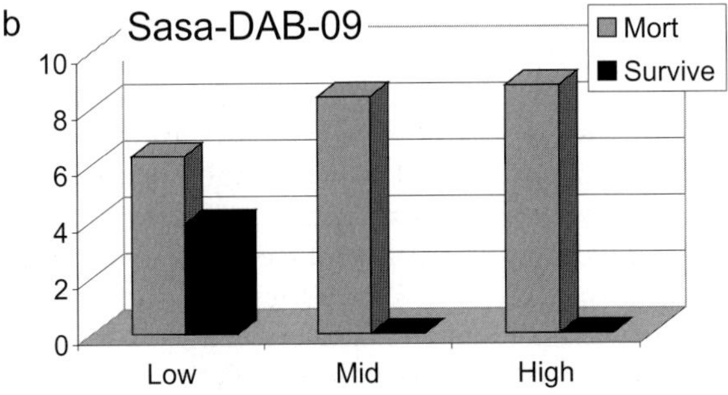

Figure 2. Frequency histograms for alleles with significant associations to IHN. Histograms show the percentage of survivors *versus* mortalities under each challenge dose, with the two replicates under each dose combined. (a) *Sasa*-B-04, highly significant association with resistance, (b) *Sasa*-DAB-09, suggestive allele not significant after p-value correction for multiple tests.

with increased susceptibility (Figure 2b). Additionally, the *Sasa*-DAB-02 allele, which was nearly significant before correction (p = 0.055, p = 0.52 after correction), was almost twice as common in the survivors as compared to the morts of the high dose (0.4 vs. 0.22) (not shown). None of the class I UBA alleles were significantly associated with susceptibility or resistance to IHN.

Discussion

Establishing associations of genetic loci with disease resistance can be daunting, given the vast number of potential gene candidates. With the sequencing of the human genome and the development of widely dispersed single nucleotide polymorphisms (SNPs), genome-wide scans can now be used to identify resistance markers to human pathogens through linkage disequilibrium mapping (Kruglyak 1999). However, this approach is not only very expensive, but requires detailed linkage maps containing numerous polymorphic markers spaced evenly throughout the genome, something that is not yet available for salmon. Another approach is to select potential candidate genes that are either mechanistically established to play a role in the disease etiology of a specific pathogen, or to simply choose genes that are associated with a number of other related diseases. We chose the latter approach, and examined three genes of the MHC in salmon. This approach seemed plausible, given the vast number of diseases at least weakly associated with MHC genes in a variety of species. The similarity of the simple MHC of salmon with that of chickens, for which MHC associations are strong, provided an even greater impetus to test this complex for associations in salmon.

The criteria to determine significance in MHC association studies is still under debate. Even with a scaled down number of candidate genes, the need for p-value corrections for multiple tests can require very large sample sizes to establish statistically significant associations (Svejgaard & Ryder 1994, Apanius et al. 1997). One approach around this is to conduct a study twice, the first time to establish a hypothesis for loci/alleles with associations, and the second to test the hypothesis (Hill 1991). In following this two-pronged approach, we present the data from the first experiments, conducted to establish a hypothesis, herein. Experience from the human HLA field suggests that associations with p-values between 0.01 and 0.05 are often not reconfirmed in subsequent studies (Svejgaard & Ryder, 1994). In this study, the *Sasa*-B-04 allele was associated with resistance to IHN and was observed at a frequency of 0.09 to 0.13 in the mid and high dose morts and 0.21 to 0.33 in the survivors. We used a conservative correction procedure but still detected a significant association for the *Sasa*-B-04 allele with a p-value less than 0.01.

A weak association of the MHC class II gene and IHNV resistance was documented in a recent study conducted on rainbow trout/cut-throat trout backcrosses (p = 0.024; Palti et al. 2001). Their study was based on the class II intron sequence between $\beta 1$ and $\beta 2$, as opposed to our analysis of the coding $\beta 1$ sequence, so unfortunately, no specific comparison of alleles is possible. Previous to their study, cutthroat trout were identified as being highly resistant to IHNV, while rainbow trout were susceptible, hence, they theorised that any allele conferring resistance should come from cutthroat trout. Contrary to their expectations, the allele associated with resistance was from the rainbow trout. As the p-values were

314

relatively low, this study requires confirmation using a greater number of families.

In most genetic associations, the MHC genes themselves do not cause the disease, they merely influence the susceptibility or resistance to the disease. Additional genes may also be involved in susceptibility to a particular disease, and may or may not interact with the MHC genes. Hence, associations of MHC alleles may depend upon the genetic background of individuals, and different strains or populations may carry different allelic associations (reviewed in Apanius et al. 1997). If MHC-based associations are to be used to enhance disease resistance for the aquaculture industry, it would be judicious not only to re-test preliminary associations on the populations or strains from which they were originally found, but also to test associations on other aquaculture strains before using them for selective breeding or vaccine development. The Atlantic salmon utilised herein have been derived over the past 30 years from multiple strains of Atlantic salmon, including two from Norway, three from the East Coast of Canada, and a single strain from Scotland. Although the strains have been extensively interbred, microsatellite analysis suggests that they most closely resemble the fish from one of the Norway strains which was added to the broodstock in the early 1990s (Withler & Miller, unpubl. data).

Associations of MHC with susceptibility to specific pathogens can also be derived through linkage disequilibrium with a susceptibility locus, and not the MHC gene itself (Bengtsson & Thomson 1981). In situations where there is not a causal relationship with disease, it can be difficult to discern whether associations are due to linkage or the associated gene. The statistical procedures used herein derives association values that are directly correlated with linkage disequilibrium values, such that the locus with the highest values should be in the closest linkage disequilibrium with the susceptibility locus, or may actually be the susceptibility locus (Svejgaard & Ryder 1994). Unlike the mammalian MHC, whereby class I and II genes reside on a single chromosome, the class II DAB/DAA, class I UBA and class I B genes of salmon each reside on separated chromosomes (Phillips et al. 2003). The *Sasa*-B locus contained the highest linkage disequilibrium values. One may surmise from this that the B locus is a susceptibility locus; however, as many other class Ib loci reside on the same chromosome and are closely linked to the B locus (ibid; Miller unpublished data), confirmation of this requires association analyses of

loci linked to B and establishment of the role of the class I B locus in the IHN disease etiology.

One cannot discount the possibility that the associations found herein are an artefact of strain or family differences in survival. Strain/population differences in susceptibility to a variety of pathogens have been documented in coho and chinook salmon (Withler & Evelyn 1990, Beacham & Evelyn 1992a,b, Balfry et al. 2001). A preliminary analysis of about half of the fish utilised herein suggested that the fish were derived from at least 30 families. This finding, and the fact that the *Sasa*-B locus was in Hardy–Weinberg equilibrium (hence no inbreeding effects), suggest that family effects may be minimal. In addition, no effort has been made by this particular hatchery to keep the strains of Atlantic salmon separate, hence, they are likely highly introgressed. However, in our next study, we will analyse the MHC composition of the three strains of Atlantic salmon used as broodstock and we will design crosses to examine both strain differences and allele-specific associations with susceptibility to IHN.

The MHC class I *Sasa*-B locus was originally isolated in gDNA (Grimholt et al. 1994, Miller and Withler 1998). There are no published cDNA sequences for the MHC class I *Sasa*-B locus, and it has not been shown to exist on northern blots, although it was not probed specifically (Aoyagi et al. 2002). For this reason, the B locus has not previously been considered a coding locus. However, recent work in three laboratories suggests that it is, in fact, present in cDNA but is expressed at lower levels than the UBA gene (Shum et al. 2002, K.M. Miller, unpubl. data, John Hansen, pers. commun., Center of Marine Biotechnology, Baltimore, Maryland). The question remains, is the B locus a classical MHC gene, or is it non-classical? Other than a potential relatively low level of expression, the B locus contains some features of a classical locus. It is highly polymorphic, with 10 alleles and a heterozygosity of 0.79 (n = 300) in Atlantic salmon α1, and the ratio of dN to dS in the PBR is greater than one for the α1 exon of Atlantic salmon (1.77 herein; unpubl. data), suggesting that selection maintains alleles with differential binding repertoires. However, a previous analysis of the B locus α2 exon in Pacific salmon did not yield a dN to dS ratio greater than one, and analysis of the key conserved N- and C-terminal peptide binding sites for classical loci suggests a lack of conformance of some residues (Shum et al. 2002). Hence, as the B locus has properties of both classical and non-classical loci, it is likely in the process of degenerating to a non-classical gene, but may still be capable of binding a reduced set of peptides, or at least still play a role in immune system function.

The information gained from this study can be used to conduct a second challenge experiment, this time using families derived from parents with known MHC genotypes, to verify linkage of the *Sasa*-B-04 allele with increased survival from IHNV challenge. In addition, we will specifically target alleles with nearly significant associations, such as *Sasa*-DAB-09 and -02. This second study will confirm the statistical relevance of the associations found herein for the mixed strain of Atlantic salmon used in this study. If associations are confirmed, further analysis of additional aquaculture strains may be in order.

Acknowledgements

The authors are grateful to Cypress Island Inc. for providing the Atlantic salmon used in this study and for helpful advice. This research was funded in part by the US Geological Survey and Fisheries and Oceans, Canada.

References

Aoyagi, K., J.M. Kijkstra, C. Xia, I. Denda, M. Ototake, K. Hashimoto & T. Nakanishi. 2002. Classical MHC class I genes composed of highly divergent sequence lineages share a single locus in rainbow trout (*Oncorhynchus mykiss*). J. Immunol. 168: 260–273.

Apanius, V., D. Penn, P. Slev, L.R. Ruff & W.K. Potts. 1997. The nature of selection on the major histocompatibility complex. Crit. Rev. Immunol. 17: 179–224.

Balfry, S.K., A.G. Maule & G.K. Iwama. 2001. Coho salmon *Oncorhynchus kisutch* strain differences in disease resistance and non-specific immunity, following immersion challenges with *Vibrio anguillarum*. Dis. Aquat. Org. 47: 39–48.

Beacham, T.D. & T.P.T Evelyn. 1992a. Population and genetic variation in resistance of chinook salmon to vibriosis, furunculosis, and bacterial kidney disease. J. Aquat. Anim. Health 4: 153–167.

Beacham, T.D. & T.P.T. Evelyn. 1992b. Genetic variation in disease resistance and growth to chinook, coho, and chum salmon with respect to vibriosis, furunculosis and bacterial kidney disease. Trans. Am. Fish. Soc. 121: 456–485.

Bengtsson, B.O. & G. Thomson. 1981. Measuring the strength of associations between HLA antigens and diseases. Tissue Antigens 18: 356–363.

Bootland, L.M. & Leong, J.C. 1999. Infectious hematopoietic necrosis virus. pp. 57–121. *In*: P.T.K. Woo & D.W. Bruno (ed.) Fish Diseases and Disorders, Vol. 3: Viral, Bacterial and Fungal Infections, CAB International, New York, NY.

Brodsky, F.M., L. Lem, A. Solache & E.M. Bennett. 1999. Human pathogen subversion of antigen presentation. Immunol. Rev. 168: 199–215.

Grimholt, U., I. Olsaker, C. de Vries Lindstrom & O. Lie. 1994. A study of variability in the MHC class II $\beta 1$ and class I $\alpha 2$ domain exons of Atlantic salmon (*Salmo salar*). Anim. Genet. 25: 147–153.

Grimholt, U., F. Drabløs, S.M. Jøorgensen, B. Høyheim & R.J.M. Stet. 2003. The major histocompatibility class I locus in Atlantic salmon (*Salmo salar* L.): polymorphism, linkage analysis and protein modelling. Immunogenetics 54: 570–581.

Hill, A.V.S. 1991. HLA associations with malaria in Africa: Some implications for MHC evolution. NATO ASI Ser, Vol. H 59. Molecular Evolution of the Major Histocompatibility Complex, Springer-Verlag, Berlin, Heidelberg.

Kaufman, J. & J. Salomonsen. 1997. The 'minimal essential MHC' revisited: Both peptide-binding and cell surface expression level of MHC molecules are polymorphisms selected by pathogens in chickens. Hereditas 127: 67–73.

Kaufman, J., H. Völk & H.J. Wallny. 1995. A 'Minimal essential MHC' and an 'unrecognized MHC': Two extremes in selection for polymorphism. Immunol. Rev. 143: 63–88.

Kent, M.L., G.S. Traxler, D. Kieser, J. Richard, S.C. Dawe, R.W. Shaw, G. Prosperi-Porta, J. Ketcheson & T.P.T. Evelyn. 1998. Survey of salmonid pathogens in ocean-caught fishes in British Columbia, Canada. J. Aquat. Anim. Health 10: 211–219.

Kono, T., M. Sakai & S. LaPatra. 2000. Expressed sequence tag analysis of kidney and gill tissues from rainbow trout (*Oncorhynchus mykiss*) infected with infectious hematopoietic necrosis virus. Mar. Biotech. 2: 493–498.

Kruglyak, L. 1999. Prospects for whole-genome linkage disequilibrium mapping of common disease genes. Nat. Genet. 22: 139–144.

Lamont, S.J. 1998. The chicken major histocompatibility complex and disease. Rev. Sci. Tech. Off. Int. Epizoot. 17: 128–142.

Malaga-Trillo E., Z. Zaleska-Rutczynska, B. McAndrew, V. Vincek, F. Figueroa, H. Sultmann & J. Klein. 1998. Linkage relationships and haplotype polymorphism among cichlid MHC class II B loci. Genetics 149: 1527–1537

Miller, K.M. & R.E. Withler. 1996. Sequence analysis of a polymorphic MHC class II gene in Pacific salmon. Immunogenetics 43: 337–351.

Miller K.M. & R.E. Withler 1998. The salmonid class I MHC: Limited diversity in a primitive teleost. Immunol. Rev. 166: 279–293

Miller, K.M., T.J. Ming, A.D. Schulze & R.E. Withler. 1999. Denaturing gradient gel electrophoresis (DGGE): A rapid and sensitive technique to screen nucleotide sequence variation in populations. BioTechniques 27: 1016–1030.

Miller, K.M., K.H. Kaukinen & A.D. Schulze. 2002. Expansion and contraction of major histocompatibility complex genes: A teleostean example. Immunogenetics 53: 941–963.

Murray B.W., P. Nilsson, Z. Zaleska-Rutczynska, H. Syltmann & J. Klein. 2000. Linkage relationships and haplotype variation of the major histocompatibility complex class I *A* genes in the cichlid fish *Oreochromis niloticus*. Mar. Biotechnol. 2: 437–448

Nei, M. & A.L. Hughes. 1991. Polymorphism and evolution of the major histocompatibility complex loci in mammals. pp. 222–247. *In*: R.K. Selander, A.G. Clark and T.S. Whittam (ed.) Evolution at the Molecular Level, Sinauer Associates Inc., Sunderland.

Palti, Y., K.M. Nichols, K.I. Waller, J.E. Parsons & G.H. Thorgaard. 2001. Association between DNA polymorphisms tightly linked to MHC class II genes and IHN virus resistance in backcrosses of rainbow and cut-throat trout. Aquaculture 194: 283–289.

Parham, P. 1994. The rise and fall of great class I genes. Seminars Immunol. 6: 373–382.

Parsons, J.E., R.A. Busch, G.H. Thorgaard & P.D. Scheerer. 1986. Increased resistance of triploid rainbow trout × coho salmon hybrids to infectious hematopoietic necrosis virus. Aquaculture 57: 337–343.

Phillips, R.B., A. Zimmerman, M.A. Noakes, Y. Palti, M.R. Morasch, L. Eiben, S.S. Ristow, G.H. Thorgaard & J.D. Hansen. 2003. Physical and genetic mapping of the rainbow trout major histocompatibility regions: evidence for duplication of the class I region. Immunogenetics 55(8): 561–569.

Potts, W.K. & P.R. Slev. 1995. Pathogen-based models favoring MHC genetic diversity. Immunol. Rev. 143: 181–197.

Raymond, M. & F. Rousset. 1995. GENEPOP (version 1.2): Population genetics software for exact tests and ecumenicism. J. Hered. 86: 248–249.

Sammut, B., L. Du Pasquier, P, Ducoroy, V, Laurens, A, Marcuz & A. Tournefier. 1999. Axolotl MHC architecture and polymorphism. Eur. J. Immunol. 29: 2897–2907.

Shum, B.P., R. Rajalingam, K.E. Magor, K. Azumi, W.H. Carr, B. Dixon, R.J.M. Stet, M.A. Adkison, R.P. Hedrick & P. Parham. 1999. A divergent non-classical class I gene conserved in salmonids. Immunogenetics 49: 479–490.

Shum, B.P., L. Guethlein, L.R. Flodin, M.A. Adkison, R.P. Hedrick, R.B. Nehring, R.J.M. Stet, C. Secombes & P. Parham. 2001. Modes of salmonid MHC class I and II evolution differ from the primate paradigm. J. Immunol. 166: 3297–3308.

Shum, B.P., P.M. Mason, K.E .Magor, L.R. Flodin, R.J.M. Stet & P. Parham. 2002. Structures of two major histocompatibility complex class I genes of rainbow trout (*Oncorhynchus mykiss*). Immunogenetics 54: 193–199.

Sidney, J., H.M. Grey, R.T. Kubo & A. Sette. 1996. Practical, biochemical and evolutionary implications of the discovery of HLA class I supermotifs. Immunol. Today 17: 261–266.

Svejgaard, A., C. Jersild, L.S. Nielson & W.F. Bodmer. 1974. HLA antigens and disease. Statistical and genetical considerations. Tissue Antigens 4: 95–105.

Svejgaard, A. & P. Ryder. 1994. HLA and disease associations: Detecting the strongest associations. Tissue Antigens 43: 18–27.

Takahata, N. 1995. MHC diversity and selection. Immunol. Rev. 143: 225–247.

Timon M., G. Elgar, S. Habu, K. Okumura & P.C.L Beverley. 1998. Molecular cloning of major histocompatibility complex class I cDNAs from the pufferfish *Fugu rubripes*. Immunogenetics 47: 170–173.

Traxler, G.S., J.R. Roome, K.A. Lauda & S. LaPatra. 1997. Appearance of infectious hematopoietic necrosis virus (IHNV) and neutralizing antibodies in sockeye salmon *Oncorhynchus nerka* during their migration and maturation period. Dis. Aquat. Org. 28: 31–38.

Winton, J.R. 1991. Recent advances in detection and control of infectious hematopoietic necrosis virus. Ann. Rev. Fish. Dis. 1: 83–93.

Winton, J.R. 1997. Immunization with viral antigens: Infectious hematopoietic necrosis. Dev. Biol. Standard. 90: 211–220.

Winton, J.R. 1998. Molecular approaches to fish vaccines. J. Appl. Ichthyol. 14: 153–158.

Withler, R.E. & T.P.T. Evelyn. 1990. Genetic variation in resistance to bacterial kidney disease within and between two strains of coho salmon from British Columbia. Trans. Am. Fish. Soc. 119: 1003–1009.

Wolf, K. 1988. Infectious hematopoietic necrosis. pp. 83–114. *In*: K. Wolf (ed.), Fish Viruses and Fish Viral Diseases, Cornell University Press, Ithaca, NY.

Environmental Biology of Fishes **69**: 317–331, 2004.
© 2004 *Kluwer Academic Publishers. Printed in the Netherlands.*

Quantitative trait loci analyses for meristic traits in *Oncorhynchus mykiss*

Krista M. Nichols[a], Paul A. Wheeler & Gary H. Thorgaard
School of Biological Sciences and Center for Reproductive Biology, Washington State University, Pullman,
WA 99164-4236, U.S.A. (e-mail: krista.nichols@noaa.gov)
[a]*Current address: National Oceanic and Atmospheric Administration, National Marine Fisheries Service,*
Northwest Fisheries Science Center, Conservation Biology Division, 2725 Montlake Blvd E,
Seattle, WA 98112, U.S.A.

Received 17 April 2003   Accepted 21 April 2003

Key words: rainbow trout, vertebrae, lateral line scales, pyloric caeca, fin rays, gill rakers, development

Synopsis

Meristic trait variation among species and populations has long been used as the basis for identification and classification of fishes. Within *Oncorhynchus mykiss*, there is considerable variation in meristic characters such as numbers of vertebrae, lateral line scales, fin rays, gill rakers, and pyloric caeca. In our laboratory the Oregon State University (OSU) rainbow trout and the Clearwater River (CW) steelhead trout clonal lines, produced by androgenesis, exhibit significant differences in values for meristic traits, making quantitative trait locus (QTL) analysis of these meristic characters possible. Our objective was to determine the number, location, and effects of QTL associated with meristic characters in order to test two hypotheses: (1) that QTL for different meristic traits co-localize to the same linkage group and (2) that meristic trait QTL co-localize to the same linkage group as a previously identified development rate QTL. Doubled haploid individuals, produced by androgenesis from sperm from an F_1 hybrid between the OSU and CW lines, were used to evaluate the joint segregation of each meristic phenotype and Amplified Fragment Length Polymorphic marker genotypes. Composite interval mapping revealed QTL for six of the seven traits analyzed. One QTL each for scales above the lateral line and for gill rakers co-localized to the same position. Only one QTL for scales above the lateral line co-localized to the same region as that for the development rate QTL, but a greater map resolution is necessary to determine if these loci are truly the same. We failed to detect pleiotropy for most meristic trait QTL. Our results suggest that different major loci are associated with variation in each meristic character and that the expression of these loci may be influenced by maternal and external environmental factors.

Introduction

Meristic characters are frequently used for the classification and identification of fishes (Kirpichnikov 1981, Behnke 1992). Counts of fin rays and spines, scales, gill rakers and pharyngeal teeth, vertebrae, and pyloric caeca are among the characters most commonly used for differentiation of species and populations. In the salmonids, scale counts have been most widely used for the differentiation of populations within species (Mottley 1934, Behnke 1992). In *Oncorhynchus mykiss* (rainbow and steelhead trout) the most notable differences among populations occur in counts of scales along the lateral line. Behnke (1992) noted consistently high numbers of scales along the lateral line in *O. mykiss* populations from the upper Sacramento River, the Oregon desert, and the upper Columbia River basin. Coastal *O. mykiss* are typically more coarse-scaled, possessing fewer pored scales along the lateral line than the fine-scaled *O. mykiss* found in interior basins (Behnke 1992). Despite the importance of these characters in the classification and identification of the fishes, the genetic basis of meristic trait variation has not been extensively studied.

Meristic trait values are often associated with differences in the rate of embryonic development

318

(Gabriel 1944, Barlow 1961, Garside 1966, MacGregor & MacCrimmon 1977a, Lindsey et al. 1984, Leary et al. 1984, Lindsey 1988, Ferguson & Liskauskas 1995). Many of these studies have involved modulation of development rate and meristic characters with changes in the environment, mostly temperature. With increasing temperature, development rate is generally accelerated and meristic trait characters generally have fewer elements. This observed negative association between development rate and number of elements has been attributed to the decrease in the time available for character differentiation before hatch or before the onset of the next stage of development (Barlow 1961). This association could be due to either pleiotropic effects of genes modulating both growth and element number or effects confounded by correlated changes due to common environment during embryogenesis. The fact that different meristic characters are similarly influenced by changes in the environment has led some to believe that the development of each of these characters is coupled to a single developmental process that affects each character similarly (Lindsey 1955, Smith & Bailey 1961, Fowler 1970). However, a genome-wide approach to study the number and overlap among loci associated with these traits has not yet been conducted to address this hypothesis.

Despite the lability of these characters with environmental changes, the heritabilities of most meristic characters are high (Kirpichnikov 1981). In rainbow trout, heritabilities of 0.84 for vertebrae, 0.67–0.93 for gill rakers on the upper first arch, and 0.23–0.93 for fin rays have been reported (Leary et al. 1985, 1992). Bergot et al. (1981) have reported a heritability of 0.43 for pyloric caeca in rainbow trout. Large amounts of additive genetic variance for meristic characters exist, but studies on the number and nature of genes involved in the differentiation and expression of meristic characters in fishes are few (Danzmann & Ferguson 1990, and references therein). A *Pgm-1* allozyme allele expressed only in the liver of one strain of rainbow trout was associated with both faster development rate and a decrease in the number of elements for meristic characters (Leary et al. 1984, 1985).

Our objective was to identify the number, position, and effects of genetic loci associated with meristic trait variation using quantitative trait loci (QTL) analyses in a cross between divergent clonal lines of *O. mykiss*. Clonal lines of *O. mykiss* facilitate QTL analyses of divergent developmental phenotypes (Young et al. 1996, 1998, Robison et al. 1999, 2001). With the identification of QTL associated with meristic trait differences in *O. mykiss*, we tested two hypotheses: (1) that genetic loci for each meristic trait are located in the same position, thus implying the same genetic loci influence more than one meristic character, and (2) that the major genetic locus associated with differences in development rate (Robison et al. 2001, Nichols unpublished) is the same as genetic loci identified for some meristic trait variation, suggesting a potential coupling of embryonic growth and differentiation.

Materials and methods

Crosses and culture

Isogenic *O. mykiss* strains produced by gynogenesis and androgenesis at Washington State University (Young et al. 1996, Robison et al. 1999) were used for characterization of meristic variation between clonal lines that were subsequently crossed for QTL studies. For clonal line characterization, common eggs were used to produce Oregon State University (OSU) and Clearwater River (CW) clones by androgenesis. Clones were produced at the same time and were reared in a common environment until 11 months of age. For QTL analyses, the OSU and CW clonal lines were crossed to produce F_1 hybrid clones. Doubled haploid (DH) progeny were produced by androgenesis using F_1 hybrid clone sperm. Briefly, outbred eggs were irradiated with gamma irradiation to destroy maternal nuclear DNA. Irradiated eggs were fertilized with F_1 hybrid sperm, and a heat shock was performed to inhibit the first cell cleavage, forming DH embryos with all-paternal nuclear DNA (Young et al. 1998, Robison et al. 2001). Doubled haploids were produced in December 1999 with eggs from two outbred females, in January 2000 with eggs from two females, and in February 2000 with eggs from three females. Fertilized eggs were incubated in vertical stack incubators in an 11°C climate-controlled chamber. The chamber was maintained in 24 h darkness except for periodic checks on the embryos and recirculation system. Embryos were maintained in incubators until absorption of the yolk-sac was complete, at which point fish were transferred to recirculating and flow-through systems to grow out for meristic trait analysis and other experiments. Doubled haploids fertilized in December 1999 and January 2000 were reared until September 2001 (herein referred to as large DH progeny), and those fertilized in February 2000 were reared until December 2000 (herein referred to as small DH progeny). At the termination of grow-out, clonal and DH fish were euthanized with a lethal dose of anesthetic. Fin clips were taken from DH and preserved in 95% ethanol for later genetic analysis. Whole fish were preserved in 10% formalin and subsequently stored in 40% isopropanol for meristic analyses. Doubled haploids were either stored individually or were tagged with identification numbers so that phenotypes and genotypes could be matched.

Meristic trait enumeration

Standard lengths (SLs) were measured on all preserved fish. Meristic trait values were enumerated only for those traits showing a significant difference between the OSU and CW clonal lines (Table 1). These traits included counts of vertebrae, lateral line scales, scales above the lateral line, pyloric caeca, anal fin rays, gill rakers on the upper first gill arch, and pectoral fin rays. Counts for all characters were enumerated as described below. Counts for OSU and CW clonal lines were made by one person (PAW). With the exception of pyloric caeca, all counts on DH progeny were also made by one person (KMN).

Vertebrae. All fish were autoradiographed at the Department of Radiology, College of Veterinary Medicine Teaching Hospital, Washington State University. Large fish were radiographed at a focal film distance (FFD) of 110 cm at 40 kVP for 20 ms; small fish were radiographed at FFD 110 cm at 40 kVP for 12.5 ms. Large fish were radiographed onto regular Kodak Lanex film, while small fish were radiographed onto Kodak Lanex Fine for better resolution of vertebrae. Vertebral counts were made on autoradiographs and were confirmed by naïve counts on all fish by the same individual a second time. Vertebrae were counted from anterior to posterior, including all elements with a definite suture from adjoining vertebrae. The most anterior vertebra counted was the first element with a definite suture separating it from the cranium. Some of the smallest fish could not be counted from autoradiographs and thus were given missing values for vertebrae number for QTL analysis.

Scales. Preserved DH were stained with Alizarin Red S (Young et al. 1995) for at least 24 h prior to counting of calcified structures such as scales, gill rakers, and fin rays. All scale counts were made on the left side of the fish. Lateral line scales were counted according to the methods of Hubbs & Lagler (1956), from the shoulder girdle to the end of the hypural plate. Lateral line scales were counted under a dissecting microscope from anterior to posterior, and then from posterior to anterior. An average of the two counts was taken for the phenotypic value used in QTL analyses. Scales above the lateral line were counted as the number of parallel rows of scales in a ventral and posterior direction from the anterior origin of the dorsal fin to, but not including, the lateral line. Scales above the lateral line were counted twice dorso-ventrally, and an average was taken for the phenotypic value. Individuals for which scales could not be reliably counted due to irregularities, interruptions, or missing scales were given missing values for QTL analyses.

Anal fin rays. Anal fin rays were counted at the base of the fin, beginning with the most anterior ray that reached the outer fin margin. This typically excluded counting of the first 2–3 short rays that did not reach the fin margin. The last two rays in the anal fin ray branch from the base, and were counted as one fin ray, as described by Hubbs & Lagler (1956). Vestigial rays

Table 1. Summary of means (±standard error of the mean (SEM)) for OSU and CW clone meristic traits. Only characters that exhibited significant differences are presented.

Trait	OSU (n = 10) (mean ± SEM)	CW (n = 7) (mean ± SEM)	Significance (ANOVA p-value)
Vertebrae	61.6 ± 0.6	65.6 ± 0.4	0.0030[KW]
Scales above the lateral line	25.1 ± 0.6	28.7 ± 0.8	0.0018
Lateral line scales	104.0 ± 3.5	132.8 ± 1.7	<0.0001
Anal fin rays	8.5 ± 0.8	10.6 ± 0.2	0.03[KW]
Average pectoral fin rays	14.1 ± 0.2	13.1 ± 0.2	0.0051
Average upper gill rakers	8.1 ± 0.1	7.6 ± 0.2	0.0377[KW]
Pyloric caeca	33.9 ± 1.4	39.7 ± 0.7	0.0027[KW]

KW = Significance by non-parametric Kruskal–Wallis test on ranked data, all other tests parametric.

could not be discerned from the first elongated ray on fins that were significantly eroded prior to preservation; individuals, in this case, were given a missing trait value for QTL analyses.

Pectoral fin rays. All elements in both the right and left pectoral fins were counted from the base of the fin. Pectoral fin ray counts used in QTL analyses included both an average and a sum of the counts from the right and left side.

Gill rakers. All ossified gill rakers on the upper part of the first gill arch on the left and right sides were counted, including vestigial rakers that stained with Alizarin Red S. Rakers were counted according to Hubbs & Lagler (1956), beginning at the top of the arch, and including all rakers to the angle where the upper and lower arches meet. If a raker was located directly on the saddle between the upper and lower arches, that raker was not included in the upper gill arch count. Both an average and a sum of counts were used in QTL analyses.

Pyloric caeca. To count pyloric caeca, the digestive tract was removed from the preserved fish and then elements were counted by dissecting and cutting each projecting caeca. In large fish, caeca could be enumerated without the aid of a dissecting scope. Caeca in small fish were enumerated under a dissecting scope. For some fish, caeca were degraded or too fragile to count reliably; these individuals were given a missing phenotype for QTL analysis.

Genotyping

Genomic DNA was extracted from fin clips with the Puregene DNA isolation kit (Gentra Systems, Minneapolis, MN, U.S.A.). Doubled haploid progeny were genotyped using Amplified Fragment Length Polymorphic (AFLP) markers as described by Vos et al. (1995) and as modified by Robison et al. (2001). Genomic DNA was digested with *Eco*RI and *Mse*I, and 20 *Eco*RI/*Mse*I primer sets were used to screen and genotype DH progeny for polymorphic markers. Marker names signify the six-cutter enzyme used, the +3 *Eco*RI selective bases, the +3 *Mse*I selective bases, followed by the fragment size (in base pairs) or arbitrary number assigned to the band, in that order.

Statistical analyses

Summary statistics. Because progeny were produced from different females, at different times, and were euthanized at different sizes, SL and meristic trait values were analyzed for significant differences among females and between large and small DH progeny using one-way analyses of variance (ANOVA). The SAS System was used for all statistical tests (SAS Statistical Institute, Cary, NC). Traits were tested for homogeneity of variance and departure from normality. Normally distributed traits were tested for significant differences among groups using parametric statistics (PROC GLM). Non-normally distributed traits were analyzed with both non-parametric tests on rank-transformed data (PROC RANK, PROC GLM) and parametric tests (PROC GLM) for comparison. Multiple comparisons of female means were made with Tukey–Kramer tests on raw and rank-transformed data. Tests for differences among fish euthanized at different times (large and small DH progeny) were made using nested females as the error term. Correlations coefficients were calculated between all pairs of traits (PROC CORR). Type I error rate was set at 0.05.

Linkage analysis. Prior to linkage analysis, AFLP marker data was pruned to include only those markers for which at least 70% of individuals were genotyped. Markers were tested for departure from expected Mendelian segregation ratios. Markers showing significant segregation distortion were checked for scoring errors and reliability, and were corrected or removed from the analysis if scoring was difficult. AFLP markers were ordered into linkage groups using Mapmaker for Mac v. 2.0 (Dr. Scott Tingey, Dupont Experimental Station, Wilmington, DE, U.S.A.) for DH data using the Kosambi map function. Markers were grouped into linkage groups at a minimum Log of the Odds (LOD) of 3.0, and maximum theta of 0.40. A framework of a maximum of eight markers in each linkage group was ordered using the 'compare' command. Any additional markers in linkage groups were added with the 'try' command. Orders of all linkage groups were checked with Mapmaker/EXP (Lander et al. 1987) using the 'ripple' command, with a window size of seven markers. The best and final linkage group order was determined by comparing all possible orders within uncertain regions identified by ripple, within the framework of correctly ordered markers given the data. Syntenies of this map with our more dense rainbow trout map (OSU × Arlee; Young et al. 1998, Nichols et al. 2003) were determined based upon shared AFLP markers.

QTL analysis. QTL analyses were performed by composite interval mapping (Zeng 1994) using QTL Cartographer (Basten 1994, 2002), to test whether additive effects were significantly different from zero at every 2 cM interval of the mapped genome. Background control markers in the analysis were selected by stepwise linear regression using forward stepwise backward elimination with Type I error rates for inserting or eliminating a marker set at 0.1. The five most significant markers in this stepwise linear regression were used as background control markers. Window size for inclusion of background markers was 10 cM. Significance thresholds for likelihood ratio (LR) tests were determined by 1000 permutations of the trait data for each analysis (Churchill & Doerge 1994, Doerge & Churchill 1996). The p = 0.05 and p = 0.01 thresholds were ascertained from the upper tails of the distributions of maximum LRs obtained from the permutations. Likelihood ratios were converted to LOD scores for visualization of results and determination of LOD support intervals. Linkage groups and QTL positions with 1 LOD support intervals were drawn with MapChart v. 2.1 (Voorrips 2002). The position and 1 LOD support intervals for each locus were compared to determine if QTL for each meristic character overlapped with other meristic QTL, or overlapped with the interval containing the development rate QTL.

Results

Summary of meristic trait counts

Seven meristic traits exhibited statistically significant differences between individuals from the OSU and CW clonal lines (Table 1) thus allowing QTL analyses of meristic trait variation in progeny of these two clonal lines. The CW line had a greater number of vertebrae, scales above the lateral line, scales along the lateral line, anal fin rays, and pyloric caeca than OSU. The OSU line had a greater number of pectoral fin rays and upper gill rakers than CW.

For the most part, DH progeny had intermediate meristic counts when compared to OSU and CW, with the exception of the number of pyloric caeca; some differences in means were observed among families (Tables 1 and 2). Within-individual coefficients of variation for counts of scales along the lateral line were 0–3.7% and for scales above the lateral line were 0–10%. Results from parametric and non-parametric tests were similar. There were no significant differences among progeny produced from different females fertilized at the same time for all traits except lateral line scales. Lateral line scale counts were significantly different among females both within and between DH families produced and euthanized at different times (data not shown). For all other traits, significant female differences were observed only between DH that were euthanized at different times (and sizes), thus DH progeny produced from different females were pooled

Table 2. Summary of meristic trait means (±standard error of the mean (SEM)) and differences between large and small DH progeny used for QTL analyses.

Trait	Small DH progeny (n = 42)		Large DH progeny (n = 57)		Significance (ANOVA p-value)
	N	Mean ± SEM	N	Mean ± SEM	
Standard length	42	72.8 ± 2.8	57	193.2 ± 6.5	<0.0001
Vertebrae	24	63.2 ± 0.3	57	63.6 ± 0.2	0.1619
Scales above the lateral line	40	25.8 ± 0.3	18	26.9 ± 0.5	0.0771
Lateral line scales*	36	120.0 ± 0.8	55	124.0 ± 0.5	<0.0001
Anal fin rays	42	10.0 ± 0.1	56	10.0 ± 0.1	0.9841
Average pectoral fin rays	42	13.7 ± 0.1	56	14.5 ± 0.1	<0.0001
Average upper gill rakers	42	7.0 ± 0.1	57	7.7 ± 0.1	<0.0001
Pyloric caeca	23	44.3 ± 2.6	49	46.1 ± 1.5	0.2632

Significance values were achieved by one-way ANOVA on ranked data, except for scales above the lateral line, which were obtained from a parametric ANOVA.

*Lateral line scale differences are attributed to differences among females within and between large and small DH groups. For all other traits, no significant differences were observed among females within the large and small DH groups.

within groups euthanized at the two different time points (referred to as the small and large DH). There was a significant difference between the large (n = 57) and small (n = 42) DH progeny in SL, as well as for average number of pectoral fin rays and average number of gill rakers (Table 2). QTL analyses of vertebrae, scales above the lateral line, anal fin rays, and pyloric caeca were conducted on pooled data from all females as well as pooled within large and small DH progeny groups since no significant differences were observed among females. QTL analyses of pectoral fin ray and gill raker numbers were conducted separately for the large and small DH. Within female, and within size-group residuals were also used for QTL analysis of traits exhibiting significant or notable differences between DH euthanized at different times. Within-female residuals from one-way ANOVA were used for lateral line scale QTL analysis.

Significant correlations were observed among some traits for the DH progeny (Tables 3 and 4). SL was significantly correlated with number of lateral line scales, pectoral fin rays, and gill rakers on the upper gill arch when large and small DH progeny were combined (Table 3). These are the same traits for which significant differences were observed between large and small DH progeny (Table 2). The significant correlations between SL and meristic element counts in combined analyses were not found in separate analyses of large and small DH (Table 4). Significant correlations observed between SL and each of these traits in the combined analysis appeared to be due to differences in meristic counts between the large and small groups. Significant correlations were observed between lateral line scale and vertebrae numbers in both combined and large DH analyses, but this correlation was not found in small DH (Tables 3 and 4).

Table 3. Spearman rank correlations between meristic trait values in DH progeny.

	Vertebrae	Lateral line scales	Scales above l.l.	Anal fin rays	Pyloric caecae	Avg. pectoral rays	Avg. upper gill rakers
SL	0.03	0.459***	0.154	−0.055	0.216	0.393***	0.585***
Vertebrae		0.359**	0.348*	0.062	0.171	0.17	0.022
Lateral line scales			0.284*	−0.153	0.031	0.206	0.155
Scales above l.l.				−0.054	−0.057	0.033	−0.028
Anal fin rays					−0.225	0.125	−0.029
Pyloric caecae						0.052	0.158
Avg. pectoral rays							0.369**

Correlations significantly different from zero are denoted with asterisks: ***p < 0.0001, **p < 0.01, *p < 0.05.
SL = standard length; l.l. = lateral line; Avg. = average.

Table 4. Spearman rank correlation coefficients between meristic trait values for small (above diagonal) and large (below diagonal) DH progeny.

Large DH	Small DH							
	SL	Vertebrae	Lateral line scales	Scales above l.l.	Anal fin rays	Pyloric caecae	Avg. pectoral rays	Avg. upper gill rakers
SL		−0.167	0.236	−0.193	−0.230	0.189	−0.109	0.182
Vertebrae	0.073		0.356**	0.074	−0.054	0.214	0.199	0.023
Lateral line scales	0.257	0.354		0.030	−0.149	0.106	0.043	−0.036
Scales above l.l.	−0.037	0.457*	0.231		0.222	−0.262	0.071	−0.098
Anal fin rays	0.250	0.280	−0.137	−0.164		−0.163	0.156	0.012
Pyloric caecae	0.047	−0.023	−0.402	0.061	−0.351		0.114	0.201
Avg. pectoral rays	0.083	−0.090	−0.051	−0.160	0.095	−0.260		0.092
Avg. gill rakers	0.292	−0.322	−0.094	−0.215	−0.083	−0.078	0.203	

Coefficients significantly different from zero are noted with asterisks: ***p < 0.0001, **p < 0.01, *p < 0.05.
SL = standard length; l.l. = lateral line; Avg. = average.

All other significant correlations among meristic characters in the combined analysis appear to be due to differences between the size groups, as most intertrait correlations disappear in separate analysis of large and small DH progeny. No significant correlation was observed between the time to hatch (embryonic development rate) and any meristic character for the large DH progeny (data not shown); time to hatch data was not collected on the small DH progeny. The fact that some correlations are dependent upon size-group further corroborates the significant differences observed between the large and small DH means, and reinforces the necessity to conduct separate and residual QTL analyses for traits exhibiting dependence on size-group.

Genetic linkage map

With a total of 238 molecular markers and the sex phenotype, linkage analyses revealed 29 linkage groups with three or more markers and six marker pairs. Only those linkage groups with significant QTL are depicted in Figures 1 and 2. The total map length is 1 233 cM with an average intermarker distance of 6.8 cM (1 233 cM/181 map intervals (non-zero cM)). Seventeen markers remained unlinked in this analysis.

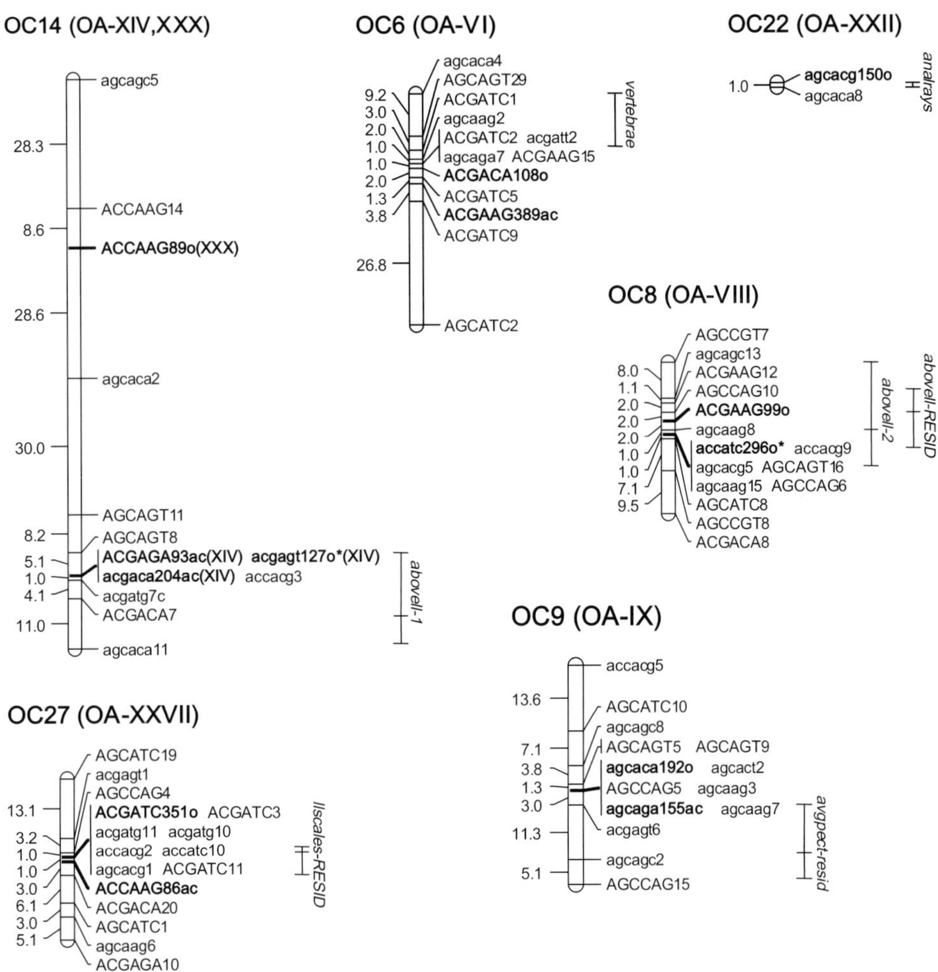

Figure 1. OSU × CW (OC) linkage groups with meristic QTL with all DH progeny. Numbers on the left are cM distances between markers. Syntenies among OC and our reference map (OSU × Arlee; OA) linkage groups are indicated in parentheses. Markers in bold are AFLP markers conferring synteny. Meristic QTL position and 1 LOD support intervals are indicated to the right of linkage groups. RESID denotes QTL identified using within-female residuals from one-way ANOVA for QTL analyses.

324

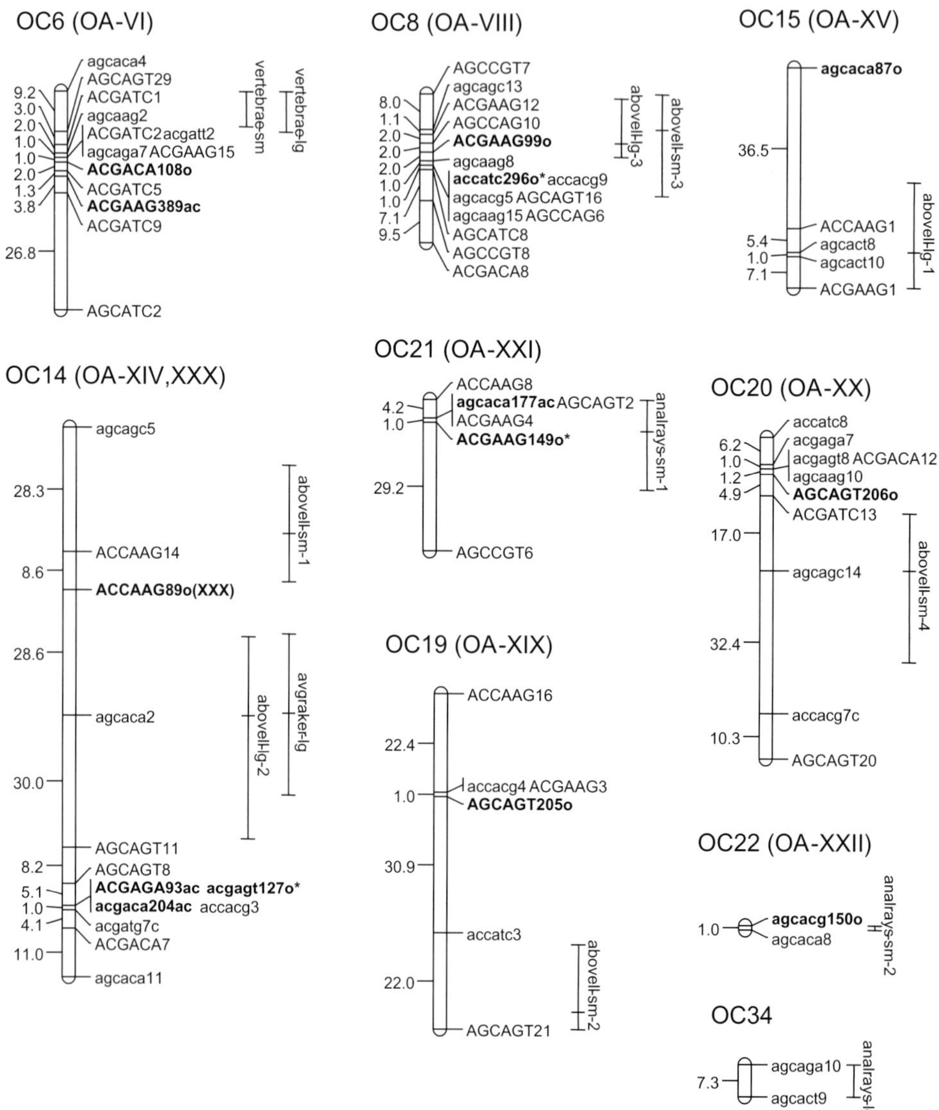

Figure 2. OSU × CW (OC) linkage groups with meristic QTL in separate analyses of small and large DH progeny. QTLs for large DH progeny are denoted with 'lg' and small with 'sm'. Separate analyses were not completed for lateral line scales. Numbers on left are cM distances. Syntenies among linkage groups in this cross and our reference OSU × Arlee (OA) map are indicated in parentheses, and are based upon shared AFLP markers shown in bold.

Since CW has 58 chromosomes (C. Ostberg, personal communication), and OSU has 60 (Ristow et al. 1998), we would expect to see 29 linkage groups, as two acrocentric chromosomes of OSU would pair with one metacentric chromosome of CW.

All but one of the linkage groups that showed significant QTL for meristic characters were matched to linkage groups in our more dense rainbow trout map (Young et al. 1998, Nichols et al. 2003) with syntenic AFLP markers. It is unclear whether this marker pair, OC34, is a smaller part of one of the larger groups identified in this study, or a small part of an independent linkage group that cannot be matched with the two markers on that group at this time. One linkage group in this cross (OC14), upon which several QTL were localized, contains markers from two linkage groups as determined by synteny with our published map. Linkage analysis in this study was unable to separate these two groups, but analysis with greater numbers of progeny and markers in another OSU × CW cross for development rate determined that these two linkage groups are separated (data not shown).

QTL analyses

At least one QTL was identified for each of six meristic characters analyzed in this cross; no QTL were identified for pyloric caeca (Tables 5 and 6). Six different QTL were revealed for five traits when all progeny were used in combined and residual analyses (Table 5, Figure 1). Separate analyses for large and small DH progeny revealed several QTL, some overlapping, and are summarized by trait below (Table 6, Figure 2).

Vertebrae. One QTL was identified for vertebrae number from combined analyses of small and large DH progeny. *Vertebrae* is located on OC6 and accounts for 23% of the variation in vertebrae number (Figure 1, Table 5). In separate analyses, *vertebrae sm* and *vertebrae lg*, for the small and large DH progeny, respectively, were localized to the same linkage group as that for the combined analysis (Figure 2). The additive effects at this locus were positive, indicating that

the substitution of a CW allele for an OSU allele at this QTL would result in an increase in the number of vertebrae, which is what would be expected based on the differences observed in the parental lines (Table 1).

Scales above the lateral line
A combined analysis of raw data for scales above the lateral line revealed two significant QTL (*abovell-1* and *abovell-2*); residual analysis (both within female and within size-group) revealed one QTL (*abovell-RESID*), which is the same locus as *abovell-2* revealed with the raw data based on overlap of LOD support intervals and similar additive effects (Figure 1). *Abovell-1* accounted for 15.9%, *abovell-2* for 27.5%, and *abovell-RESID* for 22.1% of the variation observed in above lateral line scale number (Table 5). *Abovell-1* is located on OC14 and *abovell-2* and *abovell-RESID* are located on OC8 (Figure 2). In separate analyses for scales above the lateral line in small and large DH, two QTL were identified and localized to OC8 and OC14. These are the same linkage groups associated with scales above the lateral line in the combined analyses (Figure 2). One additional QTL was identified in the large DH progeny on OC15, and two additional QTL were found in the small DH progeny on OC19 and OC20 (Figure 2). Consistent with the combined analysis, the QTL found on OC8 accounted for the largest amount of variance (32% and 42%) in scales

Table 5. QTL for DH progeny in combined QTL analyses of large and small DH groups. Horizontal lines separate QTL that are different, based on localization to different linkage groups and lack of overlap among 1 LOD support intervals.

QTL	Linkage group	Position[a]	LOD[b]	r^{2c}	a[d]
abovell-1	**OC14**	**117.9 (acgaca7)**	**3.57***	**0.159**	**−0.926**
abovell-2	**OC8**	**15.1 (agcaag8)**	**6.64****	**0.275**	**1.19**
abovell-RESID	**OC8**	**11.1 (agccag10)**	**5.39****	**0.221**	**1.00**
analrays	**OC22**	**0.01**	**4.43****	**0.134**	**0.278**
avgpect-RESID	OC9	38.8 (acgagt6)	3.47*	0.115	0.248
llscales-RESID	OC21	16.3 (agccag4)	4.73**	0.166	1.59
vertebrae	**OC6**	**0.01 (acgatc1)**	**6.15****	**0.230**	**0.637**

QTL in bold are the same as those identified in separate analyses of large and small DH. *RESID* denotes QTL detected using within-female residuals for QTL analyses.

[a]Position denotes the position from the top of the linkage group in cM, and in parentheses, the closest marker to the peak of the QTL. Note that the closest marker for the *analrays* QTL cannot be determined with current marker coverage on that linkage group.

[b]LOD = Log of the Odds that the calculated additive effects are significantly different from zero. Significance based on permutation tests is denoted as follows: *p < 0.05, **p < 0.01; p < 0.0001 were not reported as only 1000 permutations were completed.

[c]r^2 is the proportion of the variance in the trait explained by the given QTL after accounting for background markers.

[d]a = additive effect, or the effect of substituting a maternal allele (OSU) with a paternal allele (CW).

Table 6. QTL for large and small doubled haploid (DH) progeny in separate analyses. lg = QTL identified in large DH progeny, sm = QTL identified in small DH progeny. Lines separating QTL indicate significantly different loci based upon non-overlapping 1 LOD support intervals.

QTL	Linkage group	Position[a]	LOD[b]	r^{2c}	a^d
abovell-lg-1	OC15	36.5 (agcact8)	5.37*	0.114	0.898
abovell-lg-2	**OC14**	**65.5 (agcaca2)**	**6.61****	**0.171**	**−1.32**
abovell-sm-1	**OC14**	**24.0 (accaag14)**	**4.42****	**0.235**	**−1.06**
abovell-lg-3	**OC8**	**11.1 (AGCCAG10)**	**9.64****	**0.417**	**1.53**
abovell-sm-3	**OC8**	**8.0 (agcagc13)**	**6.75****	**0.320**	**1.23**
abovell-sm-2	OC19	72.3 (agcagt21)	6.02**	0.272	−1.13
abovell-sm-4	OC20	30.3 (agcagc14)	5.07**	0.216	1.29
analrays-lg	OC34	6.0 (acgatg8)	3.26*	0.169	−0.273
analrays-sm-1	OC27	7.2 (ACGAAG149o)	3.64*	0.191	0.401
analrays-sm-2	**OC22**	**0.01**	**4.27***	**0.216**	**0.433**
avgraker-lg	OC14	64.9 (agcaca2)	4.72**	0.260	0.293
vertebrae lg	**OC6**	**0.01 (AGCAGT29)**	**5.62****	**0.297**	**0.784**
vertebrae sm	**OC6**	**0.01 (AGCAGT29)**	**3.34***	**0.408**	**0.948**

QTL in bold are the same as those identified in combined and residual analyses. Lateral line scales were not analyzed in this manner, since significant female differences could not be attributed to size.

[a]Position denotes the position from the top of the linkage group in cM, and in parentheses, the closest marker to the peak of the QTL. Note that the closest marker for the *analrays-sm-2* QTL cannot be determined with current marker coverage on that linkage group.

[b]LOD = Log of the Odds that the calculated additive effects are significantly different from zero. Significance based on permutation tests is denoted as follows: *p < 0.05, **p < 0.01; ***p < 0.0001 are not reported as only 1000 permutations of the data were completed.

[c]r^2 is the proportion of the variance in the trait explained by the given QTL after accounting for background markers.

[d]a = additive effect, or the effect of substituting a maternal allele (OSU) with a paternal allele (CW) at this QTL.

above the lateral line in the separate analyses (Table 6). The QTL on OC8 had a positive additive effect, but that localized to OC14 had a negative additive effect. *Abovell-2, abovell-RESID, abovell-sm-3,* and *abovell-lg-3* – all the same locus for scales above the lateral line – are the only QTL that localized to the same linkage group and same region as that for development rate (Robison et al. 2001, K. Nichols, unpublished data).

Lateral line scales. Mapping of QTL for lateral line scales using within-female residuals revealed one QTL, *llscales-RESID* located on OC21. This QTL accounts for 16.6% of the variation observed in lateral line scale number. *Llscales-RESID* has a positive additive effect, indicating that substitution of an OSU allele with a CW allele at this locus would cause an increase in scale number as expected based on OSU and CW scale numbers.

Anal fin rays. A combined analysis of all DH progeny for anal fin ray counts revealed one significant QTL. This QTL, *analrays,* is located on OC22, a small linkage group with only two markers (Figure 1). This QTL accounted for ~13% of the variation observed in the number of anal fin rays (Table 5). In separate analyses, only one QTL for the small DH co-localized to the same position as that in the combined analysis (*analrays-sm-2*). One additional QTL was found in the large DH progeny, as well as in the small DH progeny. The negative allelic effect of *analrays-lg* is contrary to that expected by the differences in the means between the parental lines, but the combined and small DH QTL for anal fin rays indicate a positive allelic effect. The QTL identified for anal fin rays in the separate analysis were only slightly greater than the LOD threshold determined by permutation tests (0.01 < p < 0.05; Table 6).

Pectoral fin rays. One significant QTL was identified using residuals for pectoral fin rays. The QTL, *avgpect-RESID* is located on OC9 and accounts for 11.5% of the variation observed in the number of pectoral fin rays. The same result was obtained for both average and sum of pectoral rays, using residuals both within females and within size-groups for analyses. This QTL had a positive additive effect indicating that substitution of a CW allele for an OSU allele at this locus would result in an increase in the number of pectoral fin rays. This result is opposite of that expected based on OSU and CW values for this trait. No QTL were detected in separate analyses of large and small DH.

Upper gill rakers. In separate analyses of QTL for gill raker numbers, only the large DH progeny analysis revealed a significant QTL, *avgraker-lg* (Figure 2). This QTL is located on linkage group OC14 and accounts for 26% of the variation observed in gill rakers on the upper gill arch (Table 6). The additive effect at this locus is 0.293 and is opposite in sign of that expected since OSU has a greater mean number of gill rakers on the upper gill arch than CW. This QTL is located on the same linkage group as two other QTL (Figure 2), and localized to the same position as that of *abovell-lg-2*. The same result was obtained in QTL analyses conducted on the sum of the right and left gill raker numbers. No QTL were identified when residuals for average numbers of gill rakers, both within females and within size-groups, were used.

Pyloric caeca. No significant QTL were identified for pyloric caeca numbers in combined or separate analyses.

Discussion

In this first QTL analysis of meristic trait variation in any salmonid, we identified several genome regions that are associated with differences in the number of lateral line scales, scales above the lateral line, vertebrae, anal fin rays, and gill rakers. Despite significant differences in the parental lines for numbers of pyloric caeca, we failed to detect any QTL segregating in these DH progeny. These results are best discussed in light of the timing of the onset, duration, and completion of the development of these meristic characters in salmonids.

Both vertebrae and scales are associated with the early development of the anterior–posterior body axis (Lindsey 1988 and references therein). Vertebrae are among the first characters to develop. Vertebrae development follows the spacing of developed somites and is completed before hatch (Lindsey 1988). Lateral line scales are the first scales to appear during development; the appearance of scale papillae begins concomitantly with the development of lateral line organs before

hatch. The number and spacing of lateral line organs and early scale precursors are determined by the spacing of myomeres. In brown trout, Parrott (1934) reported that body scalation began at 2.4–2.8 cm in length at a focus along the lateral line just below the adipose fin (Parrott 1934). From this focus, development proceeds in all directions and fully developed scales over the entire body are not observed until well into the juvenile phase (Lindsey 1988). The fact that lateral line scale and vertebrae numbers are significantly correlated is not surprising since both arise during development of the body axis. The correlation between vertebrae and lateral line scales number in our study is consistent with the highly significant correlation among the same characters observed in natural populations (MacGregor & MacCrimmon 1977b). The significant correlation among these characters could either be due to pleiotropy or a common environment that would influence the different genes for these traits in a similar fashion. Despite this highly significant correlation, none of the QTL identified for scale numbers, either along the lateral line or above the lateral line, were localized to the same linkage groups as that for vertebrae number. This suggests that different major genes play a role in the development of each of these characters. The correlation among the traits thus may be an artifact of the similar effects of the environment on the genes for each of the traits, or the pleiotropic action of small-effect loci that were not detected in this study. In this DH design, the phenotypic correlation among these traits cannot be separated into components of genetic and environmental covariance.

Gill rakers and median and paired fin rays arise independently of the body axis (Lindsey 1988). Gill rakers arise from the dermal skeleton of the gill arches, developing initially with a cartilaginous core. Ossification of cartilaginous gill rakers proceeds from elements near the angle of the gill arch to the most dorsal and ventral elements in the upper and lower gill arches, respectively (Lindsey 1988). The completion of gill raker development and ossification does not occur until well into the juvenile phase (Lindsey 1988) and may explain the significant correlation of gill raker number with SL. In this study, only ossified structures that stained with alizarin were counted. Some of the smallest fish may not have completed ossification of gill rakers, and as a result, cartilaginous gill rakers would not have been counted. This may explain why no QTL could be detected for this trait in the small DH progeny.

Median and paired fins are controlled by internal supports that begin to develop prior to hatch. Muscles and internal supports for the rays and spines of all fins develop well before the appearance of external rays and spines. The development of fin rays is not complete until well into the juvenile phase, and in some cases, fish are preserved prematurely before ray development is completed (Lindsey et al. 1984). The significant difference in the mean number of pectoral fin rays between the large and small DHs, combined with the significant correlation among SL and pectoral fin rays in a combined correlation analysis suggests that this may be the case for the small fish. However, even in a separate QTL analysis for the large fish, in which ray development would have been completed, we failed to detect any major gene regions associated with the pectoral fin rays. One QTL was identified in the analysis of residuals for pectoral fin rays, and the lack of detection in the separate analyses may be due to small sample sizes. Size-dependent differences were not observed for anal fin rays.

The pyloric caeca are finger-like outcrops of the intestine and provide an increased surface area for the absorption of digested food (Lindsey 1988, Buddington & Diamond 1986). In *O. mykiss*, the number of caeca is fixed during the juvenile phase, by the time fish reach a length of ~40 mm (Northcote & Paterson 1960, Bergot et al. 1981). In this study, the average lengths of small (74 mm) and large (193 mm) DH were much greater than 40 mm. The pyloric caeca, unlike most of the other traits counted for this study, may have a more apparent link to fitness-related traits in rainbow trout. Bergot et al. (1981) observed that fish with greater numbers of pyloric caeca had better food conversion rates and greater weights than those fish with fewer pyloric caeca. The fact that no QTL for pyloric caeca were identified in this study suggests that pyloric caeca number is modulated by more than a few major loci in this cross, and the power for this study was not great enough to detect the potentially greater numbers of genes with smaller effects.

It is not surprising that major QTL for most of these traits do not co-localize to the same linkage group, since these meristic characters are determined at a different time during development and from different embryonic tissues. However, the disparity in the locations of QTL for the same trait in large and small DH progeny is more difficult to explain. QTL correspond to the same linkage group only for vertebrae and the largest QTL for scales above the lateral line. The remaining QTL for scales above the lateral line, anal fin rays, and lateral line scales do not correspond among large and small DH progeny. For example, for anal fin rays, there was no significant difference in the counts between large and small DH progeny, but the *analrays* QTL for the large and small groups did not localize to the same linkage groups. This disparity could be attributed to differences in the environment during the development of each of these characters, as the same genotypes may preferentially express different phenotypes under different environmental conditions. In this study, the earliest trait to complete development, vertebrae, did not show variation in position but the later developing traits did. Embryos were subjected to the same temperature, light, and water conditions prior to complete resorption of the yolk-sac and swim-up to begin exogenous feeding. During the earliest developmental periods, the only difference in environment was maternal egg source for which all nuclear DNA was destroyed. It is possible that early (large DH progeny) and late (small DH progeny) fertilized fish were produced from egg sources that had different maternal effects on embryo development. Maternal effects have been shown to play a role in the development of some meristic characters (Fowler 1970, Lindsey 1988, Ferguson & Liskauskas 1995). Other environmental factors may have also played a role in the expression of meristic trait QTL. Fish were placed in the same recirculating system immediately after swim-up, but fish from each fertilization group were ponded at different times and reared in different settings; it is possible that water temperature during post-swim-up development may not have been the same between the large and small groups. Differences between large and small fish in later developing traits, such as body scalation and fin ray development, could possibly be due to differences in rearing environment during the juvenile stage. Further studies designed to examine maternal and external environmental influence on the location and detection of QTL may clarify the disparity in QTL for large and small DH progeny for the same traits.

Previous studies suggest that the numbers of elements in meristic characters are influenced by development rate (Gabriel 1944, Garside 1966, Fowler 1970, Lindsey 1988). In these experiments, where development rate is modulated water temperature, faster-developing fish had fewer lateral line scales and vertebrae than slower-developing fish (Barlow 1961, Leary et al. 1984, Lindsey 1988). However, the negative correlation between development rate and number of meristic elements has not been surveyed in natural populations reared at the same temperature in a common garden experiment. The association between the number of meristic elements and development rate at different temperatures may not be indicative of correlated changes due to modulation by the same genes, but of correlated changes due to common environmental effects on each trait independently. In our laboratory, both the CW and Swanson River (SW) Alaska clonal lines exhibit a significantly shorter time to hatch (faster development rate) than the OSU line (Robison et al. 1999, 2001, Nichols, unpublished data). However, the faster-developing CW and SW clonal lines also have a greater number of elements than the slower-developing OSU line for most meristic characters surveyed (Table 1, Wheeler & Thorgaard, unpublished data). The fact that time to hatch is not correlated with the counts for any of the surveyed meristic characters in this study, together with the fact that most QTL identified for meristic characters do not localize to the same linkage group as that for development rate (OC8), suggest that development rate and number of meristic characters are not tightly coupled in these *O. mykiss* lines. Only one QTL – for scales above the lateral line – was localized to the same linkage group and position as that for development rate. Presently, it cannot be ascertained whether the same gene(s) on OC8 are responsible for both scales above the lateral line and rate of embryonic development, but the lack of correlation among these traits does not initially support this thought.

While the number of genes responsible for the development of many of these meristic characters is thought to be large (Kirpichnikov 1981), we have identified only a few QTL for each of these meristic traits. If the many genes suspected to play a role in the development of meristic characters were each small in their effect, the sample sizes in this study would probably have failed to detect any of these loci, as was the case for pyloric caeca. However, the fact that major genetic loci were identified for some of these traits may indicate that either only a few major genes were segregating for differences in meristic trait values, or that only the genes of large effect were detected. The distribution of QTL effects (% variance explained) for many polygenic traits has been shown to follow an L-shaped distribution whereby a few genes of large effect and a larger number genes of smaller effect influence the trait of interest (Bost et al. 2001). In studies with small sample sizes, there is a tendency to overestimate the amount of variance explained by these QTL of larger effect, when small-effect QTL are not identified (Lynch & Walsh 1998 and references therein). This tendency is exemplified by the relatively greater r^2 for loci detected in separate analyses compared to the

same locus detected in combined analyses. The power of QTL analyses is significantly affected by both the number of progeny segregating for QTL and the magnitude of the differences observed between parental lines (Lynch & Walsh 1998). We have likely detected the largest, but not all QTL for these meristic characters. Before we can truly test whether the same loci affect different meristic characters, further studies with larger sample sizes within environments and a greater number of markers are necessary to improve power.

Presently, the map developed for this study does not contain candidate genes that might offer potential hypotheses regarding the nature of the genes involved in meristic trait variation. The few major loci associated with development rate and/or meristic characters, namely *Pgm* (Leary et al. 1984) and *LDH* (Ferguson et al. 1988), have not yet been mapped in this cross. Robison et al. (2001) found no correspondence between a liver *Pgm* allele associated with rapid development rate in other studies (Leary et al. 1984) and the development rate QTL in their study. From mapped duplicated enzyme loci, microsatellites, and genes we can infer ancient duplicated or homeologous chromosome regions (Sakamoto et al. 2000, Nichols et al. 2003). Linkage groups OA-VI and OA-XXVII each contain a variant of the Wilms tumor gene, *WT-t1a* and *WT-t1b* (Brunelli et al. 2001). *WT* is a regulatory gene that plays a role in gonad and kidney development in humans. Although these linkage groups were not identified as recent homeologs by microsatellite loci (Nichols et al. 2003), the fact that the QTL for vertebrae and lateral line scale numbers fall on these linkage groups raises the hypothesis that duplicated genes may give rise to the differential regulation and development of these two traits involved in the development of the body axis.

In summary, we have identified several QTL for meristic trait variation in this *O. mykiss* cross. Evidence herein suggests that meristic trait QTL are generally not pleiotropic for more than one trait and may be influenced by maternal and external environmental factors. The lack of correlation between development rate and the meristic characters in this study, together with little evidence for co-localization of meristic QTL with the development rate QTL, suggests that the rate of development and establishment of meristic element numbers are not controlled by the same major genetic loci. While these meristic characters are important for systematics, the adaptive significance of most characters is unclear, and the forces that shape the differences among populations of *O. mykiss* are not understood. Further QTL and

330

candidate gene studies will aid in revealing the number and nature of genes influencing these meristic characters that may play a role in development of different body forms.

Acknowledgements

The authors thank Ann McEvoy and Allen Inderrieden at Washington State University for laboratory and hatchery assistance and Ana Zimmerman for assistance in counting pyloric caeca. Jerri Bartholomew and Harriet Lorz, Oregon State University kindly collected tissues and preserved the small DH for this study. This research was funded in part by a National Science Foundation grant (IBN 0082773) to GHT.

References

Barlow, G.W. 1961. Causes and significance of morphological variation in fishes. System. Zool. 10: 105–117.

Basten, C.J., B.S. Weir & Z.-B. Zeng. 1994. Zmap-a QTL cartographer. pp. 65–66. *In:* C. Smith, J.S. Gavora, B. Benkel, J. Chesnais, W. Fairfull, J.P. Gibson, B.W. Kennedy & E.B. Burnside (ed.) Proceedings of the 5th World Congress on Genetics Applied to Livestock Production: Computing Strategies and Software, Vol. 22, 5th World Congress on Genetics Applied to Livestock Production, Guelph, Ontario, Canada.

Basten, C.J., B.S. Weir & Z.-B. Zeng. 2002. QTL Cartographer, Version 1.16. Department of Statistics, North Carolina State University, Raleigh, NC.

Behnke, R.J. 1992. Native trout of western North America. American Fisheries Society Monograph 6, Bethesda, MD. 275 pp.

Bergot, P., J.M. Blanc & A.M. Escaffre. 1981. Relationship between number of pyloric caeca and growth in rainbow trout (*Salmo gairdneri* Richardson). Aquaculture 22: 81–96.

Bost, B., D. de Vienne, F. Hospital, L. Moreau & C. Dillmann, 2001. Genetic and nongenetic bases for the L-shaped distribution of quantitative trait loci effects. Genetics 157: 1773–1787.

Brunelli, J.P., B.D. Robison & G.H. Thorgaard. 2001. Ancient and recent duplications of the rainbow trout Wilms' tumor gene. Genome 44: 455–462.

Buddington, R.K. & J.M. Diamond. 1986. Aristotle revisited: The function of pyloric caeca in fish. Proc. Natl. Acad. Sci. 83: 8012–8014.

Churchill, G.A. & R.W. Doerge. 1994. Empirical threshold values for quantitative trait mapping. Genetics 138: 963–971.

Danzmann, R.G. & M.M. Ferguson. 1990. Developmental events. pp. 281–311. *In:* D.H. Whitmore (ed.) Electrophoretic and Isoelectric Focusing Techniques in Fisheries Management, CRC Press, Boca Raton, FL.

Doerge, R.W. & G.A. Churchill. 1996. Permutation tests for multiple loci affecting a quantitative character. Genetics 142: 285–294.

Ferguson, M.M., K.L. Knudsen, R.G. Danzmann & F.W. Allendorf. 1988. Developmental rate and viability of rainbow trout with a null allele at a lactate dehydrogenase locus. Biochem. Genet. 26: 177–189.

Ferguson, M.M. & A.P. Liskauskas. 1995. Heritability and evolution of meristic variation in a naturalized population of brook char (*Salvelinus fontinalis*). Nordic J. Freshw. Res. 71: 217–228.

Fowler, J.A. 1970. Control of vertebral number in teleosts – an embryological problem. Quart. Rev. Biol. 45: 148–167.

Gabriel, M.L. 1944. Factors affecting the number and form of vertebrae in *Fundulus heteroclitus*. J. Exp. Zool. 95: 105–147.

Garside, E.T. 1966. Developmental rate and vertebral number in salmonids. J. Fish. Res. Board Can. 23: 1537–1551.

Hubbs, C.L. & K.F. Lagler. 1956. Fishes of the Great Lakes Region, University of Michigan Press, Ann Arbor. 186 pp.

Kirpichnikov, V.S. 1981. Genetic Bases of Fish Selection, Springer-Verlag, New York. 410 pp.

Lander, E.S., P. Green, J. Abrahamson, A. Barlow & M.J. Daly. 1987. MAPMAKER: An interactive computer package for constructing primary genetic linkage maps of experimental and natural populations. Genomics 1: 174–181.

Leary, R.F., F.W. Allendorf & K.L. Knudsen. 1984. Major morphological effects of a regulatory gene: *pgm1-t* in rainbow trout. Mol. Biol. Evol. 1: 183–194.

Leary, R.F., F.W. Allendorf & K.L. Knudsen. 1985. Inheritance of meristic variation and the evolution of developmental stability in rainbow trout. Evolution 39: 308–314.

Leary, R.F., F.W. Allendorf & K.L. Knudsen. 1992. Genetic, environmental, and developmental causes of meristic variation in rainbow trout. Acta Zool. Fenn. 191: 79–95.

Lindsey, C.C. 1955. Evolution of meristic relations in the dorsal and anal fin supports of teleost fishes. Trans. R. Soc. Can. Sec. 5, Vol. 49, Ser. III: 35–49.

Lindsey, C.C. 1988. Factors controlling meristic variation. pp. 197–274. *In:* W.S. Hoar & D.J. Randall (ed.) Fish Physiology, Vol. 11B, Academic Press, New York.

Lindsey, C.C., A.M. Brett & D.P. Swain. 1984. Responses of vertebral numbers in rainbow trout to temperature changes during development. Can. J. Zool. 62: 391–396.

Lynch, M. & B. Walsh. 1998. Genetics and Analysis of Quantitative Traits, Sinauer Associates, Sunderland, MA. 980 pp.

MacGregor, R.B. & H.R. MacCrimmon. 1977a. Evidence of genetic and environmental influences on meristic variation in the rainbow trout, *Salmo gairdneri* Richardson. Environ. Biol. Fish. 1: 25–33.

MacGregor, R.B. & H.R. MacCrimmon. 1977b. Meristic variation among world hatchery stocks of rainbow trout, *Salmo gairdneri* Richardson. Environ. Biol. Fish. 1: 127–143.

Mottley, C.McC. 1934. The effect of temperature during development on the number of scales in the Kamloops trout, *Salmo kamloops* Jordan. Contrib. Can. Biol. 8: 253–263.

Nichols, K.M., W.P. Young, R.G. Danzmann, B.D. Robison, C. Rexroad, M. Noakes, R.B. Phillips, P. Bentzen, I. Spies, K. Knudsen, F.W. Allendorf, B.M. Cunningham, J. Brunelli, H. Zhang, S. Ristow, R. Drew, K.H. Brown, P.A. Wheeler & G.H. Thorgaard. (2003). A consolidated genetic linkage map for rainbow trout (*Oncorhynchus mykiss*). Animal Genet. 34: 102–115.

Northcote, T.G. & R.J. Paterson. 1960. Relationship between the number of pyloric caeca and length of juvenile rainbow trout. Copeia 1960: 248–250.

Parrott, A.W. 1934. The variability and growth of the scales of brown trout (*Salmo trutta*) in New Zealand. Trans. New Zeal. Inst. 63: 497–516.

Ristow, S.S., L.D. Grabowski, C. Ostberg, B. Robison & G.H. Thorgaard. 1998. Development of long-term cell lines from homozygous clones of rainbow trout. J. Aquat. Animal Health 10: 75–82.

Robison, B.D., P.A. Wheeler & G.H. Thorgaard. 1999. Variation in development rate among clonal lines of rainbow trout (*Oncorhynchus mykiss*). Aquaculture 173: 131–141.

Robison, B.D., P.A. Wheeler, K. Sundin, P. Sikka & G.H. Thorgaard. 2001. Composite interval mapping reveals a major locus influencing embryonic development rate in rainbow trout (*Oncorhynchus mykiss*). J. Hered. 92: 16–22.

Sakamoto, T., R.G. Danzmann, K. Gharbi, P. Howard, A. Ozaki, S.K. Khoo, R.A. Woram, N. Okamoto, M.M. Ferguson, L.-E. Holm, R. Guyomard & B. Hoyheim. 2000. A microsatellite linkage map of rainbow trout (*Oncorhynchus mykiss*) characterized by large sex-specific differences in recombination rates. Genetics 155: 1331–1345.

Smith, C.L. & R.M. Bailey. 1961. Evolution of dorsal fin supports of percoid fishes. Papers Mich. Acad. Sci. 46: 345–363.

Voorrips, R.E. 2002. MapChart: Software for the graphical presentation of linkage maps and QTL. J. Hered. 93: 77–78.

Vos, P., R. Hogers, M. Bleeker, M. Reijans, T. Van De Lee, M. Hornes, A. Frijters, J. Pot, J. Peleman, M. Kuiper & M. Zabeau. 1995. AFLP: A new technique for DNA fingerprinting. Nucl. Acids Res. 23: 4407–4414.

Young, W.P., P.A. Wheeler & G.H. Thorgaard. 1995. Asymmetry and variability of meristic characters and spotting in isogenic lines of rainbow trout. Aquaculture 137: 67–76.

Young, W.P., P.A. Wheeler, R.D. Fields & G.H. Thorgaard. 1996. DNA fingerprinting confirms isogenicity of androgenetically derived rainbow trout lines. J. Hered. 87: 77–81.

Young, W.P., P.A. Wheeler, V.H. Coryell, P. Keim & G.H. Thorgaard. 1998. A detailed genetic linkage map of rainbow trout produced using doubled haploids. Genetics 148: 839–850.

Zeng, Z.-B. 1994. Precision mapping of quantitative trait loci. Genetics 136: 1457–1468.

Environmental Biology of Fishes **69**: 333–344, 2004.
© 2004 *Kluwer Academic Publishers. Printed in the Netherlands.*

The mating system of steelhead, *Oncorhynchus mykiss*, inferred by molecular analysis of parents and progeny

Todd R. Seamons[a], Paul Bentzen[a,b] & Thomas P. Quinn[a]
[a]*School of Aquatic and Fishery Sciences, University of Washington, Box 355020, Seattle, WA 98195, U.S.A. (e-mail: seamonst@u.washington.edu)*
[b]*Department of Biology, Dalhousie University, Halifax, Nova Scotia, Canada B3H 4J1*

Received 17 April 2003   Accepted 27 April 2003

Key words: protandry, precocious parr, reproductive timing, size assortative mating

Synopsis

The development of molecular markers has allowed behavioral ecologists to link parents to specific offspring, providing insights into breeding systems that were not apparent from direct observations of the social system. Studies of this type in fishes have focused on species with male parental care such as centrarchids, and on salmonids, a family with little parental care. In order to gain further insight into the mating system of steelhead trout, *Oncorhynchus mykiss*, a winter-spawning species whose reproductive system is poorly known, adults returning to spawn were captured in four consecutive years in a small, unfished, wild population. Juvenile offspring were sampled by electrofishing and parentage was determined by exclusion based on a 12 locus microsatellite genotype. Both males and females mated with multiple individuals, though single pair matings were also inferred. Females and males tended to have the same number of mates (median = 1), but males were more likely to have no apparent partner (43% vs. 23% for females) and the maximum number of mates were obtained by males (range 0–10 vs. 0–5 for females). There was no difference in median arrival date by sex, but 80% of the females mated with males that had already arrived rather than males arriving with or after the females (median = 7.5, range = 1–63 days difference). Contrary to expectations, there was no evidence of size-assortative mating; larger males and larger females did not tend to mate with each other more often than would have occurred by chance. Of the juveniles with only one identified parent, most had a known mother and an unknown father rather than the reverse (88% vs. 11%). We interpret this as indirect evidence that non-anadromous males achieved a significant number of fertilizations. Thus the steelhead mating system was complex, being more strongly structured by arrival date than fish size, and including a significant genetic contribution by mature male parr.

Introduction

The relatively recent development of molecular markers suitable for discriminating genetic relationships among individuals has brought a wealth of comparisons between the observed reproductive behavior (social mating system) and genetic mating system of many organisms (Hughes 1998). Early work focused on the mating systems of birds, comparing observed patterns of paternity and offspring care with genetic patterns, often documenting extra-pair fertilizations in putatively monogamous species (e.g., Westneat 1987). The

inferences about mating systems drawn from observational and genetic studies often differed dramatically (e.g., Birkhead et al. 1990, Sillero-Zubiri et al. 1996), leading to many new theories of sexual selection, sperm competition and mate choice (Birkhead & Møller 1998).

There are now many examples of this type of comparison in fishes, mostly investigating patterns of male reproductive success in species with male parental care (e.g., DeWoody & Avise 2001, Jones & Avise 2001). Many of the other studies of genetic mating systems in fishes involve patterns of paternity, number of mates and spawning location in salmonid species (Hutchings & Myers 1988, Bentzen et al. 2001, Garant et al. 2001, Taggart et al. 2001).

The social mating system and reproductive behavior of salmonids, *Oncorhynchus*, *Salvelinus* and *Salmo* spp., is well known for most species. Males tend to arrive on the spawning grounds before females (Morbey 2000). The ability of individual males to dominate access to spawning females (and thereby maximize reproductive success) is influenced by body size, prior residence (Foote 1990), and by the operational sex ratio (OSR: ratio of sexually active females to sexually active males at any given time; Emlen & Oring 1977) which is a function of date within the spawning period (Quinn et al. 1996). Males arriving early may have first access to early arriving ripe females, whereas later in the spawning season, when OSRs may be more strongly male biased, no single male may be able to control access to females (Dickerson et al. 2002). Additionally, males that establish dominance early in the season may, regardless of size, maintain their position in the dominance hierarchy (Healey & Prince 1998).

After females arrive they immediately dig one or (less frequently) more nests, termed redds, in streambed gravel, and are courted by one or more males. They then deposit eggs fertilized by males in several discrete pockets within the redd over a period of a few hours to a few days. The female covers the fertilized eggs in each pocket with gravel before preparing the next pocket. In semelparous species, the female protects the embryos in her redd from disturbance by other females until she dies. In iteroparous species, females do not guard the completed nest; adults of both sexes migrate to feeding grounds in the ocean or fresh water habitats after spawning and may survive to return and spawn again in subsequent years.

Males may choose their mates based on size (Foote 1988), and large males may exclude smaller males from access to females through aggressive interactions (Hanson & Smith 1967, Keenleyside & Dupuis 1988, Quinn & Foote 1994). Females may also exhibit preference by delaying spawning when courted by a small or otherwise undesirable male (Foote 1989, Berejikian et al. 2000, de Gaudemar et al. 2000). One might infer that large males should be observed in dominant positions courting large females but this is not always the case (Quinn & Foote 1994).

The outcome of aggressive competition among males for access to females is generally determined by size and shape (Fleming & Gross 1994, Quinn & Foote 1994). However, many populations of anadromous salmonids have males that mature at a much smaller size than the smallest females. These males either spent a shorter period of time at sea than most males (jacks; e.g., Gross 1985) or matured without migrating to sea (precocious parr; e.g., Fleming 1998).

In contrast to other salmonids, little is known about the social or genetic mating systems of steelhead, the anadromous form of rainbow trout, *Oncorhynchus mykiss*, other than basic behaviors (Needham & Taft 1934, Shapovalov & Taft 1954, Tautz & Groot 1975). Steelhead differ significantly from other Pacific salmon in life history, so differences in mating system may be expected. Compared to most Pacific salmon, steelhead have a protracted migration and spawning season (3 months or more; Busby et al. 1996), and are generally found at much lower densities than salmon (Busby et al. 1996). Steelhead are also iteroparous and seem to produce a significant number of mature male parr (Shapovalov & Taft 1954), though the phenomenon has not been well-studied. Indeed, the breeding system of steelhead more closely resembles that of Atlantic salmon than their closer relatives, the Pacific salmon. In addition, steelhead may spawn at night, and spawn in the winter and spring when river levels are high in coastal streams, making direct observations of spawning behavior difficult. Molecular genetic tools may be especially suited to a study of steelhead mating system and reproductive behavior providing a basis for contrasts with the behavior of better known Pacific and Atlantic salmon.

The purpose of this paper is to describe the genetic mating system of steelhead, in particular the number of mates of each sex and the extent to which the breeding system is structured around the size or arrival timing of the mates in a wild population. First, we expected that both males and females would have multiple mates, but that males would be more variable in the number of males than females. Second, we predicted that male steelhead would tend to return before females and that females would mate with early arriving males. Finally, we expected that pairs of steelhead would be matched by size but that mature parr might fertilize some eggs.

Methods

Study site, tissue collection and genetic analysis

A permanent fish weir, ~0.95 km upstream of the mouth at Discovery Bay (Figure 1), has been operated continuously since 1977 on Snow Creek, Washington,

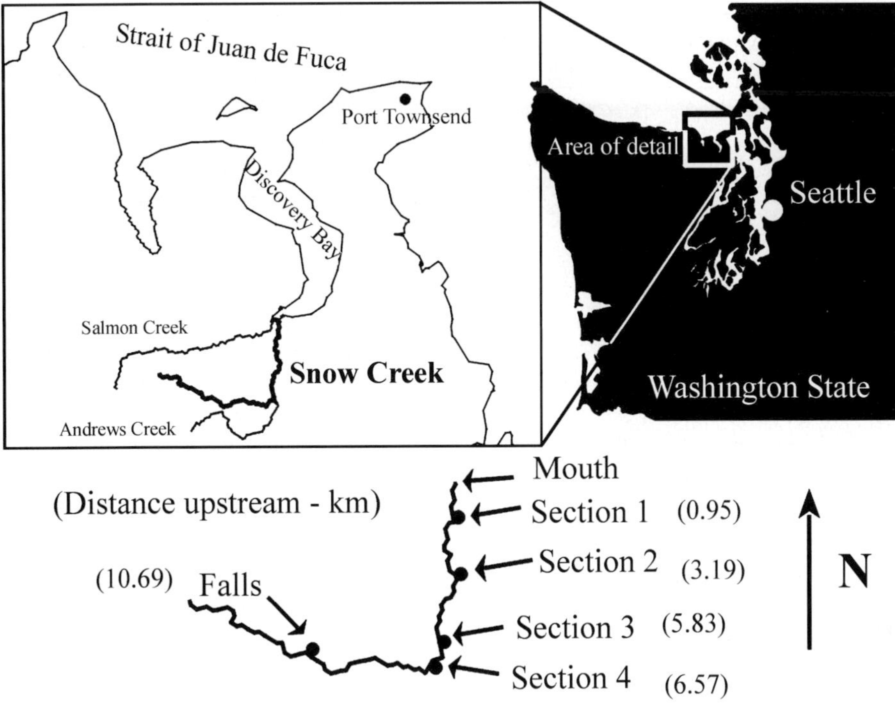

Figure 1. Map of western Washington State showing the location of Snow Creek, sampling sections within Snow Creek and their distance upstream from the mouth at Discovery Bay.

by the Washington Department of Fish and Wildlife (WDFW). Fish are unable to pass up- or downstream without being sampled except during the most extreme floods. All adult steelhead reaching the weir were sampled each spring from 1997 to 2000. Length, date of arrival and sex were recorded for each fish, and several scales were removed for age estimation. Small pieces of caudal fin were removed for DNA analysis from adults and from ∼300 juvenile steelhead that were sampled annually in October 1997–2000 by single-pass electrofishing in four different sections of Snow Creek (Figure 1). All tissue samples were stored in 95% ethanol in 1.5-ml microcentrifuge tubes at room temperature.

DNA from adults was isolated using a modified CTAB/phenol/chloroform extraction protocol (Fields et al. 1989). The DNA of juvenile fish was extracted using the DNeasy Tissue Kit (Qiagen Inc. Valencia, CA, U.S.A.) following recommended protocols. Twelve microsatellite loci (Table 1) were amplified for each individual using the following recipe: 10 μl reaction volumes containing 10 mM Tris–HCl (pH 8.3), 50 mM KCl, 1–2 mM $MgCl_2$ (see Table 1), 0.04 mM each dNTP, 1 U *Taq* DNA polymerase (Promega, Madison, WI), 0.5 μM each primer and ∼100 ng DNA template. An MJ Research PTC-200 thermal cycler (Watertown, Massachusetts) was used for all PCR amplifications. The thermal cycler profile consisted of the following conditions: (1) three cycles of 95°C (1 min) + X°C (30 s) + 70°C (1 min); (2) 22 cycles of 95°C (10 s) + X°C (30 s) + 70°C (1 min); (3) one cycle of 70°C (45 min), where X was an annealing temperature that varied among primer pairs (Table 1). Microsatellite DNA was visualized using the MegaBACE 1000 capillary electrophoresis system (Amersham Pharmacia Biotech Limited, Piscataway, NJ). One primer of each primer pair was labeled with one of three fluorescent chemical labels (FAM, NED or HEX) and electrophoresed with a 900 bp internal size standard similarly labeled (ET900-R, Amersham Biosciences).

Locus characteristics, parentage assignment and data analysis

A two-tailed exact test of Hardy–Weinberg equilibrium (HWE) was performed for each locus in each brood year (BY) combination using the Markov Chain

Table 1. Microsatellite loci used for parentage assignment, references for each locus, annealing temperature (T_A), magnesium chloride concentration in mM and repeat unit in base pairs (bp).

Locus	Source	T_A	[MgCl$_2$]	Repeat unit (bp)
Oki23	A. Spidle[a]	55	1	4
Omy1001UW	P. Bentzen[b]	55	1	4
Omy1004UW	P. Bentzen[b]	50	1	4
Omy1011UW	P. Bentzen[b]	55	1	4
Omy1191UW	P. Bentzen[b]	65	1	4
Omy1212UW	P. Bentzen[b]	65	1	4
Omy77	Morris et al. (1996)	55	1	2
One108	Olsen et al. (2000)	55	1	4
One2	Scribner et al. (1996)	55	1	2
Ots107	Nelson & Beacham (1999)	55	1.5	4
Ots108	Nelson & Beacham (1999)	55	1.5	4
Ssa85	O'Reilly et al. (1996)	60	2	2

[a]Unpublished, GenBank accession # AF272822.
[b]Unpublished, Paul Bentzen.

method implemented in Genepop 3.1 (dememorization number 1000, batches 100, 1000 iterations per batch; Raymond & Rousset 1995). Significance of probability values was adjusted for multiple tests using sequential Bonferroni correction (Rice 1989). F_{IS}, a measure of the fractional reduction in heterozygosity due to inbreeding in individuals within a subpopulation, was calculated according to Weir & Cockerham (1984) using Genepop 3.1 software (Raymond & Rousset 1995). Observed and expected heterozygosity were calculated using GENETIX 4.02 software (Belkhir et al. 2000).

Parentage was assigned through exclusion using the program WHICHPARENTS 1.0 (Will Eichert, Bodega Bay Marine Lab, California) using all 12 loci, but allowing three mismatches. The genotypes of adults identified by WHICHPARENTS as potential parents were then directly compared to juvenile genotypes; only adults that matched one allele at each locus were finally called parents. If no adults matched one allele at each locus with a juvenile it was assumed that the actual parents were not in the adult sample. Juveniles not assigned any parents were dropped from subsequent analyses. As a conservative measure, because of repeat spawning and the possibility of incomplete adult sampling in some years, all 4 years of adults were treated as potential parents of all juveniles.

In many cases only one parent was assigned to a juvenile. We calculated the probability of finding a single matching parent at random from the parental pool by using the equation: probability of a match at a single locus $= 1 - [1 - (pA + pB)]^2$, where pA is the probability of matching allele A and pB is the probability of matching allele B based on the parental population allele frequencies, multiplied across all loci. The chance of a single parent matching at random the most common possible offspring genotype (i.e., homozygous for the most frequent allele at each locus in the parent pool) for each year was between 1 and 2 in 1000 and no juvenile offspring actually had that genotype. Thus we concluded that the single assigned adult was indeed one of the parents of that juvenile.

The numbers of mates for individuals of each sex was determined by counting the number of different mates from genetically determined steelhead matings from yearly parentage data. In cases where only one parent was known it was assumed that all offspring of that parent were the offspring of only one other adult though this may underestimate the actual number of mates. Yearly male and female distributions of numbers of mates were compared using a Kolmogorov–Smirnov test (Zar 1999). To assess the patterns of arrival timing of adults we calculated an arrival score for each year of adult data using the 'protandry score' of Morbey (2000). We then calculated the frequency of males or females that arrived before their mate. We calculated the median and mean number of days between each pair of mates for those pairings where the male arrived first. To determine whether mating was size-assortative, we used Spearman's rank correlation (Zar 1999) for correlation of male and female lengths in inferred mated pairs in each year.

Results

Adult characteristics

A total of 275 adult steelhead were collected over the 4 years of this study (Table 2) ranging from 31 in 1997 to 139 in 2000. The overall sex ratio was close to 1 : 1 (135 males and 140 females) but varied considerably among years. In 1999 and 2000 the sex ratios were nearly 1 : 1 but in 1997 there were almost twice as many females as males, and in 1998 the reverse ratio was observed. Body lengths varied considerably, from 420–815 mm in males to 565–860 mm in females. The date of first arrival to Snow Creek varied from December to March among the 4 years, but most steelhead returned in March and April, and the median arrival date was in the last week of March in all 4 years.

Table 2. Number of steelhead males and females (N_M/N_F), total number of returning adults (N_T), male length size range (M_{SR}), average male length (Avg_M), female length size range (F_{SR}), average female length (Avg_F), male median return date (M_{DATE}) and female median return date (F_{DATE}) for each BY (1997–2000).

Brood year	N_M/N_F	N_T	M_{SR}	Avg_M	F_{SR}	Avg_F	M_{DATE}	F_{DATE}
1997	11/20	31	420–750	612.7	595–830	692.3	25 Mar	28 Mar
1998	35/18	53	600–735	649.6	565–795	638.6	20 Mar	21 Mar
1999	25/27	52	400–815	655.6	600–740	658.3	23 Mar	23 Mar
2000	64/75	139	430–785	645.4	565–860	640	23 Mar	25 Mar
Total	135/140	275						

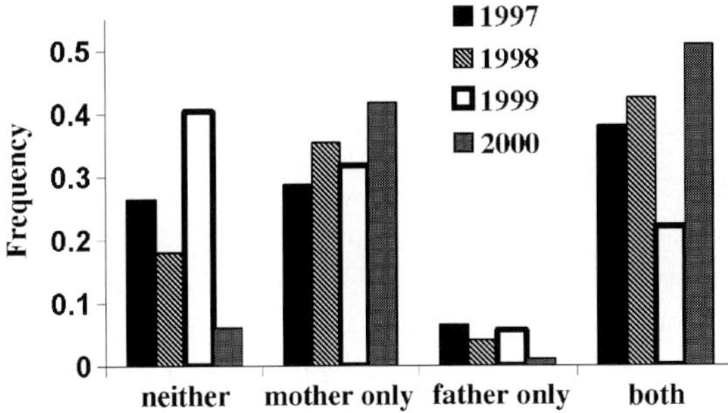

Figure 2. Frequency of the type of parental assignment – both parents, one parent (mother or father only) or neither parent – for all juvenile fish samples by BY (1997–2000).

Locus characteristics, parentage and familial relationships

Only two tests out of 48 were out of HWE after sequential Bonferroni corrections (Table 3). *Omy77*, a locus known to have null alleles in Snow Creek steelhead (Ardren et al. 1999), showed the expected deviation from HWE ($P < 0.0001$) in one of four BYs (2000), and had a significant positive F_{IS} value associated with null allele segregation ($F_{IS} = 0.209$). Locus *Ots108* was also out of HWE for BY 2000 adults ($P = 0.001$) and also showed a positive F_{IS} value (0.028).

Average heterozygosity across all 12 loci in our sample was 90%, and the probability of exclusion using all 12 loci was 0.9999999 (CERVUS 2.0; Marshall et al. 1998). We were able to assign at least one parent to 73% of all juvenile samples, and two parents to ~40%. Of those juveniles assigned only one parent, nearly 90% were assigned only a female parent (Figure 2), significantly more than expected by chance (Chi-square goodness-of-fit, $p < 0.001$,

expected $1:1$). Four percent (12) of all adult steelhead returned to spawn in Snow Creek in more than 1 year. Fifty-four juveniles appeared to be the offspring of these adults. These juveniles were assigned to a BY by one of two methods: if a male and female adult from one BY and only a single adult from another BY (the repeat year of one of the assigned adults) were assigned to a juvenile it was assumed to have been spawned in the BY of the two parents assigned (31; no juveniles appeared to be offspring of two repeat spawning adults); if only one parent was assigned (the same individual, but two BYs) the juvenile was assigned a BY by its length using juvenile age/length relationships from known age juveniles (23).

At least some male and female steelhead mated with multiple partners in all years (Figure 3). Males mated with up to 10 different females, but the median was one female (33%). Females mated with up to five different males (median $= 1$) with two mates being the mode. More males than females had no mates over all 4 years (43% vs. 23%). The male and female distribution of number of mates was significantly different for the pooled data (Kolmogorov–Smirnov test, $p < 0.001$).

The multiple matings resulted in complex patterns of relatedness. For example, in 2000 we inferred that there were eight single pair mating groups (i.e., a male and a female were each other's only mates that year; Figure 4). However, there were also two cases where one female's eggs were fertilized by two different males and three cases where one male mated with two different females. A series of progressively more complex mating groups were inferred, including one group of seven males and nine females that interbred (Figure 4). Similarly diverse patterns were inferred in the other 3 years of the study (Table 4).

There was a significant correlation between male and female size in mated pairs of steelhead only in

Table 3. Allelic variability at 12 microsatellite loci in four BYs of steelhead in Snow Creek. Number of alleles per locus (N_A), expected heterozygosity (H_E), observed heterozygosity (H_O), probability after test for goodness of fit to HWE (P_{HWE}), inbreeding coefficient (F_{IS}) and the total number of alleles across all BYs (Total N_A) are shown.

Brood year		Locus											
		Oki23	*Omy1001UW*	*Omy1004UW*	*Omy1011UW*	*Omy1191UW*	*Omy1212UW*	*Omy77*	*One108*	*One2*	*Ots107*	*Ots108*	*Ssa85*
1997	N_A	17	12	12	14	18	19	9	11	24	24	16	12
	H_E	0.89	0.87	0.74	0.88	0.92	0.92	0.8	0.88	0.92	0.94	0.91	0.86
	H_O	0.87	0.96	0.84	0.9	1	0.9	0.73	0.87	1	0.87	0.87	0.9
	P_{HWE}	0.75	0.91	0.14	0.62	0.99	0.22	0.13	0.22	0.15	0.37	0.2	0.28
	F_{IS}	0.045	−0.094	−0.119	−0.014	−0.075	0.037	0.096	0.031	−0.073	0.093	0.064	−0.036
1998	N_A	16	12	9	15	18	24	11	14	24	38	19	16
	H_E	0.88	0.86	0.83	0.89	0.9	0.94	0.82	0.88	0.93	0.96	0.91	0.86
	H_O	0.92	0.86	0.85	0.94	0.91	0.96	0.77	0.89	0.98	0.94	0.92	0.89
	P_{HWE}	0.86	0.33	0.39	0.68	0.6	0.53	0.01	0.33	0.46	0.36	0.36	0.82
	F_{IS}	−0.037	0.006	−0.015	−0.055	0.007	−0.012	0.061	0.003	−0.043	0.025	0.003	−0.025
1999	N_A	15	15	12	20	19	11	14	17	29	41	19	18
	H_E	0.87	0.89	0.83	0.91	0.91	0.93	0.84	0.9	0.94	0.93	0.9	0.87
	H_O	0.9	0.96	0.79	0.9	0.88	0.94	0.73	0.85	0.94	0.96	0.87	0.85
	P_{HWE}	0.22	0.42	0.61	0.26	0.76	0.38	0.03	0.01	0.03	0.64	0.44	0.19
	F_{IS}	−0.029	−0.069	0.056	0.012	0.036	−0.001	0.137	0.066	0.011	0	0.052	0.037
2000	N_A	18	17	11	19	22	32	16	17	33	53	22	23
	H_E	0.9	0.88	0.8	0.89	0.9	0.95	0.79	0.9	0.94	0.96	0.91	0.83
	H_O	0.95	0.9	0.78	0.92	0.93	0.96	0.63	0.87	0.99	0.96	0.88	0.83
	P_{HWE}	0.14	0.15	0.75	0.9	0.49	0.18	0.0000*	0.02	0.004	0.06	0.001*	0.37
	F_{IS}	−0.054	−0.019	0.021	−0.033	−0.038	−0.016	0.209	0.037	−0.052	0.008	0.028	0.012
Total	N_A	20	18	15	20	25	35	17	19	39	58	25	27

*Significant at $\alpha = 0.05$.

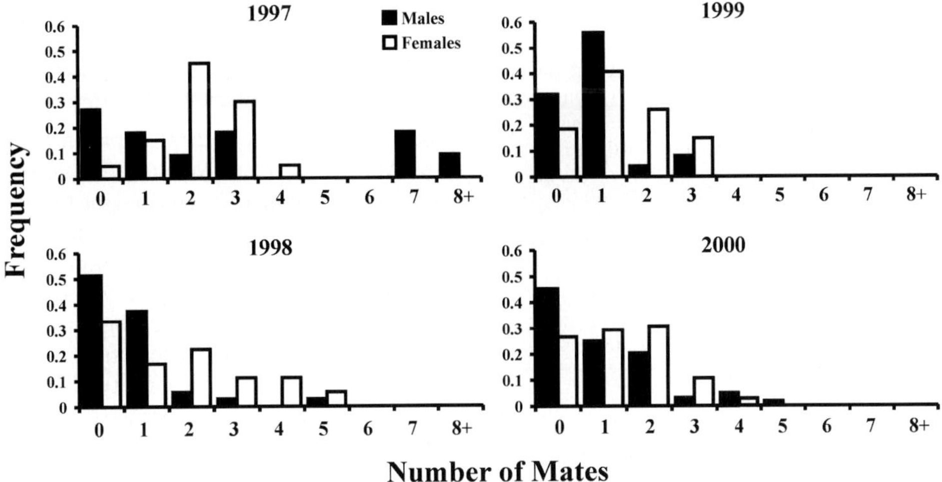

Figure 3. Estimated frequency of inferred numbers of different mates for males (black) and females (white) in four consecutive BYs (1997–2000).

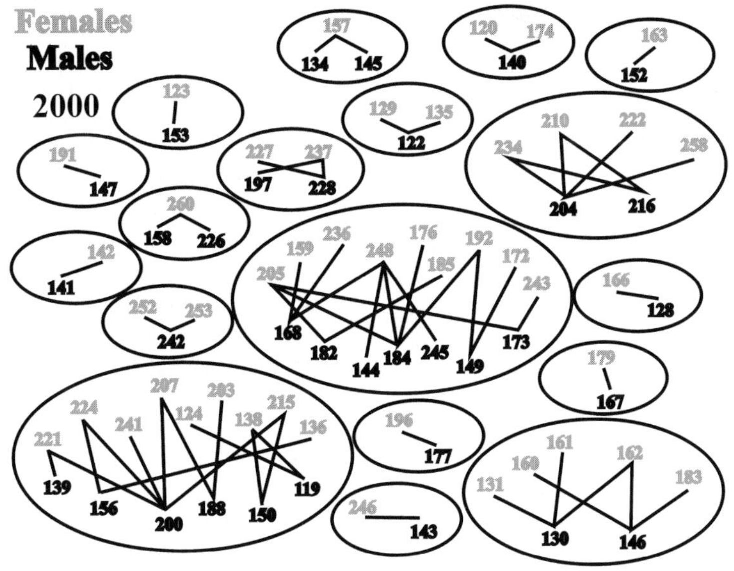

Figure 4. Inferred mating patterns of steelhead in Snow Creek from BY 2000. Each oval encompasses the individual parents (shown as identification numbers) in a mating group inferred from an extended half-sibling family of offspring. Lines connect males (solid black numbers) and females (gray numbers) with their inferred mate(s). Four types of mating were apparent: monogamy, polygyny, polyandry and polygynandry.

2000 when a slight negative correlation was found (Spearman's rank correlation, $R = -0.26$ $p = 0.046$). Jack males (i.e., those spending only 1 year at sea) may use a different mating strategy so we repeated the analysis after excluding them but there was still no relationship between the sizes of the parents.

In contrast to the lack of size-based structure in the apparent mating system, there was considerable temporal structure. Males arrived only slightly before females, with a difference in median date of 0–3 days (Table 2). The 'protandry score' (Morbey 2000), an alternative method to estimate which sex arrives first,

Table 4. Numbers of different types of mating groups (1 male (M)–1 female (F), 1 male–2 females, 2 males–1 female, multiple males–multiple females) inferred from extended half-sibling families of juvenile offspring for BYs 1997–2000. The numbers of adult males (N_M) and females (N_F) in the largest mating group with multiple individuals of both sexes are in parentheses.

Brood year	1M–1F	1M–2F	2M–1F	nM–nF (N_M/N_F)
1997	2	0	0	1 (5/14)
1998	2	0	3	1 (5/5)
1999	3	2	2	1 (3/3)
2000	8	3	2	5 (7/9)

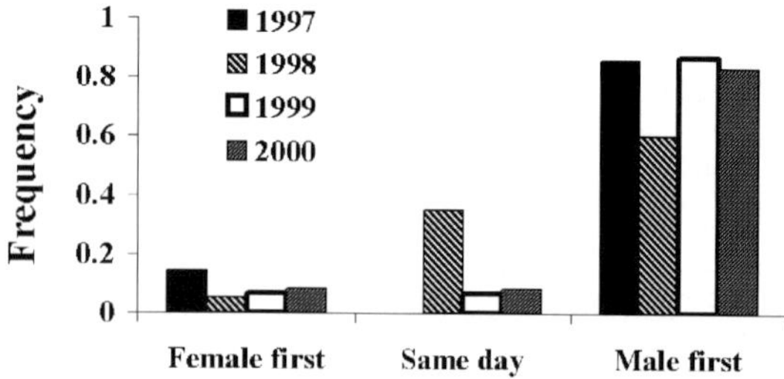

Figure 5. The order of arrival at the Snow Creek weir for inferred mated pairs of adult steelhead. Frequencies of each type of arrival order (female first, same day, male first) are shown for BYs 1997–2000.

indicated that in 1998 and 2000 males arrived before females by about 4 days, and in 1997 and 1999 females arrived before males by about 4 days.

Despite the similarity in arrival patterns, the male almost invariably arrived before the female with whom he mated (Figure 5). On average males arrived about 15 days before the females they mated with, and the yearly median difference in days ranged from 3 to 19.5 days before females. This result was not the consequence of a lack of later arriving males; an average of five males arrived on the same day or within 2 weeks after the arrival of each female, but they seldom mated with such females.

Discussion

Male steelhead were expected to have multiple mates because in other salmonids sperm production does not limit reproduction and females tend to spawn all their eggs within a few days (e.g., chum salmon, *Oncorhynchus keta* – Schroder 1981). Thus when a female has spawned all her eggs, the dominant male would be expected to seek breeding opportunities with

other females. Females may mate with multiple males as a form of mate choice, choosing better mates with each successive mating or they might benefit from diversity among their offspring in a variable environment (see Garant et al. 2001, for an in-depth discussion for Atlantic salmon). While a female might exercise some choice in mates, eggs may be fertilized by males with whom the female never intended to mate, such as satellite males, jacks or mature parr. We were unable to observe mating directly so we cannot determine whether the multiple sires resulted from female choice or male competition. Observations of salmon (e.g., Quinn et al. 1996) and paternity analysis under controlled conditions (e.g., Schroder 1981, Chebanov et al. 1983, Foote et al. 1997) suggest that male competition is probably the dominant process.

Our data showed that male and female steelhead mated with multiple partners, but the numbers of mates for each sex should be viewed as conservative estimates for several reasons. First, these numbers are biased by the numbers of offspring per parent; a male with only one offspring can only be inferred to have mated with one female. We sampled a small but unknown fraction of the offspring in the creek, and any matings that produced offspring that were not sampled or offspring that failed to survive incubation and their first summer would also be undetected. Thus fish may have had mates that we did not detect. Our conclusion that both males and females mated with multiple mates would not change with complete data but the ranges would probably increase. Second, the differences between sexes may be biased by family size. Maternal half-sib families averaged ~1 offspring more than paternal half-sib families, possibly accounting for some of the difference in the number of mates between the two sexes.

Examination of the patterns of sibship among the juveniles revealed many different types of mating, including monogamy, polygamy, polyandry and polygynandry, resulting in complex patterns of relatedness among offspring. This complexity of mating may be one reason why size-assortative mating was not revealed. Size-assortative mating was expected because there are apparent advantages for each sex to mate with large individuals. Large females are more fecund (Beacham & Murray 1993) and construct deeper redds which better protect embryos (Steen & Quinn 1999). Large males are more often able to win in competition for access to females (Quinn & Foote 1994) and may pass traits for size on to offspring (Gjerde 1986). However, large males and females mated with multiple individuals of the opposite sex, including smaller individuals. In addition, it is possible that size-assortative mating did take place, with each female mating with the largest available male at that moment, even though that male may have been smaller than ones available at other times during the season or other locations in the creek. More likely, however, a female was courted by one dominant male and other males participated in the spawning act despite his attempts to monopolize access to her. Males are also sensitive to the timing of female reproduction and will court a female about to deposit eggs rather than a larger one that is less likely to spawn shortly (Schroder 1981).

Steelhead matings were structured by arrival timing. Females almost always mated with males that had arrived before them, despite the fact that several males arrived with or soon after each female. Early male arrival timing is common in Pacific salmon (Morbey 2000), presumably to offset the skewed OSR that results from the fact that males remain sexually active until they die or leave the spawning grounds whereas females complete spawning in a few days and thereafter are not available for reproduction (Quinn et al. 1996). Male steelhead did not consistently arrive before females, and in general the breeding period was protracted and characterized by relatively few ripe females at any time. Our data showed that males need not arrive early on an absolute basis but need to arrive a week or two before a female arrives to maximize their chances of breeding with her. We have several hypotheses for this apparent pattern. First, males may have established territories within the stream and maintained them by a prior residence advantage. Assuming that most males do not stay in fresh water for the entire spawning season, which may last for up to 4 months, territories are continuously established and vacated over the spawning season. Some degree of territoriality has been reported in male sockeye salmon but it is qualitatively different from the behavior shown by females (Foote 1990). Females actively defend their redd site even in the absence of males whereas males are much less aggressive in the absence of females. Second, some amount of learning of the freshwater habitat may be necessary for males to be comfortable or established enough to spawn. Third, males may wait for females to arrive then intercept them on their upstream migration, follow them and establish a 'territory' near the female. In any of these scenarios, males arriving with or after females are less likely to mate with them than established males. However, males that arrived too early may have already left the stream or depleted their energy to the point where they cannot compete with more recent arrivals, hence the successful males tended to have arrived about 2 weeks prior to their mate.

Of all the juveniles, 39% were assigned only one parent and of those, 88% were assigned only a female parent. Several processes might prevent us from assigning parents to a juvenile trout, but we believe the contribution of mature male steelhead parr is the most likely explanation for many reasons. First, genotyping error may account for the inability to assign two parents to all juveniles, but it was unlikely to be biased toward assigning only a female parent because the DNA samples were not arranged by sex. Second, other types of laboratory or sample handling error could have caused us to be unable to assign two parents, but again this type of error was unlikely to be sex biased because tissue and DNA samples were arranged in the order that they arrived, not by sex. Third, although a few adults apparently passed the weir without being sampled, especially during floods that took place in two of the years, there was no systematic difference in timing of arrival between the sexes so this would not explain the surplus of males that were missed. Fourth, hybridization between male cutthroat and female steelhead, may have occurred. There is some evidence of steelhead hybridization with coastal cutthroat trout in Snow Creek (Young et al. 2001). However, we tested our juvenile fish with a single powerful genetic marker (Growth hormone 2 intron D; Baker et al. 2002) and found very low hybridization rates ($\sim$2%).

Mature male parr occur in steelhead populations (Needham & Taft 1934, Shapovalov & Taft 1954) as well as in many other salmonids (masu salmon, *Oncorhynchus masou* – Tsiger et al. 1994; Atlantic salmon, *Salmo salar* – Fleming 1998; chinook salmon, *Oncorhynchus tshawytscha* – Unwin et al. 1999) and in Atlantic salmon can have collectively high reproductive success as a group (Morán et al. 1996). If we are correct in our deduction that mature male parr are the cause of the difference between missing male and female parents, then they are consistently achieving considerable reproductive success. In some years of low return the mature parr may collectively be more successful than anadromous male steelhead. Most importantly for conservation, mature parr may significantly increase the effective population size (Martinez et al. 2000), thereby increasing the chances of persistence for small populations of steelhead. Ardren & Kapuscinski (2003) found that in Snow Creek the ratio of effective population size to the actual number of spawners was significantly higher in years with low spawner density. One explanation for this observed pattern may be a proportional increase in reproductive success of mature male parr when numbers of anadromous males are scarce. We currently have no estimate of their actual abundance in the population; mature parr have not been evident in our sampling operations in October.

Nearly 27% (294) of all juveniles were not assigned any parents. To account for potential errors we first assigned these juveniles to BYs using age/length histograms made from known age juveniles. Nearly 88% (258) were assigned to BY 1997 or 1999, both years in which there were floods. While it was known that high water compromised the weir in 1997 the extent to which the weir was compromised in 1999 was previously unknown. Only 23 and 13 juveniles from BY 1998 and 2000, respectively had no known parents. Tissue from these 36 juveniles and all the adults from BY 1998 and 2000 were re-extracted and re-genotyped at all loci resulting in six (two from BY 1998 and four from BY 2000) of the juveniles being assigned a single parent and one juvenile from BY 2000 that already had been assigned one parent having a second parent assigned.

The lack of parental assignment to the remaining 30 juveniles from BY 1998 or 2000 may have been due to several types of error. First, these juveniles may have been misassigned by size to BY 1998 or 2000. Some overlap in size distribution was observed, especially in 1998 YOY and 1997 yearling juveniles. Six juveniles without known parents sampled in 1998 fell within the size overlap range and could therefore be from BY 1997. Second, mutations at one or more loci could have occurred. Steinberg et al. (2002) recently calculated the mutation rate for microsatellite loci in pink salmon, *Oncorhynchus gorbushcha*, and found that two loci had mutation rates on the order of $4\text{–}9 \times 10^{-3}$ while the remaining six loci had no detected mutations. Therefore mutation does not explain a significant fraction of failed parental assignments. Third, null allele segregation could have occurred. Indeed three juvenile genotypes matched adult genotypes at all loci but *Omy77*, and at that locus both juvenile and adult were homozygous possibly indicating inheritance of a null allele. In addition, one adult individual would not amplify at locus *Omy77* despite repeated DNA extractions and PCR attempts. Since *Omy77* was known to have null alleles, it was assumed that this individual had two null alleles (or two copies of the same null allele) at this locus. Fourth, genotyping errors could have occurred. Some electrophoretic variation at dinucleotide loci caused allelic size distributions to overlap. Rather than attempting to assign alleles to bins by probabilistic methods, overlapping alleles were combined into one size bin. This conservative method reduced the power of the single locus to assign parentage, but also reduced the genotype mismatch error at that locus. The loss of power in a single locus was offset by using many different loci to assign parentage. Fifth, some of the remaining juveniles may have been the offspring of unidentified rainbow trout, the non-anadromous form of *Oncorhynchus mykiss*, though this was unlikely since no known population of rainbow trout exists in the Snow Creek watershed (Thom Johnson, WDFW, 283236 Highway 101, Port Townsend, WA, 98368 – pers. comm.). Finally, some juveniles might have been cutthroat trout (*Oncorhynchus clarki*), which are nearly impossible to distinguish from steelhead at small sizes in the field without killing the fish. Snow Creek harbors a resident cutthroat population above a barrier waterfall as well as a very small anadromous population in the lower reach.

While not substantially different from the observed mating system of other Pacific salmon, the steelhead mating system did appear to be more similar to that of Atlantic salmon in that it included multiple mates and (apparently) male parr. The complexity and flexibility of steelhead mating behavior may be important to survival in a highly variable environment and may be one reason they have been able to persist at relatively low numbers for many generations.

Acknowledgements

We gratefully acknowledge Thom Johnson, Randy Cooper and Cheri Scalf, WDFW, for providing adult phenotypic data as well as for collecting all adult tissue samples. Additional thanks go to the long list of people that helped sample juveniles each year and provided advice and assistance in the laboratory. Funding for this research was provided by the H. Mason Keeler Endowment and National Science Foundation grant DEB-9903914.

References

Ardren, W.R., S. Borer, F. Thrower, J.E. Joyce & A.R. Kapuscinski. 1999. Inheritance of 12 microsatellite loci in *Oncorhynchus mykiss*. J. Hered. 90: 529–536.

Ardren, W.R. & A.R. Kapuscinski. 2003. Demographic and genetic estimates of effective population size (Ne) reveals genetic compensation in steelhead trout. Mol. Ecol. 12: 35–49.

Baker, J., P. Bentzen & P. Moran. 2002. Molecular markers distinguish coastal cutthroat trout from coastal rainbow trout/steelhead and their hybrids. Trans. Amer. Fish. Soc. 131: 404–417.

Beacham, T.D. & C.B. Murray. 1993. Fecundity and egg size variation in North American Pacific salmon (*Oncorhynchus*). J. Fish Biol. 42: 485–508.

Belkhir, K., P. Borsa, L. Chikhi, N. Raufaste & F. Bonhomme. 2000. GENETIX 4.02, logiciel sous Windows™ pour la genetique des populations, University of Montpellier II, Laboratoire de Populations, Montpellier, France.

Bentzen, P., J.B. Olsen, J.E. McLean, T.R. Seamons & T.P. Quinn. 2001. Kinship analysis of Pacific salmon: Insights into mating, homing, and timing of reproduction. J. Hered. 92: 127–136.

Berejikian, B.A., E.P. Tezak & A.L. LaRae. 2000. Female mate choice and spawning behavior of chinook salmon under experimental conditions. J. Fish Biol. 57: 647–661.

Birkhead, T.R., T. Burke, R. Zann, F.M. Hunter & A.P. Krupa. 1990. Extra-pair paternity and intraspecific brood parasitism in wild zebra finches *Taeniopygia guttata*, revealed by DNA fingerprinting. Behav. Ecol. Sociobiol. 27: 315–324.

Birkhead, T.R. & A.P. Møller. 1998. Sperm Competition and Sexual Selection, Academic Press, London. 826 pp.

Busby, P.J., T.C. Wainwright, G.J. Bryant, L. Lierheimer, R.S. Waples, F.W. Waknitz & I.V. Lagomarsino. 1996. Status review of west coast steelhead from Washington, Idaho, Oregon, and California. NOAA Tech. Memo., U.S. Dep. Commer. NMFS-NWFSC-27. 261 pp.

Chebanov, N.A., N.V. Varnavskaya & V.S. Varnavskiy. 1983. Effectiveness of spawning of male sockeye salmon *Oncorhynchus nerka* (Salmonidae), of differing hierarchical rank by means of genetic-biochemical markers. J. Ichthyol. 23: 51–55.

de Gaudemar, B., J.M. Bonzom & E. Beall. 2000. Effects of courtship and relative mate size on sexual motivation in Atlantic salmon. J. Fish Biol. 57: 502–515.

DeWoody, J.A. & J.C. Avise. 2001. Genetic perspectives on the natural history of fish mating systems. J. Hered. 92: 167–172.

Dickerson, B.R., T.P. Quinn & M.F. Willson. 2002. Body size, arrival date, and reproductive success of pink salmon, *Oncorhynchus gorbuscha*. Ethol. Ecol. Evol. 14: 29–44.

Emlen, S.T. & L.W. Oring. 1977. Ecology, sexual selection, and the evolution of mating systems. Science 197: 215–223.

Fields, R.D., K.R. Johnson & G.H. Thorgaard. 1989. DNA fingerprints in rainbow trout detected by hybridization with DNA of bacteriophage M13. Trans. Amer. Fish. Soc. 118: 78–81.

Fleming, I. A. 1998. Pattern and variability in the breeding system of Atlantic salmon (*Salmo salar*), with comparisons to other salmonids. Can. J. Fish. Aquat. Sci. 55: 59–76.

Fleming, I.A. & M.R. Gross. 1994. Breeding competition in a Pacific salmon (coho: *Oncorhynchus kisutch*): Measures of natural and sexual selection. Evolution 48: 637–657.

Foote, C.J. 1988. Male mate choice dependent on male size in salmon. Behaviour 106: 63–80.

Foote, C.J. 1989. Female mate preference in Pacific salmon. Anim. Behav. 38: 721–722.

Foote, C.J. 1990. An experimental comparison of male and female spawning territoriality in a Pacific salmon. Behaviour 115: 283–314.

Foote, C.J., G.S. Brown & C.C. Wood. 1997. Spawning success of males using alternative mating tactics in sockeye salmon, *Oncorhynchus nerka*. Can. J. Fish. Aquat. Sci. 54: 1785–1795.

Garant, D., J.J. Dodson & L. Bernatchez. 2001. A genetic evaluation of mating system and determinants of individual reproductive success in Atlantic salmon (*Salmo salar* L.). J. Hered. 92: 137–145.

Gjerde, B. 1986. Growth and reproduction in fish and shellfish. Aquaculture 57: 37–55.

Gross, M.R. 1985. Disruptive selection for alternative life histories in salmon. Nature 313: 47–48.

Hanson, A.J. & H.D. Smith. 1967. Mate selection in a population of sockeye salmon (*Oncorhynchus nerka*) of mixed age-groups. J. Fish. Res. Board Can. 24: 1955–1977.

Healey, M.C. & A. Prince. 1998. Alternative tactics in the breeding behaviour of male coho salmon. Behaviour 135: 1099–1124.

Hughes, C. 1998. Integrating molecular techniques with field methods in studies of social behavior: A revolution results. Ecology 79: 383–399.

Hutchings, J.A. & R.A. Myers. 1988. Mating success of alternative maturation phenotypes in male Atlantic salmon, *Salmo salar*. Oecologia 75: 169–174.

Jones, A.G. & J.C. Avise. 2001. Mating systems and sexual selection in male-pregnant pipefishes and seahorses: Insights from microsatellite-based studies of maternity. J. Hered. 92: 150–158.

Keenleyside, M.H. A. & H.M.C. Dupuis. 1988. Courtship and spawning competition in pink salmon (*Oncorhynchus gorbuscha*). Can. J. Zool. 66: 262–265.

Marshall, T.C., J. Slate, L.E.B. Kruuk & J.M. Pemberton. 1998. Statistical confidence for likelihood-based paternity inference in natural populations. Mol. Ecol. 7: 639–655.

Martinez, J.L., P. Moran, J. Perez, B. De Gaudemar, E. Beall & E. Garcia-Vazquez. 2000. Multiple paternity increases effective size of southern Atlantic salmon populations. Mol. Ecol. 9: 293–298.

Morán, P., A.M. Pendás, E. Beall & E. García-Vázquez. 1996. Genetic assessment of the reproductive success of Atlantic salmon precocious parr by means of VNTR loci. Heredity 77: 655–660.

Morbey, Y. 2000. Protandry in Pacific salmon. Can. J. Fish. Aquat. Sci. 57: 1252–1257.

Needham, P.R. & A.C. Taft. 1934. Observations on the spawning of steelhead trout. Trans. Amer. Fish. Soc. 64: 333–338.

Quinn, T.P., M.D. Adkison & M.B. Ward. 1996. Behavioral tactics of male sockeye salmon (*Oncorhynchus nerka*) under varying operational sex ratios. Ethology 102: 304–322.

Quinn, T.P. & C.J. Foote. 1994. The effects of body size and sexual dimorphism on the reproductive behavior of sockeye salmon, *Oncorhynchus nerka*. Anim. Behav. 48: 751–761.

Raymond, M. & F. Rousset. 1995. An exact test for population differentiation. Evolution 49: 1280–1283.

Rice, W.R. 1989. Analyzing tables of statistical tests. Evolution 43: 223–225.

Schroder, S.L. 1981. The role of sexual selection in determining the overall mating patterns and mate choice in chum salmon. Ph.D. dissertation, University of Washington, Seattle, WA. 274 pp.

Shapovalov, L. & A.C. Taft. 1954. Life histories of the steelhead rainbow trout (*Salmo gairdneri gairdneri*) and silver salmon (*Oncorhynchus kisutch*) with special reference to Waddell Creek, California, and recommendations regarding their management. Fish. Bull. 98: 375 pp.

Sillero-Zubiri, C., D. Gottelli & D.W. Macdonald. 1996. Male philopatry, extra-pack copulations and inbreeding avoidance in Ethiopian wolves (*Canis simensis*). Behav. Ecol. Sociobiol. 38: 331–340.

Steen, R.P. & T.P. Quinn. 1999. Egg burial depth by sockeye salmon (*Oncorhynchus nerka*): Implications for survival of embryos and natural selection on female body size. Can. J. Zool. 77: 836–841.

Steinberg, E.K., K.R. Lindner, J. Gallea, A. Maxwell, J. Meng & F.W. Allendorf. 2002. Rates and patterns of microsatellite mutations in pink salmon. Mol. Biol. Evol. 19: 1198–1202.

Taggart, J.B., I.S. McLaren, D.W. Hay, J.H. Webb & A.F. Youngson. 2001. Spawning success in Atlantic salmon (*Salmo salar* L.): A long-term DNA profiling-based study conducted in a natural stream. Mol. Ecol. 10: 1047–1060.

Tautz, A.F. & C. Groot. 1975. Spawning behavior of chum salmon (*Oncorhynchus keta*) and rainbow trout (*Salmo gairdneri*). J. Fish. Res. Board Can. 32: 633–642.

Tsiger, V.V., V.L. Skirin, N.I. Krupyanko, K.A. Kashlin & A.Y. Semenchenko. 1994. Life history form of male masu salmon (*Oncorhynchus masou*) in South Primor'e, Russia. Can. J. Fish. Aquat. Sci. 51: 197–208.

Unwin, M.J., M.T. Kinnison & T.P. Quinn. 1999. Exceptions to semelparity: Postmaturation survival, morphology, and energetics of male chinook salmon (*Oncorhynchus tshawytscha*). Can. J. Fish. Aquat. Sci. 56: 1172–1181.

Weir, B.S. & C.C. Cockerham. 1984. Estimating F-statistics for the analysis of population structure. Evolution 38: 1358–1370.

Westneat, D. 1987. Extra-pair fertilizations in a predominantly monogamous birds: Genetic evidence. Anim. Behav. 35: 877–886.

Young, W.P., C.O. Ostberg, P. Keim & G.H. Thorgaard. 2001. Genetic characterization of hybridization and introgression between anadromous rainbow trout (*Oncorhynchus mykiss irideus*) and coastal cutthroat trout (*O. clarki clarki*). Mol. Ecol. 10: 921–930.

Zar, J.H. 1999. Biostatistical Analysis, Prentice Hall, Upper Saddle River, New Jersey. 663 pp.

Environmental Biology of Fishes **69**: 345–357, 2004.
© 2004 *Kluwer Academic Publishers. Printed in the Netherlands.*

Polygamous mating and high levels of genetic variation in lingcod, *Ophiodon elongatus*, of the Strait of Georgia, British Columbia

Ruth E. Withler[a], Jacquelynne R. King[a], Jeffrey B. Marliave[b], Brad Beaith[a], Shaorong Li[a],
K. Janine Supernault[a] & Kristina M. Miller[a]
[a]*Department of Fisheries and Oceans Canada, Pacific Biological Station, Nanaimo,
BC V9T 6N7, Canada (e-mail: withlerr@pac.dfo-mpo.gc.ca)*
[b]*Vancouver Aquarium Marine Science Center P.O. Box 3232, Vancouver, BC V6B 3X8, Canada*

Received 17 April 2003   Accepted 19 April 2003

Key words: polygamy, polygyny, parentage, maternity, paternity, microsatellite, mitochondrial DNA

Synopsis

Lingcod, *Ophiodon elongatus*, is a nest-guarding marine fish of western North America. Breeding occurs in late winter and early spring after males establish territories and guard nest sites therein. Eggs deposited as clutches in the nest site hatch ~7 weeks after fertilization. We evaluated the level of genetic variation in lingcod spawning in the central Strait of Georgia through analysis of microsatellite and mitochondrial D-loop variability in fertilized egg samples collected from guarded clutches. Reconstructed parental genotypes displayed a high level of allelic diversity and observed heterozygosity (83–91%) over five microsatellite loci. Progeny of a single clutch were invariably derived from a single mother and between one and five fathers. Multiple egg samples were collected from inside and outside positions on 13 lingcod egg clutches in February 2002. Fin clip samples provided microsatellite genotypes for six of the nine guardian males. Analysis of between 33 and 306 eggs from each clutch indicated that each of the 13 clutches was produced by a different mother and five of them were sired entirely by the attendant male guardian. Eight clutches were sired by multiple males, with neighboring male guardians frequently involved in clutch fertilization. Known guardian males accounted for at least 78% of observed egg fertilization, although non-territorial males were observed and may have participated in spawning. Egg fertilization by individual males was spatially heterogeneous throughout egg clutches. One male guardian failed to fertilize detectable numbers of eggs in his own or any other clutch within the study area and may have been an adoptive father. The polygynous mating structure of lingcod may help maintain genetic variation in the species.

Introduction

In higher vertebrates, and especially mammals, the greatly disparate investment in parental care between the sexes limits the fecundity of females, and hence the opportunity for female bet-hedging strategies of reproduction involving mating with multiple partners. As a result, polygyny is more common than polyandry among most vertebrate groups, with males often competing for access to females and successful males producing disproportionately large proportions of progeny (Clutton-Brock 1989, 1991). In fishes, parental care is often minimal, and more often provided by the male

than female parent (Blumer 1979). Females as well as males produce large numbers of gametes and juvenile mortality is high. Thus, females may also benefit from mating with multiple partners, especially under conditions of environmental instability (Jennions & Petrie 2000).

The lingcod, *Ophiodon elongatus*, is large carnivorous member of the family Hexagrammidae inhabiting waters off the western coast of North American from Baja California to the Aleutian Peninsula. Lingcod generally occupy waters 10–100 m in depth and are demersal spawners, locating nest sites in rocky crevices exposed to strong currents. In British Columbia, males mature as young as age 2 and acquire nest sites at depths ranging from 5 to 60 m. Mature females, aged 3 or more, remain in deeper waters. Fecundity ranges from less than 100 000 eggs in younger (smaller) females to over 500 000 eggs in large females. Females enter the nesting areas only to spawn, an activity that occurs primarily at night (Low & Beamish 1978) in January and February. Females deposit eggs in a batch or 'clutch' along with a gelatinous substance that sets the eggs in a stiff matrix.

Males usually guard nests until the young hatch, defending fertilized eggs from invertebrates and fish but exposing themselves to predation (Jewell 1968). Males occasionally guard more than one clutch in their nests, and orphaned nests may be taken over by new males (Jewell 1968, Low & Beamish 1978, LaRiviere et al. 1981). In the absence of an adoptive father, unguarded clutches are almost certainly consumed by fish or invertebrate predators (Jewell 1968, Cass et al. 1990). The incubation period ranges from 5 to 11 weeks, with an average of about 7 weeks. Hatching takes place primarily in March and April and generally occurs rapidly, within a few hours. Hatching success depends strongly on oxygenation of the eggs and varies from 5% to 95% (Giorgi & Congleton 1984).

The observation that males sometimes guard multiple clutches led us to consider the possibility that male lingcod are polygynous, as has been observed for the closely related kelp greenling (Crow et al. 1997). Males of freshwater sculpin species (Cottidae), which also acquire nest sites and guard egg clutches, tend to be highly polygynous (Goto 1993, Natsumeda 1998). It also seemed possible that the batch-spawning female lingcod might be polyandrous and allocate their eggs among the nests of multiple males. For some nest-guarding fish species, genetic analysis of parentage of fish embryos from nests has confirmed behavioral observations indicating polygyny or polyandry, although few marine species have been examined (Taborsky 2001, DeWoody & Avise 2001).

In this study, we examined levels of lingcod genetic diversity and the incidence of multiple parentage in egg clutches sampled from underwater reefs in the Strait of Georgia, British Columbia. Microsatellite loci were used to identify full and halfsib relationships among fertilized eggs within the nest, and to reconstruct genotypes of the parental fish. Polymorphism in a D-loop mitochondrial DNA sequence was used to determine the gender of the shared parent in halfsib groups. The genetic profiles of six nest guardian male

lingcod determined from fin clip samples were used to estimate the proportion of eggs under his care that each guardian fertilized.

Materials and methods

Preliminary egg sampling

In February and March 2001, divers collected samples from lingcod egg clutches at depths between 5 and 15 m at various locations in central Strait of Georgia, British Columbia. Microsatellite analysis was performed on six eggs from each of 40 clutches to enable reconstruction of 80 parental lingcod genotypes at five loci. To determine the effectiveness of these loci in parentage analysis, we calculated the number of effective alleles (Hedrick 2000) and paternal power of exclusion for each locus. Ten clutches were chosen arbitrarily for further parentage analysis. For each of two clutches, 100 fertilized eggs from a sample taken at a single location on the clutch surface were subjected to microsatellite analysis. For each of the remaining eight clutches, 12 fertilized eggs that appeared to be at a relatively early developmental stage (creamy white, no eyes visible) and 12 at a later developmental stage (grey, eyes visible) were selected from a single sample of eggs for analysis. Parental genotypes were reconstructed from the alleles observed segregating in the progeny.

Snake Island survey

In February 2002, a more thorough examination of parental contributions to lingcod nests was undertaken. The study site, Snake Island reef, is situated in the Strait of Georgia (Figure 1) and supports ongoing lingcod egg mass density surveys conducted for stock assessment purposes (King & Beaith 2001). A dive buoy was deployed to mark the location (49°12′43.7N; 123°53′04.7W) and remained in place for the duration of the study. The study site was ~10 m by 20 m with an average depth of 9 m (Figure 2). Overall, the study site was relatively flat, with several large rocks and boulders present to provide nesting sites. Much of the study site was open, or barren, without large flora, such as *Agarum* spp. present.

Egg clutches were marked with a galvanized spike and flagging tape with a number on it for identification on subsequent dives. Guardian males were attributed to an clutch if they were within 1–2 m of

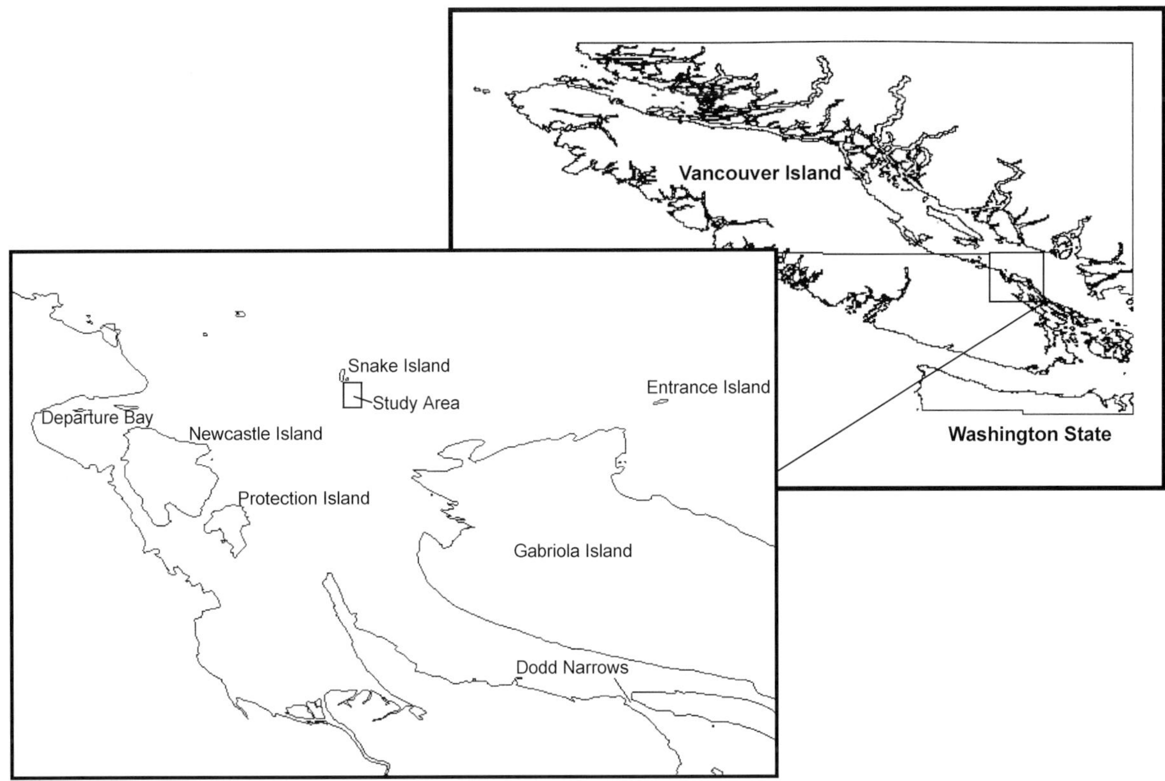

Figure 1. Snake Island reef is located in the Strait of Georgia, British Columbia. The location of the reef in proximity to Snake Island is indicated in the inset.

it and exhibited protective or territorial behavior when a diver approached the clutch. The total length (to the nearest cm) of each guardian was measured by pulling a tape measure alongside the fish as it rested on the bottom. The relative location of the nest was noted, i.e. under a rock or in a crevice, and volume of the egg clutch was estimated using length, width and height (nearest cm) measurements made underwater. Given the difficulty in measuring the dimensions of irregular shaped egg masses under rocks, measurements were made several times and the average (nearest 0.5 L) was used as the volume estimate.

Divers collected egg samples on a weekly basis from 1 to 27 February 2002. Each week, two samples were collected from each egg mass: one each from the innermost and outermost portion on the nest. Each sample consisted of ~1 cc of eggs cut from the egg mass and placed in a labelled tube filled with ambient seawater. On shore, seawater was replaced with 95% ethanol and a label with the date, clutch number and location of sample was inserted into the tube. The sampling period covered the full developmental stage of egg masses,

from creamy white (newly laid eggs) to eyed eggs just prior to hatching.

On 22 February, egg clutch 15 was removed entirely from its crevice and transported intact to the shore, weighed, and partitioned into 22 sections according to a 3-dimensional matrix of $\sim 7\,cm^3$ cubes. On 27 February, clutches 10 and 11 were removed and handled in the same manner, being cut into 12 and 21 sections, respectively. The central sections from each clutch tended to be complete cubes, whereas those from the peripheries were of various sizes due to irregular clutch shape. Each section was placed in 95% ethanol. Masses of 10 eggs were calculated five times (total of 50 eggs) to determine the average egg mass, and total clutch masses were used to estimate the total egg numbers. Microsatellite analysis was performed on eggs from each section.

Tissue samples were collected using scissors from the caudal fin of male guardians *in situ* where possible. Fin tissue samples were placed in a labelled tube filled with ambient seawater that was later replaced with ethanol.

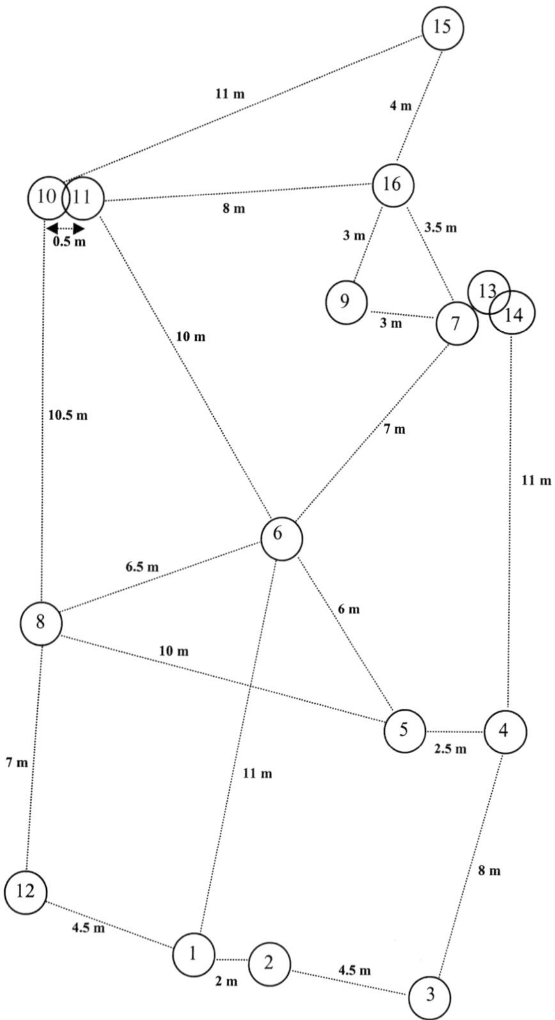

Figure 2. Measured distances (nearest 0.5 m) between egg clutches (denoted by circled numbers) at the Snake Island study site. Egg clutches 7, 9 and 14 disappeared, likely due to predation.

Microsatellite analysis

Eggs were individually detached from each clutch sample and placed in 96-well trays for DNA extraction. DNA was extracted using Qiagen DNeasy kits. Five microsatellite loci were assayed in the egg samples: *Oel* 32, *Oel* 35, *Oel* 41, *Oel* 42 and *Oel* 45 (Sewall Young, Washington Department of Fish and Wildlife, pers. comm.). Primers were labelled fluorescently and PCR products were run on gels on an ABI 377 automated DNA sequencer. Allele sizes were determined with Genescan 3.1 and Genotyper 2.5 software (PE Biosystems, Foster City, CA, U.S.A.).

For each clutch, eggs collected on different dates and locations (inside and outside) were analysed. The allelic combinations observed in the progeny at each locus were used to reconstruct two or more parental genotypes for each clutch. Chi-square tests for homogeneity (Snedecor & Cochran 1967) were used to determine if the numbers of progeny attributed to each of multiple parents were homogeneous among the various samples collected from a single clutch.

Mitochondrial DNA analysis

A 434 bp sequence of the mitochondrial D-loop was amplified from 15 adult lingcod sampled from throughout British Columbia using the primers of Crow et al. (1997). PCR of the D-loop sequence was carried out in a total volume of 100 μl containing 1 μg of DNA, 0.5 pM each primer, 200 mM each dNTP, and 4.0 units of Taq polymerase (Perkin-Elmer/Cetus, Emeryville, CA, U.S.A.). The amplification profile consisted of 35 cycles of 1 min at 94°C, 2 min at 50°C and 2 min at 68°C. DNA sequences were obtained from the ABI 377 sequencer, corrected using Sequencer 3.0 (Gene Codes Corp., Ann Arbor, MI) and aligned with ClustalX (Thompson et al. 1997). Ten distinct sequences were obtained from the 15 fish, which indicated that the lingcod D-loop was sufficiently polymorphic to serve as a maternal marker for determination of the gender of the shared parent in halfsib groups of lingcod.

The 10 D-loop sequences were used to determine denaturing gradient gel electrophoresis (DGGE) conditions that would separate them (Miller et al. 1999). DGGE was carried out using the Bio-Rad (Hercules, CA) D Code™ apparatus on 7.5% polyacrylamide gels with a 20–35% denaturing gradient. Gel temperature was maintained at 57°C for a 16-h run at 60 V. The 10 sequences were combined into three sets of standards, and all three standard sets were run in three places on each gel (right, left and center). For lingcod egg clutches in which halfsib groups of progeny were identified, the D-loop sequence from 3 to 6 progeny from each of the groups was amplified and analysed on a DGGE gel. The expectation was that if the halfsib groups from a clutch shared a mother, the mitochondrial sequence should remain constant among groups. Alternately, if the halfsib groups shared a father, the mitochondrial sequence should frequently change between the halfsib groups within a clutch.

Results

Preliminary microsatellite analysis

Observed and expected heterozygosity levels and numbers of alleles among the 80 parental genotypes developed from the 40 clutch samples in 2001 indicated a high level of genetic diversity in Strait of Georgia lingcod (Table 1). Effective allele numbers ranged from 5.3 to 11.0 over the five loci and the combined paternal probability of exclusion was 0.999. Genotypes at all five loci were consistent with those expected under conditions of Hardy–Weinberg equilibrium (p > 0.05), providing no evidence of null (non-amplifying) alleles. Reconstruction of the adult lingcod genotypes was carried out by random assignment of the two genotypes observed at each locus to the two parents. Thus, it was not possible to test for linkage equilibrium among the five loci. Analysis of adult lingcod samples from elsewhere in British Columbia has provided no evidence of linkage disequilibrium among the five loci (unpubl. data).

The progeny of all 10 nests examined more detail in 2001 could be accounted for by one parent of one

sex and between one and three parents of the other sex (Table 2). For clutches three through 10, in which eggs of two different developmental stages were examined, insufficient DNA was obtained from many of the eggs in the early development stage to provide complete microsatellite genotypes. Thus, the more developed eggs provided most of the progeny genotypes. Multiple parents of one gender were detected in 7 of the 10 nests examined, even though the sample sizes were small for most nests and thus provided minimal estimates of parental numbers.

Mitochondrial analysis

Sequencing of mtDNA obtained from 15 lingcod revealed extensive heterogeneity in the targeted D-loop sequences, providing 10 different sequences with an average pairwise proportion of variable sites of 0.01 (range 0.002–0.02) (Table 3). The DGGE conditions employed provided good separation of the mitochondrial sequences (Figure 3). Analysis of individuals from each of the seven clutches with multiple parents of one gender revealed that a single mitochondrial sequence was shared among all progeny within each clutch, but the sequence varied among clutches. Six sequences were observed among the seven clutches. Thus, the progeny from each egg clutch had the same mother but were often sired by more than one father. This result was not surprising given the very localized nature of the sample from each clutch (a small clump of eggs cut from an outside extremity). The results for clutches from which eggs of different developmental stages (but a single location on the clutch) were sampled, indicated that the apparently heterogeneous egg development rate observed within an egg mass was not due to the participation of multiple females. However, this preliminary sampling did not preclude the possibility that multiple females had laid eggs in various locations within a clutch that we failed to detect with our single small egg collection.

Table 1. Levels of allelic diversity (A = number of alleles observed), effective allele number (E), observed (H_O) and expected heterozygosity (H_E) and paternal probability of exclusion (P_E) for five microsatellite loci surveyed in Strait of Georgia lingcod.

Locus	A	E	H_O	H_E	P_E
Oel 32	10	5.6	82.5	82.9	0.67
Oel 35	8	5.3	87.8	83.1	0.66
Oel 41	13	7.7	88.6	87.0	0.73
Oel 42	19	11.0	91.1	90.9	0.81
Oel 45	13	9.5	86.1	89.5	0.78

Table 2. Percentage contributions of individual parents of one sex to lingcod eggs sampled in a preliminary survey of 10 clutches from the Strait of Georgia, British Columbia. Egg sample sizes for each clutch are given in parentheses.

Parent	Clutch									
	1 (84)	2 (93)	3 (14)	4 (22)	5 (12)	6 (11)	7 (23)	8 (24)	9 (23)	10 (13)
1	89.3	86.3	71.4	100.0	83.3	45.4	100.0	75.0	56.5	100.0
2	10.7	9.5	28.6		16.7	45.4		25.0	43.5	
3		4.2				9.1				

350

Table 3. Variable sites in a 450 bp D-loop nucleotide sequence of lingcod mitochondrial DNA. The 10 sequences (GenBank accession numbers AY466048–AY466057) were obtained from 15 lingcod sampled from British Columbia. Dots represent the nucleotide shown in sequence *Oel*-mito-1 for that site.

Sequence	Site															
	50	53	55	62	73	77	161	166	192	196	197	198	227	228	260	278
Oel-mito-1	G	C	C	G	G	G	A	C	A	G	G	G	C	T	G	C
Oel-mito-2	.	.	.	.	.	.	.	.	.	.	.	.	T	.	.	.
Oel-mito-3	.	.	.	.	.	.	G	T	G	A	A	A	.	.	.	.
Oel-mito-4	.	.	.	.	.	.	.	T	G	.	A	A	T	.	.	.
Oel-mito-5	.	.	.	.	.	A	.	.	.	.	A	A	.	.	.	.
Oel-mito-6	.	.	.	.	.	.	.	T	.	A	A	A	T	C	.	.
Oel-mito-7	.	.	.	.	A	.	.	T	.	.	A	.	T	.	.	.
Oel-mito-8	.	.	T	.	.	.	.	T	.	.	A	.	T	.	.	T
Oel-mito-9	A	T	.	A	.	A	.	T	.	A	A	A	T	.	A	.
Oel-mito-10	.	.	.	.	.	.	.	T	.	.	A	.	T	.	.	T

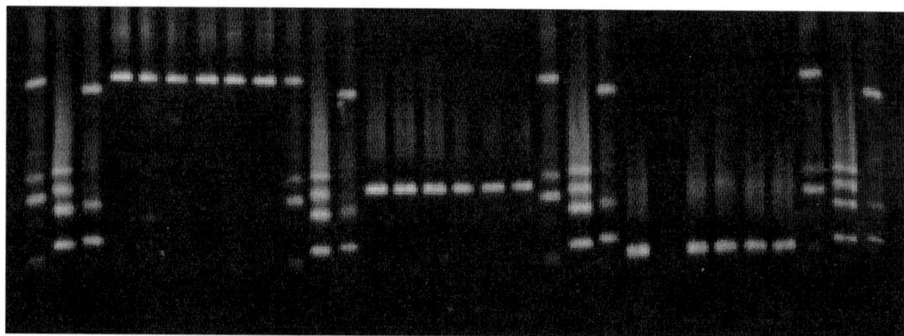

Figure 3. Denaturing gradient gel electrophoresis of lingcod D-loop mitochondrial DNA sequences provided by eggs from two clutches sampled in 2001 and one clutch sampled in 2002. Three sets of standard sequences were run on the gel in four locations and occupy lanes 1–3, 10–12, 19–21 and 28–30. Mitochondrial sequences from six eggs (three each from two halfsib groups) from clutch one (2001) are in lanes 4–9, from clutch two (2001) are in lanes 13–18, and from clutch one (2002) are in lanes 22–27. Amplification of the D-loop sequence was unsuccessful for the second egg sampled from clutch one (2002) (lane 23). Mitochondrial sequences migrated differently between clutches (indicating different mothers) but not between halfsib groups within clutches (indicating a shared mother).

2002 Snake Island survey

Sixteen lingcod egg clutches were marked with a galvanized spike and numbered flagging tape to facilitate identification in subsequent dives (Figure 2). Within 2 weeks three egg masses (clutches 7, 9 and 14) had disappeared, likely due to predation. Males remained associated with the clutch location after the egg mass disappeared. A total of 13 egg clutches were sampled, most of them under rocks. Typically, those located within a crevice occurred in pairs in close proximity to each other (e.g. clutch 10 near 11, clutch 13 near 14).

All 13 egg clutches were guarded by a male lingcod and in some instances, males guarded two or three separate but proximate clutches (Table 4). The nine guardians were situated within 1 m of their clutches and displayed varying degrees of aggressive and protective behavior so that only the six most protective males could be approached closely enough to obtain a tissue sample (Table 4). The sampled males were not consistently the largest fish nor were they guarding the largest egg clutches. Several male lingcod were observed in the study site that were not guarding nests. Typically these males were located in the area bounded by clutches 6, 7, 11 and 16 (Figure 2). On any given dive, the number of non-guarding males was estimated to be as high as four. These males were too wary to be approached, but appeared to encompass a similar size range and to be similar in coloring to guardian males.

Snake Island microsatellite analysis

Genotypes were established for between 33 and 306 progeny for each of the 13 clutches sampled within the study area. Amplification of microsatellite loci was less successful for eggs collected during the first 2 weeks of the study because of low DNA yields from eggs in early development. Nevertheless, progeny genotypes were established for each clutch from at least three samples, distinguished by time and position. Genotypes were also established for the six male guardians from which fin clips were obtained.

As in the preliminary study, progeny sampled from all clutches could be accounted for by a single mother and one to five fathers (Table 5). That the uniparental sex was female for each clutch was established either by analysis of mtDNA or by identifying one or more male parents from among the six sampled guardians, who invariably turned out to be a member of the multiparental sex. The maternal, and remaining paternal, genotypes were deduced by subtraction. A single mother and a single father that was not one of the six sampled males accounted for the progeny sampled from clutch five. A male genotype established as a parent in clutch four could account for the clutch five progeny and the genotype of the clutch five mother was deduced by subtraction. Each of the 13 clutches sampled, even the multiple clutches guarded by males 1, 4 and 10, was derived from a different female parent.

Five of the six guardian males sampled fathered at least a portion of the clutch(es) they were protecting (Table 5). Male 1 sired all the progeny sampled

Table 4. Egg clutch and male guardian numbers and associated information for the 13 lingcod egg clutches sampled in 2002.

Egg clutch	Male guardian	Male length (cm)	Estimated male age	Clutch location	Clutch volume (l)
1	1*	79	9–10	Under rock	4.5
2	1			Under rock	5.0
3	1			Under rock	4.5
4	4	67	5–6	Under rock	4.0
5	4			Under rock	3.5
6	6	71	6–7	Under rock	3.5
8	8*	69	5–6	Under rock	3.5
10	10*	62	3–4	In crevice	2.5
11	10			In crevice	4.0
12	12	62	3–4	Under rock	3.5
13	13*	65	4–5	In crevice	3.5
15	15*	58	3–4	In crevice	2.5
16	16*	58	3–4	Under rock	2.0

Asterisks denote males from which tissue samples were taken.

Table 5. Percentage contributions of individual male parents to lingcod eggs sampled from 13 clutches near Snake Island.

Male	Vol	1 (91)	2 (90)	3 (75)	4 (78)	5 (84)	6 (78)	8 (150)	10 (132)	11 (303)	12 (33)	13 (95)	15 (156)	16 (69)
1	16.0	1.1	100.0	100.0				40.7						
8	0.9							22.7		1.0				
10	5.3								100.0	68.6				
13	0.3											22.1		13.0
15	2.8												100.0	13.0
A	2.3	49.5												
B	4.2	17.6			25.6			17.3			48.5			
C	0.4	7.7												
D	1.1	24.2												
E	2.0				43.6		6.4							
F	4.7				30.8	100.0								
G	5.7						93.6			14.4	51.5			
H	0.7							19.3						
I	0.6									13.7				
J	0.1									2.3				
K	1.6											45.3		
L	0.6											17.9		
M	0.5											14.7		
N	1.0													49.3
O	0.5													24.6

Numbers of progeny sampled are shown in parentheses. Progeny were sired both by males that were sampled (males 1, 8, 13, 15 and 17) and by males not sampled in the study (males A–O). The estimated volume in litres of eggs fertilized by each male is shown (Vol).

352

from clutches two and three, but only 1 of 91 progeny sampled from clutch one. Male 1 also fathered 41% of the progeny from clutch eight, whereas male 8, the guardian, sired 23% of the progeny sampled from his clutch and 1% of the progeny sampled from clutch 11. Male 10 fertilized all the progeny sampled from clutch 10 and 69% of the progeny from clutch 11. Males 13 and 15 sired 22% and 100%, respectively, of the progeny sampled from their own clutches, as well as both contributing to the fertilization of eggs in clutch 16. However, male 16 failed to father any of the sampled progeny not only from his own but from any of the remaining 12 clutches.

Fifteen additional paternal genotypes (labelled A–O) were detected among the progeny of the 13 clutches (Table 5). Four of these males contributed to fertilization in more than one clutch. Male B, who fathered almost half the progeny of clutch 12, also made substantial contributions to clutches one, four and eight (Table 5, Figure 4). It seems likely that male B was guardian male 12 whose tissue was not sampled. Similarly, male G fertilized 94% of clutch six, 52% of clutch 12 and contributed to the orgy of clutch 11. It is possible that male G was guardian male 6. Male F, who fertilized all progeny sampled from clutch five and 31% of clutch four progeny, is a strong candidate for male 4 who guarded those two clutches. Male E, who contributed to fertilization of clutches four and six, cannot be associated with a known guarding male. However, given the positions of nests four and six, it is possible that male E was one of the two males guarding nests seven and nine who were not sampled and whose nests disappeared, precluding determination of their genotypes. Alternately, male E may have been one of the additional males frequently observed in the vicinity of nests 6, 7, 11 and 16 who were not guarding territories within the confines of the study area.

The remaining 11 males each fertilized progeny from only a single clutch within the study area, contributing between 2.3% and 49.5% of the sampled progeny (Table 5). Male E and the other 11 males not identified as male guardians fertilized ~11.4 of the 51.3 L of eggs estimated to be distributed among the 13 clutches of the study area (Table 5). Thus, non-guarding males likely fertilized a maximum of 22.2% of eggs. The actual fertilization success of non-territorial males may have been much lower, given that guardian males 7 and 9 may have been included in this group, and other group members may have held territories outside the study area.

The proportions of progeny fertilized by the various fathers within a clutch varied significantly among samples collected over time and clutch position for all clutches fertilized by two or more males (all $p < 0.05$) (Table 6). In general, samples from the outside of the clutch tended to provide better representation of the males involved, but this was not always the case.

Microsatellite analysis of clutches 10, 11 and 15

Total wet masses of the collected clutches 10, 11 and 15 were 1 173, 1 647 and 1 965 g, respectively, providing estimated egg numbers of 97 750, 137 200 and 122 800. These estimates are minimal because some eggs were lost during the removal of clutches. The volumetric estimate of clutch 15 did not indicate that it was larger than clutches 10 and 11, indicating the difficulty in estimating clutch size without collection (Table 4).

All 132 progeny sampled from the sections of clutch 10 and 156 progeny sampled from sections of clutch 15 could be attributed to the respective guardian males. The 306 progeny sampled from clutch 11 were attributed to five fathers, with the guardian siring 69% of them (Table 5). Between 11 and 24 progeny were sampled from each of 21 sections, with each section representing between 0.2% and 11.0% of the nest by weight. Male contributions to fertilization varied greatly among samples, although male 10 was identified as the principal contributor to all but two sections (Figure 5). Male 10 fertilized only 3 of 12 progeny sampled from section 3Ab and 1 of 16 progeny sampled from section 3Ac.

When male contributions to clutch 11 were weighted by the actual proportion of the nest represented by each section, the estimated contribution by the guardian male decreased by about 10% (Table 7). Examining the nest in cross-section from back to front (layers 1–3), the weighted percentage fertilization by male 10 dropped from 98% of the small back layer (presumably deposited first) to 45.4% of the front layer (presumably deposited last) (Figure 5). From bottom to top (layers a–d) his percentage fertilization remained relatively constant, but he fertilized more of the eggs on the right (B and C) than on the left (A) side of the clutch (Table 7, Figure 5). Fertilization of eggs by the remaining males was stratified by depth in the nest, with the small contributions from males 8 and J concentrated in the bottom layers, male G prominent in the midsection of the nest and male I most successful in the top layer (Table 7).

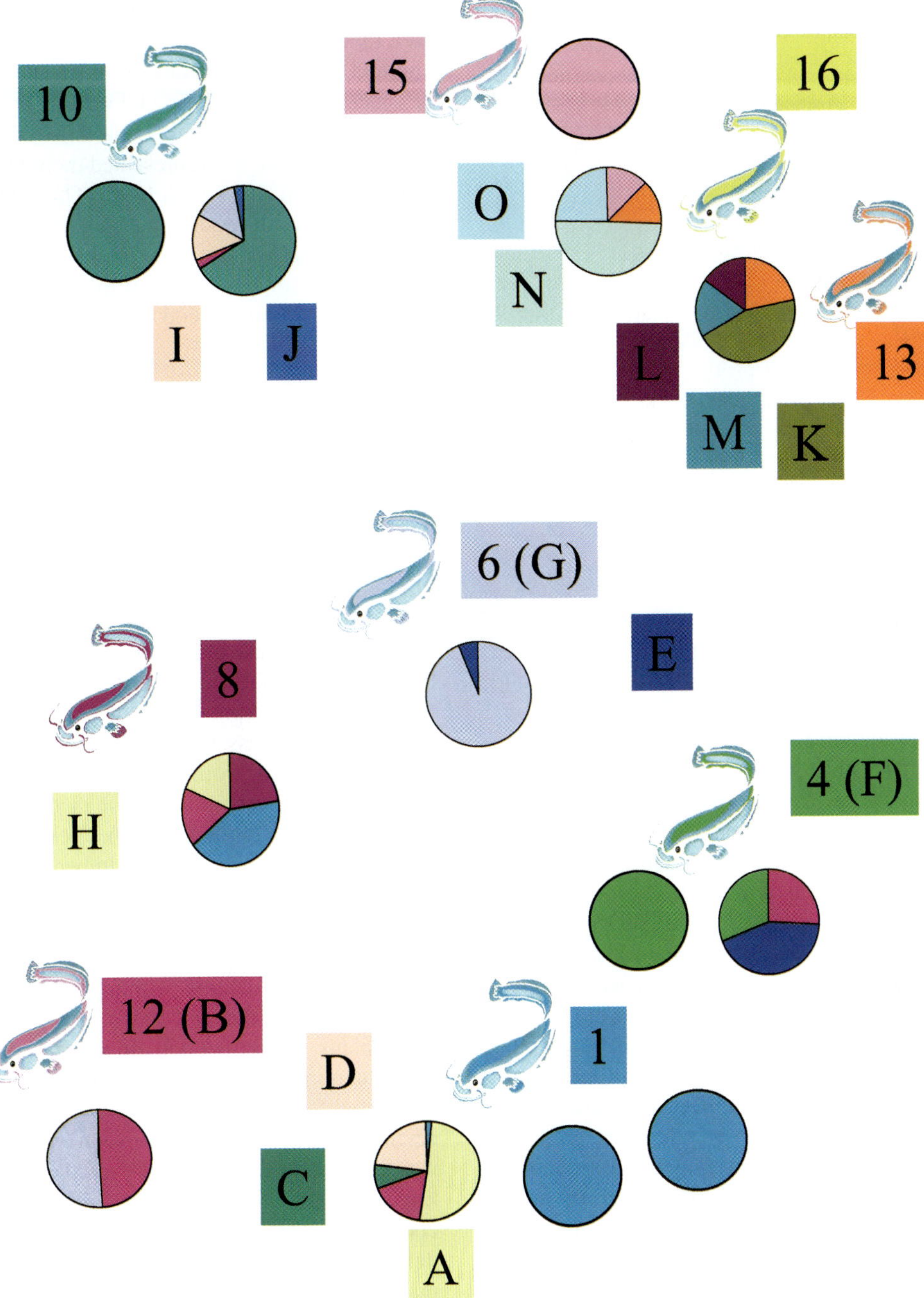

Figure 4. Pie diagrams indicating the proportion of each lingcod clutch fertilized by each of 22 male lingcod, with males differentiated by color. Clutch numbers correspond to those shown in Figure 2 except that unsampled clutches 7, 9 and 14 have been deleted. Observed guardian males are numbered and indicated by fish pictures. The sampled guardian males are indicated by number alone. Unsampled males are indicated by letters. Unsampled guardian males 4, 6 and 12 are assumed to be males F, G and B, respectively, because of the fertilizations attributed to these three males. Of the remaining unsampled males, only male E fertilized eggs in two clutches (four and six). Males A, C, D, H, I, J, K, L, M, N and O each fertilized eggs in a single clutch.

Table 6. Numbers of progeny sired by lingcod males contributing to clutches 8 and 13 in samples collected over time and from inside and outside positions on the egg clutch.

Clutch	Date	Position	Males			
8			1	8	B	H
	04/02	Inside	0	0	2	0
	04/02	Outside	0	12	0	0
	13/02	Inside	9	2	0	0
	13/02	Outside	3	6	2	1
	20/02	Outside	2	2	2	12
	26/02	Inside	45	2	0	0
	26/02	Outside	2	10	20	16
	Totals		61	34	26	29
13			13	K	L	M
	20/02	Outside	7	11	4	1
	26/02	Inside	12	10	11	3
	26/02	Outside	2	22	2	10
	Totals		21	43	17	14

Discussion

For fish species in which males guard nests, adults of both sexes may participate in multiple spawning events (DeWoody & Avise 2001, Taborsky 2001). Nests typically contain eggs from multiple females and often include eggs not fertilized by the attendant male guardian. Females may mate in succession with males who have constructed a nest or acquired the appropriate territory. The presence of eggs in the nest may increase the attractiveness of the male guardian to successive females seeking an effective and committed nest tender. Thus, the lack of evidence for polyandry among the lingcod females of this study was perhaps more surprising than the discovery of polygyny among lingcod males. A single female produced each egg clutch, and no female produced more than one clutch within the study area. If lingcod females spawn with more than one male within the breeding season, their choice of mates must be on a geographic scale that exceeds that of the current study. Moreover, the various lingcod females that spawn with a single male lay their eggs in separate clutches.

In kelp greenling, another hexagrammid in which males guard multiple clutches provided by different females, guardians have been observed with between 1 and 11 clutches (mean of four) (Crow et al. 1997). However, in that species, clutches were 'golf or tennis ball' sized, compared with the soccer ball size characteristic of lingcod clutches. The loss of one of a pair of clutches by a guardian male early in February may be indicative of some difficulty associated with defending more than one lingcod clutch, although it was in an area that contained two single clutches that also succumbed. Each of the three egg clutches collected from nest sites in this study contained $\sim$100 000 or more eggs, commensurate with females 70 cm or more in length, if the clutches represented their entire seasonal egg production (Cass et al. 1990). Thus, the laying of a single clutch each season by each female and the acquisition of a single or very few clutches by each male may be the norm for lingcod, or it may reflect the current conditions of low lingcod abundance in the Strait of Georgia.

Nest-tending males may be cuckolded by males adopting a sneaker strategy who defend no territory of their own but simply rush in and attempt to fertilize eggs while a nest guardian spawns with females (DeWoody & Avise 2001). Alternately, cuckoldry may result from related or unrelated satellite males who are tolerated by the guardian and who assist in attracting females and caring for eggs and young (Taborsky 2001). In addition, nest-tending males themselves may cuckold other nest guardians. In some fish species, these male alternative reproductive tactics (ARTs) are fixed for life or represent ontogenetic stages through which all males pass, but the most common situation is for ARTs to result from a conditional choice in which a number of factors influence the tactic adopted by an individual male (Taborsky 2001). These factors include fish density, sex ratios and number of ARTs available, as well as environmental characteristics such as availability of nesting sites and levels of predation risk.

Guardian lingcod males apparently accounted for most of the fertilization within the study area, indicating that they may be the most common adjunct participants in fertilization. In contrast, neighboring male guardian bluegill sunfish, *Lepomis macrochirus*, accounted for only 1.8% of total egg fertilization in a nesting colony in which the overall rate of cuckoldry was 21.3% (Neff 2001). However, only nine guardian lingcod males were identified in the study area but contributions from a total of 20 different males were detected in the 13 clutches. Some of the additional males may have had territories adjacent to the study area, but the continued observation of non-territorial males throughout February suggested that not all males on the breeding grounds acquired and maintained a territory.

There was no apparent phenotypic differentiation between territorial and non-territorial males in this study. Thus, it is not clear that the guardian male and sneak or satellite male are two distinct ARTs in lingcod.

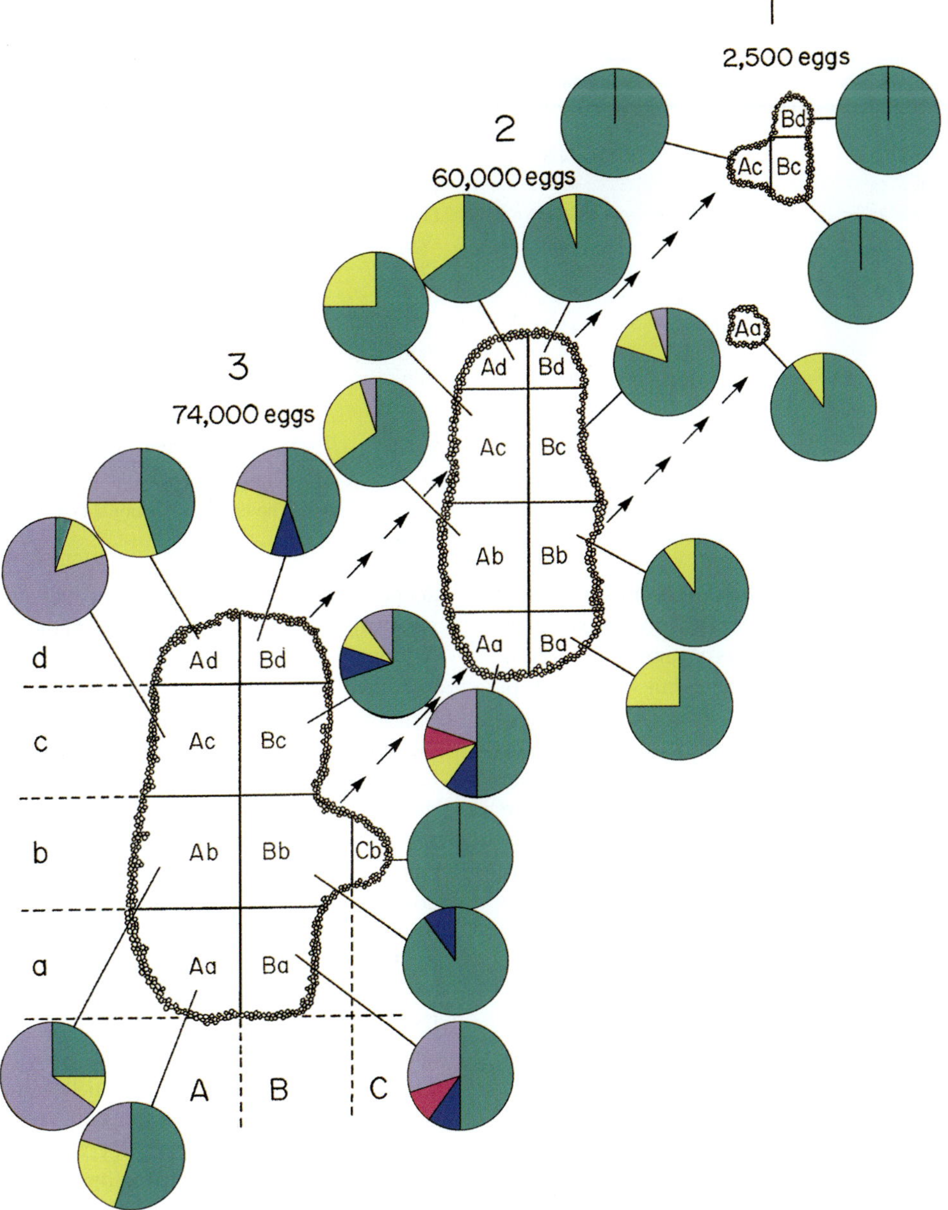

Figure 5. Pie diagrams showing unweighted percentage fertilization of egg samples from 21 sections of lingcod egg clutch 11 by five male lingcod. Fertilization by male 10 is indicated as colour teal, by male 8 as pink, by male G as purple, by male I as yellow and by male J as blue. Each section represents a portion of the nest defined on three axes from back to front as layers 1–3 (each layer shown independently), from bottom to top as layers a–d, and from left to right as layers A–C. The estimated numbers of eggs contained in each of layers 1–3 is shown above the layer.

All males may attempt to establish a territory and simultaneously participate in neighboring spawning events. Alternately, if females and/or desirable spawning sites are limited, some males may forego the expense of defending a suboptimal site entirely for the opportunity of contributing to fertilization of eggs acquired by successful territory holders. Large male lingcod tend to guard nests in deeper waters than do small males and egg survival may be greater in the deeper nests (Cass et al. 1990). Increasing size and/or experience are likely

Table 7. The weighted percentage of cross-sectional layers of clutch 11 fertilized by each of five lingcod male parents.

Layer	Clutch	Male 10	Male I	Male G	Male J	Male 8
Layer 1	0.02	98	2	0	0	0
Layer 2	0.44	74.8	19.0	5.1	0.6	0.6
Layer 3	0.54	45.4	14.0	36.8	3.5	0.4
Layer a	0.25	56.6	17.6	19.8	3.0	3.0
Layer b	0.26	63.5	9.0	24.8	2.8	0
Layer c	0.32	55.7	15.0	28.0	1.3	0
Layer d	0.17	61.3	26.1	11.2	1.5	0
Layer A	0.60	45.4	20.9	32.6	0.4	0.8
Layer B	0.39	78.4	9.4	6.7	4.8	0.6
Layer C	0.01	100.0	0	0	0	0
Total (weighted)	1.0	59.0	16.3	22.2	1.9	0.6
Total (unweighted)	1.0	68.6	13.7	14.4	2.3	1.0

Contributions of each male parent were summed over the nest sections contributing to each layer. The proportion of the nest represented by each layer is shown in the column labelled Clutch. The three-dimensional configuration of the nest axes is shown in Figure 5.

factors that increase male success in acquiring a desirable territory and persuading females to deposit eggs therein. In sculpins, male size is an important determinant of success in some species but other factors such as size of nesting area, prior deposition of eggs and degree of paternal filial cannibalism are apparently more important in others (Goto 1993, Natsumeda 1998). Two of the three males that we estimated to have fertilized the most eggs within the relatively shallow study area were the largest males but the third greatest contributor was a small male. Moreover, males may have participated in fertilizations outside the study area.

The apparently unsuccessful guardian male of this study, who fertilized none of the eggs in the clutch he was guarding, was a small/young male who may have participated ineffectively in spawning as the territory-holding or an adjunct male, or may have simply been an unrelated adoptive father. Adoption may provide unsuccessful males with a territory and the possibility of acquiring additional clutches, may provide a territory-holding male with a more attractive territory containing evidence of past success, or may simply result from frustrated paternal urges channelled into 'practice' for next year (Constantz 1985, Sargent 1989).

The spatially heterogeneous proportions of eggs fertilized by individual male parents in lingcod clutches differentiates them from the nests of some fish species, for which localized samples apparently provide accurate representation of parentage (DeWoody et al. 1998). In the dissected lingcod egg clutch, the guardian male was more successful at fertilizing eggs

deeper in the crevice perhaps because he was able to physically exclude other males from the confined area. Alternately, the adjunct males may not have attempted to spawn until egg deposition was well underway to minimize absences from their own nests. Furthermore, the guardian apparently dominated one side of the female with the adjunct males arrayed by height on the other side. There was no strong evidence that the adjunct males were present at different times during clutch deposition, with contributions from the four predominant males detected in each of both major layers of depth and width, and in all four layers of height in the clutch. Male 8, however, was either a transitory or ineffective participant, apparently successful only in fertilizing a small number of eggs in the bottom of the clutch. For studies of parentage in species with hardened egg clutches, samples from the outermost face of the clutch may provide the best representation of multiple paternity, although they would likely lead to overestimation of the contributions of adjunct males.

Lingcod are considered relatively sedentary marine fish, displaying strong adult site fidelity although females may disperse greater distances than males (Smith et al. 1990, Matthews 1992, Martell et al. 2000). These life history characteristics may have produced relatively strong population structure for a marine fish, and the evolution of one or more local populations within the Strait of Georgia (Cass et al. 1990). Nevertheless, the large number of alleles and associated high levels of heterozygosity at microsatellite loci observed in this study indicate that the effective population size(s) of lingcod in the central Strait of Georgia is large. Polygyny in mammals is often considered a factor that increases the variance in reproductive success among males and thus reduces effective population size (Storz et al. 2001). In Atlantic salmon, polygyny has been viewed as factor that might either increase the variance in male reproductive success (Garant et al. 2001) or increase the effective population size (Garcia-Vazquez et al. 2001). For iteoparous species such as lingcod, estimates of the lifetime reproductive success are required to determine if polygamy increases or decreases the interindividual variance in reproductive success. Further study may reveal that polygyny contributes to the maintenance of genetic diversity in lingcod and other nest-guarding fish species.

Acknowledgements

We thank Jim Shaklee and Sewall Young of the Washington Department of Fish and Wildlife for use of the lingcod microsatellite loci primers. Tom P. Quinn and an anonymous reviewer provided helpful comments on the manuscript.

References

Blumer, L.S. 1979. Male parental care in the bony fishes. Quart. Rev. Biol. 54: 149–161.

Cass, A.J., R.J. Beamish & G.A. McFarlane. 1990. Lingcod (*Ophiodon elongatus*). Can. Spec. Publ. Fish. Aquat. Sci. 109: 40.

Clutton-Brock, T.H. 1989. Mammalian mating systems. Proc. R. Soc. Lond (B) 236: 339–372.

Clutton-Brock, T.H. 1991. The Evolution of Parental Care, University of Princeton Press, Princeton, NJ, 368 pp.

Constantz, G.G. 1985. Allopaternal care in the tessellated darter, *Etheostoma olmstedi*. Env. Biol. Fish. 14: 175–183.

Crow, K.D., D.A. Powers & G. Bernardi. 1997. Evidence for multiple maternal contributors in nests of kelp greenling (*Hexagrammos decagrammus*, Hexagrammidae). Copeia 1997: 9–15.

DeWoody, J.A. & J.C. Avise. 2001. Genetic perspectives on the natural history of fish mating systems. J. Hered. 92: 167–172.

DeWoody, J.A., D.E. Fletcher, S.D. Wilkins, W.S. Nelson & J.C. Avise. 1998. Molecular genetic dissection of spawning, parentage and reproductive tactics in a population of redbreast sunfish, *Lepomis auritus*. Evolution 52: 1802–1810.

Garant, D., J.J. Dodson & L. Bernatchez. 2001. A genetic evaluation of mating system and determinants of individual reproductive success in Atlantic salmon. J. Hered. 92: 137–145.

Garcia-Vazquez, E., P. Moran, J.L. Martinez, J. Perez, B. de Gaudemar & E. Beall. 2001. Alternative mating strategies in Atlantic salmon and brown trout. J. Hered. 92: 146–149.

Giorgi, A.L. & J.L. Congleton. 1984. Effects of current velocity on development and survival of lingcod, *Ophiodon elongatus*, embryos. Env. Biol. Fish. 10: 15–27.

Goto, A. 1993. Male mating success and female mate choice in the river sculpin, *Cottus nozawae* (Cottidae). Env. Biol. Fish. 3: 347–353.

Hedrick, P.W. 2000. Genetics of Populations. 2nd edition, Jones and Bartlett, Portola Valley, CA, 576 pp.

Jennions, M.D. & M. Petrie. 2000. Why do females mate multiply? Biol. Rev. 75: 21–64.

Jewell, E.D. 1968. SCUBA diving observations on lingcod spawning at a Seattle breakwater. Washington Department of Fisheries Research Paper 3: 27–36.

King, J.R. & B.W. Beaith. 2001. Lingcod (*Ophiodon elongatus*) nest density SCUBA survey in the Strait of Georgia, January 16–April 26, 2001. Can. Tech. Rep. Fish. Aquat. Sci. 2374: 21.

LaRiviere, M.G., D.D. Jessup & S.B. Mathews. 1981. Lingcod, *Ophiodon elongatus*, spawning and nesting in San Juan channel, Washington. Calif. Fish. Game 67: 231–239.

Low, C.J. & R.J. Beamish. 1978. A study of the nesting behaviour of lingcod (*Ophiodon elongatus*) in the Strait of Georgia, British Columbia. Fish. Mar. Sci. Tech. Rep. No. 843, 27 p.

Martell, S.J.D., C.J. Walters & S.S. Wallace. 2000. The use of marine protected areas for conservation of lingcod (*Ophiodon elongatus*). Bull. Mar. Sci. 66: 729–743.

Matthews, K.R. 1992. A telemetric study of the home ranges and homing routes of lingcod *Ophiodon elongatus* on shallow rocky reefs off Vancouver Island, British Columbia. Fish. Bull. 90: 784–790.

Miller, K.M., T.J. Ming, A.D. Schulze & R.E. Withler. 1999. Denaturing gradient gel electrophoresis (DGGE): A rapid and sensitive technique to screen nucleotide sequence variation in populations. Biotechniques 27: 1016–1030.

Natsumeda, T. 1998. Size-assortative nest choice by the Japanese fluvial sculpin in the presence of male–male competition. J. Fish Biol. 53: 33–38.

Neff, B.D. 2001. Genetic paternity analysis and breeding success in bluegill sunfish (*Lepomis macrochirus*). J. Hered. 92: 111–119.

Sargent, R.C. 1989. Allopaternal care in the fathead minnow, *Pimephales promelas*: Stepfathers discriminate against their adopted eggs. Behav. Ecol. Sociobiol. 25: 379–385.

Smith, B.D., G.A. McFarlane & A.J. Cass. 1990. Movements and mortality of tagged male and female lingcod in the Strait of Georgia, British Columbia. Trans. Amer. Fish. Soc. 119: 813–824.

Snedecor, G.W. & W.G. Cochran. 1967. Statistical Methods. 6th edition, Iowa State University Press, Ames, Iowa, 593 pp.

Storz, J.F., H.R. Bhat & T.H. Kunz. 2001. Genetic consequences of polygyny and social structure in an Indian fruit bat, *Cynopterus sphinx*. II. Variance in male mating success and effective population size. Evolution 55: 1224–1232.

Taborsky, M. 2001. The evolution of bourgeois, parasitic and cooperative reproductive behaviors in fishes. J. Hered. 92: 100–110.

Thompson, J.D., T.J. Gibson, F. Plewniak, F. Jeanmougin & D.G. Higgins. 1997. The ClustalX windows interface: Flexible strategies for multiple sequence alignment aided by quality analysis tools. Nucl.s Acids Res. 24: 4876–4882.

Environmental Biology of Fishes **69**: 359–369, 2004.
© 2004 *Kluwer Academic Publishers. Printed in the Netherlands.*

Differential reproductive success of sympatric, naturally spawning hatchery and wild steelhead, *Oncorhynchus mykiss*

Jennifer E. McLean[a], Paul Bentzen[b] & Thomas P. Quinn
School of Aquatic and Fishery Sciences, Box 355020, University of Washington, Seattle,
WA 98195, U.S.A. (e-mail: mclean@zoology.ubc.ca)
[a]*Present address: Department of Zoology, University of British Columbia, Vancouver, British Columbia,*
Canada, V6T 1Z4
[b]*Present address: Department of Biology, Dalhousie University, Halifax, NS, Canada, B3H 4J1*

Received 17 April 2003   Accepted 24 April 2003

Key words: assignment test, fitness, hatcheries, microsatellites, salmonidae, smolt production

Synopsis

Hatchery propagation of salmonids has been practiced in western North America for over a century. However, recent declines in wild salmon abundance and efforts to mitigate these declines through hatcheries have greatly increased the relative abundance of fish produced in hatcheries. The over-harvest of wild salmon by fishing mixed hatchery and wild stocks has been of concern for many years but genetic interactions between populations, such as hybridization, introgression and outbreeding depression, may also compromise the sustainability of wild populations. Our goal was to examine whether a newly established hatchery population of steelhead trout successfully reproduced in the wild and to compare their rate of reproductive success to that of sympatrically spawning native steelhead. We used eight microsatellite loci to create allele frequency profiles for baseline hatchery and wild populations and assigned the smolt (age 2) offspring of this parental generation to a population of origin. Adults originating from a generalized hatchery stock artificially selected for early return and spawning date were successful at reproducing in Forks Creek, Washington. Although hatchery females (N = 90 and 73 in the two consecutive years of the study) produced offspring that survived to emigrate as smolts, they produced only 4.4–7.0% the number produced per wild female (N = 11 and 10). This deficit in reproductive success implies that the proportion of hatchery genes in the mixed population may diminish since deliberate releases into the river have ceased. This hypothesis is being tested in a long-term study at Forks Creek.

Introduction

Salmonid fishes have a well-documented ability to home to their natal streams for reproduction (Quinn 1993). This homing limits gene flow among populations in different locations and allows the accumulation of differences in adaptive traits such as behaviour, morphology, physiology and disease resistance among populations (Ricker 1972, Taylor 1991). Selectively neutral genetic characters, such as polymorphic proteins and DNA microsatellite loci, do not directly reflect local adaptation, but can be used to infer whether populations are reproductively isolated from one another. Microsatellite DNA techniques have been applied to examine the relationships among many natural salmonid populations (e.g. Atlantic salmon, McConnell et al. 1995; Arctic char, Brunner et al. 1998; steelhead trout, Beacham et al. 1999); here we employed these techniques to differentiate between a native wild population and a transplanted hatchery population in one stream to compare the reproductive success of these forms under natural conditions.

Local, population-specific adaptations include changes in characteristics associated with reproductive success, such as adult body size and the seasonal patterns of migration and reproduction. The body size of a female controls her capacity for egg production (number and size of eggs). The timing of spawning is also extremely important for reproductive success in salmonids, as the survival and growth of juveniles varies with the date when they emerge and begin feeding in spring (Brännäs 1995, Einum & Fleming 2000). Females that return early will have progeny that emerge earlier, and will be larger and have acquired a territory by the time later fry emerge (Chandler & Bjornn 1988). Eggs that have been deposited early, however, may be more susceptible to scour by high river flows (Cederholm[1]) and early fry may be vulnerable to predation (Brännäs 1995). The morphology and behaviour of males are shaped by the same processes of natural selection (e.g. migration, energetics and predator avoidance) but also by sexual selection for access to females on the spawning grounds (reviewed by Fleming 1996). Body size is a significant factor in these intrasexual competitive interactions. Populations evolve by culling individuals with inappropriate size, shape, breeding date or other traits, given the patterns of temperature, flow and biotic factors that characterize different rivers. Because body size and maturation timing are under partial genetic control (size in rainbow trout, Crandall & Gall 1993; timing in rainbow trout, Siitonen & Gall 1989; chinook salmon, Quinn et al. 2000; pink salmon, Smoker et al. 1998), these traits are passed on to subsequent generations.

It is widely accepted that these evolutionary processes confer a 'home court' advantage for salmon of the local population, relative to strays from other populations that might breed there (Quinn 1993). The survival advantages associated with such local adaptation are implied by the failure of virtually all transplants within the range of Pacific salmon to generate self-sustaining populations (Withler[2]), the higher survival rates of local compared to non-local salmon released from hatcheries (Reisenbichler 1988), and the evolution of local adaptations that confer survival advantages in transplanted populations (Quinn et al. 2001).

Hatcheries have artificially propagated salmon and trout in Europe and North America for more than 100 years (Nielsen 1994), and recent work has expressed concern regarding their effects on wild populations (Ryman & Laikre 1991, Waples 1991,

Utter 1998). Hatchery-produced juveniles may compete for food and space with native populations in streams (e.g. Nickelson et al. 1986, Nielsen 1994), but it is unclear what success hatchery salmon have when spawning in the wild, or what effects they have on the reproductive success of wild salmon. Hatchery fish spawning in the wild can be thought of as strays. Juveniles in hatcheries experience very different rearing conditions (e.g. feeding, use of space and predator avoidance) than do fish in streams, and this 'domestication selection' (Reisenbichler & Rubin 1999) produces differences in many behavioural traits (e.g. Berejikian 1995, Berejikian et al. 1996). This is also true for adults, who experience altered selective regimes when spawned in the hatchery (Waples 1991). Mate choice and spawning behaviour are not determined by the fish, but by the hatchery staff, who may accomplish domestication either purposely (e.g. shifting the timing of reproduction over generations by preferentially spawning early returning fish) or inadvertently, through selection of adults that would otherwise be culled by natural or sexual selection in the river.

Interactions between hatchery and wild fish are a pressing issue for salmonid conservation because hatchery production is such a large fraction of the total abundance in many areas. Light[3] estimated that approximately half of the adult steelhead trout, *Oncorhynchus mykiss*, in North America are of hatchery origin. Steelhead are an anadromous, iteroparous salmonid and spawning typically occurs during late winter and early spring (Busby et al.[4]), although dates differ among rivers. In Washington State, the great majority of hatchery steelhead are derived from a few stocks of complex ancestry. These stocks, notably one produced in the Chambers Creek Hatchery, have been selected to return and spawn earlier in the winter than wild populations (Crawford[5]).

---

[1] Cederholm, C.J. 1984. Clearwater River wild steelhead spawning timing. pp. 257–268 *In*: Proceedings of the Olympic Wild Fish Conference. Port Angeles, Washington.

[2] Withler, F.C. 1982. Transplanting Pacific salmon. Canadian Technical Report of Fisheries and Aquatic Sciences 1079, 27 pp.

[3] Light, J.T. 1989. The magnitude of artificial production of steelhead trout along the Pacific coast of North America, Fisheries Research Institute FRI-UW-8913, University of Washington, Seattle, 11 pp.

[4] Busby, P.J., T.C. Wainwright, G.J. Bryant, L. Lierhiemer, R.S. Waples, F.W. Waknitz & I.V. Lagomorsino. 1996. Status review of west coast steelhead from Washington, Idaho, Oregon and California. U.S. Dept. Commer., NOAA Tech. Memo. NMFS-NWFSC-27, 261 pp.

[5] Crawford, B.A. 1979. The origin and history of the trout brood stocks of the Washington Department of Game, Fisheries Research Report, Washington State Game Department, Olympia, 76 pp.

This project forms part of a growing literature comparing the fitness and reproductive success of wild and hatchery, and native and non-native populations, and investigating the consequences of allowing them to spawn in sympatry (steelhead, *O. mykiss*, Chilcote et al. 1986, Leider et al. 1990; coho salmon, *O. kisutch*, Fleming & Gross 1992; brown trout, *Salmo trutta*, Cagigas et al. 1999, Skaala et al. 1996; Atlantic salmon, *S. salar*, Mork 1991, Crozier 1993, Fleming et al. 1996, 2000, Crozier et al. 1997, Crozier 2000, Fleming & Petersson 2001). In Washington, as in many areas, hatcheries have been operating for many generations, making it difficult to study the genetic interactions between wild and hatchery fish. However, the initiation of steelhead production at the Forks Creek Hatchery, Washington provided a rare opportunity to investigate these interactions in the early years before the effects of introgression might occur. Here we present the initial results from our ongoing study designed to compare the reproductive success of hatchery and wild steelhead. We had three objectives at this stage of the project: (1) to characterize allele frequency profiles of the hatchery and wild populations before any interbreeding might occur and determine if they were different, (2) to determine whether hatchery steelhead successfully reproduced when permitted access to natural spawning grounds, and (3) to compare the reproductive success (in terms of smolts per female) of the hatchery and wild populations.

Methods

Study site

Forks Creek is a tributary of the Willapa River in southwest Washington (Figure 1). Forks Creek Hatchery is located ~250 m upstream of the confluence of Forks Creek and the Willapa River and has operated as a salmon hatchery since 1895. Prior to 1994, Forks Creek supported a small wild run of winter steelhead that spawns from approximately March through May (Mackey et al. 2001). In 1994, Forks Creek Hatchery received 25 000 smolts from the Bogachiel Hatchery, located to the north of the Willapa River along the west coast of Washington. This population was derived from a combination of native Bogachiel River steelhead and the Chambers Creek stock, a generalized hatchery stock artificially selected for early spawning (November to February; Ayerst 1977, Crawford[5], Mackey et al. 2001). These and all subsequent hatchery

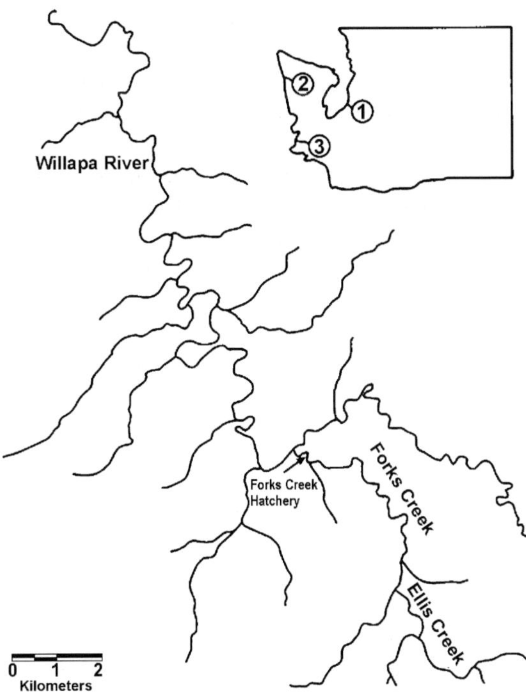

Figure 1. Location of Forks Creek Hatchery. Inset shows Washington State and the location of Chambers Creek Hatchery (1) from which the hatchery population was derived and Bogachiel Hatchery (2) from which the original hatchery smolts were taken for release in Forks Creek. Forks Creek Hatchery is designated by (3).

smolts were marked by the removal of their adipose fins. The first hatchery adults returned in the winter of 1995/1996 and we designated them as brood year ('BY') 1996. Returning salmon and steelhead are prevented from migrating upstream of the hatchery by a weir across the creek, allowing hatchery staff access to returning fish. Adult steelhead with an intact adipose fin (i.e. naturally produced) are placed upstream. Those missing an adipose fin are taken for spawning in the hatchery. In the first 2 years when hatchery adults returned, after the hatchery's capacity for steelhead eggs was met, excess hatchery fish were released upstream of the hatchery. This practice was then discontinued, and hatchery fish are no longer allowed upstream. Thus the wild population was exposed to a discrete, 2-year 'pulse' of hatchery influence.

Sampling

Our sampling in Forks Creek has been extensive. Since 1996, we have sampled all naturally produced smolts trapped at the weir, as well as all adults, both hatchery and wild, trapped at the weir. We have also subsampled naturally produced juveniles in the stream and hatchery produced juveniles in the hatchery prior to their release. Examination of the length frequency distributions of juveniles, scales collected from adults, and the general life history of steelhead in this region lead us to conclude that virtually all smolts leave the river after 2 years of residence. Here we examine adults from BY1996 and BY1997, and their offspring: smolts collected in the spring of 1998 and 1999. The 55 wild and 362 hatchery adults which returned in 1996 and 1997, before any potential mixing of the hatchery and wild gene pools, provided our genetic baseline data.

Hatchery adults were sampled weekly at the hatchery from November to February. We recorded the date of spawning, sex, fork length, weight and took scale samples for age estimation and a fin clip for DNA analysis. Fecundity was also estimated for 65 hatchery females in 1996 and 1997 by obtaining the total mass of eggs, the mass of a sample of eggs, and the number of eggs in that sample. The best relationship, a linear one: fecundity $= 13.929 \times$ length $- 4933.4$ ($r^2 = 0.414$), was used to estimate the potential egg deposition by naturally spawning females based on observed lengths.

Wild adults tended to return later in the year and were trapped on their way upstream at the hatchery weir and downstream at the smolt trap, located at the hatchery and operated from mid-late April to mid-June. Data taken from adult wild steelhead included date and direction of migration, sex, fork length, weight, a scale sample and a fin clip. Concern about the status of the wild population prevented us from sacrificing females for fecundity so we used the length–fecundity relationship from the hatchery fish to estimate the egg deposition by wild females.

Genetic analysis

Genomic DNA was extracted from fin clips stored in 95% ethanol by standard CTAB protocol (Fields et al. 1989) or by Qiagen kits. DNA was resuspended in 50–100 µl of TE solution (10 mM Tris, 1 mM EDTA in H_2O; pH 8.0) and stored at $-20°C$. Eight microsatellite loci were examined (Table 1). PCRs were carried out in 10 µl volumes (10 mM Tris–HCl (pH 8.3), 50 mM KCl, 1–2 mM $MgCl_2$ (see Table 1), 0.25 mM each dNTP, 1 U Taq DNA polymerase (Promega), 0.5 µM each primer and 100 ng DNA template) using an MJ Research PTC-200 thermocycler. Amplifications took place under the following conditions: (1) three cycles of 95°C for 1 min, X°C for 30 s, 70°C for 1 min; (2) 22 cycles of 95°C for 10 s, X°C for 30 s, 70°C for 1 min; (3) one cycle at 70°C for 45 min. X is an annealing temperature that varied among loci (see Table 1). Microsatellites were size fractioned using a 96-well capillary system Molecular Dynamics MegaBACE 1000 (Amersham Scientific). Electropherograms were analyzed using Genetic Profiler software version 1.1 (Molecular Dynamics).

Table 1. Microsatellite loci examined and PCR details. X is the PCR annealing temperature (°C), [$MgCl_2$] is the magnesium chloride concentration (mM), N_A is the number of alleles found in each population at that locus and H_E is the expected heterozygosity (%) for each population. All adults from 1996 and 1997 are included: 362 hatchery and 55 wild samples.

				Hatchery		Wild	
Locus	Reference	[$MgCl_2$]	X	N_A	H_E	N_A	H_E
Oki23	A. Spidle[a]	1	55	24	86.7	16	89.5
Omy77	Morris et al. (1996)	1	55	14	83	11	85.5
Omy1001UW	P. Bentzen[b]	1	55	21	87.7	16	90
Omy1011UW	P. Bentzen[b]	1	55	23	91.3	16	91.5
Omy1191UW	P. Bentzen[b]	1	65	26	93.3	26	94.9
Omy1212UW	P. Bentzen[b]	1	65	56	94	31	96.1
One108	Olsen et al. (2000)	1	55	26	91	16	91.4
Ssa85	O'Reilly et al. (1996)	2	60	24	82.6	15	85.6
Average				27	88.7	18	90.5

[a]A. Spidle, unpublished data. Genbank Accession # AF272822.
[b]P. Bentzen, unpublished data.

Data analysis

Heterozygosity, probability tests of Hardy–Weinberg equilibrium, tests of linkage disequilibrium, and genetic differentiation estimates among hatchery and wild populations were calculated using the GENEPOP (version 3.0) software package (Raymond & Rousset 1995). The program GENECLASS (Cornuet et al. 1999) was used both to self-classify the adult hatchery and wild samples to determine accuracy, and to assign smolts originating in the river from either hatchery- or wild-origin parents to a parental population of origin. GENECLASS uses a likelihood approach to calculate the probability of belonging to one of the baseline populations and is similar to Raanala & Mountain's (1997) method of identifying migrant individuals. We used the Bayesian likelihood algorithm option and the 'as is' procedure. A number of assignment tests were performed to (1) evaluate the extent of separation between the hatchery and wild populations, (2) determine the best baseline dataset, and (3) estimate the likelihood of misclassification of smolts. These tests included the following: 1996 and 1997 adults were self-assigned, true unknown test samples with genotypes not present in the baseline datasets were evaluated (1996 adults were assigned to 1997 and 1997 adults were assigned to 1996), and to improve the accuracy of determination we created a new baseline dataset by adjusting the sample size of the baseline hatchery population to match that of the baseline wild population. We did this by randomly selecting 55 hatchery adults from a pool of the 362 genotypes collected, and using these as the hatchery baseline samples. A list of 55 random numbers between 1 and 362 were generated in Microsoft Excel, and the DNA samples with the corresponding numbers were used. All 55 wild baseline samples were used. Multiple random samples showed consistent correct classification percentages, so we chose one random sample to use for the analyses. To estimate the likelihood of misclassification of smolts, we examined the log likelihood ratios (LLRs) of the correctly classified and misclassified individuals. Because we did not obtain 100% correct assignment of parents at an LLR of zero, we defined the ratio at which all baseline adults were correctly classified and used this ratio as a criterion for classifying smolts.

In the 'unknown' sample assignment tests, we assigned all smolts (366 from 1998 and 285 smolts from 1999) to a population of origin. Estimates of production for each population were determined by dividing the number of smolts assigned to that population by the number of females of that type that had ascended the creek to spawn in the parental year (i.e. two years before the smolts emigrated). After smolts were assigned to a population, we examined differences in date of emigration and size at emigration between the smolts produced by hatchery and wild parents with two-sample t-tests. We also assessed differences between the parent populations for these traits.

Results

Population genetics

Heterozygosities were high, ranging from 82.6% to 94.0% in the total hatchery parent population and 85.6–96.1% in the wild parent population (Table 1), 84.0–94.6% in the 1998 smolt population and 79.8–93.8% in the 1999 smolt sample (data not shown). Probability tests showed no significant departure from HWE in the wild adult population, however, with a sequential Bonferroni correction for multiple comparisons (Rice 1989), all loci in the hatchery population but one, Omy1001, significantly differed from Hardy–Weinberg proportions. The randomly generated hatchery sample used in the assignment test did not show a significant departure from Hardy–Weinberg proportions. Probability tests for genotypic linkage disequilibrium resulted in four significant values among 56 pairwise tests of eight microsatellite loci and two populations. These all occurred in the hatchery population (Omy77 × Oki23, Omy1011UW × Oki23, Ssa85 × One108, One108 × Oki23). Estimates of F_{ST} ranged from negative values to 0.013 and averaged 0.005 over all loci for the total combined adult sample (1996 and 1997, hatchery and wild). The random sub-sample of hatchery adults combined with the wild adults resulted in an F_{ST} estimate of 0.005.

Population assignment

Adult samples were used both for self-assignment and as unknowns. In the self-assignment tests, the individuals from the 1996 sample were correctly classified to their true populations, hatchery or wild, 86% of the time. The individuals from 1997 had a correct classification rate of 99%. When the 2 years were combined, correct classification was 85%. When classifying 1 year as true unknowns (genotypes not

364

found in the baseline file) to the other year as baseline data, 1996 as unknowns assigned to 1997 as baseline was 82% correct and 1997 as unknowns assigned to 1996 was 63% correct. In our created baseline data set (with the random sample from the hatchery population), self-assignment was 92% correct. Eight wild adults were incorrectly assigned to the hatchery population and one hatchery adult was incorrectly assigned to the wild population. In this, as in all of our assignment tests, there was a bias toward assigning wild fish to the hatchery population rather than the reverse, so the results will tend to over-estimate the productivity of hatchery females and underestimate the productivity of wild females.

Through self-assignment of our created baseline sample, we determined that the criterion for correct assignment of smolts was ±0.8 (Figure 2). Smolts from 1998 were assigned using the created baseline: 96 were assigned to the hatchery population and 269 to the wild population with an LLR of 0. With an LLR of ±0.8, we were unsure of the correct assignment of and therefore removed 42 hatchery-assigned smolts and 68 wild-assigned smolts (30% of the smolt sample). After this correction, 54 were assigned to the hatchery population and 201 to the wild population. Smolts from 1999 were assigned using the same created baseline dataset: 97 were assigned to the hatchery population and 188 were

assigned to the wild population with an LLR of 0. With an LLR of ±0.8, 42 individuals were removed from the hatchery population and 23 individuals were removed from the wild population (23%), leaving 55 hatchery and 165 wild individuals. The possible range, then, for 1998 was 54–164 hatchery smolts and 201–311 wild smolts. The range in 1999 was 55–120 hatchery smolts and 162–230 wild smolts.

Because the assignment test results are the same regardless of the estimate used (fewer hatchery than wild smolts produced despite many times more hatchery females than wild females), we will use the LLR = 0 estimates. Hatchery females produced an average of 1.07 smolts per capita spawning in 1996 (smolt year 1998) and 1.33 smolts per capita spawning in 1997 (smolt year 1999). Wild females produced an average of 24.50 smolts per capita in 1996 and 18.80 in 1997.

Wild and hatchery adults overlapped in size and migration date but wild fish tended to be larger and arrive later (Tables 2 and 3). Based on the length–fecundity relationship generated for the hatchery population (and assuming a similar relationship for the wild population), we estimated the mean fecundity to have been 4 547 eggs per wild female and 3 862 eggs each for hatchery females in 1996, and 4 607 eggs per wild female and 3 895 eggs each for hatchery females in

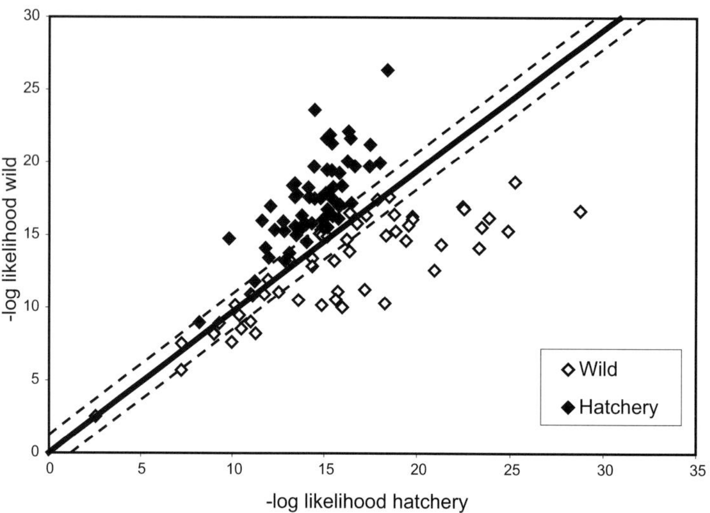

Figure 2. Separation of hatchery and wild populations based on microsatellite data. The negative log likelihood of an individual belonging to the wild population is plotted against the negative log likelihood of an individual belonging to the hatchery population, and the 1:1 line, along which an individual has an equal probability of being in either population, is a solid line. Each point is an individual; hatchery individuals are represented by black diamonds, and wild individuals are represented by white diamonds. These 110 fish include all wild fish from 1996 and 1997 and an equally sized random sample of the hatchery fish from those two years. The LLR lines, shown as dashed lines, indicate the cutoff criterion for 100% correct assignment.

Table 2. Mean fork length (mm) and estimated fecundity (# eggs per female) of hatchery- and wild-origin adult steelhead spawning naturally in 1996 and 1997 in Forks Creek.

	Females		Males	
	Wild	Hatchery	Wild	Hatchery
1996				
Mean length	681	632	686	629
Minimum length	623	440	459	406
Maximum length	756	762	910	760
Mean fecundity	4547	3862		
Total fecundity	50017	347580		
N	11	90	21	75
1997				
Mean length	685	634	619	649
Minimum length	590	530	450	433
Maximum length	780	772	810	724
Mean fecundity	4607	3895		
Total fecundity	46070	284335		
N	10	73	17	124

Table 3. Mean date naturally spawning steelhead in Forks Creek passed the weir in 1996 and 1997.

	Females		Males	
	Wild	Hatchery	Wild	Hatchery
1996	6 March	13 January	29 April	15 January
1997	23 April	31 January	20 May	30 January

Table 4. Mean date of emigration and mean fork length (mm) at emigration from Forks Creek for hatchery and wild steelhead smolts, 1998 and 1999.

	Wild	Hatchery
1998		
Mean fork length	176.8	180.9
Mean emigration date	13 May	13 May
N	96	269
1999		
Mean fork length	163.3	163.2
Mean emigration date	21 May	21 May
N	97	188

1997. We then estimated the total egg production of each population in each year to have been 50 017 wild eggs and 347 580 hatchery eggs in 1996 (12.6% wild) and 46 070 wild eggs and 284 335 hatchery eggs in 1997 (13.9% wild). Because the wild females were somewhat longer than the hatchery fish, their total egg production was closer to that of hatchery fish than might be inferred from merely the total number of females (10.9% wild in 1996 and 12.0% in 1997).

The median emigration date of the hatchery and wild smolts was the same in 1998 and 1999 (Table 4). Hatchery smolts were longer by an average of 4 mm in 1998, but the same mean length as wild smolts in 1999 (2-sample t-test p < 0.05, Table 4).

Discussion

Assignment tests and population genetics

A test of individual assignment back to population of origin with our created baseline dataset indicated that 92% of the individual assignments were correct. Correct assignment to the hatchery population was 98%, and correct assignment to the wild population was 86%. Because the hatchery fish clustered more closely than the wild fish and the percentage of correct assignment was higher for this group, it was easier to determine if a fish originated from the hatchery than the wild population. The patterns of clustering indicated a higher degree of heterogeneity among the wild fish. Although the wild population is quite small ($\sim$30 fish return on average each year), its higher diversity may be due to reproductively successful strays either from other hatchery or wild populations, some form of inbreeding avoidance occurring in the population during spawning, or a higher long-term effective population size than the hatchery population as a result of a more constant population size, genetic contributions from mature male parr (Seamons et al., this issue) and a lack of artificial selection.

The low F_{ST} value between the hatchery and wild populations (0.005) and the overlap of the two groups seen in Figure 2 can be explained by a number of features of the populations. The hatchery fish returning to Forks Creek may have clustered more closely than the wild fish due to the effects of repeated bottlenecks and small effective population size that can occur in hatcheries (Kincaid 1995, Verspoor 1988). The heterozygosities at the loci examined are high in both populations, and not very different between the two populations (Table 1). The number of alleles per locus was higher in the hatchery population because of the much higher (almost eight times) sample size. The wild adult population had allele frequencies consistent with Hardy–Weinberg proportions, but the adult hatchery population did not. This may reflect a number of population processes that have occurred extensively

366

in the history of the hatchery population but not in the wild population: non-random mating, high levels of random genetic drift from small effective population sizes, and domestication and artificial selection. All significant cases of linkage disequilibrium occurred in the hatchery population, which is also consistent with these processes acting differently on the two populations. Linkage disequilibrium can be generated by genetic drift and selection, both of which have occurred in the hatchery. Further, the rate at which linkage equilibrium re-establishes is slowed when the level of inbreeding is high, a process which may be maintaining linkage disequilibrium in the hatchery.

Smolt production

We obtained reasonable levels of correct assignment (cf. Beacham et al. 1999) and are confident in our ability to assign the offspring of naturally spawning wild and hatchery steelhead (the latter identified as adults by the absence of their adipose fins) to the correct population. However, because of the bias towards assigning to the hatchery population, reproductive success was underestimated for wild females and overestimated for hatchery females. Despite this bias, wild females outproduced hatchery females substantially in both years examined. Hatchery females produced an average of 1.07–1.33 smolts each and wild females produced an average of 18.80–24.50 smolts each.

Smolt production by steelhead varies among populations, and among years within a population. For example, estimates of smolts produced per female for Snow Creek, Washington range from 21.7 to 684 (T. Johnson, Washington Department of Fish and Wildlife, unpubl. data). Our results indicated that the production by wild fish was low but within the range observed for other populations, and the wild fish greatly outperformed the hatchery fish. A similar study on the Kalama River, Washington steelhead obtained comparable results: hatchery steelhead adults, although significantly outnumbering wild adults on the spawning grounds, produced fewer smolts per female than did wild spawners (hatchery steelhead had an average of 28% the reproductive success of wild steelhead to the smolt stage, Chilcote et al. 1986).

Why did wild steelhead outperform hatchery steelhead? In Forks Creek, differential reproductive success could be due to a number of differences between the populations but the most obvious are differential fecundity, the timing of reproduction by the parents, and domestication selection affecting adults or juveniles. The wild females were larger than the hatchery females in both years, and this may have resulted from the longer period of time spent at sea (i.e. the difference between arrival in January and April) or genetic factors. Using a relationship developed for the hatchery population, we estimated that on average each wild female would have produced 15% more eggs than each hatchery female. However, differences in number of smolts per female were 23-fold for females spawning in 1996 and 14-fold in 1997, so fecundity alone cannot explain the magnitude of the difference.

In Washington State, steelhead management policy has included the temporal segregation of hatchery and wild fish through an advancement of the spawn timing of hatchery fish. This temporal separation was originally achieved with the Chambers Creek stock, which was later taken to many hatcheries throughout the state. Such early spawning, leading to early emergence of fry, lengthens the growing period and enables the juveniles to achieve smolt transformation after 1 year, rather than 2 years as is typical of wild populations. The strategy of early spawning may work well for hatchery fish returning to hatcheries to spawn, but it may reduce their fitness under natural conditions. Forks Creek is subject to floods and scour during the early months of the year, and this may explain why few wild steelhead spawn there before March (Mackey et al. 2001). Hatchery steelhead return to Forks Creek mainly in December and January, at the time of coldest temperatures and highest discharge, which might lead to gravel scour, fine sediment transport, and embryo mortality. Cederholm[1] hypothesized that in western Washington, steelhead seldom spawn before February in order to avoid high flows that may decrease their reproductive success. In Forks Creek, the wild fish follow this strategy, and the difference in spawn timing may be one reason why wild steelhead produced many more smolts than hatchery steelhead.

If the progeny of hatchery females survived the incubation period, they would emerge much earlier than those of wild fish (estimated peak dates of 21 April as opposed to 16 June, Mackey et al. 2001). If space and food are limited, there should be an advantage for the early (hatchery) fish (Brännäs 1995, Einum & Fleming 2000). In general, the density of steelhead (and their primary interspecific competitors, coho salmon and cutthroat trout) is low in Forks Creek, as indicated by sampling of juveniles in the creek and counts of smolts leaving the creek (Quinn & McLean, unpubl. data). Thus early emergence of hatchery steelhead may not provide a substantial advantage in acquiring territories, and cold water may reduce the size advantage associated with early emergence. Unfortunately, our data do not allow us to separate patterns of mortality during the incubation stage from those in the river.

In addition to differences related to egg production and the timing of breeding and emergence, genetic differences between wild and hatchery adults or juveniles resulting from hatchery propagation may have contributed to the differential reproductive success. The hatchery fish are originally from a distant location and have undergone generations of selection for success in hatchery environments. Evidence indicates that hatchery steelhead derived from a wild population may differ from wild fish in predator avoidance (Berejikian 1995) and agonistic behaviour (Berejikian et al. 1996) of juveniles after several generations. Studies have also indicated that the reproductive behaviour of hatchery adults differs from that of wild conspecifics, and this too may have contributed to the lower productivity of hatchery fish (Jonsson et al. 1991, Fleming et al. 1996). In contrast, the wild population should be adapted for local in-stream conditions that affect fitness (e.g. redd site selection, nest preparation, etc.) and has not experienced artificial or domestication selection.

Which factor (fecundity, timing of reproduction, juvenile or adult behaviour) had the most influence on the differential reproductive success of these two groups? Population-level comparisons between groups such as these that differ in many characteristics make it difficult to determine the most important causes of unequal production. Investigation into the reproductive success of individual adults (e.g. wild and hatchery fish with the same timing and size) will be needed to disentangle the interactions among these factors.

Hybridization between hatchery and wild steelhead is perhaps the most critical issue facing the wild population in Forks Creek. Reproductive success through the smolt stage of the hatchery fish is so much lower than that of the wild fish that hybridization (if it commonly occurs and if the productivity of hybrids is intermediate between the pure forms) has the potential to extirpate the wild population. Unique locally adapted gene complexes may be lost, and a potentially severe reduction in abundance might result. If the reproductive rate does not meet replacement at the adult stage, the population will decline. The number of smolts produced by wild

females was comparatively low, suggesting that the population is not very productive and hence vulnerable to a reduction in productivity.

When hatchery fish outnumber wild fish on the spawning grounds, which was the case in the two years of our study, the two gene pools might mix despite differences in average timing (Mackey et al. 2001). Repeated releases of hatchery fish over many years (the pattern that has prevailed in most situations where hatchery and wild salmon are produced in the same river), will exaggerate this problem. In Forks Creek, hatchery steelhead are no longer deliberately released upstream of the hatchery, and this may prevent the negative effects of repeated releases. This situation is unique in that we have sampled the first generation of hatchery releases, and we will be able to monitor what happens in the future.

Acknowledgements

We thank the many people involved in sample collection, especially the Forks Creek Hatchery staff and Greg Mackey. Long Live the Kings assisted with the trapping operation, and we particularly thank Brodie Smith and Larry Sienko for help in that regard. We thank two anonymous reviewers for valuable comments on the manuscript. This work was made possible by financial support from the Weyerhaeuser Company Foundation, the National Science foundation (Grant DEB-9903914) and the Hatchery Science Reform Group in Washington State.

References

Ayerst, J.D. 1977. The role of hatcheries in rebuilding steelhead runs of the Columbia River system. pp. 84–88. *In*: E. Schweibert (ed.) Columbia River salmon and steelhead, American Fisheries Society Special Publication 10.

Beacham, T.D., S. Pollard & K.D. Le. 1999. Population structure and stock identification of steelhead (*Oncorhynchus mykiss*) in southern British Columbia, Washington, and the Columbia River based on microsatellite DNA variation. Trans. Amer. Fish. Soc. 128: 1068–1084.

Berejikian, B.A. 1995. The effects of hatchery and wild ancestry and experience on the relative ability of steelhead trout fry (*Oncorhynchus mykiss*) to avoid a benthic predator. Can. J. Fish. Aquat. Sci. 52: 2476–2482.

Berejikian, B.A., S.B. Mathews & T.P. Quinn. 1996. Effects of hatchery and wild ancestry and rearing environments on the development of agonistic behavior in steelhead trout

368

(*Oncorhynchus mykiss*) fry. Can. J. Fish. Aquat. Sci. 53: 2004–2014.

Brännäs, E. 1995. First access to territorial space and exposure to strong predation pressure: a conflict in early emerging Atlantic salmon (*Salmo salar* L.) fry. Evol. Ecol. 9: 411–420.

Brunner, C., M.R. Douglas & L. Bernatchez. 1998. Microsatellite and mitochondrial DNA assessment of population structure and stocking effects in Artic charr *Salvelinus alpinus* (Teleostei: Salmonidae) from central Alpine lakes. Mol. Ecol. 7: 209–223.

Cagigas, M.E., E. Vazquez, G. Blanco & J.A. Sanchez. 1999. Genetic effects of introduced hatchery stocks on indigenous brown trout (*Salmo trutta* L.) populations in Spain. Ecol. Fresh. Fish 8: 141: 150.

Chandler, G.L. & T.C. Bjornn. 1988. Abundance, growth, and interactions of juvenile steelhead relative to time of emergence. Trans. Amer. Fish. Soc. 117: 432–443.

Chilcote, M.W., S.A. Leider & J.J. Loch. 1986. Differential reproductive success of hatchery and wild summer-run steelhead under natural conditions. Trans. Amer. Fish. Soc. 115: 726–735.

Cornuet, J.M., S. Piry, G. Luikart, A. Estoup & M. Solignac. 1999. New methods employing multilocus genotypes to select or exclude populations as origins of individuals. Genetics 153: 1989–2000.

Crandall, P.A. & G.A.E. Gall. 1993. The genetics of body weight and its effect on early maturity based on individually tagged rainbow trout (*Oncorhynchus mykiss*). Aquaculture 117: 77–93.

Crozier, W.W. 1993. Evidence of genetic variation between escaped farmed salmon and wild Atlantic salmon (*Salmo salar* L.) in a Northern Irish river. Aquaculture 113: 19–29.

Crozier, W.W., I.J.J. Moffett & G.J.A. Kennedy. 1997. Comparative performance of native and non-native strains of Atlantic salmon (*Salmo salar* L.) ranched from the River Bush, Northern Ireland. Fish. Res. 32: 81–88.

Crozier, W.W. 2000. Escaped farmed salmon, *Salmo salar* L., in the Glenarm River, Northern Ireland: Genetic status of the wild population 7 years on. Rev. Fish Biol. Fish. 7: 437–446.

Einum, S. & I.A. Fleming. 2000. Selection against late emergence and small offspring in Atlantic salmon (*Salmo salar*). Evolution 54: 628–639.

Fields, R.D., K.R. Johnson & G.H. Thorgaard. 1989. DNA fingerprints in rainbow trout detected by hybridization with DNA of bacteriophage M13. Trans. Amer. Fish. Soc. 118: 78–81.

Fleming, I.A. 1996. Reproductive strategies of Atlantic salmon: Ecology and evolution. Rev. Fish Biol. and Fish. 6: 379–416.

Fleming, I.A. & M.R. Gross. 1992. Reproductive behaviour of hatchery and wild coho salmon (*Oncorhynchus kisutch*): Does it differ? Aquaculture 106: 101–121.

Fleming, I.A., B. Jonsson, M.R. Gross & A. Lamberg. 1996. An experimental study of the reproductive behaviour and success of farmed and wild Atlantic salmon (*Salmo salar*). J. App. Ecol. 33: 893–905.

Fleming, I.A., K. Hindar, I.B. Mjolnerod, B. Jonsson, T. Balstad & A. Lamberg. 2000. Lifetime reproductive success and interactions of farm salmon invading a native population. Proc. Royal Soc. Lond. B 267: 1517–1523.

Fleming, I.A. & E. Petersson. 2001. The ability of released hatchery salmonids to breed and contribute to the natural production of wild populations. Nordic J. Fresh. Res. 75: 71–98.

Jonsson, B., N. Jonsson & L.P. Hansen. 1991. Differences in life history and migratory behaviour between wild and hatchery-reared Atlantic salmon in nature. Aquaculture 98: 69–78.

Kincaid, H.L. 1995. An evaluation of inbreeding and effective population size in salmonid broodstocks in federal and state hatcheries. Amer. Fish. Soc. Sympo. 15: 193–204.

Leider, S.A., P.L. Hulett, J.J. Loch & M.W. Chilcote. 1990. Electrophoretic comparison of the reproductive success of naturally spawning transplanted and wild steelhead trout through the returning adult stage. Aquaculture 88: 239–252.

Mackey, G., J.E. McLean & T.P. Quinn. 2001. Comparisons of run timing, spatial distribution, and length of wild and newly-established hatchery populations of steelhead in Forks Creek, Washington. N. Amer. J. Fish. Manage. 21: 717–724.

McConnell, S., L. Hamilton, D. Morris, D. Cook, D. Paquet, P. Bentzen & J. Wright. 1995. Isolation of salmonid microsatellite loci and their application to the population genetics of Canadian east coast stocks of Atlantic salmon. Aquaculture 137: 19–30.

Mork, J. 1991. One-generation effects of farmed fish immigration on the genetic differentiation of wild Atlantic salmon in Norway. Aquaculture 98: 267–276.

Morris, D.B., K.R. Richard & J.M Wright. 1996. Microsatellites from rainbow trout and their use for genetic study of salmonids. Can. J. Fish. Aquat. Sci. 53: 120–126.

Nickelson, T.E., M.F. Solazzi & S.L. Johnson. 1986. Use of hatchery coho salmon (*Oncorhynchus kisutch*) presmolts to rebuild wild populations in Oregon coastal streams. Can. J. Fish. Aquat. Sci. 43: 2443–2449.

Nielsen, J.L. 1994. Invasive cohorts: impacts of hatchery-reared coho salmon on the trophic, developmental, and genetic ecology of wild stocks, pp. 361–385. *In*: K.L. Fresh, D.J. Stouder & R.J. Feller (ed.) Theory and Application in Fish Feeding Ecology, University of South Carolina Press, Columbia, SC.

Olsen, J.B., P. Bentzen, M.A. Banks, J.B. Shaklee & S. Young. 2000. Microsatellites reveal population identity of individual pink salmon to allow supportive breeding of a population at risk of extinction. Trans. Amer. Fish. Soc. 129: 232–242.

O'Reilly, P.T., L.C. Hamilton, S.K. McConnell & J.M. Wright. 1996. Rapid analysis of genetic variation in Atlantic salmon (*Salmo salar*) by PCR multiplexing of dinucleotide and tetranucleotide microsatellites. Can. J. Fish. Aquat. Sci. 53: 2292–2298.

Quinn, T.P. 1993. A review of homing and straying in hatchery-produced salmon. Fish. Res. 18: 29–44.

Quinn, T.P., M.J. Unwin & M.T. Kinnison. 2000. Evolution of temporal isolation in the wild: genetic divergence in timing of migration and breeding by introduced chinook salmon populations. Evolution 54: 1372–1385.

Quinn, T.P., M.T. Kinnison & M.J. Unwin. 2001. Evolution of chinook salmon (*Oncorhynchus tshawytscha*) populations in New Zealand: Pattern, rate, and process. Genetica 112/113: 493–513.

Raanala, B. & J.L. Mountain. 1997. Detecting immigration by using multilocus genotypes. Proc. Natl. Acad. Sci. USA 94: 9197–9201.

Raymond, M. & F. Rousset. 1995. GENEPOP: A population genetic software for exact tests and ecumenicism. J. Heredity 86: 248–249.

Reisenbichler, R.R. 1988. Relation between distance transferred from natal stream and recovery rate for hatchery coho salmon. N. Amer. J. Fish. Manage. 8: 172–174.

Reisenbichler, R.R. & S.P. Rubin. 1999. Genetic changes from artificial propagation of Pacific salmon affect the productivity and viability of supplemented populations. ICES J. Mar. Sci. 56: 459–466.

Rice, W.R. 1989. Analyzing tables of statistical tests. Evolution 43: 223–335.

Ricker, W.E. 1972. Hereditary and environmental factors affecting certain salmonid populations, pp. 19–160. *In*: R.C. Simon & P.A. Larkin (ed.) The Stock Concept in Pacific Salmon, H.R. MacMillan Lectures in Fisheries, University of British Columbia, Vancouver, BC.

Ryman, N. & L. Laikre. 1991. Effects of supportive breeding on the genetically effective population size. Cons. Biol. 5: 325–329.

Seamons, T.R., P. Bentzen & T.P. Quinn. 2004. The mating system of steelhead, *Oncorhynchus mykiss*, inferred by molecular analysis of parents and progeny. Environ. Biol. Fish. 69: 333–344.

Siitonen, L. & G.A.E. Gall. 1989. Response to selection for early spawn date in rainbow trout, *Salmo gairdneri*. Aquaculture 78: 153–161.

Skaala, O., K.E. Jorstad & R. Borgstrom. 1996. Genetic impact on two wild brown trout (*Salmo trutta*) populations after release of non-indigenous hatchery spawners. Can. J. Fish. Aquat. Sci. 53: 2027–2035.

Smoker, W.W., A.J. Gharrett & M.S. Stekoll. 1998. Genetic variation of return data in a population of pink salmon: A consequence of fluctuating environment and dispersive selection? Alaska Fish. Res. Bull. 5: 46–54.

Taylor, E.B. 1991. A review of local adaptation in Salmonidae, with particular reference to Pacific and Atlantic salmon. Aquaculture 98: 185–207.

Utter, F.M. 1998. Genetic problems of hatchery-reared progeny released into the wild, and how to deal with them. Bull. Mar. Sci. 62: 623–640.

Verspoor, E. 1988. Reduced genetic variability in first-generation hatchery populations of Atlantic salmon (*Salmo salar*). Can. J. Fish. Aquat. Sci. 45: 1686–1690.

Waples, R.S. 1991. Genetic interactions between hatchery and wild salmonids: Lessons from the Pacific Northwest. Can. J. Fish. Aquat. Sci 48(Suppl. 1): 124–133.

Environmental Biology of Fishes **69**: 371–378, 2004.
© 2004 *Kluwer Academic Publishers. Printed in the Netherlands.*

Genetic variation within and between domesticated chinook salmon, *Oncorhynchus tshawytscha*, strains and their progenitor populations

Ji Eun Kim[a], Ruth E. Withler[b], Carol Ritland[c] & Kimberly M. Cheng[a,d]
[a]*Department of Animal Science, University of British Columbia, Vancouver, BC, Canada*
[b]*Fisheries and Oceans Canada, Pacific Biological Station, Nanaimo, BC, Canada*
[c]*Genetic Data Center, Department of Forest Sciences, University of British Columbia, Vancouver, BC, Canada*
[d]*Corresponding author (e-mail: kmtc@interchange.ubc.ca)*

Received 17 April 2003   Accepted 19 April 2003

Key words: aquaculture, microsatellite, genetic differentiation, inbreeding, bottleneck

Synopsis

Domesticated chinook salmon strains in British Columbia (BC), Canada are believed to have originated primarily from populations of the Big Qualicum (BQ) River and Robertson Creek (RC) on Vancouver Island in the early 1980s. The number of parental fish that gave rise to the domesticated strains and their subsequent breeding history during approximately five ensuing generations of domestication were not documented. Genetic variation at 13 microsatellite loci was examined in samples from two domesticated strains and the two progenitor populations to determine the genetic relationships among them. The domesticated strains had lower allelic diversity and tended to have lower levels of expected heterozygosity than did the BQ and RC progenitor populations. Only three alleles over all 13 loci were detected in the domesticated strains that were not present in the BQ and RC samples, whereas the progenitor strains possessed over 25 (BQ) and 43 (RC) private alleles. Genetic distance and F_{ST} values also indicated a closer relationship of the domesticated strains with the BQ than the RC population. One domesticated strain had a significant excess of heterozygosity compared with that expected under conditions of mutation-drift equilibrium, indicative of a recent genetic bottleneck. Genetic differentiation between the domesticated strains was as great as that distinguishing them from the progenitor populations, indicating that the genetic base of domesticated chinook salmon could be increased by hybridization. The existence of genetically distinct domesticated strains of chinook salmon in coastal BC generates the need for an evaluation of potential genetic interactions between domesticated escapees and natural spawning populations.

Introduction

Canada is the fourth largest producer of farmed salmon in the world and British Columbia (BC) produces more than 65% of farmed salmon in Canada.[1] In recent years, production of Atlantic salmon has dominated the industry but the resistance of Pacific salmonids to endemic disease organisms is leading to increased interest in their culture. Chinook salmon, *Oncorhynchus tshawytscha*, currently maintained as four or five domesticated strains, is the most important farmed Pacific salmon species and has been cultured since the aquaculture industry began in BC in the early 1970s.

The tendency for early maturation of cultured male salmon results in small size, poor flesh pigmentation and deteriorating flesh quality prior to harvest (Hunter et al. 1982, Gobantes et al. 1998). These fish represent a significant loss of income in aquaculture operations. Female salmon can be grown consistently to optimal market size prior to sexual maturation, making the development of all-female monosex stocks for aquaculture desirable. Monosex chinook salmon

---

[1] Fisheries Statistic, British Columbia Ministry of Agriculture, Fisheries and Food. 2001. (Available online at http://www.bcfisheries.gov.bc.ca/stats/salmon-aqua.html).

372

strains have been developed in the BC aquaculture industry through use of sex-reversed males (fish carrying two X chromosomes (XX) maculinized with methyltestosterone) (Hunter et al. 1983). All sperm produced by these phenotypic males carry the X chromosome, and the mating of such males with normal females results in all-female progeny production. Each generation some of the all-female eggs are sex-reversed with testosterone during incubation in order to provide the monosex male parents for the next generation.

The BC monosex strains used in aquaculture originated from chinook salmon gametes obtained from Fisheries and Oceans Canada (DFO) hatcheries which have operated supportive breeding programs since the early 1970s to increase salmonid production on numerous river systems. In each hatchery, the local population of chinook salmon has been propagated and the hatchery production of smolts supplements those resulting from natural spawning in the river system. Adults spawned in the hatchery each year are a mixture of those produced in the wild and the hatchery. The domesticated chinook salmon strains were founded with gametes derived primarily from the Big Qualicum (BQ) River hatchery on the east coast of Vancouver Island, from which gametes were collected for several years between 1980 and 1985 (Dave Groves, Sea Spring Salmon Farm Ltd., Chemainus, BC, 2001, pers. comm.). Gametes were also obtained in at least 1 year from the Robertson Creek (RC) hatchery on the west coast of Vancouver Island. Finally, infusions of genetic material occurred sporadically in the early 1980s from other Vancouver Island hatcheries such as the ones on the Nitinat and Quinsam rivers (on the west and east coasts, respectively) and possibly from elsewhere in BC (Henrik Kreiberg, Pacific Biological Station, DFO, Nanaimo, BC, 2001, pers. comm.). The histories of various strains were not documented throughout 20 years of turbulent industry development and the extent to which the genetic material from various sources was incorporated into extant domesticated strains is not known.

In 1983, milt from a restricted number (perhaps 30 or fewer) of methyltestosterone-treated sex-reversed BQ male chinook salmon was used to create all-female (monosex) chinook salmon for industry production (Hunter et al. 1983). This represents a potential genetic bottleneck in strain development, but the subsequent isolation of genetic probes to determine the genetic sex (XX or XY) of methyltestosterone-treated salmon eliminated the need for progeny testing in the development of monosex (XX) male fish

(Devlin et al. 1991, 1994). This simplified the identification of monosex males in hormone-treated mixed-sex groups of fish and enabled industry members to independently develop monosex males from unknown, but possibly large, numbers of additional families (Robert Devlin, West Vancouver Laboratory, DFO, 2001, pers. comm.; Dave Groves, Sea Spring Salmon Farm Ltd., Chemainus, BC, 2001, pers. comm.). Broodstock numbers for the domesticated strains are currently high (200+ per generation) but the sex ratio remains biased in favor of females at ratios ranging from 20:1 to 3:1.

Highly polymorphic salmonid microsatellite DNA markers have been used to characterize and identify wild salmonid stock structure (Olsen et al. 1998, Banks et al. 2000, Nelson et al. 2001, Beacham et al. 2001) and to examine wild and hatchery-influenced salmonid populations for reduced levels of genetic variation (O'Connell et al. 1997, Hedrick et al. 2000, Hansen et al. 2000, Withler et al. 2000). Some domesticated strains of Atlantic salmon display reduced genetic variation at both microsatellite and allozyme loci compared with both ancestral and unrelated wild populations (Wilson et al. 1995, Danielsdottir et al. 1997, Norris et al. 1999), but no molecular genetic analysis of a domesticated Pacific salmonid strain has been conducted. In this study, we examine two domesticated monosex strains of chinook salmon from BC for a divergence and possible loss of genetic variability at microsatellite loci from the two primary founding (BQ and RC) populations after approximately five generations of domestication.

Methods

Samples

Adult chinook salmon were sampled from the BQ River (135 individuals in 1997) and RC (112 individuals in 1988 and 1996) hatcheries operated by DFO. In each of these hatcheries the local chinook salmon population has been supplemented through supportive breeding (i.e. hatchery spawning, egg and juvenile rearing and release of smolts to the wild for natural ocean migration and adult return) since the early 1970s. Mature females from two domesticated monosex strains, A (97 individuals) and B (120 individuals), were sampled in 1997. Both domesticated strains are currently maintained independently with large (>200) numbers of broodstock but are of uncertain provenance since the inception of domestication in the early 1980s. Genetic data for chinook salmon from the Quesnel River (100 adult fish sampled in 1996) in the interior Fraser River watershed were used to provide an outgroup in the calculation of genetic differentiation among samples (Beacham et al. in press).

Microsatellite analysis

Genomic DNA was isolated from operculum tissue stored in 95% ethanol. DNA amplification of 13 polymorphic microsatellite loci was conducted with primers from the references in Table 1. PCR reactions for all loci were carried out independently with the following exceptions: *Ots*2 and *Ots*9 were amplified together, as were *Ogo*4 and *Omy*325, and *Ogo*2 and *Oke*4. Primer concentrations in the multiplex PCRs were adjusted to ensure equal amplification of both loci. PCR products were size fractionated, 96 samples on a gel, on 4.5% denaturing polyacrylamide gels. The gels were run at 3,000 V for 2.25 h at a gel temperature of 51°C on an ABI 377 automated DNA sequencer. Allele sizes were determined with Genescan 3.1 and Genotyper 2.5 software (PE Biosystems, Foster City, CA, U.S.A.).

Statistical analysis

Allelic diversity, number of private alleles (those confined to a single sample), heterozygosity and

Table 1. Salmonid microsatellite used in survey of domesticated chinook salmon strains of BC. Annealing temperatures (°C) used in the DNA amplification of loci and size range (in base pairs) of amplified chinook salmon alleles are given for the 13 loci.

Locus	Anneal	Allele size	Reference
*Ots*2	62	118–200	Banks et al. (1999)
*Ots*9	62	80–120	Banks et al. (1999)
*Ots*100	57	210–430	Nelson & Beacham (1999)
*Ots*101	53	120–340	Nelson & Beacham (1999)
*Ots*102	50	100–400	Nelson & Beacham (1999)
*Ots*104	49	140–340	Nelson & Beacham (1999)
*Ots*107	49	150–400	Nelson & Beacham (1999)
*Ogo*2	55	200–260	Olsen et al. (1998)
*Ogo*4	58	100–180	Olsen et al. (1998)
*Oke*4	55	220–270	Buchholz et al.[1]
*Oki*100	53	160–400	Unpublished
*Omy*325	58	70–170	O'Connell et al. (1997)
*Ssa*197	55	100–350	O'Reilly et al. (1996)

[1]Buchholz, W.G., S.J. Miller & B.J. Spearman. 1999. Isolation and characterization of microsatellite loci from chum salmon and cross species amplification. US Fish Wildl. Ser. Alaska Fish. Prog. Rep. 99–1.

conformance of observed genotypic distributions with those expected under conditions of Hardy–Weinberg equilibrium (HWE) at each locus of each sample, and pairwise gametic disequilibrium between loci within samples, were estimated using GDA version 1.0.[2] Tests for HWE and linkage disequilibrium were conducted applying the Bonferroni correction (Rice 1989). F_{ST} values were also calculated for each locus and for all loci combined (Weir & Cockerham 1984) using GDA. The 95% confidence intervals for each single locus F_{ST} value were calculated by jackknifing over populations, and for the overall F_{ST} value were calculated by bootstrapping over loci. Genetic relationships among samples from the progenitor populations, domesticated strains and one unrelated sample of chinook salmon from the Quesnel River, BC population were examined with pairwise values of Nei's (1978) genetic distance and F_{ST} values. The genetic relationships among samples (F_{ST} values) were depicted in a neighbor-joining dendrogram.

The Bottleneck 1.2 program (Cornuet & Luikart 1996) was used to examine samples for a increased level of expected heterozygosity (H_e) relative to that expected under conditions of drift-mutation equilibrium (H_{eq}), as might be expected if the farmed strains experienced a genetic bottleneck during domestication. H_{eq} was estimated using the two phase model (TPM) of microsatellite evolution with an estimated 70% of single step mutation (SSM) events and a variance of 30%. The significance of heterozygosity excess was tested with a one-tailed Wilcoxin sign-rank test. The TPM apparently more realistical than either a strict SSM or infinite alleles model describes mutation processes at microsatellite loci (Estoup & Cornuet 1999, Hansen et al. 2000, Marshall & Ritland 2002).

Results

Genetic variation within samples

All 13 microsatellite loci were polymorphic in all samples of chinook salmon. Levels of expected (H_e) and observed (H_o) heterozygosity were high for all samples, with H_e ranging from 0.46 to 0.95 among loci, and from 0.73 to 0.84 among samples (Table 2). There was a significant deficit of heterozygotes in comparison with

[2]Lewis, P.O. & D. Zaykin. 1999. Genetic data analysis: Computer program for the analysis of allelic data. Version 1.0 (d1.2). Free program can be downloaded from http://lewis.eeb.uconn.edu/lewishome/software.html.

Table 2. Genetic variation within two domesticated strains of chinook salmon and their putative progenitor strains (BQ and RC) on Vancouver Island. The number of fish (N), the number of observed alleles averaged over all loci (A) and the number of private alleles (P) are shown. Also shown are the observed (H_o) and expected (H_e) levels of heterozygosity, and the estimated inbreeding coefficient (I), for each sample, calculated without the inclusion of *Ots*102, which possesses one or more non-amplifying alleles.

Sample	N	A	P	H_o	H_e	I
Progenitor						
BQ	135	18.6	25	0.82	0.84	0.02
RC	112	18.7	43	0.81	0.83	0.03
Domesticated						
A	97	9.4	0	0.77	0.73	−0.06
B	120	12.0	3	0.84	0.78	−0.08

Table 3. Number of alleles observed at each of 13 microsatellite loci in samples from two progenitor populations (BQ and RC) and two domesticated strains (A and B) of chinook salmon. The sample size is shown in parentheses.

Locus	BQ (135)	RC (112)	A (97)	B (120)
*Ots*2	17	12	9	11
*Ots*9	4	6	2	4
*Ots*100	33	31	16	18
*Ots*101	21	23	9	13
*Ots*102	28	30	11	16
*Ots*104	24	25	14	17
*Ots*107	24	25	10	11
*Ogo*2	13	9	4	5
*Ogo*4	14	14	8	12
*Oke*4	7	6	4	6
*Oki*100	23	25	16	18
*Omy*325	12	9	8	10
*Ssa*197	22	28	11	15
Total	242	243	122	156

those expected under conditions of HWE at *Ots*102, in accordance with the suspected presence of non-amplifying alleles at this locus (Nelson et al. 2001). The remaining loci tended to be in HWE in all samples, with the six exceptions (evenly split between heterozygote excess and deficiency) constituting 5% of the tests as expected. The two domesticated samples of chinook salmon were not characterized by significant heterozygote deficiency or excess. However, there was a tendency in the domesticated strains for H_o to exceed H_e, whereas the opposite was true for the BQ and RC samples. This resulted in small but negative estimated inbreeding coefficients for the domesticated strains (Table 2).

Tests for linkage disequilibrium among loci (*Ots*102 excluded) in the BQ and RC samples indicated that all 12 loci were unlinked in chinook salmon. For the BQ sample, eight of 66 pairwise comparisons were significant but they all involved the *Ots*107 locus that was out of HWE in this sample. Similarly, for the RC sample six of 66 comparisons were significant but three of these involved the *Ots*101 locus which was out of HWE. In each of the domesticated strains, three (5%) of different pairwise comparisons were significant.

Expected heterozygosity values for the domesticated strains tended be lower than, but were not significantly (P = 0.08) different from, those of the BQ and RC samples (Table 2). Over all samples, the number of alleles per locus ranged from 7 to 41, with a mean of 14.4 alleles per locus observed among samples. The domesticated strain samples possessed fewer alleles per locus (9.4 and 12.0) than did the progenitor samples (18.6 and 18.7) (p = 0.03). Loss of alleles in strain

B tended to be at the most highly polymorphic loci but was apparent at even less variable loci in strain A, which possessed only half the number of alleles as the progenitor populations (Table 3).

Many more private alleles (those confined to a single sample) were observed in the BQ (24) and RC (43) samples than in the domesticated strain samples (Table 2). Thus for the BQ sample about half of the alleles not shared with RC (53) were not found in the domesticated strains, whereas about 80% of the RC alleles not shared with BQ (54) were absent from the domesticated samples. The private BQ and RC alleles not observed in domesticated samples tended to be alleles at low-frequency in their respective strain, but the average frequency of private alleles was twice as high in the RC (0.023) than in the BQ (0.012) sample. Only three alleles were observed in the domesticated samples that were not present in one or both of BQ and RC, all in strain B and each in only one or two fish. Both the domesticated samples possessed alleles not present in the other one, many of which were at frequencies exceeding 10%.

Under the TPM of microsatellite evolution, only the strain A sample showed a significant excess of heterozygosity (H_e) compared with that expected under mutation-drift equilibrium (H_{eq}) (p = 0.01). This indicates a recent bottleneck in the development of this strain. The domesticated strains showed a decreased abundance of low-frequency alleles (those at frequencies of 0.1 or less) compared with the two progenitor

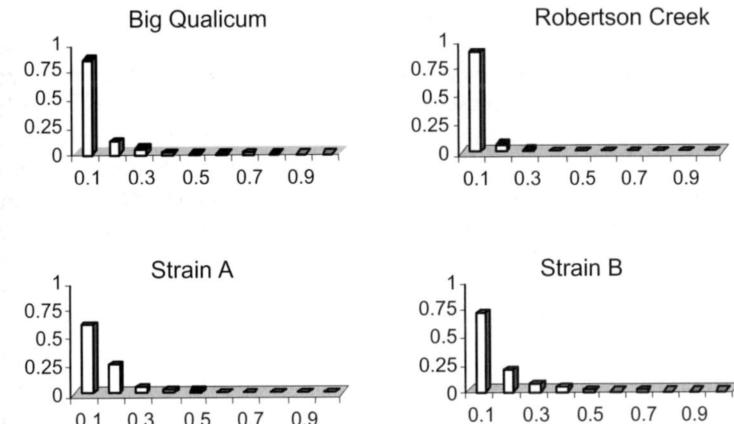

Figure 1. Allele frequency distributions for alleles at 13 microsatellite loci in samples from two populations and two domesticated strains of chinook salmon. The first column shows the proportion of alleles at frequencies between 0 and 0.1, the second column shows the proportion of alleles at frequencies between 0.1 and 0.2, etc.

Table 4. Pairwise F_{ST} (above diagonal) and Nei's (1978) genetic distance (below diagonal) values among three populations and two domesticated strains of chinook salmon from BC. The BQ and RC populations are progenitors of the domesticated strains (A and B), whereas the Quesnel population is unrelated.

	Quesnel	BQ	RC	A	B
Quesnel	–	0.056	0.040	0.109	0.082
BQ	0.40	–	0.041	0.053	0.047
RC	0.25	0.27	–	0.087	0.060
A	0.58	0.20	0.40	–	0.071
B	0.47	0.23	0.31	0.27	–

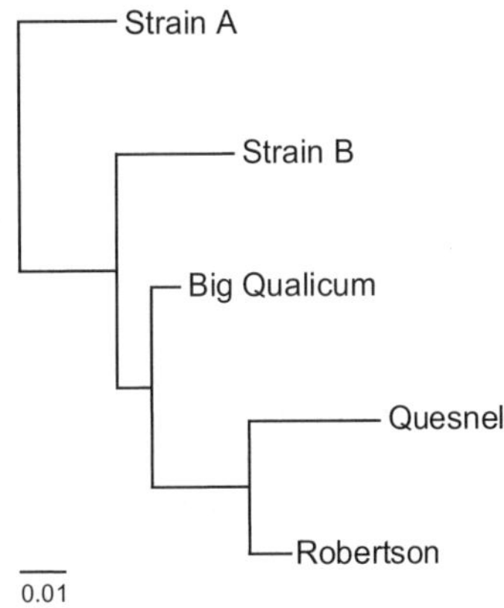

Figure 2. Neighbor-joining dendrogram of three populations and three domesticated strains of chinook salmon from BC based on pairwise F_{ST} values. The BQ and RC populations were progenitors of the domesticated strains whereas the Quesnel population was not.

strains (Figure 1), although the alteration in allele frequency distributions were not so great as to produce a 'mode shift', in which the abundance of alleles at frequencies greater than 0.1 would exceed those at frequencies of less than 0.1.

Genetic variation among samples

Locus mean F_{ST} estimates ranged from 0.026 at *Ots2* to 0.096 at *Oke4*, with an overall value of 0.058. Bootstrapping over loci indicated that the F_{ST} value among samples was significantly greater than 0, and jackknifing over samples indicated that 9 of the 13 single locus F_{ST} values were significantly greater than 0. The F_{ST} value for the comparison of the two domesticated strains (0.071) was greater than that between each of them and the BQ sample (Table 4). Strains A and B were as distinct from the BQ samples as the BQ and RC samples were from each other. Both domesticated strains were more differentiated from the RC than the BQ sample, and even more distinct from the unrelated Quesnel sample (Table 3, Figure 2). The pairwise genetic distances among samples indicated the same patterns of relationship as the F_{ST} values (Table 3).

Discussion

The loss of neutral genetic variation documented in the two domesticated chinook salmon strains examined in this study, and their genetic differentiation from founding populations, was of similar magnitude to that observed in Atlantic salmon domesticated for about the same number of generations (Mjølnerød et al. 1997, Norris et al. 1999). The domesticated strains were characterized by reduced allelic diversity and strain A showed signs of a recent genetic bottleneck. The greater loss of allelic diversity than of H_e in the domesticated strains relative to founding populations is consistent with the loss of rare alleles in small populations (Nei et al. 1975, Allendorf 1986). Nevertheless, the 10% reduction in H_e observed in strain A relative to the progenitor populations indicates a rate of reduction in heterozygosity approximately twice the 1% level below which natural selection is expected to compensate for inbreeding depression in natural populations (Franklin 1980, Frankel & Soulé 1981). However, both domesticated strains retained considerable diversity, with allele distributions that did not show a complete 'mode shift' to higher frequency alleles at the expense of lower frequency alleles (Luikart et al. 1998). Analysis of allele frequencies and private alleles both indicated that the BQ progenitor population contributed more to the genetic foundation of both domesticated strains than did the RC population.

The bottleneck detected in strain A may reflect founder effects and/or breeding practices. The use of a restricted number of males in establishing the monosex strains may have caused an initial bottleneck that was not eradicated in all strains by the subsequent incorporation of unrelated sex-reversed males. Additionally, in spite of the fact that broodstock numbers have been high in recent generations, fewer males than females are used and males in the monosex strain are obtained by hormone treatment of eggs during hatchery incubation. Eggs are generally maintained in family groups during hatchery rearing, and it is possible that relatively few families were used in the production of males each generation. The use of a non-random subset of the broodstock as males would also tend to produce different allele frequencies in the male and female parents each generation, a condition that could account for the observed slight excess of H_o compared with H_e in the domesticated strains (Luikart & Cornuet 1999).

The study supported the belief that the domesticated strains may be derived almost entirely from the BQ population with some input from RC populations but little from other enhanced populations, especially phylogenetically distinctive chinook salmon populations such as that of the Quesnel River. Only three alleles were observed in the domesticated strains, all at low-frequency, that were not detected in the BQ and RC samples. These alleles may be present in one or both of the founding populations but missed by chance in the samples analyzed in this study, may have arisen in the domesticated strains by mutation, or may reflect a low level of introgression from additional chinook salmon populations. All three 'private' alleles observed in the domesticated strains have been observed in Vancouver Island chinook salmon populations other than BQ and RC (unpubl. data). Thus, there is little evidence of significant introgression from more distant possible source populations in BC or beyond.

The ready access to gametes from the BQ and RC populations during the formative years of aquaculture in BC, the genetic predisposition of chinook salmon from Vancouver Island populations in southern BC for early seawater adaptation (Clarke et al. 1994) and the high proportion of the red-fleshed chinook salmon phenotype in these populations (Hard et al. 1989) likely led to relatively high survival and growth, and red-flesh color, under the suboptimal culture conditions that prevailed during industry development. Both BQ and RC chinook salmon survived, grew and matured well in an early strain comparison that included salmon from interior and northern populations in BC (Henrik Kreiberg, Pacific Biological Station, DFO, Nanaimo, BC, 2001, pers. comm.). In four strains of Atlantic salmon, each founded from between 8 and 24 populations, the genetic contributions of between one and three founding populations dominated each strain after three generations of selection (Gjedrem et al. 1991). As in the case of Atlantic salmon, selection for survival, growth and age of maturity in the cultured chinook salmon early in domestication likely eliminated much of the ancillary genetic material that might have been introduced from other populations.

The F_{ST} values differentiating the domesticated strains from the BQ and RC samples ranged from 0.047 to 0.087, similar to the value of 0.070 between one domesticated Norwegian strain of Atlantic salmon and the founding population believed to dominate its genetic background (Mjølnerød et al. 1997). Thus, although the chinook salmon used in BC aquaculture were generally subjected to mass selection rather than the pedigreed programs involving family and individual selection for the Norwegian Atlantic salmon, they may have acquired and retained very similar levels of genetic diversity, and become similarly differentiated from founding populations. Moreover, allele frequency differentiation between the domesticated strains was greater than that between the two founding populations. Both domesticated strains contained alleles not detected in the other. Thus, it is likely that genetic basis of domesticated chinook salmon in BC could be broadened considerably by hybridization among extant domesticated strains.

The domesticated chinook salmon of BC have become only moderately differentiated at microsatellite loci from their local progenitor populations. Nevertheless, this level of differentiation is similar to that observed for domesticated Atlantic salmon, which have been shown to possess very different growth, survival and reproductive characteristics in the wild than their progenitor populations (Fleming et al. 2000, 2002). Therefore, studies on the potential impact of interbreeding between domesticated and naturally produced chinook salmon in BC are warranted.

Acknowledgements

Funding of this project was provided by the National Research Council (Industrial Research Assistance Program), the Science Council of BC (Technology BC Program), and Fisheries and Oceans Canada. We thank Mike Wetklo for conducting the laboratory work. This paper is the part of a dissertation submitted by the first author in partial fulfillment of the requirements of a Ph.D. degree at the University of BC.

References

Allendorf, F.W. 1986. Genetic drift and the loss of alleles versus heterozygosity. Zoo. Biol. 5: 181–190.

Banks, M.A., M.S. Blouin, B.A. Baldwin, V.K. Rashbrook, H.A. Fitzgerald, S.M. Blankenship & D. Hedgecock. 1999. Isolation and inheritance of novel microsatellites in chinook salmon (*Oncorhynchus tshawytscha*). J. Hered. 90: 281–288.

Banks, M.A., V.K. Rashbrook, M.J. Calavetta, C.A. Dean & D. Hedgecock. 2000. Analysis of microsatellite DNA resolves genetic structure and diversity of chinook salmon (*Oncorhynchus tshawytscha*) in California's Central Valley. Can. J. Fish. Aquat. Sci. 57: 915–927.

Beacham, T.D., J.R. Candy, K.J. Supernault, T. Ming, B. Deagle, A. Schultz, D. Tuck, K. Kaukinen, J.R. Irvine, K.M. Miller & R.E. Withler. 2001. Evaluation and application of microsatellite and major histocompatibility complex variation for stock identification of coho salmon in British Columbia. Trans. Am. Fish. Soc. 130: 1116–1155.

Beacham, T.D., K.J. Supernault, M. Wetklo, B. Deagle, K. Labaree, J.R. Irvine, J.R. Candy, K.M. Miller, R.J. Nelson & R.E. Withler. The geographic basis of population structure in Fraser River chinook salmon, *Oncorhynchus tshawytscha*. Fish. Bull. (in press).

Clarke, W.C., R.E. Withler & J.E. Shelbourn. 1994. Inheritance of smolting phenotypes in backcrosses of hybrid stream-type × ocean-type chinook salmon (*Oncorhynchus tshawytscha*). Estuaries 17: 13–25.

Cornuet, J.-M. & G. Luikart. 1996. Description and power analysis of two tests for detecting recent population bottlenecks from allele frequency data. Genetics 144: 2001–2014.

Danielsdottir, A.K., G. Marteinsdottir, F. Arnason & S. Gudjonsson. 1997. Genetic structure of wild and reared Atlantic salmon (*Salmo salar* L.) populations in Iceland. ICES J. Mar. Sci. 54: 986–997.

Devlin, R.H., B.K. McNeil, T.D.D. Groves & E.M. Donaldson. 1991. Isolation of Y-chromosomal DNA probe capable of determining genetic sex in chinook salmon (*Oncorhynchus tshawytscha*). Can. J. Fish. Aquat. Sci. 48: 1606–1612.

Devlin, R.H., B.K. McNeil, I.I. Solar & E.M. Donaldson. 1994. A rapid PCR-based test for Y-chromosomal DNA allows simple production of all-female strains of chinook salmon. Aquaculture 128: 211–220.

Estoup, A. & J.-M. Cornuet. 1999. Microsatellite evolution: Inferences from population data. pp. 45–65. *In*: D.B. Goldstein & C. Schlötterer (ed.) Microsatellites. Evolution and Applications, Oxford University Press, Oxford.

Fleming, I.A., K. Hindar, I.B. Mjølnerød, B. Jonsson, T. Balstad & A. Lamberg. 2000. Lifetime success and interactions of farm salmon invading a native population. Proc. R. Soc. Lond. B 267: 1517–1523.

Fleming, I.A., T. Agustsson, B. Finstad, J.I. Johnsson & B.T. Björnsson. 2002. Effects of domestication on growth physiology and endocrinology of Atlantic salmon (*Salmo salar*). Can. J. Fish. Aquat. Sci. 59: 1323–1330.

Frankel, O.H. & M.E. Soulé. 1981. Conservation and Evolution, Cambridge University Press, Cambridge. 327 pp.

Franklin, I.A. 1980. Evolutionary change in small populations. pp. 131–150. *In*: M.E. Soulé & B.A. Wilcox (ed.) Conservation Biology: An Evolutionary-Ecological Perspective, Sinauer Associates, Sunderland, MA.

Gjedrem, T., H.M. Gjøen & B. Gjerde. 1991. Genetic origin of Norwegian farmed Atlantic salmon. Aquaculture 98: 41–50.

Gobantes, I., G. Choubert, J.C.G. Milicua & R. Gomez. 1998. Serum carotenoid concentration changes during sexual maturation in farmed rainbow trout (*Oncorhynchus mykiss*). J. Agric. Food. Chem. 46: 383–387.

Hansen, M.M., E.E. Nielsen, D.E. Ruzzante, C. Bouza & K.-L.D. Mensberg. 2000. Genetic monitoring of supportive breeding in brown trout (*Salmo trutta* L.), using microsatellite DNA markers. Can. J. Fish. Aquat. Sci. 57: 2130–2139.

Hard, J.J., A.C. Wertheimer & W.F. Johnson. 1989. Geographic variation in the occurrence of red- and white-fleshed chinook salmon (*Oncorhynchus tshawytscha*) in western North America. Can. J. Fish. Aquat. Sci. 46: 1107–1113.

Hedrick, P.W., V.K. Rashbrook & D. Hedgecock. 2000. Effective population size of winter-run chinook salmon based on microsatellite analysis of returning spawners. Can. J. Fish. Aquat. Sci. 57: 2368–2373.

Hunter, G.A., E.M. Donaldson, F.W. Goetz & P.R. Edgell. 1982. Production of all-female and sterile coho salmon, and experimental evidence for male heterogamety. Trans. Am. Fish. Soc. 111: 367–372.

Hunter, G.A., E.M. Donaldson, J. Stoss & I. Baker. 1983. Production of monosex female groups of chinook salmon (*Oncorhynchus tshawytscha*) by the fertilization of normal ova with sperm from sex-reversed females. Aquaculture 33: 355–364.

Luikart, G., F.W. Allendorf, J.-M. Cornuet & W.B. Sherwin. 1998. Distortion of allele frequency distributions provided a test for recent population bottlenecks. J. Hered. 89: 238–247.

Luikart, G. & J.-M. Cornuet. 1999. Estimating the effective number of breeders from heterozygote excess in progeny. Genetics 151: 1211–1216.

Marshall, H.D. & K. Ritland. 2002. Genetic diversity and differentiation of Kermode bear populations. Mol. Ecol. 11: 685–697.

Mjølnerød, I.B., U.H. Refseth, E. Karlsen, T. Balstad, K.S. Jakobsen & K. Hindar. 1997. Genetic differences between two wild and one farmed population of Atlantic salmon (*Salmo salar*) revealed by three classes of genetic markers. Hereditas 127: 239–248.

Nei, M. 1978. Estimation of average heterozygosity and genetic distance from a small number of individuals. Genetics 89: 583–590.

Nei, M., T. Maruyama & R. Chakraborty. 1975. The bottleneck effect and genetic variability in populations. Evolution 29: 1–10.

Nelson, R.J. & T.D. Beacham. 1999. Isolation and cross species amplification of microsatellite loci useful for study of Pacific salmon. Anim. Genet. 30: 228–229.

Nelson, R.J., M.P. Small, T.D. Beacham & K.J. Supernault. 2001. Population structure of Fraser River chinook salmon (*Oncorhynchus tshawytscha*): An analysis using microsatellite DNA markers. Fish. Bull. 99: 94–107.

Norris, A.T., D.G. Bradley & E.P. Cunningham. 1999. Microsatellite genetic variation between and within farmed and wild Atlantic salmon (*Salmo salar*) populations. Aquaculture 180: 247–264.

O'Connell, M., R.G. Danzmann, J.-M. Cornuet, J.M. Wright & M.M. Ferguson. 1997. Differentiation of rainbow trout (*Oncorhynchus mykiss*) populations in Lake Ontario and the evaluation of the stepwise mutation and infinite allele mutation models using microsatellite variability. Can. J. Fish. Aquat. Sci. 54: 1391–1399.

Olsen, J.B., P. Bentzen & J.E. Seeb. 1998. Characterization of seven microsatellite loci derived from pink salmon. Mol. Ecol. 7: 1083–1090.

O'Reilly, P.T., L.C. Hamilton, S.K. McConnell & J.M. Wright. 1996. Rapid analysis of genetic variation in Atlantic salmon (*Salmo salar*) by PCR multiplexing of dinucleotide and tetranucleotide microsatellites. Can. J. Fish. Aquat. Sci. 53: 2292–2298.

Rice, W.R. 1989. Analyzing tables of statistical tests. Evolution 43: 223–225.

Weir, B.S. & C.C. Cockerham. 1984. Estimating F-statistics for the analysis of population structure. Evolution 38: 1358–1370.

Wilson, I.F., E.A. Bourke & T.F. Cross. 1995. Genetic variation at traditional and novel allozyme loci, applied to interactions between wild and reared *Salmo salar* L. (Atlantic salmon). Hereditas 75: 578–588.

Withler, R.E., K.D. Le, R.J. Nelson, K.M. Miller & T.D. Beacham. 2000. Intact genetic structure and high levels of genetic diversity in bottlenecked sockeye salmon (*Oncorhynchus nerka*) populations of the Fraser River, British Columbia, Canada. Can. J. Fish. Aquat. Sci. 57: 1985–1998.

Environmental Biology of Fishes **69**: 379–393, 2004.
© 2004 *Kluwer Academic Publishers. Printed in the Netherlands.*

Lopsided fish in the Snake River Basin – fluctuating asymmetry as a way of assessing impact of hatchery supplementation in chinook salmon, *Oncorhynchus tshawytscha*

Orlay Johnson, Kathleen Neely & Robin Waples
National Marine Fisheries Service, Northwest Fisheries Science Center, 2725 Montlake Blvd. East,
Seattle, WA 98112, U.S.A. (e-mail: orlay.johnson@noaa.gov)

Received 17 April 2003   Accepted 19 April 2003

Key words: developmental instability, Pacific salmon, meristics

Synopsis

The use of developmental instability (an individual's failure to produce a consistent phenotype in a given environment) was evaluated to detect the effects of outplanting hatchery fish on wild salmon. Juvenile chinook salmon were collected in 1989, 1990, and 1991 from five drainages in the Snake River Basin. In each drainage we attempted to collect fish from streams with no hatchery supplementation (wild), naturally spawning fish from streams with hatchery supplementation (natural), and fish collected at a hatchery. Forty fish were collected per site and the number of elements in bilateral characters were counted on each side of the fish. Indices of fluctuating asymmetry (FA), a measure of minor, random deviations in perfect symmetry of bilateral counts, were calculated as an estimator of developmental instability. Analysis of character counts from seven paired characters revealed normal distributions. Only one of the characters displayed counts that were statistically larger on one side than the other, indicating that directional asymmetry (DA) or antisymmetry was not a major bias of FA. However, the means of all individual characters revealed a non-statistically significant left side bias. We analyzed our data using two indices of FA (FA1 and FA5) with different levels of sensitivity to DA. Differences in both FA indices were found among years, with collection sites in 1989 having significantly larger FA values than in 1991 (FA p < 0.01). Levels of FA among wild, natural, and hatchery fish were comparatively small (FA1 p = 0.17). This suggests developmental conditions were different in the first year of the study than in the last. The cause of these differences may be linked to either genetic or environmental variation or to gene–environment interactions, but the general population declines of salmon that occurred during this time obscures more specific conclusions.

Introduction

The abundance of most naturally spawning Pacific salmon species has been substantially below historical levels in recent years (Nehlsen et al. 1991, McClure et al. 2003). The cause of these low levels has been attributed to a multitude of impacts such as poor forestry and agriculture practices, overgrazing, over-fishing, industrialization, urbanization, dams, hatcheries, variation in ocean conditions, and global climate change (Magnuson 1996).

To stem the decline in the Columbia River Basin, a wide variety of recovery and restoration programs have been proposed, including hatchery supplementation of wild populations. Hatchery supplementation programs have been evaluated in numerous studies in order to improve their effectiveness (Allendorf & Ryman 1987, Reisenbichler & Rubin 1999, Waples & Drake 2003), but there are still substantial gaps in our knowledge of how to supplement natural populations effectively. Among the most important factors to consider are the genetic consequences of releasing hatchery-reared fish into the wild (e.g., Waples 1991, Hindar et al. 1991, Cuenco et al. 1993, Campton 1995, Lynch & O'Hely 2001, Ford 2002) and the resulting genetic introgression that presumably occurs with wild populations.

This is an important consideration because the genetic makeup of native wild stocks has presumably been shaped by many years of adaptation to local conditions (Ricker 1972, Taylor 1991), and transplanted fish (especially if not of local origin) may be less well suited to local conditions (Reisenbichler & Rubin 1999).

Assessment of changes in the population structure or stability resulting from these population declines and recovery efforts have been widely acknowledged as both important and difficult (Magnuson 1996, Ham & Pearsons 2000). For thousands of years, salmon have adapted to local conditions (e.g., evolved co-adapted gene complexes) that allow them to home to natal streams and survive in a variety of changing and often harsh conditions. Crosses of genetically distinct parents (intraspecific hybridization) may result in disruption of these adaptive mechanisms, and even small changes in this structure may cause extinctions of locally adapted populations. However, our understanding of what these changes are, or how they occur, are often too little for managers to implement effective remediation action until it is too late (Magnuson 1996, Ham & Pearsons 2000).

Ecological parameters, such as productivity, survivorships, or fecundity are often used as indicators of population well being. However, these parameters are often lagging indicators of stress, documenting problems that have already occurred (Freeman[1]). Estimation of changes in a population's developmental instability has been suggested as a leading indicator of problems that could signal trouble before demographic declines occur (Freeman[1]).

Developmental instability represents the inability of an embryo to produce a consistent phenotype in a given environment (Waddington 1957, Møller & Swaddle 1997, Polak 2003) (Figure 1). Estimators of developmental instability have been used to access the well being of populations in a variety of species and ecosystems (Graham et al. 1993, Clarke 1993, 1995a,b, Freeman et al. 1996, Lens et al. 2002) including salmonids (e.g., Leary et al. 1983, 1985a, 1993, Wilkins et al. 1995, Gharrett et al. 1999). These estimators are responsive to a wide range of biotic and physical stresses (Figure 1) including chemical stressors (Valentine & Soule 1973, Jagoe & Haines 1985, Lindsey 1988, Kieser 1992, Graham et al. 1993),

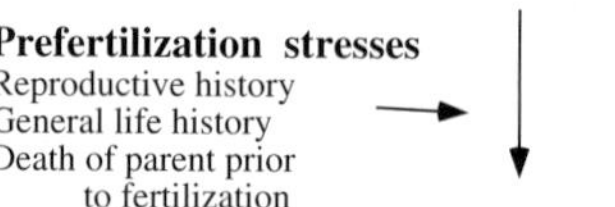

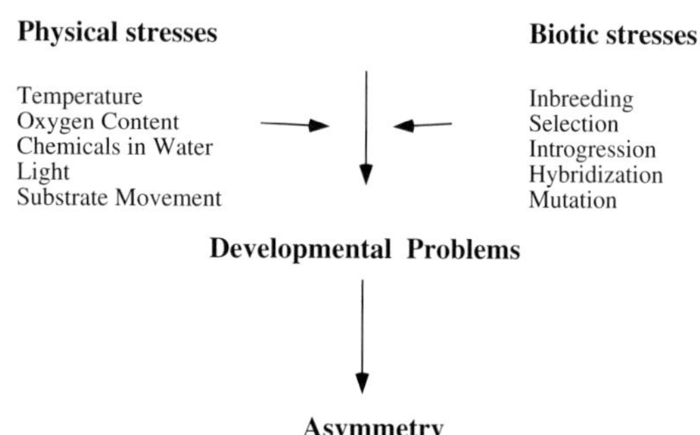

Figure 1. Conceptual model of how pre- and post-fertilization biological and physical stresses may result in meristic variation and asymmetry.

parasites (Alados et al. 1993, Polak 1994, Escos et al. 1995, Mara 1995, Perez-Tris et al. 2002), inbreeding (Leary et al. 1985c, 1987, Markow & Martin 1993), and hybridization (Leamy 1984, Leary et al. 1985b, Gharrett & Smoker 1991, Graham 1992, Gharrett et al. 1999, Alibert & Auffray 2003).

Development of corresponding morphological characters (such as pectoral fins) on opposite sides of an organism are controlled by the same genes and in theory, should show perfect bilateral symmetry (Van Valen 1962). However, stresses (Figure 1) during morphogenesis can cause random changes in an organism's developmental program and result in asymmetry. Asymmetrical development, unlike changes in morphometric characters, can only be modified during early development. Once meristic characters become fixed during embryogenesis, they remain unchanged, regardless of subsequent changes in the environment or an organism's body size and shape (Waddington 1940, review in Møller & Swaddle 1997). Instability during development is assumed to result in lower fitness due to reduced performance during natural and sexual selection (Palmer & Strobeck 1986, Møller & Swaddle 1997, Møller 1997). However, not all studies have found a negative relation between developmental instability and fitness (Markow & Ricker 1992, Ueno 1994, Martin & Hosken 2002).

---

[1] Freeman, D.C., J.M. Emlen, J.H. Graham, R.L. Mara & M. Tracy. 1996. Developmental instability as a bioindicator of ecosystem health. pp. 170–177. *In*: G.R. Barrow, E.D. McArthur, R. Sosebee & R. Tausch (ed.) Proceedings: Shrubland Ecosystem Dynamics in a Changing Environment, Las Cruces, NM.

In the literature, a common estimator of developmental instability is fluctuating asymmetry (FA), the random fluctuations in meristic counts from the left and right sides of an organism. FA is believed to arise from the organism's inability to buffer development against biological or environmental disturbances (Van Valen 1962, Palmer & Strobeck 1986, 1992). Other kinds of asymmetry, such as directional asymmetry (DA) and antisymmetry, can also occur when deviations from perfect symmetry are not random, but develop in a particular direction. The mean of the left and right deviations under fluctuating asymmetry should be zero, but with DA, the individuals are either left or right dominant. Antisymmetry is a pattern of left (L) and right (R) counts where the counts are directional, but not consistently so, and the pattern is bimodal. An example of antisymmetry is claw size in fiddler's crabs, where one claw is always larger than the other, but is not consistently the right or left (Neville 1976). Other examples of antisymmetry include the direction, the sail is set in wind blown jellyfish such as *Vellela vellela* or the direction of crossing in the beak of crossbill birds (*Loxia curvirostrea*) (Neville 1976).

Supplementation of wild salmon populations with hatchery stocks involves the intentional integration of hatchery and wild populations. Supplementation should result in the introgression of genes from hatchery fish into the naturalized (supplemented) populations once these hatchery fish return and spawn with the wild fish. The nature and magnitude of the consequences of this interbreeding will depend on two different types of effects. The first effect is one of stock origin: if the hatchery stock was derived from a different population than the one being supplemented, progeny of hatchery–wild matings might exhibit outbreeding depression due to different local adaptations of the source and supplemented populations. Outbreeding depression is the reduction in fitness caused by hybridization among genetically distinct parents and is caused by two mechanisms, the 'dilution' of locally adapted genotypes or through the loss of co-adapted gene complexes (Alibert & Auffray 2003). The second effect arises from domestication, or genetic change due to selective differences in the hatchery and wild environments. Even a locally derived hatchery population can diverge over time from its local source population. These changes could also lead to outbreeding depression if the hatchery population was subsequently used for supplementation. If the hatchery stock is recently derived from a local population and the hatchery and wild populations are well mixed

each generation, substantial divergence between the hatchery and wild populations would not be expected. However, because of the cumulative effects of domestication, the hatchery–wild system as a whole would be expected to genetically diverge over time from the original wild population, with the most likely outcome being a reduction in average fitness in the wild environment (Reisenbichler & Rubin 1999, Lynch & O'Hely 2001, Ford 2002).

Our hypothesis is that supplemented populations will show changes in levels of FA compared to their parent wild or hatchery populations. Soule (1967) suggested that populations in hybrid zones should show increased developmental instability due to the breakdown in genomic co-adaptation and outbreeding depression. However, the direction of this change may be difficult to predict, as developmental stability is theoretically expected to be enhanced by moderate outbreeding before decreasing in highly outbred groups (Vrijenhoek & Lerman 1982). Actual studies have given conflicting results. As an example, Felley (1980) investigated a zone of secondary species contact in blue gill sunfish and found no evidence of lowered developmental homeostasis, but Graham and Felley (1985) found that more recently introgressed populations of centrarchids had higher FA levels than either of the parental populations.

In a survey of 47 studies of developmental instability in hybrids among genus, species, subspecies, and populations, Alibert & Auffray (2003) found instability decreased in 11 studies, increased in 25, and did not change in 11. One of these studies, Freeman et al. (1995) is recorded twice as they found two traits with increased instability and two with decreased instability. Seven of these studies were of crosses among salmonids. In these studies, three reported increased levels of FA (Leary et al. 1985b, Gharrett & Smoker 1991, Wilkins et al. 1995), two showed decreased levels of FA (Ferguson 1986, 1988), and one found no change in FA (Gharrett et al. 1999). In studies of hybridization below the species level, five studies found increased levels of instability, five had no change, and four showed decreased instability. These conflicting results are not surprising, as we know little about the genetic and/or environmental forces that mediate development in populations.

Differences in the genetic and/or environmental forces among groups of hatchery, wild, and natural chinook are at least equally murky. Compared to streams, hatcheries provide a protected rearing habitat for salmonids, but the two environments may be dramatically different in many other ways, including differences in mate selection criteria, pre-spawning mortality, egg densities, water temperature, oxygen content, and pH. Because hatcheries usually rear salmon eggs in far denser conditions than in the wild, prophylactic chemical baths are often used to prevent or treat infections, diseases, and other problems (Piper et al. 1982). Although these chemicals increase embryo and juvenile survival, they may result in increased levels of stress and developmental instability in hatchery fish (Smith & Piper 1972, Piper et al. 1982).

In this study, we attempted to estimate developmental instability in chinook salmon using two indices of FA (FA1 and FA5). We collected juvenile salmon over 3 years from streams with no hatchery planting records (wild), from streams previously supplemented with hatchery fish (natural), and from associated hatcheries. We calculated levels of FA from bilateral characters and tested to see if they might reveal changes attributable to hatchery supplementation, years of collection, or drainage of collection site. More generally, we attempted to determine whether changes in FA levels across all populations in the basin could be used as leading indicators of demographic change prior to its occurrence.

Materials and methods

Study design

Juvenile chinook salmon (*Oncorhynchus tshawytscha*) were collected in 1989–1991 (1988–1990 broodyears) from five drainages in the Snake River Basin in Idaho and Oregon (Figure 2, Table 1). Whenever possible, yearly samples were collected in each basin from (1) a hatchery population; (2) a hatchery supplemented, but naturally reproducing population; and (3) a wild population considered by state fishery agencies to be uninfluenced by hatchery releases (Table 1). However, samples from these different types of management scenarios were not always available each year. For example, samples from the Upper Salmon River (supplemented stream) were available only in 1989, and samples from Catherine Creek (supplemented stream) on the Grande Ronde River were available only in 1990. An isolated wild population in Marsh Creek on the Middle Fork of the Salmon River was sampled in all years to allow a comparison with basins that have been influenced by supplementation and hatchery programs. Hatchery samples were collected from two facilities that use broodstock derived from local chinook salmon

populations (Sawtooth Hatchery and Imnaha facility) and two that use broodstock originally collected outside the local watershed (Looking glass and McCall hatcheries) (Table 1).

Sample collections

Juvenile chinook salmon from streams managed as wild or natural (supplemented) were collected in August and September by seine and electroshocker along stream reaches of ~0.5–1 km. Efforts were made in supplemented streams to avoid sampling planted fish that were not the result of natural spawning. Juvenile hatchery fish were sampled from August to February of each year (primarily later in the season) with dip nets from raceways containing progeny of the targeted population. All fish were killed by an overdose of MS-222 and placed on dry ice for transport to and storage in the laboratory at $-80°C$. Detailed collection of information is provided for 1989 and 1990 in Waples et al. (1993). All fish were 1-year-old juveniles, so fish collected in 1989 were 1988 broodstock.

Character counts

Eight bilateral meristic characters previously shown to exhibit asymmetry in salmonid fishes (Landrum 1966, Leary et al. 1983, 1984) were counted under a binocular dissecting microscope (Figure 3): pectoral fin rays (P1), pelvic fin rays (P2), mandibular pores (MP), lower first branchial arch gill rakers (LGR1), upper first branchial arch gill rakers (UGR1), lower second branchial arch gill rakers (LGR2), upper second branchial arch gill rakers (UGR2), and branchiostegal rays (BR). Counts were made on ~40 randomly selected chinook salmon from each site in each year. Data were pooled from males and females because no differences between sexes in the variance and means of these characters have been reported in salmonids (Landrum 1966, Leary et al. 1983, 1984). BR counts were not used in the analysis of FA indices as they were only collected in 1989 and 1991 and the character displayed strong directionality that has been associated with an anatomical advantage to the organism (Landrum 1966, Lagler et al. 1977).

Measurement error

The estimation of individual FA is subject to large measurement or counting errors and several authors have discussed methods to increase repeatability (R) in FA

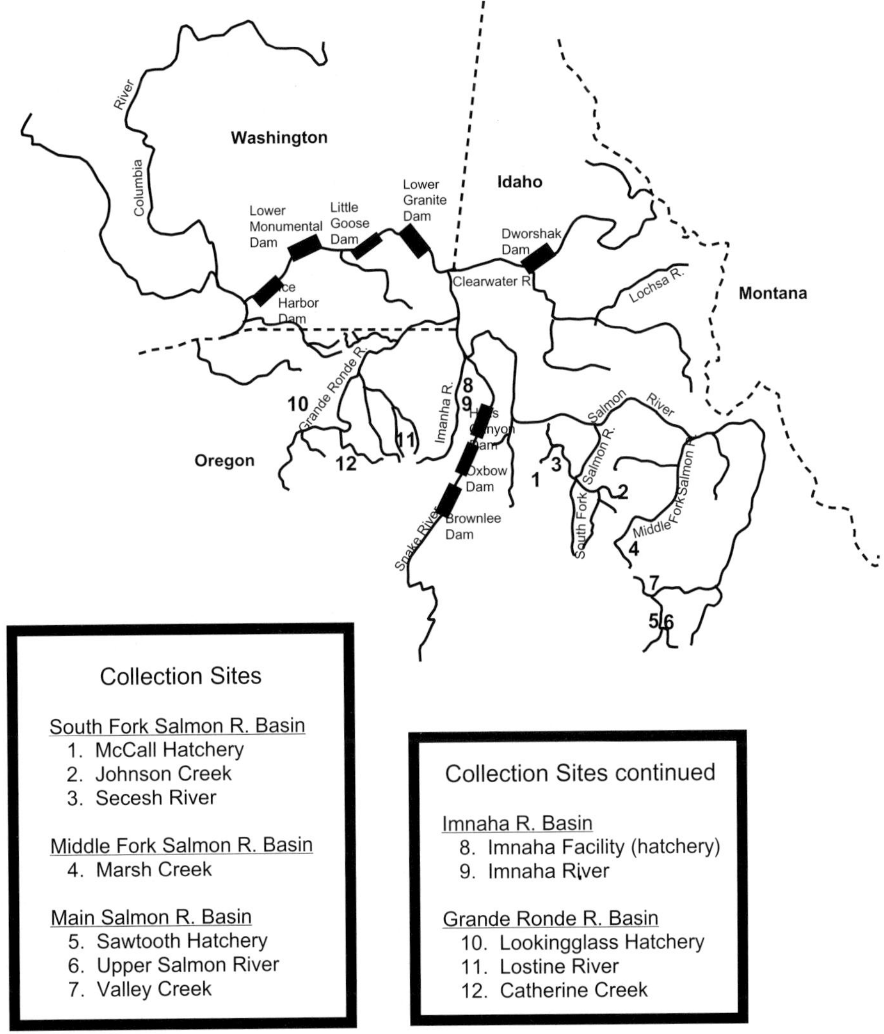

Figure 2. Map of lower Snake River Basin with collection sites numbered.

samples (Palmer & Strobeck 1986, 1992, Whitlock & Fowler 1997, Palmer 1994, Van Dongen 1999). In our study, each character was counted three times and repeatability of character counts was checked through periodic tests of within and between-counter variability on counts from the same fish. Differences among repeated character counts by the same technicians were near zero. However, when we totaled R and L counts from different technicians of selected characters on the same fish, differences among counters initially averaged almost 10%. This was because some technicians counted more elements in a character than others did. However, there were no significant differences in FA values between technicians; the differences were in the total number of character elements counted. As an example, some counters included gill rakers that were visible under the skin, but had not yet erupted to the surface of the gill arch. Others did not. After further training and continual communication among counters, these differences were reduced to zero.

We tested for the significance of non-DA relative to counting error by using a two-way analysis of variance (ANOVA) (sides × individuals). We found no statistical differences in counts of the same samples between the two counters. However, as pointed out by Palmer & Strobeck (1986) and Palmer (1994), an ANOVA may miss actual differences when sides differ by only one or two (as with these bilateral character counts). Therefore, a low level of undetected scoring error may exist.

Table 1. Collection sites for chinook salmon in 1989–1991 in Snake River basin (see Figure 1 for locations on map).

Drainage	Collection site	Run-time	Type	Hatchery broodstock Years	
				Origin	Sampled
South Fork Salmon River		Summer			
	McCall Hatchery		Hatchery	Collected at Little Goose Dam	All years
	Johnson Creek		Natural		All years
	Secesh River		Wild		All years
Middle Fork Salmon River		Spring			
	Marsh Creek		Wild		All years
Main Fork Salmon River		Spring			
	Sawtooth Hatchery		Hatchery	In basin	All years
	Upper Salmon River		Natural		1989
	Valley Creek		Wild		All years
Imnaha River		Summer			
	Imnaha Facility		Hatchery	In basin	All years
	Imnaha River		Natural		1989/1990
Grande Ronde River		Spring			
	Looking glass Hatchery		Hatchery	Stock – Rapid R. Hatchery, Idaho	All years
	Lostine River		Wild		All years
	Minam River		Wild		1990
	Catherine Creek		Natural		1990

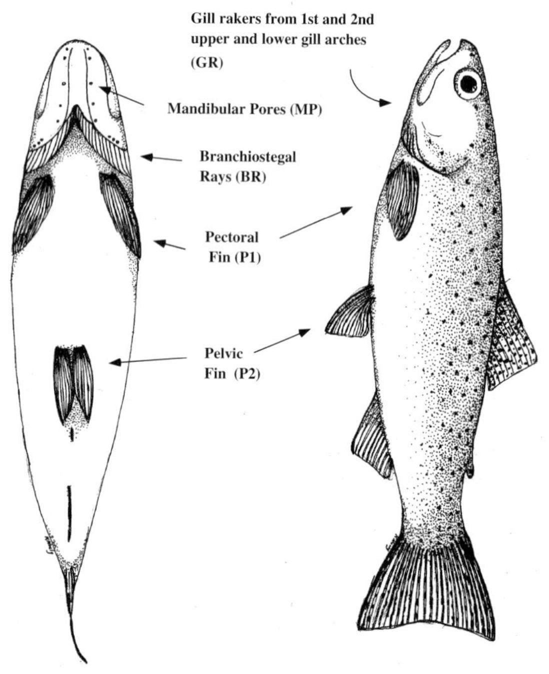

Figure 3. Bilateral characters that were counted on each fish for the study (abbreviations in parentheses).

To increase reliability of counts within a collection site, counts that differed by more than two from the mean count for that site were recounted by an additional counter. In no case did the second counter find different counts than the first. To determine if counting errors were related to fish size (bigger fish being easier to count than smaller fish), a sub-sample from the same collection site of fish greater and smaller than 80 mm were counted by two technicians and differences in R–L counts analyzed. Of five characters (P1, P2, MP, LGR1, and UGR1) enumerated in 20 fish (10 small, 10 large), only two fish showed any differences among counts between technicians and this only in MP. Although, more characters were omitted as uncountable in small than large fish (3 out of 10 smaller fish had one or more uncountable characters compared to zero in the larger fish), there were significant differences ($p > 0.05$) in FA1 or FA5 between large and small fish.

Data analysis

Correlation matrices (Sokal & Rohlf 1981) were used to test for independence among counts of meristic characters and to determine whether a simple additive approach for combining data was valid for this study. Presence of antisymmetry or DA in data from individual characters was tested using a factorial ANOVA procedure (Palmer 1994) and corrected with sequential Bonferroni inequality tests for multiple samples. The normality of the meristic data was evaluated for skewness and kurtosis. High values of these results would suggest that characters might not exhibit the pattern of random bilateral variation expected during development (Palmer & Strobeck 1992, Palmer 1994).

Statistical analyses followed the procedure of Palmer & Strobeck (1986) and Palmer (1994) for small samples where there are not large differences in counts between characters. An asymmetry count (A_{ij}) was obtained from each fish and character when the count taken from the left side (L_{ij}) was subtracted from the corresponding count on the right side (R_{ij}) to obtain the signed difference:

$$A_{ij} = (R_{ij} - L_{ij})$$

where i = individual fish and j = individual character counted.

DA for a particular character was the mean of A_i.

$$DA_j = \text{Mean } A_{ij}$$

We evaluated several indices of FA and choose two (FA1 and FA5) that met the characteristics of our dataset, as recommended in Palmer (1994). FA1 is the absolute or unsigned value of DA (Palmer & Strobeck 1986, 1992):

$$FA1_j = \text{Mean } |A_{ij}|$$

FA1 is easy to compute and is intuitively easy to understand, because it is simply the unbiased estimator of the sample standard deviation, and it is relatively insensitive to outliers. This FA estimator will be biased if either DA or antisymmetry is strongly present, and its statistical power is less than some other estimators. It is also sensitive to size of the difference between R and L counts.

FA5 is a population estimator of FA for a particular character and is the mean squared A_i divided by number of fish counted (N_j) (Palmer & Strobeck 1986):

$$FA5_j = \frac{\sum (A_i)^2}{N_j}$$

FA5 is an estimator of the between-sides variance of the data and allows an added degree of freedom over the standard variance as the mean is assumed to be non-zero. Thus, this index has added statistical power, particularly for samples <40, but it is also biased by either DA or antisymmetry and it is more sensitive to outliers than FA1. It is also sensitive to the size dependence of |R–L|.

Statistical procedures (regression and ANOVA) were used to determine the effect of length, site of collection, drainage of collection site, and type of fish (wild, hatchery, or natural) on the meristic variables. When multiple significance tests were conducted (e.g., number of statistically significant differences among FA values at the different collection sites), we used a sequential Bonferroni inequality test to reduce the occurrence of false positive results to 5% or less (Rice 1989, Palmer 1994, Anderson & Finn 1996).

Results

Independence of character counts

A correlation matrix of A_i for the seven characters from 1,123 chinook (241 fish had one or more characters missing) showed only two characters, P1 and UGR2, to be significantly correlated (p = 0.046) with a correlation of 0.059 (Table 2), after adjusting for multiple tests.

Length and character counts or measurement error

Mean length of hatchery fish (104.6 mm $\pm$ 0.8 SE) was significantly greater (p < 0.001) greater than either wild (68.61 $\pm$ 0.4 mm) or natural fish (65.2 $\pm$ 0.5 mm). Wild fish were also longer than fish from supplemented sites (p = 0.026), although the average length was

Table 2. Correlation matrix of asymmetry indices (A_i) for all values, excluding BR, from chinook salmon in all years from all sites.

Characters	P1	P2	MP	LGR1	UGR1	LGR2	UGR2
P1	1.00	0.03	−0.03	0.00	0.01	−0.02	0.06[*]
P2		1.00	−0.04	−0.04	0.02	0.01	−0.02
MP			1.00	0.03	0.02	0.03	−0.01
LGR1				1.00	0.04	0.00	0.00
UGR1					1.00	0.00	0.01
LGR2						1.00	0.03
UGR2							1.00

There were 1,123 observations used in the matrix and 231 cases omitted due to missing values. Asterisk (*) indicates value is statistically significant (p < 0.05).

only 2.2 mm greater. There were no significant differences in length between fish collected in different years.

Since it is easier to count characters in larger fish than smaller fish, there was a significant positive correlation between fish length and total number of elements per character counted ($R = 0.447$, $R^2 = 0.200$, and $p < 0.001$) for all characters except LG2. Even when counts were restricted to fish less than 80 mm in length (61 hatchery, 244 natural, and 447 wild fish), length was still significantly correlated to character counts ($p = 0.0004$).

However, using regression analysis, we found no significant correlation of length with A_{ij} ($p = 0.2550$, $R^2 = 0.001$), FA1 ($p = 0.3218$, $R^2 = 0.001$), or FA5 ($p = 0.5832$, $R^2 = 0.010$). We concluded that although more characters could be seen and counted in larger fish, the increases were equal on both sides of the fish.

Directional asymmetry pattern of counts for individual bilateral characters

Following Palmer (1994), we tested our data to determine whether individual characters showed consistent directionality in counts of the L and R sides. Student's t-tests comparing raw L and R character counts indicated only one character (1LGR) had statistically different ($p = 0.02$) bilateral counts. The mean DA values for all characters in the study were slightly negative (Table 3). However, after sequential Bonferroni correction for multiple tests, factorial ANOVA tests on counts from left and right sides of all characters (Palmer 1994) indicated means were not significantly different from zero ($p < 0.01$).

Normality of data – kurtosis and skewness

Kurtosis and skewness were calculated for each bilateral character following Palmer (1994) (Table 3). Skewness of counts from all characters is low with no statistical differences among characters ($p = 0.08$). The kurtosis index for all characters (especially P1, P2) was positive, indicating a leptokurtotic distribution for the data (Table 3). These results are within a range similar to that calculated from asymmetry indices of the same characters in other studies (e.g., Leary et al. 1984, 1985). Kurtosis of the mean values for all characters across fish was less than one (0.97), further indicating the data were normally distributed.

Asymmetry in years, drainages, sites, and type of management (wild, natural, or hatchery)

The data were analyzed to determine if differences in asymmetry could be partitioned by collection, year, site, drainage, or fisheries management type (wild, hatchery, or natural). There were 13 collection sites: 11 in 1989, 12 in 1990, and 10 in 1991. Nine sites were replicated in all 3 years, so we grouped all sites by year for analysis. Summary of data for collections sites, including mean DA, FA1, and FA5, is presented in Table 4.

There were significant differences in FA among years of collection. ANOVA of both FA1 and FA5 indices showed significant ($p = 0.001$) declines from 1989 to 1991 (Figure 4). FA5 indices were significantly greater in 1989 than in both 1990 (mean 0.31 ± 0.013, $p = 0.04$) and 1991 (mean 0.27 ± 0.013, $p = 0.003$), but not between 1990 and 1991 ($p = 0.1525$). Indices of FA1 from fish collected in 1989 (mean $= 0.31 \pm 0.01$) were significantly greater than

Table 3. Character summaries of asymmetry indices for DA and FA $\pm$ SE. N_t is the number of fish counted per character.

Character	DA		N_t	Skewness	Kurtosis	FA1		FA5	
	Mean	SE				Mean	SE	Mean	SE
P1	−0.01	0.02	1235	0.52	7.78	0.24	0.01	0.28	0.04
P2	−0.01	0.01	1209	−0.55	11.13	0.13	0.01	0.14	0.03
MP	−0.03	0.02	1189	0.13	1.05	0.40	0.02	0.43	0.03
LGR1	−0.02	0.02	1255	−0.09	1.81	0.35	0.02	0.38	0.02
UGR1	−0.01	0.02	1258	−0.10	1.66	0.39	0.02	0.43	0.03
LGR2	−0.03	0.01	1248	−0.33	2.21	0.24	0.01	0.25	0.02
UGR2	−0.01	0.02	1245	−0.07	2.62	0.26	0.01	0.28	0.02
Mean	−0.02	0.01	1284	0.01	0.97	0.29	0.01	0.05	0.00
BR	−1.14	0.04	719	0.04	7.73	1.23	0.03	1.32	0.24

Mean skewness ± 0.07. Mean kurtosis ± 4.04.

Table 4. Values of mean DA, FA1, and FA5 ± SE for collection sites pooled by years.

Collection site	Type	Years sampled	No. of fish counted	DA ± SE	FA1 ± SE	FA5 ± SE
Catherine Creek	N	1	40	−0.04 ± 0.04	0.24 ± 0.02	0.26 ± 0.02
Imnaha Hatchery	H	3	119	−0.03 ± 0.02	0.29 ± 0.02	0.31 ± 0.02
Imnaha River	N	2	80	0.02 ± 0.03	0.28 ± 0.02	0.31 ± 0.01
Johnson Creek	N	3	120	−0.02 ± 0.02	0.28 ± 0.02	0.33 ± 0.05
Looking glass Hatchery	H	3	120	0.01 ± 0.02	0.28 ± 0.02	0.28 ± 0.01
Lostine River	W	3	120	0.02 ± 0.02	0.29 ± 0.02	0.29 ± 0.01
Marsh Creek	W	3	126	−0.02 ± 0.02	0.30 ± 0.02	0.34 ± 0.08
McCall Hatchery	H	3	117	−0.09 ± 0.02	0.32 ± 0.02	0.35 ± 0.04
Minam River	W	1	40	0.01 ± 0.03	0.27 ± 0.03	0.33 ± 0.04
Sawtooth Hatchery	H	3	121	−0.04 ± 0.02	0.32 ± 0.02	0.38 ± 0.07
Secesh River	W	3	121	0.01 ± 0.02	0.27 ± 0.02	0.28 ± 0.04
Upper Salmon River	N	1	40	0.02 ± 0.03	0.27 ± 0.02	0.29 ± 0.03
Valley Creek	W	3	120	−0.01 ± 0.02	0.28 ± 0.02	0.29 ± 0.03

Because data for BR were not collected in 1990, data from character was not included in calculations of DA or FA. Type is management type (N = natural, W = wild, and H = hatchery).

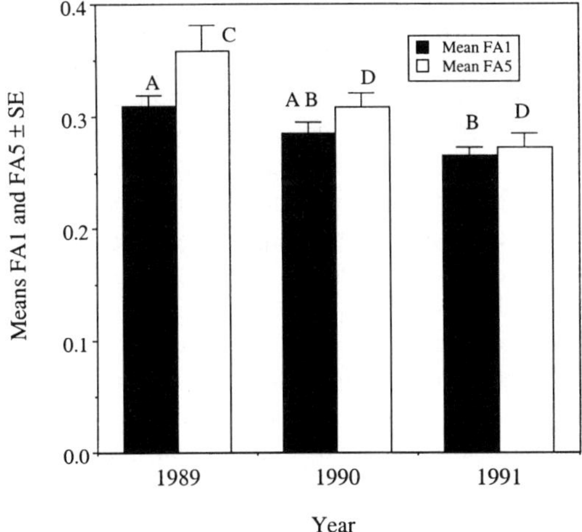

Figure 4. Fluctuating asymmetry (FA1 and FA5) indices from fish collected in different years (i.e., different broodyears). Both indices had statistically significant differences among years (p = 0.01 for FA1 and 0.009 for FA 5). FA1 indices of fish collected in 1989 were significantly greater (indicated with A) than fish collected 1991 (p < 0.001) (B), but not 1990 (p > 0.05) (AB). FA5 indices in 1989 were significantly greater (indicated with C) than fish collected both in 1990 (p < 0.05) (D) and 1991 (p < 0.001) (D). Fish collected in 1990 were not significantly different from fish collected in 1991 (p > 0.10).

Catherine Creek (only sampled in 1990) had the lowest mean FA values for both FA1 (0.24 ± 0.02) and FA5 (0.26 ± 0.02). Sawtooth Hatchery demonstrated the highest mean FA5 mean value (0.38±0.07). McCall Hatchery had the highest mean FA1 value (0.32±0.02) and the second highest mean FA5 (0.35 ± 0.04) value (Table 4). There were no significant differences in FA among management types (FA1 p = 0.18 and FA5 p = 0.65).

Counter measurement error

Four scientists counted all chinook salmon characters in the study. In general, two counters had primary responsibilities for 1989 and 1990 and a two had primary responsibility for 1991. There were significant differences in FA among years, and there is the potential that these yearly differences represent differences in counts among technicians rather than differences among fish. However, there was overlap among years for all technicians and there are no significant differences among technicians in overall character counts (p = 0.55), in DA (p = 0.56), or in FA1 (p = 0.33).

Discussion

Lack of significant differences among hatchery, wild, and natural fish

The purpose of the study was to evaluate the use of FA to detect changes in developmental instability in

1991 (mean 0.37 ± 0.01, p = 0.0012), but indices in 1990 (mean 0.29 ± 0.01) were not different from 1989 (p = 0.068) or 1991 (p = 0.1232).

We found no significant differences in either FA1 (p = 0.48) or FA5 (p = 0.89) among collection sites.

388

juvenile fish from streams where hatchery-reared fish are planted onto wild populations. We analyzed two FA indices (FA1 and FA5) and found no statistical differences among hatchery fish, naturally spawning fish in streams with previous hatchery plants (natural), or naturally spawning fish from streams without known hatchery supplementation (wild). Nor did we find significant differences (after corrections for multiple tests) among collection sites or drainage basins (over all collection years). However, we did detect significant differences in overall and character means of both FA1 and FA5 among fish collected in different years, regardless of management type, collection site, or drainage (Figure 4).

Several hypotheses can be proposed for the lack of difference in FA indices among hatchery, wild, and natural populations.

(1) Developmental instability was sufficiently low across the region buffering developmental problems during embryogenesis regardless of wild, hatchery, or natural origin.
(2) Hatchery fish planted into streams that contained wild fish did not survive, or if they did survive, their reproductive success with wild fish was negligible.
(3) Hatchery fish survived and produced hybrid offspring that did not experience increased developmental instability.
(4) Large region-wide environmental perturbations caused increased stress and developmental instability in most Snake River chinook populations and obscured any differences detected by FA among management types.

All these hypotheses may have occurred and may be contributors to the lack of significant differences in mean FA among hatchery, wild, and natural fish. However, the first hypothesis, that instability was low across all populations, has little support. Salmon populations in the region have experienced repeated population declines and environmental alterations (reviewed in Magnuson 1996) at a level that, in other species, has resulted in detectable changes in developmental instability (Freeman et al. 1995, Zakharov 2003). The second hypothesis, that planted hatchery fish had low rates of survival and/or poor reproductive success, has general support in the literature (e.g., Chilcote et al. 1989, Leider 1990). A genetic model has been proposed (Emlen 1991) that shows interbreeding between genetically   distinct hatchery and wild populations may substantially reduce productivity for more than 10 generations after interbreeding. However, although a number of studies on hatchery–wild reproductive success among chinook salmon are in progress, none have yet been published.

The third hypothesis, that hybrid offspring did not experience increased developmental instability, or that changes were insufficient to be detected in our analysis, is possible. The relationships of genomic co-adaptation, outbreeding depression, and developmental instability during hybridization are complex, confusing, and often unpredictable (Alibert & Auffray 2003). Detectable changes would partially depend on the differences among hatchery and wild stocks, and the differential mortality or reproductive success that hatchery and wild fish might experience during their life history. However, differences in FA from other studies on salmonid hybrid crosses, using a variety of indices (e.g., Leary et al. 1985b, Ferguson 1986, 1988, Gharrett & Smoker 1991, Wilkins et al. 1995, Gharrett et al. 1999) would suggest the technique is adequately sensitive to detect any changes in developmental instability that did occur among these hatchery–wild crosses.

The fourth hypothesis, that large scale environmental and other changes overwhelmed differences related due to wild–hatchery interactions also seems plausible. Throughout the Pacific Northwest during the late 1980s and early 1990s, salmon experienced freshwater and ocean conditions which resulted in changes in population structure and significant declines in abundance in many salmon stocks (Magnuson 1996). An example of other changes that may have altered population structure include the widespread straying of supplemented chinook salmon which have been documented to have occurred in the Grande Ronde during these years (Crateau[2]).

The significant differences detected in FA among fish collected in 1989–1991 (broodyears 1988–1990, respectively) occur across management types, collection sites, and river drainages, suggesting a basin-wide cause. It has been shown in previous studies (Allenbach et al. 1999) that FA is increased in organisms that develop in a stressful and/or marginal environment. The mid-1980s to early 1990s were drought years across the Snake River Basin. Precipitation was lowest in 1988,

---

[2] Crateau, E. 1997. Straying of hatchery origin spring/summer-run chinook salmon in the Grande Ronde Basin. *In*: W.S. Grant (ed.) Genetic Effects of Straying of Non-Native Hatchery Fish into Natural Populations. Proceedings of the Workshop, June 1–2, 1995, Seattle, WA.

increased slightly in 1989, but declined again from 1990–1993.[3] This drought may have added stress both to spawners prior to fertilization and to embryos during incubation in 1988 through increased water temperatures, low spawner concentrations, increased predation, and other factors. The slightly improved conditions in 1989 and 1990 may be reflected in the higher FA values from some sites in 1990 and 1991.

It is interesting that the highest FA values from 1989 were from Marsh Creek in the Salmon River. This creek is located in the Middle Fork of the Salmon River, a drainage within the Salmon River that does not have any hatcheries or hatchery-enhanced streams (Figure 2). However, the next highest FA values were from fish reared at two hatcheries (McCall and Sawtooth), also from the greater Salmon River Basin. These three sites also had the highest mean FA values in 1990 (along with wild fish from the Secesh River, a tributary of the South Fork Salmon River). However, in 1991, Marsh Creek and Secesh River had the lowest FA values (i.e., developmental instability), while the two Salmon River hatcheries had FA values in the middle of the FA distribution for that year. As noted in Palmer (1994), outliers or miscounts in a dataset can exaggerate FA indices, but the values for these sites appear to be true counts of bilateral asymmetry and not the result of injury or counting error.

Did FA vary with size of fish?

We did find significant differences in fork length between hatchery and naturally spawning fish, which might have resulted in a bias in our results. The mean length of hatchery fish was nearly 65% larger than the mean length of either wild or natural fish. The significant differences in length between wild and natural fish (even though small) are more surprising as both types of fish are from naturally spawning parents, and would be expected to have experienced natural rearing environments. However, streams are usually supplemented with hatchery fish because the natural fish populations have declined. The causes of this decline may still be impacting the supplemented populations and could result in a reduction in fish size.

We also found that more characters could be counted in larger fish than smaller fish, so the mean of characters counted in hatchery fish was significantly larger

(p > 0.05) for most traits than in wild or natural fish. There were no statistical differences among naturally spawning fish (p > 0.05). Palmer (1994) has pointed out that apparent asymmetry may be influenced by trait size and can cause spurious results in studies of development stability. However, we did not find significant differences (p < 0.05) in DA, FA1, or FA5 among fish of different sizes in our study, even though larger fish had more countable characters than smaller fish. Therefore, in our study, the question, 'Does FA vary with size' can be answered, 'No.'

Several authors have suggested that size of organisms within a population may bias FA analysis because the size may reflect differences in general 'condition' of the organism (reviewed in Møller 1997). This concept is rooted in the view of developmental instability as an easily measured surrogate for fitness, even though it has been shown that developmental processes are not linear but rather complex, dynamic, and non-linear (Graham et al. 1993, 1998, Møller 1997). In this study, the mean length of hatchery fish was significantly larger, by over 65%, than either of the naturally spawned groups. However, it is unlikely the size differential represented increased condition or fitness, but more likely reflected collection timing and rearing conditions. Wild and natural fish were all collected during August and early September, while hatchery fish were usually collected November through January, giving hatchery fish 3–5 months longer to grow. In addition, hatchery fish are fed to satiation in a protected environment with stable temperatures and no predators and would be expected to be larger than stream fish at the same age. Conversely, wild or natural fish that are unfit, sick, or in poor condition would probably be smaller than fish that are fit, healthy, and robust. However, these small fish would also be more likely to die or be eaten than bigger fish and so not be included in this study. More study would be needed to determine if size differences among fish from different management types correlates to increases in their condition or fitness.

Why use FA and not DA?

Fluctuating Asymmetry measures developmental errors during morphogenesis (i.e., the number of times the organism is unable to develop equally numbered bilateral elements), whereas DA and antisymmetry are believed to measure patterns of morphological development, which may have an adaptive purpose and an inherited (genetic) component (e.g., human heart and

---

[3] Western Regional Climate Center. 2002. 2215 Raggio Parkway, Reno, NV 89512.

fiddler crab claws) (Graham et al. 1998). However, there has been an extensive debate in the literature of the values of using deviations from antisymmetry and DA as measures of developmental instability (Palmer & Strobeck 1992, Graham et al. 1993, 1998, McKenzie & O'Farrell 1993).

Studies have shown that if FA occurs in an organism, it is possible to induce an evolutionary change to DA in response to intense directional selection (e.g., Mather 1953, McKenzie & Clarke 1988, Leary & Allendorf 1989, Graham et al. 1993). These findings would support claims that DA, antisymmetry, and FA are 'dynamically interrelated' (Graham et al. 1993b), can be co-produced by stressed environments, and all represent a form of developmental instability. Further, Graham et al. (1998) have presented a method to estimate developmental instability from DA data using residual variance from either a major axis regression or a general structural model.

In our study, only two individual characters displayed statistically significant directionality (i.e., more counts consistently to one side than the other). The first, BR, showed significantly more counts on the left than the right ($p < 0.0001$), but was excluded from the study because samples were not counted in 1990. In addition, BR has been associated with an anatomical advantage to the organism (Landrum 1966, Lagler et al. 1977). Only one other character in the study had a statistically significant directionality ($p = 0.02$) across all samples and that was LGR1 (mean number of rakers on right was 8.66 ± 0.91 SE, and on left 8.65 ± 0.90 SE). The reason for this was unclear. However, certain groupings of collection sites also revealed DA, especially among management groups. When sub-groups of larger normally distributed asymmetry datasets reveal directionality, this deviation from perfect symmetry is less likely to be adaptive or inherited through natural selection. Further, the DA displayed by BR counts also demonstrated patterns related to different treatment groups, and these differences may indicate an embryogenic environmental insult. Graham et al. (1998, 2003) demonstrate that FA, DA, and antisymmetry are dynamically interrelated and that all three components of asymmetry must be calculated to properly estimate developmental instability. These relationships will be further investigated in future work.

Conclusions

Were the apparent high levels of population instability in 1989 and 1990 an early warning of decreased fish abundance? This is difficult to determine, as during this time, chinook salmon populations were taking a precipitous decline throughout all management groups and drainages in the Snake River Basin. In addition, during the years of this study, the region was experiencing one of the most severe droughts of the century, during which water temperatures in the region were high and flow diminished (Western Regional Climate Center[3]). Although precipitation increased beginning in 1989, the drought persisted until the mid-1990s (Western Regional Climate Center[3]).

Analysis of FA indices was envisioned as a possible indicator of future population instability prior to actual declines in abundance. As this is a relatively non-invasive technique (only 40 lethal samples of juvenile fish from a population are required), and does not require extensive technology (binocular microscope and calipers are sufficient), a region-wide meristic database could be developed to monitor changes in FA among index populations. Changes in FA levels might alert managers that problems at the population level were occurring and this would allow preventative steps to be taken prior to demographic declines. Although we detected changes in FA among different years, the use of FA by itself did not seem to provide sufficient information to be useful in predicting future population declines. In future studies, we will include estimates of deviations from DA and antisymmetry following Graham et al. (1998), who proposed that using all these indices are crucial to properly evaluating the significance of developmental instability.

Acknowledgements

By far the most difficult task in this type of study is the consistent and accurate counting of meristic characters and the authors wish to extend a special thanks to Stacy Jones, Kathleen Neely, Amy Cook, Tami Peperell, and Ruth Levine for this challenging task. Selection of study sites and stock histories was carried out with assistance from Richard Carmichael, Oregon Department of Fish and Wildlife (ODFW), and Steve Yundt, Idaho Department of Fish and Game (IDFG). Assistance with field collections was provided by numerous field and hatchery biologists from ODFW and IDFG, along with Steve Achord, Paul Moran, and others on the National Marine Fisheries Service (NMFS) collection crews. Steve Smith and Ben Alexander of NMFS provided statistical consulting. Funding was partially provided by the Division of Fish and Wildlife, Bonneville Power Administration (Project No. 89-096).

References

Alados, C.L., J. Escos & J.M. Emlen. 1993. Developmental instability as an indicator of environmental-stress in the Pacific hake (*Merluccius productus*). Fish. Bull. 91: 587–593.

Alibert, P. & J. Auffray. 2003. Genomic coadaptation, outbreeding depression, and developmental instability. pp. 116–134. *In*: M. Polak (ed.) Developmental Instability Causes and Consequences, Oxford University Press, New York.

Allenbach, D.M., K.B. Sullivan & M.J. Lydy. 1999. Higher fluctuating asymmetry as a measure of susceptibility to pesticides in fishes. Environ. Toxicol. Chem. 18: 899–905.

Allendorf, F.W. & N. Ryman. 1987. Genetic management of hatchery stocks. pp. 141–159. *In*: N. Ryman & F. Utters (ed.) Population Genetics and Fishery Management, University of Washington Press, Seattle, WA.

Anderson, T.W. & J.D. Finn. 1996. The New Statistical Analysis of Data. Springer-Verlag, New York.

Campton, D.E. 1995. Genetic effects of hatchery fish on wild populations of Pacific salmon and steelhead: What do we really know? pp. 337–353. *In*: H.L. Schramm Jr. & R.G. Piper (ed.) Uses and Effects of Cultured Fishes in Aquatic Ecosystems. Am. Fish. Soc. Symp. 15.

Clarke, G.M. 1993. The genetic basis of developmental stability. I. Relationships between stability, heterozygosity and genomic coadaptation. Genetica 89: 15–23.

Clarke, G.M. 1995a. The genetic basis of developmental stability. II. Asymmetry of extreme phenotypes revisited. Am. Nat. 146: 708–725.

Clarke, G.M. 1995b. Relationships between developmental stability and fitness: Applications for conservation biology. Conserv. Biol. 9: 18–25.

Cuenco, M.L., T.H. Backman & P.R. Mundy. 1993. The use of supplementation to aid in natural stock restoration. pp. 269–293. *In*: J.G. Cloud & G.H. Thorgaard (ed.) Genetic Conservation of Salmonid Fishes. Proceedings of a NATO advanced Study Institute, Plenum Press, New York.

Emlen, J.M. 1991. Heterosis and outbreeding depression: A multilocus model and an application to salmon production. Fish. Res. 12: 187–212.

Escos, J., C.L. Alados, J.M. Emlen & S. Alderstein. 1995. Developmental instability in the Pacific hake parasitized by myxosporeans *Kudoa* spp. Trans. Amer. Fish. Soc. 124: 943–945.

Felley, J. 1980. Analysis of morphology and asymmetry in blue gill sunfish (*Lepomis macrochirus*) in the southeastern United states. Copeia 1: 1980, 18–29.

Ferguson, M. 1986. Developmental stability of rainbow trout hybrids: Genomic coadaptation or heterozygosity. Evolution 40: 323–330.

Ford, M.J. 2002. Selection in captivity during supportive breeding may reduce fitness in the wild. Conserv. Biol. 16: 815–825.

Freeman, D.C., J.H. Graham, D.W. Byrd, E.D. McArthur & W.A. Turner. 1995. Narrow hybrid zone between two subspecies of big sagebrush, *Artemisia tridentate* (Asteraceae). III. Developmental instability. Amer. J. Bot. 82: 1144–1152.

Gharrett, A.J. & W.W. Smoker. 1991. Two generations of hybrids between even- and odd-year pink salmon (*Oncorhynchus gorbuscha*): a test for outbreeding depression? Can. J. Fish. Aq. Sci. 48: 1744–1749.

Gharrett, A.J., W.W. Smoker, R.R. Reisenbichler & S.G. Taylor. 1999. Outbreeding depression in hybrids between odd- and even-broodyear pink salmon. Aquaculture 173: 117–129.

Graham, J.H. 1992. Genomic coadaptation and developmental stability in hybrid zones. Acta Zool. Fennica 191: 121–131.

Graham, J.H. & J.D. Felley. 1985. Genomic coadaptation and developmental stability within introgressed populations of *Enneacanthus gloriosus* and *E. obesus* (Pisces, Centrarchidae). Evolution 391: 104–114.

Graham, J.H., D.C. Freeman & J.M. Emlen. 1993. Antisymmetry, directional asymmetry, and dynamic morphogenesis. Genetica 89: 121–137.

Graham, J.H., J.M. Emlen, D.C. Freeman, L.J. Leamy & J.A. Kieser. 1998. Directional asymmetry and the measurement of developmental instability. Biol. J. Linn. Soc. 64: 1–16.

Graham, J.H., J.M. Emlen & D.C. Freeman. 2003. Nonlinear dynamics and developmental instability. pp. 35–50. *In*: M. Polak (ed.). Developmental Instability Causes and Consequences. Oxford University Press, New York.

Ham, K.D. & T.N. Pearsons. 2000. Can reduced salmonid population abundance be detected in time to limit management impacts? Can. J. Fish. Aquat. Sci. 57: 17–24

Hindar, K., N. Ryman & F. Utter. 1991. Genetic effects of cultured fish on natural fish populations. Can. J. Fish. Aquat. Sci. 48: 945–957.

Jagoe, C.H. & T.A. Haines. 1985. Fluctuating asymmetry in fishes inhabiting acidified and unacidified lakes. Can. J. Zool. 63: 130–138.

Kieser, J.A. 1992. Fluctuating odontometric asymmetry and maternal alcohol consumption. Ann. Hum. Biol. 19: 513–520.

Lagler, K.F., J.E. Bardach, R.R. Miller & D.R. Passion. 1977. Ichthyology, 2nd edition. John Wiley & Sons, New York, 506 pp.

Landrum, B.J. 1966. Bilateral asymmetry in paired meristic characters of Pacific salmon. Pac. Sci. 20: 193–202.

Leamy, L. 1984. Morphometric studies in inbred and hybrid house mice. V. Directional and fluctuating asymmetry. Am. Nat. 123: 579–593.

Leary, R. & F. Allendorf. 1989. Fluctuating asymmetry as an indicator of stress: implications for conservation biology. Trends in Ecology and Evolution. 4(7): 214–217.

Leary, R.F., F.W. Allendorf & K.L. Knudsen. 1983. Developmental stability and enzyme heterozygosity in rainbow trout. Nature 301: 71–72.

Leary, R.F., F.W. Allendorf & K.L. Knudsen. 1984. Superior developmental stability of heterozygotes at enzyme loci in salmonid fishes. Am. Nat. 124: 540–551.

Leary, R.F., F.W. Allendorf & K.L. Knudsen. 1985a. Inheritance of meristic variation and the developmental stability in rainbow trout. Evolution 39: 308–314.

Leary, R.F., F.W. Allendorf & K.L. Knudsen. 1985b. Developmental instability and high meristic counts in interspecific hybrids of salmonid fishes. Evolution 39: 1318–1326.

Leary, R.F., F.W. Allendorf & K.L. Knudsen. 1985c. Developmental instability as an indicator of reduced genetic variation in hatchery trout. Trans. Amer. Fish. Soc. 114: 230–235.

Leary, R.F., F.W. Allendorf & K.L. Knudsen. 1987. Differences in inbreeding coefficients do not explain the association between heterozygosity at allozyme loci and developmental stability in rainbow trout. Evolution 41: 1413–1415.

Leary, R.F., F.W. Allendorf & K.L. Knudsen. 1993. Null alleles at two lactate dehydrogenase loci in rainbow trout are associated with decreased developmental stability. Genetica 89: 3–14.

Lens, L., S. Van Dongen & E. Matthysen. 2002. Fluctuating asymmetry as an early warning system in the critically endangered Taita thrush. Conserv. Biol. 16: 479–487.

Lerner, I.M. 1954. Genetic Homeostasis. Wiley, New York. 134 pp.

Leung, B. & M.R. Forbes. 1997. Modeling fluctuating asymmetry in relation to stress and fitness. OKIOS 78: 397–405.

Lindsey, C.C. 1988. Factors controlling meristic variation. pp. 197–274. *In*: W.S. Hoar & D.J. Randall (ed.) Fish Physiology. Vol. XIB, Academic Press, New York.

Lynch, M. & M. O'Hely. 2001. Captive breeding and the genetic fitness of natural populations. Conserv. Genet. 2: 363–378.

Mara, R.L. 1995. Developmental Stability in *Acer rubrum*. M.S. Thesis, Wayne State University, Detroit, Michigan. 28 pp.

Magnuson, J.J. 1996. Upstream: Salmon and Society in the Pacific Northwest. National Academy of Sciences, Washington D.C. 452 pp.

Markow, T.A. & J.P. Ricker. 1992. Male size, developmental stability, and mating success in natural populations of three *Drosophila* species. Heredity 69: 122–127.

Markow, T.A. & J. Martin. 1993. Inbreeding and developmental stability in a small human population. Ann. Hum. Biol. 20: 389–394.

Martin, O.Y. & D.J. Hosken. 2002. Asymmetry and fitness in female yellow dung flies. Biol. J. Linn. Soc. Lond. 76: 557–563.

Mather, K. 1953. Genetical control of stability in development. Heredity 7: 297–336.

McClure, M.M., E.E Holmes, B.L. Sanderson & C.E. Jordan. 2003. A large–scale, multi-species assessment: Anadromous salmonids in the Columbia River Basin. Ecol. Appl. 13(4): 964–989.

McKenzie, J.A. & G.M. Clarke. 1988. Diazinon resistance, fluctuating asymmetry, and fitness in the Australian sheep blowfly. Genetics 120: 213–220.

McKenzie, J.A. & K. O'Farrell. 1993. Modification of developmental instability and fitness: Malathion-resistance in the Australian sheep blowfly, *Lucilia cuprina*. Genetica 89: 67–76.

Møller, A.P. 1997. Developmental stability and fitness: a review. Am. Nat. 149: 916–932.

Møller, A.P. & J.P. Swaddle. 1997. Asymmetry, Developmental Stability, and Evolution. Oxford University Press Inc., New York. 291 pp.

Nehlsen, W., J.E. Williams & J.A. Lichatowich. 1991. Pacific salmon at the crossroads: Stocks at risk from California, Oregon, Idaho, and Washington. Fisheries 16: 4–21.

Neville, A.C. 1976. Animal Asymmetry. Edward Arnold Publishers Ltd., London.

Palmer, A.R. 1994. Fluctuating asymmetry: A primer. pp. 335–364. *In*: T.A. Markow (ed.) Developmental Instability: Its Origin and Evolutionary Implications, Kluwer Academic Publishers, Dordrecht, the Netherlands.

Palmer, A.R. & C. Strobeck. 1992. Fluctuating Asymmetry as a measure of developmental stability: Implications of non-normal distributions and power of statistical tests. Acta Zool. Fennica 191: 57–72.

Palmer, A.R. & C. Strobeck. 1986. Fluctuating asymmetry: Measurement, analysis and pattern. Annu. Rev. Ecol. Syst. 17: 391–421.

Perez-Tris, J., R. Carbonell & J.L. Telleria. 2002. Parasites and the blackcap's tail: Implications for the evolution of feather ornaments. Biol. J. Linn. Soc. 76: 481–492.

Piper, R.G., L.E. McElwain, J.P. McCaren, L.G. Fowler & J.R. Leonard. 1982. Fish Hatchery Management. U.S. Department of the Interior, Fish and Wildlife Service, Washington, D.C., 517 pp.

Polak, M. 2003. Introduction. pp. xix–xxiii. *In*: M. Polak (ed.). Developmental Instability Causes and Consequences. Oxford University Press, New York.

Polak, M. 1994. Parasites increase fluctuating asymmetry of male *Drosophila nigrospiraculla*: Implication for sexual selection. pp. 257–268. *In*: T. Markow (ed.) Developmental Instability: Its Origins and Evolutionary Implications, Kluwer Academic Publishers, Dordrecht, the Netherlands.

Reisenbichler, R.R. & S.P. Rubin. 1999. Genetic changes from artificial propagation of Pacific salmon affect the productivity and viability of supplemented populations. ICES J. Mar. Sci. 56: 459–466.

Rice, W.R. 1989. Analyzing tables of statistical tests. Evolution 43: 223–225.

Ricker, W.E. 1972. Hereditary and environmental factors affecting certain salmonid populations. pp. 19–160. *In*: R.C. Simon & P.A. Larkin (ed.) The Stock Concept in Pacific Salmon, Macmillan Lectures in Fisheries, University of British Columbia, Vancouver, B.C.

Sokal, R.R. & F.J. Rohlf. 1981. Biometry. W.H. Freeman & Co., New York, 859 pp.

Soule, M.E. 1967. Phenetics of natural populations. II. Asymmetry and evolution in a lizard. Amer. Nat. 101: 142–159.

Smith, C.E. & R.G. Piper. 1972. Pathological effects in formalin-treated rainbow trout (*Salmo gairdneri*). J. Fish. Res. Board. 29: 328–329.

Taylor, E.B. 1991. A review of local adaptation in Salmonidae, with particular reference to Pacific and Atlantic salmon. Aquaculture 98: 185–207.

Ueno, H. 1994. Fluctuating asymmetry in relation to two fitness components, adult longevity and male mating success in a ladybird beetle, *Harmonia axyridis* (Coleoptera: Coccinellidae). Ecol. Entomol. 19: 87–88.

Valentine, D.W. & M.E. Soule. 1973. Effect of p,p-DDT on developmental stability of pectoral fin rays in the grunion, *Leuresthes tenuis*. Fish. Bull. 71: 921–926.

Van Dongen, S. 1999. Accuracy and power in fluctuating asymmetry studies: Effects of sample size and number of within-subject repeats. J. Evol. Biol. 12: 547–550.

Van Valen, L. 1962. A study of fluctuating asymmetry. Evolution 16: 125–142.

Vrijenhoek, R.C. & S. Lerman. 1982. Heterozygosity and developmental stability under sexual and asexual breeding systems. Evolution 36: 768–776.

Waddington, C.H. 1940. Organisers & Genes. Cambridge University Press, Cambridge, England, 160 pp.

Waddington, C.H. 1957. The Strategy of the Genes. George Allen & Unwin, London.

Waples, R.S. 1991. Genetic interactions between hatchery and wild salmonids: Lessons from the Pacific Northwest. Can. J. Fish. Aquat. Sci. 48(Suppl. 1): 124–133.

Waples, R.S. & J. Drake. 2003. Risk/benefit considerations for marine stock enhancement: A Pacific salmon perspective. pp. 260–306. *In*: K.M. Leber, S. Kitada, T. Svåsand & H.L. Blankenship (ed.) Proceedings of Second International Symposium on Stock Enhancement. Kobe, Japan, January 2002, Blackwell Publishing Ltd, Oxford, U.K.

Whitlock, M.C. & K. Fowler. 1997. The instability of studies of instability. J. Evol. Biol. 10: 63–67.

Wilkins, N.P., E. Gosling, A. Curatolo, A. Linnane, C. Jordan & H.P. Courtney. 1995. Fluctuating asymmetry in Atlantic salmon, European trout and their hybrids, including triploids. Aquaculture 137: 77–85.

Zakharov, V.M. 2003. Linking developmental stability and environmental stress; A whole organism approach. pp. 402–414. *In*: M. Polak (ed.) Developmental Instability Causes and Consequences. Oxford University Press, New York.

Environmental Biology of Fishes **69**: 395–407, 2004.
© 2004 *Kluwer Academic Publishers. Printed in the Netherlands.*

Temporal comparisons of genetic diversity in Lake Michigan steelhead, *Oncorhynchus mykiss*, populations: effects of hatchery supplementation

Meredith L. Bartron[1,2] & Kim T. Scribner[1]
[1]*Department of Fisheries and Wildlife, Michigan State University, 13 Natural Resources Building, East Lansing, MI 48824, U.S.A.*
[2]*U.S. Fish and Wildlife Service, Northeast Fishery Center, PO Box 75, Lamar, PA 16848, U.S.A.*
(e-mail: meredith_bartron@fws.gov)

Received 21 April 2003   Accepted 24 April 2003

Key words: microsatellites, outbreeding, salmon, stocking, temporal variation

Synopsis

Steelhead, *Oncorhynchus mykiss*, were first introduced into the Great Lakes in the late 1800s. Subsequently, natural recruitment across the Lake Michigan basin has been regularly supplemented by primarily one hatchery strain. Recently, multiple strains derived from locations across the species native range along the west coast of the United States have also been stocked by different management agencies. Prior to 1983, hatchery supplementation of Lake Michigan steelhead populations in Michigan was largely unsuccessful due to low smolting rates of small (<120 mm) hatchery yearlings (estimated survival 0.01%). Accordingly, contributions of hatchery fish to historical adult spawning runs in Michigan tributaries were low (0–30%) across six major drainages. Large yearlings of different hatchery strains (>150 mm) have been stocked exclusively since 1983, increasing estimates of survival to smolting (90%). Consequently, the proportion of hatchery adults in spawning runs increased to 13–79%. We examined the effects of changes in stocking practices on straying rates of hatchery steelhead and to temporal changes in levels of genetic diversity and relationships among populations. We used microsatellite loci to estimate allele frequencies for six populations sampled for two time periods (1983–1984 and 1998–1999). Measures of inter-population divergence (mean F_{ST}) were not significant for either time period. However, spatial genetic relationships among historical and contemporary populations were significantly correlated with geographic distance; a result not expected if gene flow (natural straying) among populations was mediated solely by hatchery supplementation. Increased numbers of alleles in spawning adults from populations can be attributed to alleles specific to recently introduced hatchery strains.

Introduction

Molecular markers have been used widely in studies of Pacific salmon species, *Oncorhynchus* spp., to document interactions between natural populations and between fish from natural and hatchery origins. For steelhead, *Oncorhynchus mykiss*, molecular markers have identified genetic population structure at micro- and macro-geographic spatial scales (Parkinson 1984, Reisenbichler & Phelps 1989, Reisenbichler et al. 1992, Beacham et al. 1999, Nielsen 1999, Nielsen & Fountain 1999, Osterberg & Thorgaard 1999). Changes in population estimates of allele frequency and genetic

diversity have been used to examine how levels of genetic variation within and among populations have changed over time (Heath et al. 2002), and have provided insight into factors (both natural and anthropogenic) contributing to temporal change (Waples 1998, Garant et al. 2000).

Hatchery supplementation has been used widely for salmon populations. Concerns regarding the potential for interactions between individuals of wild and hatchery-origin have increased the awareness of the need to re-evaluate hatchery management practices (Ryman 1991, Waples 1991, Lynch & O'Hely 2001). Theoretical and empirical evidence

(Washington & Koziol 1993, Gharrett 1994, Lynch[1], Reisenbichler & Rubin 1999) has focused on potential consequences of introgression and breakdown of physiological or biochemical compatibilities between genes (co-adapted gene complexes) in wild and/or naturalized populations (Lynch & O'Hely 2001). One hypothesized mechanism for outbreeding and potential fitness reduction in salmonids relates to increased levels of gene flow (straying) between hatchery and native or naturalized populations leading to the breakdown of locally adapted gene complexes.

Pacific salmon generally exhibit low levels of straying to non-natal spawning grounds (Quinn 1993). However, increased levels of gene flow between hatchery and native or naturalized populations may occur as hatchery fish exhibit higher rates of straying from rivers into which they were stocked (see Waples 1991). While co-occurrence of hatchery and native fishes has been widely reported (largely from direct observations of tags or fin clips), evidence for introgression and consequences to population viability are not widely known. Assessments of the magnitude of genetic change within and among populations would profit from knowledge of historical benchmarks of genetic characteristics prior to outbreeding events. Reisenbichler & Phelps (1989) hypothesized that introgression between native steelhead and hatchery steelhead widely stocked across large areas led to lower inter-population levels of genetic diversity. However, Reisenbichler & Rubin (1999) suggest that although interbreeding may occur between hatchery and wild individuals, the potential reproductive contribution of hatchery fish to wild spawning populations could be quite low (due to lower reproductive success of hatchery individuals).

Steelhead were introduced into the Great Lakes in the late 1800s (MacCrimmon & Gots[2]). Natural reproduction primarily occurs in Michigan rivers due to the lack of suitable spawning habitat elsewhere in the Lake Michigan basin (MacCrimmon & Gots[2]). Widespread natural reproduction in the Great Lakes led to development of self-sustaining populations. However, natural recruitment has been supplemented by hatchery-production (Biette et al. 1981, Seelbach 1987a). Management agencies around the Lake Michigan basin stock multiple steelhead strains to take advantage of strain-specific variation in life history characteristics (i.e. run timing) that are valued by recreational anglers. Because naturalized steelhead spawn in drainages in only a portion of the basin, stocking acts to increase the number of fish present in rivers during spawning runs, and to increase the distribution of steelhead in open-water areas throughout the basin. One hatchery strain (the Michigan strain, derived from the Little Manistee River) has historically been used across the basin, and is presently used for supplementation by the state of Michigan. Recently released hatchery strains include the Chambers Creek and Skamania strains originating from Washington, and the Ganaraska strain established from the naturalized population in the Ganaraska River in Ontario. Currently, genetic differences between the four hatchery strains represent the major component of genetic diversity in Lake Michigan steelhead (Bartron et al.[3]). Introgression among hatchery strains stocked by other agencies across the Great Lakes basin may lead to increasing levels of straying, outbreeding, and loss of strain-specific hatchery traits.

Changes in hatchery management practices have increased the relative contribution of hatchery steelhead to naturalized populations in Michigan. The state of Michigan changed the size and age of stocked juvenile steelhead in 1983, from stocking fall fingerlings (<110 mm) to large yearlings (>150 mm; Seelbach 1987a). Larger juveniles had higher rates of smolting (mean 48.2% vs. 0.5–2.9%; Seelbach 1987a) and survival to smolt stage (90% vs. 0.01%; Rand et al. 1993) compared to juveniles stocked at smaller sizes. Prior to changes in age and size at stocking, the average contribution of hatchery fish to spawning runs in six Michigan rivers was generally low (0–30%; Seelbach & Whelan 1988). Following the change in stocking practices, the contribution of hatchery fish in the same six rivers during the 1998–1999 spawning run substantially increased (13–79%; Bartron et al.[3]).

In this study, we compared allele frequencies and levels of genetic diversity for two different time periods (1983–1984 and 1998–1999) for each of six

[1] Lynch, M. 1997. Inbreeding depression and outbreeding depression. pp. 59–67. *In*: Grant W.S. (ed.) Genetic Effects of Straying of Non-native Hatchery Fish into Natural Populations: Proceedings of the Workshop. U.S. Dept. Comm., NOAA Tech Memo. NMFS-NWFSC-30.

[2] MacCrimmon, H.R. & B.L. Gots. 1972. Rainbow trout in the Great Lakes. Ontario Ministry of Natural Resources, Sport Fisheries Branch, Toronto, Ontario, 66 pp.

[3] Bartron, M.L., D.R. Swank, E. Rutherford & K.T. Scribner. Methodological bias in estimates of strain composition and straying rates of hatchery-produced steelhead in Lake Michigan tributaries. N. Am. J. Fish. Manag. submitted.

populations before and after changes in juvenile age at stocking. Comparisons should show whether increased survival of hatchery juveniles and abundance of adult hatchery steelhead led to increased levels of introgression as evidenced by altered genetic characteristics within, and inter-relationships among naturalized populations. Our two main objectives were (1) to determine if time series data on survival and stocking histories for steelhead in the Lake Michigan basin were suggestive of increasing threats of introgression between hatchery and wild individuals, and (2) using time-series data of population allele frequencies and estimates of genetic diversity, to determine if the increased presence of hatchery individuals during the spawning period had demonstrable effects on population genetic characteristics of naturalized populations.

Methods

Stocking history

Original introductions of steelhead into the Great Lakes in the late 1800s utilized gametes taken from rivers in California (primarily McCloud River, Klamath River, and Redwood Creek), and the Willamette and Rogue rivers in Oregon (MacCrimmon & Gots[2]). Within Lake Michigan specifically, steelhead (or non-anadromous rainbow trout) from the McCloud River were first stocked in 1896 (MacCrimmon & Gots[2]). Naturalized populations were well established and widely distributed around the Lake Michigan basin by the 1920s although primary reproduction occurred in Michigan due to the abundance of spawning habitat (MacCrimmon & Gots[2]). Supplemental stocking within Lake Michigan did not occur again until the mid-1950s following a decline in the lake-wide steelhead population (MacCrimmon & Gots[2]). Until the mid-1980s, stocking primarily utilized one hatchery strain, derived from returning adults captured at the Little Manistee weir in Michigan's Little Manistee River (Figure 1). Starting in the mid-1980s, additional hatchery strains have been used for stocking by states around the Lake Michigan basin. These additional hatchery strains include: the Chambers Creek strain from the Puget Sound region of Washington; the Skamania strain, from the Washougal River in Washington; the Ganaraska strain, from the Ganaraska River in Ontario, Canada. The Ganaraska and Michigan strains originate from naturalized populations within the Great Lakes. Each of the four hatchery strains stocked into Lake

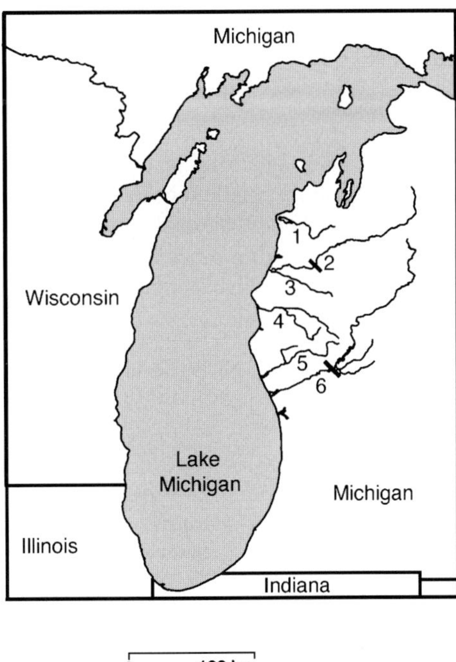

Figure 1. Geographic locations where historical and contemporary populations of steelhead were sampled from Michigan tributaries of Lake Michigan. Samples were obtained from throughout the river drainages. Numbers on maps corresponding to river names are as follows: (1) Betsie River, (2) Manistee River, (3) Little Manistee River, (4) Pere Marquette River, (5) White River, (6) Muskegon River. The bars that intersect both the Manistee River and Muskegon River correspond to dams that prevent upstream migration of adult steelhead. Sampling locations for these rivers were distributed throughout the portion of the rivers downstream from the dams.

Michigan are maintained by gametes taken from adults returning to weirs located on four rivers around the Lake Michigan basin each year. No specific broodstocks are maintained. Thus, unintentional straying and introgression of fish of multiple strains could compromise the genetic integrity of each strain.

Stocking records of Michigan hatchery strain steelhead for the Lake Michigan basin (Table 1) and specifically for the Lake Michigan tributaries examined in this study (Table 2) were obtained from the Michigan Department of Natural Resources (Lansing, MI) for two time periods (1979–1982 and 1993–1997). Time periods represent years preceding the two sampling events of adult steelhead spawning runs, and reflect cohorts that may contribute to the spawning runs sampled. Smolt-equivalents were calculated from estimates of survival rates of juvenile steelhead stocked into Lake Michigan based on age, size, and stocking location (Rand et al. 1993).

Sample collection

Samples from adult steelhead spawning populations were obtained from six populations for each of two time periods, before (1983–1984) and after (1998–1999, 2000) changes in stocking practices. Adult steelhead from both time periods were sampled from the Betsie River, Manistee River, Little Manistee River, Pere Marquette River, White River, and Muskegon River (Figure 1). Historical samples were based on archived scale samples held at the Michigan Department of Natural Resources Institute for Fisheries Research (Ann Arbor, MI) that had been obtained by creel surveys and hook and line sampling of adult steelhead. Archived scales were sampled from adults during the fall 1983 and spring 1984 spawning runs in each river (Seelbach & Whelan 1988). Among historical populations, fall 1983 and spring 1984 samples were obtained for the Bestie, Little Manistee, and White rivers. The Manistee, Pere Marquette, and Muskegon rivers were only sampled in the fall of 1983. Contemporary populations in the Manistee, Little Manistee, Pere Marquette, and Muskegon rivers were sampled during both fall 1998 and spring 1999. The White and Betsie rivers were sampled in the spring of 2000. Among historical populations, the Betsie (fall run $N = 17$, spring run $N = 33$), Little Manistee (fall run $N = 29$, spring run $N = 29$), and White (fall run $N = 7$, spring run $N = 6$) rivers were sampled during both fall and spring runs. Among contemporary populations, the Manistee (fall run $N = 53$, spring run $N = 105$), Little Manistee (fall run $N = 57$, spring run $N = 60$), Pere Marquette (fall run $N = 29$, spring run $N = 73$), and Muskegon (fall run $N = 37$, spring run $N = 39$) rivers were sampled during both fall and spring runs. Contemporary populations were sampled by electrofishing and creel surveys of adult steelhead returning to Michigan rivers throughout the fall of 1998, spring 1999, and spring of 2000, to include samples from the entire spawning period.

Table 1. Number of juvenile Michigan strain steelhead annually stocked into Michigan rivers of the Lake Michigan basin, estimates of the number of juveniles that would be expected to survive to smolt stage, and percentage of the total number stocked based on estimated survival.

Year	Number stocked	Smolt-equivalent[1]	Estimated percent survival
1978	Data not available	—	—
1979	1 092 273	21 091	1.9
1980	1 325 177	19 942	1.5
1981	674 343	77 508	11.5
1982	1 091 154	15 493	1.4
Mean	1 045 737	33 509	4.1
1993	475 139	388 765	81.8
1994	532 688	439 297	82.3
1995	544 530	448 243	82.3
1996	527 329	426 517	80.9
1997	544 791	482 485	88.6
Mean	524 895	437 061	83.2

[1]Smolt equivalents were calculated as described by Rand et al. (1993). Estimated survival to smolt stage was based on juvenile size and age at stocking, and stocking location.

Table 2. Estimates of the mean ($\pm$95% CI) number of juvenile steelhead stocked into Michigan tributaries of Lake Michigan that were expected to survive to smolt (i.e. number of smolt-equivalents), and the proportion of wild-origin adults in the spawning runs of six rivers in Michigan estimated from two time periods prior to and following changes in juvenile stocking practices.

River	1983–1984		1998–1999	
	Mean number stocked per year	Proportion wild[1]	Mean number stocked per year	Proportion wild[2]
Betsie	2 588	0.82 ($\pm$0.10)	44 635	0.65 ($\pm$0.18)
Manistee	826	0.88 ($\pm$0.08)	42 781	0.65 ($\pm$0.06)
Little Manistee	1 400	0.98 ($\pm$0.03)	90	0.69 ($\pm$0.01)
Pere Marquette	113	1.00 ($\pm$0.00)	9 252	0.87 ($\pm$0.06)
White	4 880	0.88 ($\pm$0.16)	20 178	0.55 ($\pm$0.17)
Muskegon	8 514	0.70 ($\pm$0.28)	52 683	0.21 ($\pm$0.04)
Mean	3 054	0.88	28 269	0.60

[1]Data from Seelbach & Whelan (1988).
[2]Data from Bartron et al.[3].

Scale pattern analysis

Scale pattern analysis (SPA) was used to determine hatchery or natural origin of each adult steelhead (Seelbach & Whelan 1988). SPA for fish sampled from both historic and contemporary time periods was performed using the ratio 23 method (Seelbach & Whelan 1988). Ratio 23 quantifies differential winter and spring growth rates in juvenile steelhead through comparison of the width of the five intercirculus spaces prior to the first annulus to the width of the five intercirculus spaces that follow the first annulus (Seelbach & Whelan 1988). For contemporary populations, juvenile stream resident time (Bartron et al.[3]) was also used to identify hatchery-origin individuals. Juvenile stream resident time was used when the ratio 23 value was between 0.7 and 0.8 (0.7 being the threshold for determination of hatchery or natural-origin). Hatchery yearlings migrate downstream to Lake Michigan within a year after stocking, whereas juvenile steelhead produced in the wild may reside in the stream for more than one year prior to smolting (Seelbach 1987a). The use of an additional technique to identify hatchery-origin individuals was necessary due to changes in growth patterns resulting from the increased size and age-at-stocking of hatchery fish. The likelihood of classification errors resulting from the ratio 23 technique is discussed in Seelbach & Whelan (1988). All steelhead identified as originating in hatcheries based on SPA were removed from the genetic analysis.

Genetic analysis

After removal of hatchery individuals from each population, sample sizes for historical and contemporary populations for each river were: Betsie River ($N = 49$ and $N = 25$), Manistee River ($N = 52$ and $N = 107$), Little Manistee River ($N = 57$ and $N = 116$), Pere Marquette River ($N = 10$ and $N = 102$), White River ($N = 15$ and $N = 18$), and Muskegon River ($N = 7$ and $N = 75$). DNA was obtained from scales and fin clips. DNA was extracted from scales (Qiagen DNeasy[®] Tissue Kit, Qiagen Inc.), and from fin clips (PURGENE[®], Gentra Inc.). Individuals were genotyped at six microsatellite loci, including Ogo1a and Ogo4 (Olsen et al. 1998), Omy77 (Morris et al. 1996), Oneμ10 (Scribner et al. 1996), Ots103 (Beacham et al. 1998), and Oki200 (Beacham et al. 1999). PCR reactions for Ogo1a, Omy77, Oneμ10, and Ogo4 followed protocols described in Bartron et al.[3]. PCR reactions

for Ots103 and Oki200 were conducted in $10\,\mu$l volumes using $40\,$ng DNA, $1\,\mu$l 10x PCR Buffer ($0.1\,$M Tris-HCl, pH 8.3, $0.015\,$M MgCl$_2$, $0.5\,$M KCl, 0.1% gelatin, 0.1% NP-40, 0.1% Triton-X 100), $0.2\,$mM dNTP's, $0.4\,\mu$M fluorescent labeled forward primer, $0.4\,\mu$M unlabeled reverse primer, and 0.3 units Taq Polymerase. PCR reactions for all loci utilized an initial denaturing step at $94°$C for $2\,$min, followed by 30 cycles for tissue-derived DNA and 40 cycles for scale-derived DNA of $94°$C for $1\,$min, annealing temperature for $1\,$min, and extension at $72°$C for $1\,$min, and a final extension period of $2.5\,$min. The annealing temperature for Ots103 and Oki200 was $50°$C. Genotypes for Ogo1a, Ogo4, Omy77, and Oneμ10 were visualized on a Hitachi FM-BIO[®] II scanner. Genotypes for Ots103 and Oki200 were visualized on a LI-COR[®]IR^2 Global Edition DNA Sequencer. Molecular weight standards and individuals of known genotype were run on each gel to standardize scoring.

Statistical analysis

Samples were taken over two portions of the spawning run (fall and spring) from most populations. We used Fisher's exact test (GENMOD procedure) in SAS software (SAS Institute 1999) to determine if proportions of hatchery and river-origin individuals differed between fall and spring runs, and between historical and contemporary time periods.

Estimates of observed and expected heterozygosity and allele frequencies for each population were calculated using BIOSYS-1 (Swofford & Selander 1981). Significance of deviations of genotypic frequencies from Hardy–Weinberg expectations were determined for each population using Markov chain methods of Guo & Thompson (1992) implemented in program GENEPOP (Raymond & Rousset 1995). Allelic richness values for each population for each locus (Petit et al. 1998) were calculated using FSTAT (Goudet 1995) v2.9.3.1 to standardize population measures of allelic diversity for differences in sample size. Mean allelic richness estimates for each population were presented as the sum of the allelic richness values assayed.

Pair-wise F_{ST} (Weir & Cockerham 1984) comparisons were used to determine if genetic differences existed between fall and spring runs for those rivers that were sampled for both runs. In the absence of differences, fall and spring samples were combined for analyses of population differences. Estimates of pair-wise and overall population differentiation were summarized using F-statistics, implemented in the program FSTAT (Goudet 1995) v.2.9.3.1. Nominal alpha values were corrected for multiple pair-wise comparisons using Bonferroni corrections (Rice 1989). Cavalli-Sforza & Edwards (1967) chord distances were estimated for all population comparisons during both time periods using BIOSYS-1 (Swofford & Selander 1981) and were used to construct neighbor-joining (Saitou & Nei 1987) dendrograms. Neighbor-joining trees and associated bootstraps (500 iterations) were generated using PHYLIP v3.5 (Felsenstein 1993). The resulting trees were displayed using TREEVIEW (Page 1996).

Generalized Mantel tests (Smouse et al. 1986) were used to test for correlations between genetic relationships (inter-population genetic distance) and population geographic proximity, providing measures of spatial autocorrelation in allele frequency. Geographic distances (in kilometers) were estimated between river mouths along the Lake Michigan shoreline.

Results

Hatchery contributions to spawning runs

Stocking histories for steelhead of the Michigan hatchery strain into rivers in Michigan revealed a decline in the number of steelhead stocked over time (Table 1). On average, 1 045 737 Michigan-strain steelhead were stocked into Lake Michigan each year between 1979 and 1982, whereas an average of 524 895 were stocked each year between 1993 and 1997 (Table 1). Following the change in age and size at stocking, the estimated number of stocked juvenile steelhead that survive to smolting increased from 4.1% to 83.2% between the two time periods (Table 1). Survival estimates were comparable to those of river-origin steelhead in Lake Michigan, which ranged from 13% to 90% for presmolt winter survival (Seelbach 1987b). Between 1978 and 1982, the estimated yearly mean number of Michigan strain juveniles stocked into Lake Michigan surviving to smolt stage was 33 509 (Table 1), compared to a mean of 437 061 between 1993 and 1997 (Table 1). Between 1983–1984 and 1998–1999, the mean number of Michigan strain steelhead stocked into the six rivers examined in this study that survived to smolt stage increased from 3 054 to 28 269 (Table 2). Estimates of natural recruitment are not available for the six rivers, however the 10-fold increase in hatchery contributions

that survive to the smolt stage suggests an increased potential for straying of adult hatchery fish into natural spawning populations across the Lake Michigan basin.

We observed an increase in the proportion of hatchery-origin individuals found in adult steelhead spawning runs for six rivers in Michigan between the 1983–1984 and 1998–1999 spawning runs (Table 2). The contribution of hatchery steelhead in the six rivers examined averaged 12% during 1983–1984 spawning run, and 40% during the 1998–1999 spawning run. Significant differences were observed in hatchery contributions to the spawning run of each river between the historical and contemporary populations for the Little Manistee River ($X^2_{df=1} = 37.86$, p < 0.0001), Manistee River ($X^2_{df=1} = 15.54$, p < 0.0001), Muskegon River ($X^2_{df=1} = 10.49$, p < 0.0012), and White River ($X^2_{df=1} = 5.38$, p < 0.02).

Genetic analysis

The mean number of alleles per locus across all loci increased over time (comparisons of historical and contemporary populations; Table 3). The mean number of alleles per locus ranged from 4.2 to 4.7 for the historical populations and 5.3–7.8 for the contemporary populations (Table 3; specific allele frequencies are provided in Appendix 1). When measured by allelic richness to correct for sample size, the mean number of alleles per locus increased between time periods for three of the six rivers (Table 3). Alleles not found in historic populations were observed in contemporary populations, half of which are found in one or all of the Skamania, Chambers Creek, or Ganaraska hatchery strains (Appendix 1). Five alleles (in loci Ogo4 and Omy77) were found in the contemporary populations, and in three of the hatchery strains, but not in the historic populations (Appendix 1). One allele (in locus Ogo1a) was absent from historic populations except for the Little Manistee, but was found in three of the other five contemporary naturalized populations (Appendix 1). Given that the majority of common alleles are shared across strains and naturalized populations, estimates of introgression based solely on the appearance of rare alleles are likely under-estimates. Mean observed heterozygosity for the historical populations ranged from 0.563 to 0.632 and from 0.555 to 0.686 for the contemporary populations (Table 3). Deviations from Hardy–Weinberg equilibrium were not observed in any population sampled during either time period (p > 0.05).

Table 3. Sample sizes (N), number of alleles per locus (A), allelic richness (A_r), and observed and expected (H_o and H_e) heterozygosity for each steelhead population during each time period (1983–1984 and 1998–1999 or 2000), and for each hatchery strain.

| | Population | | | | | | | | | | Hatchery strains | | | | |
| | Betsie | | Manistee | | Pere Marquette | | White | | Muskegon | | Little Manistee | | Skamania | Chambers Creek | Ganaraska |
Locus	1983–1984	2000	1983–1984	1998–1999	1983–1984	1998–1999	1983–1984	2000	1983–1984	1998–1999	1983–1984	1998–1999			
Ogo1a															
N	47	23	52	105	10	101	15	17	7	75	57	114	103	59	60
A	3	3	3	4	3	5	3	3	3	5	4	4	3	3	3
A_r	2.981	2.990	2.993	3.307	3.000	3.483	2.934	2.993	3.00	3.339	3.115	3.206	2.052	2.178	2.924
H_o	0.596	0.739	0.654	0.629	0.600	0.683	0.600	0.294	0.714	0.720	0.544	0.614	0.271	0.271	0.483
H_e	0.657	0.666	0.672	0.648	0.679	0.681	0.618	0.670	0.560	0.687	0.677	0.667	0.333	0.333	0.585
Ogo4															
N	49	25	52	106	10	100	14	16	7	72	56	115	111	58	60
A	4	6	7	10	4	11	7	7	5	7	7	8	8	9	9
A_r	4.525	4.785	4.297	4.585	3.621	4.758	4.989	4.917	5.000	4.261	4.359	4.073	5.629	5.192	5.053
H_o	0.673	0.760	0.654	0.670	0.500	0.680	0.857	0.563	0.714	0.722	0.679	0.652	0.802	0.776	0.700
H_e	0.754	0.765	0.673	0.697	0.626	0.738	0.759	0.730	0.780	0.717	0.874	0.702	0.817	0.761	0.659
Omy77															
N	49	25	48	104	10	100	15	16	7	75	56	116	97	59	59
A	13	15	15	16	11	16	10	11	9	13	16	17	6	13	13
A_r	7.549	8.516	7.479	7.762	9.164	7.315	7.144	6.902	9.000	7.292	7.817	7.247	4.807	7.005	7.630
H_o	0.857	0.920	0.854	0.827	1.00	0.770	0.867	0.938	0.857	0.747	0.875	0.784	0.707	0.712	0.915
H_e	0.864	0.898	0.870	0.874	0.932	0.855	0.871	0.821	0.912	0.872	0.874	0.832	0.722	0.866	0.892
Oneµ10															
N	49	24	51	107	10	95	15	16	7	75	57	115	103	60	60
A	5	5	7	7	3	6	5	4	3	6	6	7	3	6	6
A_r	3.490	3.632	4.044	4.143	2.956	3.992	4.122	3.432	3.000	3.728	4.038	3.549	2.720	4.209	4.613
H_o	0.673	0.708	0.804	0.692	0.556	0.705	0.733	0.813	0.857	0.680	0.649	0.696	0.417	0.583	0.817
H_e	0.641	0.650	0.675	0.677	0.464	0.672	0.678	0.688	0.648	0.685	0.688	0.665	0.443	0.653	0.775
Oki200															
N	49	24	46	89	10	100	15	18	7	75	57	113	106	57	60
A	5	6	5	6	2	6	2	4	4	6	6	6	5	7	6
A_r	2.854	4.034	3.107	3.390	2.000	2.948	1.999	2.767	4.000	3.635	3.010	3.257	3.595	4.389	3.366
H_o	0.510	0.708	0.500	0.506	0.667	0.450	0.333	0.556	0.429	0.480	0.439	0.522	0.670	0.847	0.567
H_e	0.456	0.559	0.475	0.555	0.523	0.453	0.434	0.440	0.396	0.533	0.481	0.523	0.650	0.770	0.543
Ots103															
N	49	25	52	96	10	102	15	18	7	73	57	115	108	43	58
A	3	3	3	4	2	3	4	3	2	3	4	3	4	3	5
A_r	2.268	2.488	2.265	2.303	1.921	2.032	2.867	2.022	2.000	2.107	2.221	2.112	2.913	2.155	2.794
H_o	0.184	0.280	0.231	0.208	0.200	0.196	0.400	0.167	0.000	0.164	0.193	0.209	0.417	0.217	0.362
H_e	0.241	0.287	0.244	0.229	0.189	0.183	0.352	0.160	0.264	0.202	0.211	0.214	0.383	0.251	0.367
Mean															
A	5.8	6.3	6.7	7.8	4.2	7.8	5.2	5.3	4.3	6.7	7.2	7.5	4.8	6.8	7.0
A_r	3.945	4.408	4.031	4.248	3.777	4.088	4.009	3.836	4.333	4.060	4.093	3.907	3.619	4.188	4.397
H_o	0.582	0.686	0.616	0.588	0.587	0.581	0.632	0.555	0.595	0.586	0.563	0.580	0.540	0.568	0.641
H_e	0.602	0.637	0.601	0.613	0.569	0.597	0.619	0.585	0.593	0.616	0.610	0.601	0.540	0.606	0.637

Hatchery individuals were excluded from river samples.

We found no evidence for significant differences in allele frequency between fall and spring portions of spawning runs within rivers. Variance in allele frequency as estimated using F_{ST} values for comparisons between fall and spring runs sampled within each river for each time period did not indicate evidence for significant genetic differentiation ($p > 0.05$ for mean F_{ST} values across all pair-wise comparisons). Consequently, samples from each run were combined for those populations that included samples from the fall and spring runs. Mean F_{ST} values when estimated across all six populations for each time period were not statistically significant (mean F_{ST} for historical and contemporary populations were 0.006 and 0.002, respectively; Table 4).

Mantel tests were conducted using genetic distances among populations from each time period and for correlations between inter-population genetic distances among populations. The Mantel tests indicated that genetic relationships among populations were consistent between the two time periods ($R = 0.65$; $p < 0.01$; Figure 2). Correlations between the geographic and genetic distances between each spawning population were significant for historic populations ($R = 0.55$; $p < 0.01$), indicating significant spatial autocorrelation in allele frequencies between population (isolation by distance) relationships. Relationships between genetic and geographic distance were also evident for contemporary populations, though the relationship was less strongly supported ($R = 0.39$;

Table 4. Pair-wise estimates of F_{ST} between historic (above diagonal) and contemporary (below diagonal) populations.

Population	Population					
	Betsie	Manistee	Little Manistee	Pere Marq.	White	Muskegon
Betsie	*0.000*	0.007	0.000	0.021	0.001	0.000
Manistee	0.002	*0.001*	0.000	0.002	0.002	0.000
Little Manistee	0.000	0.007	*0.000*	0.013	0.000	0.000
Pere Marquette	0.003	0.005	0.004	*0.007*	0.016	0.032
White	0.000	0.000	0.000	0.000	*0.000*	0.000
Muskegon	0.000	0.007	0.001	0.004	0.000	*0.000*

Pair-wise F_{ST} values along the diagonal (italicized) represent comparisons between time periods for the same population. Comparisons are not statistically significant.

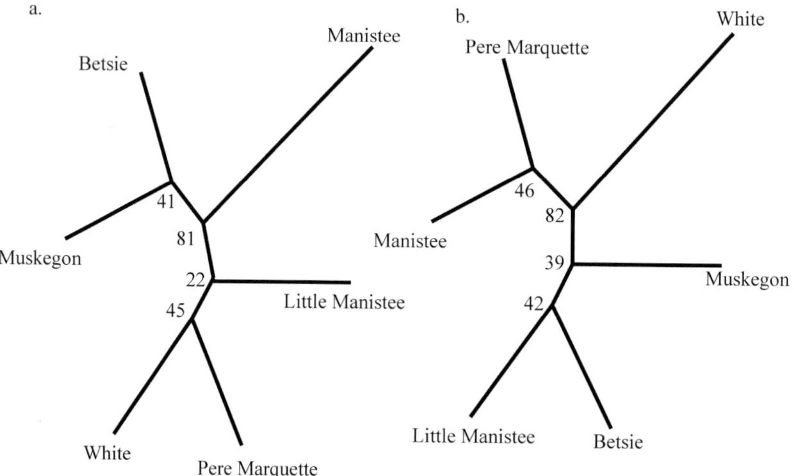

Figure 2. Unrooted neighbor-joining trees based on Cavalli-Sforza & Edwards (1967) chord distance, demonstrating the genetic relationships among each of the six populations for two time periods (a) 1983–1984 and (b) 1998–2000. Bootstrap values at nodes represent the percentage of 500 trees where the populations past the nodes were grouped together.

p < 0.02). Neighbor-joining tree topology representing inter-population variation in allele frequency across loci as viewed in the Cavalli-Sforza & Edwards (1967) distances among historic populations generally grouped populations with geographic neighbors, but topology may have been biased by populations with small sample sizes (i.e. Muskegon, Pere Marquette, and White river populations). Among contemporary populations grouping was more random, and again may be biased by populations with small sample sizes (i.e. Betsie and White rivers; Figure 2). Nodes for both trees were weakly supported by bootstrap values (Figure 2).

Discussion

Widespread stocking of hatchery fish has raised concerns regarding the potential for interactions between hatchery and wild populations of many fish species. Particular emphasis has been placed on salmonid populations in their native range due to their socio-economic importance and to dramatic declines in population numbers and distribution. Individuals of hatchery origin commonly co-occur with wild individuals, which pose increased risks of introgression. Outbreeding can lead to the loss of genetic adaptations within populations (Lynch & O'Hely 2001). Loss of naturally evolved co-adapted gene complexes may not be a pressing issue in introduced and artificially maintained systems, such as those that currently exist in the Great Lakes. However, phenotypic differences (e.g. variation in run timing) are often exploited by managers to provide increased recreational fishing opportunities for anglers. In Lake Michigan, multiple hatchery strains with evolved tendencies to enter rivers at different times of the year are stocked to extend the duration of once seasonal fisheries. Concurrent stocking of multiple strains increase risks of introgression among strains and between individuals of hatchery and natural origin. Variation among strains in phenotype or life history, whether adaptive or not, is useful to managers and is at risk of loss through outbreeding.

Although large numbers of juvenile steelhead are stocked yearly (Table 1), natural reproduction provides a substantial component of the steelhead production to Lake Michigan, and hence to the basin-wide recreational fishery. Rand et al. (1993) estimated that steelhead smolts emigrating from the Little Manistee River contribute 13–21% of the total basin-wide smolt yield. Although the total number of hatchery steelhead stocked in Michigan has decreased since the early 1980s (Table 1), survival of hatchery fish to smolt-stage has increased (Table 2). As a consequence, the proportion of hatchery fish present in spawning runs increased significantly between the two time periods (Table 2). Clearly, increased abundance of hatchery adults in spawning runs increases the potential for introgression between the hatchery and naturalized populations of steelhead in tributaries across the Lake Michigan basin. Previously, there was a lack of information regarding how the threat of increased introgression translated to actual outbreeding.

We did not observe strong evidence of increased levels of gene flow on the basis of changes in magnitude of inter-population variance in allele frequency as seen from F-statistics. However, our ability to detect effects of potential introgression if occurring (i.e. statistical power) was low, as historical populations were not significantly diverged genetically. We did however observe significant correlations between genetic distance and geographic distance among populations, suggesting that steelhead populations in Michigan (both historical and contemporary) do not represent a single panmictic population, but rather are structured as a function of geographic distance among populations. Decreased correlations between genetic and geographic distance in contemporary populations as compared to historical populations are suggestive of some level of changes in patterns of gene flow or homogenization of allele frequencies due to increased introgression as would be expected between hatchery and naturalized individuals.

Recent introduction of new genetically differentiated hatchery strains by states across the Lake Michigan basin has let to increasing incidence of these strains in Michigan rivers (Bartron et al.[3]). Introgression most likely has contributed to increased levels of within-population diversity for most contemporary populations surveyed in this study. Spatial variance in allele frequency represents a small component of the total genetic variation apportioned across the Lake Michigan basin. Genetic diversity within populations was higher in contemporary populations than in historical populations due to increased mean number of alleles per locus. The appearance of alleles not found in historical populations is due to the recently introduced hatchery strains (Appendix 1). Since all fish surveyed are of wild origin, the formerly unseen alleles can only be attributed to past introgression.

The absence of significant genetic structure among historical populations was not unexpected because of the proximity of natural populations to each other, and the relatively short time steelhead have been present in Lake Michigan. Differentiation over relatively short geographic distances was detected among steelhead populations within the Skeena River (Beacham et al. 2000). On a larger spatial scale, Krueger & May (1987) analyzed steelhead populations within the Lake Superior basin and found greater levels of population differentiation ($F_{ST} = 0.026$). Observation of significant spatial structure by Krueger & May (1987) was potentially due to the sampling of juveniles, and spatial structure could reflect family groups rather than population differences (Allendorf & Phelps 1981).

Despite lack of significant spatial genetic differentiation, genetic characteristics observed within steelhead populations in Lake Michigan (such as the lack of significant genetic differences between fall and spring run steelhead) are consistent with comparisons of winter and summer-run steelhead in their native range (Chilcote et al. 1980, Leider et al. 1984). Heath et al. (2002) found temporal comparisons of genetic diversity among steelhead populations to be much greater over a shorter time period than was found among Michigan populations. The lack of greater temporal genetic divergence among Michigan populations could be attributed in part to disparities in sample size obtained for populations in each time period. However, three of the historic populations had large sample sizes for both time periods and pairwise comparisons between those populations and their contemporary counterpart did not indicate significant differences.

Within regional spatial contexts such as the Great Lakes, gene flow among fish from naturalized populations may be high due to natural levels of straying. However, the Great Lakes basins share a number of characteristics of other multi-jurisdictional fisheries resources that are increasingly at risk to accelerated levels of outbreeding due to increased reliance on hatchery supplementation. In the Great Lakes, different agencies stock fish of different genetic backgrounds. Due to higher rates of straying by hatchery fish (Waples 1991), the potential to interbreed with native or naturalized fish from locally adapted gene pools increases as seen by levels of gene diversity within and among steelhead populations in Lake Michigan. Concomitantly, apportionment of genetic diversity may be altered within and among populations. Findings of significant correlations of geographic and genetic distance in contemporary populations suggest that contributions to recruitment from naturalized fish continue to be proportionally higher than is realized by fish of hatchery origin despite changes in management practices used for hatchery supplementation.

Acknowledgements

The authors wish to thank Ed Rutherford for sampling assistance, Jan Sapak and Dave Swank for sampling assistance and scale reading, and Pat DeHaan and Kristine Bennett for lab assistance. We are grateful to Paul Seelbach for access to historical samples and comments on earlier drafts of the paper. Support was provided by Michigan Department of Natural Resources (MDNR) through the Federal Aid Program, Great Lakes Restoration Act through the U.S. Fish and Wildlife Service, and the Partnership for Ecosystem Research and Management (PERM) program through the MDNR and the Department of Fisheries and Wildlife at Michigan State University.

References

Allendorf, F.W. & S.R. Phelps. 1981. Use of allelic frequencies to describe population structure. Can. J. Fish. Aquat. Sci. 38: 1507–1514.

Beacham, T.D., L. Margolis & R.J. Nelson. 1998. A comparison of methods of stock identification for sockeye salmon (*Oncorhynchus nerka*) in Barkley Sound, British Columbia. N. Pac. Anadromous Fish Comm. Bull. 1: 227–239.

Beacham, T.D., S. Pollard & K.D. Le. 1999. Population structure and stock identification of steelhead in southern British Columbia, Washington, and the Columbia River based on microsatellite DNA variation. Trans. Am. Fish. Soc. 128: 1068–1084.

Beacham, T.D., S. Pollard & K.D. Le. 2000. Microsatellite DNA population structure and stock identification of steelhead trout (*Oncorhynchus mykiss*) in the Nass and Skeena Rivers in northern British Columbia. Mar. Biotech. 2: 587–600.

Biette, R.M., D.P. Dodge, R.L. Hassinger & T.M. Stauffer. 1981. Life history and timing of migrations and spawning behavior of rainbow trout (*Salmo gairdneri*) populations of the Great Lakes. Can. J. Fish. Aquat. Sci. 38: 1759–1771.

Cavalli-Sforza, L.L. & A.W.F. Edwards. 1967. Phylogenetic analysis: Models and estimation procedures. Evolution 21: 550–570.

Chilcote, M.W., B.A. Crawford & S.A. Leider. 1980. A genetic comparison of sympatric populations of summer and winter steelheads. Trans. Am. Fish. Soc. 109: 203–206.

Felsenstein, J. 1993. Phylip: Phylogeny inference package. Ver 3.5c. Department of Genetics, University of Washington, Seattle.

Gharrett, A.J. 1994. Genetic dynamics of a small population system: A model applicable to interactions between hatchery and wild fish. Aquacult. Fish. Manage. 25 (Suppl. 2): 79–92.

Garant, D., J.J. Dodson & L. Bernatchez. 2000. Ecological determinants and temporal stability of the within-river population structure in Atlantic salmon (*Salmo salar* L.). Mol. Ecol. 9: 615–628.

Goudet, J. 1995. F-STAT v. 1.2: A computer program to calculate F-Statistics. J. Hered. 86: 485–486.

Guo, S.W. & E.A. Thompson. 1992. Performing the exact test of Hardy–Weinberg proportions for multiple alleles. Biomet. 48: 361–372.

Heath, D.D, C. Busch, J. Kelly & D.Y. Atagi. 2002. Temporal change in genetic structure and effective population size in steelhead trout (*Oncorhynchus mykiss*). Mol. Ecol. 11: 197–214.

Krueger, C.C. & B. May. 1987. Genetic comparison of naturalized rainbow trout populations among Lake Superior tributaries: Differentiation based on allozyme data. Trans. Am. Fish. Soc. 116: 795–806

Leider, S.A., M.W. Chilcote & J.J. Loch. 1984. Spawning characteristics of sympatric populations of steelhead trout (*Salmo gairdneri*): Evidence for partial reproductive isolation. Can. J. Fish. Aquat. Sci. 41: 1454–1462.

Lynch, M. & M. O'Hely. 2001. Captive breeding and the genetic fitness of natural populations. Cons. Gen. 2: 363–378.

Morris, D.B., K.R. Richard, & J.M. Wright. 1996. Microsatellites from rainbow trout (*Oncorhynchus mykiss*) and their use for genetic study of salmonids. Can. J. Fish. Aquat. Sci. 53: 120–126.

Nielsen, J.L. 1999. The evolutionary history of steelhead (*Oncorhynchus mykiss*) along the US Pacific Coast: Developing a conservation strategy using genetic diversity. ICES J. Mar. Sci. 56: 449–458.

Nielsen, J.L. & M.C. Fountain. 1999. Microsatelite diversity in sympatric reproductive ecotypes of Pacific steelhead (*Oncorhynchus mykiss*) from the Middle Fork Eel River, California. Ecol. Freshw. Fish 8: 159–168.

Olsen, J.B., P. Bentzen & J.E. Seeb. 1998. Characterization of seven microsatellite loci derived from pink salmon. Mol. Ecol. 7: 1087–1089.

Osterberg, C.O. & G.H. Thorgaard. 1999. Geographic distribution of chromosome and microsatellite DNA polymorphisms in *Oncorhynchus mykiss* native to Western Washington. Copeia 1999: 287–298.

Page, K.S. 1996. TreeView: An application to display phylogenetic trees on personal computers. Comp. Applic. Biol. Sci. 12: 357–358.

Parkinson, E.A. 1984. Genetic variation in populations of steelhead trout (*Salmo gairdneri*) in British Columbia. Can. J. Fish. Aquat. Sci. 41: 1412–1420.

Petit, R.J., A. El Mousadik & O. Pons. 1998. Identifying populations for conservation on the basis of genetic markers. Cons. Biol. 12: 844–855.

Quinn, T.P. 1993. A review of homing and straying of wild and hatchery-produced salmon. Fish. Res. 18: 29–44.

Rand, P.S., D.J. Stewart, P.W. Seelbach, M.L. Jones & L.R. Wedge. 1993. Modeling steelhead population energetics in Lakes Michigan and Ontario. Trans. Am. Fish. Soc. 122: 977–1001.

Raymond, M. & F. Rousset. 1995. GENEPOP (version 1.2): Population genetics software for exact tests and ecumenicism. J. Hered. 86: 248–249.

Reisenbichler, R.R. & S.R. Phelps. 1989. Genetic variation in steelhead (*Salmo gairdneri*) from the north coast of Washington. Can. J. Fish. Aquat. Sci. 46: 66–73.

Reisenbichler, R.R. & S.P. Rubin. 1999. Genetic changes from artificial propagation of Pacific salmon affect the productivity and viability of supplemented populations. ICES J. Mar. Sci. 56: 459–466.

Reisenbichler, R.R., J.D. McIntyre, M.F. Solazzi, & S.W. Landino. 1992. Genetic variation of steelhead of Oregon and northern California. Trans. Am. Fish. Soc. 131: 158–169.

Rice, W.R. 1989. Analyzing tables of statistical tests. Evolution 43: 223–225.

Ryman, N. 1991. Conservation genetics considerations in fishery management. J. Fish Biol. 39: 211–224.

Saitou, N. & M. Nei. 1987. The neighbor-joining method: A new method for reconstructing phylogenetic trees. Mol. Biol. Evol. 4: 406–425.

SAS Institute. 1999. SAS/STAT Version 8.0. SAS Institute, Cary, North Carolina.

Scribner, K.T., J.R. Gust & R.L. Fields. 1996. Isolation and characterization of novel salmon microsatellite loci: Cross-species amplification and population genetic applications. Can. J. Fish. Aquat. Sci. 53: 833–841.

Seelbach, P.W. 1987a. Smolting success of hatchery-raised steelhead planted in a Michigan tributary of Lake Michigan. N. Am. J. Fish. Manag. 7: 223–231.

Seelbach, P.W. 1987b. Effect of winter severity on steelhead smolt yield in Michigan: An example of the importance of environmental factors in determining smolt yield. Am. Fish. Soc. Symp. 1: 411–450.

Seelbach, P.W. & G.E. Whelan. 1988. Identification and contribution of wild and hatchery steelhead stocks in Lake Michigan tributaries. Trans. Am. Fish. Soc. 117: 444–451.

Smouse, P.E., J. Neel & R.R. Sokal. 1986. Multiple regression and correlation extensions of the Mantel test of matrix correspondence. Syst. Zool. 35: 627–632.

Swofford, D.L. & R.B. Selander. 1981. BIOSYS-1: A FORTRAN program for the comprehensive analysis of electrophoretic data in population genetics and systematics. J. Hered. 72: 281–283.

Waples, R.S. 1991. Genetic interactions between hatchery and wild salmonids: Lessons from the Pacific Northwest. Can. J. Fish. Aquat. Sci. 48(Suppl. 1): 124–133.

Waples, R.S. 1998. Separating the wheat from the chaff: patterns of genetic differentiation in high gene flow species. J. Hered. 89: 438–450.

Washington, P.M. & A.M. Koziol. 1993. Overview of the interactions and environmental impacts of hatchery practices on natural and artificial stocks of salmonids. Fish. Res. 18: 105–122.

Weir, B.S. & C.C. Cockerham. 1984. Estimating F-Statistics for the analysis of population structure. Evolution 38: 1358–1370.

Appendix 1. Sample sizes (N) and allele frequencies at each of the six microsatellite loci for six Michigan steelhead populations sampled during each time period (1983–1984 and 1998–1999), and each hatchery strain.

| | Population | | | | | | | | | | Hatchery Strains | | | | |
| | Betsie | | Manistee | | Pere Marquette | | White | | Muskegon | | Little Manistee | | Skamania | Chambers Creek | Ganaraska |
Locus	1983–1984	2000	1983–1984	1998–1999	1983–1984	1998–1999	1983–1984	2000	1983–1984	1998–1999	1983–1984	1998–1999			
Ogo1a															
N	47	23	52	105	10	101	15	17	7	75	57	114	103	59	60
115	0.000	0.000	0.000	0.000	0.000	0.005	0.000	0.000	0.000	0.000	0.000	0.000	0.000	0.000	0.000
125	0.000	0.000	0.000	0.029	0.000	0.040	0.000	0.000	0.000	0.020	0.009	0.018	0.000	0.000	0.000
139	0.415	0.413	0.356	0.486	0.300	0.405	0.500	0.412	0.643	0.353	0.360	0.417	0.874	0.797	0.566
141	0.000	0.000	0.000	0.000	0.000	0.000	0.000	0.000	0.000	0.007	0.000	0.000	0.000	0.000	0.000
143	0.234	0.239	0.327	0.247	0.250	0.223	0.367	0.239	0.143	0.293	0.325	0.254	0.107	0.186	0.267
161	0.351	0.348	0.317	0.238	0.450	0.327	0.133	0.349	0.214	0.327	0.306	0.311	0.019	0.017	0.167
Ogo4															
N	49	25	52	106	10	100	14	16	7	72	56	115	111	58	60
118	0.328	0.360	0.490	0.473	0.550	0.375	0.393	0.438	0.429	0.375	0.348	0.370	0.148	0.387	0.558
120	0.296	0.280	0.279	0.250	0.300	0.305	0.250	0.281	0.214	0.340	0.330	0.365	0.135	0.147	0.100
122	0.020	0.000	0.010	0.000	0.000	0.005	0.000	0.000	0.000	0.014	0.000	0.004	0.000	0.009	0.050
124	0.061	0.100	0.038	0.028	0.000	0.030	0.036	0.063	0.071	0.035	0.045	0.052	0.000	0.000	0.075
126	0.000	0.000	0.000	0.005	0.000	0.000	0.000	0.000	0.000	0.000	0.000	0.000	0.000	0.000	0.000
128	0.000	0.000	0.000	0.000	0.000	0.000	0.000	0.000	0.000	0.007	0.000	0.000	0.000	0.000	0.000
130	0.000	0.000	0.000	0.000	0.000	0.010	0.000	0.000	0.000	0.000	0.000	0.000	0.032	0.052	0.000
132	0.224	0.160	0.096	0.085	0.100	0.165	0.213	0.125	0.143	0.160	0.205	0.166	0.297	0.250	0.117
134	0.071	0.080	0.058	0.108	0.000	0.050	0.036	0.031	0.143	0.069	0.027	0.026	0.171	0.078	0.033
136	0.000	0.020	0.029	0.028	0.050	0.030	0.036	0.031	0.000	0.000	0.027	0.013	0.162	0.034	0.017
138	0.000	0.000	0.000	0.005	0.000	0.005	0.000	0.000	0.000	0.000	0.000	0.000	0.041	0.009	0.000
140	0.000	0.000	0.000	0.009	0.000	0.020	0.000	0.000	0.000	0.000	0.000	0.000	0.014	0.034	0.008
142	0.000	0.000	0.000	0.009	0.000	0.005	0.036	0.031	0.000	0.000	0.018	0.004	0.000	0.000	0.042
Omy77															
N	49	25	48	104	10	100	15	16	7	75	56	116	97	59	59
82	0.061	0.060	0.021	0.005	0.050	0.020	0.000	0.031	0.000	0.000	0.063	0.034	0.000	0.000	0.144
98	0.000	0.020	0.021	0.019	0.100	0.005	0.000	0.000	0.071	0.000	0.009	0.009	0.000	0.000	0.017
100	0.305	0.260	0.270	0.279	0.200	0.305	0.233	0.375	0.286	0.240	0.285	0.362	0.000	0.000	0.085
102	0.000	0.000	0.000	0.000	0.000	0.000	0.000	0.031	0.000	0.000	0.000	0.000	0.000	0.000	0.000
104	0.082	0.020	0.010	0.058	0.050	0.020	0.033	0.031	0.000	0.047	0.054	0.017	0.082	0.186	0.034
106	0.082	0.040	0.094	0.091	0.000	0.055	0.200	0.063	0.143	0.053	0.089	0.056	0.000	0.000	0.169
108	0.000	0.020	0.000	0.014	0.000	0.000	0.000	0.000	0.000	0.000	0.000	0.004	0.000	0.017	0.008
110	0.041	0.000	0.115	0.043	0.100	0.110	0.100	0.031	0.000	0.033	0.045	0.056	0.000	0.017	0.153
112	0.082	0.120	0.156	0.120	0.150	0.095	0.067	0.188	0.072	0.133	0.080	0.082	0.000	0.059	0.127

	Population										Hatchery Strains				
	Betsie		Manistee		Pere Marquette		White		Muskegon		Little Manistee		Skamania	Chambers Creek	Ganaraska
Locus	1983–1984	2000	1983–1984	1998–1999	1983–1984	1998–1999	1983–1984	2000	1983–1984	1998–1999	1983–1984	1998–1999			
114	0.000	0.000	0.000	0.024	0.000	0.015	0.033	0.031	0.000	0.013	0.009	0.013	0.000	0.008	0.000
116	0.082	0.040	0.063	0.077	0.050	0.060	0.033	0.031	0.072	0.047	0.035	0.052	0.160	0.068	0.034
120	0.000	0.040	0.000	0.024	0.000	0.000	0.000	0.000	0.000	0.000	0.000	0.004	0.160	0.220	0.000
122	0.031	0.100	0.042	0.034	0.100	0.045	0.000	0.000	0.143	0.093	0.080	0.099	0.000	0.000	0.059
124	0.000	0.000	0.000	0.000	0.000	0.000	0.000	0.000	0.000	0.000	0.000	0.000	0.041	0.144	0.000
126	0.000	0.000	0.000	0.000	0.000	0.000	0.000	0.000	0.000	0.000	0.000	0.000	0.000	0.017	0.000
128	0.092	0.060	0.094	0.068	0.050	0.050	0.067	0.000	0.071	0.154	0.134	0.065	0.134	0.034	0.093
130	0.092	0.080	0.052	0.101	0.100	0.145	0.201	0.125	0.071	0.127	0.036	0.086	0.464	0.154	0.068
132	0.020	0.080	0.021	0.029	0.000	0.020	0.000	0.063	0.000	0.027	0.027	0.039	0.119	0.017	0.000
134	0.020	0.040	0.021	0.000	0.050	0.040	0.033	0.000	0.000	0.020	0.036	0.009	0.000	0.059	0.008
136	0.000	0.000	0.010	0.000	0.000	0.010	0.000	0.000	0.000	0.000	0.009	0.000	0.000	0.000	0.000
142	0.010	0.020	0.010	0.014	0.000	0.005	0.000	0.000	0.071	0.013	0.009	0.013	0.000	0.000	0.000
Oneμ10															
N	49	24	51	107	10	95	15	16	7	75	57	115	103	60	60
121	0.000	0.000	0.020	0.009	0.000	0.000	0.000	0.000	0.000	0.000	0.000	0.000	0.000	0.000	0.000
125	0.051	0.021	0.020	0.051	0.000	0.047	0.033	0.000	0.000	0.020	0.018	0.004	0.098	0.142	0.200
127	0.102	0.104	0.196	0.178	0.167	0.237	0.133	0.250	0.143	0.200	0.149	0.170	0.184	0.142	0.192
129	0.449	0.458	0.490	0.491	0.722	0.489	0.500	0.438	0.500	0.420	0.438	0.439	0.718	0.541	0.300
131	0.388	0.375	0.225	0.224	0.111	0.179	0.267	0.281	0.357	0.320	0.316	0.339	0.000	0.017	0.250
133	0.000	0.000	0.010	0.014	0.000	0.011	0.000	0.000	0.000	0.020	0.000	0.005	0.000	0.133	0.033
135	0.010	0.042	0.039	0.033	0.000	0.037	0.067	0.031	0.000	0.020	0.035	0.039	0.000	0.025	0.025
137	0.000	0.000	0.000	0.000	0.000	0.000	0.000	0.000	0.000	0.000	0.044	0.004	0.000	0.000	0.000
Oki200															
N	49	24	46	89	10	100	15	18	7	75	57	113	106	57	60
86	0.021	0.021	0.022	0.022	0.000	0.020	0.000	0.028	0.000	0.047	0.018	0.035	0.000	0.018	0.042
90	0.010	0.083	0.000	0.006	0.000	0.010	0.000	0.028	0.071	0.013	0.009	0.009	0.009	0.009	0.017
92	0.041	0.083	0.054	0.096	0.000	0.050	0.000	0.000	0.000	0.073	0.053	0.058	0.317	0.298	0.042
94	0.704	0.646	0.696	0.618	0.556	0.710	0.700	0.722	0.787	0.653	0.683	0.646	0.132	0.219	0.625
96	0.000	0.000	0.000	0.000	0.000	0.000	0.000	0.000	0.000	0.000	0.000	0.000	0.000	0.009	0.000
98	0.224	0.146	0.206	0.236	0.444	0.205	0.300	0.222	0.071	0.187	0.228	0.239	0.481	0.219	0.257
100	0.000	0.021	0.022	0.022	0.000	0.005	0.000	0.000	0.071	0.027	0.009	0.013	0.061	0.228	0.017
Ots103															
N	49	25	52	96	10	102	15	18	7	73	57	115	108	43	58
55	0.000	0.000	0.000	0.000	0.000	0.000	0.000	0.000	0.000	0.000	0.000	0.000	0.000	0.000	0.009
57	0.082	0.080	0.048	0.042	0.100	0.054	0.133	0.028	0.000	0.068	0.044	0.035	0.000	0.000	0.121
73	0.000	0.000	0.000	0.000	0.000	0.000	0.000	0.000	0.000	0.000	0.000	0.000	0.046	0.023	0.000
77	0.051	0.080	0.087	0.068	0.000	0.044	0.033	0.056	0.143	0.041	0.061	0.083	0.125	0.117	0.078
81	0.867	0.840	0.865	0.875	0.900	0.902	0.801	0.916	0.857	0.891	0.886	0.882	0.773	0.860	0.783
85	0.000	0.000	0.000	0.015	0.000	0.000	0.033	0.000	0.000	0.000	0.009	0.000	0.056	0.000	0.009

Hatchery individuals were excluded from river samples.

Environmental Biology of Fishes **69**: 409–418, 2004.
© 2004 *Kluwer Academic Publishers. Printed in the Netherlands.*

Genetic selection and molecular analysis of domesticated rainbow trout for enhanced growth on alternative diet sources

Ken Overturf[a], Dan Bullock[a], Scott LaPatra[b] & Ron Hardy[c]
[a]*USDA/ARS, Hagerman Fish Culture Experiment Station, 3059-F National Fish Hatchery Road, Hagerman, ID 83332 U.S.A. (e-mail: kennetho@uidaho.edu)*
[b]*University of Idaho, Hagerman Fish Culture Experiment Station, Hagerman, ID 83332 U.S.A.*
[c]*Clear Springs Foods Inc., Buhl, ID 83316 U.S.A.*

Received 17 April 2003   Accepted 16 June 2003

Key words: real-time PCR, immunology, growth

Synopsis

Real-time polymerase chain reaction was used to monitor the expression of specific genes and confirm that an alteration in level of expression correlates with changes in diet, metabolism, or immune response. To test these probe and primer sets for their ability to monitor changes in gene expression, groups of fish were experimentally challenged either via daylength, vitamin availability, or with pathogenic microorganisms. The probes and primers sets were then used to analyze expression levels of specific gene products from isolated RNA. The expression of certain genes such as myosin in muscle correlated significantly with protein intake and seasonal daylength. Other genes involved with growth and metabolism such as insulin-like growth factor and pyruvate kinase showed lower levels of significance in correlating with planes of nutrition and varied with diet and treatment. However differences for pyruvate kinase were found in early tests done with animals receiving a vitamin-reduced diet. For factors relating immunological status after infection with microbiological pathogens, several factors such as MX-1 and CD-8 correlated with dose of the viral pathogen infectious hematopoeitic necrosis virus. Association of pathogen dose with immunological expression level was much less pronounced when fish were tested with bacterial microorganisms. This work will aid in evaluating the effects of diet on fish health and nutrient utilization and for evaluating the effects of selection of trout strains.

Introduction

Aquaculture is a rapidly developing agricultural economic resource with room for great expansion. Of increasing importance for the continued growth of the salmonid aquaculture industry is the replacement of fish meal with alternative protein sources (Hardy & Roberts 1998). Cereal grains and their products are promising candidates as renewable replacements for fish meal. Replacement of protein in fish diets with that of cereal grains would significantly reduce our dependence on fish protein sources, and also reduce the content of enriching nutrients, such as phosphorus in facility effluents.

At the Hagerman Fish Culture Experiment Station (HFCES), work has progressed in the study of diets containing cereal grains for rainbow trout (Pfeffer & Henrichfreise 1994, Sugiura et al. 1999, Arndt et al. 1999). Studies have shown that partial replacement of fishmeal with cereal grain proteins can be effective (Gomes et al. 1995, Skonberg et al. 1998), but that high replacement levels reduce growth rates and negatively affect the health of fish (Burrells et al. 1999). These negative effects include loss of appetite, intestinal damage, weakened immune function leading to disease outbreaks, and noticeable physical alterations (splotchy discoloration and fin erosion) (Ketola 1983, Rumsey et al. 1994, Gomes et al. 1995). Before new diet formulations can be widely recommended, they must be evaluated for their effects on fish physiology and health.

In order to achieve the maximal benefit of alternative diets, it is necessary to screen and genetically select for fish from existing strains that flourish on these diets. We have begun a genetic enhancement program for the selection of rainbow trout families fed a 40% barley replacement diet, based upon fish weight gain and feed conversion. Selection for growth has been shown to be a highly heritable quantitative trait (Fishback et al. 2002). However, selection for growth alone may lead to undesirable phenotypic side effects. For example selection for large, fast growing fish may result in fish that produce and store fat instead of muscle (Kause et al. 2002). Factors related to growth and growth control have also been linked to the immune system of trout (Yada et al. 2001). Opportunities now exist to identify the genes that play a role in the growth process. Molecular examination of genes involved with such quantitative genetic traits as growth can steer a genetic enhancement program toward a more clearly defined phenotype. Depending upon the contribution of the gene in question for the trait under selection, knowledge of the profile of that particular gene's expression pattern may lead to more rapid phenotypic gains when selection emphasis is placed on that specific gene, rather than on growth alone.

In our studies, the initial selection of rainbow trout families and individuals within families was based on weight gain and feed conversion ratio (FCR). Subsequently we expect to evaluate selected families by quantifying the expression of genes related to the phenotype under selection. Identification and analyses of the expression of appropriate genes related to growth, health, and nutrient utilization have been undertaken to determine methods and techniques to analyze gene expression related to physical characteristics (Overturf & Hardy 2001). One method is through the use of fluorescent-labeled probes and real-time PCR. Similar to reverse-transcribed polymerase chain reaction (RT-PCR) for quantification, this method uses two primers, one that acts to prime the reverse transcription reaction for the generation of complimentary DNA (cDNA) and also functions in concert with the other primer during the polymerase chain reaction (PCR) amplification steps. This standard method of quantitative PCR has been used in measuring TGF-beta mRNA in teleost fish (Harms et al. 2000). Typically, in quantitative PCR, an aliquot of the reaction product is run on agarose gels and the gel is analyzed by an external imaging method that compares the intensity of stained or radioactively labeled bands to that of a standard. Recently this method has been improved to provide an internal measurement of expansion of individual PCR product during the log phase of amplification for each individual reaction. This is accomplished by the addition of a sequence-specific, complementary fluorescent-labeled reporter probe that binds to the region being amplified between the two PCR primers. Upon cleavage of the sequence-bound probe by the 5′ nuclease activity of the *Taq* polymerase, the fluorescent reporter component of the probe is released from a terminal-quenching dye attached to the probe, emitting a detectable fluorescent signal. Accumulation of PCR products is detected directly by monitoring the increase in fluorescence of the reporter dye. When the probe is intact, the proximity of the reporter dye to the quencher dye results in suppression of the reporter fluorescence (Lakowicz 1983). Normalization of the amplification reactions is accomplished by dividing the emission intensity of the reporter dye by the emission intensity of a passive reference dye that is included in all experimental and control reactions.

Prior work from this laboratory has demonstrated that myosin expression correlates directly with nutritional intake in rainbow trout (Overturf & Hardy 2001). The expression of this gene is now being studied in selected lines of rainbow trout to determine if it can be used for the analysis of growth as related to muscle deposition and feed usage throughout the year. This same myosin probe was tested on steelhead reared in outdoor raceways under seasonal photoperiod to determine if a relationship exists between day length and myosin expression in fish fed at a constant level.

We are also initiating efforts to develop other probe and primer sets that can be used in genetic enhancement programs to evaluate and select for rainbow trout that grow on alternative protein sources. Specific probes for immune factors have been developed to generate an expression profile of 'healthy' fish and to look at the fish response on varied formulated diets. Here we report on the ability of the recently designed probes to measure changes in expression occurring in animals exposed to different disease-causing microorganisms. These probes might also be used for the selection of disease resistant stocks of trout.

Probe and primer sequences have also been generated to amplify gene sequences for trout metabolic and growth factors. These types of factors should prove useful in monitoring the effectiveness of different diet formulations, and lead to the evaluation and identification of metabolic enzymes or pathways that are specific for energy or nutrient utilization. Knowledge regarding the stimulation of growth factor expression will be useful in selecting for fish that grow rapidly while converting energy to muscle instead of depositing it as fat. With an understanding of gene expression levels and their relationships to changes for trait attributes using real-time PCR to quantify expression of specific genes can be used to enhance selection for specialized phenotypes in rainbow trout.

Materials and methods

Care and feeding of the trout strains

Steelhead myosin study. Steelhead, *Oncorhynchus mykiss*, (80 g) were reared at the U.S. Fish and Wildlife National Hatchery in Hagerman, Idaho in outdoor cement raceways receiving a constant flow (1 212 gal min^{-1}) of 58°C water during the entire trial from September to March. They were fed a standard commercial trout feed containing 44% crude protein and 16% fat (Nelson and Sons, Inc., Murray, UT) at a constant feed rate per average weight of fish each day. Steelhead were sampled every 2 weeks for 7 months (five fish for each sample time point).

Myosin strain study. After known crosses (four strains, 2×2 factorial crosses) were performed, 16 families of rainbow trout were obtained from suppliers as eggs or swim-up young and reared to 3.0 g prior to the study. Each family for each strain was placed in triplicate 145 l tanks at a density of 100 fish per tank. When the fish reached 100 g, they were transferred to 450 l tanks for the remainder of the study. The fish were randomly assigned to tanks to avoid confounding environmental effects on growth. The fish were reared in pathogen-free, 14.5°C spring water at a flow rate of 4–8 l min^{-1} per tank, depending on the size of the tank. Photoperiod was controlled by timers at 14 h light and 10 h of dark. Control fish families (three tanks of 100 randomly selected fish for each strain from all generated families of that strain) were reared on a commercial trout feed containing 44% crude protein and 16% fat (Nelson and Sons, Inc., Murray, UT). For experimental selected trout strain and family studies analyzed, the standard diet was modified to contain 40% Waxbar barley. Each group of fish was counted and bulk weighed every

2 weeks, and weight gain and FCR were calculated for a 4-month period. After 4 months, the top 10% of fish from the two highest performing families of each strain, determined by a combination of best average weight gain and FCR levels, and the bottom 15% of fish from the poorest performing families were kept for further studies. Samples of fish were overdosed with MS-222 (tricaine methanosulfonate, Argent Laboratories, Redmond, WA) and RNA was extracted from tissue.

Immunology factor study. Specific-pathogen-free (SPF) rainbow trout (mean initial weight, 14.0 g) were supplied by Clear Springs Foods, Inc. (Buhl, ID). Fish were held in indoor research 378 l tanks supplied with SPF 15°C spring water treated by ultraviolet light. During the pathogen infection phase of these studies the fish were transferred to 19 l tanks for use in challenge studies. Fish were fed 1% body weight per day of a pelleted trout feed (Clear Springs Foods) while being held in the large tanks, and *ad libitum* after pathogen infection.

Vitamin deficiency study. A standard fish meal, fish oil, and ground wheat diet (Table 1) was manufactured by compression pelleting at the HFCES (Hagerman, Id.). Fish oil and ground wheat levels were varied to maintain isocaloric equilibrium within the diets. Four experimental diets were produced, varying only in the level of vitamin premix (Table 1). The vitamin premix was made to provide the NRC (1993) requirements when added at 4% to the diet. This vitamin premix was fed at 1%, 2%, 4%, and 8% of the diet to provide 25%, 50%, 100%, and 200% of the NRC requirements for rainbow trout. Each diet was fed to three replicate tanks of trout, starting weight 10 g, and the arrangement of the diets among tanks was according to a completely randomized design. Every 4 weeks fish were bulk weighed and two fish from each tank were removed for mRNA expression analysis. Fish were handled and treated in accordance with the guidelines approved by the Animal Care and Use Committee of the University of Idaho.

Bacterial pathogen infection

A virulent strain of *Flavobacterium psychrophilum* (CSF-259-93) was used. The isolate was cultured in tryptone yeast extract salts (TYES) broth and maintained at 15°C (0.4% tryptone, 0.04% yeast extract,

Table 1. Composition of diets fed to rainbow trout for 12 weeks in the vitamin deficiency study.

Ingredients	Diet 1	Diet 2	Diet 3	Diet 4
Fish meal (LT aqua-grade)	600	600	600	600
Ground wheat	259	249	229	189
Fish oil	123	130	137	144
Ascorbic acid (phosphate ester) 35% active	1.0	1.0	1.0	1.0
Choline chloride (50% dry)	6	6	6	6
Trace mineral premix	1	1	1	1
Vitamin premix*	10	20	40	80

Vitamin	Dose (per kg/feed)
A	2 500 IU
D	2 400 IU
E	50 IU
K	10 g
Thiamin	1 g
Riboflavin	4 g
Pyridoxine	3 g
Pantothenic acid	20 g
Niacin	10 g
Biotin	0.150 g
Folic acid	1 g
Vitamin B$_{12}$	0.01 g

*Vitamin premix formula for purified diets.

0.05% calcium chloride, 0.05% magnesium sulfate, and pH 7.2). For challenge trials, *F. psychrophilum* was grown on TYES agar at 15°C for 72 h. Bacteria was harvested by gentle swabbing with a cotton applicator stick and re-suspended in 0.85% saline to an optical density (OD) of 0.2, 0.4, and 0.6 at 525 nm. Twenty rainbow trout were inoculated subcutaneously with 25 µl of each OD suspension, or 0.85% saline alone as a negative control. Plate counts were done on TYES agar at the time of subcutaneous injection of the bacteria.

A virulent strain of *Aeromonas salmonicida* was also tested in rainbow trout. Stock 1% (w/v) bacterial suspensions were prepared in 0.5% (w/v) tryptone after growth on TS agar plates. Twenty rainbow trout were inoculated intraperitoneally with 100 µl of a 1 : 10, 1 : 100, and 1 : 1 000 dilutions of the 1% stocks, or tryptone alone as a negative control. Plate counts were done on TS agar at the time of subcutaneous injection of the bacteria.

Virus infection

A virulent strain of infectious hematopoietic necrosis virus (IHNV; CSF-220-90) was propagated in EPC cells using standard methods (LaPatra et al. 1994). Twenty rainbow trout were inoculated intraperitoneally with 100 µl of a 1 : 10 000, 1 : 100 000, and 1 : 1 000 000 dilutions of the stock virus. Concentrations of viruses that were inoculated or detected in rainbow trout tissues after infection were determined by plaque assay as previously described (LaPatra et al. 1994).

Tissue sampling – At 1 and 5 day post-infection the kidneys, spleens, and livers of five fish were sampled for pathogen detection. Tissue samples were processed using standard methods (Drolet et al. 1994).

RNA isolation from tissues

Either muscle removed from the left mid section or tissue isolated from the brain, spleen, and kidney of freshly sacrificed control (sham infected) or experimental fish was immediately placed in TRIZol or frozen in liquid nitrogen in 2 ml cryo-vials or in aluminum foil. If the isolation was not performed onsite, the tissue samples were transported to the HFCES on dry ice where they were kept at −80°C until the time of isolation. Within a week the frozen tissue samples were thawed and total RNA was isolated using the TRIZol extraction method (GIBCO, Lifetechnologies, Grand Island, New York). The quantity and purity of the RNA was determined by analysis on a spectrophotometer at 260 nM.

In vitro *transcription of standard mRNAs*

PCR primers with *BamHI* 3′ and *EcoRI* 5′ restriction sites were engineered to amplify a region slightly larger than the amplicon used for sequence detection for each of the immunological sequences as standards. The PCR primers used for amplification of control standards are as follows: IL-8 (interleukin-8), accession #AJ279069 5′ forward primer 161–181 and 3′ reverse primer 314–335; CD-8 (cell surface antigen) accession #AF178054 5′forward primer 413–433 and 3′ reverse primer 504–524; C-3 (complement component C3) accession #L24433 5′ forward primer 2022–2054 and 3′ reverse primer 2136–2156R; MX-1 (interferon induced protein) accession #U30253 5′ forward primer 964–986 and 3′ reverse primer 1139–1160;

pyruvate kinase accession #AF246146 5′ forward primer 37–56 and 3′ reverse primer 323–342; ILGF-II (insulin-like growth factor-II) accession #M95184 5′ forward primer 366–386 3′ reverse primer 526–547; and the control β-actin accession #AF254414 5′ forward primer 353–373 and 3′ reverse primer 480–500. After PCR the amplified DNA was gel isolated and restriction digested with *BamHI* and *EcoRI* and cloned into a similarly digested pBluescript plasmid (Stratagene, La Jolla, CA). From these clones control standards were generated by *in vitro* transcription off the T7 priming site using the Riboprobe *in vitro* transcription system (Promega, Madison, WI). The myosin probe and primer sets have been previously described (Overturf & Hardy 2001).

Transcripts were run on formaldehyde/MOPS gels to ensure the presence of a single band corresponding to the correct size. Quantification of isolated transcripts was performed by two methods, using a spectrophotometer (Eppendorf, Westbury, NY) and fluorometer (Turner Designs, Sunnyvale, CA) with RiboGreen (Molecular Probes, Eugene, OR). These transcripts were then used as quantitative standards in the analysis of experimental samples. Standardization of samples and determination of copy number has been previously reported (Overturf & Hardy 2001, Overturf et al. 2001).

Sequence detection

After quantification, 75 ng of total isolated RNA from each sample were added to a microcentrifuge tube containing the following; 1X TaqMan buffer, 3 mM MnOAc, 0.3 mM dNTPs except dTTP, 0.6 mM dUTP, 0.3 μM forward primer, 0.3 μM reverse primer, 0.2 μM FAM-6 (6-carboxyfluorescein) labeled probe, 5 units *rTH* DNA polymerase (reverse transcriptase and DNA polymerase), and 0.5 units AmpErase UNG enzyme (to prevent reamplification cDNA amplified products) (PE Biosystems, Foster City, CA). The primers and fluorescent-labeled probes used for reverse transcription of mRNA and then in the subsequent sequence detection reaction are as follows: IL-8 forward primer 189–208, reverse primer 287–311, and the probe 209–237; CD-8 forward primer 445–469, reverse primer 498–518, and the probe 472–499; C-3 forward primer 2060–2084, reverse primer 2124–2143, and the probe 2085–2109; MX-1 forward primer 1044–1064, reverse primer 1105–1127, and the probe 1079–1096; ILGF-II forward primer 421–444, reverse

primer 495–514, and the probe 461–485; pyruvate kinase forward primer 97–122, reverse primer 182–205, and the probe 164–183; and the housekeeping control probe and primer set β-actin was forward primer 372–390, reverse primer 428–451R, and the probe 399–319. All probes were labeled with the fluorescent tag 6-FAM (PE Biosystems, Foster City, CA). Fifty microliters from each sample was then pipeted into individual wells of a 96 well optical plate, capped, and placed into an ABI-Prism model 7700 sequence detector. To ensure consistency of reagent handling between each reaction all liquid manipulation was done on a Qiagen 8000 liquid handling robot (Valencia, CA). A serial dilution of six duplicate standards was run with each probe for quantification. RT-PCR conditions were as follows; 2 min @ 50°C, 30 min @ 60°C, 5 min @ 95°C and then 40 cycles of PCR consisting of 20 s @ 92°C followed by 1 min @ 62°C. The fluorescence output for each cycle of the polymerase reaction was measured and downloaded to a Macintosh G3 computer after completion of the run. Accumulated data were analyzed using the computer program Sequence Detector version 1.7 (Applied Biosystems, Foster City, CA).

Data analysis

Average fish weight gain and FCRs were calculated monthly to determine effects of dietary changes. Statistical significance was calculated for the standardized expression levels between experimental groups. Data were transformed as necessary and analyzed using Student's t-tests, correlation or statistical analysis of variance using Sigma Stat 2.0 (Jandel Scientific, San Rafael, CA). A significance level of $P < 0.05$ was used and tank mean values were considered units of observation for statistical analysis of growth performance data and expression studies.

Results

Myosin expression relative to daylength and growth

From September to March, steelhead trout held in outdoor raceways were sampled every 2 weeks for muscle myosin expression levels. A significant correlation ($r = 0.8$) was found between the level of myosin expression and day length. The lowest

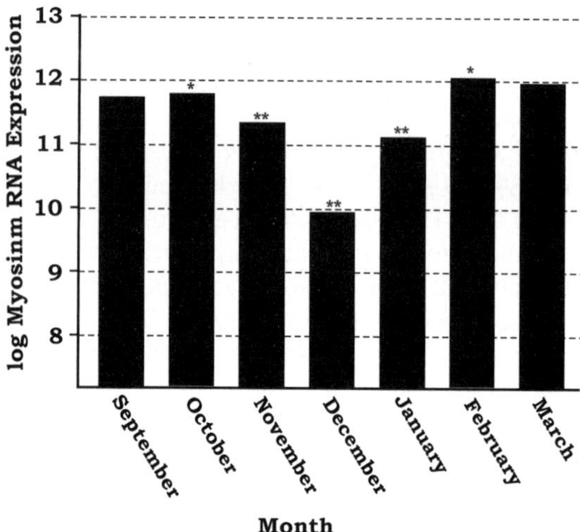

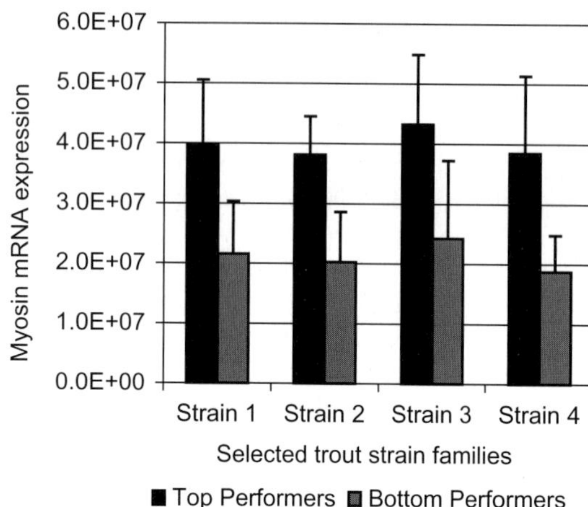

Figure 1. Analysis of myosin expression in families from four different trout strains reared under selection for weight and FCR for 7 months. *Significantly different from the next following month for October or the preceding month for February ($p < 0.05$). **Significantly different from the preceding and following month ($P < 0.05$).

Figure 2. Myosin expression levels in muscle RNA from steelhead maintained on constant feed reared in outdoor raceways. *Significant difference between top and bottom performers for each family ($p < 0.05$).

myosin expression levels occurred during the month of December (Figure 1).

Myosin expression was measured in families from four strains of trout fed a diet containing 40% barley. Higher levels of myosin expression were observed in the top 10% from the highest performing families than from their smaller counterparts (Figure 2) ($p < 0.05$).

Expression of immunological factors

To determine the effectiveness of the probes and primer sets used to analyze health and immunological status, initial investigations were carried out on fish infected with pathogenic microorganisms. Groups of fish received three different doses of bacterial pathogens or virus; their organs (spleen, kidney, and liver) were harvested 1 and 5 days post-infection to determine the presence of pathogens and expression levels of immunofactors. *F. psychrofilum* infection resulted in elevated changes of IL-8 and CD-8 in the spleen at 5 days post-infection and a slight elevation with a low dose for C-3 at 1 day post-infection. No significant changes for C-3, MX-1, IL-8, and CD-8 were detected in the liver. At 5 days post-infection significant increases were found in the kidney for the expression of CD-8, IL-8, and C-3 in animals that

received the mid-range dose of *F. psychrofilum* (data not shown).

Fish exposed to *A. salmonicida* showed elevation in the spleen for MX-1 at 1 day post-infection but not at 5 days. In the spleen there was an increase in the expression of IL-8 with dose after 1 day post-infection ($r = 0.71$). However, only at the high dose of *A. salmonicida* was the expression level significantly higher than that of control animals. This trend was also observed with the 5 days post-infected fish; unfortunately none of the high level dosed fish survived to the 5 days post-infection sampling. MX-1 levels were elevated at 1 day post-infection in the spleen but not at 5 days post-infection. In the liver, levels of C-3 and IL-8 were significantly increased in fish receiving the high dose of *A. salmonicida* at 1 day post-infection but not after 5 days, again because none of these fish survived (data not shown). The kidneys of fish infected with *A. salmonicida* showed elevated levels of IL-8 at high pathogen dose 1 day after infection and elevated levels of C-3 at 5 days after infection at all doses. However, there was no correlation between dose and expression level (Figure 3a). CD-8 levels were elevated at 1 day post-infection for the mid- and high-dosed animals, while MX-1 expression appeared to be elevated for only mid-dosed fish at this time.

Infectious hematopoietic necrosis virus infected fish showed elevated expression for CD-8 and IL-8 in the spleen at 1 day post-infection and CD-8 was correlative with pathogen dose ($r = 0.88$). At 5 days

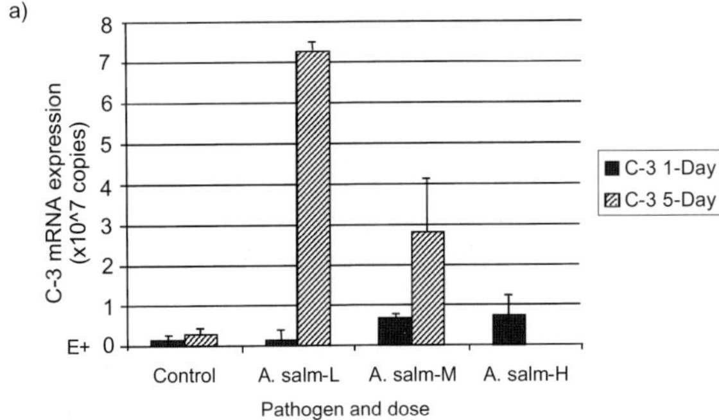

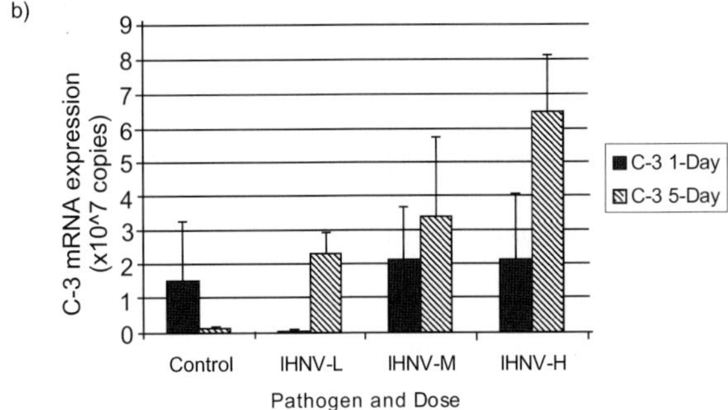

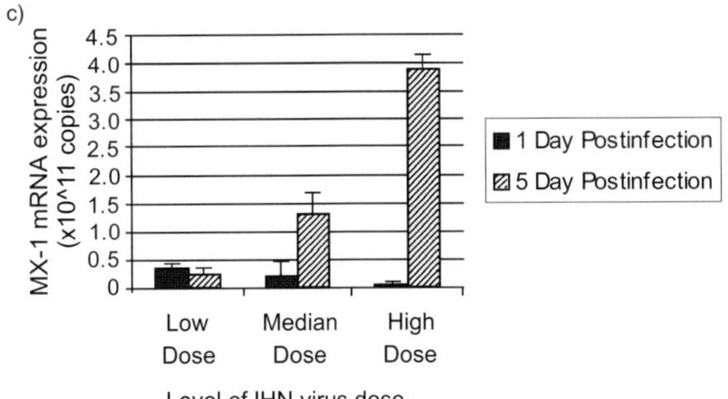

Figure 3. Expression levels of immunological factors from injected fish. (a) Expression level of C-3 from the kidney of fish infected with *A. salmonicida*. (b) Expression level of C-3 from the spleen of fish infected with *IHNV*. (c) Expression level of MX-1 from the liver of fish infected with *IHNV*.

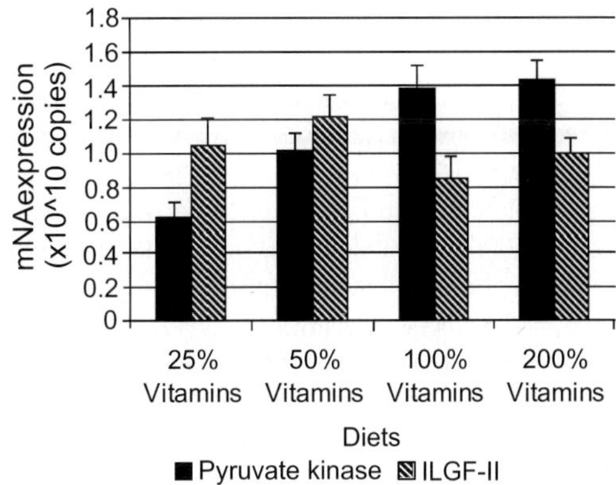

Figure 4. Expression levels of pyruvate kinase and ILGF-II in fish maintained on four diets varying in vitamin concentrations.

Expression of metabolic factors in fish on a vitamin limited diet

Groups of fish fed diets containing 200%, 100%, 50%, and 25% of the recommended vitamin levels for rainbow trout were tested for expression of pyruvate kinase and ILGF-II in their livers. The fish had been fed the diets for 12 weeks and sampled three times during the study (once every 4 weeks). Figure 4 shows the results from fish sampled at 8 and 12 weeks during the experiment. Pyruvate kinase displayed a positive correlation with dietary vitamin level (r = 0.89) from 25% to 100%; expression differed significantly at these values (p < 0.05). Increasing the vitamin level to 200% did not significantly increase (p > 0.1) pyruvate kinase expression. ILGF-II mRNA levels in fish were not significantly different between any of the vitamin diets.

Discussion

These results demonstrate that changes in mRNA transcript levels are measurable and correlate with alterations in the diet, environment, or health of the fish. Previous research examining myosin expression showed that it was highly correlated with fish growth and energy intake (Overturf & Hardy 2001). Federal and state hatcheries raising steelhead for mitigation are concerned with the size and health of smolts at the time of release. Strict feeding regimes are used to produce fish of a specific size for a specified release date. To gain a better understanding of feed utilization in steelhead during hatchery rearing, myosin expression

post-infection, elevated C-3 levels correlated with dose (r = 0.91) (Figure 3b). In the liver C-3, CD-8, and MX-1 were all elevated in a manner that correlated with dose at 5 days post-infection, r = 0.865, 0.94, and 0.98, respectively. Expression of MX-1 was greatly increased at all doses over controls (Figure 3c). Within the kidney MX-1 was elevated at 5 days post-infection with a median dose of virus, and C-3 and IL-8 appeared to also be elevated at 5 days but only significantly in animals that received the lowest dose of virus.

was monitored in a group of fish maintained on a constant feeding regime from young until they were released, a period of 8 months. Even though these fish were fed at a constant rate, their level of myosin expression significantly decreased with declining day length during the late fall and increased after the winter solstice. Wild fish grow when food is plentiful during the warmer months, and accumulate energy stores to survive during the winter months when food is scarce (Rand-Weaver et al. 1995). Hence, during the fall and winter months of short day length, hatchery fish with continued access to feed accumulate fat instead of producing somatic tissue. These findings suggest that it might be possible to select for fish that express myosin at continual high levels even with decreased daylength. Commercial hatcheries would benefit from fish stocks that grow more evenly or continually year-round.

Myosin expression was also examined in families of trout that had been positively and negatively selected for growth. In the early stages of our genetic selection program, families generated by 2×2 factorial crosses from four different strains were placed under selection for growth based upon weight gain and FCRs. In all strains, myosin was expressed at higher levels in fish from families that exhibited more rapid weight gain and lower FCRs. Monitoring myosin expression might prove useful in determining the potential of different diets for growth when weight differences or other physical determinants between control and experimental groups is not significant. For selection programs seeking strains with enhanced growth, incorporating selection for individuals that express higher myosin on the same diet may select for fish with a higher propensity to produce muscle instead of fat.

Knowledge of the factors involved in the expression of immunological factors can have an impact on a number of issues relating to aquaculture. These include genetic selection for disease resistance and animal health, as well as those that involve stress and effects of the environment and essential dietary nutrient levels. The findings from this study demonstrate that changes in expression of genes involved in immunological responses can be detected. In some instances, such as for IHNV exposure, gene expression levels and pathogen dosage were positively correlated for certain immune related factors. Following IHNV injection, expression of C-3, IL-8, CD-8, and MX-1 increased at 5 days post-infection with a high correlation for almost

all tissues analyzed. In contrast, although exposure to bacterial pathogens showed changes in expression level in various tissues for some of the factors, increased microorganism dosage only infrequently correlated with mRNA expression. IL-8 expression in the spleen of fish infected with *A. salmonicida* is an example of a correlative dose-response that occurred after bacterial infection. Except for elevated IL-8 and CD-8 expression with *F. psychrofilum* infection, almost all changes detected with bacterial exposure occurred at 1 day post-infection. This could be due to problems with the injection, dose or condition of the bacteria and thus not a true dose-response. At low pathogen doses certain immune factors may not be activated to combat the pathogen and at higher doses the infection may overwhelm the animal and specific immune responses. Studies are necessary to determine the optimum dose and range of specific bacterial pathogens to generate elevated immune response to increased dosage challenge. Nevertheless, this work demonstrates that immune responses can be measured with gene expression using RT-PCR. Further, under conditions whereupon either a pathogen possesses a strong antigen, or proper dose of pathogen is used, gene expression shows a strong correlation with dose. As our understanding of the fish health and the immune system increases, selection of disease resistant fish based upon evaluation of immune factor expression may become important for stock enhancement programs. The goal at this facility is to attempt to generate a profile of the expression of immune- and health-related genes and use this to evaluate immunostimulatory or antinutritional stress related components in formulated feeds.

Another area of interest for testing the effects of diet on fish is how different protein sources or other dietary components affect metabolic and regulatory growth pathways. Fish were fed diets with reduced level of essential vitamins, and the expression levels of pyruvate kinase and ILGF-II were studied. The expression level of pyruvate kinase correlated well with the expression level of the enzyme, decreasing as the vitamin concentration of the diet was reduced. Increases in vitamin levels over NRC recommended levels did not significantly increase pyruvate kinase gene expression. This work was done on relatively young trout, it will be interesting to see if the same effect occurs in older larger fish. It may be that as vitamins become limiting in the diet, key factors necessary for coenzymes and other cofactors become reduced. This could either alter the kinetics of enzymes in the pathway or trigger a negative feedback pathway for reducing the rate of transcription of genes involved directly or indirectly with the pathway. Differing vitamin concentrations did not significantly change the expression of ILGF-II. It might be more useful to measure the expression of other factors related to the growth hormone pathway, such as ILGF binding proteins that appear to be more tightly regulated in controlling growth (Moriyama et al. 2000).

Most of the recent work with real-time PCR involving fish or aquaculture has focused on disease detection and analysis of immunological factors related to disease (Overturf et al. 2001). The research presented here demonstrates how the use of real-time PCR can be expanded to evaluate change in expression as a means of quantifying a physiological change that cannot be easily measured by other methods. If expression patterns of specific genes vary in a manner that correlates with experimental treatment, the mRNA expression of that gene can be used to monitor that trait even if the product of the gene itself is not directly involved with the physical or physiological change of interest, but is only indirectly acted upon by the pathway of actual interest. It is possible to speculate about the function of the gene under study but the function cannot be completely understood until research is performed on the translated protein. Nevertheless, this approach can be effective in analyzing and tracking specific traits through populations and during selection.

We have demonstrated that expression of a metabolic enzyme correlates with dietary vitamin level consistent with previous reports showing relationships between vitamin levels and growth and health of fish (Kitamura et al. 1967, Morito et al. 1986, Halver 2002). Analyses of expression of immunological factors are likely to be beneficial in diet studies as it has been previously shown that high protein replacement with some cereal crops can actually be detrimental and generate an intestinal immune response in fish (Rumsey et al. 1995). Analysis of expression of immunological factors that are related to other health related factors, such as a cortisol, will likely become more important for understanding and selecting for growth and health of fish on new diet formulations. In fact there is evidence from other researchers documenting an association between stress resistance, immunity and growth performance (Fevolden et al. 2002). The use of RT-PCR will be of great assistance in evaluating expression changes of genes involved with these factors and determining the interactions between them.

Acknowledgements

We thank the aquaculture operations responsible for providing us with the test fish strains, and Jerry Jones and Bill Shewmaker from Clear Springs Foods for their excellent technical assistance during the immunological studies. The work of Ken Overturf was funded by the USDA-Agricultural Research Service. Mention of trade names or commercial products in this article is solely for the purpose of providing specific information and does not imply recommendation or endorsement by the U.S. Department of Agriculture.

References

Arndt, R.E., R.W. Hardy, S.H. Sugiura & F.M. Dong. 1999. Effects of heat treatment and substitution level on palatability and nutritional value of soy defatted flour in feeds for Coho Salmon, *Onchorhychus kisutch*. Aquaculture 180: 129–145.

Burrells, C., P.D. Williams, P.J. Southgate & V.O. Crampton. 1999. Immunological, physiological and pathological responses of rainbow trout (*Oncorhynchus mykiss*) to increasing dietary concentrations of soybean proteins. Vet. Immunol. Immunopathol. 72: 277–288.

Drolet, B.S., J.S. Rohovec & J.C. Leong. 1994. The route of entry and progression of infectious haematopoietic necrosis virus in *Oncorhynchus mykiss* (Walbaum): A sequential immunohistochemical study. J. Fish Dis. 17: 337–347.

Fevolden, S.E., K.H. Roed & K.T. Fjalestad. 2002. Selection response of cortisol and lysozyme in rainbow trout and correlation to growth. Aquaculture 205: 61–75.

Fishback, A.G., R.G. Danzmann, M.M. Ferguson & J.P. Gibson. 2002. Estimates of genetic parameters and genotype by environment interactions for growth traits of rainbow trout (*Oncorhynchus mykiss*) as inferred using molecular pedigrees. Aquaculture 206: 137–150.

Gomes, E.F., P. Rema & S.J. Kaushik. 1995. Replacement of fish meal by plant proteins in the diet of rainbow trout (*Oncorhynchus mykiss*): Digestibility and growth performance. Aquaculture 130: 177–186.

Halver, J.E. 2002. The Vitamins. pp. 62–143. *In*: J.E. Halver & R.W. Hardy (ed.) Fish Nutrition, 3rd edition, Academic Press, San Diego, CA.

Hardy, R.W. & R.J. Roberts. 1998. Atlantic salmon. The feed requirements and market for Atlantic salmon. Int. Aqua Feed 2: 21–24.

Harms, C.A., S. Kennedy-Stoskopf, W.A. Horne, F.J. Fuller & W.A. Tompkins. 2000. Cloning and sequencing hybrid striped bass (*Morone saxatilis* × *M. chrysops*) transforming growth factor-beta (TGF-beta), and development of a reverse transcription quantitative competitive polymerase chain reaction (TR-qcPCR) assay to measure TGF-beta mRNA of teleost fish. Fish Shellfish Immunol. 10: 61–85.

Kause, A., O. Ritola, T. Paananen, E. Mantysaari & U. Eskelinen. 2002. Coupling body weight and its composition: A quantitative genetic analysis in rainbow trout. Aquaculture 211: 65–79.

Ketola, H.G. 1983. Requirement for dietary lysine and arginine by fry of rainbow trout. J. Anim. Sci. 56: 101–107.

Kitamura, S., T. Suwa, S. Ohara & K. Nakagawa. 1967. Studies on vitamin requirements of rainbow trout. III. Requirements for vitamin A and deficiency symptoms. Bull. Jpn. Soc. Sci. Fish. 33: 1126–1131.

Lakowicz, J.R. 1983. Energy transfer. In: Principles of Fluorescent Spectroscopy, Plenum Press, N.Y. pp. 303–339.

LaPatra, S.E., K.A. Lauda & G.R. Jones. 1994. Antigenic variants of infectious hematopoietic necrosis virus and implications for vaccine development. Dis. Aquat. Org. 20: 119–126.

Morito, C.L.H., D.H. Conrad & J.W. Hilton. 1986. The thiamin deficiency signs and requirement of rainbow trout (*Salmo gairdneri* Richardson). Fish Physiol. Biochem. 1: 93–104.

Moriyama, S., F.G. Ayson & H. Kawauchi. 2000. Growth regulation by insulin-like growth factor-I in fish. Biosci. Biotechnol. Biochem. 64: 1553–1562.

National Research Council (NRC). 1993. Nutrient Requirements of Fish, National Research Council, National Academy Press, Washington, D.C. 114 pp.

Overturf, K. & R.W. Hardy. 2001. Myosin expression levels in trout muscle: A new method for monitoring specific growth rates for rainbow trout *Oncorhynchus mykiss* (Walbaum) on varied planes of nutrition. Aqua. Res. 12: 315–322.

Overturf, K., S. LaPatra & M. Powell. 2001. Real-time PCR for the detection and quantitative analysis of IHNV in salmonids. J. Fish Dis. 24: 325–333.

Pfeffer, E. & B. Henrichfreise. 1994. Evaluation of potential sources of protein in diets for rainbow trout (*Oncorhynchus mykiss*). Arch. Anim. Nutr. 45: 371–377.

Rand-Weaver, M., T.G. Pottinger & J.P. Sumpter. 1995. Pronounced seasonal rhythms in plasma somatolactin levels in rainbow trout. J. Endocrinol. 146: 113–119.

Rumsey, G.L., A.K. Siwicki, D.P. Anderson & P.R. Bowser. 1994. Effect of soybean protein on serological response, non-specific defense mechanisms, growth, and protein utilization in rainbow trout. Vet. Immunol. Immunopathol. 41: 323–339.

Rumsey, G.L., J.G. Endres, P.R. Bowser, K.A. Earnest-Koons, D.P Anderson & A.K. Siwicki. 1995. Soy protein in diets of rainbow trout: Effects on growth, protein absorption, gastrointestinal histology, and nonspecific serologic and immune response. pp. 166–188. *In*: C.E. Lim & D.J. Sessa (ed.) Nutrition and Utilization Technology in Aquaculture, AOCS Press, Champaign, IL.

Skonberg, D.I., R.W. Hardy, F.T. Barrows & F.M. Dong. 1998. Color and flavor analyses of fillets from farm-raised rainbow trout (*Oncorhynchus mykiss*) fed low-phosphorus feeds containing corn or wheat gluten. Aquaculture 166: 269–277.

Sugiura, S.H., V. Raboy, K.A. Young, F.M. Dong & R.W. Hardy. 1999. Availability of phosphorus and trace elements in low-phytate varieties of barley and corn for rainbow trout (*Oncorhynchus mykiss*). Aquaculture 170: 285–296.

Yada, T., T. Azuma & Y. Takagi. 2001. Stimulation of non-specific immune functions in seawater-acclimated rainbow trout, *Oncorhynchus mykiss*, with reference to growth hormone. Comp. Biochem. Physiol. B. Biochem. Mol. Biol. 129: 695–701.

Environmental Biology of Fishes **69**: 419–425, 2004.
© 2004 *Kluwer Academic Publishers. Printed in the Netherlands.*

Improvement of sperm motility of sex-reversed male rainbow trout, *Oncorhynchus mykiss*, by incubation in high-pH artificial seminal plasma

Toru Kobayashi[a], Shozo Fushiki[b] & Koichi Ueno[a]
[a]*Laboratory for Aquatic Biology, Department of Fisheries, Faculty of Agriculture, Kinki University, Nakamachi, Nara 631-8505, Japan (e-mail: kobayasi@nara.kindai.ac.jp)*
[b]*Fisheries Laboratory of Kinki University, Takada, Shingu, Wakayama 647-1101, Japan*

Received 17 April 2003   Accepted 24 April 2003

Key words: testicular spermatozoa, artificial seminal plasma

Synopsis

Changes in the motility time of spermatozoa collected from the testes and the sperm duct of normal and sex-reversed male (XX) rainbow trout in physiological balanced salt solution were examined after incubation in artificial seminal plasmas of various pHs. Although untreated spermatozoa from the sperm duct retained motility for 60–90 s in the balanced salt solution, the spermatozoa collected from the testes were immotile. During the incubation in artificial seminal plasma of pH 7.0, the spermatozoa from the sperm duct hardly moved, similar to the testicular spermatozoa in the balanced salt solution. By suspending and incubating the testicular spermatozoa in artificial seminal plasma of pH 9.9 for 2 h at 4°C, the percentage of motile spermatozoa increased from 0–5% to 80%. The spermatozoa remained motile for at least 2 min after long-term incubation (12 h). When the full-sib eggs were inseminated with untreated testicular spermatozoa or testicular sperm treated for 2 h at high pH, the percentage survival increased from 5.5% to 53.8% at the eyed stage due to the high-pH treatment. The incubation of the spermatozoa in high-pH artificial seminal plasma improved the motility of the spermatozoa from the testes of the sex-reversed male that had lost its sperm duct. By this treatment, it is possible to markedly increase the mass production efficiency of all-female or all-female triploid sterile progenies.

Introduction

All-female or sterile fish are desired for the purpose of preventing the adverse effects of maturity in salmon and trout culture (Donaldson & Hunter 1982, Lincoln & Scott 1983, Lincoln & Bye 1984, Bye & Lincoln 1986, Benfey et al. 1988, Quillet et al. 1988, Olito & Brock 1991, Johnstone et al. 1991, McGeachy et al. 1995, Ojolick et al. 1995). The economic ramifications of such adverse effects on fish breeding management are remarkable. Sex determination models of most species of salmonid fish are of the XY type (Johnstone et al. 1978, Donaldson & Hunter 1982, Yamazaki 1983, Devlin & Nagahama 2002). Therefore, to produce large amounts of all-female progeny, eggs should be fertilized using only spermatozoa with the X chromosome (Donaldson et al. 1993). For this purpose, methods have

been developed to secure the sex-reversed male (XX) by individual selection from both genetic sex mixture groups, which were changed to functional males that produce only genetic females in the next generation by test cross-breeding, or by obtaining genetic females and reversing their sex (Purdom 1983, Thorgaard 1983, 1986, Tsumura et al. 1991). Male sex hormones such as 17α-methyltestosterone are used to change the sex in these cases (Okada 1979, Nakamura 1994, Piferrer et al. 1994). However, structural and functional abnormalities occur during testicular development when the steroid concentration and water temperature during the oral administration of steroids for the fish are inappropriate. The breeding water temperature in conventional progeny production, especially when the breeding water is spring water, does not fall below 12.0°C even in winter, which is the sex differentiation period of

420

salmonid fish. The frequency of fish that have their testes inclined toward the anterior region and no sperm duct increased markedly from the case of common males (Bye & Lincoln 1986), although the testes are able to undergo spermatogenesis and spermiogenesis when 17α-methyltestosterone treatment is performed at such a high water temperature. To produce progeny using sex-reversed males with abnormal testes, the testes must be picked out from the open abdomen of sex-reversed males, and the eggs have to be inseminated with the X-sperms that flow out from the seminal lobules in the testes that are broken artificially. However, the spermatozoa that have not been spermiated in the testes are not motile, the flagella of these spermatozoa are entirely inactive, and the fertilization ratio is extremely low.

With such a background, in this study, we attempt to improve the production efficiency of all-female progenies by investigating the acquisition and improvement of motility *in vitro* of testicular spermatozoa of sex-reversed males.

Materials and methods

Testicular spermatozoa were collected from testicular cysts mainly of sex-reversed males, which grew on oral administration of 17α-methyltestosterone, and partly of common males by opening their abdomen of five fish. Spermatozoa collected from the sperm duct by pressing the abdomen were used as control. Artificial seminal plasma (ASP)[1] for salmonidae, which consisted of 7.60 g of NaCl, 2.98 g of KCl, 0.37 g of $CaCl_2 \cdot 2H_2O$, 0.31 g of $MgCl_2 \cdot 6H_2O$, 0.21 g of $NaHCO_3$, and 1000-ml of distilled water, was used. Its pH was adjusted by adding drops of 1 N-NaOH. To compare the motilities between testicular spermatozoa and sperm duct spermatozoa, and to check the effects of the difference in pH on each spermatozoon, the testicular spermatozoa and sperm duct spermatozoa were diluted 100 times with ASPs of pHs 8.0 and 10.5 (ASP 8.0, ASP 10.5). After incubation for about 10 min, 0.02 ml of the suspension was transferred to a glass slide and mixed with 0.05 ml of physiological balanced salt solution (BSS; 7.5 g of NaCl, 0.2 g of KCl, 0.2 g of $CaCl_2 \cdot 2H_2O$, 0.02 g of $NaHCO_3$, and 1000 ml of distilled water). The motility time of the spermatozoa was immediately observed under a microscope.

The instant when BSS was added was designated as zero time, and the motility termination time was when all spermatozoa in one microscopic view field stopped moving.

Then, testicular spermatozoa were suspended in ASPs of three different pHs (ASP 7.4, ASP 8.4, ASP 9.9) and stored at 4°C. These suspensions were checked for motility time and the percentage of motile sperm using a microscope to determine individual spermatozoon motility immediately after dilution, and after incubation for 10, 20, 40, 60, 90, 120, 150 min. The motility of the spermatozoa stored for 12 h in ASP of a certain pH was checked to determine the influence of long-term storage. To determine the influence of pH of ASP with which spermatozoa were incubated on the survival ratio in the eyed stage (eyed ratio), the spermatozoa that were stored for 2 h in ASPs of pHs 7.0 and 9.9 (ASP 7.0, ASP 9.9) were inseminated into full-sib eggs, and the eyed ratios and the survival ratios in the swim-up stage (swim-up ratio) were compared.

All quantities are presented as mean ± SEM. The *t*-test was used to compare the mean values on demand.

Results

Comparison of influences of pH on motility between testicular spermatozoa and sperm duct spermatozoa from the same fish

The motility times of testicular spermatozoa and sperm duct spermatozoa that were incubated in ASP 8.0 and ASP 10.5 in BSS are shown in Figure 1. Although

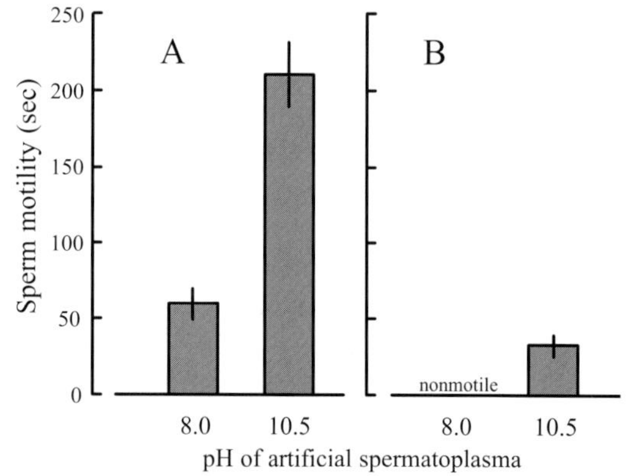

Figure 1. Comparison of motilities between testicular spermatozoa (A) and sperm duct spermatozoa (B) suspended in artificial spermatoplasmas of different pHs. Vertical bars represent the standard errors of the mean.

---

[1] Takahashi, K., Y. Omori, T. Inoda, & M. Morisawa: Abst. Meeting Jpn. Soc. Sci. Fish., April, 1986, p. 89 (in Japanese).

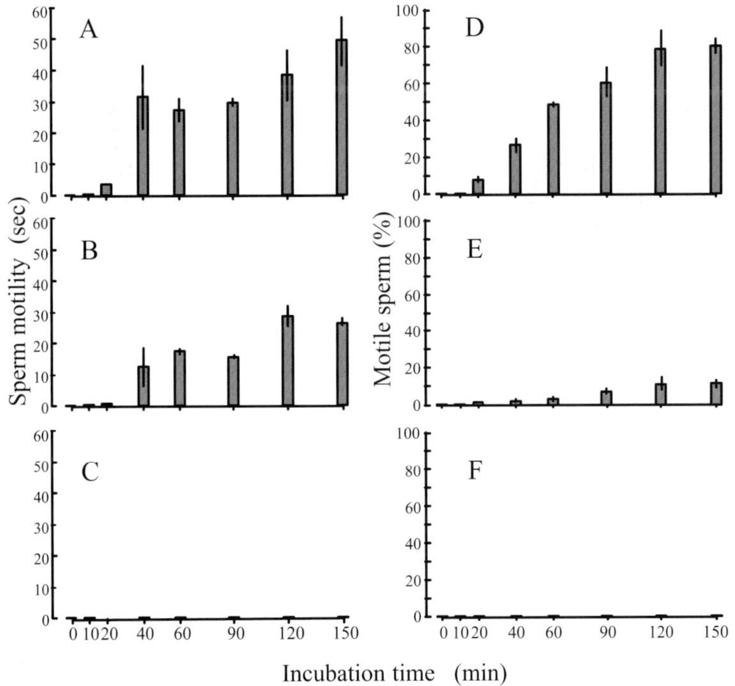

Figure 2. Influence of the difference in pH of artificial spermatoplasmas on the motility acquisition of testicular spermatozoa. A–C, changes in sperm motility; D–F, changes in percentage of motile sperm. The pHs of artificial spermatoplasmas for incubation of sperm are as follows: A, D, pH 9.9; B, E, pH 8.4; C, F, pH 7.4. Vertical bars represent the standard errors of the mean.

testicular spermatozoa incubated in ASP 8.0 were not motile in BSS, those incubated in ASP 10.5 moved with a motility time of 35 s. In contrast, sperm duct spermatozoa showed long-time motility for 3 min and 33 s when suspended in ASP 10.5, although a motility time of about 1 min is observed when the spermatozoa were suspended in ASP 8.0.

Storage time and sperm motility

The conditions for the acquisition of motility by the testicular spermatozoa that were incubated in ASPs of various pHs are shown in Figure 2. The testicular spermatozoa incubated in ASP 7.4 were not motile at any time period from the start of incubation up to 150 min after the start of incubation. The testicular spermatozoa incubated in ASP 8.4 showed a motility time of about 10 s at 40 min after the start of incubation and 28 s at 2 h after the start of incubation, although they did not move at all until 20 min after the start of incubation. The percentage of motile spermswas 11.5%. The testicular spermatozoa that were preserved in ASP 9.9 did not move until 10 min after the start of incubation.However, 20 min later, the spermatozoa

moved for 3.2 s and became motile at this time. Furthermore, 150 min later, the motility time of these spermatozoa was extended to 48.8 s and 80% of the spermatozoa were motile. Therefore, it was determined that ASP 9.9 is more suitable than ASP 8.4 for incubating testicular spermatozoa, increasing their motility.

Results of similar experiments that use testicular spermatozoa at the end of the spermiation stage of a sex-reversed male parent fish are shown in Figure 3. These testicular spermatozoa continued to move for 20–60 s after placing in each ASP. Optimal testis condition was determined to be the condition under which spermiation was induced producing the greatest number of cysts and the environment of the spermatozoa inside the testis was very similar to that in the sperm duct. Such results are often obtained for testicular spermatozoa at the end of the spermiation stage. There was no significant difference in sperm motility among ASP 9.9, 8.4, and 7.4 at 0 incubation time. At the incubation time of 120 min in ASP 8.4, sperm motility time become stable at 30–40 s, and no significant change was observed. However, the spermatozoa incubated in ASP 9.9 moved for a much longer time than those in ASP 8.4 within an incubation time of 40 min. Similar

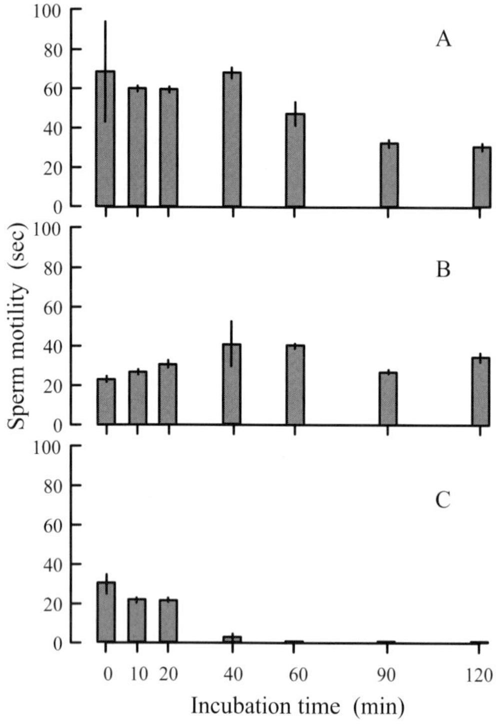

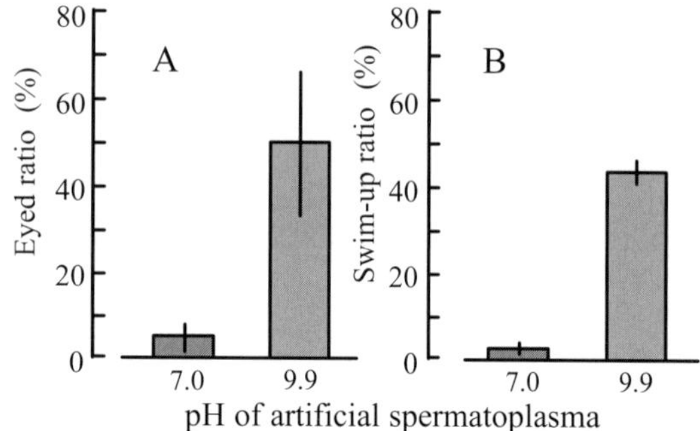

Figure 4. Influence of insemination of spermatozoa incubated in artificial spermatoplasmas of various pHs on the eyed ratio (A) and the ratio of number of swim-up juveniles to number of eyed eggs (B). Vertical bars represent the standard errors of the mean.

The eyed ratio and the swim-up ratio for cultivation in ASP 7.0 were 5.5% and 3.0%, respectively, and for cultivation in ASP 9.9, they were 53.8% and 47.5%, respectively. The pH of the ASP used in spermatozoon cultivation markedly influenced the results of progeny production. Accordingly, it was determined that sperm motility markedly influences production efficiency as well as egg quality.

Discussion

Generally, the motile factor of fish spermatozoa differs by the species. The spermatozoa of most of the marine or freshwater teleosts are not active in the solutions with or without electrolytes that the osmolality is iso-smotic to that of seminal plasma (Morisawa 1985, 1997, Billard & Cosson 1992). However, the spermatozoon starts the movement when it was diluted to the hyposmotic solution in the freshwater teleosts, or hyperosmotic solution in the marine teleosts (Lahnsteiner et al. 1992, 1996, Ohta & Tsuji 1998, Detweiler & Thomas 1998). The spermatozoa of the salmonid fishes starts its flagellar movement right away when the milt is diluted to the solution of low K^+ ion concentration, although it does not begin to move in the case that milt was diluted to the medium containing K^+ (Lahnsteiner et al. 1993, 1997, 1998). The same phenomenon is observed in shishamo smelt *Spirinchus lanceolatus* (Ohta et al. 1995), zebrafish *Brachydanio rerio* (Takai & Morisawa 1995), Japanese eel *Anguilla japonica* (Ohta et al. 1997) and ayu *Plecoglossus altivelis* (Ohta et al. 2001). However, the spermatozoa immediately after completed the spermiogenesis have no potential for motility (Morisawa & Morisawa 1986, Miura et al. 1992, Lahnsteiner et al. 1999). After

Figure 3. Influence of the difference in pH of artificial spermatoplasmas (A, pH 9.9; B, pH 8.4; C, pH 7.4) on the motility acquisition of testicular spermatozoa. Vertical bars represent the standard errors of the mean.

motility time was observed in the case of incubation in ASP 8.4. Spermatozoa incubated in ASP 7.4 showed a sharp decrease in motility time with incubation time and stopped moving at 60 min after the start of incubation.

Influence of long-term preservation on sperm motility

The spermatozoa preserved for 12 h in ASP 8.4 or ASP 9.9 had a motility time of 2 min and 70% spermatozoa moved after BSS was added. Spermatozoa preserved in ASP 7.4 did not move at all even 12 h after the start of incubation.

Influence of sperm motility on improvement of progeny production results from the eyed ratio and swim-up ratio

The difference in results of the eyed and swim-up ratios due to the difference in pH of ASP in which testicular spermatozoa are cultivated is shown in Figure 4.

the onset of spermatogenesis in the testis and spermiogenesis is completed, spermatozoa move to the sperm duct by spermiation. The potential for motility of spermatozoa develops during storage in the lobular lumen and sperm duct (Morisawa & Morisawa 1986). In the semen that is used in conventional production, the spermatozoa have already acquired potential for motility. These semen exhibit active flagellar movement in river water or BSS after they have been collected from the sperm duct by pressing the abdomen of the male parent fish (Morisawa 1985, Miura et al. 1992; Lahnsteiner et al. 1993, 1998). However, there are many cases in which the testes of the sex-reversed male rainbow trout show various structural abnormalities, such as having only one gonad, the absence of sperm duct, or testes inclination toward the anterior region of the abdominal cavity (Bye & Lincoln 1986, Tsumura et al. 1991, Piferrer et al. 1994). In individuals that lack the sperm duct, the collection of semen by squeezing becomes physically impossible, even if secondary sexual characteristics are distinct, and most of the germ cells of the testes reach spermiogenesis. In the testicular spermatozoa of the sex-reversed male that lacks the sperm duct, the eyed ratio becomes significantly low at 20–30%, because semen was collected by picking out testes, clipping them, and squeezing out the semen. The spermatozoa that were used in this experiment have not been spermiated from the cyst in the testes, and therefore, they have not yet acquired motility. It is known that the pH and HCO_3^- concentration of seminal plasma are higher in the sperm duct than in the testes (Morisawa & Morisawa 1986, Morisawa 1987, Boitano & Omoto 1991, Miura et al. 1992). Particularly, it is suggested that 17α, 20β-dihydroxy-4-pregnen-3-one (17,20-DP), the final maturation hormone of salmonid fish, is involved in the increase in pH (Miura et al. 1992, Nagahama, 1994). 17,20-DP elevates the pH of seminal plasma directly or indirectly, and the CAMP quantity in the spermatozoa is increased markedly by this phenomenon. CAMP acts as a direct trigger that induces motility of the spermatozoa and its quantity influences the motility time of the spermatozoa (Morisawa 1985, Morisawa & Ishida 1987, Cosson et al. 1995). In the acquisition of potential for motility by the spermatozoa, various effectors, such as the forward-motility protein (FMP) and 15-kDa protein, are involved (Morisawa & Hayashi 1985, Hayashi et al. 1987). Accordingly, it is possible with the high potassium ion concentration and the high pH of ASP that the treatment confers motility to the spermatozoa and extends motility time. The eyed

ratio of the egg fertilized by the untreated testicular spermatozoa was 20–30%. However, the eyed ratio was successfully improved to 70–80% by subjecting the testicular spermatozoa to the above treatment. This treatment is expected to be useful for not only the sex-reversed male in the all-female progeny production, but also the improvement of production efficiency of fish species such as the Japanese char, *Salverinus pluvius*, in which only a small quantity of milt can be collected by squeezing the abdomen. We have introduced this method to the production of char in recent years and obtained a high eyed ratio of 70% (unpubl. data). It was demonstrated that the adjustment of the motility of the spermatozoa influences both the production results and egg quality even in conventional progeny production. Moreover, such longer sperm storages were also examined (Stoss & Holtz 1983), it is necessary to investigate formulae of seminal plasma as external environmental factor of sperm, minutely.

Morisawa (1987) showed that the pH in the sperm duct of a common rainbow trout is 7.8–8.0. Miura et al. (1992) reported that the highest motility was obtained when ASPs of pH 8.5–9.0 were used, when they confer motility artificially to the cherry salmon, *Oncorhynchus masou,* testicular spermatozoa. The environment occupied by the spermatozoa in the seminal lobule, lobular lumen, and sperm duct at each stage of testicular maturity, especially in terms of pH, is very different. Accordingly, the optimal pH of ASP that confers motility to the sperm must be maintained. It might also be necessary to mitigate damage caused by abrupt pH changes on the sperm using buffer solutions such as HEPES.

Recently, the river environments are gradually turning acid due to the acid rain by the increase of carbon dioxide, nitrogen oxide density in the atmosphere, or some pollutants (Rurangwa et al. 2002). It is reported that the acidified external environment influences pH of internal environment of the fishes and changes the organization and pH of its body fluid (Wood et al. 1998). If pH of the seminal plasma that influences to the potential acquisition for motility of spermatozoa shifted more acidly than common condition, it was considerable that the potential for motility of the spermatozoa gotten in such environment degenerates.

In this study, we confirm that the incubation of spermatozoa in ASP of high pH results in motility increase or the improvement of the potential for motility of testicular spermatozoa of sex-reversed males. However, the influences of various factors such as the presence of glucose, trigriceride, and FMP on the acquisition of

424

motility by the spermatozoa, in addition to the increase in pH, cannot be disregarded (Lahnsteiner et al. 1999). It will be possible to freely control the motility of spermatozoa during fertilization, if further analyses of these factors are carried out.

References

Benfey, T.J., P.G. Bosa, N.L. Richardson & E.M. Donaldson. 1988. Effectiveness of a commercial-scale pressure shocking device for producing triploid salmonids. Aquacult. Eng. 7: 146–154.

Billard, R. & M.P. Cosson. 1992. Some problems related to the assessment of sperm motility in freshwater fish. J. Exp. Zool. 261: 122–131.

Boitano, S. & C. Omoto. 1991. Membrane hyperpolarisation activates trout sperm without an increase in intracellular pH. J. Cell. Sci. 98: 343–349.

Bye, V.J. & R.F. Lincoln. 1986. Commercial methods for the control of sexual maturation in rainbow trout (*Salmo gairdneri* R.). Aquaculture 57: 299–309.

Cosson, M.P., J. Cosson, F. André, & R. Billard. 1995. CAMP/AMP relationship in the activation of trout sperm motility: Their interaction in membrane-deprived models and in live spermatozoa. Cell. Mot. Cytoskel. 31: 159–176.

Detweiler, C. & P. Thomas. 1998. Role of ions and ion channels in the regulation of Atlantic croaker sperm motility. J. Exp. Zool. 281: 139–148.

Devlin, R.H., & Y. Nagahama. 2002. Sex determination and sex differentiation in fish: An overview of genetic, physiological, and environmental influences. Aquaculture 208: 191–364.

Donaldson, E.M. & G.A. Hunter. 1982. Sex control in fish with particular reference to salmonids. Can. J. Fish. Aquat. Sci. 39: 99–110.

Donaldson, E.M., R.H. Devlin, I.I. Solar & F. Piferrer. 1993. The reproductive containment of genetically altered salmonids. pp. 113–129. *In*: J.G. Cloud & G.H. Thorgaard (ed.) Genetic Conservation of Salmonid Fishes, Plenum Press, New York.

Hayashi, H., K. Yamamoto, H. Yonekawa & M. Morisawa. 1987. Involvement of tyrosine protein kinase in the initiation of flagellar movement in rainbow trout spermatozoa. J. Biol. Chem. 262: 16692–16698.

Johnstone, R., T.H. Simpson & A.F. Youngson. 1978. Sex reversal in salmonid culture. Aquaculture 13: 115–134.

Johnstone, R., H.A. McLay & M.V. Walsingham. 1991. Production and performance of triploid Atlantic salmon in Scotland. Can. Tech. Rep. Fish. Aquat. Sci. 1789: 15–35.

Lahnsteiner, F., B. Berger, T. Weismann & R.A. Patzner. 1996. Motility of spermatozoa of *Alburnus alburnus* (Cyprinidae) and its relationship to seminal plasma composition and sperm metabolism. Fish. Physiol. Biochem. 15: 167–179.

Lahnsteiner, F., R.A. Patzner & T. Weismann. 1992. Monosaccharids as energy resources during motility of spermatozoa in *Leuciscus cephalus* (Cyprinidae Teleostei). Fish. Physiol. Biochem. 10: 283–289.

Lahnsteiner, F., R.A. Patzner & T. Weismann. 1993. Energy resources of spermatozoa of the rainbow trout (*Oncorhynchus mykiss*) (Pisces, Teleostei). Reprod. Nutr. Dev. 33: 349–360.

Lahnsteiner, F., T. Weismann & R.A. Patzner. 1997. Aging processes in semen of the rainbow trout, *Oncorhynchus mykiss*. Progr. Fish. Cult. 58: 149–159.

Lahnsteiner, F., T. Weismann & R.A. Patzner. 1998. Evaluation of the semen quality of the rainbow trout, *Oncorhynchus mykiss*, by sperm motility, seminal plasma parameters, and spermatozoal metabolism. Aquaculture 163: 163–181.

Lahnsteiner, F., B. Berger & T. Weismann. 1999. Sperm metabolism of the teleost fishes *Chalcalburnus chalcoides* and *Oncorhynchus mykiss* and its relation to motility and viability. J. Exp. Zool. 284: 454–465.

Lincoln, R.F. & A.P. Scott. 1983. Production of all-female triploid rainbow trout. Aquaculture 30: 375–380.

Lincoln, R.F. & V. Bye. 1984. Triploid rainbows show commercial potential. Fish Farmer 7: 30–32.

McGeachy, S.A., T.J. Benfey & G.W. Friars. 1995. Fresh water performance of triploid Atlantic salmon (*Salmo salar*) in New Brunswick aquaculture. Aquaculture 137: 333–341.

Miura, T., K. Yamauchi, H. Takahashi & Y. Nagahama. 1992. The role of hormones in the acquisition of sperm motility in salmonid fish. J. Exp. Zool. 261: 359–363.

Morisawa, M. 1985. Initiation mechanism of sperm motility at sperm motility at spawning in teleosts. Zool. Sci. 2: 605–615.

Morisawa, M. 1987. The process of initiation of sperm motility at spawning and ejaculation. pp. 137–157. *In*: H. Mohri (ed.) New Horizons in Sperm Cell Research, Japan Science Society Press, Tokyo/Gordon and Breach Science Publishers New York.

Morisawa, M. & H. Hayashi. 1985. Phosphorylation of a 15 K axonemal protein is the trigger initiating trout sperm motility. Biochem. Res. 6: 181–184.

Morisawa, S. & M. Morisawa. 1986. Acquisition of potential for sperm motility in rainbow trout and chum salmon. J. Exp. Biol. 126: 89–96.

Morisawa, M. & K. Ishida. 1987. Short-term change in levels of cyclic AMP, adenylate cyclase, and phosphodiesterase during the initiation of sperm motility in rainbow trout. J. Exp. Zool. 242: 199–204.

Nagahama, Y. 1994. Endocrine regulation of gametegenesis in fish. Int. J. Dev. Biol. 38: 217–229.

Nakamura, M. 1994. A study of susceptibility of sex reversal after a single 2-hour treatment of androgen in amago salmon. Fish. Sci. 60: 483–484.

Ojolick, E.J., R. Cusack, T.J. Benfey & S.R. Kerr. 1995. Survival and growth of all-female diploid and triploid rainbow trout (*Oncorhynchus mykiss*) reared at chronic high temperature. Aquaculture 131: 177–187.

Okada, H. 1979. Functional masculinization of genetic females in rainbow trout. Bull. Jpn. Soc. Sci. Fish. 45: 413–419.

Ohta, H., S. Kusuda & S. Kudo. 1995. Motility of testicular spermatozoa in the shishamo smelt *Spirinchus lanceolatus*. Nippon Suisan Gakkaishi 61: 7–12.

Ohta, H., K. Ikeda & T. Izawa. 1997. Increases in concentrations of potassium and bicarbonate ions promote acquisition of motility *in vitro by* Japanese eel spermatozoa. J. Exp. Zool. 277: 171–180.

Ohta, H. & M. Tshuji. 1998. Ionic environment necessary for maintenance of potential motility in the common carp spermatozoa during *in vitro* storage. Fish. Sci. 64: 547–552.

Ohta, H., T. Unuma, M. Tshuji, M. Yoshioka & M. Kashiwagi. 2001. Effects of bicarbonate ions and pH on acquisition and maintenance of potential for motility in ayu, *Plecoglossus altivelis* Temminck et Schlegel. Aquac. Res. 32: 385–392.

Olito, C. & I. Brock. 1991. Sex reversal of rainbow trout: Creating an all-female population. Prog. Fish-Culturist 53: 41–44.

Piferrer, F., T.J. Benfey & E.M. Donaldson. 1994. Gonadal morphology of normal and sex-reversed triploid and gynogenetic diploid coho salmon (*Oncorhynchus kisutch*). J. Fish Biol. 45: 541–553.

Purdom, C.E. 1983. Genetic engineering by the manipulation of chromosome. Aquaculture 33: 287–300.

Quillet, E., B. Chevassus, J-M. Blanc, F. Krieg & D. Chourrout. 1988. Performance of auto and allotriploids in salmonids I. Survival and growth in fresh water farming. Aquat. Living Resour. 1: 29–43.

Rurangwa, E., A. Biegniewska, E. Slominska, E.F. Skorkowski & F. Ollevier. 2002. Effect of tributyltin on adenylate content and enzyme activities of teleost sperm: A biochemical approach to study the mechanisms of toxicant reduced spermatozoa motility. Comp. Biochem. Physiol. C Toxicol. Pharmacol. 131: 335–344.

Stoss, J. & W. Holtz. 1983. Successful storage of chilled rainbow trout (*Salmo gairdneri*) spermatozoa for up to 34 days. Aquaculture 31: 269–274.

Takai, H. & M. Morisawa. 1995. Change in intracellular K^+ concentration caused by external osmolality change regulates sperm motility of marine and freshwater teleosts. J. Cell. Sci. 108: 1175–1181.

Thorgaard, G.H. 1983. Chromosome set manipulation and sex control in fish. pp. 405–434. *In*: W.S. Hoar, D.J. Randall & E.M. Donaldson (ed.) Fish Physiology, Volume 9B, Academic Press, New York.

Thorgaard, G.H. 1986. Ploidy manipulation and performance. Aquaculture 57: 57–64.

Tsumura, K., V.E. Blann & C.A. Lamony. 1991. Progeny test of masculinized female rainbow trout having functional gonoducts. Prog. Fish-Culturist 53: 45–47.

Wood, C.M., R.W. Wilson, R.J. Gonzalez, M.L. Patrick, H.L. Bergman, A. Narahara & A.L. Val. 1998. Responses of an Amazonian teleost, the tambagui (*Colossoma macropomum*), to low pH in extremely soft water. Physiol. Zool. 71: 658–670.

Yamazaki, F. 1983. Sex control and manipulation in fish. Aquaculture 33: 329–354.

Environmental Biology of Fishes **69**: 427–432, 2004.
© 2004 *Kluwer Academic Publishers. Printed in the Netherlands.*

Temporal and spatial occurrence of female chinook salmon carrying a male-specific genetic marker in the Columbia River watershed

Trevor R. Chowen & James J. Nagler
*Department of Biological Sciences, Center for Reproductive Biology, University of Idaho, Moscow,
ID 83844-3051, U.S.A. (e-mail: jamesn@uidaho.edu)*

Received 17 April 2003   Accepted 24 April 2003

Key words: genotype, *OtY1*, phenotype, sex chromosomes, sex-linked marker

Synopsis

Recent declines in many chinook salmon, *Oncorhynchus tshawytscha*, populations within the Columbia River watershed have prompted an examination of their reproductive biology. In a previous study many female fall chinook salmon collected in 1999 from the Hanford Reach of the Columbia River tested positive for a male-specific DNA marker (*OtY1*) found on the Y chromosome. The purpose of this study was to determine if females testing positive for the *OtY1* marker could be found in other populations of fall chinook salmon from the Columbia River, and to assess the prevalence of *OtY1* incidence in different female cohorts. Post-spawned male and female fall chinook salmon from three different naturally spawning populations (Hanford Reach, Yakima River and Ives Island) and one hatchery population (Priest Rapids Hatchery) on the Columbia River were tested in 2000 and 2001 for the *OtY1* marker. Among naturally spawning populations, 57.4% of the females tested positive from the Hanford Reach, 33.3% tested positive from the Yakima River, and 32.5% tested positive from Ives Island. Of the Priest Rapids Hatchery fish, 62.5% of the females tested positive, and significant differences were detected between the 1995–1996 and 1997–1998 female cohorts from this population. No significant differences were detected between any of the female cohorts from the naturally spawning populations. All male chinook salmon samples, tested positive for *OtY1*.

Introduction

Declining stocks of chinook salmon in the northwestern United States have been the focus of concern and conservation efforts for many decades. While overharvest and habitat degradation typically have been recognized as primary causes, recent attention has turned towards examining the reproductive biology of these fish and how this may be impacted by human activities. A possible contributing factor in these declines may be inappropriate expression of genotypic sex.

Chinook salmon are a gonochoristic species in which determination of phenotypic sex occurs through an XX/XY chromosome system, such that males are heterogametic (XY) and females are homogametic (XX). A molecular marker based upon the polymerase chain reaction (PCR) (Devlin et al. 1994) exists that is capable of identifying genetic male (XY) chinook salmon via the amplification of a 209 bp sequence of DNA (*OtY1*) that is unique to the Y chromosome and, therefore, found only in male chinook salmon (Devlin et al. 1991). Nagler et al. (2001) reported that a majority of adult female fall chinook salmon sampled in 1999 from the Hanford Reach of the Columbia River tested positive for the *OtY1* marker. A similar observation has been made in female chinook salmon populations from the Sacramento-San Joaquin River basin in California (Williamson & May 2001).

This study follows the finding of Nagler et al. (2001) and sought to study the spatial and temporal occurrence of Columbia River female fall chinook salmon bearing the *OtY1* marker. Our first objective was to expand

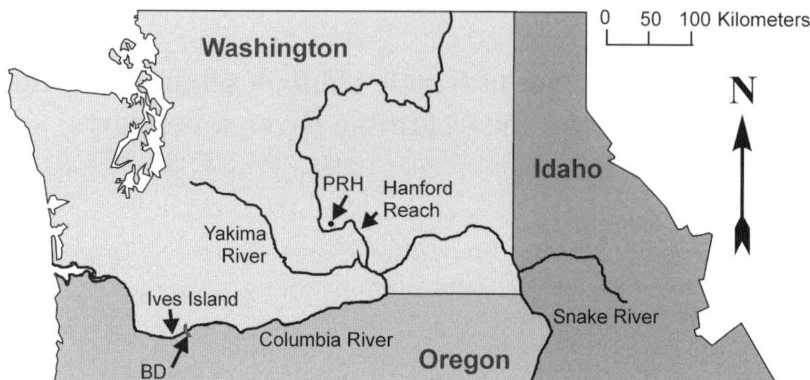

Figure 1. Map of the northwestern United States showing the Columbia River watershed and sites where chinook salmon were sampled in this study. Abbreviations: BD, Bonneville Dam; PRH, Priest Rapids Hatchery.

the spatial aspect by sampling other naturally spawning populations of fall chinook salmon within the Columbia River watershed outside the Hanford Reach. Two additional sites were chosen for study based upon their population size and feasibility of obtaining samples: (1) the Yakima River, a tributary of the Columbia River entering the Hanford Reach, and (2) the vicinity of Ives Island, ~4 km downstream from Bonneville Dam on the mainstem of the Columbia River (Figure 1). The second objective was to sample populations from each site over two consecutive years (2000 and 2001) and age individuals to assess the presence of the *OtY1* marker in each female cohort and determine whether any differences existed.

Materials and methods

Our aim was to sample additional populations of fall chinook salmon (*Oncorhynchus tshawytscha*) within the Columbia River watershed, geographically distinct from the Hanford Reach. Few populations in accessible locations still exist with sizes sufficient to permit feasible sampling. Two naturally spawning populations were identified that met these criteria – one in the lower Yakima River, and another at Ives Island (Figure 1). Sampling took place at these sites and the Hanford Reach during November 2000 and 2001 in collaboration with Washington Department of Fish and Wildlife (WDFW) salmon carcass survey crews. Samples were also collected from the Priest Rapids Hatchery (situated on the Hanford Reach) (Figure 1) in November 2000 and 2001 from fall chinook salmon that had returned to the hatchery and were being killed for collection of gametes. Approximately 50 female and 50 male

fish were sought at each of the four sites. Pieces of pectoral fin tissue (~5–10 cm^2) were removed from post-spawned adults using scissors and forceps that had been previously cleaned with either 95% ethanol or 10% bleach. Samples were preserved in the field in individual 20 ml vials of 95% ethanol. At the time of fin tissue collection, the phenotypic sex of each fish sampled was determined by gross visual examination of the gonads (i.e. testes or ovaries), and scales were collected for analysis of age.

Genomic DNA was isolated from 5–10 mg of each of the fin tissue samples using the Puregene DNA Isolation Kit (Gentra Systems, Minneapolis, MN) and a protocol supplied by the manufacturer for solid tissue. PCRs were prepared using a *Taq* DNA Polymerase kit (no. 18038-042; Gibco BRL, Rockville, MD) and primers Y1 and Y2 as per Devlin et al. (1994). Reactions (25 μl) were performed with slight modification to the thermal profile in Devlin et al. (1994), beginning with an initial denaturation at 94°C for 5 min, followed by 30 cycles consisting of 30 s denaturation at 94°C, 30 s annealing at 60°C and 30 s polymerization at 72°C. This protocol resulted in the amplification of a 209 bp male-specific DNA sequence (i.e. *OtY1*), which allowed for the comparison of the sex-linked genotype of each individual (presence or absence of *OtY1*) with its sexual phenotype (presence of testes in males or ovaries in females).

Cohorts of overlapping age classes from the two sequential sampling seasons were combined for each of the four sites and labeled with a cohort classification corresponding to their year of fertilization, as inferred from their age. For example, age 3 females sampled in 2000 and age 4 females sampled in 2001 were both of the 1997 cohort. Percentages of females testing positive for *OtY1* were compared among cohorts within each sampling site with χ^2-tests (SigmaStat Version 2.0, Jandel Scientific, Inc.). Due to small sample sizes, it was necessary in some cases to pool two sequential cohorts for statistical testing. Significant differences were concluded at the $\alpha = 0.05$ level.

Results

Returning adult fall chinook salmon were sampled from four different sites over a period of two consecutive years within the Columbia River watershed and tested for the presence of the *OtY1* marker. Results show that 171 out of a total of 320 adult females sampled during 2000 and 2001, or 53.4%, were positive for the *OtY1* marker. Among naturally spawning females sampled, 57.4% tested positive from the Hanford Reach, 33.3% tested positive from the Yakima River, and 32.5% tested positive from Ives Island. In addition, 62.5% of the females from the Priest Rapids Hatchery also tested positive for *OtY1*. Females that were found to bear the *OtY1* marker had all been confirmed in the field as phenotypic females based upon the presence of either intact or spawned-out ovaries. All male chinook salmon sampled from each site in both years (23–99 males per site) consistently tested positive for *OtY1*.

Age analysis from scale patterns revealed that at all four sites *OtY1*-positive females were comprised of multiple age classes in both 2000 and 2001 samples. Overlapping age classes from the two sampling years were combined to show that a total of four cohorts were represented at each site. These cohorts were categorized by their inferred fertilization years as either 1995, 1996, 1997, or 1998 cohorts. χ^2-tests among the cohorts from the Hanford Reach, Yakima River, and Ives Island, individually, did not indicate any statistically significant differences ($p > 0.05$) in percentage of positive females (Figure 2). However, a highly significant difference ($p < 0.001$) in percentage of positive females was detected between cohorts from Priest Rapids Hatchery (Figure 2). The incidence of *OtY1* in the 1995–1996 cohorts was much higher than that found in the 1997–1998 cohorts.

Discussion

Similar to female fall chinook salmon sampled in 1999 (Nagler et al. 2001) from the Hanford Reach

of the Columbia River, we found numerous females from the Hanford Reach in 2000 and 2001 bearing the *OtY1* marker. Further, two other naturally spawning populations of fall chinook salmon also have been identified to contain females bearing *OtY1*; one from the lower Yakima River, a tributary of the Columbia River, and another in the vicinity of Ives Island, ~325 river kilometers downstream of the Hanford Reach. These data support the finding of Nagler et al. (2001) and demonstrate that the phenomenon is not unique to the Hanford Reach, but also exists in other geographically distinct naturally spawning populations within the Columbia River watershed.

In addition, this study also discovered the *OtY1* marker in numerous female fall chinook salmon from Priest Rapids Hatchery, and therefore marks the first record of this phenomenon occurring in fish from a Columbia River hatchery. This is a significant finding since *OtY1* was not found in females sampled previously from Priest Rapids Hatchery in 1999 (Nagler et al. 2001). The basis for this discrepancy is not readily apparent, since it is assumed that the same cohorts returning in consecutive years were sampled in both 1999 and 2000. However, the fact that age data was not collected in 1999 (Nagler et al. 2001), as it was in 2000 and 2001, weakens this assumption and makes an unqualified comparison of these two studies impossible. Nevertheless, the observation reported here of *OtY1*-positive females at Priest Rapids Hatchery between 2000 and 2001 is consistent with the recent finding of other *OtY1*-positive females at several hatcheries within the Sacramento and San Joaquin River drainages in California (Williamson & May 2001).

Age at sexual maturity in chinook salmon is typically variable (Healey 1998), and females of the Hanford Reach and Priest Rapids Hatchery populations in particular may return to spawn anywhere between three and seven years of age (Dauble & Watson 1997). In this study, as many as four cohorts of *OtY1*-positive females were detected at each of the sampling sites over the two-year study period. There were no significant differences in percentage of *OtY1*-positive females among the cohorts from the naturally spawning populations sampled. In contrast, a highly significant difference ($p < 0.001$) was detected between the 1995 and 1996 cohorts versus the 1997 and 1998 cohorts from Priest Rapids Hatchery (Figure 2). The incidence of *OtY1*-positive females declined dramatically between the 1996 cohorts and the 1997 cohorts, such that the 1995 and 1996 cohorts both had much higher incidence

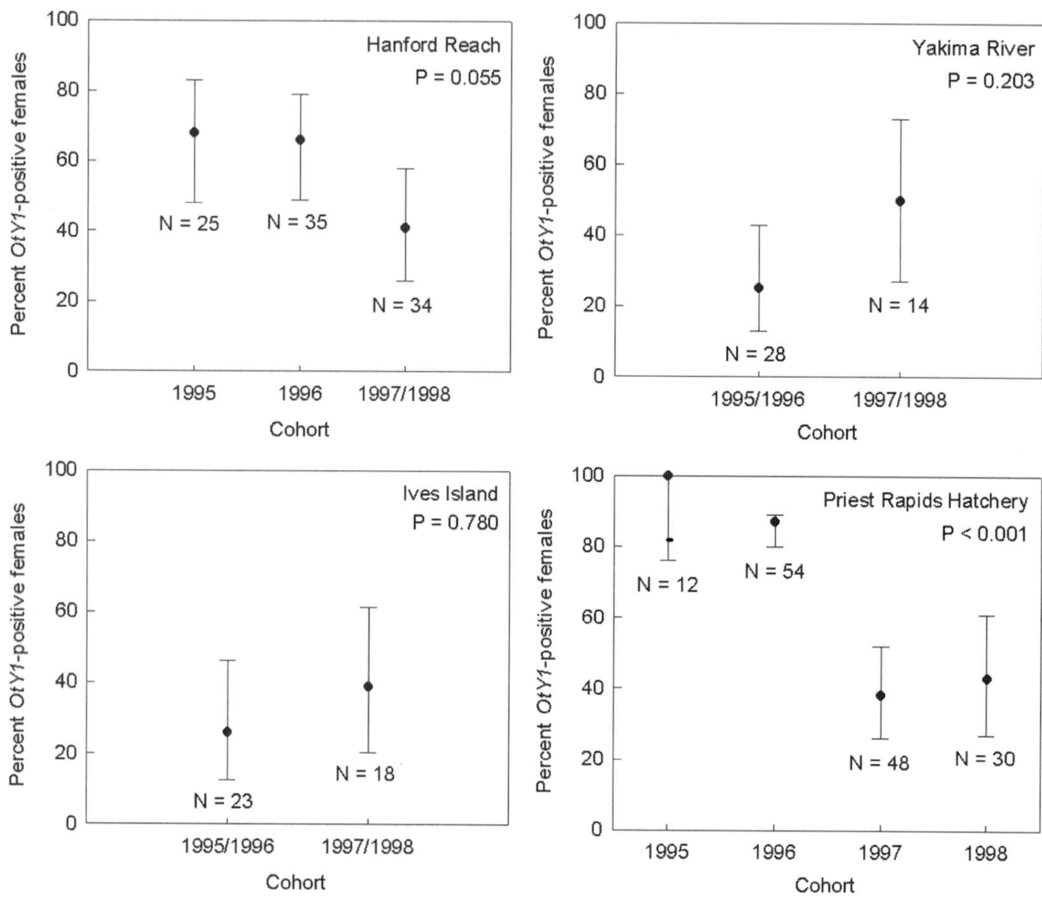

Figure 2. Percentages of female fall chinook salmon cohorts that tested positive for *OtY1* from each of the four populations sampled from the Columbia River watershed. P-values correspond to χ^2-tests among cohorts within each population. Bars depict 95% confidence intervals. Numbers below bars represent the total number of females assigned to each cohort group.

levels of *OtY1*-positive females than either the 1997 or 1998 cohorts. A similar trend, although not significant, was also apparent between the same cohorts from the Hanford Reach. However, results from the Hanford Reach should not be viewed exclusive from Priest Rapids Hatchery data, since strays from Priest Rapids Hatchery influence the population sampled at the Hanford Reach. Returning fall chinook salmon of Priest Rapids Hatchery origin have been shown to stray by an average rate of 29.8%, contributing an estimated average of 8.63% to the census of fish spawning naturally on the Hanford Reach (Evenson et al. 2002[1]).

---

[1] Evenson, D.F., D.R. Hatch & A.J. Talbot. 2002. Hatchery contribution to a natural population of chinook in the Hanford Reach of the Columbia River. Technical Report, Columbia River Inter-Tribal Fish Commission.

The coincidence that the only significant difference between cohorts was found in the one population examined that is entirely the product of artificial propagation should not be overlooked. Indeed, all stages of adult spawner selection and egg fertilization at Priest Rapids Hatchery involve a large degree of artificial manipulation, and therefore open the possibility for non-random selection of genotypes, even if unintentional.

In 2001, Nagler et al. put forward several possible explanations, including hypotheses of either sex reversal of genetic males to develop as phenotypic females or translocation of the region of the Y chromosome containing the *OtY1* sequence to another chromosome. A large number of xenobiotics that exist as pollutants from agricultural, industrial and domestic activities have been identified to have estrogenic properties in salmonid fish bioassays (Petit et al. 1997), and many of these so-called 'environmental estrogens' have been found to exist at low levels in the waters of the Columbia (USGS-NASQAN[2]) and Yakima rivers (Kolpin et al. 2002). The concentrations of chemicals examined in these two surveys are probably all individually too low to cause such an extreme effect as sex reversal; however, very little is known about the potential for complex mixtures of these chemicals to act in an additive manner. Furthermore, a recent laboratory study conducted by Afonso et al. (2002) demonstrated both development of an intersex condition and complete feminization of genetically male chinook salmon following exposure to varying supernormal concentrations of either 17β-estradiol, bleached kraft mill effluent, or secondary sewage effluent. Despite this, the scenario of sex reversal via endocrine disruption as investigated in these studies is difficult to reconcile with observations in our study, since none of our four sampling locations are suspected of receiving concentrated effluent from either pulp mills or sewage treatment facilities.

Another possibility that also involves the scenario of sex reversal may be that gonadal differentiation in these fall chinook salmon has been affected by fluctuating water temperature at an early life stage. Varying incubation temperature of eggs has been implicated in altering sex ratios of a few gonochoristic fish species (e.g. Kitano et al. 1999, Pavlidis et al. 2000, D'Cotta et al. 2001, Wang & Tsai 2000). However, this does not most likely explain our observations within the Columbia River drainage. For example, a preliminary study (Nagler & Thorgaard, unpublished data) subjected fertilized chinook salmon eggs to a fluctuating temperature regime modeled after the intragravel water temperatures observed for the Hanford Reach (Chapman et al. 1983), which fluctuate daily as a result of upstream hydroelectric dam activities. The experimental treatment continued for three months through the period of gonadal differentiation, but failed to result in a sex ratio significantly different from the expected 1 : 1 ratio. To our knowledge, there is no experimental data to indicate that sex differentiation in chinook salmon is temperature labile.

The other possibility is that female chinook salmon that test positive for *OtY1* are not phenotypically sex reversed males, but are in fact genetic females, and the inconsistent expression of *OtY1* that has been observed in the Columbia River is evidence of past genetic rearrangement involving the Y chromosome. Evidence for close physical linkage of *OtY1* with the sex-determining locus on the Y chromosome comes from analyses of both inheritance patterns of *OtY1* within experimental families, and the inability to detect recombinant genotypes in individual male chinook salmon sampled from naturally spawning populations (Devlin et al. 1991, 2001). However, these studies were conducted with chinook salmon from populations in British Columbia. No inheritance studies have been done for populations of chinook salmon south of British Columbia, where expression of *OtY1* in females has been observed independently for the Columbia River in Washington State (Nagler et al. 2001, and this study) and the Sacramento and San Joaquin Rivers in California (Williamson & May 2001). Thus, the potential remains that *OtY1* is an inconsistent molecular marker for genotypic sex in these more southern populations of chinook salmon – i.e. in some populations of chinook salmon, the *OtY1* DNA sequence may occur on the X chromosome or an autosome. In male chinook salmon in which *OtY1* has been characterized (Devlin et al. 1998), the 209 bp sequence is found within a larger 8 kb tandem repeat (*OtY8*), present ∼300 times on the Y chromosome and organized into at least six clusters spanning a total of 3.7 Mb. Fluorescent *in situ* hybridization in male chinook salmon from British Columbia has localized *OtY1* to the terminal portion of the short arm of the Y chromosome (Stein et al. 2001). However, it is possible that in populations of chinook salmon in the Columbia, Sacramento and San Joaquin Rivers, some or all of the 300 copies of the *OtY1* repeat have been involved in either meiotic crossover between the X and Y chromosomes or translocation between the Y chromosome and an autosome. The ancestral chromosomal location of *OtY1* is unknown, so it is uncertain whether this hypothetical genetic rearrangement might have involved the translocation of *OtY1* to the Y chromosome or from the Y chromosome. However, the fact that expression of *OtY1* continues to be strongly male-biased, even in those populations with female incidence, suggests that at least some of the copies of *OtY1* remain associated with the Y chromosome.

Incorporating multiple DNA markers that are unrelated and physically distant from one another may help to elucidate whether genetic rearrangement involving the Y chromosome could account for the non-sex-specific patterns of *OtY1* expression observed in this study. Work is currently underway in our laboratory to include such additional male-specific markers in

[2] United States Geological Survey – National Stream Quality Accounting Network. http://water.usgs.gov/nasqan/data/finaldata/vernita.html. Cited 17 May 2002.

further studies of these and future chinook salmon samples from the Columbia River.

Acknowledgements

The authors thank Tim Cavileer for assistance with sample processing, Washington Department of Fish and Wildlife for assistance with sample collection, and John Sneva for the scale age analysis. This study was funded by grant no. 2001-008-00 from the Bonneville Power Administration.

References

Afonso, L.O.B., J.L. Smith, M.G. Ikonomura & R.H. Devlin. 2002. Y-chromosomal DNA markers for discrimination of chemical substance and effluent effects on sexual differentiation in salmon. Environ. Health Perspect. 110: 881–887.

Chapman, D.W., D.E. Weitkamp, T.L. Welsh & T.H. Schadt. 1983. Effects of minimum flow regimes on fall Chinook spawning at Vernita Bar, 1972–1982. Grant County Public Utility District No. 2., Ephrata, WA.

Dauble, D.D. & D.G. Watson. 1997. Status of fall chinook salmon populations in the mid-Columbia River, 1948–1992. N. Amer. J. Fish. Mgmt. 17: 283–300.

D'Cotta, H., A. Fostier, Y. Guiguen, M. Govoroun & J.-F. Baroiller. 2001. Aromatase plays a key role during normal and temperature-induced sex differentiation of tilapia *Oreochromis niloticus*. Mol. Reprod. Dev. 59: 265–276.

Devlin, R.H., B.K. McNeil, T.D.D. Groves & E.M. Donaldson. 1991. Isolation of a Y-chromosomal DNA probe capable of determining genetic sex in chinook salmon (*Oncorhychus tshawytscha*). Can. J. Fish. Aquat. Sci. 48: 1606–1612.

Devlin, R.H., B.K. McNeil, I.I. Solar & E.M. Donaldson. 1994. A rapid PCR-based test for Y-chromosomal DNA allows simple production of all-female strains of chinook salmon. Aquaculture 128: 211–220.

Devlin, R.H., G.W. Stone & D.E. Smailus. 1998. Extensive direct-tandem organization of a long repeat DNA sequence on the Y chromosome of chinook salmon. J. Mol. Evol. 46: 277–287.

Devlin, R.H., C.A. Biagi & D.E. Smailus. 2001. Genetic mapping of Y-chromosomal DNA markers in Pacific salmon. Genetica 111: 43–58.

Healey, M.C. 1998. Life history of chinook salmon (*Oncorhynchus tshawytscha*). pp. 311–394. *In:* C. Groot & L. Margolis (ed.) Pacific Salmon Life Histories. University of British Columbia Press, Vancouver, BC.

Kitano, T., K. Takamune, T. Kobayashi, Y. Nagahama & S.-I. Abe. 1999. Suppresion of P450 aromatase gene expression in sex reversed males produced by rearing genetically female larvae at a high water temperature during a period of sex differentiation in the Japanese flounder (*Paralichthys olivaceus*). J. Mol. Endocrinol. 23: 167–176.

Kolpin, D.W., E.T. Furlong, M.T. Meyer, E.M. Thurman, S.D. Zaugg, L.B. Barber & H.T. Buxton. 2002. Pharmaceuticals, hormones, and other organic wastewater contaminants in U.S. streams, 1999–2000: a national reconnaissance. Environ. Sci. Technol. 36: 1202–1211.

Nagler, J.J., J. Bouma, G.H. Thorgaard & D.D. Dauble. 2001. High incidence of a male-specific genetic marker in phenotypic female chinook salmon from the Columbia River. Environ. Health. Perspect. 109: 67–69.

Pavlidis, M., G. Koumoundours, A. Sterioti, S. Somarakis, P. Divanach & M. Kentouri. 2000. Evidence of temperature-dependent sex determination in the European sea bass (*Dicentrarchus labrax*). J. Exp. Zool. 287: 225–232.

Petit, F., P. Le Goff, J.-P. Cravedi, Y. Valotaire & F. Pakdel. 1997. Two complementary bioassays for screening the estrogenic potency of xenobiotics: recombinant yeast for trout estrogen receptor and trout hepatocyte cultures. J. Mol. Endocrinol. 19: 321–335.

Stein, J., R.B. Phillips & R.H. Devlin. 2001. Identification of the Y chromosome in chinook salmon (*Oncorhynchus tshawytscha*). Cytogenet. Cell Genet. 92: 108–110.

Wang, L.-H. & C.-L. Tsai. 2000. Effects of temperature on the deformity and sex differentiation of tilapia, *Oreochromis mossambicus*. J. Exp. Zool. 286: 534–537.

Williamson, K.S. & B. May. 2001. Incidence of phenotypic female chinook salmon positive for the male Y-chromosome-specific marker *OtY1* in the Central Valley, California. J. Aquat. Animal Health 14: 176–183.

Environmental Biology of Fishes **69**: 433–440, 2004.
© 2004 *Kluwer Academic Publishers. Printed in the Netherlands.*

Molecular systematics and evolution of the growth hormone introns in the Salmoninae

Ruth B. Phillips[a], Makoto P. Matsuoka[b], Nick R. Konkol[c] & Sheldon McKay[d]

[a]*School of Biological Sciences, Washington State University, 14204 NE Salmon Creek Ave., Vancouver, WA 98686-9600, U.S.A. (e-mail: phllipsr@vancouver.wsu.edu)*
[b]*Present address: Institute for Marine Biosciences National Research Council, 1411 Oxford St., Halifax, NS, Canada B3H 3Z1*
[c]*Present address: Department of Biology, Marquette University, Milwaukee, WI 53201-1881, U.S.A.*
[d]*Present address: Genome Sciences Centre, British Columbia Cancer Agency, Vancouver, BC, Canada V5Z 4E6*

Received 17 April 2003   Accepted 16 June 2003

Key words: phylogenetics, growth hormone introns, selection, salmonid, estrogen response elements

Synopsis

DNA sequence data was collected for the C and D introns in the duplicate growth hormone loci (GH1 and GH2) from *Brachmystax lenok*, two subspecies of *Hucho hucho*, *Hucho (Parahucho) perryi*, *Salmo salar*, *Salmo trutta*, *Acantholingua ohridana* (*Salmothymus*), six species of *Salvelinus*, eight species of *Oncorhynchus* including *O. masou*, and three outgroups including *Thymallus thymallus*, *Coregonus artedi*, and *Coregonus clupeaformis*. Phylogenetic analyses were performed using maximum parsimony and maximum likelihood (PAUP, version 4.08beta) with gaps as missing data and as a fifth base. *B. lenok* was basal in all of the trees and all of the other genera were monophyletic with the exception that *A. ohridana* always placed within *Salmo*, and *H. hucho* sp. often placed with *B. lenok*. The GH1 introns supported a sister relationship between *Oncorhynchus* and *Salvelinus*, while the combined GH2 introns were ambiguous at this node. This result contrasts with trees based on morphology and the ribosomal ITS1 sequences that support a sister relationship between *Salmo* and *Oncorhynchus*. The only estrogen response element (ERE) in the gene is found in the C intron and has mutated in GH2 in all of the species except *B. lenok*. The ERE element in GH1 has undergone another mutation in all of the species except for *B. lenok*, and members of the two genera *Salvelinus* and *Oncorhynchus*. Thus these latter two genera are the only ones with a difference in expression of GH1 and GH2 in the presence of estrogen. Differences in selective pressure on the introns in the duplicate genes in different taxa could account for the conflicting results obtained in the phylogenetic analysis.

Introduction

As a result of diploidization following an autotetraploidization event (Allendorf & Thorgaard 1984), salmonid fishes have two unlinked growth hormone genes, GH1 and GH2. There are 6 exons and 5 introns in the GH genes. The two largest introns are intron C which averages 805 bp for GH1C and 489 bp for GH2C, and intron D which averages 1010 bp for GH1D and 1048 bp for GH2D in Pacific salmon (Devlin 1993, Blackhall 1994, McKay et al. 1996). Phylogenetic analysis of the GH introns has been used for examination of relationships of Pacific salmonids (Devlin 1993, Blackhall 1994, McKay et al. 1996), charrs (Westrich et al. 2002), and for intergeneric relationships (Oakley & Phillips 1999).

Relationships among the genera in the subfamily Salmoninae have been controversial. Previous analyses of intergeneric relationships utilizing the growth hormone introns (Oakley & Phillips 1999, McKay 1997) showed that phylogenies based on the GH1 introns strongly supported a sister relationship between the genus *Salvelinus* and *Oncorhynchus*, instead of *Salmo* and *Oncorhynchus* as expected from morphology. Phylogenies based on GH2 introns did not resolve intergeneric relationships. Trees based on the ribosomal ITS2 likewise do not resolve this node, but trees based on ribosomal ITS1 sequences (Phillips & Oakley 1997, Phillips et al. unpubl.) support the conventional hypothesis.

Previous phylogenetic analysis of the GH introns included 12 taxa for GH1C, 17 taxa for GH2C, 8 taxa for GH1D and 9 taxa for GH2D (McKay 1997). Only one species in the genus *Salmo* was included in the GHC data sets and *Brachymystax* and *Hucho* sp. were not included in the GHD data sets. In this paper, we present a phylogenetic analysis of the subfamily Salmoninae based on data from the sequences of GH1C, GH1D, GH2C and GH2D in 20 species of the Salmoninae including three species from *Salmo* and three outgroup taxa.

Duplicate genes can diverge in function as a result of mutations in regulatory elements (reviewed in Force et al. 1999, Prince et al. 2002). The 'subfunctionalization' model proposes that the two duplicate genes acquire complementary loss of function mutations in regulatory elements, so each locus carries out a different subset of functions provided by the ancestral locus. The duplicates may experience differences in selective pressures as a result of this divergence that could affect their usefulness for phylogenetic analysis.

The duplicated GH genes in rainbow trout have a variety of hormone response elements in the noncoding regions including the 5′UTR, 3′UTR and the five introns (Yang et al. 1997). All of these are present in multiple copies, except one, the estrogen response element (ERE) which is found only once in the C intron. This ERE has mutated in GH2, resulting in preferential expression of GH1 in females (Yang et al. 1997). In this paper we examine the pattern of evolution of sequence of this element in the Salmoninae to determine if it might help explain the differences in phylogenetic signal in the data sets based on the GH1 and GH2 introns.

Methods

Table 1 shows collection location for fish used in this project and GenBank accession numbers or citations for sequences. Total genomic DNA extractions were done from either liver or finclips. The previously

sequenced GH1C and GH2C for *S. salar* were obtained from Genbank (accession numbers X61938 and M21573) (Johansen et al. 1989).

Amplification of introns C and D from one fish from each species was done using the polymerase chain reaction (PCR). Primers used to amplify GH1C were GH1L and GH2CR (Oakley & Phillips 1999), those used to amplify GH2C were GH2CFB (5′ATCGTGAGCCC-AATCGACAAGCAG3′) and GH2CR, and those for GH1D and GH2D were GH7 (5′CTTATGCATGTCC-TTCTTGAA3′) and GH56 (5′AAGCTCAGCGACCT-CAAAGT3′). Internal primers were used for several of the GHD introns. For example, internal primers for GH1D of *S. namaycush* were as follows: GH1DFint (5′GGTAACTTCACAGACACTTCAC3′) and GH1-DRint (5′CCTGTGAAGTGTCTGTGAAG3′). PCR reactions for GHC and GHD contained ∼500 or 200 ng of template DNA, respectively, 30 pmol of each primer, 1× thermal buffer, 200 μM of each deoxynucleotide triphosphate, 25 mM MgCl$_2$, and 2.5 units Taq polymerase in 100 μl total volume. For GHC, the reactions were done in an Ericomp DeltaCycler II System thermal cycler as follows: one initial denaturation at 94°C for 3 min, 35 cycles of denaturation at 94°C for 30 s, annealing at 55°C for 1 min, extension at 72°C for 2 min, followed by one final extension at 72°C for 5 min. For GHD, the reactions were done in a MJ PTC 100 thermocycler as follows: one initial denaturation for at 94°C for 5 min, 35 cycles of 94°C for 1 min, annealing at 55°C for 1 min, extension at 72°C for 1 min, followed by a final extension at 72°C for 5 min. The PCR products were electrophoresed in 2% low melting point gel for 3 h at 55 V and stained with ethidium bromide. DNA was excised from the gels and purified using Wizard PCR columns (Promega, WI) according to the manufacturer's instructions. There was no ambiguity in determining the GH1 introns from the GH2 introns as the sequences are quite diverged and the size of GH1C was always larger than GH2C. For the GHD introns there were several indels of 15–20 bp that were diagnostic for GH1D or GH2D (McKay et al. 1997).

PCR products from several individuals including those from GHD had to be cloned into the pGEM-T Easy Vector (Promega, WI) before sequencing. Sequencing was done in both directions using an ABI Model 373 automated sequencer. Internal primers were used for some of the D introns.

Sequence chromatograms were edited with Edit View (Perkin Elmer) and corresponding forward and reverse sequences were aligned with Sequencer 3.1

Table 1. Location and GenBank accession numbers for growth hormone sequences used in this study*.

Species	Location	Accession numbers			
		GH1C	GH1D	GH2C	GH2D
B. lenok	River Kohr, RU	AF005918	AY125095	AF005916	AY125097
H. h. hucho	Turiec River, SLV	AY125124	AY125119	AF005906	AY125121
H. h. taimen	Amur River, RU	AY125129	AY125125	AF005907	AY125127
P. perryi	Hokkaido Fish Hat. Hokkaido, Japan	AF005920	AY125113	AF005909	AY125115
S. alpinus	Norwegian stock Freshwater Institute, Manitoba, Canada	AF005921	AY125164	AF005909	AY125163
S. namaycush	Green Lake stock Iron River, WI	AF005922	AY125192	AF005910	U29954
S. confluentus	Arrow Lake, BC	AY125172	AY125168	AF005911	AY125173
S. fontinalis	Nevin Hatchery, Madison,WI	AY125178	AY125179	AY125174	AY125175
S. malma	Fox River, AK	AY125129	AY125144	AY125145	AY125146
S. leucomaenis	Hokkaido Fish Hat. Hokkaido, Japan	AY125182	AY125184	AY125180	AY125183
S. salar	Norway	X61938	AY125198	M21573	AY125200
S. trutta	Scotland	AY125208	AY125204	AF005912	AY125206
A. ohridana	Lake Ohrid, Bosnia-Herzegovina	AY125143	AY125141	AF005915	AY125139
O. mykiss	Evergreen Hat. Pound, WI	AF005913	AY125131	J3797	J3797
O. clarki	coastal stock, Seattle, WA	AF005924	AY124131	AF005913	AY125133
O. kisutch	Kettle Moraine Hatchery, WI	AF005925	AF541851	U04930	AF541852
O. tshawytscha	Milwaukee River, Milwaukee, WI	Du et al. 1993	AY125160	AF005914	AY125157
O. gorbuscha	Juneau, Alaska	AF005926	AY125094	AY125089	AY125090
O. keta	Yukon River, AK	AF005927	AF541853	L04688	AF541854
O. nerka	Bristol Bay, AK	U14551	U14551	U14535	U14535
O. masou	Hokkaido Fish Hat Hokkaido, Japan	AF541855	AF541856	AF541857	AF541858
C. clupeaformis	Lake Michigan, WI	AY125111	AY125112	AY125107	AY125108
C. artedi	Lake Michigan, WI	AY125105	AY125101	AY125102	AY125103
T. arcticus	Bozeman, MT	AY125215	AY125213	AY125212	AY125210

to produce a composite file of the amplified product for each individual sequenced. The alignment of the composite files was done by eye and the character matrix generated by Sequencer was analyzed by PAUP (version 4.08b, Swofford, 2001). The composite sequences were deposited in GenBank (see Table 1 for Accession numbers).

Phylogenetic trees were constructed based on separately aligned GH1C, GH2C, GH1D and GH2D sequences. Combined trees were then constructed for various combinations of these sequences including data previously obtained from ITS1 and ITS2.

The branch and bound option of PAUP 4.0 d65 was used for maximum parsimony analysis and for the nonparametric bootstrap analysis (Felsenstein 1985) using the heuristic search option. Alignment gaps were either considered as missing, or were recoded as binary characters and added to the dataset. Including gap characters did not significantly affect tree topology (data not shown). Trees were rooted with the two *Coregonus* species as outgroups. To determine if the data sets could be combined, the CHARPARTITION function in PAUP was used.

436

Results

Phylogenetic analysis of the GH1 introns

GH1 intron C sequences ranged from 536 to 1047 bp (except for *T. thymallus* at 335 bp and *O. kisutch* at 1311 bp) and the alignment was 1445 characters of which 302 were variable and 160 parsimony informative. A branch and bound search produced five trees. These trees differed only in rearrangements among species in the genus *Salvelinus*. All genera were supported by high bootstraps and *Brachymystax* and *H. h. taimen* were basal taxa. There was support for a clade with *P. perryi* and *Salmo* (83% bootstrap) and another clade with *Salvelinus* and *Oncorhynchus* (94% bootstrap).

GH1 intron D sequences ranged from 831 to 1157 bp and the alignment of sequences was 1242 of which 439 were variable and 206 parsimony informative. A branch and bound search produced 15 trees that differed in rearrangements among species in *Salvelinus* and *Oncorhynchus*. *B. lenok* was in the basal position followed by *Salmo*, then *Parahucho*, followed by a clade with *Salvelinus* and *Oncorhynchus* (94% bootstrap).

In the tree based on the combined GH1C and GH1D data, there was one clade with *Salvelinus* and *Oncorhynchus* (100% bootstrap) and a weakly supported clade with *Salmo*, *Parahucho*, and *H. hucho* sp. (59% bootstrap) (Figure 1). There was strong support for *B. lenok* as the basal genus and for monophyly of *Salmo*, *Salvelinus* and *Oncorhynchus*.

Phylogenetic analysis of the GH2 introns

GH2 intron C sequences ranged from 443 to 624 bp (except *T. thymallus* which was only 286 bp), and the alignment was 794 characters of which 195 were variable and 111 parsimony informative. A branch and bound search produced 12 trees, and there was no resolution of intergeneric relationships in the consensus tree. All of the trees supported monophyly of each of the more derived genera and *Brachymystax* or *Hucho* was always the basal genus.

GH2 intron D sequences ranged from 846 to 1177 bp (except *O. masou* which was 435 bp), and the alignment was 1253 characters of which 424 were variable and 195 parsimony informative. A branch and bound search produced 764 trees and there was no resolution of intergeneric relationships. All of the trees supported monophyly of each of the more derived genera and *Brachymystax* or *Hucho* was always the basal genus. The branch and bound search based on the combined GH2C and GH2D data set gave 16 trees. In the bootstrapped consensus tree there was no resolution of intergeneric relationships (Figure 2). There was strong support for monophyly of each of the more derived genera and for *Brachymystax* or *Hucho* as the basal genera.

Sequence of the ERE element in the C intron

The sequence of the ERE element in the GH C introns is shown in Table 2. The sequence in *Brachymystax lenok* (GGG CAA GCA GAC C) is the same in both GH1C and GH2C and is only one base pair different from the mammalian consensus sequence. This is the only taxon in which both sequences are the same. ERE elements in both genes are mutated in members of *Hucho*, *Parahucho* and *Salmo*. In *Salvelinus* and *Oncorhynchus* the ERE element is intact in GH1C, but is mutated in GH2C.

Discussion

Phylogenetic analysis of the data sets from all four introns (GH1C, GH1D, GH2C, GH2D) gave strong support to the monophyly of *Salmo*, *Salvelinus* and *Oncorhynchus* and to the basal position of either *Brachymystax* or a clade with *Brachymystax/Hucho*. The position of the monotypic genus *P. perryi* was variable.

The analysis also gives strong support to the inclusion of *A. ohridana* in *Salmo*, which is consistent with previous work in our laboratory using both mitochondrial and nuclear sequences (Phillips et al. 2000). This species has previously been considered a monotypic genus (Hadzisce 1961), a subgenus of *Salmo* (Behnke 1968) or a subgenus of *Salmothymus* (Svetovidov 1975). *Salmothymus* was originally proposed to include *A. ohridana* and a more widely distributed species, *S. obtusirostris* (Svetovidov 1975), but Stearley & Smith (1993) suggested it might include up to five species. Recently Snoj et al. (2002) obtained mitchondrial and nuclear sequences from *S. obtusirostris* and their phylogenetic analysis produced a tree supporting these two species as sister taxa closely related to *S. trutta*. This work supports reclassification of both species as *Salmo*.

The consensus tree based on GH1 intron sequences also gave strong support to the sister relationship

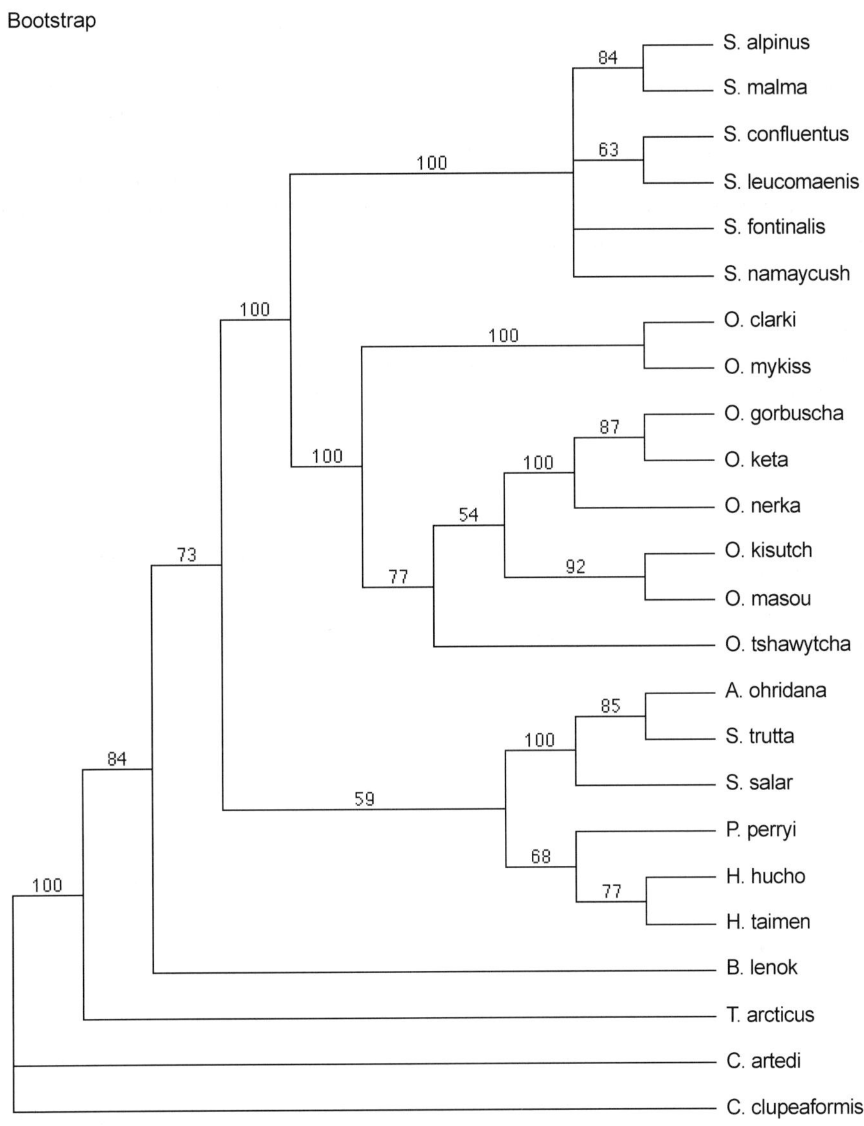

Figure 1. Results of phylogenetic analysis of sequences of the GH1C and GH1D introns from 24 taxa using maximum parsimony. The consensus tree is shown obtained treating gaps as missing with outgroups *C. artedi* and *C. clupeaformis*. Numbers above nodes are nonparametric bootstrap percentages based on 1 000 replications.

between *Salvelinus* and *Oncorhynchus*. Strong support for this relationship was obtained previously with smaller datasets from GH1C (Oakley & Phillips 1999) and GH1D (McKay 1997). There was no resolution of intergeneric relationships in the tree based on the GH2 intron sequences.

The data from the GH1 introns conflicts with data from the ribosomal ITS1. The tree based on ITS1 sequences supported a clade with *Oncorhynchus* and *Salmo* while the tree based on data from the ITS2 was unresolved with respect to intergeneric relationships (Oakley & Phillips 1999, Phillips et al. unpubl.). In summary, phylogenetic analysis of data sets obtained from four different genes produced two trees in which intergeneric relationships were not resolved (GH2, ITS2), one which supported a clade between *Oncorhynchus* and *Salmo* (ITS1), and the other which supported a clade between *Oncorhynchus* and *Salvelinus* (GH1). The placement of *Parahucho* was ambiguous in the datasets based on ITS2 and GH2,

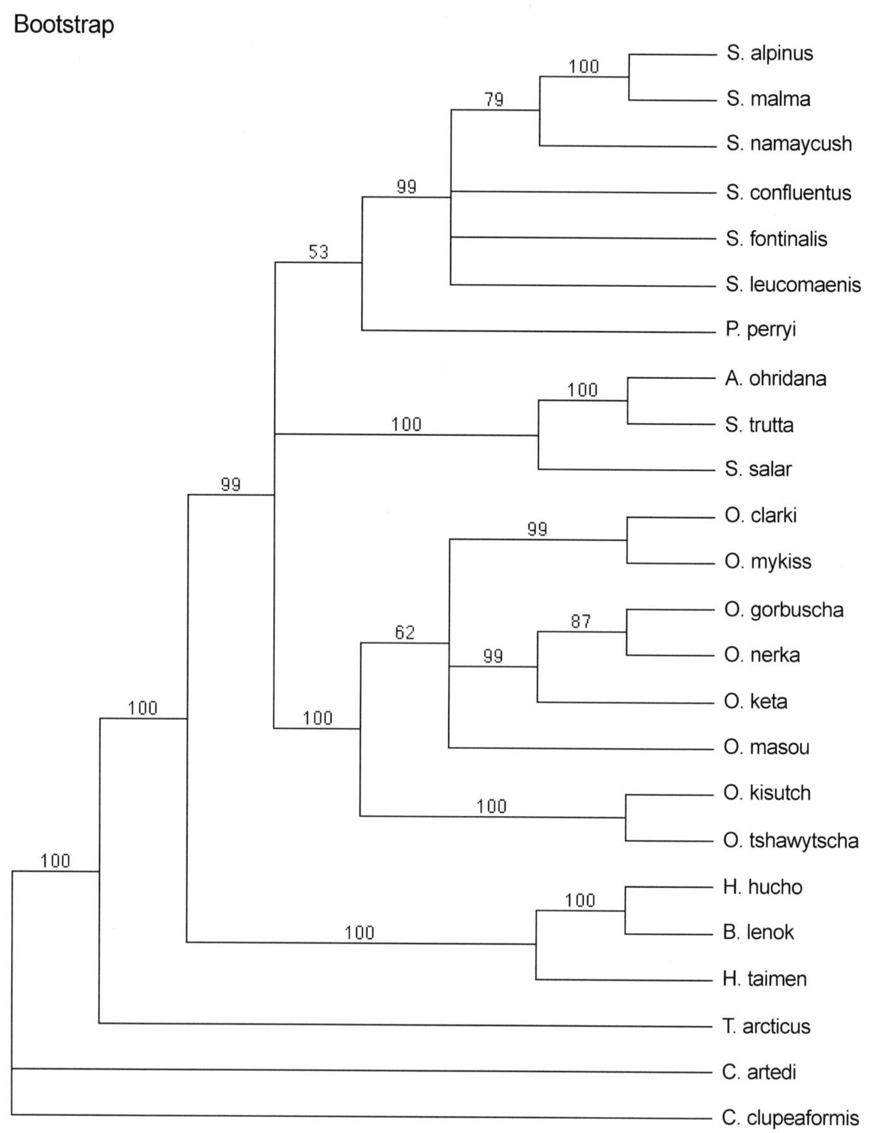

Figure 2. Results of phylogenetic analysis of sequences of the GH2C and GH2D introns from 24 taxa using maximum parsimony. Consensus of 16 trees obtained treating gaps as missing with outgroups *C. artedi* and *C. clupeaformis*. Numbers above nodes are nonparametric bootstrap percentages based on 1 000 replications.

but ITS1 and GH1C were in agreement in its position as a sister clade with *Salmo*. *Parahucho* was between *Salmo* and the clade with *Salvelinus* and *Oncorhynchus* in the dataset based on GH1D. These conflicting results are typical of a starburst adaptive radiation in which taxa branch off from each other closely in time. In that case, some gene trees will not correspond with species trees.

The sequence analysis of the ERE in the duplicate C introns suggests another explanation for the lack of congruence between the trees based on GH1 introns and ITS1 data. The two GH genes may be diverging more rapidly in certain taxa compared with others as a result of 'subfunctionalization'. The GH1 C intron in rainbow trout retains a functional ERE that has been mutated in the GH2 C intron with the result that the GH1 protein is preferentially expressed in rainbow trout females (Yang et al. 1997).

We have shown in this study that this sequence difference in the ERE between GH1C and GH2C

439

Table 2. Sequence of the Estrogen Response Element in Intron C of Growth Hormone Genes in Salmoninae. (Mammalian consensus is same as *B. lenok* except for a T in position # 9).

	Growth Hormone 1					Growth Hormone 2				
Brachmystax lenok	G G G	C A A	G C A	G A C	C	G G G	C A A	G C A	G A C	C
H. h. hucho	. T .	A . .	. . .	. . .	.	. . .	. . .	. . .	. . .	.
H. h. taimen	. T .	A . .	. . .	. . .	.	. . .	. . .	T . .	. . .	.
P. perryi	. T .	A . .	. . .	. . .	.	. . .	. . .	. . .	. . T	.
S. salar	. . .	A . .	. . .	. . .	.	. . .	. . .	. . .	. . T	.
S. trutta	. . .	A . .	. . .	. . .	.	. . .	. . .	. . .	. . T	.
*Acantho. ohridana**	. . .	A . .	. . .	. . .	.	. . .	. . .	. . .	. . T	.
Salvelinus alpinus	. . .	. . .	. . .	. . .	.	. . .	. . .	. . .	. . T	.
S. malma	. . .	. . .	. . .	. . .	.	. . .	. . .	. . .	. . T	.
S. confluentus	. . .	. . .	. . .	. . .	.	. . .	. . .	. . .	. . T	.
S. namaycush	. . .	. . .	. . .	. . .	.	. . .	. . .	. . .	. . T	.
S. leucomaenis	. . .	. . .	. . .	. . .	.	. . .	. . .	. . .	. . T	.
S. fontinalis	. . .	. . .	. . .	. . .	.	. . .	. . .	. . .	. . T	.
Oncorhynchus masou	. . .	. . .	. . .	. . .	.	. . .	. . .	. . .	. . T	.
O. mykiss	. . .	. . .	. . .	. . .	.	. . .	. . .	. . .	. . T	.
O. clarki	. . .	. . .	. . .	. . .	.	. . .	. . .	. . .	. . T	. .
O. kisutsch	. . .	. . .	. . .	. . .	.	. . .	. . .	. . .	. . T	.
O. tshawytscha	. . .	. . .	. . .	. . .	.	. . .	. . .	. . .	. . T	.
O. nerka	. . .	. . .	. . .	. . .	.	. . .	. . .	. . .	. . T	.
O. keta	. . .	. . .	. . .	. . .	.	. . .	. . .	. . .	. . T	.
O. gorbuscha	. . .	. . .	. . .	. . .	.	. . .	. . .	. . .	. . T	.

*Full name is *Acantholingua ohridana*.
Base pair deletions are denoted by -.

is only present in taxa of the genus *Oncorhynchus* and *Salvelinus*, the two genera which exhibit a sister relationship based on this data. Both EREs are intact in *B. lenok* and both have been mutated in *Hucho, Parahucho* and *Salmo*. Thus 'subfunctionalization' of expression of the duplicate GH genes in response to estrogen has occurred only in *Salvelinus* and *Oncorhynchus*. Thus the two duplicate genes may be evolving rapidly in a similar fashion in *Salvelinus* and *Oncorhynchus* where they are diverging in function, compared to taxa where either both or neither of the duplicate genes respond to estrogen.

In conclusion a 'starburst phylogeny' and presence of duplicate genes which are diverging in function along different paths in different taxa may explain the lack of consistency in the intergeneric relationships among the Salmoninae based on different genes.

Acknowledgements

We thank the large number of people who assisted us with collection of samples. We would also like to thank Marc Noakes for submitting the additional sequences to GenBank. He also assisted with the sequencing of the GH-D introns in *H. h. taimen*.

References

Allendorf, F.W. & G.H. Thorgaard. 1984. Tetraploidy and the evolution of salmonid fishes. pp. 1–53. *In*: B.J. Turner (ed.) Evolutionary Biology of Fishes, Plenum Publishing Co., New York.

Behnke, R.J. 1968. A new subgenus and species of trout: *Platysalmo platycephalus* from southcentral Turkey with comments on the classification of the subfamily Salmoninae. Mitteilungen Hamburgischen Zool. Mus. Instit. 66: 1–15.

Blackhall, W.J. 1994. A molecular study of introgression between westslope cutthroat trout and rainbow trout. MS Thesis. University of Alberta, Edmonton, Canada. 80 pp.

Devlin, R.H. 1993. Sequence of sockeye type 1 and type 2 growth hormone genes and the relationships of rainbow trout with Atlantic and Pacific salmon. Can. J. Fish. Aquat. Sci. 50: 1738–1748.

Felsenstein, J. 1985. Confidence limits on phylogenies: An approach using the bootstrap. Evol. 39: 783–791.

Force, A., M. Lynch, F.B. Pickett, A. Amores, Y.L. Yan & J. Postlethwait. 1999. Preservation of duplicate genes by complementary degenerative mutations. Genetics 151: 1531–45.

Hadzisce, S. 1961. Zur Kenntnis des *Salmothymus orhidanus* (Steindachner) (Pisces, Salmonidae). Intl Vereinigung theoretische angewandte Limnol. Verhandlungen 14: 785–791.

Johansen, B., O.C. Johnsen & S. Valla. 1989. The complete nucleotide sequence of the growth-hormone gene from Atlantic salmon (*Salmo salar*). Gene (Amsterdam) 77: 317–324.

McKay, S.J. 1997. Evolutionary Genetic Analysis of Pacific Salmon and Trout, PhD. Thesis, Simon Fraser University, Burnaby, Canada. 140 pp.

McKay, S.J., R.H. Devlin & M.J. Smith. 1996. Phylogeny of Pacific salmon and trout based on growth hormone type-2 (GH2) and mitochondrial NADH dehydrogenase subunit 3 (ND3) DNA sequences. Can. J. Fish. Aquat. Sci. 53: 1165–1176.

McKay, S.J., M.J. Smith & R.H. Devlin. 1997. Polymerase chain reaction-based species identification of salmon and coastal trout in British Columbia. Mol. Mar. Biol. Biotech. 6: 131–140.

Oakley, T.H. & R.B. Phillips. 1999. Phylogeny of salmonine fish based on growth hormone introns: Atlantic (*Salmo*) and Pacific (*Oncorhynchus*) salmon are not sister taxa. Mol. Phylogen. Evol. 11: 381–393.

Phillips, R.B. & T.H. Oakley. 1997. Evolutionary relationships among salmonid fishes inferred from nuclear and mitochondrial sequences. Chapter 10 pp. 145–162. *In*: T. Kocher & C. Stepien (ed.) Molecular Systematics of Fishes, Academic Press, New York.

Phillips, R.B., M.P. Matsuoka, I. Konon & K.M. Reed. 2000. Phylogenetic analysis of mitochondrial and nuclear sequences supports inclusion of *Acantholingua ohridana* in *Salmo*. Copeia 2000: 546–550.

Prince, V.E. & F.B. Pickett. 2002. Splitting pairs: The diverging fates of duplicated genes. Nat. Rev. Genet. 3: 827–837.

Snoj, A.E. Melkic, S. Susnik, S. Muhamedagic & P. Dovc. DNA phylogeny supports revised classification of *Salmothymus obtusirostris*. 2002. Biol. J. Linn. Soc. 77: 399–411.

Stearley, R.F. & G.R. Smith. 1993. Phylogeny of the Pacific trouts and salmons (*Oncorhynchus*) and genera of the family Salmonidae. Trans. Amer. Fish. Soc. 122: 1–33.

Svetovidov, A. 1975. Comparative osteological study of the Balkan endemic genus *Salmothymus* in relation to its classification. Zool. Zhurnal 54: 1174–1190.

Swofford, D. 2001. PAUP: Phylogenetic analysis using parsimony (beta ver. 4.08b). Illinois Natural History Association, Champaign, Illinois.

Westrich, K.M., N.R. Konkol, M.P. Matsuoka & R.B. Phillips. 2002. Interspecific relationships among charrs based on phylogenetic analysis of nuclear growth hormone introns. Environ. Biol. Fish. 64: 217–222.

Yang, B.-Y., K-M Chan, C.-M. Lin & Chen, T.T. 1997. Characterization of rainbow trout (*Oncorhynchus mykiss*) growth hormone 1 gene and the promoter region of growth hormone 2 gene. Arch. Biochem. Biophy. 340: 359–368.

Environmental Biology of Fishes **69**: 441–447, 2004.
© 2004 *Kluwer Academic Publishers. Printed in the Netherlands.*

Karyological differentiation of northern Dolly Varden and sympatric chars of the genus *Salvelinus* in northeastern Russia

Sergei V. Frolov & Valentina N. Frolova
*Institute of Marine Biology, Far East Branch, Russian Academy of Sciences,
Vladivostok 690041, Russia (e-mail: sergeifrolov@hotmail.com)*

Received 21 April 2003   Accepted 24 April 2003

Key words: chars, sympatry, karyotype structure, NOR, species differentiation

Synopsis

In many cases the taxonomic status of sympatric chars (with exception of white-spotted char *Salvelinus leucomaenis*) is not clear and is actively debated. We karyotyped three pairs of sympatric chars inhabiting the Russian Far East-northern Dolly Varden, *S. malma malma*, and Lavanidov's char, *S. levanidovi*, from the Yama River (northern coast of the Sea of Okhotsk, Magadan Region), northern Dolly Varden and Taranetz's char, *S. taranetzi*, from Lake Achchen (east Chukotka), and northern Dolly Varden and white char, *S. albus*, from the Kamchatka River (Kamchatka Peninsula). Three of them had similar chromosome numbers. But all chars studied had an individual and discrete set of karyotypic characters, which enabled reliable identification each of them by chromosome number, chromosome arm number, and number and location of active nuclear organizer regions.

Introduction

The chars of the genus *Salvelinus* are characterized by high morphological plasticity. As a result there are some difficulties in their identification on the basis of traditional morphological characters. The problem of char differentiation is critical in zones of sympatric habitation of several forms. In many cases, the taxonomic status of sympatric chars (with the exception of white-spotted char, *Salvelinus leucomaenis*) is not clear and actively debated.

There are many places in the north where two or more anadromous chars live and spawn in the basin of one river or lake. In the Russian Far East these sympatric chars are northern Dolly Varden, *S. malma malma*, and Lavanidov's char, *S. levanidovi*, in the Yama River (coast of the Sea of Okhotsk, Magadan Region), northern Dolly Varden and Taranetz's char, *S. taranetzi*, in Lake Achchen (eastern Chukotka), and northern Dolly Varden and white char, *S. albus*, in the Kamchatka River (Kamchatka Peninsula).

Some of the Russian ichthyologists believe that most of morphological characters of sympatric chars (such as a number of gill rakers, number of pyloric caeca, etc.) are overlapping. Therefore sympatric chars in northeastern Asia cannot be identified on the basis of morphological criteria alone and should be classified as *S. alpinus* (Barsukov 1960, Volobuev et al. 1979, Savvaitova et al. 1988, Savvaitova 1989). At the same time these authors mentioned that they 'never had problems with identification of each form of chars due to their differences in the shape of body and coloration' (Savvaitova 1989), or 'all studied chars may be identified . . . on the basis of body shape and coloration' (Barsukov 1960).

In contrast, I. Chereshnev and his colleagues have shown that sympatric chars in the first two pairs could be easily differentiated on the basis of external coloration, coloration of the coelom, and the form of the caudal peduncle. They believe that Dolly Varden and Levanidov's char, and Dolly Varden and Taranetz's char could be identified in each river where they are sympatric using the number of gill rakers and the mean number of pyloric caeca, and must be considered as distinct species (Chereshnev 1982, Chereshnev et al. 1989). Additionally, protein markers also demonstrated

442

significant differentiation of sympatric Dolly Varden –
Levanidov's char and Dolly Varden – Taranetz's char
pairs in sympatric habitats (Kartavtsev et al. 1983,
Omelchenko et al. 1986, 1988). White char in the
Kamchatka River could be differentiated from sym-
patric Dolly Varden by morphology of skull bones
(Glubokovsky 1977b) and proteins, and are also con-
sidered a distinct species (Glubokovsky 1977b).

We karyotyped these pairs of sympatric chars to
ascertain if their karyological differences are stable and
to clarify char taxonomy. We believe the data obtained
will be useful for differentiation of sympatric chars in
northeast Asia and for determining of their taxonomic
status.

Materials and methods

We karyotyped northern Dolly Varden and Levanidov's
char from the Yama River; northern Dolly Varden
and Taranetz's char from Lake Achchen; and northern
Dolly Varden and white char from the Kamchatka
River (Figure 1, Table 1). Chars from the Yama River
and Lake Achchen were identified by external coloration, coloration of the coelom and by the caudal
peduncle as described by Chereshnev (Chereshnev
1982, Chereshnev et al. 1989, I.A. Chereshnev pers.
commun.). Some chars from the Kamchatka River were
identified by M.K. Glubokovsky and others on the basis
of external coloration according his recommendation.

Chromosomal slides were prepared by the air-dry
method from kidney tissue after colchicine injection
(Frolov 1989), and chromosomes were stained with
Giemsa (Sigma, 4% solution in phosphate buffer pH 6.8
or in distilled water). To reveal active nuclear organizer regions (NORs) we used the Ag–NOR-staining
procedure (Howell & Black 1980).

Chromosomes were classified as metacentric (M),
submetacentric (SM), subtelocentric (ST), and acrocentric (A) with the boundaries between these chromosome types at $r = 1.67$, $r = 3.0$, and $r = 7.00$.
This classification is similar to Levan et al. (1964),
except for the last type of chromosome. We classified
chromosomes with a very short and rare visible second
arms as 'A' instead of 't' and 'T' (telocentric, namely
chromosomes with a very short or without second arm).

Some chars (Dolly Varden from Lake Achchen and
the Kamchatka River, white char from the Kamchatka
River) had one pair of the middle-sized SM or ST chromosome with a satellite-like structure and an active
NOR in their short arms. Due to variable length of this

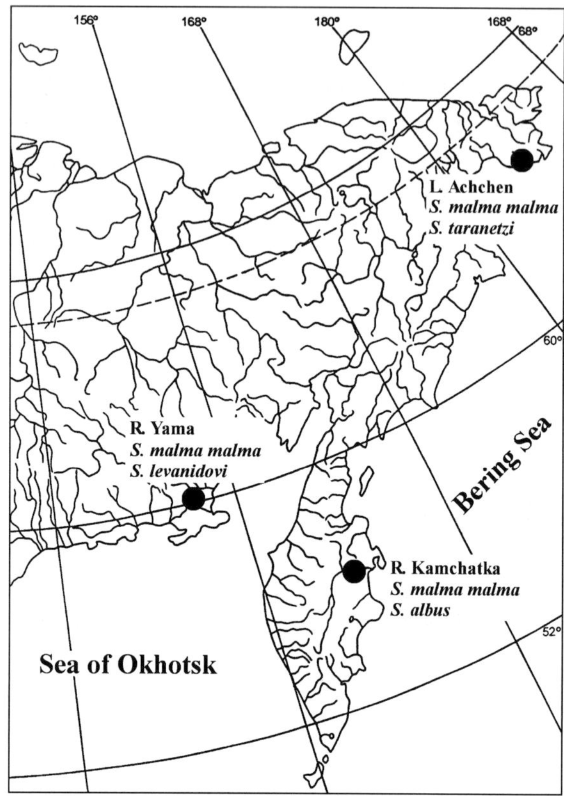

Figure 1. Some places of sympatry of northern Dolly Varden and
other chars in the Far East of Russia.

structure, NOR-bearing chromosomes of these chars
looked as M, SM, or ST in different cells (arm ratio
$r = 1.25$–4.0) and were frequently heteromorphic. So,
we classified these chromosomes as SM–ST and noted
NF = 98 + 2 for karyotypes of these chars.

Results and Discussion

Dolly Varden and Levanidov's char

Differences of karyotypes of Dolly Varden and
Levanidov's char from the the Yama River were very
much expressed in both chromosome number and chromosome arm number: $2n = 78$, NF = 98 and $2n = 80$,
NF = 100, respectively (Table 1, Figure 2). Both
species had invariable 2n and NF and could be easily
differentiated by these characters of their karyotypes.
Rows of M and SM chromosomes in karyotypes of
Dolly Varden and Levanidov's char were very similar

Table 1. Species studied and their karyotypes.

Locality, species	Giemsa staining				Ag-NOR staining		
	Number of fish	Number of cells	Karyotype		Number of fish	Number of cells	Number of active NORs in cell
			2n	NF			
R. Yama							
S. levanidovi	15	231	80	100	11	150	1–4 (5 pairs)
S. malma malma	17	319	78	98	14	209	1–6 (5 pairs)
L. Achchen							
S. taranetzi	17	309	76–79	98	10	223	1–3 (2 pairs)
S. malma malma	13	202	78	98 + 2	6	135	1–4 (5 pairs)
R. Kamchatka							
S. albus	12	121	76–79	98 + 2	5	40	1–3 (2 pairs)
S. malma malma	11	173	78	98 + 2	5	78	1–2 (1 pair)

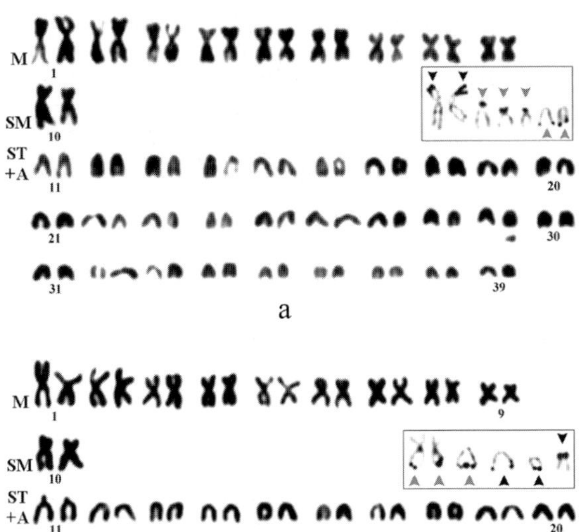

a

b

Figure 2. Karyograms of northern Dolly Varden (a) and Levanidov's char (b) from the Yama River, with NOR-bearing chromosomes in frames (black arrows indicate major NOR sites).

(if not identical, see Figure 2). But Levanidov's char had one additional pair of large ST chromosome (# 11 in Figure 2), which added two additional chromosome arms in karyotype of Levanidov's char.

Most karyologists believe that primitive characters states in char's karyotypes are 2n = 84 and NF = 100, and reduction in 2n and in NF are derived states (Viktorovsky 1978, Cavender & Kimura 1989, Phillips et al. 1989a, Frolov 1995). If so, Levanidov's char could be considered as a phylogenetically more ancient

Table 2. Frequency of active NORs at different chromosome location in Dolly Varden and Levanidov's char from the Yama River.

NOR location	NOR frequency	
	S. malma malma (N = 14)	*S. levanidovi* (N = 11)
M_t	0.043	0.267
SM_t^*	**1.071**	0.220
ST_{sa}	0.445	**0.513**
A_c	0.024	0.260
A_t	0.311	**0.627**

t = telomere, sa = shortarm, c = centromere, * = short arm in Dolly Varden and long arm in Levanidov's char. Bold numbers are frequencies of major NOR sites.

species, and Dolly Varden as a phylogenetically young species (Frolov & Frolova 1997). We have shown earlier, that heterochromatin in C-banded karyotypes of Levanidov's char and Dolly Varden was located only in small centromeric regions of chromosomes in both species (Frolov 2000). Therefore differences between these chars in 2n and NF could not be explained by additional heterochromatic chromosome pair or by additional heterochromatic chromosome arms in karyotype of Levanidov's char. In all probability, karyotype of Dolly Varden was formed from Levanidov's char-like karyotype by tandem translocation of two small A with formation of one large A. Both 2n and NF were reduced in Dolly Varden karyotype.

Both species had multiple NOR (Table 1), but major NOR sites were located in different chromosomes (Table 2). Some NORs were located in different positions on similar chromosomes or in chromosomes of the same morphology which had different size (see NOR in SM and A in Figure 2), and were also species specific. At the same time, NORs also showed phylogenetic primitiveness of Levanidov's char because their location were very similar to those in karyotypes of lake trout *S. namaycush* and brook trout *S. fontinalis* (Frolov & Frolova 1999).

Levanodov's char could be differentiated from Dolly Varden using common morphological characters (Chereshnev et al. 1989). They also may be rather easily differentiated by external coloration, coloration of a stomatic cavity and form of a basis of a tail stalk (Chereshnev et al. 1989), and we primarily differentiated them by these three characters. According to these authors, external coloration of Levanidov's char is more similar to those of lake trout (*S. namaycush*) and brook trout (*S. fontinalis*) than to sympatric Dolly Varden. Our results were in accordance with these observations: major NOR sites of Levanidov's char (see Table 2) are more similar to those in primitive karyotypes of lake and brook trouts (Phillips et al. 1989a,b, Frolov & Frolova 1999).

Although Levanidov's char and Dolly Varden move for spawning in the Yama River at the same time, Levanidov's char prefers main-stream spawning in the Yama River and other rivers which it inhabits, but Dolly Varden prefers upper stream tributaries of these rivers (Chereshnev et al. 1989). This suggests reproductive isolation of Levanidov's char and Dolly Varden due to differential spawning location. Our karyological data and results of biochemical study of proteins of the same specimens of Levanidov's char and Dolly Varden from the Yama River (Omelchenko et al. 1986) showed that hybrids between Levanidov's char and Dolly Varden were absent.

Dolly Varden and Taranetz's char

First views of routinely stained karyotypes of Dolly Varden and Taranetz's char showed that chromosome numbers were very similar in most cells of the specimens studied. But Dolly Varden had stable counts $2n = 78$, where as Taranetz's char had a mosaic $2n$ in most specimens due to Robertsonian translocation (Table 1). Six fishes had $2n = 76–78$, six fishes had $2n = 77–78$, and five fishes had $2n = 76–79$, $2n = 76–77$, $2n = 77–79$, $2n = 78–79$ or $2n = 78$ (Frolov 1997, 2000). Dolly Varden had one pair of marker SM–ST chromosomes (SM–ST, # 11, Figure 3) and NF $= 98 + 2$, but Taranetz's char had no similar chromosomes in their karyotype and NF $= 98$ (Figure 3).

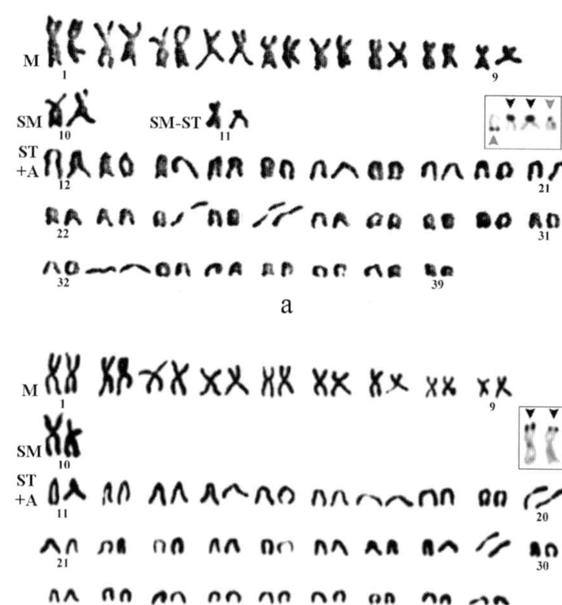

Figure 3. Karyograms of northern Dolly Varden (a) and Taranetz's char (b) from Lake Achchen with NOR-bearing chromosomes in frames (black arrows indicate major NOR sites).

Table 3. Frequency of active NORs at different chromosome locations in Dolly Varden and Taranetz's char from Lake Achchen.

NOR location	NOR frequency	
	S. malma malma (N = 6)	*S. taranetzi* (N = 10)
M$_t$	0.04	—
SM$_t$	—	**1.80**
ST$_{sa}$	**1.90**	0.05
A$_t$	0.19	—

t = telomere, sa = short arm. Bold numbers are frequencies of major NOR sites.

These characters were sufficient to identify any specimens of Dolly Varden or Taranetz's char. But the number and location of active NORs were much more informative for differentiation of these chars. Dolly Varden had multiple NORs with the major NOR site in small ST, while Taranetz's char usual had one NOR and the major NOR site was located on the large SM (Figure 3, Tables 1 and 3).

According to common opinion of char karyologists, single or multiple NORs are opposite states (primitive or diverged and vice versa) in char karyotypes

(Cavender & Kimura 1989, Phillips et al. 1989c, Frolov 1995). If so, Dolly Varden and Taranetz's char from the Lake Achchen belong to different phylogenetic lines of *Salvelinus* and may be considered as distinct species. Moreover, NORs of Dolly Varden from Lake Achchen are more similar to those of north American lake and brook trouts, and of Arctic char (*S. alpinus*) from Europe, which have multiple NORs located in the same chromosomes (Phillips et al. 1988, 1989a,c). On the contrary, NOR-bearing large SM chromosomes of Taranetz's char (Figure 3) are typical for karyotypes of Arctic char from Northwest Territory (2n = 78) (Phillips 1988, 1989a,c) and for southern American Dolly Varden (*S. malma lordi*, 2n = 82) (Cavender & Kimura 1989, Ueda et al. 1991, Alonso et al. 1999).

Some Russian ichthyologists believe morphological characters are overlapped when comparing Dolly Varden and Taranetz's char from many Chukotka rivers, and thus these chars cannot be strongly differentiated one from another (Barsukov 1960, Volobuev et al. 1979, Savvaitova et al. 1988). On the contrary, our data are in agreement with separation of Dolly Varden and Taranetz's char taxonomy. I. Chereshnev and M. Glubokovsky showed that these chars could be differentiated in all Chukotka basins by mean numbers of gill rakers, pyloric caeca and vertebrae, skull bones, spawning coloration, coloration of a stomatic cavity, etc. (Chereshnev 1978, 1982, Glubokovsky & Chereshnev 1981). Both chars spawn in Lake Achchen, but Dolly Varden spawns in tributaries of this lake, and Taranetz's char spawns in the lake not far from the shore (Chereshnev 1982, Gudkov 1994).

Protein systems study have also shown reproductive isolation of Dolly Varden and Taranetz's char in Lake Achchen and species status of these chars (Omelchenko et al. 1988). Moreover, according to the analysis of repetitive nuclear DNA sequences northern Dolly Varden and Taranetz's char belong to different phylogenetic groups of the genus *Salvelinus* (Oleinik & Skurikhina 1999).

Dolly Varden and white char

Karyotypes of Dolly Varden and white char from the Kamchatka River are more similar in comparison with other sympatric chars (Figure 4). Both char had identical pair of SM–ST chromosomes (# 11, Figure 4) and identical chromosome arm numbers NF = 98 + 2. But karyotypes of these chars could be differentiated by chromosome number, which is stable

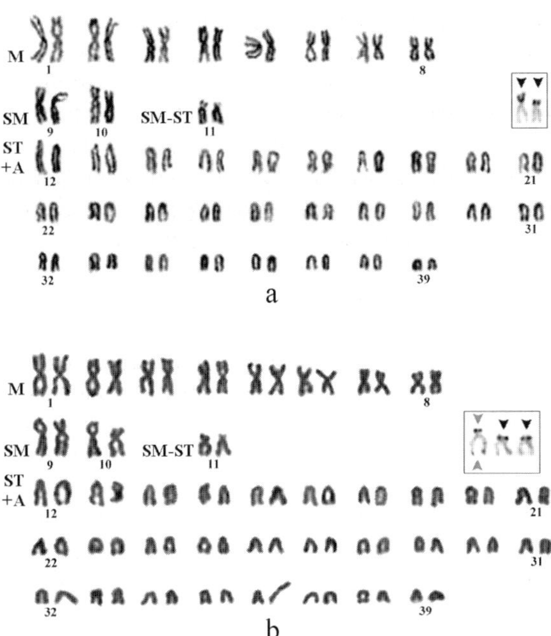

Figure 4. Karyograms of northern Dolly Varden (a) and white char (b) from the Kamchatka River, with NOR-bearing chromosomes in frames (black arrows indicate major NOR sites).

2n = 78 in Dolly Varden, and mosaic/polymorphic 2n = 76–79 due to Robertsonian translocation in white char (Table 1). Seven specimens of white char had 2n = 78, three had 2n = 78–79, one had 2n = 76–79, and one had 2n = 77 (Frolov 2001).

One active NOR was revealed in Dolly Varden, while white char had multiple NORs (Table 1). NORs were located in the telomeric region of the marker SM–ST chromosome in Dolly Varden, but were detected in the same location and additionally in telomeric regions of the short and long arms of the large ST chromosome in white char (Figure 4). In Dolly Varden, NORs were usually active in both homologous chromosomes (78% of cells examined). In white char, one or two marker SM–ST chromosomes had active NORs in 85% of cells, and the large ST chromosomes had active NORs in one or two sites in 48% of cells (Frolov 2001).

We would like to note that northern Dolly Varden invariable has 2n = 78 in any population studied in the far east of Russia (see upper), and in North America (Phillips et al. 1999). But white char from L. Kronotskoye (Kamchatka) (Viktorovsky 1978) and from the Kamchatka River basin (Frolov 2001), and melanistic form of white char – 'stone char' from the Kamchatka River (Frolov 1991) had

446

polymorphic karyotypes. Moreover, karyotype and single NOR-bearing chromosome in Dolly Varden from the Kamchatka River and from R. Noatak (northern Alaska – Phillips et al. 1999) may be identical. These data indicated that Dolly Varden and white char from the Kamchatka River belong to the same phylogenetic line of *Salvelinus* and are close related. On the other hand, 2n and NOR variations in white char may provide evidence for their reproductive isolation.

Morphologically Dolly Varden and white char are very similar. Savvaitova (1989) wrote that there were no sufficient differences between populations of chars in the Kamchatka River. On the contrary, Glubokovsky (1977b) described a new char species from the Kamchatka River – white char (*S. albus*), which was differentiated from Dolly Varden on the basis of skull bones. He believes that the Kamchatka River is inhabited by three char species – *S. leucomaenis*, *S. malma* and *S. albus* (Glubokovsky 1977a). This assumption is consistent with results of numerical taxonomy studies of chars of the Kamchatka River basin (Bagyryan 1981). On the other hand, Vasilieva (1980) noted that craniological characters are not useful for diagnostic of char forms in the Kamchatka River.

First attempts to differentiate Dolly Varden and white char in the Kamchatka River by protein markers were unsuccessful (Efremov 1991). Osinov[1] noted that Dolly Varden had some rare alleles which were absent in white char, and he 'cannot reject the possibility of reproductive isolation between local populations or ecological forms of chars inhabiting … Kamchatka River', but he concluded that the level of genetic divergence between these chars are very low. Very low differences of Dolly Varden and white char (comparable with differences of northern Dolly Varden populations) was also shown in the studies of their nuclear and mitochondrial DNA (Oleinik & Polyakova 1994, Oleinik & Skurikhina 1999).

Conclusions

The degree of karyotype differences of sympatric chars varies and reflects (on our opinion) their origin and distribution. According to karyotypes, as northern Dolly Varden and Levanidov's char from the Yama River, and northern Dolly Varden and Taranetz's char from Lake Achchen belong to different phylogenetic lines,

whereas northern Dolly Varden and white char from the Kamchatka River are very close related, but may be reproductively isolated. Karyological data obtained here are in good agreement with other genetic data of sympatric chars of the Russian Far East – proteins and nuclear and mitochondrial DNA. Thus, we propose that differences in karyotypes of sympatric chars studied reflect real level of genetic differences of these pairs of species, and that karyotyping is a reliable method for differentiation of sympatric chars and may be widely used to resolve the problems of differentiation, evolution and phylogeny of char.

In conclusion we would like to note that differences in karyotypes of Dolly Varden from different localities were also revealed by this research. First, SM–ST chromosomes varied among chars with NF = 98 or NF = 98 + 2. Second, Dolly Varden had different numbers and locations of NORs in different populations (Figures 2–4). In our opinion, these differences are extremely interesting and should be a matter of further research.

Aknowledgements

We thank M. Glubokovsky (Institute of Marine Biology FEB RAS, Vladivostok) and I. Chereshnev (Institute of Biological Problems of the North FEB RAS, Magadan) for recommendations on identification of char specimens, P. Gudkov (Institute of Biological Problems of the North FEB RAS, Magadan) for his help in catching chars from Lake Achchen, and sport-fishing firm 'Talan' (Magadan) for technical and financial support of our research of chars from the Yama River. Comments of anonymous reviewers and editorial suggestions were extremely helpful for manuscript revision. Our study of char's karyotypes was supported by grants of the Russian Foundation for Basic Research (## 94-04-11129, 98-04-48919, 00-04-63057).

References

Alonso, M., A. Fujiwara, E. Yamaha, S. Kimuira & S. Abe. 1999. Ribosomal RNA gene loci and silver-stained nucleolar organizer regions associated with heterochromatin in Alaskan char *Salvelinus malma* and chum salmon *Oncorhynchus keta*. Hereditas 131: 221–225.

Bagyryan, S. Sh. 1981. Using the methods of numerical taxonomy for differentiation of two sympatric groups of chars in Lake Azabachye. Biologicheskiye Nauki 1: 107–110 (in Russian).

Barsukov, V.V. 1960. On the taxonomy of Chukotsk chars of the genus *Salvelinus*. Voprosy Ikhtiologii 14: 3–17 (in Russian).

---

[1] Osinov, A.G. 1999. The Dolly Varden (*Salvelinus malma*) of Eurasia and chars of the Kamchatka River basin: Data from allozyme analysis. ISACF Information Series 7: 173–181.

Cavender, T.M. & S. Kimura. 1989. Cytotaxonomy and interrelationships of the Pacific basin *Salvelinus*. Physiol. Ecol. Jpn. 1: 49–68.

Chereshnev, I.A. 1978. Taxonomy of the chars *Salvelinus* (Nilson) Richardson from the Bering Sea coast of the Chukotski Peninsula. Biologiya Morya 1: 36–46 (in Russian with English summary).

Chereshnev, I.A. 1982. On the taxonomic status of sympatric anadromous chars of the genus *Salvelinus* (Salmonidae) from Eastern Chukotka. Voprosy Ikhtiologii 22: 922–936 (in Russian).

Chereshnev, I.A., M.B. Skopetz & P.K. Gudkov. 1989. New species of char, *Salvelinus levanidovi* sp. nov. from the basin of Okhotsk Sea. Voprosy Ikhtiologii 29: 691–704 (in Russian).

Efremov, V.V. 1991. Electrophoretic variability of chars of the genus *Salvelinus* (Salmonidae) in the Kamchatka River basin. pp. 94–102. *In*: I.A. Chereshnev & M.K. Glubokovsky (ed.) Biology of Chars in the Far East, FEB Acad. Sci. USSR, Vladivostok (in Russian).

Frolov, S.V. 1989. Differentiation of sex chromosomes in the Salmonidae. 1. Karyotype and sex chromosomes in *Parasalmo mykiss*. Tsitologia 31: 1391–1394 (in Russian with English summary).

Frolov, S.V. 1991. Karyotypes of white-spotted char *Salvelinus leucomaenis* (Pallas) and white char *S. albus* Glubokovsky from the basin of River Kamchatka. pp. 103–111. *In*: I.A. Chereshnev & M.K. Glubokovsky (ed.) Biology of the Chars of the Far East, FEB Acad. Sci. USSR , Vladivostok (in Russian).

Frolov, S.V. 1995. Comparative karyology and karyotype evolution of chars. J. Ichthyol. 35: 14–23.

Frolov, S.V. 1997. Divergence of karyotypes of Chukotkan chars. Doklady Akademii Nauk 357: 703–705 (in Russian).

Frolov, S.V. 2000. Karyotype variability and evolution in Salmonidae. Dalnauka, Vladivostok. 229 pp (in Russian).

Frolov, S.V. 2001. Karyological differences between northern Dolly Varden *Salvelinus malma malma* and white char *S. albus* from the Kamchatka River basin. Russ. J. Genet. 37: 269–275.

Frolov, S.V. & V.N. Frolova. 1997. Karyological differentiation of *Salvelinus levanidovi* and *S. malma* from the Yama River. Russ. J. Mar. Biol. 23: 328–331.

Frolov, S.V. & V.N. Frolova. 1999. Nuclear organizer regions of chromosomes as a phylogenetic marker of karyotypes in northern malma trout and Levanidov's char. Russ. J. Mar. Biol. 25: 438–440.

Glubokovsky, M.K. 1977a. Taxonomic relationships of chars from the genus *Salvelinus* in the basin of the Kamchatka River. Biol. Morya 3: 24–35 (in Russian with English summary).

Glubokovsky, M.K. 1977b. *Salvelinus albus* sp. n. from the basin of River Kamchatka. Biol. Morya 4: 48–56 (in Russian).

Glubokovsky, M.K. & I.A. Chereshnev. 1981. Debatable questions on the Holarctic chars of the genus *Salvelinus*. 1. A study of anadromous chars from the basin of the East Siberian Sea. Voprosy Ikhtiologii 21: 771–786 (in Russian).

Gudkov, P.K. 1994. On biology of Taranetz's char *Salvelinus taranetzi* from the Lake Achchen. Voprosy Ikhtiologii 34: 58–63 (in Russian).

Howell, W.M. & D.A. Black. 1980. Controlled silver-staining of nucleolus organizer regions with a protective colloidal developer: A 1-step method. Experientia 36: 1014–1015.

Kartavtsev, Yu.F., M.K. Glubokovsky & I.A. Chereshnev. 1983. Genetic differentiation and variability of two sympatric species of chars (*Salvelinus*, Salmonidae) of Chukotka. Genetika 19: 584–593 (in Russian with English summary).

Levan, A., K. Fredga & A.A. Sandberg. 1964. Nomenclature for centromeric position on chromosomes. Hereditas 52: 201–220.

Oleinik, A.G. & N.E. Polyakova. 1994. Restriction analysis of the mitochondrial genome in the family Salmonidae. Russ. J. Genet. 30: 1043–1054.

Oleinik, A.G. & L.A. Skurikhina. 1999. Phylogenetic relationships among anadromous chars of the genus *Salvelinus* based on restriction endonuclease analysis of nuclear DNA. Russ. J. Genet. 35: 1082–1091.

Omelchenko, V.T., D.V. Polytov, E.A. Salmenkova, T.V. Malinina & S.V. Frolov. 1986. Genetic differentiation of sympatric chars of the genus *Salvelinus* from the Yama River. Russ. J. Genet. 32: 1358–1364.

Omelchenko, V.T., E.A. Salmenkova, T.V. Malinina & S.V. Frolov. 1988. Genetic differentiation of sympatric populations of chars of the genus *Salvelinus* from the Achchen Lake (Chukotka). Russ. J. Genet. 34: 311–316.

Phillips, R.B., K.A. Pleyte & S.E. Hartley. 1988. Stock-specific differences in the number and chromosomal position of the nucleolar organizer regions (NORs) in Arctic char (*Salvelinus alpinus*). Cytogenet. Cell Genet. 48: 9–12.

Phillips, R.B., K.A. Pleyte & P.E. Ihssen. 1989a. Patterns of chromosomal nucleolar organizer region (NOR) variation in fishes of the genus *Salvelinus*. Copeia 1: 47–53.

Phillips, R.B., K.D. Zajicek & P.E. Ihssen. 1989b. Population differences in chromosome-banding polymorphisms in lake trout. Trans. Amer. Fish. Soc. 118: 64–73.

Phillips, R.B., K.A. Pleyte, L.M. Ert & S.E. Hartley. 1989c. Evolution of nucleolar organizer regions and ribosomal RNA genes in *Salvelinus*. Physiol. Ecol. Jpn. 1: 429–447.

Phillips, R.B., L.I. Gudex, K.M. Westrich & A.L. DeCicco. 1999. Combined phylogenetic analysis of ribosomal ITS1 sequences and new chromosome data supports three subgroups of Dolly Varden char (*Salvelinus malma*). Can. J. Fish. Aquat. Sci. 56: 1504–1511.

Savvaitova, K.A. 1989. Arctic chars (structure of population systems, perspectives of using of reserves in economy), Agropromizdat, Moscow. 223 pp (in Russian).

Savvaitova, K.A., V.A. Maximov & V.V. Volobuev. 1988. On the relations between anadromous forms of chars of the genus *Salvelinus* (Salmonidae, Salmoniformes) at the Chukot Peninsula. Zoologicheskii Zurnal 67: 1498–1508 (in Russian with English summary).

Ueda, T., H. Fukuda & J. Kobayashi. 1991. Karyological study of the Dolly Varden *Salvelinus malma* from Alaska. Jpn. J. Ichthyol. 37: 354–357.

Vasilieva, E.D. 1980. An experience of utilization of osteological characters in taxonomy of chars of the genus *Salvelinus* (Salmoniformes, Salmonidae). Zoologicheskii Zurnal 59: 1671–1682 (in Russian with English summary).

Viktorovsky, R.M. 1978. Mechanisms of speciation of the chars of Kronotskoye Lake, Nauka, Moscow. 106 pp (in Russian).

Volobuev, V.V., E.D. Vasilieva & K.A. Savvaitova. 1979. On taxonomic status of Chukotkan anadromous chars of the genus *Salvelinus*. Voprosy Ikhthiologii 19: 408–418 (in Russian).

Environmental Biology of Fishes **69**: 449–459, 2004.
© 2004 *Kluwer Academic Publishers. Printed in the Netherlands.*

Differences between two subspecies of Dolly Varden, *Salvelinus malma*, revealed by RFLP–PCR analysis of mitochondrial DNA

Alla G. Oleinik[a], Lubov A. Skurikhina[a], Sergei V. Frolov[a], Vladimir A. Brykov[a] & Igor A. Chereshnev[b]
[a]*Institute of Marine Biology, Far East Branch Russian Academy of Science,*
Vladivostok 690041, Russia (e-mail: inmarbio@mail.primorye.ru)
[b]*Institute of Biological Problems of the North Far East Branch Russian Academy of Science, Magadan, Russia*

Received 21 April 2003   Accepted 16 June 2003

Key words: char, PCR, RFLP-analysis

Synopsis

Genetic differences between two subspecies of Dolly Varden, northern *Salvelinus malma malma* and southern *Salvelinus malma krascheninnikovi*, from rivers of eastern Russia were studied. Mitochondrial DNA was analyzed by restriction fragment length polymorphisms (RFLP) performed on products amplified with polymerase chain reaction. Three adjacent segments (approximately 7670 bp), comprising 47% of the mitochondrial genome were used: two encoding the five complete NADH dehydrogenase subunits and the other the cytochrome *b* gene and the control region (D-loop). Total composite haplotypes 46 were found among 136 fishes using RFLP analysis with 14 restriction enzymes. The amount of nucleotide divergence between haplotypes of two subspecies of Dolly Varden was estimated to be approximately 4%. The differences in the level of nucleotide diversity, mismatch distribution between haplotypes, and population-genetic structure of two subspecies of Dolly Varden suggest that these two forms have existed separately for a long time.

Introduction

Fishes of the genus *Salvelinus* have presented a great challenge to taxonomists and evolutionary biologists who are interested in resolving inter-relationships and understanding the evolution of diversity within salmonid genera (Behnke 1980). Char exhibit a complex mosaic of variability in morphology, coloration, and ecology throughout their range. Many species of char also show varying degrees of anadromy, with both resident freshwater and anadromous populations, often exist sympatrically. Because taxonomic recognition is based primarily on all these characters, systematic and evolutionary scenarios of the origin of the different forms in this species complex have been controversial (Behnke 1980).

The monophyly of the genus is supported both morphologically (Glubokovsky 1995) and genetically (Grewe et al. 1990), although taxonomic confusion persists within the genus, especially in the eastern part of Russia (Chereshnev 1983, Glubokovsky 1995). The phylogenetic relationships in *Salvelinus* species have been obscured by the lack of comparative genetic studies throughout this area and by the use of genetic markers with low resolving power (Osinov & Pavlov 1998, Salmenkova et al. 2000). The evolution of molecular genetic techniques during the last two decades, particularly analysis of mitochondrial DNA (mtDNA), has substantially contributed to an understanding of natural genetic diversity, biogeography, and speciation issues (Avise 2000). This is especially helpful in such a puzzling group as Salmonidae, which show a mosaic of morphological and ecological traits.

Behnke (1980) postulated an early Pleistocene divergence into Arctic char (*S. alpinus*) and Dolly Varden (*S. malma*). Arctic char from the Chukchi (*S. alpinus*) and Bering (*S. malma*) seas has been considered a distinct form, based on their peculiar coloration

450

(Behnke 1980). The latter lives on the eastern and western sides of the Pacific Ocean. It is the only widely accepted subspecies from Central and Eastern Asia, whereas systematic relationships among numerous others nominates species and subspecies from this region remain controversial. Two subspecies of *S. malma* are suggested to be present in North America: the northern form, *S. m. malma*, occurring north of the Alaska Peninsula and the southern form, *S. m. lordi*, occurring throughout south central and southeastern Alaska and south to Washington (McPhail 1961, Behnke 1980). In Asia, a northern form and southern form also occur. The area of the northern form, probably identical to *S. m. malma*, stretches from the Bering Strait to the southern Kamchatka Peninsula and reaches the delta of the Amur River (Berg 1948). The southern subspecies, *S. m. krascheninnikovi*, inhabits the area south to the Amur River mouth to the Peter the Great Gulf, including Primorye, Sakhalin Island, southern Kuril Islands and northern Japan (Berg 1948).

In this investigation our primary goal was to define the differences between southern *S. m. kraschen-innikovi* and northern *S. m. malma* subspecies of Dolly Varden from eastern Russia using restriction fragment length polymorphisms (RFLP) analysis of the polymerase chain reaction (PCR) amplified fragments of mtDNA. Mitochondrial DNA is especially useful for recognizing phylogenies because it evolves 5–10 times faster than many nuclear loci, is primarily clonally inherited, and does not undergo recombination (Brown et al. 1979, Avise 2000).

Materials and methods

Sample collection

Samples of Dolly Varden were collected, 1997–2000, from northern and southern localities of eastern Russia (Table 1, Figure 1). Samples of DNA from white-spotted char *S. leucomaenis* Pallas were used for comparison and as the outgroup in phylogenetic analysis. Samples of heart tissue from each specimen were preserved in 95% ethanol or a solution of 20% dimethyl sulfoxide (DMSO) and 0.25 M ethylenediaminetetra-acetic acid (EDTA) at pH 8, saturated with NaCl.

DNA isolation, PCR-amplification, and RFLP-analysis

Total genomic DNA was isolated using standard methods (Sambrook et al. 1989). Target sequences were PCR-amplified from total genomic DNA using primers developed for coho salmon mtDNA (Gharrett et al. 2001). Target sequences were amplified by heating to 94°C for 5 min, followed by 30 cycles of 1 min at 94°C, 1 min at 55°C, and 3 min at 72°C using Taq polymerase from Perkin Elmer (Norwalk, CT).

We surveyed mtDNA variation by examining restriction site variation in three amplified portions of the mtDNA (Table 2), which together comprise about 50% of the mitochondrial genome. Single digests of sub-samples of the PCR-amplified mtDNA

Table 1. Geographical localization and sample sizes of char.

Number of population	Species	Locality	Number of specimens
1	*S. leucomaenis*	Izmena Bay, south coast of Kunashir, Kuril Islands	10
2	*S. leucomaenis*	Kamchatka R., Kamchatka Peninsula	2
3	*S. m. malma*	Kamchatka R., Kamchatka Peninsula	25
4	*S. m. malma*	Paratunka R., Kamchatka Peninsula	25
5	*S. m. malma*	Achchon Lake, Chukchi Peninsula	8
6	*S. m. malma*	Yama R., North Coast of Okhotsk Sea	13
7	*S. m. krascheninnikovi*	Aniva Creek, Sakhalin Island	31
8	*S. m. krascheninnikovi*	Teplyi Creek, Sakhalin Island	10
9	*S. m. krascheninnikovi*	Val R., Sakhalin Island	24

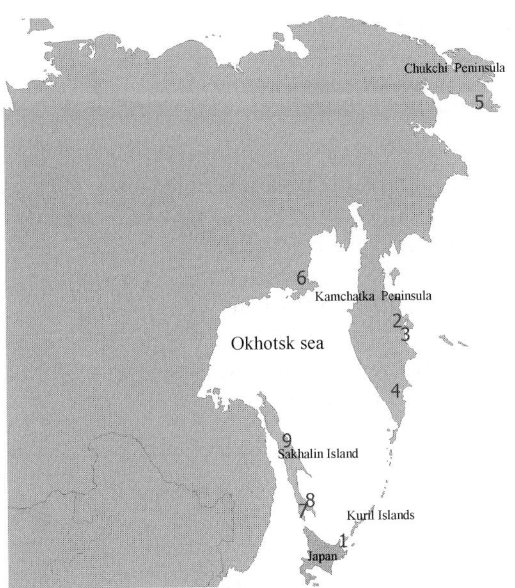

Figure 1. Map of the study area showing the location of sample points. Locality numbers as in Table 1.

Table 2. Primers for PCR amplification (Gharrett et al. 2001) and PCR fragment sizes of char mtDNA (bp).

Region	Sequence	Frag. size
ND1/	5′ACCTCGATGTTGGATCAGG 3′	2689
ND2	5′ATTAAAGTGNTTGA(T/G)TTGCATTC 3′	
ND5/	5′AACAGCTCATCCATTGGTCTTAGG 3′	2488
ND6	5′TTACAACGATGGTTTTTCATGTCA 3′	
Cyt *b*/	5′TGAA(G/A)ACCACCGTTGTTATTCAA 3′	2795
D-loop	5′TAGGGCCTCTCGTATAACCG 3′	

regions were made using digestions with 14 restriction enzymes each: *FnuDII (BstUI), DdeI, EcoRV, RsaI, MspI, Sau96I, HinfI, HhaI, AccBII (HgiCI), AvaII, BstNI, MboI, AvaI, StyI* ('Fermentas', Lithuania and 'Sibenzyme', Novosibirsk). Resulting fragments were separated by electrophoresis performed in 1.5% agarose (a mixture composed of one part regular agarose [Sigma] and two parts Synergel™ [Diversified Biotech Inc., Boston MA]) in 0.5 × TBE buffer (TBE is 90 mM Tris–boric acid, and 2 mM EDTA, pH 7.5). DNA in the gel was stained with ethidium bromide and photographed on an ultraviolet light transilluminator. Digests that produced fragments that could not be adequately detected were resolved on 12% polyacrylamide gel. *HinfI* digested $\phi\chi$ *174* RF phage DNA and PstI λ phage DNA was used as molecular weight markers to estimate restriction fragment sizes.

Statistical analysis

Restriction sites were inferred from fragment patterns that could be related to each other by the gain or loss of one single or more sites. The restriction site matrix of presence/absence was constructed for each enzyme and each haplotype was given a single letter code. Composite haplotypes were constructed from restriction fragment patterns of all restriction enzymes across three mtDNA PCR-amplified regions. Thus, each fish was characterized by the 42 letter composite haplotype code. A site matrix for each of the composite haplotypes resolved was constructed using the software package REAP (McElroy et al. 1992). Networks of minimum restriction site changes and mismatch distributions (distribution of the number of nucleotide differences between all haplotypes) were made with the use of software package arlequin2000 (Schneider et al. 1999). The differences between the mismatch distributions was estimated with the use χ^2 test.

Proportion of nucleotide substitutions (π) and their standard errors (Nei & Tajima 1981, Nei 1987, Nei & Miller 1990) and haplotype (nucleon) diversities (h) (Nei & Tajima 1981, Nei 1987) were estimated with REAP. As a measure of genetic diversity among haplotypes, we computed the average numbers of nucleotide substitutions per site between all haplotypes (nucleotide divergence) (Nei 1987) using the DA program in REAP. The UPGMA algorithm was used to cluster haplotypes according to nucleotide divergence in the program NTSYS (Rohlf 1990). Haplotype frequencies were calculated for each population sample. A number of contingency χ^2 tests (Roff & Bentzen 1989) for differences in haplotype frequencies among populations and different groups of populations were conducted using REAP. The F_{ST} values and significance F_{ST} p values (significance level = 0.0500) among all samples were estimate with the use of software package Arlequin2000.

Results

mtDNA diversity

Our analysis surveyed 277 restriction sites over 7 670 bp part of char mtDNA, 46 composite haplotypes were resolved among 136 samples of Dolly Varden and 6 haplotypes among 12 samples of white-spotted char *S. leucomaenis* (Table 3). The distribution of mtDNA haplotypes in Dolly Varden populations is typical for

Table 3. Composite haplotypes mtDNA detected among the population of *S. leucomaenis u S. malma*.

Haplotype	Composite haplotype mtDNA														Population								
	BstUI	DdeI	EcoRV	RsaI	MspI	AsuI	HinfI	HhaI	HgiCI	AvaII	BstNI	MboI	AvaI	StyI	1	2	3	4	5	6	7	8	9
KAA	AKK	KKK	DAA	KKK	KAB	KKK	KKK	ABK	KAA	KKK	KKK	KKK	KAB	KAK	7	0	0	0	0	0	0	0	0
KAB	AKK	LKK	DAA	KKK	KAB	KKK	KLK	ABK	KAA	KKK	KKK	KKK	KAB	LAK	1	0	0	0	0	0	0	0	0
KAC	AKK	KKL	DAA	KKK	KAB	KKK	KKK	ABK	KAA	KKK	KKK	KKK	KAB	KAK	1	0	0	0	0	0	0	0	0
KAD	AKK	KKK	DAA	KKK	KAB	KKK	KKK	ABK	KAA	KKK	LKK	KKK	KAB	KAK	1	0	0	0	0	0	0	0	0
KAE	AKK	LKK	DAA	KKK	KAB	KKK	KLK	ABK	KAA	KKK	KKK	KKK	KAB	AEK	0	1	0	0	0	0	0	0	0
KAF	AKK	LKK	DAA	KKK	KAB	KKK	KLK	ABK	KAA	KKK	KKK	KKK	KAB	AAK	0	1	0	0	0	0	0	0	0
NAA	AAA	AAA	AAA	AAA	ACA	AAA	AAA	AAA	AAA	AAA	AAA	BAA	AAA	BAA	0	0	13	0	0	0	0	0	0
NAB	ABA	AAB	BAA	AAA	GCA	AAA	AAC	ABA	AAA	AAA	AAA	BAA	AAA	BAA	0	0	1	0	0	0	0	0	0
NAC	AAA	AAB	AAA	AAA	ACA	AAC	AAA	AAA	AAA	AAA	AAD	BAA	AAA	BAA	0	0	2	0	0	0	0	0	0
NAD	AAA	AAA	AAA	AAA	ACA	AAA	ABA	AAA	AAA	AAA	AAA	BAA	AAA	BAA	0	0	1	0	0	0	0	0	0
NAE	AAA	AAD	AAA	AAA	ACA	AAC	AAC	AAA	AAA	AAA	AAD	BAA	AAA	BAA	0	0	1	0	0	0	0	0	0
NAF	ABA	AAB	AAA	ADA	ACA	AAA	AAC	ABA	AAA	AAA	AAA	BAA	AAA	BDA	0	0	1	0	0	0	0	0	0
NAG	AAA	AAD	AAA	AAA	ACA	AAA	AAA	AAA	AAA	AAA	AAA	BAA	AAA	BAA	0	0	1	0	0	1	0	0	0
NAH	AAA	AAD	AAA	AAA	ACA	AAA	ACA	AAA	AAA	AAA	AAA	CAA	AAA	BAA	0	0	1	0	0	0	0	0	0
NAI	AAA	AAB	AAA	AAA	CCA	AAA	AAA	AAA	AAA	AAA	AAA	BAA	AAA	BAA	0	0	1	0	3	0	0	0	0
NAJ	AAA	AAB	AAA	AAA	ACA	AAA	AAA	AAA	AAA	AAA	AAA	BAA	AAA	BAA	0	0	2	23	0	11	4	0	5
NAK	AAA	AAA	AAA	AAA	FCA	AAA	AAA	BAA	AAA	AAA	AAA	BAA	AAA	BAA	0	0	1	0	0	0	0	0	0
NAL	AAA	AAB	AAA	AAA	ACA	AAA	AAA	AAA	AAA	AAA	AAA	CAA	AAA	BAA	0	0	0	0	1	0	0	1	0
NAM	ABA	AAB	BAA	AAA	ACA	AAA	AAC	ABA	AAA	AAA	CAA	BAA	AAA	BAA	0	0	0	0	1	0	0	0	0
NAN	ABA	AAB	AAA	AAA	ACA	AAA	AAC	AAA	AAA	AAA	CAA	BAA	AAA	AAA	0	0	0	0	1	0	0	0	0
NAO	AAA	AAA	AAA	AAA	ACA	AAA	AAA	AAA	CAA	AAA	AAA	BAA	AAA	BAA	0	0	0	0	1	0	0	0	0
NAP	AAA	AAB	AAA	AAA	ACA	AAA	AAA	AAA	AAA	AAA	AAA	BAA	KAA	BAA	0	0	0	0	1	0	0	0	0
NAQ	AAA	AAB	AAA	AAA	ACA	AAA	AAA	AAA	AAA	AAA	AAA	BAA	AAA	BAE	0	0	0	1	0	0	0	0	0

NAR	AAA	AAB	AAA	AAA	ACA	AAA	AAE	AAA	AAA	AAA	AAA	BAA	AAA	BAA	0	0	0	1	0	0	0	0	0
NAS	AAA	AAB	AAA	AAA	ACA	AAA	AAA	AAA	AAA	AAA	AAA	BAA	AAA	AAA	0	0	0	0	0	0	1	0	0
SAA	ABA	BBD	CAA	BBB	BAB	ABB	AAB	ABB	BAA	AAA	BBB	AAB	BAA	BBB	0	0	0	0	0	1	0	0	0
SAB	ABA	DBB	CAA	BBB	BAB	ABB	AAB	ABB	BAA	AAA	BBB	AAB	BAA	ABB	0	0	0	0	0	0	15	3	1
SAC	ABA	DBB	CAA	BBB	BAB	ACB	AAB	ABB	BAA	AAA	BBB	AAB	BAA	ABB	0	0	0	0	0	0	1	0	0
SAD	ABA	DBB	CAA	BBB	BAB	ACB	AAB	ABB	BAA	AAA	BBB	EAB	BAA	ABB	0	0	0	0	0	0	1	2	0
SAE	ABA	DBB	CAA	BBB	DBB	ABB	ADB	ABB	CAA	AAA	BAB	EAB	BAA	BBB	0	0	0	0	0	0	1	0	0
SAF	ABA	DBI	CAA	BBB	DAB	ABB	AEB	ABB	CAA	AAA	BBB	EAB	BAA	BBB	0	0	0	0	0	0	1	0	0
SAG	ABA	BBB	CAA	BBB	DBB	CBB	ADB	ABB	CAA	AAA	BCB	EAB	BAA	BBB	0	0	0	0	0	0	1	0	0
SAH	ABA	BBB	CAA	BBB	BAB	ABB	AAB	ABB	BAA	AAB	BBB	AAB	BAA	BBB	0	0	0	0	0	0	1	0	0
SAI	ABA	BBB	CAA	BBB	BAB	ABD	AAB	ABB	BAA	AAB	BBC	AAB	BAA	BBB	0	0	0	0	0	0	1	0	0
SAJ	ABA	BBB	CAA	BBB	DBB	ABB	ADB	ABB	BAA	AAA	BCB	EAB	BAA	BBB	0	0	0	0	0	0	1	0	0
SAK	ABA	DBB	CAA	BBB	BAB	ABB	AAB	ABB	BAA	AAA	BBB	EAB	BAA	ABB	0	0	0	0	0	0	1	0	0
SAL	ABA	BBB	CAA	BBB	BAB	ABB	AAB	ABB	BAA	DAB	BBB	AAB	BAA	BBB	0	0	0	0	0	0	1	0	0
SAM	ABA	BBA	CAA	BBB	DBB	ABB	ADB	ABB	CAA	AAA	BCB	EAB	BAA	BBD	0	0	0	0	0	0	1	0	0
SAN	ABA	BBD	CAA	BBB	BAA	ABB	AAB	ABB	BAA	AAB	BBB	EAB	BAA	BBB	0	0	0	0	0	0	0	1	0
SAO	ABA	BBB	CAA	BBB	DBB	ABB	ADB	ABB	BAA	AAA	BCB	AAB	BAA	BBB	0	0	0	0	0	0	0	1	0
SAQ	ABA	DBD	CAA	BBB	BAB	ABB	AAB	ABB	BAA	AAA	BBB	AAB	BAA	ABB	0	0	0	0	0	0	0	0	2
SAS	ABA	BBB	CAA	BBB	DBB	ABB	ADB	ABB	CAA	AAA	BCB	EAB	BAA	BBB	0	0	0	0	0	0	0	0	4
SAT	ABA	BBI	CAA	BBB	DAB	ABB	AEB	ABB	CAA	AAA	BBB	EAB	BAA	BBB	0	0	0	0	0	0	0	0	2
SAU	ABA	BBD	CAA	BBB	BAB	ABB	AAB	ABB	BAA	AAB	BBB	AAB	BAA	BBB	0	0	0	0	0	0	0	0	1
SAV	ABA	BBD	CAA	BBB	BAB	ABB	AAB	ABB	BAA	AAB	BBB	AAB	BAA	BBB	0	0	0	0	0	0	0	0	3
SAW	ABA	DBD	CAA	BBB	BAB	ABD	AAB	ABB	BAA	AAA	BBB	AAB	BAA	ABB	0	0	0	0	0	0	0	0	1
SAX	ABA	DBB	CAA	BBB	BAB	ABB	AAB	ABB	CAA	AAA	BBB	AAB	BAA	ABB	0	0	0	0	0	0	0	0	1
SAY	ABA	DBD	CAA	BBB	DAB	ABB	AAB	ABB	BAA	AAA	BBB	AAB	BAA	ABB	0	0	0	0	0	0	0	0	1
SAZ	ABA	BBF	CAA	BBB	DAB	ABB	AEB	ABB	CAA	AAA	BBB	EAB	BAA	BBB	0	0	0	0	0	0	0	0	2
SBA	ABA	BBD	CAA	BBB	BAB	ABD	AAB	ABB	CAA	AAB	BBC	AAB	BAA	BBB	0	0	0	0	0	0	0	0	1

Population abbreviations as Table 1 and Figure 1.

454

most marine fish species, when there are a few haplotypes with high frequencies and a substantial number of rare haplotypes, which are their mutational derivatives. The mean haplotype diversity among Dolly Varden individuals was 0.6430 ± 0.01102, nucleotide diversity was $1.010017 \pm 0.00075\%$.

mtDNA haplotype genealogy of Dolly Varden

Clustering of the Dolly Varden haplotype sequence divergence resolved two broad groupings of haplotypes: group N ('northern') included 21 haplotypes, and group S ('southern') included the remaining 25 haplotypes (Figures 2 and 3). There were large breaks

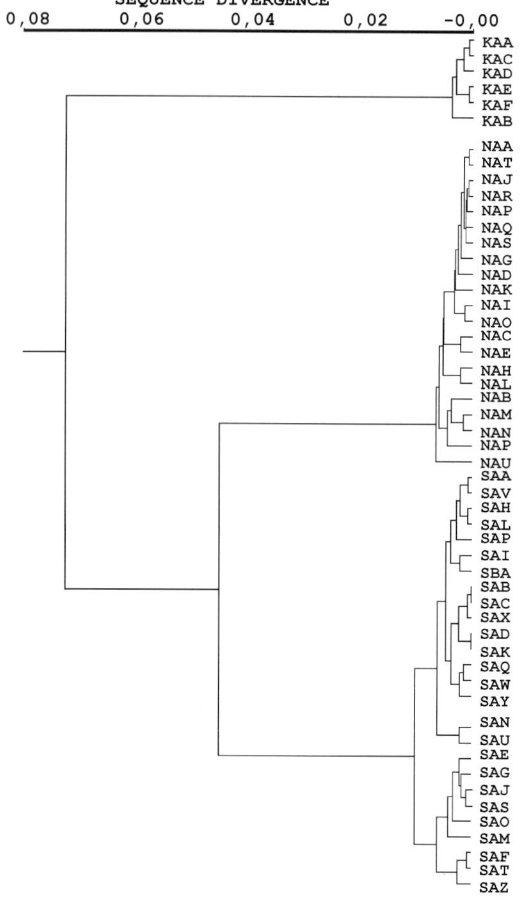

Figure 2. UPGMA phenogram based on sequence divergence estimates observed among the haplotypes of Dolly Varden. The cophenetic correlation has the value of 0.99149. The codes of haplotypes are as in Table 3.

in the genealogy between these two groups and the minimal number of mutational steps between haplotypes NAH and SAI is 50. These two clades of Dolly Varden mtDNA differed by about 4% in sequence divergence. In comparison, the minimum difference between haplotype KAE of the char *S. leucomaenis* and haplotype SAJ from southern clade of the Dolly Varden is 79 mutational steps (Figure 3). Average sequence divergence between haplotypes of the Dolly Varden and *S. leucomanis* was 7.0%.

The 'northern' clade of the haplotypes exhibited a star-like pattern, branching from the central haplotype (NAJ) and separated from the neighboring haplotypes by one or more mutational steps. The 'southern' clade of mtDNA haplotypes had more complex structure, with at least two groups of haplotypes (S1 and S2), which are separated by at least eight mutational steps. These two groups (sub-clades) of southern Dolly Varden mtDNA differed by about 1% in sequence (Figures 1 and 2).

The numbers of variable restriction sites and haplotypes were greater in the southern than in the northern population of Dolly Varden. In common, the southern samples of the Dolly Varden had significantly higher haplotype and nucleotide diversity than the northern samples (Table 4).

Mismatch distributions into two clades of mtDNA haplotypes are presented in Figure 4. The mean value of the differences between all haplotypes are higher in the southern clade (x = 10.4 ± 0.115) as compared with the northern clade (x = 6.4 ± 0.104) and the differences between two distributions was significant ($\chi^2 = 62.6$, df = 9, p < 0.05?).

Geographical distribution of mtDNA haplotypes and population structure analysis

The occurrence of mtDNA haplotypes in the seven populations is shown in Table 3. All specimens from northern drainage basins (Kamchatka, Paratunka, Yama, Achchon) had the 'northern' haplotypes excluding one fish from Yama River (Table 3). Most of the individuals (80–90%) from the southern drainage (Sakhalin Island) belonged to the 'southern' clade of haplotypes. From 10% to 20% of the individuals from southern populations revealed the 'northern' variants of mtDNA haplotypes. In this paper we presented the data for only Sakhalin Island rivers. Our preliminarily data obtained from other regions, South Kuril Island and Primorye, suggest that Dolly Varden from these

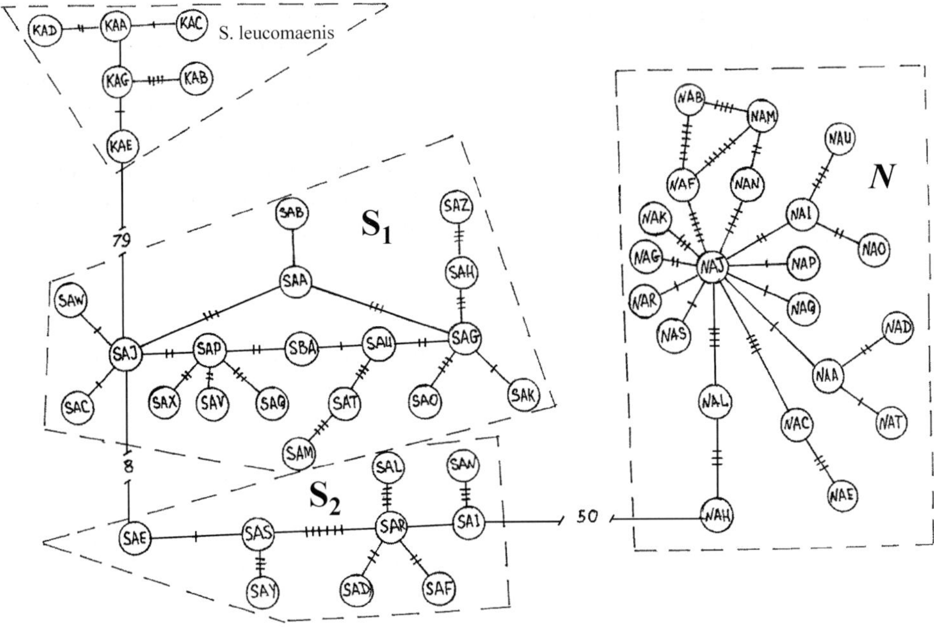

Figure 3. Minimum spanning haplotype network of Dolly Varden. Tick marks on the branches indicate mutational changes. The codes of haplotypes are as in Table 3.

Table 4. Values for haplotype diversity (h) and nucleotide diversity (π) of Dolly Varden populations.

Species	Population	Haplotype diversity ($\pm$SE) (h)	Nucleotide diversity (π)
S. m. malma	Kamchatka R.	0.7333 ± 0.09516	0.002712
S. m. malma	Paratunka R.	0.1567 ± 0.09574	0.000114
S. m. malma	Achchon L.	0.8929 ± 0.11127	0.003727
S. m. malma	Yama R.	0.2949 ± 0.15577	0.006627
S. m. krascheninnikovi	Aniva Creek	0.7613 ± 0.07887	0.015001
S. m. krascheninnikovi	Teplyi Creek	0.9111 ± 0.07734	0.021959
S. m. krascheninnikovi	Val R.	0.9203 ± 0.03150	0.019647
S. m. malma (sum data)		0.6902 ± 0.03817	0.001650
S. m. krascheninnikovi (sum data)		0.8514 ± 0.03145	0.006301

regions also have the 'southern' variants of the mtDNA haplotypes.

There are large differences in the haplotype composition between northern and southern populations of Dolly Varden ($\chi^2 = 781$; df $= 329$; p $= 0$). Haplotype composition differs among northern populations ($\chi^2 = 129$; df $= 54$; p $= 0$), as well as among southern populations are ($\chi^2 = 88$; df $= 56$; p < 0.01). Pairwise comparison among southern populations revealed the significant differences between Val Rivers samples and other samples ($\chi^2 = 49.6$; p $= 0.$), but no differences among the samples from Teplyi Crick and Aniva Crick ($\chi^2 = 14.5$; p $= 0.5428$). The pairwise comparisons of populations from north and south regions resulted in large F_{ST} values (0.3878–0.76861). F_{ST} values were significant for all pairwise comparisons of northern populations except Yama–Paratunka ($F_{ST} = 0.05874$) and Yama–Achchon populations (0.02857). Pairwise comparisons for southern populations of Dolly Varden resulted in small (0.01603–0.01393) insignificant F_{ST} values among all three populations.

Discussion

An integrated analysis of the mtDNA genealogical tree and the geographical distribution of haplotypes and haplotype clades provide insight into the

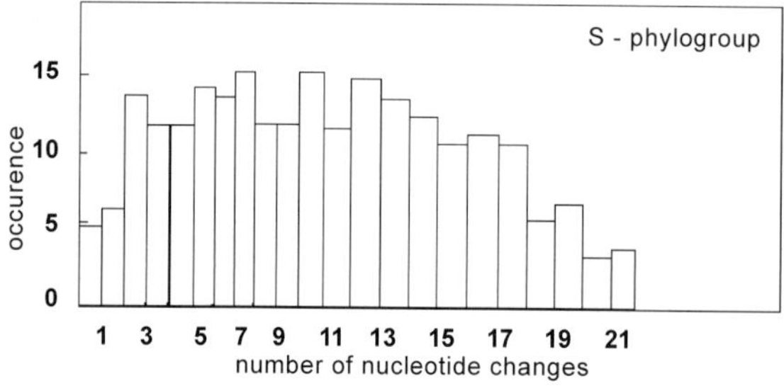

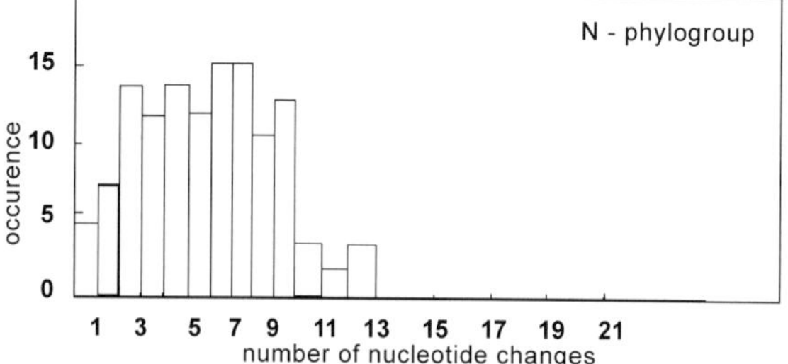

Figure 4. The mismatch distribution between haplotypes in southern (S) and northern (N) phylogroups.

evolutionary history of the species. Phylogeographic patterns depend on both historical population events and current population processes (Avise 2000). Our analysis of mtDNA sequence variation in Dolly Varden over its native range along the Asian coast of the Pacific Ocean demonstrated the presence of at least two phylogenetic groupings. Haplotypes between these differed by about 4% in nucleotide sequence divergence. Based on an approximate molecular clock of 2% sequence divergence per million years (Brown et al. 1979) this divergence can be translated into a separation of about 2 million years ago. The time of the divergence between two mtDNA phylogroups of Dolly Varden correspond to the border between Pleistocene and Pliocene periods, which is characterized by vertebrate speciations (Avise et al. 1998). The great differences between clades of the haplotypes indicate that these two forms of Dolly Varden have existed separately for a long time. The extent of divergence in mtDNA sequence between these forms is close to that observed between valid species of char. For example, the difference between *S. malma* and *S. leucomaenis* as estimated between 4.78% (Oleinik & Polyakova 1994) and 7.0% (these data), between arctic char *S. alpinus* and *S. namaycush* – 4% (Gylensten & Wilson 1987, Brunner et al. 2001), *S. alpinus complex* and *S. fontinalis* – 5.82% (Brunner et al. 2001), *S. alpinus complex* and *S. leucomaenis* – 3.99% (Brunner et al. 2001). It should be noted, that our data may be overestimated, compared with the data

of other investigators, because we used the relatively high variable fragments of the mtDNA (Churikov et al. 2001). The karyological investigations support species level of differences between southern and northern forms of Dolly Varden in Asiatic populations (Frolov et al. 1997). In contrast, the comparative study of the nucleotide sequences of the nuclear ribosomal genes (Phyllips et al. 1999) and allozyme data (Crane et al. 1994, Osinov & Pavlov 1998, Salmenkova et al. 2000) has revealed a lower level of divergence between *S. m. krascheninnikovi* and *S. m. malma*. This is probably due to the higher rate sequence divergence in mtDNA and to the greater effect of drift on the mtDNA genome when compared to that of nuclear sequences (Avise 2000).

The different genealogy of haplotypes, the shape of the mismatch distribution, and the levels of diversity within two clades of mtDNA (Slatkin & Hudson 1991, Rodgers & Harpending 1992, Nee et al. 1995) suggest the different histories of 'northern' and 'southern' subspecies of the Dolly Varden. MtDNA from southern populations were more diverse and the haplotype tree more complex, showing differentiation into two sub-clades. The mismatch distribution between haplotypes in southern populations also revealed relatively high level of dispersion when compared to that in northern populations. All these observation indicate that 'southern' clade of the Dolly Varden haplotypes is older. Moreover, the complex structure of the haplotypes genealogies separated by more than the average number of mutational steps (S1 and S2 sub-clades) apparently suggest fragmentation of the species range during the Pleistocene period. Sakhalin Island was never cover by ice sheet, but the periodical lowering of the level of the ocean during the glaciation periods probably had great effects on the distribution of this species in the past.

The 'northern' clade of haplotypes showed a typical star-like genealogy of haplotypes. The northern populations are also characterized by low diversity and a relative sharp wave in the mismatch distribution. These properties can be explained by reduced genetic variability resulting from recent recolonization of northern populations and/or founder effect. These indicate that the northern clade of haplotypes and northern populations of Dolly Varden are younger than southern populations.

The Dolly Varden from southern and northern regions exhibit different population structures. *S. m. malma* from northern regions is characterized by distinct differences among most of the populations.

The absence of differences between the population from Yama River and other northern populations is possibly due to the small number of specimen analyzed from Yama River, that is, sample size effect. *S. m. krascheninnikovi* from southern region is characterized by little changes in haplotype frequencies across the investigated area and the level of interpopulation diversity is weak compared to intrapopulation variability, as is evidently by F_{ST} values. We suggest that the observed differences between two subspecies maybe due to differences in the history of Northern Pacific region during the Pleistocene period. Periodical glaciations and/or lowering of the ocean level caused extirpation of Dolly Varden populations from the north regions except for a few small areas that remained ice-free. When glaciers retreated and freshwater habitat became available for spawning, the populations were gradually reestablished. The star-like structure of the haplotypes and mismatch distribution from northern populations support this scenario. Probably, the southern populations of Dolly Varden were not directly influenced by glaciation Pleistocene events to the same extent. The relative low level of significant differences between populations in southern regions suggested greater gene flow due to ancient or ongoing exchange of individuals.

From 10% to 20% of fish in the each southern population possessed mtDNA haplotypes that matched the common 'northern' variants (Table 3). Although the presence of the northern haplotypes in the southern populations could potentially be accounted for by the paraphyletic retention of ancestral haplotypes or convergent evolution, these explanations seem improbable. Other explanations include the possibility of the migration of Dolly Varden between northern and southern regions. High levels of 'homing' characterize Dolly Varden, but the long-distance migrations of immature or non-spawning mature have been observed (McPhail 1961, Behnke 1972, Morrow 1980, DeCicco 1992, Wilson et al. 1996). In this case it is necessary to assume very high levels of migrations between regions, possibly millions of specimens, and consequently, this assumption seems unlikely too. Moreover, the allozyme data on the char from these regions contradict this assumption (Salmenkova et al. 2000).

One potential explanation involves introgressive hybridization between northern and southern subspecies of Dolly Varden (Behnke 1980). Wilson & Hebert (1993) reported natural interspecies hybridization between lake trout, *S. namaycush*, and arctic char, *S. alpinus*, using molecular data. Later Bernatchez et al. (1995) found a population of brook trout, *S. fontinalis*, in southern Quebec that was completely introgressed with arctic char mtDNA. A similar instance of interspecific mitochondrial replacement in a population of lake trout, *S. namaycush*, that was fixed for *S. alpinus* mtDNA, was reported by Wilson & Bernatchez (1998). Wilson & Bernatchez (1998) supposed that hybrids between two species, *S. namaycush* and *S. alpinus*, may have been favored under periglacial conditions due to selective advantages of 'northern' (*S. alpinus*) mitochondrial types or associate nuclear genes.

The investigation of two north American Pacific species over abroad range, Dolly Varden *S. malma* and bull trout *S. confluentus* (Taylor et al. 2001), revealed some parallelism in divergent evolution, similar to what we found in populations in eastern Russia. Both species were subdivided into two major mtDNA lineages, 'southern' and 'northern'. The analysis of mtDNA indicated a paraphyletic relationship for both species, at the same time Dolly Varden and bull trout show reciprocal monophyly at the nuclear genome. Taylor et al. (2001) suggested that the presence of two divergent mtDNA clades in Dolly Varden was attributable to introgression of bull trout mtDNA into Dolly Varden.

In our case the most probable explanation for the presence of the 'northern' variants of mtDNA haplotypes in populations of southern subspecies would also involve asymmetric introgressive hybridization between northern and southern subspecies of Dolly Varden, with subsequent multiple backcrossing between hybrids and the southern forms. It seems probably that this took place during the Pleistocene glacial advances, when northern forms were displaced into the southern region by the ice sheet and/or the lowering of ocean level.

Our data and similar observations (Wilson & Hebert 1993, Bernatchez et al. 1995, Wilson & Bernatchez 1998, Taylor et al. 2001) suggest that introgressive hybridizations between char species or forms could have occurred very often during the evolution of this group. If it is so it seems possible that reticulate evolution may determine the great plasticity and produced complex mosaic of variability in morphology and ecology that char species exhibit throughout their range in the Northern Pacific Ocean.

Acknowledgements

The manuscript benefited from critical reading by D.Yu. Churikov. We would like to thank two anonymous reviewers for their excellent reviews. This work supported by grants RFBR numbers 94-04-11129, 98-04-48919 and 02-04-49480.

References

Avise, J.C. 2000. Phylogeography. The history and formation of species. Harvard University Press, Cambridge, 447 pp.

Avise, J.C., D. Walker & G.C. Johns. 1998. Speciation durations and Pleistocene effects on vertebrate phylogeography. Proc. R. Soc. Lond. B. 265: 1707–1712.

Behnke, R.J. 1972. The systematics of salmonid fishes of recently glaciated lakes. J. Fish. Res. Board Can. 29: 639–671.

Behnke, J.R. 1980. A systematic review of genus *Salvelinus*. pp. 441–480. *In*: E.K. Balon (ed.) Char: Salmonid Fishes of the Genus *Salvelinus*, Dr W. Junk, The Hague.

Berg, L.S. 1948. Freshwater Fishes of the USSR. and Adjacent Countries, Vol. 1, Acad. Sci. USSR. Press, Moscow, 466 pp (in Russian).

Bernatchez, L., H. Glement, C.C. Wilson & A.G. Danzmann. 1995. Introgression and fixation of Arctic char (*Salvelinus alpinus*) mitochondrial genome in an allopatric population of brook trout (*Salvelinus fontinalis*). Can. J. Fish. Aquat. Sci. 52: 179–185.

Brown, W.M., M. George Jr. & A.C. Wilson. 1979. Rapid evolution of animal mitochondrial DNA. Proc. Natl Acad. Sci. U.S.A. 76: 1967–1971.

Brunner, P.C., M.R. Douglas, A. Osinov, C.C. Wilson & L. Bernatchez. 2001. Holarctic phylogeography of arctic char (*Salvelinus alpinus* L.) inferred from mitochondrial DNA sequences. Evolution 55: 573–586.

Chereshnev, I.A. 1983. The fauna, systematics and relations of the freshwater fishes of the east Chukotka. pp. 89–108. *In*: Ecology and Systematics of the Freshwater Organisms of the Far East, Far East Science Center Press, Vladivostok (in Russian).

Churikov, D., M. Matsuoka, X. Luan, A.K. Gray, VI.A. Brykov & A.J. Gharrett. 2001. Assessment of concordance among genealogical reconstruction from various mtDNA segments in three species of Pacific salmon (genus *Oncorhynchus*). Mol. Ecol. 10: 2329–2339.

Crane, P.A., L.W. Seeb & J.E. Seeb. 1994. Genetic relationships among *Salvelinus* species inferred from allozyme data. Can. J. Fish. Aquat. Sci. 51(Suppl. 1): 182–197.

DeCicco, A.L. 1992. Long-distance movements of anadromous Dolly Varden between Alaska and the U.S.S.R. Arctic 45: 120–123.

Frolov, S.V., V.V. Frolova & A.V. Molodichenko. 1997. Karyotype of the char *Salvelinus malma* from Yama River (Magadan area) and the taxonomic status of the northern and the southern, char. Russ. J. Mar. Biol. 23: 269–272.

Gharrett, A.J., A.K. Gray & V.A. Brykov. 2001. Phylogeographical analysis of mitochondrial DNA variation in Alaskan coho salmon, *Oncorhynchus kisutch*. Fish. Bull. US 99: 528–544.

Glubokovsky, M.K. 1995. The Evolutionary Biology of Salmonid Fishes, Nauka, Moscow, 343 pp (in Russian).

Grewe, M.P., N. Billington & P.D.H. Hebert. 1990. Phylogenetic relationships among members of *Salvelinus* inferred from mitochondrial DNA divergence. Can. J. Fish. Aquat. Sci. 47: 984–991.

Gyllensten, U.B. & A.C. Wilson. 1987. Mitochondrial DNA of salmonids: inter- and intraspecific variability detected with restriction enzymes. pp. 301–317. *In*: F. Utter & N. Ryman (ed.) Population Genetics and Fishery Management, University of Washington Press, Seattle.

McElroy, D., P. Moran, E. Bermingham & I. Kornfield. 1992. REAP: An integrated environment for the manipulation and phylogenetic analysis of restriction data. J. Hered. 83: 153–158.

McPhail, J.D. 1961. A systematic study of the *Salvelinus alpinus* complex in North America. J. Fish. Res. Board Can. 18: 793–816.

Morrow, J.E. 1980. Analysis of the Dolly Varden char, *Salvelinus malma*, of northwestern North America and northeastern Siberia. pp. 323–338. *In*: E.K. Balon (ed.) Char: Salmonid Fishes of the Genus *Salvelinus*, Dr W. Junk, The Hague.

Nee, S., Holmes, E.C. & P.H. Harvey. 1995. Inferring population history from molecular phylogenies. Philos. Trans. R. Soc. Lond. B. 349: 25–31.

Nei, M. 1987. Molecular Evolutionary Genetics, Columbia University Press, New York, NY. 512 pp.

Nei, M. & J.C. Miller. 1990. A simple method for estimating average number of nucleotide substitution within and between populations from restriction data. Genetics 125: 837–879.

Nei, M. & F. Tajima. 1981. DNA polymorphism detectable by restriction endonucleases. Genetics 105: 207–217.

Oleinik, A.G. & N.E. Polyakova. 1994. Restriction analysis of the mitochondrial genome in the family Salmonidae. Russ. J. Genet. 30: 1034–1054.

Osinov, A.G. & S.D. Pavlov. 1998. Allozyme variation and genetic divergence between populations of Arctic char and Dolly Varden (*Salvelinus alpinus–S. malma* complex). J. Ichthyol. 38: 42–55.

Phyllips, R.V., L.I. Gudex, K.M. Westrich & A.L. DeCicco. 1999. Combined phylogenetic analysis of ribosomal ITS1 sequences and new chromosome data supports three subgroups of Dolly Varden char (*Salvelinus malma*). Can. J. Fish. Aquat. Sci. 56: 1504–1511.

Rodgers, A.R. & H. Harpending. 1992. Population growth makes waves in the distribution of pairwise genetic differences. Mol. Biol. Evol. 9: 552–569.

Roff, D.A. & P. Bentzen. 1989. The statistical analysis of mitochondrial DNA polymorphism: χ^2 and the problem of small samples. Mol. Biol. Evol. 6: 539–545.

Rohlf, F.J. 1990. NTSYS-pc: Numerical Taxonomy and Multivariate Analysis System, Version 1.60, Exeter Publishing Ltd., Setauket, New York, 117 pp.

Salmenkova, E.A., V.T. Omel'chenko, A.A. Kolesnikov & T.V. Malinina. 2000. Genetic differentiation of chars in the Russian north and Far East. J. Fish Biol. 57(Suppl. A): 136–157.

Sambrook, J., E.F. Fritsch & T. Maniatis. 1989. Molecular Cloning: A Laboratory Manual, Cold Spring Harbor Lab. Press, New York, 1,659 pp.

Schneider, S., D. Roessli & L. Excoffier. 2000. ARLEQUIN 2000: A Software for Population Genetics Analysis, Genetics and Biometry Lab., University of Geneva, Geneva, 111 pp.

Slatkin, M. & R.R. Hudson. 1991. Pairwise comparisons of mitochondrial DNA sequences in stable and exponential growing populations. Genetics 129: 555–562.

Taylor, E.B., Z. Redenbach, A.B. Costello, S.M. Pollard & C.J. Pacas. 2001. Nested analysis of genetic diversity in northwestern North American char, Dolly Varden (*Salvelinus malma*) and bull trout (*Salvelinus confluentus*). Can. J. Fish. Aquat. Sci. 58: 406–420.

Wilson, C.C. & L. Bernatchez. 1998. The ghost of hybrids past: Fixation of Arctic char (*Salvelinus alpinus)* mitochondrial DNA in an introgressed population of lake trout (*S. namaycush*). Mol. Ecol. 7: 127–132.

Wilson, C.C. & P.D.N. Hebert. 1993. Natural hybridization between Arctic char (*Salvelinus alpinus*) and like trout (*S. namaycush*) in Canadian Arctic. Can. J. Fish. Aquat. Sci. 50: 2652–2658.

Wilson, C.C., P.D.N. Hebert, J.D. Reist & J.B. Dempson. 1996. Phylogeography and postglacial dispersal of arctic char *Salvelinus alpinus* in North America. Mol. Evol. 5: 187–197.

Environmental Biology of Fishes **69**: 461–470, 2004.
© 2004 *Kluwer Academic Publishers. Printed in the Netherlands.*

Use of microsatellite locus flanking regions for phylogenetic analysis?
A preliminary study of *Sebastes* subgenera

Takashi Asahida[a], Andrew K. Gray[b,c] & Anthony J. Gharrett[b]
[a]*School of Fisheries Sciences, Kitasato University, Sanriku, Ofunato, Iwate 022-0101, Japan
(e-mail: asahida@kitasato-u.ac.jp)*
[b]*Fisheries Division, School of Fisheries and Ocean Sciences, University of Alaska, Fairbanks,
11120 Glacier Highway, Juneau, Alaska 99801 U.S.A. (e-mail: ffajg@uaf.edu)*
[c]*Current address: National Marine Fisheries Service, Auke Bay Laboratory, 11305 Glacier Highway,
Juneau, Alaska 99801 U.S.A. (e-mail: andy.gray@noaa.gov)*

Received 21 April 2003   Accepted 24 April 2003

Key words: rockfish, Scorpaenidae, *Hozukius*, systematics, PCR, DNA sequence

Synopsis

The systematics of the species-rich genus of *Sebastes* rockfish has not been resolved, and is unlikely to be resolved using morphological criteria. Using an alternative approach based on DNA sequence variation, we sampled the genome for random, presumably neutral, nuclear DNA sequences, by sequencing microsatellite flanking regions of 15 *Sebastes* species (representing 15 of 22 extant subgenera) and *Hozukius emblemarius*. We aligned sequences of flanking regions of eight of the 13 loci that amplified, which presumably are homologous. In aggregate, the aligned sequences included 848 bp, 53 bp of which were polymorphic. The base changes among species included 27 transitions, 20 transversions, three multiple substitutions, and three deletions or insertions. The flanking regions at different loci had different base substitution rates. Nucleotide divergence among *Sebastes* species ranged from 0.0012 to 0.0216, whereas nucleotide divergence between *Sebastes* and *Hozukius* species ranged from 0.0095 to 0.0240. We used maximum parsimony, neighbor-joining, and maximum likelihood methods to examine relationships among the species. *H. emblemarius* formed a separate group, and five of the western Pacific Ocean species of rockfish clustered together, suggesting that they may have shared an evolutionary history distinct from many North American species. The flanking regions of some microsatellite loci contain DNA sequence variation that distinguishes rockfish species and may prove useful for clarifying relationships among rockfish, or other closely related species.

Introduction

Of the approximately 100 *Sebastes* species occurring in the northern Pacific Ocean, the majority is distributed along the North American coast, about 22 species are found exclusively in Asian waters, and about five trans-Pacific species occur along the northern Pacific Rim (Kendall 2000). Five additional very closely related *Sebastes* occur in the North Atlantic Ocean and two (possibly three) species occur in the southern hemisphere (Rocha-Olivares et al. 1999a). The species in the North Atlantic probably originated from a trans-Arctic colonization from the North Pacific Ocean during an interglacial period; whereas, the species in the southern hemisphere probably originated from movements from the northern Pacific through equatorial waters cooled during a glacial maximun (Love et al. 2002). Although the genus emerged roughly five million years ago (Johns & Avise 1998), many of its species are quite closely related; and within some subgenera, species probably diverged within the past two million years (Rocha-Olivares et al. 1999b).

The abundance of *Sebastes* species created a chaotic taxonomic history, during which numerous genera and subgenera were established and re-established (Kendall 2000). Although *Sebastes* presently includes about 22 subgenera, as well as many subspecies, neither the relationships among subgenera nor the systematic validity of most of them have been determined (Ishida 1984, Kendall 1991, 2000). Only the subgenus *Sebastosomus* has been rigorously investigated recently (e.g., Chen 1971, 1975, Rocha-Olivares et al. 1999b); and although broader studies have been conducted (Ishida 1984, Seeb 1986, Johns & Avise 1998, Rocha-Olivares et al. 1999c), a thorough investigation of the entire genus has yet to be completed and many systematic questions remain unresolved. Consequently, no phylogeny for the *Sebastes* subgenera exists because efforts to develop a robust phylogeny from morphological criteria have failed;[1] and previous studies based on molecular genetic variation did not include both Asian and North American species (e.g., Seeb 1986, Johns & Avise 1998, Rocha-Olivares et al. 1999c).

Identification of closely related taxa and their phylogenetic relationships have been advanced substantially by DNA sequence analyses (e.g., Hillis et al. 1996). Several such studies investigated specific genes or assemblages of genes (such as mtDNA) to identify *Sebastes* species (Rocha-Olivares 1998, Roques et al. 1999, Gharrett et al. 2001) or to investigate relationships among species (e.g., Bentzen et al. 1998, Rocha-Olivares et al. 1999a, Rocha-Olivares & Vetter 1999, Gharrett et al. 2001). Although structural genes may have the advantage that some of the divergence results from natural selection, regions of those genes are usually constrained from diverging. Moreover, focusing on one or a few loci generally restricts the samples of nucleotides to a small, and not necessarily representative, portion of the genome (Karl & Avise 1993). To examine species that have diverged recently requires sequences that evolve rapidly must be included. Our approach here is to examine the flanking sequences of microsatellite loci. Microsatellites are distributed broadly throughout the genome, generally do not code for structural genes, and most are probably not subject to strong selection pressures (but see Streelman et al. 1998, Streelman & Kocher 2002). It is likely that the sequences flanking most microsatellite loci are relatively unconstrained by selection and can mutate freely. Recently, differences in the flanking regions of a microsatellite locus shared by two sturgeon species

---

[1] Kendall, A.W., Jr. Personal communication. U.S. National Marine Fisheries Service, retired; 635 Wanapum Drive, La Conner, WA 98257 U.S.A.

Acipenser oxyrinchus and *A. sturio* contributed to documentation of the movement of the North American species, *A. oxyrinchus*, into the Baltic sea during the Middle Ages (Ludwig et al. 2002). In concept, data from a number of these regions should serve as a sample of relatively neutral sequences from the entire nuclear genome.

Primers that can be used to PCR amplify a number of *Sebastes* microsatellite loci have been developed (Wimberger et al. 1999, Westerman et al. GenBank no. AF269052-269061, Seeb et al. GenBank nos. AF142587 and AF142483-142496, Roques et al. 1999, Miller et al. 2000, Sekino et al. 2001). We tested amplification of 13 microsatellite loci and chose eight loci to sequence for *Hozukius emblemarius* and 15 *Sebastes* species. Sequence divergence among species provides a means to address a variety of questions. Although there is no phylogeny available for *Sebastes* subgenera, we expect that our results will provide an initial step in developing one. Other specific questions that we examine are relationship between *Sebastes* and a closely related taxon, *Hozukius*, which has not been investigated (Kendall 2000). Also, the disjunct distributions of many Pacific Ocean *Sebastes* species evoke questions of the origin and distribution of the genus, in particular the relationships between the eastern and western Pacific species, some of which have been assigned to common subgenera (*Pteropodus*, *Mebarus*, and *Zalopyr*) (Kendall 2000).

Materials and methods

Adult specimens of 15 different species of *Sebastes* rockfish (three subspecies of *S. pachycephalus* were analyzed) and *H. emblemarius* were collected from coastal waters of Southeast Alaska, Oregon, Southern California, Baja California, and Japan (Table 1). Samples of heart or muscle tissue from each specimen were placed in a preservative solution (Seutin et al. 1991, 20% dimethyl sulfoxide (DMSO), 0.25 M EDTA, pH 8.0, saturated with NaCl). Preserved samples were stored at $-20°C$ until DNA isolation. Total cellular DNA was isolated using Puregene™ DNA isolation kits (Gentra Systems Inc., U.S.A.).

We tested amplification of 13 microsatellite loci (Table 2). Polymerase chain reaction (PCR) amplification was done as follows: (1) one cycle of melting at 94°C for 3 min 45 s, annealing for 30 s at temperatures ranging from 52 to 58°C (Table 2), and extension for 45 s at 72°C; (2) one cycle at 90 s at 94°C, 30 s

Table 1. Rockfish, *Sebastes* and *Hozukius* spp., sampling locations and dates, and number of specimens analyzed organized by subgenus.

Species	Subgenus	Locality	Date	N
S. zacentrus	*Allosebastes*	NE Pacific, Southeast Alaska	Jul 1995	10
S. glaucus	*Emmelas*	NW Pacific, Hokkaido, Japan	Nov 1999	2
S. aurora	*Eosebastes*	NE Pacific, Oregon	Sep 1995	9
S. elongatus	*Hispaniscus*	NE Pacific, California	Sep 1995	10
S. taczanowskii	*Mebarus*	NW Pacific, Hokkaido, Japan	Dec 1998	14
S. pachycephalus nigricans	*Murasoius*	NW Pacific, Chiba, Japan	Feb 1999	3
S. p. nudus	*Murasoius*	NW Pacific, Iwate, Japan	Feb 1999	3
S. p. pachycephalus	*Murasoius*	NW Pacific, Iwate, Japan	Feb 1999	3
S. vulpes	*Neohispaniscus*	NW Pacific, Iwate, Japan	Jan 1999	5
S. hubbsi	*Pteropodus*	NW Pacific, Iwate, Japan	Feb 1999	5
S. babcocki	*Rosicola*	NE Pacific, Oregon	Jul 1994	10
S. serriceps	*Sebastocarus*	NE Pacific, California and Baja California	Jun 1995	10
S. steindachneri	*Sebastodes*	NW Pacific, Iwate, Japan	Feb 1999	6
S. helvomaculatus	*Sebastomus*	NE Pacific, California and Oregon	Jul 1994	10
S. ruberrimus	*Sebastropyr*	NE Pacific, Southeast Alaska	Jul 1994	6
S. oblongus	*Takenokius*	NW Pacific, Iwate, Japan	Jun 1999	3
S. matsubarae	*Zalopyr*	NW Pacific, Iwate, Japan	May 1999	6
H. emblemarius		NW Pacific, Iwate, Japan	Feb 1999	9

at the annealing temperature, and 45 sec extension at 72°C; (3) 30 cycles of 30 s at 94°C, 30 s at the annealing temperature, and 45 s at 72°C; and (4) a final extension step of 1 min at 68°C. We used *Taq* DNA polymerase (Promega, U.S.A.) according to the manufacturer's directions. Amplified microsatellite fragments were analyzed with a LI-COR DNA sequencing and genetic analysis system (LI-COR, U.S.A.). Typically, 0.5 µl of each PCR product labeled with IRDye™ 700 or 800 (LI-COR, U.S.A.) and 0.5 µl of loading buffer (95% formamide, 20 mM EDTA, 0.1% bromophenol blue) were loaded in a 25 cm, 6% denaturing polyacrylamide gel. Fragments were electrophoresed for approximately 2 h at 1500 V. We detected and scored alleles at each locus with Gene ImagIR™ software (LI-COR, U.S.A.). The molecular weight markers used to estimate the microsatellite allele sizes were 50–350 bp or 50–700 bp size standards (LI-COR, U.S.A.).

After verifying microsatellite amplification and determining the extent of polymorphism for each locus, we sequenced microsatellites that amplified successfully in all species. We obtained sequence data for all species from eight loci: *Sma* 3, *Sma* 5, *Sma* 7, *Sma* 10, and *Sma* 11 isolated from *S. maliger* (Wimberger et al. 1999), and *SR* 7-7, *SR* 15-8, and *SR* 16-5 isolated from *S. rastrelliger* (Westerman et al. GenBank nos. AF269055, 269059, 269061).

Sequencing the short (about 100–300 bp) fragments was challenging because it is difficult to purify the short, easily lost amplification fragments for the subsequent sequencing reaction. We succeeded by treating the PCR products with ExoSAP-IT (Exonuclease I and shrimp alkaline phosphatase in buffer (USB Corp., U.S.A.)) to remove unincorporated dNTPs and single stranded DNA (e.g., primers) that remained with the PCR product and might interfere with the sequencing reaction. We sequenced both strands of the PCR product using IRDye 800 Termination mixes (LI-COR, U.S.A.) and Thermo Sequenase Cycle Sequencing kits (USB Corp., U.S.A.), amplifying with the forward and reverse primers that we originally used to amplify each locus. The amplification and labeling reaction involved: (1) heating the mixture to 95°C for 2 min; (2) 30 cycles of 30 s at 95°C, 30 s at the appropriate annealing temperature (Table 2) for each primer pair, and 45 s at 72°C; and (3) a final extension for 5 min at 72°C. The sequence reaction products were analyzed on a LI-COR DNA sequencer (LI-COR, U.S.A.). Sequence data were aligned by hand using sequence editor Align IR (LI-COR, U.S.A.) and DNASIS (Hitachi, Japan).

The GenBank accession numbers of the microsatellite flanking sequences of *Sma* 3, *Sma* 5, *Sma* 7, *Sma* 10, *SR* 7-7, *SR* 15-8, and *SR* 16-5 are AY132151–AY132290. Flanking sequences of the *Sma* 11 locus are available from T. Asahida on request (GenBank does not accept sequences shorter than 50 bp).

Differences in the extent of divergence of the flanking regions among the loci were examined

Table 2. Rockfish microsatellites, sizes of alleles and flanking sequences, and annealing temperatures.

Locus	Primers	Total allele size (bp)	Flanking sequences (bp)	Annealing temperature	Reference
Sma 3	5'-GCAGACTTACAGCGGTTTCAC-3' 5'-ACCATCCAGTCATACGAGCAC-3'	138–187	88	58	Wimberger et al. 1999
Sma 5	5'-CGAGTAACACCAGTGCCAAC-3' 5'-GAGATTTCTGGAGTCCACGC-3'	125–178	117	58	Wimberger et al. 1999
Sma 6	5'-ATGATGAAGTGTCGGTTGCTC-3' 5'-AAGGGAGGGCACCCCAAAAC-3'	172–187	NS	58	Wimberger et al. 1999
Sma 7	5'-CATAGGTCATTTCTCAAAGGTGTG-3' 5'-GAGAAACAAACAGGAACTGAGAGAG-3'	166–190	151	58	Wimberger et al. 1999
Sma 10	5'-TGTCAGTTGTCATAGAAAATCCTCTG-3' 5'-CACCAGTGGAACACACGCAC-3'	66–138	46	58	Wimberger et al. 1999
Sma 11	5'-AATAGAGGACGGGCAACG-3' 5'-ATCGTACCAGCTGACAACCTG-3'	106–200	19	56	Wimberger et al. 1999
SR 5-9	5'-CTTGCTACTGCAGAGTGACTAC-3' 5'-CCTCATAATAGAGCTTGTAATAACG-3'	94–102	NS	57	Westerman et al. GenBank no. AF269052
SR 7-2	5'-GAACATCCCTCCTTCCGACGC-3' 5'-GTCAAACAACTGCAGAATGTTCG-3'	130–209	NS	57	Westerman et al. GenBank no. AF269054
SR 7-7	5'-GCATGAAAGTGTATGAAAGGC-3' 5'-CATGTGATTCTGTGTCTAACTGAG-3'	176–228	111	57	Westerman et al. GenBank no. AF269055
SR 7-25	5'-GACCTTTCCCTGAACACACTCG-3' 5'-CAAGAGGCGGTGGTGCTGATGG-3'	162–234	NS	57	Westerman et al. GenBank no. AF269056
SR 11-103	5'-CTTGCAGGTAACGGGAAGG-3' 5'-GGCTGATGACATTGCAACCTTG-3'	257–289	NS	57	Westerman et al. GenBank no. AF269058
SR 15-8	5'-GGAGATGTGCGTGGCTCGTCTGG-3' 5'-GGGTTTACTCATTGTAGAC-3'	289–349	222	52	Westerman et al. GenBank no. AF269059
SR 16-5	5'-CCATCTGTGCTGAGCTGTCACTG-3' 5'-GAGAAGAGGCCTACAAGTACC-3'	177–277	106	52	Westerman et al. GenBank no. AF269061

NS indicates not sequenced.

by testing the homogeneity in the proportions of nucleotide positions at which base substitutions or indels were observed using log-likelihood (*G*-tests) analysis (Sokal & Rohlf 1995).

We examined the phylogenetic relationships among the species using maximum parsimony (MP), neighbor-joining (NJ), and maximum likelihood (ML) methods (PAUP 4.0 b, Swofford 2001). The MP analyses were performed using a heuristic search [the parameters of the process were: equal weighting ($1:1 =$ ns : nv), gaps were treated as a 'fifth base', sequences were added randomly, zero length branches were collapsed, the steepest descent was not enforced, and branch swapping (TBR) was used]; and all minimal trees were saved. A NJ tree was generated using HKY85 (Hasegawa et al. 1985) distance metric. ML heuristic searches were performed using the HKY85 model (Hasegawa et al. 1985) [the parameters of the process were: unequal base frequencies were estimated empirically, the ns : nv ratio was estimated from the data set, rate heterogeneity among sites was considered, and the shape parameter of the gamma distribution was estimated from the data]. Multiple MP trees were combined to produce a majority-rule consensus tree. We bootstrapped the NJ and ML trees with 1000 iterations using both variable and invariant sites. Multiple bootstrapped trees were combined to produce bootstrap majority-rule consensus trees.

Results

Eight of the 13 microsatellite loci we tested amplified, and their banding patterns were interpretable in all the species tested. Two of the eight loci exhibited variation in all of the species tested; and only three loci were invariant in more than seven species, even though some species were represented by only two or three individuals (Table 1, Appendix). The other five loci (*Sma* 10, *SR* 5-9, *SR* 7-2, *SR* 7-25, and *SR* 11-103) failed to amplify or their banding patterns could not be interpreted in one or two of the species (Appendix).

We sequenced the flanking regions of eight loci in *H. emblemarius* and 15 species of *Sebastes* as well as two additional subspecies of *S. pachycephalus*. We were able to align the sequences for each locus, which suggests that the loci are homologous among species. Flanking sequences of the eight microsatellite loci totaled 848 nucleotides. Variation among the aligned sequences resulted from 27 transitions, 20 transversions, three multiple substitutions, and three insertions or deletions (indels) (Figure 1). Substitution rates differed among loci (p = 0.0051) ranging from 1.0% at *Sma*5 to 11.3% at *Sma* 7. Indels were observed at the *Sma* 3 and *Sma* 10 loci.

Overall, nucleotide divergences between *Sebastes* species ranged from 0.0012 to 0.0216 substitutions per nucleotide (Table 3). In comparison the intraspecific divergence was 0.0036 for *S. serriceps* and 0.0012 for *H. emblemarius*. The intergeneric nucleotide divergence between *Sebastes* species and *H. emblemarius*, which generally exceeded the divergence between *Sebastes* species pairs (except for comparisons of *S. steindachneri* with some other *Sebastes* species), ranged from 0.0095 to 0.0240 substitutions (Table 3). The intraspecific variation we observed in *S. pachycephalus* and *S. serriceps* did not complicate the analysis of microsatellite flanking sequences.

Figure 2 shows two of the trees, the NJ tree overlaid with bootstrap values for the nodes and a majority-rule consensus MP tree based on the 922 most parsimonious trees. The maximum parsimony and bootstrapped ML and NJ analyses produced nearly identical trees (Figure 2; the ML tree is not shown). The sample of species examined was disproportionately represented by Asian species, but it is notable that five of the Asian species (*S. pachycephalus*, *S. oblongus*, *S. hubbsi*, *S. taczanowski*, and *S. glaucus*), which belong to different subgenera (Table 1), formed a branch (Figure 2) that included no North American species. In all the trees, one of the main branches included four North American species (*S. aurora*, *S. babcocki*, *S. serriceps*, and *S. elongatus*) and one Asian species (*S. vulpes*), but inclusion of *S. vulpes* was not supported by high bootstrap values (Figure 2). Also, the North American *S. ruberrimus* and the Asian *S. matsubarae* share a branch. Because differences between many of the species were based on one or two nucleotide changes, support was weak at many of the nodes.

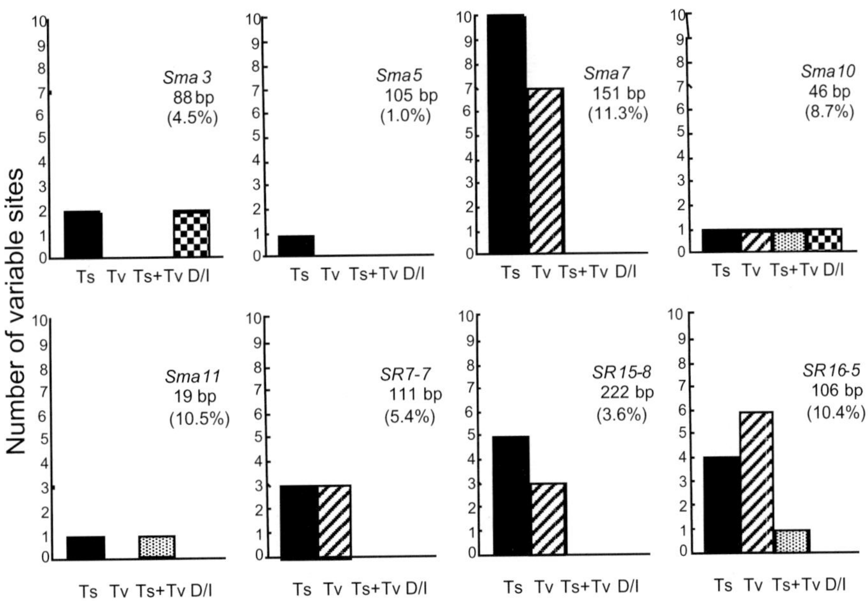

Figure 1. The base substitution pattern in rockfish, *Sebastes* and *Hozukius* spp., for each microsatellite locus (Table 2). Ts. = Transition. Tv. = Transversion. Ts. + Tv. = Transition or Transversion. D/I = Indels.

Table 3. Sequence divergence (Hasegawa et al. 1985) between samples of rockfish (*Sebastes* and *Hozukius*).

Species	Saur	Sbab	Selo	Sgla	Shel	Shub	Smat	Sobl	Spni	Spnu	Sppa	Srub	Sser 1	Sser 2	Sste	Stac	Svul	Szac	Hemb 1
Sbab	0.0036																		
Selo	0.0012	0.0048																	
Sgla	0.0132	0.0144	0.0120																
Shel	0.0048	0.0059	0.0036	0.0083															
Shub	0.0131	0.0168	0.0119	0.0168	0.0131														
Smat	0.0072	0.0107	0.0060	0.0132	0.0048	0.0156													
Sobl	0.0119	0.0156	0.0107	0.0131	0.0095	0.0131	0.0095												
Spni	0.0120	0.0156	0.0108	0.0131	0.0095	0.0131	0.0120	0.0095											
Spnu	0.0144	0.0180	0.0132	0.0156	0.0119	0.0156	0.0144	0.0095	0.0024										
Sppa	0.0120	0.0156	0.0108	0.0131	0.0095	0.0131	0.0120	0.0095	0.0000	0.0024									
Srub	0.0036	0.0071	0.0024	0.0119	0.0036	0.0119	0.0036	0.0083	0.0107	0.0132	0.0107								
Sser 1	0.0048	0.0059	0.0059	0.0156	0.0071	0.0156	0.0120	0.0168	0.0168	0.0192	0.0168	0.0083							
Sser 2	0.0036	0.0048	0.0048	0.0144	0.0060	0.0168	0.0107	0.0156	0.0156	0.0180	0.0156	0.0071	0.0036						
Sste	0.0143	0.0156	0.0131	0.0180	0.0095	0.0180	0.0143	0.0192	0.0192	0.0216	0.0192	0.0131	0.0168	0.0156					
Stac	0.0108	0.0144	0.0096	0.0119	0.0083	0.0143	0.0108	0.0107	0.0107	0.0132	0.0107	0.0095	0.0156	0.0144	0.0180				
Svul	0.0048	0.0059	0.0036	0.0132	0.0048	0.0156	0.0095	0.0144	0.0120	0.0144	0.0120	0.0059	0.0071	0.0060	0.0143	0.0132			
Szac	0.0071	0.0083	0.0059	0.0156	0.0071	0.0180	0.0119	0.0168	0.0156	0.0180	0.0156	0.0083	0.0095	0.0083	0.0119	0.0156	0.0071		
Hemb 1	0.0155	0.0168	0.0143	0.0192	0.0107	0.0240	0.0143	0.0192	0.0204	0.0229	0.0204	0.0131	0.0179	0.0168	0.0155	0.0168	0.0156	0.0120	
Hemb 2	0.0143	0.0155	0.0131	0.0180	0.0095	0.0228	0.0131	0.0180	0.0192	0.0217	0.0192	0.0119	0.0167	0.0155	0.0143	0.0156	0.0143	0.0108	0.0012

Key to species: *Saur – S. aurora*; *Sbab – S. babcocki*; *Selo – S. elongatus*; *Sgla – S. glaucus*; *Shel – S. helvomaculatus*; *Shub – S. hubbsi*; *Smat – S. matsubarae*; *Sobl – S. oblongus*; *Spni – S. p. nigricans*; *Spnu – S. p. nudus*; *Sppa – S. p. pachycephalus*; *Srub – S. ruberrimus*; *Sser – S. serriceps*; *Sste – S. steindachneri*; *Stac – S. taczanowskii*; *Svul – S. vulpes*; *Szac – S. zacentrus*; *Hemb – H. emblemarius*.

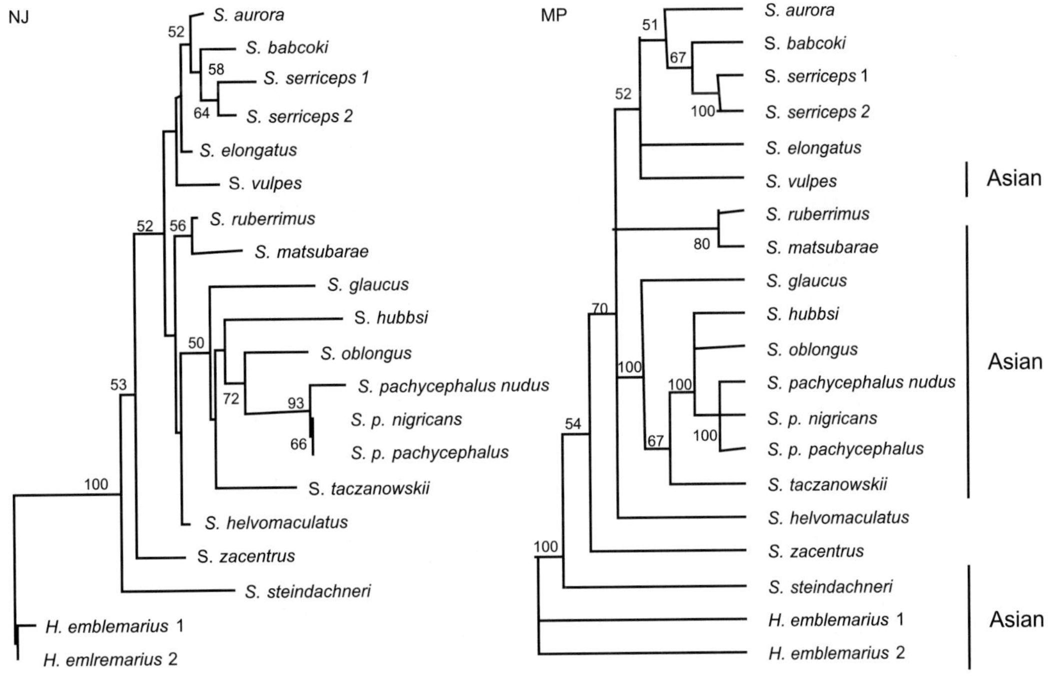

Figure 2. Trees of rockfish, *Sebastes* and *Hozukius*, species based on microsatellite flanking region sequence data. MP is a majority-rule consensus tree of 922 equally parsimonius trees, and NJ is a neighbor-joining tree on which bootstrap (1 000 iterations) values are superimposed. Nodes supported by less than 50% of the trees included are not labeled.

Discussion

The microsatellite locus primers we tested amplified broadly across the range of *Sebastes* species, including 15 of 22 currently recognized subgenera and species from both the eastern and western Pacific Ocean. Not only does each of these microsatellite loci occur across the range of species, but the flanking regions are also sufficiently conserved to support amplification. These loci must predate the divergence of the *Sebastes* and *Hozukius* genera from the ancestral scorpaenid lineage and may serve as characters for deducing their relationships to other scorpaenid taxa. Our results demonstrate the success of cross-species amplification of microsatellite primers among closely related (congeneric) species (Primmer et al. 1996, Fields & Scribner 1997). Many of the microsatellite loci we tested exhibited variability in small samples (2–14 individuals; Appendix), which suggests that they may be useful (barring the existence of null alleles) for population genetics analyses for numerous species (e.g., Matala et al. 2003; Appendix).

Most of the flanking regions of the microsatellite loci that we sequenced are short sequences (19–222 bp), each of which yields relatively little information. However, in aggregate they included 848 nucleotides that represent presumably neutral or nearly neutral sequences sampled throughout the genome. This sample of the genome can be increased by surveying sequences at additional microsatellite loci.

The range of intraspecific divergence observed, including *S. pachycephalus* subspecies (0–0.0024) and two *S. serriceps* haplotypes (0.0036), as well as two *H. emblemarius* haplotypes (0.0012) is of the same order of magnitude as the average intraspecific divergence (0.0024) reported for the mitochondrial NADH-dehydrogenase-3 and -4 subunit genes and 12S and 16S ribosomal RNA genes in *Sebastes* species (Gharrett et al. 2001). That study estimated divergences from restriction site data taken from a different set of *Sebastes* species, but included several closely related species (Gharrett et al. 2001). Some of the interspecific differences were less than intraspecific differences, which may be attributed either to stochastic differences resulting from small samples or may reflect very recent divergences. The interspecific differences observed from microsatellite flanking sequences (0.0012–0.0216) were similar to the interspecific divergences (0.0015–0.0384) observed by Gharrett et al. (2001). The intergeneric nucleotide divergence between *Sebastes* species and *H. emblemarius* (0.0095–0.0240) is much lower than the intergeneric mitochondrial gene comparisons between *Sebastes* and *Sebastolobus* and between *Sebastes* and *Helicolenus*, which were 0.1073 and 0.0805, respectively (Gharrett et al. 2001). Although it is difficult to compare estimates of divergences obtained by different methods, the similarities suggest that the mutation rate is of the same order of magnitude. Also, the relatively low divergence between *H. emblemarius* and *Sebastes* in this study suggests a much closer relationship than between *Sebastes* and the *Sebastolobus* or *Helicolinus* genera.

In this preliminary survey we used MP, NJ, and ML methods to estimate phylogenetic relationships among the species included. Because we broadly examined the subgenera within *Sebastes*, we had no expectation of constructing phylogenies that would allow us to test the coherence of species within subgenera, and many of the species we examined were of Asian origin and have not been previously included in molecular phylogenetic surveys (e.g., Rocha-Olivares et al. 1999a,b,c). Because the relationships among *Sebastes* subgenera based on morphological data or molecular data have not been described (Kendall 2000), we have no standard with which to compare our results. However, it appears that the subgenera may group on a geographic basis because many of the Asian species clustered separately from North American species (Figure 2), although additional North American species must be studied to be conclusive. Another pair of species, the Asian *S. matsuabrae* and northeastern Pacific *S. ruberrimus*, clustered, suggesting that they diverged earlier and independently of the other Asian cluster. The similarities among North Atlantic *Sebastes* species suggest a single colonization followed by radiation and the divergence among Southern Hemisphere species reflects the same pattern. In contrast, the patterns of divergence among Asian species suggests a much more complicated history that may involve much longer time spans and more than one wave of contact from the eastern Pacific Ocean.

The issue of whether *Sebastes* and *Hozukius* are sister taxa or if *Hozukius* species should be included in *Sebastes* remains open. The genus *Hozukius*, which is endemic to Asian waters, is quite similar to *Sebastes*, separated only by characteristics of the second suborbital bone (Matsubara 1943). In future studies, *Helicolenus hilgendorfi* may be an appropriate outgroup for examining the monophyly of *Sebastes* and *Hozukius* because *Helicolenus* has remarkably different characters (e.g., the absence of an air-bladder) from *Sebastes* and *Hozukius*.

These results demonstrate that the flanking sequences of some microsatellite loci may contain information useful for identifying *Sebastes* species and potentially contribute to addressing phylogenetic questions. By sampling flanking sequences of additional loci (e.g., Seeb et al. GenBank nos. AF142587 and AF142483-142496, Roques et al. 1999, Miller et al. 2000, Sekino et al. 2001), the number of nucleotides can be increased substantially and, presumably, the genome will be more broadly sampled. It will be necessary to increase the number of each species sampled in order to verify that intraspecific variation does not interfere with phylogenetic inferences. To better elucidate the phylogeny of rockfish and the nature of the geographic distribution of species, flanking regions of many additional species must be sequenced. In particular, both Asian and North American species are included in the subgenera *Pteropodus*, *Mebarus*, and *Zalopyr*. It will be interesting to see if those assignments are correct or if their phylogeny more strongly reflects their geographic distribution across the Pacific Ocean.

Our comparison of *Sebastes* microsatellite flanking regions resolved differences among all species included in the study. The species, which were chosen from different subgenera to ensure that we would find divergence, may provide a preliminary glimpse of relationships among subgenera. Two observations were of particular interest. First, many of the Asian subgenera clustered together, which suggests that their species may have radiated subsequent to isolation from most eastern Pacific species, or that the subgeneric assignments are not warranted. Also, *H. emblemarius* appears to be closely related to *Sebastes*. Although our results are preliminary, they warrant further effort. By including flanking sequences of additional microsatellite loci and more species, particularly ones that have been studied in other molecular studies (e.g., Johns & Avise 1998, Rocha-Olivares et al. 1999b), it may be possible to describe the relationships among *Sebastes* species and rigorously

Appendix Amplification and variability of microsatellites for *Sebastes* and *Hozukius* species.

Species	N	Sma 3	Sma 5	Sma 6	Sma 7	Sma 10	Sma 11	SR 5-9	SR 7-2	SR 7-7	SR 7-25	SR 11-103	SR 15-8	SR 16-5
S. zacentrus	10	++	+	+	++	++	++	+	++	++	++	++	++	++
S. glaucus	2	++	++	++	+	+	+	−	+−	++	++	+−	++	++
S. aurora	9	++	++	+	+	++	++	+	++	++	++	+	++	++
S. elongatus	10	++	++	+	+	++	++	+	++	++	++	+	++	++
S. taczanowskii	14	++	++	+	++	++	++	+	++	++	++	++	++	++
S. pachcephalus nigricans	3	+	+	+	+	++	+	+	++	+	++	+	++	++
S. p. nudus	3	+	++	+	+	+	++	+	++	++	++	+	++	++
S. p. pachycephalus	3	++	++	+	+	++	+	+	++	++	++	+	++	++
S. vulpes	5	++	++	+	++	+	++	+	++	++	++	+	++	++
S. hubbsi	5	++	+	+	++	++	++	+	++	++	++	++	++	++
S. babcocki	10	++	++	++	++	++	++	+	++	++	++	++	++	++
S. serriceps	10	++	++	+	+	++	++	+−	++	++	−	++	++	++
S. steindachneri	6	++	++	++	++	++	++	+	+−	++	++	++	++	++
S. helvomaculatus	10	++	++	++	++	++	++	+	++	++	++	++	++	++
S. ruberrimus	6	++	++	+	++	++	++	++	++	++	++	+	++	++
S. oblongus	3	+	+	++	+	+−	+	+	++	+	++	++	++	++
S. matsubarae	6	++	++	+	+	+	++	++	++	++	++	−	++	++
H. emblemarius	9	++	++	++	++	+−	++	+	++	++	++	+	++	++

++ indicates amplification, + indicates amplification but no variability for the species, and − indicates failure to amplify in any of the samples tested for the species. +− indicates amplification but difficulty in scoring, interpretation, or other problems.

evaluate the validity of the existing subgenera and their membership.

Acknowledgements

We thank M. Love (University of California Santa Barbara), J. Heifetz (NMFS Auke Bay Laboratory), and H. Ida (Kitasato University) for collecting fish samples used in this study. We are grateful to K. Hayashizaki and Y. Shinotsuka (Kitasato University) for help with the data analysis. J. Seeb (Department of Fish and Game, State of Alaska) and three anonymous reviewers provided constructive comments. This project was supported by a grant from the United States Geological Survey Biological Resources Division to AJG and by the Promotion and Mutual Aid Corporation for Private Schools of Japan to TA.

References

Bentzen, P., J.M. Wright, L.T. Bryden, M. Sargent & C.T. Zwanenburg. 1998. Tandem repeat polymorphism and heteroplasmy in the mitochondrial control region of Redfishes (*Sebastes*: Scorpaenidae). J. Heredity 89: 1–7.

Chen, L.-C. 1971. Systematics, variation, distribution, and biology of rockfishes of the subgenus *Sebastomus* (Pisces, Scorpaenidae, *Sebastes*), Bull. Scripps Inst. Ocean., Univ. Cal. 115 pp.

Chen, L.-C. 1975. The rockfishes, genus *Sebastes* (Scorpaenidae), of the Gulf of California, including three new species, with a discussion of their origin. Proc. Calif. Acad. Sci. XL: 109–141.

Fields, R.L. & K.T. Scribner. 1997. Isolation and characterization of novel waterfowl microsatellite loci: Cross-species comparisons and research applications. Mol. Ecol. 6: 199–202.

Gharrett, A.J., A.K. Gray & J. Heifetz. 2001. Identification of rockfish (*Sebastes* spp.) from restriction site analysis of the mitochondorial ND-3/ND-4 and 12S/16S rRNA gene regions. U.S. Fish. Bull. 99: 49–62.

Hasegawa, M., H. Kishino & T. Yano. 1985. Dating of the human–ape splitting by a molecular clock of mitochondrial DNA. J. Mol. Evol. 21: 160–174.

Hillis, D.M., B.K. Mable, A. Larson, S.K. Davis & E. Zimmer. 1996. Nucleic acids IV: Sequencing and cloning. pp. 321–381. *In*: D.M. Hillis, C. Moritz & B.K. Mable (ed.) Molecular Systematics, Sinauer Assoc., Sunderland, MA.

Ishida, M. 1984. Taxonomic study of the sebastine fishes in Japan and its adjacent waters, M.S. thesis, Hokkaido University, Hakodate, Hokkaido, Japan. 267 pp.

Johns, G.C. & J.C. Avise. 1998. Tests for ancient species flocks based on molecular phylogenetic appraisals of *Sebastes* rockfishes and other marine fish. Evolution 52: 1135–1146.

Karl, S.A. & J.C. Avise. 1993. PCR-based assays of Mendelian polymorphisms from anonymous single-copy nuclear DNA: Techniques and applications for populaton genetics. Mol. Biol. Evol. 10: 342–361.

Kendall, A.W., Jr. 1991. Systematics and identification of larvae and juveniles of the genus *Sebastes*. Environ. Biol. Fish. 30: 173–190.

Kendall, A.W., Jr. 2000. An historical review of *Sebastes* taxonomy and systematics. Mar. Fish. Rev. 62: 1–23.

Love, M.S., M. Yoklavich and L. Thorsteinson. 2002. The Rockfishes of the Northeast Pacific, University of California Press, Berkeley, CA. 405 pp.

Ludwig, A., L. Debus, D. Lieckfeldt, I. Wirgin, N. Benecke, I. Jenneckens, P. Williot, J.R. Waldman & C. Pitra. 2002. When American sea sturgeon swam east. Nature 419: 447–448.

Matala, A.P., A.K. Gray, J. Heifetz & A.J. Gharrett. 2003. Population Structure of Alaskan Shortraker rockfish, *Sebastes borealis*, inferred from microsatellite variation. Environ. Biol. Fish.

Matsubara, K. 1943. Studies on the Scorpaenoid fishes of Japan: Anatomy, phylogeny and taxonomy (1 and 2). Trans. Singenkagaku kenkyusyo 1, 2: 1–485.

Miller, K.M., A.D. Schulze & R.E. Withler. 2000. Characterization of microsatellite loci in *Sebastes alutus* and their conservation in congeneric rockfish species. Mol. Ecol. 9: 240–242.

Primmer, C.R., A.P. Moller & H. Ellegren. 1996. A wide-range survey of cross-species microsatellite amplification in birds. Mol. Ecol. 5: 365–378.

Rocha-Olivares, A. 1998. Multiplex haplotype-specific PCR: A new approach for species identification of the early life stages of rockfishes of the species-rich genus *Sebastes* Cuvier. J. Exp. Mar. Biol. Ecol. 231: 279–290.

Rocha-Olivares, A. & R.D. Vetter. 1999. Effects of oceanographic circulation on the gene flow, genetic structure, and phylogeography of the rosethorn rockfish (*Sebastes helvomaculatus*). Can. J. Fish. Aquat. Sci. 56: 803–813.

Rocha-Olivares, A., R.H. Rosenblatt & R.D. Vetter. 1999a. Cryptic species of rockfishes (*Sebastes*: Scorpaenidae) in the Southern Hemisphere inferred from mitochondrial lineages. J. Hered. 90: 404–411.

Rocha-Olivares, A., R.H. Rosenblatt & R.D. Vetter. 1999b. Molecular evolution, systematics, and zoogeography of the rockfish subgenus *Sebastomus* (*Sebastes*, Scorpaenidae) based on mitochondrial cytochrome *b* and control region sequences. Mol. Phylogenet. Evol. 11: 441–458.

Rocha-Olivares, A., C.A. Kimbrell, B.J. Eitner & R.D. Vetter. 1999c. Evolution of mitochondrial cytochrome *b* gene sequence in the species rich genus *Sebastes* (Teleostei, Scorpaenidae) and its utility in testing the monophyly of the subgenus *Sebastomus*. Mol. Phylogenet. Evol. 11: 426–440.

Roques, S.D. Pallotta, J.-M. Sevigny & L. Bernatchez. 1999. Isolation and characterization of polymorphic microsatellite markers in the North Atlantic redfish (Teleostei: Scorpaenidae, genus *Sebastes*). Mol. Ecol. 8: 685–702.

Seeb, L.W. 1986. Biochemical systematics and evolution of the scorpaenid genus *Sebastes*, Ph.D. dissertation, University of Washington, Seattle. 176 pp.

Sekino, M., N. Takagi, M. Hara & H. Takahashi. 2001. Analysis of microsatellite DNA polymorphisms in rockfish *Sebastes thompsoni* and application to population genetics studies. Mar. Biotechnol. 3: 45–52.

Seutin, G., B.N. White & P.T. Boag 1991. Preservation of avian blood and tissue samples for DNA analysis. Can. J. Zool. 69: 82–90.

Sokal, R.R. & F.J. Rohlf. 1995. Biometry, 3rd edition, Freeman, San Francisco, CA. 573 pp.

Streelman, J.T. & T.D. Kocher. 2002. A gene for salinity tolerance in tilapia: Microsatellite variation associated with prolactin expression and growth of salt-challenged tilapia. Physiol. Genomics 9: 1–4.

Streelman, J.T., R. Zardoya, A. Meyer & S.A. Karl. 1998. Multilocus phylogeny of cichlid fishes (*Pisces: Perciformes*): Evolutionary comparison of microsatellite and single copy nuclear loci. Mol. Biol. Evol. 15: 798–808.

Swofford, D.L. 2001. PAUP. Phylogenetic Analysis Using Parsimony, Sinauer, Sunderland, MA.

Wimberger, P., J. Burr, A. Gray, A. Lopez & P. Bentzen. 1999. Isolation and characterization of twelve microsatellite loci for rockfish (*Sebastes*). Mar. Biotechnol. 1: 311–315.

Authors' index